Linear System Theory

The State Space Approach

McGraw-Hill Series in System Science

Editorial Consultants

George Dantzig
William Linvill
Lotfi Zadeh

Zadeh and Desoer Linear System Theory: The State Space Approach

Linear System Theory

The State Space Approach

Lotfi A. Zadeh & Charles A. Desoer

Department of Electrical Engineering
University of California
Berkeley, California

McGraw-Hill Book Company, Inc.

New York / San Francisco / Toronto / London

Linear System Theory: The State Space Approach

 Library of Congress Catalog Card Number 63-14581

72746

To Fay and Claudine

Preface

This book was born out of the authors' conviction that the scientific and technological developments of the last two decades have set the stage for an approach to linear system theory that differs radically from those currently taught in the classroom.

By far the most important of these developments is the advent of large-scale general-purpose digital computers. By providing the means for the analysis and design of highly complex systems, such computers have generated a need for new approaches and concepts bearing on the characterization, identification, and control of systems of various types. Indeed, not long ago an engineer could rely on his intuition, empirical know-how, and a collection of classical analytical and numerical methods for the solution of most of the problems encountered in engineering design and analysis. Today he is frequently called upon to devise a program or an algorithm that in conjunction with a digital computer leads to an optimal configuration for a complex system subjected to a large number of inputs and constraints. It hardly needs arguing that to design a large-scale system in this fashion requires a much greater degree of mathematical sophistication and, more particularly, much more rigorous and extensive training in circuit theory, information theory, control theory, optimization techniques, and computer programming than were necessary in the past.

In addition to its direct impact on system design and analysis, the development of digital computers has stimulated interest in finite-state systems and automata, which in turn has led to many new concepts and techniques centering on the notion of state. Parallel developments in control theory, information processing, and related fields have reawakened interest in the classical phase-space techniques of analysis of systems characterized by first-order vector differential or difference equations. Within the last several years, in particular, the phase-space or, equivalently, state-space approach has come to play a central role in the theories of optimal control of constrained as well as unconstrained systems.

The rapidly growing interest in state-space techniques is reflected in the spirit as well as the choice of subject matter in the book. Instead of following the conventional pattern of starting with the discussion of differential equations and proceeding to the Laplace and Fourier transformations, we introduce the notion of state at the outset and construct a general system-theoretic framework for both linear and nonlinear, nondifferential as well as differential systems. In this way, such notions as object, system, and state equivalence are given precise meaning, and a solid foundation is laid for the state-space approach to linear differential systems which is developed in succeeding chapters.

In order to attain a high level of precision and clarity we have freely used modern mathematical notations and terminology and have stated all significant results and conclusions in the form of theorems, lemmas, or assertions. This should not be construed to imply that the book is addressed to the mathematician. Its mathematical flavor notwithstanding, our exposition is directed primarily at an engineering audience, and it has no pretense at having the level of rigor which prevails in the mathematical literature. Essentially, we assume that the reader is a graduate student or a practicing engineer having the usual background in calculus, complex variable, ordinary differential equations, Laplace and Fourier transformations, and linear circuit theory. Some familiarity with the elements of linear algebra is also taken for granted. For completeness, self-contained expositions of delta functions and distributions, the Fourier and Laplace transformations, and linear vector spaces are included in appendixes.

To aid in the understanding of the subject matter as well as to facilitate the use of the book as a reference, we have numbered and cross-referenced all comments, remarks, assertions, theorems, examples, exercises, etc. The numbering system employed is explained in the Note to the Reader on p. xi.

A fairly good idea of the contents and scope of the book can be obtained by scanning the Contents. However, a few specific comments concerning the subject matter of various chapters may be of help in orienting the reader.

To begin with, the material presented in the first three chapters is essentially an attempt at constructing a conceptual framework for nonlinear as well as linear system theory in which the central notions are those of object, input-ouput relation, input-output-state relation, state, state and system equivalence, state equations, etc. Since much of this material is new, our exposition of it has no claim at definitiveness or completeness. This applies, in particular, to the discussion of such notions as the state space of an interconnection, state and system equivalence in the case of systems whose state space changes with time, input-output-state relations for systems with infinite-dimensional state space, etc.

The material in Chaps. 4 through 11 is concerned in the main with the application of state-space techniques in the analysis of systems characterized by differential equations with time-varying as well as constant coefficients. Since these applications range over a very wide field, we have not attempted to present an exhaustive coverage of them; instead, we have concentrated on a relatively few basic topics which we consider to be of particular relevance to applications of linear system theory. These include the study of modes in time-invariant systems, the association of state vectors and state equations with time-varying systems, the analysis of periodically varying systems, the stability of linear systems, and the notions of controllability and observability.

For convenience of the reader, each chapter includes a list of books under the heading of References and Suggested Reading. These lists are intended to direct the reader to basic sources of information on subjects related to those discussed in the text. To make the book substantially self-contained, compact expositions of the mathematical techniques which are of frequent use in the text are included in the four appendixes which follow Chap. 11.

The reader may want to know how the order of the authors' names was arrived at. The facts are these: one of us (L.A.Z.) wrote the first four chapters and Appendix *B*; the other (C.A.D.) wrote the remaining seven chapters and the other three appendixes. We saw no reason for following the tradition of using the alphabetical order since it puts a perpetual burden on those whose initial happens to be at the end of the alphabet. Therefore, we flipped a fair coin and tails won.

Acknowledgments

The authors thank Glenn Bacon, Jack Wing, and Albert Chang for reading parts of the manuscript and suggesting improvements; and Jocelyn Harleston, LaDelia Gilmore, and Kim Teramoto for typing the manuscript. Finally, the book could not have been written without the generous research support provided by the University of California, the National Science Foundation, and the Department of Defense, in particular the Office of Naval Research, the Office of Scientific Research, and the Army Research Office.

Lotfi A. Zadeh
Charles A. Desoer

Note to the Reader

The following system of numbering and cross-referencing is used in this book: At the top of each page in the outer margin appear chapter and section numbers in boldface italic type; for example, ***3.8*** at the top of a page means that the discussion on that page is part of Chapter *3*, Section *8*. In addition, each item (definition, theorem, example, comment, remark, etc.) is given a number that appears in the left-hand margin; such items are numbered consecutively within each section. Item numbers and all cross references in the text are in italic type. Cross references are of the form "by Definition *5.6.3*"; this means "by the definition which is item *3* of Section *6* in Chapter *5*." Since the four appendixes are labeled *A*, *B*, *C*, and *D*, the reference "*C.17.3*" is to "item *3* in Section *17*, Appendix *C*." When we refer in a section to an item within the same section, only the item number is given; thus "substituting in *7*" means "substituting in Eq. *7* of this section." Also, "*7 et seq.*" means "Eq. *7* and what follows Eq. *7*."

Preceding the index is a glossary of symbols which describes notational conventions and contains brief definitions and references to the principal symbols used in the book. The reader is advised to study this glossary before reading the book. It may be helpful to note a few of the symbols here. The symbol $\triangleq$ stands for "is defined to be" or "represents" or "denotes." For example, F $\triangleq$ force, or F($\triangleq$ force), means "F denotes force." The symbol $\Rightarrow$ stands for "implies." The symbols $\forall$ and $\exists$ mean, respectively, "for all" and "there exists." For example, $\forall t$ means "for all t." The symbol $\triangleleft$ indicates the end of discussion, and the abbreviations R.H.S. and Q.E.D. stand, respectively, for "right-hand side" and "*quod erat demonstrandum.*"

Contents

Language, notation and basic concepts of system theory

1

1 *Introduction*

This is an introductory chapter. As such, it contains few concrete results in the form of theorems, formulae, etc., which abound in the rest of this book. Essentially, the main function of this chapter is to provide a conceptual and notational framework in terms of which the techniques of linear system theory can be presented in a clear and concise fashion.

In the initial stages of our presentation, there will of necessity be some concepts which cannot be defined precisely—indeed, not even explained in completely unequivocal terms. This applies, for example, to the concepts of measurable attribute, physical object, and experimenter. Such concepts, and, in particular, the notion of an "object," constitute the primitive concepts of system theory. In later stages, most of the concepts introduced will be nonprimitive in nature, and they will be defined—though not always in a formal fashion—in terms of the primitive concepts introduced earlier.

In the interest of attaining maximum clarity, we shall not hesitate to depart—whenever it is expedient—from traditional modes of exposition. In particular, we shall frequently interrupt the presentation with comments, notes, remarks, recapitulations, digressions, examples, and exercises. We shall also frequently resort to the pedagogically helpful technique of preceding the formal definition of a term X with an informal explanation whose function is to provide an insight into the significance of X rather than to give X a mathematically precise meaning.

We begin with an informal discussion of the central notion in this book, the notion of a system. A more precise definition of the term "system" will be given in Sec. *10* at the end of this chapter.

2 System analysis: an example

According to Webster's dictionary, a system is "a collection of objects united by some form of interaction or interdependence." Since we shall be concerned only with the quantitative aspects of system behavior, the notion of a system will have for us a narrower meaning. Specifically, we shall assume that each object which is part of a system is characterized by a finite number of measurable attributes[1] and that the interaction between such objects as well as the interdependence between the attributes of each object can be expressed in some well-defined mathematical form.

To introduce in an informal way some of the basic concepts of system theory, let us consider a very simple mechanical system such as that shown in Fig. *1.2.1*, which comprises two particles M_1 and M_2, together with a spring S_3 linking them. M_1 and M_2 are constrained to move along the OX axis, and M_1 is subjected to an external force F which varies with time in a manner specified by a given time function f.

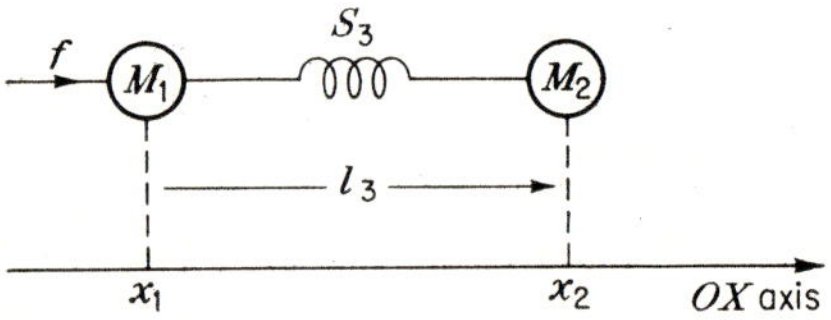

Fig. 1.2.1 Example of a system.

Let us assume for a moment that the analysis of this system, call it $\mathcal{S}$, is presented as a problem to one who is unfamiliar with the laws governing the motion of mechanical systems. What are the questions which will arise in his mind? Among them should be:

What attributes of the objects of which $\mathcal{S}$ is comprised need be considered?

What are the mathematical relations between the relevant attributes of each object in $\mathcal{S}$?

What are the mathematical relations between the attributes of different objects in $\mathcal{S}$; in other words, what are the relations representing the *interactions* of objects in $\mathcal{S}$?

These are essentially the three basic questions which arise in the analysis of most types of system, and learning how to answer them constitutes a large part of the training one receives as an engineer or a

[1] By a *measurable attribute* we mean a characteristic which can be expressed in terms of one or more real or complex numbers. For example, height (a real number), velocity (a real vector), voltage (a real or complex number), impedance (a complex number or a function of frequency), and impedance matrix (a set of complex numbers or complex functions of frequency) are measurable attributes. On the other hand, such attributes as smell, taste, intonation, and beauty are much too subjective and fuzzy to admit of measurement.

physicist. For example, one who is conversant with the analysis of mechanical systems would know that in the case of the system $\mathcal{S}$ the relevant attributes of a particle M are its mass, position, velocity, acceleration, and applied force, while the relevant attributes of a spring are its spring constant, force-free length, and length under tension (or compression). Thus, if the masses are denoted by m (with appropriate subscripts, for example, m_1 is the mass of M_1), the forces by F, the positions by x, the velocities by $\dot{x}$ ($\triangleq dx/dt$), and the spring constant by k, then we have the following three relations:

Relation between the attributes of M_1:

1 $$F_1 = m_1\ddot{x}_1$$

Here F_1 is the total force acting on M_1, with F_1 positive along the positive direction of the OX axis; m_1 is the mass of M_1; and $\ddot{x}_1$ is the second derivative of x_1. Note that *1* relates F_1, m_1, and x_1 through $\ddot{x}_1$ rather than directly.

Relation between the attributes of M_2:

2 $$F_2 = m_2\ddot{x}_2$$

Here F_2 represents the total force acting on M_2, m_2 is the mass of M_2, and $\ddot{x}_2$ is the acceleration of M_2.

Relation between the attributes of S_3:

3 $$F_3 = k(l_3^0 - l_3)$$

Here l_3^0 denotes the length of the spring S_3 when no force is applied to S_3, l_3 is the length of S_3 when an external force F_3 (tending to reduce the length of S_3) is applied, and k is the spring constant ($k > 0$).

4 *Comment* The relation between the applied force and the length of a physical spring is in reality much more complicated than *3*. In effect, *3* may be regarded as a first approximation to the true relation between F_3 and l_3; or alternatively, one may regard S_3 as an idealized spring for which *3* is an exact relation between F_3 and l_3. The results of an analysis in which *3* is used will be approximate or exact depending on whether the first or the second viewpoint is used.

Next, we have the relations representing interactions. These are

5 $$l_3 = x_2 - x_1$$

which means that the length of S_3 is determined by the positions of M_1 and M_2;

6 $$F_1 = f - F_3$$

which means that the total force acting on M_1 is the difference between the externally applied force f and the force exerted on M_1 by the spring S_3; and

7 $$F_2 = F_3$$

which means that the total force acting on M_2 is the force exerted on M_2 by S_3.

In sum, we have six equations involving nine attributes of objects of which $\mathcal{S}$ is comprised, namely, the mass of M_1, the force acting on M_1, the acceleration of M_1, the mass of M_2, the force acting on M_2, the acceleration of M_2, the force acting on S_3, the spring constant, and the length of S_3. These equations can be divided into two groups: Eqs. *1* to *3*, which represent the relations between the attributes of the same object, and Eqs. *5* to *7*, which represent the interactions, i.e., the relations between the attributes of different objects.

A typical problem in connection with the system under consideration might be the determination of the position of M_1 ($\triangleq x_1$) as a function of time, given the external force acting on M_1 ($\triangleq f$) as a function of time. Another way of stating this problem is to say that F is the input to $\mathcal{S}$, that is, F is an attribute of $\mathcal{S}$ which is varied directly by an "experimenter," and that x_1 is the corresponding output, that is, x_1 is an attribute which varies as a result of variation in F and which is observed but not varied directly by the experimenter. To determine x_1 in terms of F, it is necessary to solve Eqs. *1* to *3* and *5* to *7* for x_1 as a function of time, given f as a function of time. This entails the elimination of unwanted attributes $\ddot{x}_1$, x_2, $\ddot{x}_2$, and l_3 from the equations in question. These unwanted variables are what will later be called the *suppressed outputs* of the system (see Sec. *4*).

This brief analysis of the simple mechanical system of Fig. *1.2.1* clearly reveals certain basic problems which appear in one guise or another in the analysis of any system, regardless of its nature, composition, and degree of complexity. These are:

The identification of the relevant attributes of the objects of which the system is comprised

The characterization of the relations between these attributes

The representation of interactions between different objects in terms of interrelations between their attributes

The formation of a system of relations between the attributes of the system as a whole

The determination of relations between the attributes varied by the experimenter (the inputs) and the attributes which are observed but not varied directly (the outputs)

These problems—as well as certain other aspects of system analysis and synthesis which are not placed in evidence by the example just considered—are dealt with in system theory on an abstract rather than a physical level. Thus, what matters in system theory is not the physical identity of the attributes associated with a system but the mathematical relations between the attributes. This is the viewpoint which we shall adopt in this book.

Although the present text is concerned in the main with the theory of linear systems, it will be helpful to begin our exposition on a more general level and define the basic concepts of system theory in a way which makes them applicable to nonlinear as well as linear systems. This is particularly true of such basic concepts as state, input, output, input-output relation, state and system equivalence, and interconnection.

In the sections that follow we shall concern ourselves primarily with terminology and notation. Generally, our exposition will have the pattern of an informal discussion terminating in formal definitions which serve to summarize the preceding discussion. For convenient reference, the definitions, comments, notations, equations, etc., are numbered consecutively within each section.

We begin our discussion with time functions, since in general the measurable attributes of an object are time-dependent quantities.

3 *Time functions*

Notation for time functions

Let v denote a measurable attribute of an object $\mathfrak{A}$. Unless stated to the contrary, v will be understood to represent a time function, with the value of v at time t denoted by $v(t)$. Furthermore, it will be understood—unless otherwise indicated—that t ranges over the entire real line $(-\infty, \infty)$, which will be referred to as the *time axis*. (Note that the real line does not include $-\infty$ and ∞.)

If the values which t can assume are restricted to a specified subset T of the time axis, then we write (in the usual mathematical notation) $t \in T$, where $\in$ stands for "belongs to." Typically, T may be a semi-infinite interval $[t_0, \infty)$, where t_0 denotes a particular value of t, or a finite interval $[a,b]$, or a set of equispaced instants of time $(0,h,2h,3h, \ldots)$, where h is the interval between two successive values of t. In the latter case t is a discrete variable, whereas in the former examples t is continuous.

1 *Note* In the sequel we shall frequently be making statements which are subject to the qualification "for all values of t belonging to the set T." To simplify such statements, we shall say "for all t," meaning "for all t in T." When no confusion is likely to arise, we shall use the same convention for variables other than t. For example, "for all x" will mean "for all values of x in the range of x." Furthermore, we shall use the standard symbol $\forall$ (universal quantifier) for "for all" and $\exists$

(existential quantifier) for "there exists." When it is expedient to do so and there is no danger of misinterpretation, we shall place the quantifiers on the right side rather than the left side of an equation.

2 *Example* $f(t) = g(t) \quad \forall t$ has the same meaning as $\forall t[f(t) = g(t)]$, which in turn means "for all values of t in T, $f(t)$ is equal to $g(t)$." $\exists t[f(t) = g(t)]$ means "there exists a value of t in T for which $f(t) = g(t)$." $\forall \alpha \exists t[f(t,\alpha) = g(t,\alpha)]$ means "for all values of α there exists a value of t (*depending on* α) such that $f(t,\alpha) = g(t,\alpha)$." ◁

3 **Remark** In dealing with a time function—or for that matter any function—it is important to distinguish between the time function as such and the value which it assumes at some particular instant of time. More specifically, when we say "a time function v" or, more particularly, "a time function v defined on a set T," we mean the entire curve (graph) of v versus t as t runs through the set T, or, more precisely, the set of pairs $\{(t,v(t))\}$, $t \in T$. On the other hand, when we write $v(t)$, we mean the value which v assumes at the instant t. Unfortunately, it is standard practice to write $v(t)$ when one actually means v, and this gives rise to considerable confusion. [The distinction between v and $v(t)$ is illustrated in Fig. *1.3.1*, where for simplicity it is assumed that $v(t)$ is a real-valued variable for each t in $(-\infty, \infty)$.]

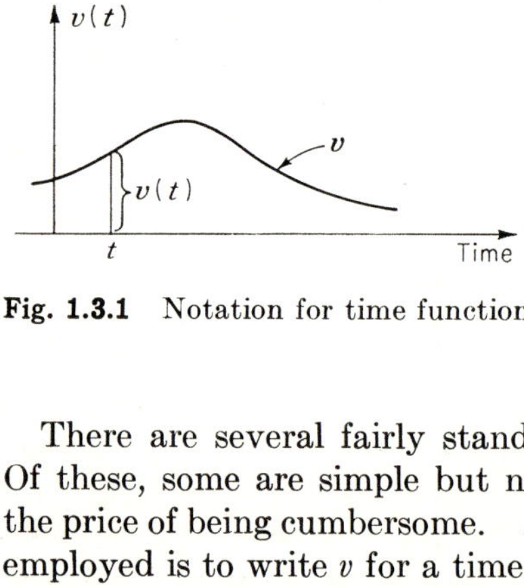

Fig. 1.3.1 Notation for time functions.

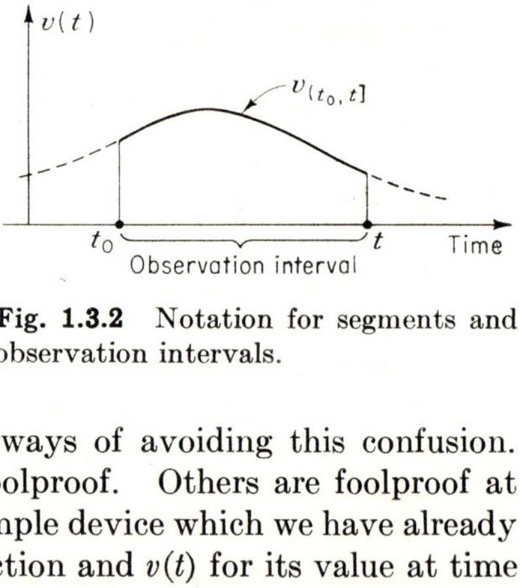

Fig. 1.3.2 Notation for segments and observation intervals.

There are several fairly standard ways of avoiding this confusion. Of these, some are simple but not foolproof. Others are foolproof at the price of being cumbersome. A simple device which we have already employed is to write v for a time function and $v(t)$ for its value at time t. Another is to use a dot in place of t to denote a time function; for example, $v(\cdot)$ is a time function and $v(t)$ is its value at time t. To denote a segment of a time function u over an interval, say, $[t_0,t_1]$, we shall use the symbol $u_{[t_0,t_1]}$, meaning that $u_{[t_0,t_1]}$ is the totality of pairs $\{(t,u(t)\}$, with $t_0 \le t \le t_1$. The interval $[t_0,t_1]$ will be referred to as the *observation interval*. Usually, the observation interval will be a semiclosed interval $(t_0,t_1]$ or $[t_0,t_1)$. (See Fig. *1.3.2* for illustration.) Strongly entrenched traditions will force us on occasion to depart from this agreement. For example, following standard practice, a sine function will be written as sin t rather than sin $(\cdot)$ or sin. Similarly, a delta function will be written as $\delta(t)$ rather than δ or $\delta(\cdot)$. Also, the Laplace transform of a time function $f(\cdot)$ will be written as $F(s)$ rather than $F(\cdot)$ or F. Indeed, in sections in which delta functions or Laplace transforms are used extensively, we shall have to rely on the context to indicate whether a symbol such as $v(t)$ should be interpreted as the time function v or as the value of v at time t.

Range of a time function

In general, our concern will be not so much with a single time function v as with a class of time functions $\{v\}$ of which v is an element. (The braces denote a class or a set.) When this is the case, v may be regarded as a variable whose range is the class $\{v\}$. For convenience, the range of v will be denoted by $R[v]$, where R stands for range. Thus, $R[v]$ is a class (space, set) of time functions to which v is constrained to belong.

When v varies over $R[v]$, the values of v at a fixed time t vary over a set which is the *range* of the variable $v(t)$. The range of $v(t)$ will be denoted by $R[v(t)]$.

In general, the range of $v(t)$ may be different from the range of $v(t')$, $t' \neq t$. However, in all of the cases with which we shall be concerned, $R[v(t)]$ is independent of t. Thus, it will be understood, unless stated to the contrary, that the range of $v(t)$ is the same for all values of t in T.

Typically, we shall deal with the following special cases.

$v(t)$ is a real scalar ranging over the real line $(-\infty, \infty)$. (Scalars are identified by lightface type.) In this case, $R[v(t)]$ is the real line; that is, $R[v(t)] = \mathcal{R}^1$ ($\mathcal{R}^1 \triangleq$ space of real numbers).

$v(t)$ is a complex scalar, with the range of $v(t)$ being $\mathcal{C}^1$, the space of complex numbers.

$\mathbf{v}(t)$ is an n-vector (identified by boldface type), with the range of $\mathbf{v}(t)$ being $\mathcal{R}^n$, the space of n-tuples of real numbers. [That is, $\mathcal{R}^n$ is the set of all ordered n-tuples $(\rho_1, \ldots ,\rho_n)$, where $\rho_1, \ldots, \rho_n$ are real numbers.]

$\mathbf{v}(t)$ is an n-vector, with the range of $\mathbf{v}(t)$ being $\mathcal{C}^n$, the space of n-tuples of complex numbers. [$\mathcal{C}^n$ is the set of all ordered n-tuples $(x_1, \ldots ,x_n)$, where $x_1, \ldots, x_n$ are complex numbers.]

$\mathbf{V}(t)$ is a matrix, with the elements of $\mathbf{V}(t)$ ranging over $\mathcal{R}^1$ or $\mathcal{C}^1$. (Matrices are denoted by boldface uppercase symbols.)

$v(t)$ ranges over a finite set of numbers or letters. For example, $v(t)$ can be any one of the letters of the English alphabet or an integer between, say, -10 and $+10$. In this case, $R[v(t)]$ is a finite set. (When $R[v(t)]$ is a finite set, it is customary to call it an *alphabet* even though its elements may not be letters.)

4 **Remark** In later stages of our discussion, where $\mathbf{v}(t)$ will have a concrete meaning such as input, output, or state, $R[\mathbf{v}(t)]$ will be referred to in a way which will place in evidence the significance of $\mathbf{v}(t)$. Thus, if $\mathbf{v}(t)$ is the input at time t, then $R[\mathbf{v}(t)]$ will be called the *input space.* Similarly, $R[\mathbf{v}(t)]$ will be called the *output space* if $\mathbf{v}(t)$ is the output at time t, and $R[\mathbf{v}(t)]$ will be referred to as the *state space* when $\mathbf{v}(t)$ is the state at time t. ◁

It is important to have a clear understanding of the distinction between the range of $\mathbf{v}(t)$, $R[\mathbf{v}(t)]$, and the range of $\mathbf{v}$, $R[\mathbf{v}]$. The former is a set such as $\mathcal{R}^n$, whereas the latter is a set of time functions, i.e., a function space. This distinction is illustrated by the following examples.

5 *Example* Suppose that t ranges over the interval [0,1] and that $\mathbf{v}(t)$ is a 2-vector with components $v_1(t)$ and $v_2(t)$. Assume that the constraints on v_1 and v_2 are (I) $|v_1(t) + v_2(t)| \leq 1$ for all t in [0,1] and (II) $\int_0^1 |v_1(t)|\, dt \leq 2$. In this case, $R[\mathbf{v}(t)]$ is a subset of $\mathcal{R}^2$ (space of pairs of real numbers) defined by (I) and $R[\mathbf{v}]$ is the space of all vector-valued time functions $\mathbf{v}$ defined on [0,1] which satisfy the constraints (I) and (II).

6 *Example* Suppose that T is the set $\{0,1,2, \ldots\}$ and $R[v(t)]$ is the set $\{0,1\}$ (for example, v is a binary sequence). Assume that the successive values of $v(t)$ are constrained by the condition $v(t) + v(t + 1) \leq 1$ for all t. Then $R[v]$ is the set of all binary sequences starting at $t = 0$ in which 1 does not occur twice in a row.

7 **Remark** In most of the problems with which we shall deal in this text, a typical $\mathbf{v}$ will be a vector-valued time function which (I) may contain at most a finite number of delta functions[1] of various orders over any finite interval and (II) is piecewise continuous and has piecewise continuous derivatives of all finite orders on every finite interval over which it has no delta functions. Such a function will be said to be *finitely differentiable in the distribution sense.* Thus, it will be assumed, unless otherwise specified, that $R[\mathbf{v}(t)]$ is $\mathcal{R}^n$ (space of n-tuples of real numbers) and that $R[\mathbf{v}]$ is the space of all vector-valued time functions which are finitely differentiable in the distribution sense.

Usually, a symbol such as $\mathbf{v}$ will be used to denote not the entire time function but a segment of it over an observation interval such as $[t_0,t_1]$. Correspondingly, $R[\mathbf{v}]$ or, more explicitly, $R[\mathbf{v}_{[t_0,t_1]}]$ will denote the set (space) of such segments.

[1] We assume that the reader is familiar with the elementary properties of delta functions, Laplace transforms, and linear vector spaces. Brief treatments of these topics are presented in Appendixes *A*, *B*, and *C*, respectively. The notion of a derivative in the distribution sense is discussed in Appendix *A* and, more sketchily, in Chap. *3*, Sec. *8*.

In most practical problems, neither $R[\mathbf{v}(t)]$ nor $R[\mathbf{v}]$ can be specified in precise terms. For example, suppose that v represents the input to an electronic amplifier. In this case the restrictions on v are rather fuzzy. For instance, $R[v]$ might be specified to be a class of time functions whose amplitude is not much in excess of 10 volts and whose bandwidth is not much in excess of 20,000 cps. This implies that $R[v(t)]$ is a set of numbers which are not much greater than 10 in magnitude. Note that neither $R[v]$ nor $R[v(t)]$ is a well-defined set.

4 *Physical and abstract objects*

Terminal variables and relations

We proceed to introduce in an informal fashion some of the basic notions which are common to systems of all types. The first such notion is that of a *physical object,* by which is usually meant a physical device associated with a number of attributes which are relevant to the purpose for which the device is used. Thus, in the example considered in Sec. *2*, the spring S_3 is an object whose relevant attributes are its spring constant k, its length l_3, its length in the absence of external forces l_3^0, and the external force F_3 applied to S_3. The cost of the spring, its chemical composition, etc., are attributes which are not relevant to the performance of the system $\mathcal{S}$ considered in the example. In the sequel, we shall always assume—unless explicitly stated to the contrary—that we have a priori knowledge of what the relevant attributes of the objects under consideration are and that we have means of measuring them.

As was pointed out in Sec. *1*, what matters in system theory is not the *physical,* but the *mathematical* identity of the attributes of an object and the relations between them. Thus, to a system theorist an object $\mathcal{A}$ is essentially an abstract entity associated with a set of attributes $v_1, v_2, \ldots$ —which in general are numbers or arrays of numbers— which is characterized by the relations between these attributes. Thus, in effect, an *abstract object,* or simply an *object,* is a set of variables together with a set of relations between them.

Before proceeding to enlarge on this informal definition and illustrate its meaning with examples, we shall have to introduce some auxiliary terminology and notations. Specifically, the variables associated with an object $\mathcal{A}$ will be called the *terminal variables* of $\mathcal{A}$ and will be denoted by $v_1, v_2, \ldots$ or other lowercase Latin or Greek letters, with or without subscripts or superscripts, as the need may dictate. The relations between the terminal variables will be referred to as the *terminal variable relations* or simply *terminal relations* for $\mathcal{A}$. An object $\mathcal{A}$ will be said to be *characterized* by its terminal relations, which will be written in the symbolic form

1
$$\begin{aligned} \mathcal{A}^{(1)}(v_1, \ldots, v_n) &= 0 \\ \mathcal{A}^{(2)}(v_1, \ldots, v_n) &= 0 \\ \cdots\cdots\cdots \\ \mathcal{A}^{(m)}(v_1, \ldots, v_n) &= 0 \end{aligned}$$

where each $\mathcal{A}^{(j)}$, $j = 1, \ldots, m$, represents a relation between the v_i, $i = 1, \ldots, n$. In the sequel, our attention will be restricted to objects for which both n and m are finite numbers. To denote an object, we shall usually use letters such as $\mathcal{A}$, $\mathcal{B}$, $\mathcal{C}$, often with subscripts and sometimes with superscripts.

The terminal variables are assumed to vary with time, or, for short, to be *time functions*. Thus, unless stated to the contrary,

$$v_i \qquad i = 1, \ldots, n$$

will represent a time function, with the value of v_i at time t denoted by $v_i(t)$. For simplicity, we assume that the $v_i(t)$ are scalar-valued. For convenience, the $v_i(t)$ are identified with the components of an n-vector $\mathbf{v}(t) = (v_1(t), \ldots, v_n(t))$ which will be referred to as the *terminal variable vector* or the *terminal function vector*.

2 *Example* Let $\mathcal{A}$ be an abstract object associated with four terminal variables v_1, v_2, v_3, and v_4, which are related to one another by the differential equations

$$v_3 = a_{11}\frac{dv_1}{dt} + a_{12}\frac{dv_2}{dt}$$

$$v_4 = a_{21}\frac{dv_1}{dt} + a_{22}\frac{dv_2}{dt}$$

where $a_{11}, \ldots, a_{22}$ are constants. These equations constitute the terminal relations which characterize $\mathcal{A}$.

Oriented and nonoriented objects

So far we have not made any mention of inputs and outputs. If an abstract object $\mathcal{A}$ is characterized by relations of the form *1* without any specification as to which variables are the inputs (causes) and which are the outputs (effects), then $\mathcal{A}$ will be said to be *nonoriented*. If, on the other hand, the terminal variables are divided into two categories—input variables and output variables—playing the respective roles of independent and dependent variables, then $\mathcal{A}$ will be said to be *oriented*. Such a division can be made arbitrarily (subject to conditions discussed in Sec. *6*) and carries no implication of physical causality between the inputs and outputs.

The terminal relations *1* can arise from two sources. First, they can be postulated without any reference to physical objects; second, they can be the result of measurements (experiments) performed on a physical object $\mathcal{P}$.

More specifically, consider a physical object $\mathcal{P}$ which is associated with a set of variable attributes $v_1, \ldots, v_n$. Suppose that an experimenter selects a subset of these attributes, say, $v_1, \ldots, v_k$, and proceeds to vary them with time, starting at $t = t_0$ and ending at $t = t_1$ in a manner specified by the time functions $u_1(\cdot), \ldots, u_k(\cdot)$, with $v_i(t) = u_i(t)$, $i = 1, \ldots, k$, $t_0 < t \leq t_1$. Furthermore, suppose that the experimenter observes the resulting variations $y_1(\cdot), \ldots, y_m(\cdot)$ in, say, the attributes $v_{k+1}, \ldots, v_{k+m}$, respectively, and disregards the remaining attributes $v_{k+m+1}, \ldots, v_n$. In such an experiment, the time functions $u_1, \ldots, u_k$ (starting at t_0 and ending at t_1) are the *inputs*, the time functions $y_1, \ldots, y_m$ (starting at t_0 and ending at t_1) are the *outputs* or the *responses*, and the variables $v_{k+m+1}, \ldots, v_n$ are the *suppressed output variables.* For convenience, the k-tuple $(u_1, \ldots, u_k)$ will be referred to as the *input vector* and will be denoted by $\mathbf{u}$. Similarly, the m-tuple $(y_1, \ldots, y_m)$ will be referred to as the *output vector* and will be denoted by $\mathbf{y}$. The $(n\text{-}k\text{-}m)$-tuple $v_{k+m+1}, \ldots, v_n$ will be referred to as the *suppressed output vector* and will be denoted by $\tilde{\mathbf{y}}$. Note that $\mathbf{u}$ and $\mathbf{y}$ are abbreviations for the segments $\mathbf{u}_{(t_0,t_1]}$ and $\mathbf{y}_{(t_0,t_1]}$, respectively. (In later discussions, the observation interval will sometimes be taken to be $[t_0,t_1]$ or $[t_0,t_1)$ (see *1.5.5*).)

3 **Remark** Since $\mathbf{u}$ represents the input to $\mathcal{P}$, the range of $\mathbf{u}(t)$ (see 1.3.4) will be referred to as the *input space* of $\mathcal{P}$ and will be denoted by $R[\mathbf{u}(t)]$. (Unless stated otherwise, it will be assumed that $R[\mathbf{u}(t)]$ is independent of t.) Correspondingly, the range of the time function $\mathbf{u} \triangleq \mathbf{u}_{(t_0,t_1]}$ will be referred to as the *input segment space* of $\mathcal{P}$ and will be denoted by $R[\mathbf{u}]$. Note that $R[\mathbf{u}]$ depends on two parameters, t_0 and t_1, with t_0 and t_1 varying over T. In effect, $\mathcal{P}$ is associated not with a single input segment space $R[\mathbf{u}]$, but with a family $\{R[\mathbf{u}]\}$ which is generated by the parameters t_0 and t_1. The same comment applies to $R[\mathbf{y}]$. Thus, it should be understood that we are using somewhat loose language when we refer to $R[\mathbf{u}]$ as *the* input segment space of $\mathcal{P}$.

It should also be understood that, in cases in which v is a time function defined on, say, $(-\infty, \infty)$, a constraint on v of the form $\int_{-\infty}^{\infty} v^2(t)\, dt \leq 1$ does not induce a corresponding constraint on a segment $v_{[t_0,t_1]}$, whereas a constraint of the form $|v(t)| \leq 1$ for $-\infty < t < \infty$ induces the same constraint on $v_{[t_0,t_1]}$. Thus, in the latter case $R[v_{[t_0,t_1]}]$ is the space of all time functions on $[t_0,t_1]$ satisfying $|v(t)| \leq 1$ for $t_0 \leq t \leq t_1$, whereas in the former case $R[v_{[t_0,t_1]}]$ is merely the space of segments $v_{[t_0,t_1]}$ of time functions v defined on $(-\infty,\infty)$ which satisfy the condition

$$\int_{-\infty}^{\infty} v^2(t)\, dt \leq 1$$

When $R[\mathbf{u}]$ and $R[\mathbf{y}]$ are not specified in explicit terms, it will be understood that they are spaces of vector-valued time functions (defined over a finite or infinite observation interval) which are finitely differentiable in the distribution sense (see *1.3.7*). ◁

By varying the input vector $\mathbf{u}$ over all the values which it is allowed to assume, i.e., over $R[\mathbf{u}]$, and observing the resulting output vector $\mathbf{y}$, the experimenter can derive, at least in principle, a set of relations between the y_j and the u_i which can be expressed in the symbolic form

4
$$\begin{aligned} \mathcal{P}^{(1)}(u_1, \ldots, u_k, y_1, \ldots, y_m) &= 0 \\ \cdots\cdots\cdots\cdots\cdots\cdots \\ \mathcal{P}^{(m)}(u_1, \ldots, u_k, y_1, \ldots, y_m) &= 0 \end{aligned}$$

or more compactly,

5
$$\mathcal{P}(\mathbf{u},\mathbf{y}) = 0$$

where $\mathbf{u}$ and $\mathbf{y}$ range over $R[\mathbf{y}]$ and $R[\mathbf{y}]$, respectively. In the sequel, a relation such as *5* will be referred to as an *input-output relation* for $\mathcal{P}$. Note that the main difference between an *input-output relation* and the *terminal relations* (*1*) of $\mathcal{P}$ is that in the former some of the terminal variables are designated as inputs and others as outputs, whereas in the latter no such designations are made.

6 *Example* Suppose that an experiment of the kind described above is performed on an electrical capacitor $\mathcal{P}$, with the voltage v across the capacitor being the output and the current i through the capacitor being the input. The result of the experiment would be the empirical input-output relation

7
$$C \frac{dv}{dt} = i$$

where C is a constant parameter. In symbolic form, this relation would be written as

8
$$\mathcal{P}(i,v) = 0$$

Observe that *7* does not define v uniquely as a function of i. Rather, v is determined to within an additive constant (constant of integration) which represents the voltage across $\mathcal{P}$ at the beginning of the experiment. Furthermore, *7* remains unchanged when the roles of i and v as input and output are interchanged. Thus *7* may also be regarded as a terminal relation for $\mathcal{P}$, with $\mathcal{P}$ playing the role of a nonoriented object.

Abstract model and physical realization

With reference to *4* and *5*, consider an abstract object $\mathcal{A}$ whose terminal variables are $u_1, \ldots, u_k, y_1, \ldots, y_m$ and which is characterized by an input-output relation

9 $$\mathcal{P}(\mathbf{u},\mathbf{y}) = 0$$

in which $\mathbf{u} \triangleq \mathbf{u}_{(t_0,t_1]}$ and $\mathbf{y} \triangleq {}_{(t_0,t_1]}$, with t_0 and t_1 ranging over T. Such an object $\mathcal{A}$ will be called an *abstract oriented model* of $\mathcal{P}$ and will be denoted by $\mathcal{A}(\mathcal{P})$. Conversely, $\mathcal{P}$ will be called a *physical realization* of $\mathcal{A}$ and will be denoted by $\mathcal{P}(\mathcal{A})$. In effect, an abstract model of $\mathcal{P}$ is merely an abstract object $\mathcal{A}$ whose input-output relation coincides with that of $\mathcal{P}$.

It is obvious that every physical object has an abstract oriented model. However, the converse is not always true, which implies that there exist abstract oriented as well as nonoriented objects which do not have a physical realization. If $\mathcal{A}$ is physically realizable, i.e., if $\mathcal{A}$ has a physical realization, then, in general, it can be realized in a variety of physical forms $\mathcal{P}_1, \mathcal{P}_2, \ldots$. In this case, $\mathcal{A}(\mathcal{P}_1)$, $\mathcal{A}(\mathcal{P}_2)$, . . . will all represent the same abstract object $\mathcal{A}$.

We proceed to illustrate these points by simple examples.

10 *Example* Suppose that $\mathcal{A}$ is a nonoriented object with terminal variables v_1 and v_2 connected by the terminal relation

$$v_1(t) = jv_2(t) \qquad \text{for all } t,\ j = \sqrt{-1}$$

It is clear that $\mathcal{A}$ does not admit of physical realization, since either $v_1(t)$ or $v_2(t)$ or both must be complex-valued and hence cannot be identified with the attributes of a physical object. (This implies that we stipulate that the attributes of a physical realization of $\mathcal{A}$ be real-valued.)

11 *Example* Let $\mathcal{P}$ be a particle with mass m. Assume that the variable attributes of $\mathcal{P}$ are the applied force F (along the OX axis), the position of $\mathcal{P}$ ($\triangleq x$) on the OX axis, the velocity of $\mathcal{P}$ ($\triangleq \dot{x}$), and the acceleration of $\mathcal{P}$ ($\triangleq \ddot{x}$). Suppose that the experimenter varies F, observes $\ddot{x}$, and suppresses x and $\dot{x}$. Let $F \triangleq u$, $\ddot{x} \triangleq y$, and $\mathcal{A} \triangleq$ abstract oriented model of $\mathcal{P} \triangleq \mathcal{A}(\mathcal{P})$.

The input-output relation for $\mathcal{P}$ is

12 $$m\ddot{x} = F$$

Correspondingly, the input-output relation for $\mathcal{A}(\mathcal{P})$ is

13 $$my = u$$

Now, if we start with an abstract nonoriented object $\mathcal{A}$ which is characterized by *13*, then $\mathcal{P}$ is a physical realization of $\mathcal{A}$, with $\ddot{x}$ and

F identified with y and u, respectively. Another physical realization is yielded by a resistor in which y is identified with the current through the resistor, u is identified with the voltage across the resistor, and $m = R$ ($\triangleq$ resistance). Still another realization (this time an approximate one) is an amplifier with input voltage y, output voltage u, and amplification factor m. The point of the latter realization is that it shows that a terminal variable of $\mathcal{A}$, say, u, can correspond to an input in one realization (e.g., the force applied to a particle $\mathcal{P}$) and to an output in another realization (e.g., the output voltage of an amplifier). In other words, whether a particular terminal variable v_i of a nonoriented abstract object $\mathcal{A}$ is designated as an input variable or an output variable does not depend on any intrinsic properties of $\mathcal{A}$.

14 *Example* $\mathcal{A}$ (a nonoriented object) is associated with two terminal variables v_1 and v_2 which range over real-valued time functions and which are related by the differential equation

15 $$\frac{dv_2}{dt} = \frac{d^2v_1}{dt^2} + v_1$$

In this case, $\mathcal{A}$ can be realized in a variety of ways. A realization involving an electrical network is shown in Fig. *1.4.1*. Here v_1 plays the role of input voltage and v_2 is the current flowing into the network.

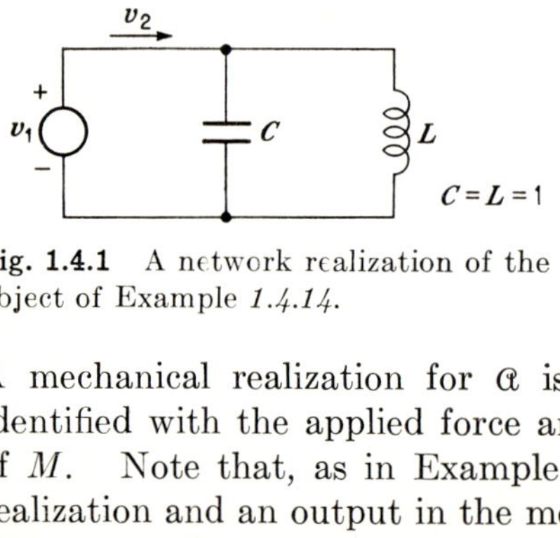

Fig. 1.4.1 A network realization of the object of Example *1.4.14*.

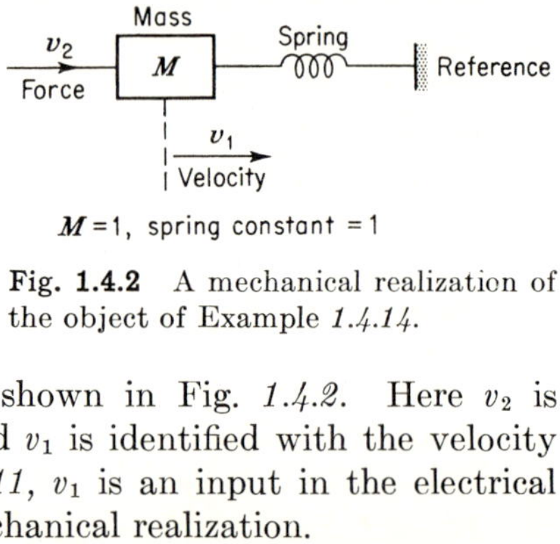

Fig. 1.4.2 A mechanical realization of the object of Example *1.4.14*.

A mechanical realization for $\mathcal{A}$ is shown in Fig. *1.4.2*. Here v_2 is identified with the applied force and v_1 is identified with the velocity of M. Note that, as in Example *11*, v_1 is an input in the electrical realization and an output in the mechanical realization.

16 *Comment* The notion of a nonoriented abstract object helps to clear away some of the confusion surrounding the notion of causality. Specifically, much of this confusion stems from the lack of distinction between a physical object $\mathcal{P}$ and its abstract model $\mathcal{A}(\mathcal{P})$. We saw that, in general, there is no implication of physical causality in the designation of a subset of terminal variables of $\mathcal{A}$ as inputs and the rest as outputs. Furthermore, we have noted that if $\mathcal{A}$ is nonoriented, then a terminal variable of $\mathcal{A}$, say, v_1, may be an input variable in one realization of $\mathcal{A}$ and an output variable in another realization. Also, $\mathcal{A}$ may be physically realizable as a nonoriented object and yet may not

have a physical realization corresponding to a particular orientation. For example, if $\mathcal{A}$ is characterized by the relation $v_2(t) = v_1(t + 1)$ for all t, then $\mathcal{A}$ is realizable as a "perfect delay line" with unity delay if v_2 is identified with the input and v_1 with the output of the delay line. On the other hand, if $\mathcal{A}$ is oriented, with v_1 designated as the input and v_2 as the output, then $\mathcal{A}$ has no physical realization, since $\mathcal{P}(\mathcal{A})$ would be a perfect predictor of the future—a physical impossibility.

Formal definition of an abstract object

Up to this point, our discussion of the notion of an abstract object has been informal in nature, its intent being that of preparing the way for the formal definition given in the sequel. Specifically, consider an observation interval such as $[t_0,t_1]$ (or $(t_0,t_1]$ or (t_0,t_1)). Let $(\mathbf{u},\mathbf{y})$, $\mathbf{u} \triangleq \mathbf{u}_{[t_0,t_1]}$, $\mathbf{y} \triangleq \mathbf{y}_{[t_0,t_1]}$, be an ordered pair of time functions defined on this interval and let $R_{[t_0,t_1]} \triangleq \{(\mathbf{u},\mathbf{y})\}$ be a set (space) of such pairs for fixed t_0 and t_1, with $t_0,t_1 \in T$ [$T \triangleq$ range of t (Sec. *3*)]. Furthermore, let $\{R_{[t_0,t_1]}\}$ be a family of such sets generated by varying t_0 and t_1 over T, with $t_1 \geq t_0$. In terms of these symbols, the definition of an oriented abstract object may be formulated as follows:

17 **Definition** An *oriented abstract object* $\mathcal{A}$ is a family $\{R_{[t_0,t_1]}\}$, $t_0,t_1 \in T$, of sets of ordered pairs of time functions, with a generic pair denoted by $(\mathbf{u},\mathbf{y})$, in which the first component, $\mathbf{u} \triangleq \mathbf{u}_{[t_0,t_1]}$, is called an *input segment,* or, simply, an *input,* and the second component, $\mathbf{y} \triangleq \mathbf{y}_{[t_0,t_1]}$, is called an *output segment,* or, simply, an *output.* A pair of time functions $(\mathbf{u},\mathbf{y})$, $\mathbf{u} \triangleq \mathbf{u}_{[t_0,t_1]}$, $\mathbf{y} \triangleq \mathbf{y}_{[t_0,t_1]}$, is an *input-output pair belonging to* $\mathcal{A}$, written as $(\mathbf{u},\mathbf{y}) \in \mathcal{A}$, if $(\mathbf{u},\mathbf{y}) \in R_{[t_0,t_1]}$ for some t_0,t_1 in T. Thus, an oriented abstract object $\mathcal{A}$ can be identified with the totality of input-output pairs which belong to $\mathcal{A}$.

The members of the family $\{R_{[t_0,t_1]}\}$ are required to satisfy the consistency condition: If $(\mathbf{u}_{[t_0,t_1]},\mathbf{y}_{[t_0,t_1]})$ is an input-output pair belonging to $\mathcal{A}$, then so is any section of this pair, i.e., any pair of the form $(\mathbf{u}_{[\tau_0,\tau_1]},\mathbf{y}_{[\tau_0,\tau_1]})$ in which $t_0 \leq \tau_0 \leq t_1$, $\tau_0 \leq \tau_1 \leq t_1$, and $\mathbf{u}_{[\tau_0,\tau_1]} = \mathbf{u}_{[t_0,t_1]}$, $\mathbf{y}_{[\tau_0,\tau_1]} = \mathbf{y}_{[t_0,t_1]}$ over $[\tau_0,\tau_1]$. In plain words, this condition requires that every section of an input-output pair of $\mathcal{A}$ be in itself an input-output pair for $\mathcal{A}$. The purpose of this condition is to ensure that the family $\{R_{[t_0,t_1]}\}$, $t_0,t_1 \in T$, is defined in a consistent fashion.

The set of all segments $\mathbf{u}$ over $[t_0,t_1]$ such that $(\mathbf{u},\mathbf{y}) \in \mathcal{A}$ will be referred to as an *input segment space* of $\mathcal{A}$ and will be denoted by $R[\mathbf{u}]$. Similarly, the set of all segments $\mathbf{y}$ such that $(\mathbf{u},\mathbf{y}) \in \mathcal{A}$ will be referred to as an *output segment space* of $\mathcal{A}$ and will be denoted by $R[\mathbf{y}]$. This implies that $R_{[t_0,t_1]}$, which is the set of all pairs $(\mathbf{u}_{[t_0,t_1]},\mathbf{y}_{[t_0,t_1]})$ belonging to $\mathcal{A}$, is a subset of the product space $R[\mathbf{u}] \times R[\mathbf{y}]$. [By the *product space* of n spaces $X_1, \ldots, X_n$ with generic elements $x_1, \ldots, x_n$ is meant the set of all n-tuples $(x_1, \ldots, x_n)$ with $x_i \in X_i$, $i = 1, \ldots, n$ (see Ref. 2).] The relationship between $R_{[t_0,t_1]}$ and $R[\mathbf{u}] \times R[\mathbf{y}]$ is depicted in Fig. 1.4.3, in which $\delta\beta$ and $\gamma\alpha$ represent $R[\mathbf{u}]$ and $R[\mathbf{y}]$,

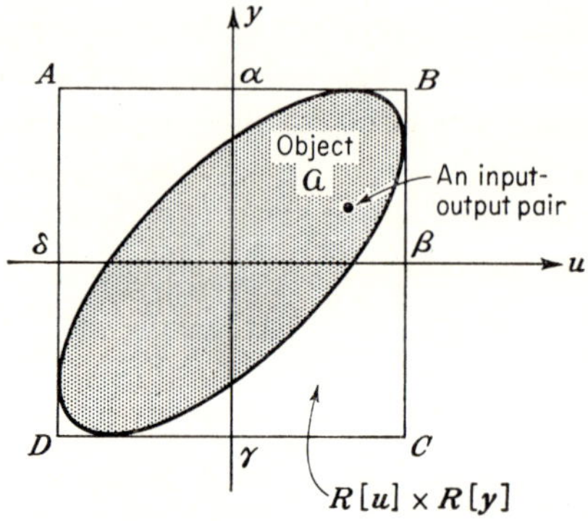

Fig. 1.4.3 Geometrical representation of an abstract object. The segments $\alpha\gamma$ and $\beta\delta$ represent $R[\mathbf{u}]$ and $R[\mathbf{y}]$, respectively; the set bounded by the rectangle $ABCD$ represents the product space $R[\mathbf{u}] \times R[\mathbf{y}]$.

respectively, the rectangle $ABCD$ represents the set $R[\mathbf{u}] \times R[\mathbf{y}]$, and the set labeled $\mathcal{A}$ denotes the set of input-output pairs $\{\mathbf{u},\mathbf{y}\}$.

Finally, a relation—written as $\mathcal{A}(\mathbf{u},\mathbf{y}) = 0$—which defines the set of all input-output pairs belonging to $\mathcal{A}$ will be called an *input-output relation* for $\mathcal{A}$, and $\mathcal{A}$ will be said to be *characterized by* its input-output relation. ◁

18 *Example* Let $\mathcal{A}$ be an object characterized by the input-output relation

19
$$\frac{dy}{dt} = u$$

in which $T = (-\infty, \infty)$ and u ranges over the space of continuous real-valued time functions. In this case, $R_{[t_0,t_1]}$ (which is the space of input-output pairs over a finite interval $[t_0,t_1]$) is the set of all ordered pairs of time functions of the form $\left(u(t), \alpha + \int_{t_0}^{t} u(\xi)\, d\xi\right)$, $t_0 \le t \le t_1$, where α is an arbitrary real constant. For example, $(1,t)$, $t_0 \le t \le t_1$; $(1, 1 + t)$, $t_0 \le t \le t_1$; and $\left(t, \frac{t^2}{2} + 1\right)$, $t_0 \le t \le t_1$ are input-output pairs for $\mathcal{A}$.

20 *Example* If in the above example the input-output relation is replaced by

21
$$\frac{dy}{dt} + y = u$$

the corresponding expression for an input-output pair belonging to $\mathcal{A}$

can be found by solving *21* for y in terms of u. Thus, in this case, the expression for an input-output pair over $[t_0,t_1]$ reads

22 $$\left(u(t),\ e^{-(t-t_0)}\alpha + \int_{t_0}^{t} e^{-(t-\xi)}u(\xi)\, d\xi\right),\ t_0 \leq t \leq t_1$$

where u ranges over time functions for which $\int_{t_0}^{t} e^{-(t-\xi)}u(\xi)\, d\xi$ is defined for all $t \in [t_0,t_1]$ and α is an arbitrary real number. Note that to each input segment $u_{[t_0,t_1]}$ there corresponds a family of output segments $\{y_{[t_0,t_1]}\}$ given by *22*, with α playing the role of a parameter which generates the family.

Uniform objects

The definition of an oriented object given above is complicated somewhat by the fact that, although every segment of an input-output pair which belongs to $\mathcal{A}$ is likewise an input-output pair belonging to $\mathcal{A}$, the converse is not necessarily true; i.e., in general there can exist input-output pairs which are not segments of other input-output pairs. As a simple example, consider an object $\mathcal{A}$ which, up to time $t = 0$, is characterized by the equation

23 $$y = u$$

and thereafter (for $t \geq 0$) is characterized by

24 $$\frac{dy}{dt} = \frac{du}{dt}$$

In this case, an input-output pair such as $(1,2)$, $t \geq 0$, is not a segment of any input-output pair which is defined on an observation interval which starts at a negative value of t_0. [All such pairs are of the form $(u(t),u(t))$, $t \geq t_0$, $t_0 < 0$.]

It is for this reason that $\mathcal{A}$ must be defined as a family of sets of input-output pairs, rather than a single set of input-output pairs. However, for most applications it is sufficient to restrict one's attention to objects which can be characterized by a single set of input-output pairs defined over T. Such objects will be said to be *uniform*. More formally, they are defined as follows:

25 **Definition** An oriented object $\mathcal{A}$ is *uniform* if every input-output pair $(\mathbf{u}_{[t_0,t_1]},\mathbf{y}_{[t_0,t_1]})$ belonging to $\mathcal{A}$ is a segment of an input-output pair $(\mathbf{u}_T,\mathbf{y}_T)$ defined over the set T. $\lhd$

An immediate consequence of this definition is that a uniform oriented object $\mathcal{A}$ can be characterized by a single set of input-output pairs $\{(\mathbf{u}_T,\mathbf{y}_T)\}$. For example, if $T = [0,\infty)$, then $\mathcal{A}$ can be characterized by the set (space) of input-output pairs $\{(\mathbf{u}_{[0,\infty)},\mathbf{y}_{[0,\infty)})\}$ defined over the interval $[0, \infty)$.

Henceforth we shall assume, unless explicitly stated to the contrary, that $\mathfrak{A}$ is a uniform oriented object. Since the notion of a uniform oriented object will play a central role in the sequel, it will be helpful to restate its definition in a more complete form as well as to recapitulate the definitions of such related notions as input and output which were given in Definition *17*.

26 **Definition** A *uniform oriented abstract object*, or, simply, an *object*, $\mathfrak{A}$, is a set, written as $\mathfrak{A} = \{(\mathbf{u}_T,\mathbf{y}_T)\}$, of ordered pairs of time functions $(\mathbf{u}_T,\mathbf{y}_T)$ defined on a set T of the time axis $(-\infty, \infty)$, with the generic values of these time functions at time t denoted by $\mathbf{u}(t)$ and $\mathbf{y}(t)$, respectively. [Thus, $\mathbf{u}_T \triangleq \{(t,\mathbf{u}(t))|t \in T\}$, and similarly for $\mathbf{y}_T$. Unless stated otherwise, it will be understood that $T = (-\infty, \infty)$.]

The set of all $\mathbf{u}_T$ such that $\mathbf{u}_T$ is the first component of a pair belonging to the set $\mathfrak{A}$ constitutes the *domain* of $\mathfrak{A}$, written as $R[\mathbf{u}_T]$. In symbols

27 $$R[\mathbf{u}_T] = \{\mathbf{u}_T|(\mathbf{u}_T,\mathbf{y}_T) \in \mathfrak{A}\} \qquad \text{for some } \mathbf{y}_T$$

Similarly the *range* of $\mathfrak{A}$, denoted by $R[\mathbf{y}_T]$, is defined by

28 $$R[\mathbf{y}_T] = \{\mathbf{y}_T|(\mathbf{u}_T,\mathbf{y}_T) \in \mathfrak{A}\} \qquad \text{for some } \mathbf{u}_T$$

(In Fig. *1.4.3*, the sets $\delta\beta$ and $\gamma\alpha$ represent $R[\mathbf{u}_T]$ and $R[\mathbf{y}_T]$, respectively.) A segment $\mathbf{u} \triangleq \mathbf{u}_{[t_0,t_1]} \triangleq \{(t,\mathbf{u}(t))|t_0 \leq t \leq t_1\}$ of $\mathbf{u}_T$ is called an *input segment*, or simply an *input*, over the *observation interval* $[t_0,t_1]$. Similarly, $\mathbf{y} \triangleq \mathbf{y}_{[t_0,t_1]}$ is called an *output segment*, or, simply, an *output*, over $[t_0,t_1]$. A pair $(\mathbf{u},\mathbf{y})$, in which $\mathbf{u}$ and $\mathbf{y}$ are defined over the same observation interval, will be said to be an *input-output pair* belonging to $\mathfrak{A}$, or, simply, an *input-output pair*. The set of all $\mathbf{u}$ (over a fixed observation interval) such that, for some $\mathbf{y}$, $(\mathbf{u},\mathbf{y})$ belongs to $\mathfrak{A}$ constitutes an *input segment space* of $\mathfrak{A}$, written as $R[\mathbf{u}]$. Similarly, $R[\mathbf{y}] = \{\mathbf{y}|(\mathbf{u},\mathbf{y}) \in \mathfrak{A}\}$ constitutes an *output segment space* of $\mathfrak{A}$. The sets $R[\mathbf{u}(t)] \triangleq \{\mathbf{u}(t)|\mathbf{u}_T \in R[\mathbf{u}_T]\}$ and $R[\mathbf{y}(t)] \triangleq \{\mathbf{y}(t)|\mathbf{y}_T \in R[\mathbf{y}_T]\}$ will be referred to, respectively, as the *input* and *output spaces* of $\mathfrak{A}$ and, unless stated to the contrary, will be assumed to be independent of t. $\lhd$

29 *Note* For simplicity, we shall usually refer to both $R[\mathbf{y}_T]$ and $R[\mathbf{y}]$ as the *range* of $\mathfrak{A}$, relying on the context for indication of whether it should be interpreted as $R[\mathbf{y}_T]$ or $R[\mathbf{y}]$. The same comment applies to $R[\mathbf{u}_T]$ and $R[\mathbf{u}]$. [Note that $R[\mathbf{u}_T]$ is a set of time functions defined on $(-\infty, \infty)$—assuming that $T = (-\infty, \infty)$—whereas $R[\mathbf{u}]$ is a set of segments of these time functions over a fixed observation interval (see *1.4.3* and *1.4.17*).] On occasion, to differentiate between $R[\mathbf{u}]$ and $R[\mathbf{u}_T]$ the latter will be called the *input-function space* of $\mathfrak{A}$.

30 *Comment* It is important to note that, under Definition *26,* to a given input **u** is associated, in general, not a unique output **y**, but a set of distinct **y**'s (see Fig. *1.4.3*) which correspond to different "initial conditions," or, equivalently, "initial states" of $\mathcal{A}$. As we shall see presently, the "state of $\mathcal{A}$" is essentially a label associated with each input-output pair of $\mathcal{A}$ in such a way that **u** ($\triangleq \mathbf{u}_{(t_0,t]}$) and the state of $\mathcal{A}$ at time t_0 uniquely determine $\mathbf{y}_{(t_0,t]}$.

31 *Comment* In mathematical terminology, an abstract object as defined by *26* is a *relation* rather than a function (or an operator). [A relation is a set of ordered pairs (**u**,**y**), whereas a function is a relation in which to every **u** corresponds a unique **y** (see Ref. *11*).] In defining an abstract object as a relation we depart from the conventional and, in our opinion, unsatisfactory definitions found in the literature in which the notion of a system—which has essentially the same meaning as an oriented abstract object—is identified with a function (e.g., a transfer function) or an operator which associates a unique output with each input.

32 *Comment* In the foregoing discussion we have focused our attention on oriented objects, since most of the objects treated in later chapters are of this type and, furthermore, it is a simple matter to extend a definition concerning an oriented object to a nonoriented object. For example, by deleting from *26* the definitions of input, output, and associated concepts, we obtain the definition of a nonoriented object. We shall not formulate this definition, since the reader can easily do it himself.

Parametrization of the space of input-output pairs

As pointed out above in Comment *30,* a central feature of an oriented object $\mathcal{A}$ is the nonuniqueness of the output. More specifically, to each input segment $\mathbf{u} \triangleq \mathbf{u}_{[t_0,t_1]}$ there corresponds, in general, a set of output segments $\mathbf{y} \triangleq \mathbf{y}_{[t_0,t_1]}$ such that **u** and each **y** in the set constitute an input-output pair belonging to $\mathcal{A}$. In a similar fashion, to each output segment **y** there will, in general, correspond a set of input segments **u**.

One way of associating a unique **y** with each **u** consists in attaching to each input-output pair (**u**,**y**) a label (parameter) $\mathbf{s}(t_0)$, ranging over a space Σ (which for simplicity is assumed to be independent of t_0), such that **y** is uniquely determined by **u** and $\mathbf{s}(t_0)$. We shall refer to this process as the *parametrization of the space of input-output pairs*, and we shall call $\mathbf{s}(t_0)$ the *state* of $\mathcal{A}$ at time t_0. As an aid in the visualization of what is involved in this process, assume that $\mathcal{A}$ is represented by a catalog each page of which shows an input **u** and a corresponding output **y**, with (**u**,**y**) constituting an input-output pair belonging to $\mathcal{A}$. The catalog, then, represents the space of input-output pairs, and one natural way of parametrizing this space is to number those pages of the catalog on which the same $\mathbf{u}$ appears. For example, assuming for simplicity that the number of inputs is finite and that a particular $\mathbf{u}$, say $\mathbf{u}^i$, $i = 1, \ldots, k$, appears on pages $1, \ldots, N_i$, then given the page number and $\mathbf{u}^i$ one can uniquely determine the corresponding $\mathbf{y}$. In this case, the state space Σ consists of the page numbers $1, \ldots, N$, where $N = \max [N_1, \ldots, N_k]$.

This informal explanation serves merely as a brief introduction to the discussion of the notion of state in Secs. *5* to *7*. As we shall see in the sequel, the notion of state is by no means trivial to define in a satisfactory fashion. We shall also see that the notion of state will play a key role in the development of the theory presented in the following chapters.

5 *An introduction to the notion of state*

1 *Note* In the following discussion the observation interval will usually be denoted by $(t_0,t]$ rather than $[t_0,t_1]$ as in Sec. *4*. ◁

As pointed out above, the state of $\mathcal{A}$ at time t_0 may be regarded as a label $\mathbf{s}(t_0)$ attached to each input-output pair $(\mathbf{u}_{(t_0,t]},\mathbf{y}_{(t_0,t]})$ in such a way that $\mathbf{y}_{(t_0,t]}$ is uniquely determined by $\mathbf{u}_{(t_0,t]}$ and $\mathbf{s}(t_0)$. In this sense, the state of $\mathcal{A}$ at time t_0 serves essentially the purpose of parametrizing the space of input-output pairs $R_{(t_0,t]}$ in such a way that $\mathbf{y}$ becomes a function of $\mathbf{u}$ and $\mathbf{s}(t_0)$.

Stated less formally, the state of $\mathcal{A}$ at time t_0 is a collection of numbers, represented by, say, a vector $\boldsymbol{\alpha}$ ranging over a space Σ, such that the knowledge of (I) $\boldsymbol{\alpha}$, (II) input-output relation of $\mathcal{A}$, and (III) input segment $\mathbf{u}_{(t_0,t]}$ is sufficient to determine the output segment $\mathbf{y}_{(t_0,t]}$ uniquely. The vector $\boldsymbol{\alpha}$ plays the role of a value of $\mathbf{s}(t_0)$, the state of $\mathcal{A}$ at time t_0. (In the sequel, in some cases $\boldsymbol{\alpha}$ will stand for a fixed point in Σ; in others, $\boldsymbol{\alpha}$ will be a variable ranging over Σ. The sense in which $\boldsymbol{\alpha}$ is used will be clear in most cases from the context.)

This, however, is not the only property that the state of $\mathcal{A}$ should have. Thus, suppose that we associate a label $\mathbf{s}(t_0)$ with an input-output pair $(\mathbf{u}_{(t_0,\infty)},\mathbf{y}_{(t_0,\infty)})$. Then, if we consider a section of this pair, $(\mathbf{u}_{(t_1,\infty)},\mathbf{y}_{(t_1,\infty)})$, starting at some later time t_1, it is reasonable to stipulate that the label $\mathbf{s}(t_1)$ associated with the input-output pair $(\mathbf{u}_{(t_1,\infty)},\mathbf{y}_{(t_1,\infty)})$ be uniquely determined by $\mathbf{s}(t_0)$ and the input segment $\mathbf{u}_{(t_0,t_1]}$. This property of the state of $\mathcal{A}$ will be discussed in greater detail in Sec. *6*, where it will manifest itself in the form of a consistency condition (*1.6.35*) entering into the definition of the state of $\mathcal{A}$.

2 *Note* We are tacitly assuming here, and will continue to do so in the

sequel, that $\mathcal{A}$ contains no random elements. Otherwise we would have to say "the knowledge of (I), (II), and (III) conveys as much information about $\mathbf{y}_{(t_0,t]}$ as does the knowledge of (I), (II), and (III) *plus* the knowledge of all the values of $\mathbf{u}(t)$, $\mathbf{y}(t)$, and $\mathbf{s}(t)$ for $t < t_0$." In any case, the basic property of the state of $\mathcal{A}$ at time t_0 is that it "separates" the future ($t > t_0$) from the past ($t < t_0$) by providing all the information about the past of $\mathcal{A}$ that is relevant to the determination of the response of $\mathcal{A}$ to any input starting at t_0. ◁

Accepting for the moment this informal explanation of the notion of state, let us consider the following basic question. Suppose that we have an oriented object $\mathcal{A}$ which is characterized by a family of sets of input-output pairs or, equivalently, by an input-output relation (see *1.4.17*)

3 $$\mathcal{A}(\mathbf{u},\mathbf{y}) = 0$$

How can we associate a state $\mathbf{s}(t)$ with $\mathcal{A}$? Or, in other words, how can we parametrize the space of input-output pairs of $\mathcal{A}$ in such a way that $\mathbf{y}_{(t_0,t]}$ would be uniquely determined by $\mathbf{s}(t_0)$ and $\mathbf{u}_{(t_0,t]}$ for all $\mathbf{u}_{(t_0,t]}$ in $R[\mathbf{u}]$ and all $\mathbf{s}(t_0)$ in Σ?

A clue to the solution of this problem can be obtained in many cases by raising the question: What do we have to know about the history of $\mathcal{A}$ up to time t_0 in order to be able to determine $\mathbf{y}_{(t_0,t]}$ uniquely when $\mathbf{u}_{(t_0,t]}$ is given? As we shall see in Chap. *4*, in the case of objects characterized by differential equations, the knowledge of a certain finite number of the derivatives of the input and output at time t_0 provides all of the information about the past history of $\mathcal{A}$ that is relevant to the determination of $\mathbf{y}_{(t_0,t]}$. Thus, in the case of such objects, the state of $\mathcal{A}$ at time t_0 is a vector whose components are the derivatives of various orders (or their linear combinations) of the input and output of $\mathcal{A}$ evaluated at time t_0. However, as we have already pointed out, the mere fact that $\mathbf{y}_{(t_0,t]}$ is uniquely determined by $\mathbf{s}(t_0)$ and $\mathbf{u}_{(t_0,t]}$ does not in itself qualify $\mathbf{s}(t_0)$ as the state of $\mathcal{A}$ at time t_0.

4 *Example* In the case of Example *1.4.20* the state of $\mathcal{A}$ at time t_0 is given by $y(t_0)$, which is the value of the output of $\mathcal{A}$ at time t_0. [Here $y(t_0)$ is the zero-order derivative of the output evaluated at t_0.] Note that $y(t)$ (for $t \geq t_0$) is uniquely determined by $y(t_0)$ and the values of u over the interval $[t_0,t]$. Clearly, this is equivalent to saying that $y_{[t_0,t]}$ is uniquely determined by $y(t_0)$ and $u_{[t_0,t]}$.

5 **Remark** Our notation for observation intervals may appear to be somewhat lacking in consistency because in some places we write $\mathbf{u}_{(t_0,t]}$, in others $\mathbf{u}_{[t_0,t]}$, and in still others $\mathbf{u}_{[t_0,t)}$. The explanation for this variability is as follows. In the first place, we generally use semi-closed intervals $(t_0,t]$ or $[t_0,t)$ rather than a closed interval $[t_0,t]$ because of the need to be able to adjoin a segment $\mathbf{u}$ (ending at t) to a segment

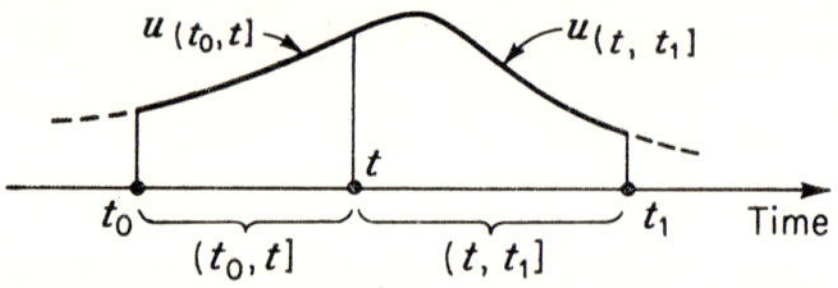

Fig. 1.5.1 Notation for observation intervals and segments.

v (starting at t) (Fig. *1.5.1*) without creating ambiguity at the common point t when the value of **u** at t is different from that of **v** at t. So far as this particular consideration is concerned, there is no difference between using $(t_0,t]$ and $[t_0,t)$ so long as we use one or the other consistently.

In the second place, when the state of $\mathcal{A}$ is specified at time t_0, we denote the input segment by $\mathbf{u}_{(t_0,t]}$ in order to indicate that the input **u** is applied to $\mathcal{A}$ "slightly" after the state of $\mathcal{A}$ is established. In many cases, however, and particularly in the case of objects characterized by differential equations, it is more convenient to convey the same idea by specifying the state at time t_0- (that is, at time $t_0 - \varepsilon$, where ε is an arbitrarily small positive number) and denoting the input by $\mathbf{u}_{[t_0,t)}$. In this connection, it should be noted that it does not make a difference whether the state is specified at t_0 or t_0- or whether the input is defined over $(t_0,t]$ or $[t_0,t]$ so long as **u** is sufficiently "smooth" at $t = t_0$. Thus, in situations in which this is tacitly or explicitly assumed to be true, we may use $\mathbf{s}(t_0)$ and $\mathbf{s}(t_0-)$ interchangeably, as well as $\mathbf{u}_{(t_0,t]}$, $\mathbf{u}_{[t_0,t)}$, and $\mathbf{u}_{[t_0,t]}$, without affecting the results of the analysis. The same comment applies to $\mathbf{y}_{(t_0,t]}$, $\mathbf{y}_{[t_0,t)}$, and $\mathbf{y}_{[t_0,t]}$. $\triangleleft$

With these remarks and observations in the background, we are in a position to formulate a definition which attaches well-defined meaning to the notion of state. Since the definition in question is rather long and replete with conditions, it will be helpful to explain the idea behind it first.

Let $\mathcal{A}$ be an oriented object which is derived from a nonoriented object $\mathcal{A}'$ by arbitrarily designating a subset of the terminal variables of $\mathcal{A}'$ as the inputs $u_1, \ldots, u_k$ and the remainder as the outputs $y_1, \ldots, y_m$ of $\mathcal{A}$, with the terminal relations of $\mathcal{A}'$ becoming the input-output relation for $\mathcal{A}$.

As we shall see in Chap. *4*, there are many ways in which a state vector $\mathbf{s}(t_0)$ can be associated with $\mathcal{A}$. Thus, if we start with $\mathcal{A}'$, then there is considerable arbitrariness in the way $\mathcal{A}'$ is oriented and, once $\mathcal{A}'$ is oriented, there is further arbitrariness in the way in which a state vector is associated with $\mathcal{A}$. Because of this arbitrariness, the definition given below is essentially a set of qualifying conditions which, if satisfied, enable one to call $\mathbf{s}(t_0)$ a *state vector* or simply *state* of $\mathcal{A}$ at time t_0.

We proceed to formulate the definition in question in the following section.

6 *Definition of state, input, and output*

Let $\mathcal{A}$ be an oriented abstract object with **u** and **y** denoting, respectively, the input and output of $\mathcal{A}$ over an observation interval $(t_0,t]$, with the terms "input" and "output" interpreted in the sense of *1.4.17*. Let $R[\mathbf{u}]$ and $R[\mathbf{y}]$ be the corresponding input and output segment spaces of $\mathcal{A}$ (*1.4.17*), and let $\mathcal{A}$ be characterized by its input-output relation

1 $$\mathcal{A}(\mathbf{u},\mathbf{y}) = 0$$

or, equivalently, by the family $\{R_{(t_0,t]}\}$ of spaces of input-output pairs of $\mathcal{A}$. Furthermore, let $\boldsymbol{\alpha}$ be a variable ranging over a space Σ. (In most of the applications with which we shall be concerned later, Σ will be an n-dimensional linear vector space and $\boldsymbol{\alpha}$ will be an n-vector.)

In these terms, a definition of state, input-output-state relation, etc., can be formulated in the following manner.

2 **Definition** If the spaces of input-output pairs of $\mathcal{A}$ admit of parametrization in the form of a relation

3 $$\mathbf{y}(t) = \mathbf{A}(\boldsymbol{\alpha};\mathbf{u}_{(t_0,t]}) \qquad \forall t > t_0, \qquad \forall t_0$$

where **A** denotes a function of $\boldsymbol{\alpha}$ and $\mathbf{u}_{(t_0,t]}$, with t_0 and t ranging over T, $\boldsymbol{\alpha}$ ranging over Σ, **u** ranging over $R[\mathbf{u}]$ and **y** ranging over $R[\mathbf{y}]$, which satisfies the four mutual- and self-consistency conditions set forth below, then *3* qualifies to be called an *input-output-state relation* for $\mathcal{A}$ and Σ qualifies to be called the *state space* of $\mathcal{A}$, with the elements of Σ being the *states* of $\mathcal{A}$ and $\boldsymbol{\alpha}$ being the *state of* $\mathcal{A}$ *at time* t_0. Furthermore, if these conditions are satisfied, we shall say that (I) $\mathcal{A}$ is *completely characterized* by its input-output-state relation *3*, (II) the segment $\mathbf{y}_{(t_0,t]}$ is the *response segment* of $\mathcal{A}$ to $\mathbf{u}_{(t_0,t]}$ starting in state $\boldsymbol{\alpha}$, and (III) $\mathbf{u}_{(t_0,t]}$ and $\mathbf{y}_{(t_0,t]}$ constitute an *input-output pair associated with, or with respect to,* $\boldsymbol{\alpha}$. The relation between $\mathbf{y}_{(t_0,t]}$ and $\mathbf{u}_{(t_0,t]}$ will be expressed symbolically as

4 $$\mathbf{y}_{(t_0,t]} = \bar{\mathbf{A}}(\boldsymbol{\alpha};\mathbf{u}_{(t_0,t]})$$

or for short

$$\mathbf{y} = \bar{\mathbf{A}}(\boldsymbol{\alpha};\mathbf{u})$$

where the bar over **A** serves to differentiate between $\mathbf{y}(t)$ and $\mathbf{y}_{(t_0,t]}$. (Note that $\bar{\mathbf{A}}$ is a function on the product space $\Sigma \times R[\mathbf{u}]$ with values in $R[\mathbf{y}]$, whereas **A** is a function on $\Sigma \times R[\mathbf{u}]$ with values in $R[\mathbf{y}(t)]$. In the sequel, both *3* and *4* will be referred to as input-output-state relations for $\mathcal{A}$, since *3* is derivable from *4* and vice versa.) Finally, a pair $(\mathbf{u}_{(t_0,t]},\mathbf{y}_{(t_0,t]})$ will be said to *satisfy* the input-output-state relation *4* (or, equivalently, *3*) if $\mathbf{u}_{(t_0,t]}$ and $\mathbf{y}_{(t_0,t]}$ constitute an input-output pair with respect to some $\boldsymbol{\alpha}$ in Σ. (Note that in terms of *4* we can write $R[\mathbf{y}] = \{\bar{\mathbf{A}}(\boldsymbol{\alpha};\mathbf{u}) | \boldsymbol{\alpha} \in \Sigma, \mathbf{u} \in R[\mathbf{u}]\}$.) ◁

The consistency conditions in question read as follows:

5 **Mutual-consistency condition** Every input-output pair for $\mathcal{A}$ satisfies the input-output-state relation *4*, and vice versa. More specifically: (I) if $(\mathbf{u}_{(t_0,t]},\mathbf{y}_{(t_0,t]})$ or, more simply, $(\mathbf{u},\mathbf{y})$ is any pair of time functions (with $\mathbf{u} \in R[\mathbf{u}]$ and $\mathbf{y} \in R[\mathbf{y}]$) satisfying *1*, then $(\mathbf{u},\mathbf{y})$ also satisfies *4* in the sense that there exists an $\boldsymbol{\alpha}$ in Σ, say, $\boldsymbol{\alpha}_0$ such that

6 $$\mathbf{y}_{(t_0,t]} = \bar{\mathbf{A}}(\boldsymbol{\alpha}_0;\mathbf{u}_{(t_0,t]})$$

and (II) any pair of time functions $(\mathbf{u},\mathbf{y})$ satisfying *4* for some $\boldsymbol{\alpha} \in \Sigma$ over an observation interval $(t_0,t]$ is an input-output pair for $\mathcal{A}$. (This must hold for all t_0,t in T and all $\mathbf{u}$ in $R[\mathbf{u}]$.) ◁

7 *Example* Suppose that $\mathcal{A}$ is characterized by the input-output relation

8 $$\frac{dy}{dt} = u$$

where u and y are real-valued time functions. Consider the relation

9 $$y(t) = \alpha + \int_{t_0}^{t} u(\xi)\, d\xi$$

where ξ is the variable of integration and α is a real-valued variable ranging over the real line $(-\infty,\infty)$. [That is, $\Sigma = \mathcal{R}^1 \triangleq (-\infty,\infty)$.] Clearly, the mutual-consistency condition is satisfied in this case for the following reasons. (I) Every input-output pair for $\mathcal{A}$ can be represented as

10 $$(u(t),y(t)) = \left(u(t),\, c + \int_{t_0}^{t} u(\xi)\, d\xi\right) \qquad t > t_0$$

where c is a real constant. Hence, *10* satisfies *9* with $\alpha = c$. (II) By direct substitution of *9* into *8* it follows at once that every pair of time functions (u,y) satisfying *9* also satisfies *8*. Note that the mutual-consistency condition would not be satisfied if Σ, the range of α, were restricted to the interval $[0,\infty)$; for in this case there is no α in $[0,\infty)$ with which, for example, the pair $(t, -1 + t^2/2)$—an input-output pair for $\mathcal{A}$—can satisfy *9* over $[0,\infty)$.

Essentially, the purpose of the mutual-consistency condition is to ensure that the input-output relation *1* and the input-output-state relation *3* (or, equivalently, *4*) represent the same object $\mathcal{A}$.

11 **First self-consistency condition** For all t_0, the response $\mathbf{y}(t)$ at any time $t > t_0$ is uniquely determined by $\boldsymbol{\alpha}$ and $\mathbf{u}_{(t_0,t]}$. In other words, in order to qualify as a state space of $\mathcal{A}$, Σ must have the property that, given

any point $\boldsymbol{\alpha}$ in Σ (which we call the state of $\mathcal{A}$ at time t_0) and any input $\mathbf{u}_{(t_0,t]}$ in the input segment space of $\mathcal{A}$, the output of $\mathcal{A}$ at time t is uniquely determined by $\boldsymbol{\alpha}$ and $\mathbf{u}_{(t_0,t]}$ and hence is independent of the values of $\mathbf{u}$ and $\mathbf{y}$ prior to t_0. As was stressed in our informal discussion in Sec. *5*, this is the key property of the notion of state. ◁

12 *Example* It is obvious that the input-output-state relation *9* satisfies the first self-consistency condition. Note that this is true regardless of the range of α.

As another example, consider a nonoriented object $\mathcal{A}$ which is characterized by the terminal relation

13 $$v_3 = v_1 + v_2$$

Assume that $\mathcal{A}$ is oriented by designating v_3 as the input variable and v_1 and v_2 as the output variables. That is, $u \triangleq v_3$ and $\mathbf{y} \triangleq (v_1,v_2)$. Furthermore, assume that Σ is identified with $\mathcal{R}^1$ and that

14 $$y_1 = \alpha u \qquad y_1 \triangleq v_1,\ y_2 \triangleq v_2,\ u \triangleq v_3$$
$$y_2 = (1 - \alpha)u$$

in which $\alpha \in \mathcal{R}^1$. Then it is evident that u, y, Σ, and the input-output-state relation *14* satisfy the first self-consistency condition. On the other hand, the mutual-consistency condition is not satisfied, because while it is true that every pair (u,y) which satisfies *14* also satisfies *13*, the converse is clearly untrue. ◁

A simple example of an input-output-state relation which does not satisfy the first self-consistency condition because of nonuniqueness is

$$y^2 = \alpha u \qquad \alpha \in \mathcal{R}^1$$

where u and y are real-valued time functions. Another example is

$$y(t) = \alpha + \int_{t_0}^{t+1} u(\xi)\, d\xi \qquad \alpha \in \mathcal{R}^1,\ t > t_0$$

Here the condition is violated because the output at time t cannot be determined without the knowledge of the values of the input between t and $t + 1$.

15 **Second self-consistency condition** If the input-output-state relation *4* is satisfied by an input-output pair $(\mathbf{u}_{(t_0,t_1]},\mathbf{y}_{(t_0,t_1]})$, then it is also satisfied by all input-output pairs of the form $(\mathbf{u}_{(t,t_1]},\mathbf{y}_{(t,t_1]})$, where $t_0 < t \leq t_1$ and $\mathbf{u}_{(t,t_1]}$ and $\mathbf{y}_{(t,t_1]}$ are sections of $\mathbf{u}_{(t_0,t_1]}$ and $\mathbf{y}_{(t_0,t_1]}$, respectively. This must hold true for all $\boldsymbol{\alpha}$ in Σ and all input-output pairs associated with $\boldsymbol{\alpha}$.

16 **Remark** Before we begin to discuss the significance of this condition it will be helpful to simplify the notation by introducing the following abbreviations.

(I) A segment of the input such as $\mathbf{u}_{(t_0,t]}$ defined over an observation interval $(t_0,t]$ will be denoted simply by $\mathbf{u}$.

(II) A segment of the input defined over $(t,t_1]$ will be denoted by $\mathbf{u}'$.

(III) The segment $\mathbf{u}_{(t_0,t_1]}$, which consists of $\mathbf{u}_{(t_0,t]}$ followed by $\mathbf{u}_{(t,t_1]}$, will be denoted by $\mathbf{uu}'$.

(IV) The response of $\mathcal{A}$ at time t to $\mathbf{u}_{(t_0,t]}$ starting in state $\boldsymbol{\alpha}_0$ will be denoted by

17 $$\mathbf{y}(t) \triangleq \mathbf{A}(\boldsymbol{\alpha}_0;\mathbf{u})$$

and the corresponding response segment $\mathbf{y}_{(t_0,t]}$ will be expressed as

18 $$\mathbf{y}_{(t_0,t]} \triangleq \bar{\mathbf{A}}(\boldsymbol{\alpha}_0;\mathbf{u})$$

with the bar in $\bar{\mathbf{A}}$ signifying that $\bar{\mathbf{A}}(\boldsymbol{\alpha}_0;\mathbf{u})$ is a segment of $\mathbf{y}$ and not just the value of $\mathbf{y}$ at time t.

(V) The symbol $\bar{\mathbf{A}}(\boldsymbol{\alpha}_1;\mathbf{u}')$ will stand for $\mathbf{y}_{(t,t_1]}$, that is, the response segment of $\mathcal{A}$ to $\mathbf{u}'$ ($\triangleq \mathbf{u}_{(t,t_1]}$) with $\mathcal{A}$ starting in state $\boldsymbol{\alpha}_1$ at time t_1. ◁

Using the above notation, let $(\mathbf{uu}',\mathbf{yy}')$ be an input-output pair satisfying the input-output-state relation *4* with $\boldsymbol{\alpha} = \boldsymbol{\alpha}_0$. That is,

19 $$\mathbf{yy}' = \bar{\mathbf{A}}(\boldsymbol{\alpha}_0;\mathbf{uu}')$$

Now, to say that $(\mathbf{u}',\mathbf{y}')$ satisfies *4* is equivalent to saying that there exists a nonempty set Q of values of $\boldsymbol{\alpha}$ in Σ such that

20 $$\mathbf{y}' = \bar{\mathbf{A}}(\boldsymbol{\alpha};\mathbf{u}')$$

is satisfied for all $\boldsymbol{\alpha}$ in Q. In symbols,

21 $$Q \triangleq \{\boldsymbol{\alpha}|\mathbf{y}' = \bar{\mathbf{A}}(\boldsymbol{\alpha};\mathbf{u}')\} \qquad \boldsymbol{\alpha} \in \Sigma$$

that is, Q is the set of all $\boldsymbol{\alpha}$ in Σ which satisfy *20*. Thus, in terms of Q the second self-consistency condition can be expressed as follows.

The set Q is nonempty for all $\mathbf{u}$, $\mathbf{u}'$, *and* $\boldsymbol{\alpha}_0$. This statement is merely an abbreviation of the equivalent statement, *if* $(\mathbf{uu}',\mathbf{yy}')$ *is any input-output pair satisfying the input-output-state relation 4, then* $(\mathbf{u}',\mathbf{y}')$ *also satisfies 4; that is,*

22 $$\mathbf{y}' = \bar{\mathbf{A}}(\boldsymbol{\alpha};\mathbf{u}')$$

for some $\boldsymbol{\alpha}$ *in* Σ (see Fig. *1.6.1*). It is understood that this must hold true for all $\boldsymbol{\alpha} \in \Sigma$, all $\mathbf{uu}'$ in the input segment space of $\mathcal{A}$, and all t_0, t, and t_1.

Essentially, the second self-consistency condition is designed to ensure that the state space Σ is sufficiently "rich" to include all possible initial conditions in $\mathcal{A}$. To illustrate this point, consider the following

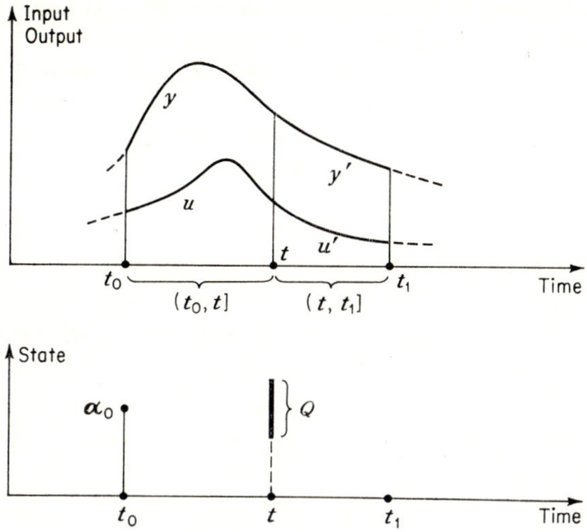

Fig. 1.6.1 Significance of the second self-consistency condition.

23 *Example* Let $\mathcal{A}$ be an object characterized by the input-output-state relation

24 $$y(t) = \alpha^2 + \int_{t_0}^{t} u(\xi)\, d\xi$$

where u and y are real-valued time functions and α ranges over the real line. Suppose that initially (at $t_0 = 0$) α is 1 and $u(t) = -2t$ over the observation interval [0,3]. Then $y(t) = 1 - t^2$ over [0,3], so that the input-output pair associated with α_0 ($\alpha_0 = 1$) is $(-2t, 1 - t^2)$ over [0,3].

Let the intermediate time t be 2. Then, it is easy to verify that the only value of α^2 for which *24* can be satisfied by the pair $(-2t, 1 - t^2)$ over [2,3] is given by

25 $$\begin{aligned}\alpha^2 &= 1 + \int_0^2 (-2\xi)\, d\xi \\ &= -3\end{aligned}$$

Since the equation $\alpha^2 = -3$ does not have a solution in Σ, it follows that Q is empty and hence the second self-consistency condition is not satisfied.

The condition in question would be satisfied if, instead of *24*, we had

26 $$y(t) = \alpha + \int_{t_0}^{t} u(\xi)\, d\xi$$

In this case, for any α_0 in Σ and any τ in the interval $(t_0,t]$ *26* yields

27 $$y(t) = \alpha_0 + \int_{t_0}^{\tau} u(\xi)\, d\xi + \int_{\tau}^{t} u(\xi)\, d\xi \qquad t_0 < \tau \leq t$$

which indicates that for every τ in $(t_0,t]$ the pair $(u_{(\tau,t]}, y_{(\tau,t]})$ is an input-output pair with respect to the state

28 $$\alpha_1 = \alpha_0 + \int_{t_0}^{\tau} u(\xi)\, d\xi \quad \triangleleft$$

As a preliminary to stating the third self-consistency condition, it will be helpful to analyze briefly what happens to the set Q in the second self-consistency condition when the segment $\mathbf{u}'$ ($\triangleq \mathbf{u}_{(t,t_1]}$) is varied over the input segment space of $\mathcal{A}$. Specifically, consider an input-output pair $(\mathbf{u}_{(t_0,t_1]},\mathbf{y}_{(t_0,t_1]})$ which satisfies *4* with respect to a state $\boldsymbol{\alpha}_0$ in Σ, and let t be an intermediate time between t_0 and t_1. Assuming that the second self-consistency condition is satisfied, the set Q of values of $\boldsymbol{\alpha}$ with respect to which $(\mathbf{u}',\mathbf{y}')$ is an input-output pair is nonempty. Since the only variables in the problem are $\boldsymbol{\alpha}_0$, t, $\mathbf{u}_{(t_0,t_1]}$, and $\mathbf{y}_{(t_0,t_1]}$, Q can depend only on these variables. If, in addition, we assume that the first self-consistency condition is satisfied, then $\mathbf{y}_{(t_0,t_1]}$ is uniquely determined by $\boldsymbol{\alpha}_0$ and $\mathbf{u}_{(t_0,t_1]}$. In this case, Q is determined solely by $\boldsymbol{\alpha}_0$, t, and $\mathbf{u}_{(t_0,t_1]}$ or, if t is held fixed, by $\boldsymbol{\alpha}_0$, $\mathbf{u}$, and $\mathbf{u}'$ (since $\mathbf{u}_{(t_0,t_1]} = \mathbf{u}\mathbf{u}'$). To make this more explicit, we shall write Q as $Q(\boldsymbol{\alpha}_0;\mathbf{u},\mathbf{u}')$. Thus

29 $$Q(\boldsymbol{\alpha}_0;\mathbf{u},\mathbf{u}') \triangleq \{\boldsymbol{\alpha} | \mathbf{y}\mathbf{y}' = \bar{\mathbf{A}}(\boldsymbol{\alpha}_0;\mathbf{u}\mathbf{u}') \text{ and } \mathbf{y}' = \bar{\mathbf{A}}(\boldsymbol{\alpha};\mathbf{u}')\}$$

Consider now a series of experiments in which $\boldsymbol{\alpha}_0$ and $\mathbf{u}$ ($\triangleq \mathbf{u}_{(t_0,t]}$) are the same for all experiments but $\mathbf{u}'$ ($\triangleq \mathbf{u}_{(t,t_1]}$) is varied from experiment to experiment. (It is understood that not only the values of $\mathbf{u}'$ over $(t,t_1]$ but also the value of t_1 are varied.) For simplicity, assume that two experiments are performed on $\mathcal{A}$ and that in one of them the input $\mathbf{u}'$ is equal to, say, $\mathbf{v}$ (that is, $\mathbf{u}_{(t,t_1]} = \mathbf{v}_{(t,t_1]}$) and in the other $\mathbf{u}'$ is equal to $\mathbf{w}$, with $\mathbf{v} \neq \mathbf{w}$. This is illustrated in Fig. *1.6.2*, where $\mathbf{y}_{\mathbf{v}}$ and $\mathbf{y}_{\mathbf{w}}$ denote the corresponding response segments of $\mathcal{A}$.

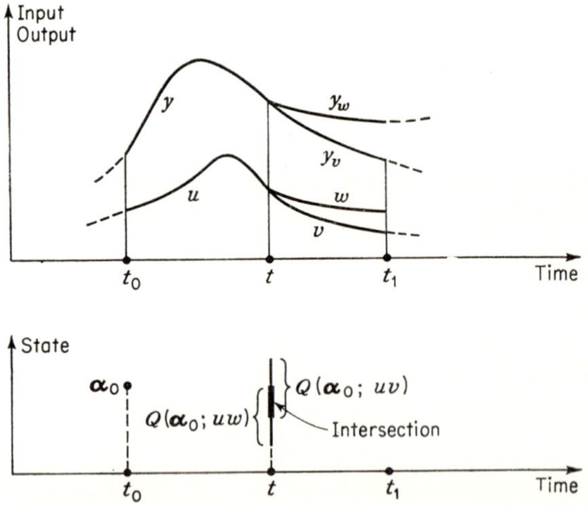

Fig. 1.6.2 Significance of the third self-consistency condition.

Now the set Q for the first experiment is $Q(\boldsymbol{\alpha}_0;\mathbf{u},\mathbf{v})$, and for the second experiment it is $Q(\boldsymbol{\alpha}_0;\mathbf{u},\mathbf{w})$. By definition, $Q(\boldsymbol{\alpha}_0;\mathbf{u},\mathbf{v})$ is the set of all points in Σ with respect to which $(\mathbf{v},\mathbf{y}_\mathbf{v})$ is an input-output pair satisfying

30 $$\mathbf{y}_\mathbf{v} = \bar{\mathbf{A}}(\boldsymbol{\alpha};\mathbf{v})$$

and

31 $$\mathbf{y}\mathbf{y}_\mathbf{v} = \bar{\mathbf{A}}(\boldsymbol{\alpha}_0;\mathbf{u}\mathbf{v})$$

Likewise, $Q(\boldsymbol{\alpha}_0;\mathbf{u},\mathbf{w})$ is the set of all points in Σ with respect to which $(\mathbf{w},\mathbf{y}_\mathbf{w})$ is an input-output pair satisfying

32 $$\mathbf{y}_\mathbf{w} = \bar{\mathbf{A}}(\boldsymbol{\alpha};\mathbf{w})$$

and

33 $$\mathbf{y}\mathbf{y}_\mathbf{w} = \bar{\mathbf{A}}(\boldsymbol{\alpha}_0;\mathbf{u}\mathbf{w})$$

Thus, the set of all points in Σ with respect to which *both* $(\mathbf{v},\mathbf{y}_\mathbf{v})$ and $(\mathbf{w},\mathbf{y}_\mathbf{w})$ are input-output pairs satisfying both *30*, *31* and *32*, *33* is the *intersection* of $Q(\boldsymbol{\alpha}_0;\mathbf{u},\mathbf{v})$ and $Q(\boldsymbol{\alpha}_0;\mathbf{u},\mathbf{w})$, that is, $Q(\boldsymbol{\alpha}_0;\mathbf{u},\mathbf{v}) \cap Q(\boldsymbol{\alpha}_0;\mathbf{u},\mathbf{w})$.

More generally, if we consider the totality of experiments in which $\boldsymbol{\alpha}_0$ and $\mathbf{u}$ are held fixed and $\mathbf{u}'$ is varied over the input segment space of $\mathcal{A}$, and if we form the intersection of all the Q sets corresponding to these experiments, then the intersection (if it is not empty) will constitute the set of all points in Σ with respect to which every pair $(\mathbf{u}',\mathbf{y}')$ satisfying the input-output-state relation

34 $$\mathbf{y}\mathbf{y}' = \bar{\mathbf{A}}(\boldsymbol{\alpha}_0;\mathbf{u}\mathbf{u}')$$

satisfies

$$\mathbf{y}' = \bar{\mathbf{A}}(\boldsymbol{\alpha};\mathbf{u}')$$

for all $\boldsymbol{\alpha}$ in the intersection. Clearly, it would not be meaningful to speak of the "state of $\mathcal{A}$ at time t" if the intersection in question were empty; for to do so would imply that there does not exist a point (state) in Σ which can account for all possible input-output pairs starting at time t.

With these observations as a background, we are prepared to express the third self-consistency condition in a concise fashion.

35 **Third self-consistency condition** Let $(\mathbf{u}\mathbf{u}',\mathbf{y}\mathbf{y}')$ (where $\mathbf{u}\mathbf{u}' \triangleq \mathbf{u}_{(t_0,t_1]}$, $\mathbf{y}\mathbf{y}' \triangleq \mathbf{y}_{(t_0,t_1]}$, $\mathbf{u} \triangleq \mathbf{u}_{(t_0,t]}$, $\mathbf{u}' \triangleq \mathbf{u}_{(t,t_1]}$) be an input-output pair satisfying *4* (or, equivalently, *3*) with respect to some value of $\boldsymbol{\alpha}$, say, $\boldsymbol{\alpha}_0$, in Σ. In symbols,

36 $$\mathbf{y}\mathbf{y}' = \bar{\mathbf{A}}(\boldsymbol{\alpha}_0;\mathbf{u}\mathbf{u}')$$

which is merely a restatement of *34*. Furthermore, let $Q(\boldsymbol{\alpha}_0;\mathbf{u},\mathbf{u}')$ be the set of all $\boldsymbol{\alpha}$ in Σ with respect to which $(\mathbf{u}',\mathbf{y}')$ is an input-output pair satisfying *4*. That is,

37 $$Q(\boldsymbol{\alpha}_0;\mathbf{u},\mathbf{u}') \triangleq \{\boldsymbol{\alpha}|\mathbf{y}' = \bar{\mathbf{A}}(\boldsymbol{\alpha};\mathbf{u}') \text{ and } \mathbf{y}\mathbf{y}' = \bar{\mathbf{A}}(\boldsymbol{\alpha}_0;\mathbf{u}\mathbf{u}')\}$$

which is a restatement of *29*. Then, the third self-consistency condition requires that the intersection

38 $$Q_t \triangleq \bigcap_{\mathbf{u}'} Q(\boldsymbol{\alpha}_0;\mathbf{u},\mathbf{u}')$$

taken over all $\mathbf{u}'(\triangleq \mathbf{u}_{(t,t_1]})$ in the input segment space of $\mathcal{A}$ be nonempty for all t_0, all $\boldsymbol{\alpha}_0 \in \Sigma$, and all segments $\mathbf{u}$ in the input segment space of $\mathcal{A}$. Since Q_t as defined by *38* depends only on $\boldsymbol{\alpha}_0$ and $\mathbf{u}$, we shall denote it by $Q_t(\boldsymbol{\alpha}_0;\mathbf{u})$, where t identifies the right end point of the observation interval on which $\mathbf{u}$ is defined.

39 **Remark** Note that it follows at once from the way in which the third self-consistency condition is formulated that it cannot be satisfied unless all the $Q(\boldsymbol{\alpha}_0;\mathbf{u},\mathbf{u}')$ sets are nonempty, which implies that if the third self-consistency condition is satisfied, then the second self-consistency condition is also satisfied. In this sense, the third self-consistency condition *implies* the second self-consistency condition. The converse, however, is not true. Indeed, it is not difficult to construct examples of relations of the form $\mathbf{y} = \bar{\mathbf{A}}(\boldsymbol{\alpha};\mathbf{u})$ which satisfy the first and second self-consistency conditions but not the third self-consistency condition. (*Exercise:* Construct such an example.)

Essentially, the purpose of the third self-consistency condition is to make it meaningful to speak of "the state of $\mathcal{A}$ at time t." This basic notion will be defined in Sec. *7*. ◁

To illustrate the third self-consistency condition, we shall use a simple

40 *Example* Consider the input-output-state relation

41 $$y(t) = \alpha + \int_{t_0}^{t} u(\xi)\, d\xi$$

where u and y are real-valued time functions and Σ is the real line. For concreteness, let $\alpha_0 = -3$, $t_0 = 0$, $t = 1$, $t_1 > 1$, and $u(\xi) = 2 - 2\xi$ for $0 < \xi \leq t$. The values of $u(\xi)$ for ξ between 1 and t_1 are assumed to vary from experiment to experiment (see Fig. *1.6.2*).

In order to avoid confusion with t (which in our example is fixed at 1), let τ denote the running time in the interval $(1,t_1]$. Then, on dividing the range of integration in *41* into two parts, from 0 to 1 and from 1 to τ, we have for the assumed α_0, t_0, t, and $u_{(0,\tau]}$

42 $$y(\tau) = -3 + \int_0^1 (2 - 2\xi)\, d\xi + \int_1^{\tau} u(\xi)\, d\xi \qquad 1 < \tau \leq t_1$$

or

43 $$y(\tau) = -2 + \int_1^{\tau} u(\xi)\, d\xi \qquad 1 < \tau \leq t_1$$

Clearly, this relation is of the same form as *41*, which implies that $(u_{(1,t_1]},y_{(1,t_1]})$, where $y_{(1,t_1]}$ is defined by *43*, satisfies the input-output-state relation *41* with $\alpha = -2$. Consequently, the set Q (as defined by *37*) is nonempty, since it contains $\alpha = -2$ as an element. (Note that $u_{(1,t_1]}$ and $y_{(1,t_1]}$ play the roles of $\mathbf{u}'$ and $\mathbf{y}'$, respectively.) Furthermore, no matter what $u_{(1,t_1]}$ is, the pair $(u_{(1,t_1]},y_{(1,t_1]})$ will satisfy *41* with $\alpha = -2$. Consequently, all the Q sets corresponding to different $u_{(1,t_1]}$ contain the point $\alpha = -2$ and hence so does their intersection Q_1 (defined by *38*). This verifies that Q_1 is nonempty when $t = 1$ and $u(\xi) = 2 - 2\xi$ for $0 < \xi \leq t$. Obviously, the same conclusion holds for arbitrary t between t_0 and t_1 as well as arbitrary $u_{(t_0,t]}$. Consequently, *41* satisfies the third self-consistency condition.

44 **Remark** Suppose that we have established that an input-output-state relation $\mathbf{y} = \mathbf{A}(\boldsymbol{\alpha};\mathbf{u})$, in which $\boldsymbol{\alpha} \in \Sigma$, satisfies the four consistency conditions. Then it is evident that the input-output-state relation $\mathbf{y} = \bar{\mathbf{A}}(\boldsymbol{\alpha}',\mathbf{u})$, where $\boldsymbol{\alpha}'$ is a variable which is in one-one correspondence with $\boldsymbol{\alpha}$, will also satisfy the four consistency conditions. In other words, if $\boldsymbol{\alpha}' = \boldsymbol{\mu}(\boldsymbol{\alpha})$, where $\boldsymbol{\mu}$ denotes a one-one mapping of Σ onto itself, then the values of $\boldsymbol{\alpha}'$ also qualify as states of $\mathcal{A}$. Note that such one-one transformations on the state space amount merely to a relabeling of the elements of Σ and that relabeling does not affect the consistency conditions or the form of the input-output-state relation. However, a relabeling will, in general, change the form of the so-called state equations of $\mathcal{A}$ (see Sec. *8*). For this reason, it may be advantageous in specific cases to use $\boldsymbol{\alpha}'$—where $\boldsymbol{\alpha}'$ is in one-one correspondence with $\boldsymbol{\alpha}$—rather than $\boldsymbol{\alpha}$ itself as a description of a state of $\mathcal{A}$ in order to simplify the form of state equations. In the cases which we shall consider in later chapters, the transformations in question are linear and amount essentially to a change in the coordinate system of Σ. ◁

This concludes our formulation and discussion of Definition *2*. Additional comments and examples clarifying the significance of the various concepts introduced in this definition are presented in the following two sections.

7 *More on the notion of state and oriented objects*

As will be shown in Chap. *2* (*2.2.19*), the set Q_t defined by *1.6.38* comprises values of $\boldsymbol{\alpha}$ which are *equivalent* in the sense that (I) the response of $\mathcal{A}$ to $\mathbf{u}'$ is the same for all starting states in Q_t and (II) this holds true for all $\mathbf{u}'$ in the input segment space of $\mathcal{A}$. In this sense, any $\boldsymbol{\alpha}$ in Q_t can be regarded as a representative of the entire set Q_t.

This fact permits us to define the state of $\mathcal{A}$ at time t as follows.

1 **Definition** Consider an object $\mathcal{A}$ which is characterized by the input-output-state relation (see *1.6.3* and *1.6.4*)

2 $$\mathbf{y}(t) = \mathbf{A}(\boldsymbol{\alpha}_0;\mathbf{u}_{(t_0,t]})$$

or equivalently

$$\mathbf{y}_{(t_0,t]} = \bar{\mathbf{A}}(\boldsymbol{\alpha}_0;\mathbf{u}_{(t_0,t]})$$

where $\boldsymbol{\alpha}_0$ ranges over the state space Σ. Let $\mathbf{u} \triangleq \mathbf{u}_{(t_0,t]}$, $\mathbf{y} \triangleq \mathbf{y}_{(t_0,t]}$, and let $\mathbf{uu}'$ denote the segment $\mathbf{u}$ followed by a segment $\mathbf{u}'$ (see *1.6.16*) and $\mathbf{yy}'$ denote the response of $\mathcal{A}$ to $\mathbf{uu}'$ starting in state $\boldsymbol{\alpha}_0$. Thus

$$\mathbf{yy}' = \bar{\mathbf{A}}(\boldsymbol{\alpha}_0;\mathbf{uu}')$$

As in *1.6.29* and *1.6.38*, define the sets Q and Q_t by

3 $$Q \triangleq Q(\boldsymbol{\alpha}_0;\mathbf{u},\mathbf{u}') \triangleq \{\boldsymbol{\alpha} | \mathbf{yy}' = \bar{\mathbf{A}}(\boldsymbol{\alpha}_0;\mathbf{uu}') \text{ and } \mathbf{y}' = \bar{\mathbf{A}}(\boldsymbol{\alpha};\mathbf{u}')\}$$

4 $$Q_t \triangleq Q_t(\boldsymbol{\alpha}_0;\mathbf{u}) \triangleq \bigcap_{\mathbf{u}'} Q(\boldsymbol{\alpha}_0;\mathbf{u},\mathbf{u}')$$

Then the *state of* $\mathcal{A}$ *at time* t is defined to be any state $\boldsymbol{\alpha}$ in $Q_t(\boldsymbol{\alpha}_0;\mathbf{u})$, with $\boldsymbol{\alpha}_0$ being the state of $\mathcal{A}$ at time t_0, or equivalently, the starting (initial) state of $\mathcal{A}$. [Note that in this definition we could not have said "any state $\boldsymbol{\alpha}$ in Q_t" were it not for the fact that all the elements of Q_t are equivalent to one another (*2.2.19*).] ◁

The state of $\mathcal{A}$ at time t will be denoted by $\mathbf{s}(t)$,[1] with the understanding that $\mathbf{s}(t)$ is merely a label which identifies an element of $Q_t(\boldsymbol{\alpha}_0;\mathbf{u})$, say, $\boldsymbol{\alpha}$, as the state of $\mathcal{A}$ at time t. Consistent with this notation, the initial state of $\mathcal{A}$ at time t_0 will be denoted by $\mathbf{s}(t_0)$. When we wish to place in evidence that $\mathbf{s}(t_0)$ is the initial state of $\mathcal{A}$, the input-output-state relation *2* will be written as

5 $$\mathbf{y}(t) = \mathbf{A}(\mathbf{s}(t_0);\mathbf{u}_{(t_0,t]})$$

Note that $\mathbf{s}(t_0)$ and, more generally, $\mathbf{s}(t)$ range over the state space Σ; that is, for each fixed t, $R[\mathbf{s}(t)] = \Sigma$ (see *1.3.4*). The components of $\mathbf{s}(t)$ will be called the *state variables* of $\mathcal{A}$. ◁

Consider an object $\mathcal{A}$ which is initially in state $\mathbf{s}(t_0)$ and to which an input $\mathbf{u}_{(t_0,t_1]}$ is applied. The state of $\mathcal{A}$ at time t_1, $\mathbf{s}(t_1)$, is called the *terminal state of* $\mathcal{A}$ relative to the input $\mathbf{u}_{(t_0,t_1]}$. Clearly, if the input segment $\mathbf{u}_{(t_0,t_1]}$ is followed by another input segment $\mathbf{u}_{(t_1,t_2]}$, then $\mathbf{s}(t_1)$ plays the dual role of the terminal state for $\mathbf{u}_{(t_0,t_1]}$ and the initial state for $\mathbf{u}_{(t_1,t_2]}$. Note also that the state of $\mathcal{A}$ at time t may be regarded as

[1] In Chap. *3* we shall switch to the symbol $\mathbf{x}(t)$ to denote the state of $\mathcal{A}$ at time t in order to avoid confusion with the complex variable s and to conform more closely to the notation commonly employed in the theory of ordinary differential equations.

the terminal state of $\mathcal{A}$ relative to the input $\mathbf{u}_{(t_0,t]}$. On occasion we shall find it convenient to use the notation $\boldsymbol{\alpha}_0^*$ for the terminal state relative to $\mathbf{u}_{(t_0,t]}$ when the initial state is $\boldsymbol{\alpha}_0$.

6 *Comment* It should be recalled (see *1.3.3*) that the segment $\mathbf{u}_{(t_0,t]}$ is defined as the set of pairs $\{(\tau,\mathbf{u}(\tau))\}$, with $t_0 < \tau \leq t$. Thus, when we write $\mathbf{y}(t) = \mathbf{A}(\mathbf{s}(t_0);\mathbf{u}_{(t_0,t]})$, it is understood that $\mathbf{y}(t)$ depends not only on $\mathbf{s}(t_0)$ and the values of $\mathbf{u}(\tau)$ for $t_0 < \tau \leq t$ but also on the end points t_0 and t. If this is not the case, i.e., if $\mathbf{y}(t)$ depends only on $\mathbf{s}(t_0)$ and $\mathbf{u}(\tau)$, $t_0 < \tau \leq t$, then $\mathcal{A}$ is a *time-invariant* object. The notion of time invariance will be discussed in greater detail in Chap. *3*, Sec. *2*.

7 *Comment* As was pointed out in *1.5.5*, we use $\mathbf{u}_{(t_0,t]}$ rather than $\mathbf{u}_{[t_0,t]}$ in the input-output-state relation for two reasons: (I) to avoid overlap between two adjoining segments such as $\mathbf{u}_{(t_0,t]}$ and $\mathbf{u}_{(t,t_1]}$ and (II) to resolve the ambiguity arising when $\mathbf{u}$ is discontinuous or has a delta function at $t = t_0$. Furthermore, to write the input as $\mathbf{u}_{(t_0,t]}$ is consistent with our intuitive feeling that the input is applied "slightly" after the initial state is established. In later chapters, we shall frequently find it more convenient to resolve the ambiguity in question by identifying the initial state with the state of $\mathcal{A}$ at time t_0-, that is, the instant immediately preceding t_0. In this case, the input will be assumed to start at t_0 rather than at t_0+ (the instant immediately following t_0) as in *5*. (This point will be elaborated upon in later chapters in the course of our discussion of linear systems.) Needless to say, when $\mathbf{u}$ is sufficiently smooth at t_0, it does not matter whether $\mathbf{u}_{(t_0,t]}$ or $\mathbf{u}_{[t_0,t]}$ is used in *5*.

8 *Comment* In dealing with an object $\mathcal{A}$ which is characterized by an input-output relation $\mathcal{A}(\mathbf{u},\mathbf{y}) = 0$ (*1.4.17*), it will be our standard practice to associate a state vector $\mathbf{s}(t)$ with $\mathcal{A}$ by defining $\mathbf{s}(t)$ in terms of $\mathbf{u}_{(t_0,t]}$ and $\mathbf{y}_{(t_0,t]}$ and then proceed to verify that it satisfies the mutual- and self-consistency conditions. As we shall see later in Chap. 4, in cases in which the input-output relation $\mathcal{A}(\mathbf{u},\mathbf{y}) = 0$ is a differential equation, $\mathbf{s}(t)$ can be defined in terms of $\mathbf{u}(t)$, $\mathbf{y}(t)$, and a finite number of their derivatives at t. As was indicated previously (*1.6.44*), a variety of other expressions for $\mathbf{s}(t)$ in terms of $\mathbf{u}_{(t_0,t]}$ and $\mathbf{y}_{(t_0,t]}$ can be obtained by changing the coordinate system in the state space. The important point to bear in mind is that there are many ways in which a state vector $\mathbf{s}(t)$ can be associated with an object characterized by an input-output relation; so that when we speak of $\mathbf{s}(t)$ as *the* state of $\mathcal{A}$ at time t, we do not mean that there is just one way of defining $\mathbf{s}(t)$ in terms of the input and output of $\mathcal{A}$. This point will be elaborated upon in Chap. *4*, Sec. *2*.

9 *Comment* Throughout the foregoing discussion we have tacitly assumed that the time function $\mathbf{s}$ is defined for all t. This assumption is usually satisfied in practice. It is not difficult, however, to envisage situations in which s may have discontinuities or, worse, may contain delta functions of various orders. In such cases s may be undefined for one or more values of t and, strictly speaking, should be regarded as a distribution (see *A.4.1 et seq.*) rather than a function of time in the classical sense of the term.

We proceed to illustrate the notions introduced above by several informal examples. In the discussion of these examples we shall not attempt to be completely rigorous, since we have not yet developed the necessary techniques for dealing with the types of objects which are used as illustrations. Such techniques will be developed in later chapters.

10 *Example* Consider first a very simple type of object, namely, an electrical capacitor. The abstract model $\mathcal{A}$ of such a capacitor is characterized by the terminal relation

$$i = C \frac{dv}{dt}$$

where $i \triangleq$ current through the capacitor, $v \triangleq$ voltage across the capacitor, and $C \triangleq$ capacitance, which is assumed to be a positive constant.

Let $\mathcal{A}$ be oriented by designating i to be the input and v the output. Now let us ask the question: What does one have to know about $\mathcal{A}$ at time t_0 in order to be able to find the value of the output v at any time $t > t_0$, given the values of the input ($\triangleq i$) between t_0 and t? The answer is clearly $v(t_0)$, for by integrating between t_0 and t, we have

11 $$v(t) = v(t_0) + \frac{1}{C} \int_{t_0}^{t} i(\xi)\, d\xi$$

which in effect is the input-output-state relation for $\mathcal{A}$, with $v(t_0)$ playing the role of $s(t_0)$. Here the state space Σ is the real line $(-\infty, \infty)$, as are the input and output spaces $R[i(t)]$ and $R[v(t)]$.

We have already verified that *11* satisfies the consistency conditions (see *1.6.12* and *1.6.40*), with the state of $\mathcal{A}$ at time t given by $v(t)$. Note that if there is a possibility that i may contain a delta function $\delta(t - t_0)$ at t_0, then *11* should be written as

$$v(t) = v(t_0-) + \frac{1}{C} \int_{t_0}^{t} i(\xi)\, d\xi \qquad t \geq t_0$$

with the understanding that $\delta(t - t_0)$ is included in the range of integration $[t_0,t]$. For example, if $i(t) = \delta(t)$ and $t_0 < 0 < t$, then (see Appendix *A*)

$$\begin{aligned} v(t) &= v(t_0-) + \frac{1}{C} \int_{t_0}^{t} \delta(\xi)\, d\xi \\ &= v(t_0-) + \frac{1}{C} \qquad t > 0 \end{aligned}$$

which shows that there is a jump in v at 0 equal to $1/C$.

It is instructive to consider also the case where $\mathcal{A}$ is oriented by designating v to be the input and i to be the output. In this case, care must be exercised because of the possibility that the input may be discontinuous at t_0. On defining the initial state to be $s(t_0-)$ rather than $s(t_0)$ (see Comment *7*), the input-output-state relation is found to be (see *1.10.18*)

$$i(t) = C\frac{dv}{dt} - Cv(t_0-)\delta(t - t_0)$$

where $\delta(t - t_0)$ is a delta function at $t = t_0$, $v(t_0-)$ represents the state of $\mathcal{A}$ at time t_0-, $v(t)$ is assumed to vanish for $t < t_0$, and dv/dt is the derivative of v in the distribution sense. [For example, if $v(t)$ is the unit step function, $v(t) = 1(t)$, then $dv/dt = \delta(t)$. See Appendix *A* for a fuller discussion.] Note that the state at time t is simply $v(t)$.

12 *Example* Consider the abstract model of a particle $\mathcal{P}$ in one-dimensional space to be an oriented object $\mathcal{A}$ with applied force f as the input and the position of the particle ($\triangleq x$) as the output. In this case, the input-output relation for the object is

13 $$m\ddot{x} = f$$

where $m \triangleq$ mass and $\ddot{x} \triangleq$ acceleration.

We observe that (I) knowing the position and velocity of $\mathcal{A}$ at any time t and (II) knowing $f_{(t_0,t]}$, we can find the position at any time $t > t_0$ by integrating *13* and using the position and velocity at time t_0 as initial conditions. This suggests that the vector

$$\mathbf{s}(t_0) = (x(t_0),\dot{x}(t_0))$$

with components $x(t_0)$ and $\dot{x}(t_0)$, is a likely candidate for the state of $\mathcal{A}$ at time t_0. The corresponding input-output-state relation can readily be found by integrating both sides of *13* twice and substituting $x(t_0)$ and $\dot{x}(t_0)$ for the initial conditions at t_0. The result reads:

14 $$x(t) = x(t_0) + \dot{x}(t_0)(t - t_0) + \frac{1}{m}\int_{t_0}^{t} d\eta \int_{t_0}^{\eta} d\xi\, f(\xi)$$

or in vector form

15 $$x(t) = \langle \mathbf{s}(t_0),\boldsymbol{\tau}(t - t_0)\rangle + \frac{1}{m}\int_{t_0}^{t} d\eta \int_{t_0}^{\eta} d\xi\, f(\xi)$$

where $\langle \mathbf{a},\mathbf{b}\rangle$ denotes the scalar product of two vectors $\mathbf{a}$ and $\mathbf{b}$ and $\boldsymbol{\tau}(t - t_0)$ is the vector

$$\boldsymbol{\tau}(t - t_0) = (1, t - t_0)$$

If we had assumed that the point in question moves in three-dimensional rather than one-dimensional space, then $\mathbf{x}(t)$ and $\dot{\mathbf{x}}(t)$ would be 3-vectors and the state $\mathbf{s}(t_0)$ would have $\mathbf{x}(t_0)$ and $\dot{\mathbf{x}}(t_0)$ as vector components:

16 $$\mathbf{s}(t_0) = (\mathbf{x}(t_0), \dot{\mathbf{x}}(t_0))$$

In this case, the input-output-state relation has exactly the same form as *15* with $x(t)$ and $\dot{x}(t)$ replaced by $\mathbf{x}(t)$ and $\dot{\mathbf{x}}(t)$, respectively. Note that the state space in this case is six-dimensional ($\Sigma = \mathcal{R}^6$), whereas the input and output spaces are three-dimensional ($R[\mathbf{x}(t)] = R[\mathbf{f}(t)] = \mathcal{R}^3$). ($\mathcal{R}^n \triangleq$ space of ordered n-tuples of real numbers.)

It is intuitively obvious and it will be proved in a general way in Chap. *4* that *14* satisfies the four consistency conditions and that the state of $\mathcal{a}$ at time t is given by the vector $\mathsf{s}(t) = (x(t),\dot{x}(t))$. By differentiating *14* with respect to t and forming the vector $(x(t),\dot{x}(t))$, the expression for $\mathsf{s}(t)$ in terms of $\mathsf{s}(t_0)$ and $\mathbf{u}_{(t_0,t]}$ is found to be (in vector form)

17
$$\mathsf{s}(t) = \mathbf{\Phi}(t - t_0)\mathsf{s}(t_0) + \int_{t_0}^{t} \mathbf{h}(t - \xi) f(\xi)\, d\xi \qquad t \geq t_0$$

where

$$\mathbf{\Phi}(t) = \begin{bmatrix} 1 & t \\ 0 & 1 \end{bmatrix} \qquad \mathbf{h}(t) = \begin{bmatrix} t \\ 1 \end{bmatrix} \qquad t \geq 0$$

This is a special instance of a general expression for $\mathsf{s}(t)$ which will be derived in Chap. *3*, Sec. *7*.

Note that the output of $\mathcal{a}$ at time t is expressed in terms of $\mathsf{s}(t)$ by

18
$$x(t) = \langle \mathsf{c},\mathsf{s}(t) \rangle$$

where $\mathsf{c} = (1,0)$. On substituting *17* in *18*, we obtain the input-output-state relation *14*. Thus, the pair of equations *17* and *18* can be deduced from *14*, and *14* in turn is deducible from *17* and *18*. As we shall see at a later point, this is a general property of the so-called *state equations* of $\mathcal{a}$ (see *1.8.2 et seq.*).

19 *Example* Consider a clock with just one hand, the hour hand. Suppose that the position of the hand is represented by a variable x which can take on the values 0, 1, . . . , 11, with the understanding that any time between 0 and 1 and including 1 is regarded as the first hour ($x = 1$), any time between 1 and 2 and including 2 is regarded as the second hour ($x = 2$), etc. Now assume that at the end of each hour an experimenter tosses a die and advances the hour hand manually by the number of hours appearing on the face of the die. For example, if at time $t = 3$ ($t = 0, 1, 2, 3, \ldots$ hr), $x(3)$ is 4 and the number appearing on the face of the die is 6, then the hand is advanced to $x = 4 + 6 = 10$; that is, $x(4) = 10$.

The number appearing on the face of the die at time t is regarded as the input to the clock. As for the output, it is taken to be 0 at time t if $x(t)$ is even and 1 if $x(t)$ is odd. Now as a guide in deciding on how to identify the states of the object in question, let us ask the question: What do we have to know at time t_0 in order to be able to determine the response to any input starting at t_0? The answer is clearly $x(t_0)$. To verify that this is indeed the case, it will suffice to derive the input-output-state relation for the object under consideration and to verify that it satisfies the self-consistency conditions. (Note that the mutual-consistency condition would also have to be verified if the object were initially characterized by an input-output relation. Since this is not the case, only the self-consistency conditions need be satisfied.) To this end, let n mod p denote the least nonnegative residue of an integer n which results when multiples of the integer p are subtracted from or added to n. (For example, 5 mod 3 = 2, 2 mod 3 = 2, 10 mod 5 = 0, 7 mod 2 = 1.) Thus, if the initial position (at $t = t_0$) of the hour hand is denoted by $x(t_0)$ and the successive numbers appearing on the face of the die are denoted by $u(t_0)$, $u(t_0 + 1)$, . . . , we can write

$$x(t) = [x(t_0) + u(t_0) + u(t_0 + 1) + \cdots + u(t - 1)] \bmod 12$$

and since the output is related to $x(t)$ by

20 $$y(t) = x(t) \bmod 2$$

the desired input-output-state relation is

$$y(t) = \{[x(t_0) + u(t_0) + \cdots + u(t - 1)] \bmod 12\} \bmod 2$$

or more simply

21 $$y(t) = [x(t_0) + u(t_0) + u(t_0 + 1) + \cdots + u(t - 1)] \bmod 2$$

since 2 is a divisor of 12.

In this example, the state space is a finite set $\{0,1, \ldots ,11\}$. The input space is likewise a finite set, $R[u(t)] = \{1,2, \ldots ,6\}$. The output space is $R[y(t)] = \{0,1\}$.

To verify that *21* satisfies the consistency conditions, it is expedient to put *21* into the form

22 $$y(t) = [x(t_0) + u(t_0) + \cdots + u(t' - 1) + u(t') + \cdots + u(t - 1)] \bmod 2$$

where t' is an arbitrary time between t_0 and $t - 1$. Then, regarding $u(t'), \ldots ,u(t - 1)$ as an input sequence starting at t', *22* can be written as

23 $$y(t) = [x(t') + u(t') + \cdots + u(t - 1)] \bmod 2 \qquad t' > t$$

where

24 $$x(t') = x(t_0) + u(t_0) + \cdots + u(t' - 1)$$

Since *23* is of the same form as *21*, with $x(t')$ independent of $u(t')$, . . . $,u(t - 1)$, it follows at once that the third self-consistency condition is satisfied and that $x(t')$, as given by *24*, represents the state of

$\mathcal{A}$ at time t'. Note that the arguments used here are similar to those employed in the case of Example *1.6.40*. Note in particular that the first consistency condition is satisfied trivially and that the second self-consistency condition is implied by the third condition (see Remark *1.6.39*).

It is easy to verify that the second self-consistency condition would not be satisfied if the state space were restricted to, say, the set $\{1,3,5,7\}$. In that case, $x(t')$ as determined by *24* would not always be a member of this set, and hence for some u the set $Q_{t'}$ would be empty. Specifically, set $t_0 = 0$, $t' = 4$, $x(t_0) = 7$, $u(t_0 + 1) = u(1) = 1$, $u(t_0 + 2) = u(2) = 4$, $u(t' - 1) = u(3) = 2$. Then $x(t') = (7 + 1 + 4 + 2) \bmod 2 = 0$, and $x(t')$ is not an element of the state space $\{1,3,5,7\}$.

25 *Example* Consider a so-called "perfect delay line" with delay δ ($\delta \triangleq$ constant > 0). The abstract model of such a device is an oriented object $\mathcal{A}$ characterized by the input-output relation

26 $$y(t) = u(t - \delta) \qquad t \geq t_0 + \delta,\ t_0 \in T$$

In other words, the output of $\mathcal{A}$ at time $t \geq t_0 + \delta$ is equal to the value of the input of $\mathcal{A}$ at time $t - \delta$. The state of $\mathcal{A}$ at time t is the input "stored" in the delay line, that is, $u_{[t-\delta,t)}$. The input-output-state relation is expressible as

27 $$\begin{aligned} y(t) &= \text{value of } \mathbf{s}(t_0) \text{ at } 2t_0 - t && \text{for } t_0 \leq t < t_0 + \delta \\ &= u(t - \delta) && \text{for } t \geq t_0 + \delta \end{aligned}$$

where $\mathbf{s}(t_0) \triangleq u_{[t_0-\delta,t_0)}$.

Note that in all of the preceding examples the state space is finite-dimensional. In the present example, it is not. Specifically, the state space is the set of all segments of the form $u_{[t-\delta,t)}$, where u ranges over time functions in the input function space of $\mathcal{A}$.

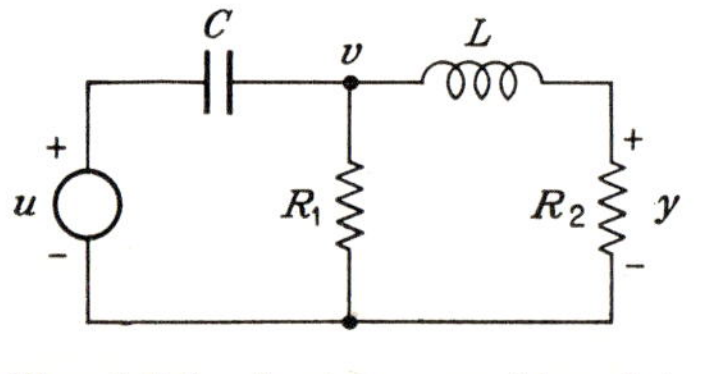

Fig. 1.7.1 System considered in Example *1.7.28*.

28 *Example* Consider the electrical network $\mathcal{A}$ shown in Fig. *1.7.1*. This is an oriented object in which u is the input voltage, y is the output voltage, and v is a suppressed output variable (see Sec. *4*).

The input-output relations for this object (derived by the rules of circuit analysis, see Refs. *3* to *5*) are

29 $$y + \frac{R_2}{R_1} v + R_2 C \dot{v} = R_2 C \dot{u}$$
$$\dot{y} + \frac{R_2}{L} y = \frac{R_2}{L} v$$

As will be shown in a general way in Chap. *4*, Sec. *6*, a vector $\mathbf{s}(t)$ which has as its components the voltages across the capacitors and the currents through the inductors of an *RLC* network $\mathcal{A}$ at time t qualifies as a state vector for $\mathcal{A}$. For the network under consideration this yields

30 $$\mathbf{s}(t) = \left(\frac{y(t)}{R_2}, u(t) - v(t)\right)$$

as the expression for the state vector. The corresponding input-output-state relation can be found from *29* and *30* by using techniques described in Chap. *4*. Employing these techniques and assuming for simplicity that $R_1 = R_2 = C = L = 1$, we obtain

31 $$y(t) = \langle \boldsymbol{\phi}(t - t_0), \mathbf{s}(t_0) \rangle + \int_{t_0}^{t} e^{-(t-\xi)}[\cos(t - \xi) - \sin(t - \xi)]u(\xi)\, d\xi \qquad t \geq t_0$$

where $\langle \mathbf{a},\mathbf{b} \rangle$ denotes the scalar product of $\mathbf{a}$ and $\mathbf{b}$ and the vector $\boldsymbol{\phi}(t)$ is given by

32 $$\boldsymbol{\phi}(t) = (e^{-t} \cos t, -e^{-t} \sin t) \qquad t \geq 0$$

Our purpose in considering this example is merely that of illustrating the form of the input-output-state relation in the case of an electrical network such as the one shown in Fig. *1.7.1*, rather than that of justifying the choice of $\mathbf{s}(t)$ as expressed by *30* as the state vector or showing how to derive the input-output-state relation *31* from the circuit equations *29*. This we shall do in a general way in Chap. *4*.

8 *Miscellaneous concepts*

In this and the following two sections, we shall discuss briefly and in quick succession several miscellaneous concepts which will be needed in later chapters. We begin with

State equations

In the course of analyzing Examples *1.7.10*, *1.7.12*, *1.7.19*, and *1.7.28* in the preceding section, we have obtained explicit expressions for the state of $\mathcal{A}$ at time t in terms of the state of $\mathcal{A}$ at time t_0 and the input to $\mathcal{A}$ between t_0 and t. More generally, if the consistency conditions are satisfied, then it can be asserted that $\mathbf{s}(t)$, the state at time t, is uniquely determined by $\mathbf{s}(t_0)$, the state at time t_0, and $\mathbf{u}_{(t_0,t]}$, and that the functional dependence of $\mathbf{s}(t)$ on $\mathbf{s}(t_0)$ and $\mathbf{u}_{(t_0,t]}$ can be derived, at least in principle, from the input-output-state relation

1 $$\mathbf{y}(t) = \mathbf{A}(\mathbf{s}(t_0); \mathbf{u}_{(t_0,t]})$$

This assertion is an immediate consequence of the definition of $\mathbf{s}(t)$ (*1.7.1*). Specifically, we have defined $\mathbf{s}(t)$ to be any element of the set $Q_t(\boldsymbol{\alpha}_0;\mathbf{u}_{(t_0,t]})$, where $\boldsymbol{\alpha}_0 = \mathbf{s}(t_0)$. Clearly, since Q_t depends only on $\boldsymbol{\alpha}_0$ and $\mathbf{u}_{(t_0,t]}$, so must $\mathbf{s}(t)$. Furthermore, from the way in which Q_t is defined (*1.6.38*), it is evident that, given the input-output-state relation *1* and given $\boldsymbol{\alpha}_0$ and $\mathbf{u}_{(t_0,t]}$, one can determine the sets $Q(\boldsymbol{\alpha}_0;\mathbf{u},\mathbf{u}')$, with $\mathbf{u}'$ varying over the input segment space of $\mathcal{A}$, and then determine their intersection, which is Q_t. Thus, in principle at least, Q_t and hence $\mathbf{s}(t)$ can be deduced from $\boldsymbol{\alpha}_0$, $\mathbf{u}_{(t_0,t]}$, and the input-output-state relation *1*. As Examples *1.7.10*, *1.7.12*, and *1.7.19* demonstrate, this can be done quite easily in cases in which it can be shown directly that there exists a state, say, $\boldsymbol{\alpha}$, which belongs to all $Q(\boldsymbol{\alpha}_0;\mathbf{u},\mathbf{u}')$ sets and hence lies in the intersection Q_t.

For convenience in reference, the expression for $\mathbf{s}(t)$ as a function of $\mathbf{s}(t_0)$ and $\mathbf{u}_{(t_0,t]}$ will be called the *state equation* for $\mathcal{A}$ and it will be written in the general form

2 $$\mathbf{s}(t) = \mathbf{s}(\mathbf{s}(t_0);\mathbf{u}_{(t_0,t]})$$

where $\mathbf{s}$ represents a function with values in Σ, defined on the product space (*1.4.17*) of Σ and $R[\mathbf{u}]$. To emphasize that *2* is derivable from *1*, we shall say that the state equation *2* is *induced* by the input-output-state relation *1*. A typical example of a state equation is *1.7.17*.

Now if we let t_0 approach t from the left in both *1* and *2*, then, under suitable regularity assumptions on $\mathbf{s}$ and $\mathbf{u}$, *1* and *2* will tend in the limit to differential equations of the form

3 $$\dot{\mathbf{s}}(t) = \mathbf{f}(\mathbf{s}(t),\mathbf{u}(t), \ldots ,\mathbf{u}^{(k)}(t),t)$$

4 $$\mathbf{y}(t) = \mathbf{g}(\mathbf{s}(t),\mathbf{u}(t), \ldots ,\mathbf{u}^{(l)}(t),t)$$

where $\mathbf{f}$ and $\mathbf{g}$ are ordinary (point) functions of their arguments and $u^{(k)} \triangleq k$th derivative of $\mathbf{u}$. If this holds true for $\mathcal{A}$, then $\mathcal{A}$ will be said to be of *differential type*, or, simply, a *differential object*, and *3* and *4* will be referred to as its *state equations in differential form*. Differential objects and their state equations are of particular interest to us, since most of the systems considered in later chapters will be of differential type.

A somewhat more general type of object is one for which *4* but not necessarily *3* holds true. In this case, the state equations are of the form

5 $$\mathbf{s}(t) = \mathbf{s}(\mathbf{s}(t_0);\mathbf{u}_{(t_0,t]})$$

6 $$\mathbf{y}(t) = \mathbf{g}(\mathbf{s}(t),\mathbf{u}(t), \ldots ,\mathbf{u}^{(k)}(t),t)$$

In most cases, the expression for $\mathbf{y}(t)$ does not involve the derivatives of the input or can be put into such form by a redefinition of $\mathbf{s}(t)$. Thus, the usual form of state equations is

7 $$\mathbf{s}(t) = \mathbf{s}(\mathbf{s}(t_0);\mathbf{u}_{(t_0,t]})$$

8 $$\mathbf{y}(t) = \mathbf{g}(\mathbf{s}(t),\mathbf{u}(t),t)$$

A simple example of an object which admits of the representation *7* and *8* but not *3* and *4* is the "perfect delay line" of Example *1.7.25*. Note that in this case the output at time t depends only on $\mathbf{s}(t)$ and not on $u(t)$.

When there is a need to differentiate between *3* and *5*, we shall say that *5* is a state equation in *explicit form*, whereas *3* is a state equation in differential form. In effect, if $\mathfrak{A}$ is characterized by *3* and *4*, then its state equations in explicit form (*5* and *6*) can be obtained, in principle, by solving the vector differential equation *3* for $\mathbf{s}(t)$ and expressing $\mathbf{s}(t)$ as a function of $\mathbf{s}(t_0)$ and $\mathbf{u}_{(t_0,t]}$. On the other hand, if $\mathfrak{A}$ is characterized by *5* and *6*, then it may not be possible to put *5* into the differential form *3*.

When we say that $\mathfrak{A}$ is *characterized* by *5* and *6*, what we have in mind is that on substituting $\mathbf{s}(t)$ from *5* into *6* one obtains the input-output-state relation for $\mathfrak{A}$. Consequently, not only are *5* and *6* deducible (under suitable regularity assumptions on $\mathbf{s}$ and $\mathbf{u}$) from the input-output-state relation *1* but also the input-output-state relation *1* is deducible from the state equations. Consequently, the state equations of $\mathfrak{A}$ provide a complete characterization of $\mathfrak{A}$ in the sense of Definition *1.6.2*. Note that this is true also of the differential state equations *3* and *4* provided *3* has a unique solution.

In the case of linear differential systems—which are the main concern of this book—the differential state equations assume a particularly simple form, namely,

9 $$\dot{\mathbf{s}}(t) = \mathbf{A}(t)\mathbf{s}(t) + \mathbf{B}(t)\mathbf{u}(t)$$

10 $$\mathbf{y}(t) = \mathbf{C}(t)\mathbf{s}(t) + \mathbf{D}_0(t)\mathbf{u}(t) + \mathbf{D}_1(t)\mathbf{u}^{(1)}(t) + \cdots + \mathbf{D}_k(t)\mathbf{u}^{(k)}(t)$$

where the coefficients are (possibly time-dependent) matrices. Such state equations will be discussed at length in Chaps. *4* through *7*.

Continuous-state, discrete-state, and finite-state objects

The nature of the state space Σ of an object $\mathfrak{A}$ is one of its more important characteristics. For most of the objects considered in this book, Σ is a finite-dimensional space or, more specifically, the set of all ordered n-tuples of real or complex numbers. [One simple exception is the state space of a perfect delay line (*1.7.25*), which is infinite-dimensional.] More generally, if Σ is a continuum, we shall say that $\mathfrak{A}$ is a *continuous-state object*. If Σ is a countable set, $\mathfrak{A}$ will be said to be a *discrete-state object*. And, finally, if Σ is a finite set, we shall say that $\mathfrak{A}$ is a *finite-state object*. An example of the latter is the object discussed in *1.7.19*.

Deterministic and probabilistic (stochastic) objects

In defining an abstract object and related concepts we have made the tacit assumption that the behavior of $\mathcal{A}$ does not exhibit any randomness. In effect, we have defined via *1.6.2* an oriented abstract object which is *deterministic* in the sense that $\mathbf{y}(t)$ is uniquely determined by $\mathbf{s}(t_0)$ and $\mathbf{u}_{(t_0,t]}$ for all $t > t_0$. However, it is not difficult to extend the definition in question to *probabilistic* or *stochastic* objects in which, for each t, $\mathbf{y}(t)$ is a random variable. For such objects, the input-output-state relation *1* which characterizes a deterministic object is replaced by an input-output distribution-state relation which expresses the probability measure on the space of response segments $\mathbf{y}_{(t_0,t]}$ as a function of the initial state $\mathbf{s}(t_0)$ and the input segment $\mathbf{u}_{(t_0,t]}$. Correspondingly, the state equation $\mathbf{s}(t) = \mathbf{s}(\mathbf{s}(t_0);\mathbf{u}_{(t_0,t]})$ is replaced by an equation which expresses the conditional probability distribution of $\mathbf{s}(t)$ for each t, given $\mathbf{s}(t_0)$ and $\mathbf{u}_{(t_0,t]}$. We shall not dwell further on the properties of stochastic objects, since we shall have no occasion to treat them in this book.

Single and multiple experiments and measurability of states

Underlying the notions of input and output of an abstract object $\mathcal{A}$ is the informal understanding (see Sec. *4*) that the inputs are those terminal variables of a physical realization $\mathcal{P}(\mathcal{A})$ of $\mathcal{A}$ (if one exists) which are varied directly by an experimenter and that the outputs are the terminal variables which are observed but not varied directly. On the other hand, in our definition of state (*1.7.1*) there was no mention of whether, or to what extent, the initial state $\mathbf{s}(t_0)$ is under the control of the experimenter. The reason for this is that the question can be answered in a number of ways depending on the type of experiment which is performed on $\mathcal{P}(\mathcal{A})$. The following discussion merely touches on some of the aspects of this many-faceted problem.

To begin with, one has to distinguish between a so-called *single experiment,* where it is assumed that the experimenter has at his disposal just one copy of $\mathcal{P}(\mathcal{A})$, and a *multiple experiment,* where the experimenter has access to an unlimited number of identical copies of $\mathcal{P}(\mathcal{A})$, all in the same initial state $\mathbf{s}(t_0)$.[1] In the sequel, the term "experiment" will be used in the sense of "single experiment," unless otherwise noted. It should be noted, however, that any property of $\mathcal{A}$ whose definition involves qualifying conditions such as "for all $\mathbf{u}$" or "for all $\mathbf{s}(t_0)$" cannot be verified, in general, without performing a

[1] The notions of "single experiment" and "multiple experiment" have been introduced by E. F. Moore in his basic work on the theory of sequential machines (see Ref. *6*). Moore uses the term *simple experiment* rather than *single experiment,* as we do here.

multiple experiment on $\mathcal{A}$. For example, the property of linearity (*3.4.25*) involves such conditions; and thus it is not possible, in general, to determine whether $\mathcal{A}$ is linear or not by experimenting with a single copy of $\mathcal{A}$.

In a single or a multiple experiment starting at time t_0, the initial state $\mathbf{s}(t_0)$ is *settable* if, at any initial time t_0, $\mathbf{s}(t_0)$ can be chosen at will within the state space Σ. $\mathbf{s}(t_0)$ is *nonsettable* if the experimenter cannot assign to $\mathbf{s}(t_0)$ an arbitrarily selected value in Σ at the beginning of the experiment.

11 *Example* Consider the clock described in Example *1.7.19*. Here it is conceivable that the experimenter would be permitted to adjust the position of the hour hand at will at the beginning of the experiment. In this case, $\mathbf{s}(t_0)$ would be settable.

On the other hand, consider the network of Fig. *1.7.1* and suppose that it is encapsulated in plastic, with only the input and output terminals accessible to the experimenter. In that case, the initia' state of the network would not be settable. A different kind of restriction on the choice of the initial state is discussed in *2.6.24*. ◁

Another notion of importance is that of *measurability*. We shall say that the state of $\mathcal{A}$ is *measurable* if the experimenter can determine the value of $\mathbf{s}(t)$ for each t either directly or from the knowledge of the input segment $\mathbf{u}_{(t_0,t]}$ and the output segment $\mathbf{y}_{(t_0,t]}$, without knowing the initial state $\mathbf{s}(t_0)$. In this connection it should be noted that, in the case of a deterministic object, the knowledge of $\mathbf{s}(t_0)$ and $\mathbf{u}_{(t_0,t]}$ suffices to determine $\mathbf{s}(t)$, but this is not true of stochastic objects. We shall also say that the state is *intermittently measurable* if the experimenter can determine $\mathbf{s}(t)$ at some but not all instants of time in the course of the experiment. The so-called *interrupted control processes* defined by Bellman and Kalaba (see Ref. *7*) fall into this category.

A basic problem that is encountered in many different contexts is that of the determination of the initial state of $\mathcal{A}$. In one variant of this problem, the experimenter is given $\mathbf{u}_{(t_0,t]}$ and $\mathbf{y}_{(t_0,t]}$ and is asked to find $\mathbf{s}(t_0)$. In effect, this amounts to solving the input-output-state relation

12 $$\mathbf{y}_{(t_0,t]} = \bar{\mathbf{A}}(\mathbf{s}(t_0);\mathbf{u}_{(t_0,t]})$$

for $\mathbf{s}(t_0)$ in terms of $\mathbf{u}_{(t_0,t]}$ and $\mathbf{y}_{(t_0,t]}$. If this can be done for all $\mathbf{s}(t_0)$ and all $\mathbf{u}$ and $\mathbf{y}$ within $R[\mathbf{u}]$ and $R[\mathbf{y}]$, respectively (see *1.6.2*), and if the solution is unique (to within equivalent states, see *2.2.1*), then $\mathcal{A}$ will be said to be *initial state determinable in strict sense.*[1] [As will be

[1] A closely related notion in the case of linear systems is that of *observability*, which will be discussed in detail in Chap. *11*, Sec. *4*.

seen later (*3.7.55*), this property is possessed by reduced linear systems.] More generally, $\mathcal{a}$ will be said to be *initial state determinable in wide sense* if no matter what $\mathbf{s}(t_0)$ is, the experimenter can always find an input $\mathbf{u}_{(t_0,t]}$ such that $\mathbf{s}(t_0)$ can be determined uniquely (to within equivalent states) from the knowledge of $\mathbf{u}_{(t_0,t]}$ and $\mathbf{y}_{(t_0,t]}$. Note that the difference between the two definitions is that in the former every $\mathbf{s}(t_0)$ can be determined from *any* $\mathbf{u}_{(t_0,t]}$ and $\mathbf{y}_{(t_0,t]}$ associated with $\mathbf{s}(t_0)$, whereas in the latter *there exists* a $\mathbf{u}_{(t_0,t]}$ [dependent on $\mathbf{s}(t_0)$] such that from this $\mathbf{u}_{(t_0,t]}$ and the response segment $\mathbf{y}_{(t_0,t]}$ the experimenter can determine $\mathbf{s}(t_0)$.

In a similar fashion, one can define *terminal state determinability*, i.e., the determinability of $\mathbf{s}(t)$ from the knowledge of $\mathbf{u}_{(t_0,t]}$ and $\mathbf{y}_{(t_0,t]}$, under the assumption that $\mathbf{s}(t_0)$ is not known. For a discussion of this problem in the context of finite-state systems see Refs. *6* and *8*.

Nonanticipative and anticipative objects

Although we have not stated so explicitly, Definition *1.6.2* defines an object $\mathcal{a}$ that is *nonanticipative*, meaning, roughly, that $\mathcal{a}$ acts only on the present ($t = t_0$) and past ($t < t_0$) values of its input and not on its future ($t > t_0$) values. The tacit assumption that $\mathcal{a}$ is nonanticipative enters into the formulation of the third self-consistency condition (*1.6.35*). To place it in evidence, we make the

13 **Definition** An abstract object $\mathcal{a}$ is *nonanticipative* if and only if for all t_0 in T, all initial states, and all inputs $\mathcal{a}$ has the following property. Using the notation of *1.6.16*, let $\mathbf{uu}'$ be an input segment (consisting of $\mathbf{u}$ followed by $\mathbf{u}'$) and let $\mathbf{yy}'$ be the corresponding output with $\mathcal{a}$ starting in state $\mathbf{s}(t_0)$. Then, if $\mathbf{u}$ is kept fixed and $\mathbf{u}'$ is varied over $R[\mathbf{u}']$, $\mathbf{y}$ remains fixed and $\mathbf{y}'$ varies over $R[\mathbf{y}']$. (In plain words, this means that the input segment $\mathbf{u}'$ has no influence on a prior output segment $\mathbf{y}$.)

An abstract object which is not nonanticipative will be said to be *anticipative*. More particularly, if $\mathcal{a}$ acts only on the present and future values of its input, and not on its past values, it will be said to be *purely anticipative*. In the case of a purely anticipative object, the input-output-state relation has exactly the same form *2* as for a nonanticipative object, with the only difference being that it is valid for $t < t_0$ rather than $t > t_0$. In a sense, a purely anticipative object is a mirror image in time of its nonanticipative counterpart.

14 *Example* An object characterized by the input-output relation

$$y(t) = u(t-1) \qquad t > t_0 + 1,\ t_0 \in T$$

where u and y are real-valued time functions, is nonanticipative. (One physical realization of such an object is a perfect delay line.)

15 *Example* An object characterized by the input-output relation

$$y(t) = u(t + 1) \qquad t > t_0,\ t_0 \in T$$

is purely anticipative.

16 *Example* An object characterized by the input-output relation

$$y(t) = u(t - 1) + u(t + 1) \qquad t > t_0 + 1,\ t < t_0 - 1,\ t_0 \in T$$

is anticipative. *Question:* Why is this relation valid over the intervals indicated? What is the state of this object at time t?

Memoryless and finite-memory objects

17 **Definition** An object $\mathcal{A}$ is *memoryless* if, for each t, $\mathbf{y}(t)$ depends only on $\mathbf{u}(t)$, that is, the input-output relation is of the form

$$\mathbf{y}(t) = \mathbf{g}(\mathbf{u}(t),t)$$

where $\mathbf{g}$ is a function defined on $R[\mathbf{u}(t)] \times T$. For example, a *squarer*, which is an object characterized by the input-output relation

18 $$y(t) = [u(t)]^2$$

is memoryless. On the other hand, a *differentiator*, which is an object characterized by the input-output relation

$$\mathbf{y}(t) = \frac{d\mathbf{u}}{dt}$$

is not memoryless, since $\mathbf{y}(t)$ is determined not only by $\mathbf{u}(t)$ but also by the values of $\mathbf{u}$ in the immediate vicinity of t.

To say that $\mathbf{y}(t)$ depends only on $\mathbf{u}(t)$ and nothing else is equivalent to saying that the state space Σ consists of a single point. In later discussions we shall find it convenient to denote the only state of a memoryless object by a special symbol such as (1).

19 **Definition** An object $\mathcal{A}$ is said to have a *finite memory of length L* if the input-output relation of $\mathcal{A}$ admits of the representation

20 $$\mathbf{y}(t) = \mathbf{g}(\mathbf{u}_{[t-L,t]}) \qquad t > t_0 + L,\ t_0 \in T$$

where $\mathbf{g}$ is a function defined on the space of segments $\{\mathbf{u}_{[t-L,t]}\}$ and L is a finite real number. Essentially, *20* means that for each t the output of $\mathcal{A}$ at time t depends only on the values of the input in the interval $[t - L, t]$. A simple example of a finite memory object is a perfect delay line characterized by the input-output relation

$$y(t) = u(t - L) \qquad t > t_0 + L,\ t_0 \in T$$

Another example is a so-called *moving averager*, which is characterized by the input-output relation

21
$$y(t) = \frac{1}{L}\int_{t-L}^{t} u(\xi)\, d\xi \qquad t > t_0 + L,\ t_0 \in T$$

Sources

Up to this point, we have concerned ourselves with oriented objects which have an input $\mathbf{u}$ ranging over $R[\mathbf{u}]$ and an output $\mathbf{y}$ ranging over $R[\mathbf{y}]$. There is, however, an important class of objects, namely, *sources*, which have an output $\mathbf{y}$ but no input. We have not treated such objects as a separate class because a source $\mathcal{S}$ can be regarded as a degenerate form of an oriented object $\mathcal{A}$ whose input segment space $R[\mathbf{u}]$ comprises a single time function, say, $\mathbf{u}^0$. Based on this viewpoint, a source can be defined as follows.

22 **Definition** A source $\mathcal{S}$ is an object whose output $\mathbf{y}(t)$ admits the representation

23
$$\mathbf{y}(t) = \mathbf{S}(\mathbf{s}(t_0); \mathbf{u}^0_{(t_0, t]}) \qquad \mathbf{s}(t_0) \in \Sigma$$

where $\mathbf{u}^0$ is a fixed time function and *23* is an input-output-state relation. A source is *memoryless* if $\mathcal{S}$ is a memoryless object in the sense of Definition *17*.

24 *Example* A source whose output is expressed by $y(t) = \sin t,\ t \geq t_0$, is a memoryless source. On the other hand, a source whose output is defined by *1.7.25* with, say, $u(t) = \sin t,\ t \geq t_0$, is a source with memory, since the output over the interval $t_0 \leq t \leq t_0 + \delta$ depends on the initial state of $\mathcal{A}$ and is not determined by the values of $u(t)$ for $t \geq t_0$.

9 *Graphical representation and system elements*

So far our discussion has been centered on the notion of a single abstract object $\mathcal{A}$ which is associated with a set of terminal variables and is characterized by the relations between them. Our basic concern, however, is with *systems* of objects rather than single objects, i.e., with the collective behavior of a set of objects which interact with one another. Until we define the term "system" more formally in Sec. *10*, it will be used by us in its usual not well-defined sense.

As a preliminary, we shall dwell briefly on a few topics which involve largely matters of terminology and notation. We begin with a discussion of

Graphical symbols for abstract objects

To denote an oriented abstract object $\mathcal{A}$ in a graphical form, we shall use for the most part the conventional rectangular-block representation illustrated in Fig. *1.9.1*. Here the leads represent the terminal variables and the inputs and outputs are identified by arrowheads which point toward $\mathcal{A}$ in the case of inputs and away from $\mathcal{A}$ in the case of outputs.

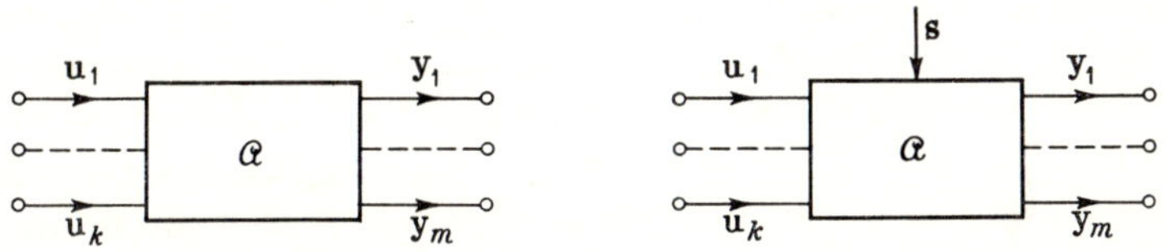

Fig. 1.9.1 Block-diagram representations of an object.

1 **Remark** Note that it is customary, but not really necessary, to associate a scalar rather than a vector variable with each terminal. The only point that matters is that a terminal variable—whatever the nature of its range—must be treated as a single indivisible entity when the object in question is interconnected with other objects. Thus, if an output variable, say, $\mathbf{y}^i$, is a vector with components $y_1^i, \ldots, y_l^i$, then these components cannot be treated as terminal variables in their own right when it comes to connecting $\mathcal{A}$ to other objects in a system (see Sec. *10*). Rather, they must be treated as a single "bundle" which can be connected only to matching input bundles of one or more other objects. ◁

We assume that $\mathcal{A}$ is characterized or is characterizable by an input-output-state relation of the form (*1.6.3*)

2 $$\mathbf{y}(t) = \mathbf{A}(\boldsymbol{\alpha};\mathbf{u})$$

where $\mathbf{u}$ is an input segment $\mathbf{u}_{(t_0,t]}$, $\boldsymbol{\alpha}$ is the state at time t_0, $\mathbf{y}(t)$ is the output at time t, and $\mathbf{A}$ is a function defined on the product of the state space Σ and the input segment space $R[\mathbf{u}]$. Equation *2* implies that each of the components of $\mathbf{y}(t)$ is, like $\mathbf{y}(t)$, a function of $\boldsymbol{\alpha}$ and $\mathbf{u}$. We shall express this by the notation

3 $$y_i(t) = A_i(\boldsymbol{\alpha};\mathbf{u}) \qquad i = 1, \ldots, m$$

where $y_i(t)$ is the ith component of $\mathbf{y}(t)$, $i = 1, \ldots, m$. This implies —as illustrated in Fig. *1.9.2*—that an object $\mathfrak{A}$ having k input variables and m output variables can always be decomposed into m objects $\mathfrak{A}_1, \ldots, \mathfrak{A}_m$ each of which has k input variables and one output variable. Thus, in the case of a system of oriented objects there is no loss in generality in assuming that the system is comprised of oriented

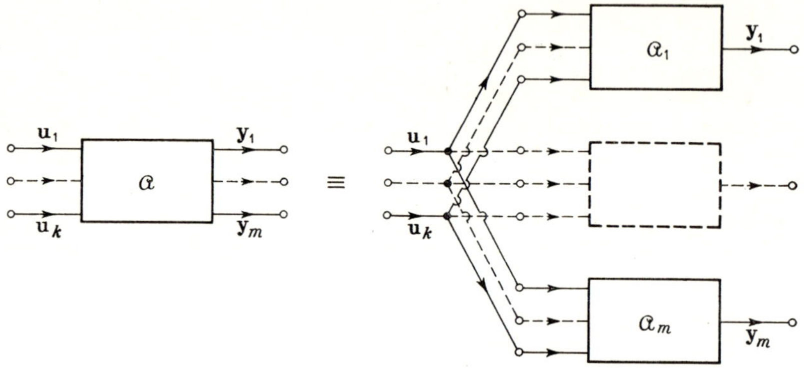

Fig. 1.9.2 Decomposition into objects having a single output terminal.

objects each having just one output variable. To take advantage of this, it is expedient to adopt a simplified notation for oriented objects having k input variables $u_1, \ldots, u_k$ and one output variable y. Two examples of such notation are shown in Fig. *1.9.3*. Here (*a*) depicts the conventional signal-flow-graph type of representation (see Chap. *9*, Sec. *15*) in which the nodes are associated with variables and the

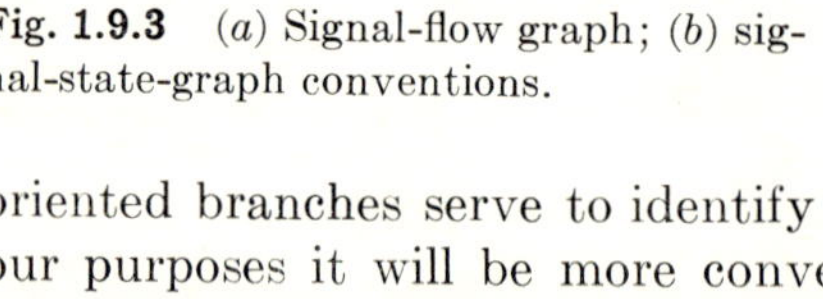

Fig. 1.9.3 (*a*) Signal-flow graph; (*b*) signal-state-graph conventions.

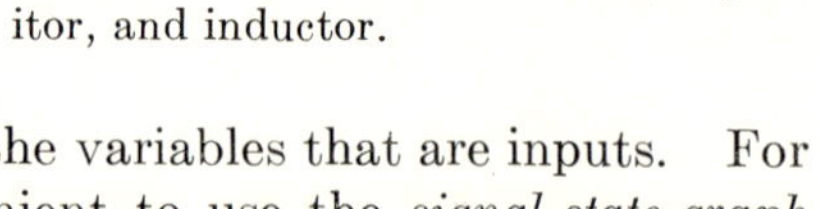

Fig. 1.9.4 Symbols for resistor, capacitor, and inductor.

oriented branches serve to identify the variables that are inputs. For our purposes it will be more convenient to use the *signal-state-graph* representation depicted in (*b*). Here the branches are associated with the input and output variables and the node is associated with state. In this way, it is made clear that the output y depends not only on the inputs $u_1, \ldots, u_k$ but also on the state $\mathbf{s}$.

It should be noted that the diagrammatic representations of $\mathfrak{A}$

depicted in Figs. *1.9.1* to *1.9.3* convey information only concerning the number of terminal variables of $\mathcal{A}$ and which of these terminal variables are inputs and which are outputs; they do not provide quantitative information on how the input, output, and state variables are related to one another—which is the information contained in the input-output and input-output-state relations of $\mathcal{A}$. Such information can be supplied, of course, by writing the input-output relation or its equivalent inside or adjacent to the block representing $\mathcal{A}$. Another obvious way of conveying information concerning the input-output relation of $\mathcal{A}$ is to use a set of specialized graphical symbols each of which is associated with a particular form of input-output relation. This is what is customarily done in the case of electric circuits, where the symbols shown in Fig. *1.9.4* represent nonoriented objects with terminal variables v and i. These objects, called resistor, capacitor, and inductor, are characterized by the terminal relations $v = Ri, i = C\,dv/dt$, and $v = L\,di/dt$, where R, C, and L are constant nonnegative parameters called resistance, capacitance, and inductance, respectively. (The plus sign specifies the positive polarity for v and the arrow indicates the positive sense of current flow for i.)

In the sequel, we shall on occasion use terms that are synonymous with "object" but give it somewhat specialized or different shades of meaning. The terms in question and the reasons for their use are listed below.

Black box To stress that the physical object of which $\mathcal{A}$ is an abstract model is dealt with through its terminal variables. We shall try to avoid using this term even though it is a very suggestive one, largely because its meaning has become fuzzy through wide and careless usage.

Multipole To indicate that $\mathcal{A}$ has two or more terminal variables.

n-pole To indicate that $\mathcal{A}$ has n terminal variables.

2-pole or two-pole To indicate that $\mathcal{A}$ has one input and one output variable. On occasion we shall find the longer name "single-input single-output system" preferable for our purposes.

System element To stress that $\mathcal{A}$ is one of a set of objects of which a system is comprised, or to indicate that $\mathcal{A}$ can be used as a "building block" for other systems.

System To imply that the abstract object in question is a combination of other abstract objects.

There are several types of system element that play a particularly important role in system theory, both because of their frequent occurrence as components of complex systems and because they can be used as building blocks in the synthesis of almost any physically realizable system. A set of such elements is defined in the next subsection.

Adders, multipliers, scalors, delayors, and integrators

The five elements defined below have two significant properties. First, they form an *independent* set, in the sense that none can be constructed out of a finite number of the other four types of element. Second, they form a *complete* set, in the sense that almost any physically realizable system can be synthesized—to an arbitrarily high degree of accuracy—out of a sufficiently large number of such elements. We shall not attempt at this stage to state this assertion in a more precise form. In any case, its proof is outside the scope of our treatment.

The five elements in question are the *adder, multiplier, scalor, delayor,* and *integrator*. We proceed to give brief definitions of these terms.

4 **Definition** An *adder* is a memoryless (*1.8.17*) $(n + 1)$-pole $(n = 2, 3, \ldots)$ having n inputs $u_1, \ldots, u_n$ and one output y. The defining input-output relation for the adder is

5 $$y = u_1 + u_2 + \cdots + u_n$$

An adder is usually represented as a circle with a plus inside (Fig. *1.9.5a*).

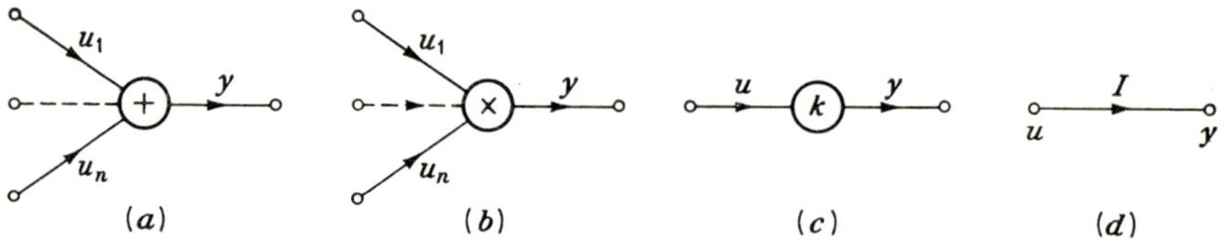

Fig. 1.9.5 Symbols for (*a*) adder, (*b*) multiplier, (*c*) scalor, and (*d*) unitor.

6 *Comment* Equation *5* reads: The output time function is the sum of input time functions. In other words, *5* implies that

$$y(t) = u_1(t) + u_2(t) + \cdots + u_n(t)$$

for all t. The point to note here is that whenever the terms in a relation are time functions (rather than their values at time t, as above), it should be understood that the relation holds for all t. ◁

For each t, the $u_i(t)$ are usually understood to be scalars ranging over the real line $\mathcal{R}^1$. More generally, the range $R[u_i]$ of each u_i can be a function space of the type defined in *1.3.7*. For our purposes, it will suffice to assume that, for each t and each i, $u_i(t)$ is a real or complex scalar. The same comment applies to the multiplier, scalor, and delayor defined below.

7 **Definition** A *multiplier* is an $(n + 1)$-pole $(n = 2, 3, \ldots)$ having n inputs $u_1, \ldots, u_n$ and one output y. The defining input-output relation for the multiplier is

$$y = u_1 u_2 \ . \ . \ . \ u_n$$

A multiplier is represented as a circle with a cross inside (Fig. *1.9.5b*).

8 **Definition** A *scalor*[1] is a 2-pole characterized by the input-output relation

$$y = ku$$

where k is a constant (not necessarily real or positive). A scalor may be regarded as an "amplifier" having a constant gain k. A scalor is represented as shown in Fig. *1.9.5c*. The constant k will be referred to as the *scale factor* or *gain* of a scalor. A scalor with a unity scale factor will be called a *unitor*, or the *identity operator*, and it will be denoted by 1 or **I** or I. (Fig. *1.9.5d*).

9 **Definition** A *delayor* is a 2-pole characterized by the input-output relation (see *1.7.25*)

$$y(t) = u(t - \delta) \qquad t > t_0 + \delta,\ t_0 \in T$$

where δ (δ = constant > 0) is the amount of delay. In words, the output of the delayor at time t is equal to the value of its input at time $t - \delta$. A delayor is represented as shown in Fig. *1.9.6a*. A delayor

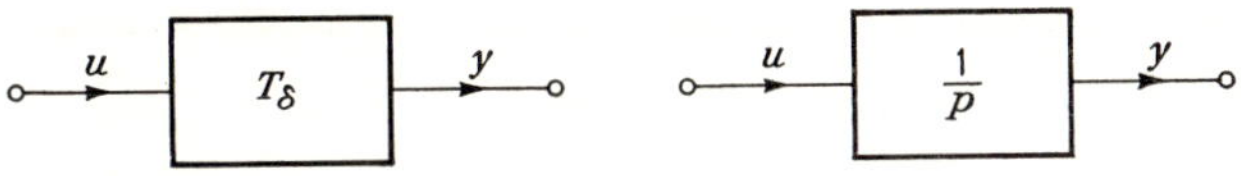

(*a*) (*b*)

Fig. 1.9.6 Symbols for (*a*) delayor and (*b*) integrator.

with $\delta = 1$ will be called a *unit delayor*, and will be denoted as in Fig. *1.9.6a*, with D replacing T_δ.

10 **Definition** An *integrator* is a 2-pole defined by the input-output-state relation

11 $$y(t) = y(t_0-) + \int_{t_0}^{t} u(\xi)\, d\xi \qquad t \geq t_0,\ t_0 \in T$$

$y(t_0-)$ [$\triangleq s(t_0-)$] is the output of the integrator at time t_0- (see *1.7.10*). An integrator is represented as shown in Fig. *1.9.6b*, where the symbol $1/p$ stands for the operation of integration. (A slightly more general definition of an integrator is given in *4.3.48*. The notion of a *differentiator* is defined in *4.4.37*.) ◁

We are now ready to consider the basic question of how two or more objects can be combined with each other to form a *system* of objects, or

[1] The terms "scalor" and "delayor" are due to Prof. W. H. Huggins, Jr., of the Johns Hopkins University, Baltimore, Md.

simply a *system*. In this connection, the notion of an *interconnection* plays a central role. We focus our attention on this and related notions in the following section.

10 *Interconnection of objects*

Roughly speaking, a system is a collection of interacting objects. To concretize the notion of a system, the interactions between the elements $\mathcal{A}_1, \ldots, \mathcal{A}_N$ of a system $\mathcal{S}$ are represented as a set of constraints on terminal variables, of which a typical one reads "μth terminal variable of object $\mathcal{A}_i$ is equal to νth terminal variable of object $\mathcal{A}_j$ for all t in T ($T \triangleq$ range of t)."[1] A collection of objects $\mathcal{A}_1, \ldots, \mathcal{A}_N$ subjected to such constraints is said to be an *interconnection* of $\mathcal{A}_1, \ldots, \mathcal{A}_N$ or simply an *interconnection*. An interconnection is usually represented in the form of a *block diagram* (Fig. *1.10.1*), in which the joining of a

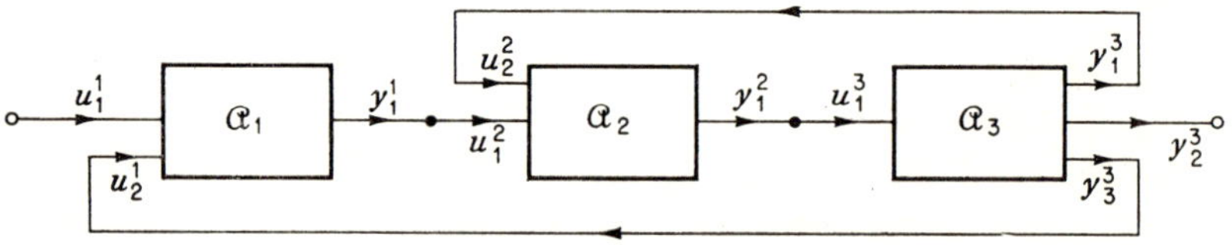

Fig. 1.10.1 An interconnection of $\mathcal{A}_1$, $\mathcal{A}_2$, $\mathcal{A}_3$.

terminal of $\mathcal{A}_i$ to a terminal of $\mathcal{A}_j$ signifies that the corresponding terminal variables are constrained to be equal for all t.

Suppose that $\mathcal{A}_1, \ldots, \mathcal{A}_N$ are characterized by their input-output-state relations and that they are interconnected in a specified manner, forming an object which we call $\mathcal{A}$. There are several questions concerning $\mathcal{A}$ that immediately suggest themselves, namely: What is the input-output-state relation for $\mathcal{A}$? What is the state space of $\mathcal{A}$? What is the output function space of $\mathcal{A}$? To provide a basis for answering these questions, it is convenient to introduce the notion of an *initially free interconnection*, denoted by $\mathcal{A}(t_0)$, in which the connections between the $\mathcal{A}_i$ in $\mathcal{A}$ are broken with switches which remain open till $t = t_0$ (Fig. *1.10.2*). As we shall see in the sequel, the artifice of regarding $\mathcal{A}$ as a limiting form, as $t_0 \to -\infty$, of an initially free interconnection $\mathcal{A}(t_0)$ makes it possible to relate the properties of $\mathcal{A}$ to those of $\mathcal{A}(t_0)$ and, in particular, give a characterization of the state space of

[1] It may appear that this does not apply to electrical networks, in which the constraints on terminal variables have the form of Kirchhoff's current and voltage relations. However, such relations can be put into the form stated in the text by regarding an adder as a network element. See Example *37*.

$\mathcal{C}$—a problem that is much less simple than it may appear to be on the surface.

In general, the questions posed above are by no means easy to answer, for we do not as yet possess a well-developed body of results bearing on the behavior of an interconnection of objects each of which is characterized by an input-output-state relation. Indeed, some of the concepts introduced in this section should be regarded as tentative in nature.

In order to facilitate the understanding of the role played by the notion of an initially free interconnection, we shall introduce it in the

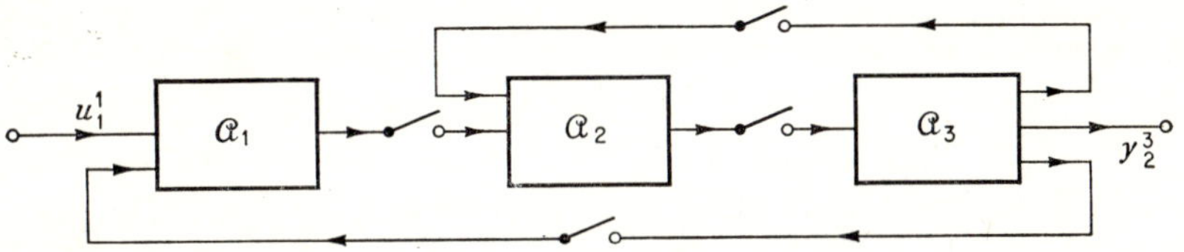

Fig. 1.10.2 Initially free interconnection of $\mathcal{C}_1$, $\mathcal{C}_2$, $\mathcal{C}_3$.

simple context of a tandem combination of two objects. Furthermore, in order to conserve space, most of our definitions will be formulated in terms of oriented objects, since it will usually be quite clear how analogous definitions can be made for nonoriented objects.

We begin our discussion with a degenerate type of interconnection, namely the

Direct product

Consider a collection of objects $\mathcal{C}_1, \ldots, \mathcal{C}_N$ each of which is characterized by an input-output-state relation of the form

1 $$\mathbf{y}^i = \bar{\mathbf{A}}_i(\boldsymbol{\alpha}^i;\mathbf{u}^i) \qquad \boldsymbol{\alpha}^i \in \Sigma_i,\ i = 1, \ldots, N$$

where $\mathbf{u}^i \triangleq \mathbf{u}^i_{(t_0,t]}$ and $\mathbf{y}^i = \mathbf{y}^i_{(t_0,t]}$, with $\mathbf{u}^i$, $\mathbf{y}^i$, and $\boldsymbol{\alpha}^i$ ranging over $R[\mathbf{u}^i]$, $R[\mathbf{y}^i]$, and Σ_i, respectively.

When there is no interaction at all between the $\mathcal{C}_i$, $i = 1, \ldots, N$, the collection $\mathcal{C}_1, \ldots, \mathcal{C}_N$ constitutes the *direct product* of $\mathcal{C}_1, \ldots, \mathcal{C}_N$. More formally, we have the

2 **Definition** The *direct product* of $\mathcal{C}_1, \ldots, \mathcal{C}_N$ is an object $\mathcal{C}$ comprising $\mathcal{C}_1, \ldots, \mathcal{C}_N$ and denoted by $\mathcal{C}_1 \times \cdots \times \mathcal{C}_N$, with the input to and the state of each $\mathcal{C}_i$, $i = 1, \ldots, N$, being independent of the inputs, outputs, and states of other objects in the collection. The input and output of $\mathcal{C}$ are defined to be the composite vectors

3 $$\mathbf{u} = (\mathbf{u}^1, \ldots, \mathbf{u}^N)$$

and

4 $$\mathbf{y} = (\mathbf{y}^1, \ldots, \mathbf{y}^N)$$

respectively. Correspondingly, the input and output segment spaces of $\mathcal{A}$ are the product spaces (*1.4.17*)

5 $$R[\mathbf{u}] = R[\mathbf{u}^1] \times \cdots \times R[\mathbf{u}^N]$$

and

6 $$R[\mathbf{y}] = R[\mathbf{y}^1] \times \cdots \times R[\mathbf{y}^N]$$

Furthermore, as will be shown in *2.6.10*, the composite vector

$$\mathbf{s}(t) = (\mathbf{s}^1(t), \ldots, \mathbf{s}^N(t))$$

qualifies as the state of $\mathcal{A}$ at time t, with the state space of $\mathcal{A}$ being the product space

7 $$\Sigma = \Sigma_1 \times \cdots \times \Sigma_N$$

8 *Example* The object of Fig. *1.10.5b* is the direct product of $\mathcal{A}$ and $\mathcal{B}$, with the input and output of $\mathcal{A} \times \mathcal{B}$ being $\mathbf{u} = (u_1,u_2)$ and $\mathbf{y} = (y_1,y_2)$, respectively. ◁

We turn next to

Initially free tandem combinations and products

Roughly speaking, $\mathcal{A}_1$ and $\mathcal{A}_2$ are connected in *tandem*, with $\mathcal{A}_1$ preceding $\mathcal{A}_2$, if the input to $\mathcal{A}_2$ ($\triangleq \mathbf{u}^2$) is constrained to be equal to the output of $\mathcal{A}_1$ ($\triangleq \mathbf{y}^1$), as illustrated in Fig. *1.10.3*. Here we have to distinguish

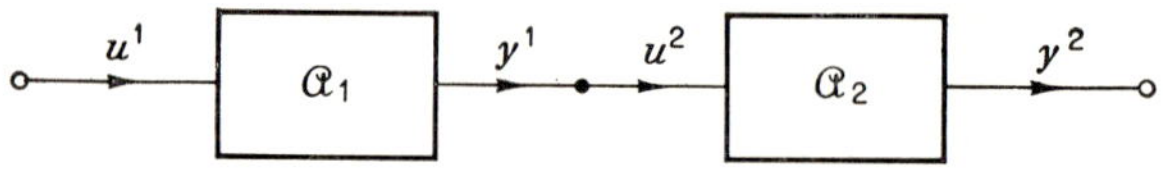

Fig. 1.10.3 Tandem combination of $\mathcal{A}_1$ and $\mathcal{A}_2$, with $\mathcal{A}_1$ preceding $\mathcal{A}_2$.

between two cases: (I) $\mathbf{u}^2 = \mathbf{y}^1$ for $t > t_0$, where t_0 is the instant at which $\mathbf{u}^1$ (input to $\mathcal{A}_1$) starts, and (II) $\mathbf{u}^2 = \mathbf{y}^1$ for all t. In the former case, the tandem combination of $\mathcal{A}_1$ and $\mathcal{A}_2$ will be said to be *initially free* or simply *free*. In the latter case, the tandem combination will be said to be initially *constrained*. Essentially, in case I, the combination of $\mathcal{A}_1$ and $\mathcal{A}_2$ acts like the direct product of $\mathcal{A}_1$ and $\mathcal{A}_2$ for $t < t_0$. This is illustrated in Fig. *1.10.4*, in which the switch K signifies that from time t_0 on (but not before) $\mathbf{u}^2$ is constrained to be equal to $\mathbf{y}^1$.

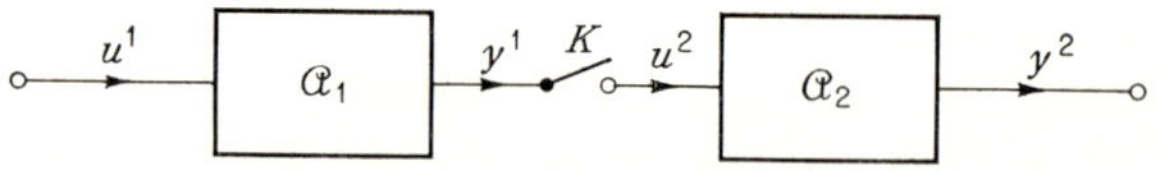

Fig. 1.10.4 Initially free tandem combination of $\mathcal{A}_1$ and $\mathcal{A}_2$.

We assume that $\mathcal{A}_1$ and $\mathcal{A}_2$ are characterized by input-output-state relations of the form *1*, $i = 1, 2$, with the output segment space of $\mathcal{A}_1$

contained in the input segment space of $\mathcal{A}_2$ in order to allow the use of the output of $\mathcal{A}_1$ as input to $\mathcal{A}_2$. Furthermore, we assume that the states of $\mathcal{A}_1$ and $\mathcal{A}_2$ at time t are given by the respective state equations ($i = 1, 2$)

9 $$\mathbf{s}^i(t) = \mathbf{s}^i(\boldsymbol{\alpha}^i;\mathbf{u}^i) \qquad \boldsymbol{\alpha}^i \triangleq \mathbf{s}^i(t_0),\ \mathbf{u}^i \triangleq \mathbf{u}^i_{(t_0,t]}$$

which are induced by the input-output-state relations *1* (see *1.8.2 et seq.*).

We focus our attention first on case I. In this case, the initial states $\boldsymbol{\alpha}^1$ and $\boldsymbol{\alpha}^2$ can be chosen arbitrarily in Σ_1 and Σ_2, respectively, or, equivalently, the pair $(\boldsymbol{\alpha}^1, \boldsymbol{\alpha}^2)$ can be chosen arbitrarily in the product space $\Sigma_1 \times \Sigma_2$. Clearly, given $\boldsymbol{\alpha}^1$, $\boldsymbol{\alpha}^2$, and $\mathbf{u}^1 \triangleq \mathbf{u}^1_{(t_0,t]}$, we can find from *1* and *9* the output $\mathbf{y}^2 \triangleq \mathbf{y}^2_{(t_0,t]}$ and the states $\mathbf{s}^1(t)$ and $\mathbf{s}^2(t)$ for any $t > t_0$. On the basis of this observation, we make the following

10 **Definition** The *free tandem combination* of $\mathcal{A}_1$ and $\mathcal{A}_2$ with $\mathcal{A}_1$ preceding $\mathcal{A}_2$, or, equivalently, the *free product* of $\mathcal{A}_2$ and $\mathcal{A}_1$, is an object $\mathcal{A}(t_0)$ characterized by the input-output-state relation

11 $$\mathbf{y}^2 = \bar{\mathbf{A}}_2[\boldsymbol{\alpha}^2;\bar{\mathbf{A}}_1(\boldsymbol{\alpha}^1;\mathbf{u}^1)]$$

in which the composite state $(\boldsymbol{\alpha}^1,\boldsymbol{\alpha}^2)$ plays the role of the initial state of $\mathcal{A}(t_0)$. Correspondingly, the state of $\mathcal{A}(t_0)$ at time t is given by the pair $(\mathbf{s}^1(t),\mathbf{s}^2(t))$, where $\mathbf{s}^1(t)$ and $\mathbf{s}^2(t)$ are expressed by (see *9*)

12 $$\begin{aligned}\mathbf{s}^1(t) &= \mathbf{s}^1(\boldsymbol{\alpha}^1;\mathbf{u}^1_{(t_0,t]})\\ \mathbf{s}^2(t) &= \mathbf{s}^2(\boldsymbol{\alpha}^2;\bar{\mathbf{A}}_1(\boldsymbol{\alpha}^1;\mathbf{u}^1_{(t_0,t]}))\end{aligned}$$

and the output segment space of $\mathcal{A}(t_0)$ is given by (see *1.6.2*)

13 $$R[\mathbf{y}^2] = \{\mathbf{y}^2|\mathbf{y}^2 = \bar{\mathbf{A}}_2[\boldsymbol{\alpha}^2;\bar{\mathbf{A}}_1(\boldsymbol{\alpha}^1;\mathbf{u}^1)]\} \qquad \mathbf{u}^1 \in R[\mathbf{u}^1],\ \boldsymbol{\alpha}^1 \in \Sigma_1,\ \boldsymbol{\alpha}^2 \in \Sigma_2$$

14 *Comment* It is evident that *11* satisfies the first self-consistency condition *1.6.11*. This, however, is not sufficient to qualify *11* as an input-output-state relation. The justification for referring to *11* as an input-output-state relation is provided by results which will be established in Chap. *2* (see *2.6.10*).

15 **Remark** It is important to observe that, unlike the initial state $(\boldsymbol{\alpha}^1,\boldsymbol{\alpha}^2)$ which ranges over the product space $\Sigma_1 \times \Sigma_2$, the pair $(\mathbf{s}^1(t),\mathbf{s}^2(t))$, which plays the role of the state of $\mathcal{A}(t_0)$ at time t, ranges over a subset of $\Sigma_1 \times \Sigma_2$ rather than $\Sigma_1 \times \Sigma_2$ itself. The subset in question is defined by

16 $$\Sigma(t_0,t) = \{(\mathbf{s}^1(t),\mathbf{s}^2(t))|\boldsymbol{\alpha}^1 \in \Sigma_1,\boldsymbol{\alpha}^2 \in \Sigma_2,\mathbf{u}^1 \in R[\mathbf{u}^1]\}$$

where $\mathbf{s}^1(t)$ and $\mathbf{s}^2(t)$ are given by *12*. This subset constitutes the state space of $\mathcal{A}(t_0)$ at time t, and the arguments t_0, t serve to indicate that it depends on both t_0 and t.

17 *Example* Let $\mathcal{A}_1$ be an integrator with input-output-state relation *1.9.11* and let $\mathcal{A}_2$ be a differentiator with input-output-state relation (see *4.4.39*)

18
$$y(t) = -u(t_0-)\delta(t - t_0) + \frac{d}{dt}[1(t - t_0)u(t)]$$

Then the input-output-state relation for the free product of $\mathcal{A}_2$ and $\mathcal{A}_1$ is given by

$$y^2(t) = u^1(t) + (\alpha^1 - \alpha^2)\delta(t - t_0-) \qquad t \geq t_0$$

where $\alpha^1 = y^1(t_0-)$ and $\alpha^2 = u^2(t_0-)$. Since $y^1(t) = u^2(t)$ for $t \geq t_0$, the state space $\Sigma(t,t_0)$ is $\mathcal{R}^2$ for $t = t_0-$ and $\Sigma(t_0,t) = \mathcal{R}^1$ for $t \geq t_0$ (see *22*). Thus, $\Sigma(t_0,t)$ changes from $\mathcal{R}^2$ to $\mathcal{R}^1$ at time t_0 and remains constant thereafter. Later on we shall see that this is characteristic of linear systems (Remark *2.6.8*).

19 **Remark** The definition of the free product of $\mathcal{A}_1$ and $\mathcal{A}_2$ can be extended in an obvious fashion to the free product of $\mathcal{A}_1, \ldots, \mathcal{A}_N$. For example, the input-output-state relation for the free product of $\mathcal{A}_3$, $\mathcal{A}_2$, and $\mathcal{A}_1$ (with $\mathcal{A}_1$, $\mathcal{A}_2$, and $\mathcal{A}_3$ separated by switches which close at t_0) is given by the nested expression

20
$$\mathbf{y}^3 = \bar{\mathbf{A}}_3\{\boldsymbol{\alpha}^3;\bar{\mathbf{A}}_2[\boldsymbol{\alpha}^2;\bar{\mathbf{A}}_1(\boldsymbol{\alpha}^1;\mathbf{u}^1)]\}$$

Note that this expression implies that the free product is associative in the sense that $(\mathcal{A}_3\mathcal{A}_2)\mathcal{A}_1$ and $\mathcal{A}_3(\mathcal{A}_2\mathcal{A}_1)$ are characterized by the same input-output-state relation. This is a special instance of the *associative principle* which is stated in *34*.

Constrained tandem combinations and products

The notion of an initially free tandem combination of $\mathcal{A}_1$ and $\mathcal{A}_2$ provides a convenient way of defining the tandem combination of $\mathcal{A}_1$ and $\mathcal{A}_2$ in which the input to $\mathcal{A}_2$ is constrained to be equal to the output of $\mathcal{A}_1$ for all t rather than $t > t_0$. Such a combination, call it $\mathcal{A}$, is represented diagrammatically as in Fig. *1.10.3*, with the connection of the output of $\mathcal{A}_1$ to the input of $\mathcal{A}_2$ signifying that $\mathbf{y}^1(t) = \mathbf{u}^2(t)$ for all t.

Essentially, we regard a constraint of the form "$\mathbf{y}^1(t) = \mathbf{u}^2(t)$ for all t" as a limiting form (as $t_0 \to -\infty$) of the constraint "$\mathbf{y}^1(t) = \mathbf{u}^2(t)$ for $t > t_0$." This implies that the state space of $\mathcal{A}$ is formally characterized by the limit, if it exists,

21
$$\Sigma(t) = \lim_{t_0 \to -\infty} \Sigma(t_0,t)$$

For general state spaces, the precise meaning which should be attached to *21* and the conditions under which the limit exists are neither simple nor even completely settled questions. For our purposes, it will suffice to take *21* to mean that, for each t, any point in $\Sigma(t)$ can be approached arbitrarily closely (in terms of a metric defined on the product space $\Sigma_1 \times \Sigma_2$) by a point in $\Sigma(t_0,t)$. Actually, the precise meaning of *21* will not play a material role in the following discussion, since our use of *21* on a general level will be largely formal in nature. (See also Remark *2.6.8*.)

22 *Comment* In later chapters we shall be concerned for the most part with differential objects (see *1.8.3 et seq.*) for which the state at time t is defined by a finite number of the derivatives of the input and output at time t. For such systems, $\Sigma(t)$, if it exists, can be determined by a direct process of elimination of identical variables from the expressions for the states of $\mathcal{A}_1$ and $\mathcal{A}_2$ at time t (see *2.6.8*). For example, suppose that $\mathbf{s}^1(t)$ and $\mathbf{s}^2(t)$ are real-valued and are respectively equal to the output of $\mathcal{A}_1$ and input of $\mathcal{A}_2$ at time t, that is, $s^1(t) = y^1(t)$ and $s^2(t) = u^2(t)$. Then the constraint "$y^1(t) = u^2(t)$ for all t" implies "$s^1(t) = s^2(t)$ for all t," and consequently if $y^1(t)$ ranges over $\mathcal{R}^1$ (for all t), so does the pair $(s^1(t),s^2(t))$. Thus, in this case Σ is a subspace ($= \mathcal{R}^1$) of the product space $\Sigma_1 \times \Sigma_2$ ($= \mathcal{R}^2$). This conclusion is consistent with that of Example *17*.

Note that $\Sigma(t)$ is independent of t by virtue of the assumption that (I) the dependence of $s^1(t)$ and $s^2(t)$ on the input and output does not involve t and (II) the range of $y^1(t)$ is independent of t. Such conditions are satisfied by most of the types of system with which we shall be concerned in later chapters. For this reason, we shall abbreviate $\Sigma(t)$ to Σ, with the tacit understanding that $\Sigma(t)$ exists and is independent of t.

Based on these observations, we make the following

23 **Definition** An object $\mathcal{A}$ is an *initially constrained tandem* (*series*) *combination*, or, more simply, *tandem combination* of $\mathcal{A}_1$ and $\mathcal{A}_2$, with $\mathcal{A}_1$ preceding $\mathcal{A}_2$, if the input to $\mathcal{A}_2$ is constrained to be equal to the output of $\mathcal{A}_1$ for all t. The tandem combination of $\mathcal{A}_1$ and $\mathcal{A}_2$ will also be referred to as an *initially constrained product*, or simply *product*, of $\mathcal{A}_2$ and $\mathcal{A}_1$, and it will be denoted by $\mathcal{A}_2\mathcal{A}_1$. $\lhd$

We regard $\mathcal{A}$ as a limiting form of the free product $\mathcal{A}(t_0)$ as $t_0 \to -\infty$. Then, the state space of $\mathcal{A}$ is a subset Σ of the product space $\Sigma_1 \times \Sigma_2$, with Σ formally defined by

$$\Sigma = \lim_{t_0 \to -\infty} \{(\mathbf{s}^1(t),\mathbf{s}^2(t)) | \boldsymbol{\alpha}^1 \in \Sigma_1, \boldsymbol{\alpha}^2 \in \Sigma_2, \mathbf{u}^1 \in R[\mathbf{u}^1]\}$$

where $\mathbf{s}^1(t)$ and $\mathbf{s}^2(t)$ are expressed by *12*. Correspondingly, the input-output-state relation of $\mathcal{A}$ is given by *20*, with $(\boldsymbol{\alpha}^1,\boldsymbol{\alpha}^2)$ ranging over Σ.

24 *Comment* As in the case of the free product, the product of $\mathcal{A}_2$ and $\mathcal{A}_1$ has the associative property, and consequently the product of $\mathcal{A}_N, \ldots, \mathcal{A}_1$ can be written as $\mathcal{A}_N \cdots \mathcal{A}_1$ without the use of parentheses.

25 *Comment* Throughout the foregoing discussion we have assumed that $\mathcal{A}_1$ and $\mathcal{A}_2$ are characterized by their respective input-output-state relations, and our aim was that of obtaining a similar characterization for $\mathcal{A}_2\mathcal{A}_1$. It should be noted, however, that alternatively we could have started with the characterization of $\mathcal{A}_1$ and $\mathcal{A}_2$ in terms of their spaces of input-output pairs (see *1.4.26*) and asked merely for a definition of the space of input-output pairs of $\mathcal{A}_2\mathcal{A}_1$. Specifically, this can be done quite easily as follows. Let $\mathcal{A}_1$ and $\mathcal{A}_2$ be defined symbolically by $\mathcal{A}_1 \triangleq \{(\mathbf{u}^1,\mathbf{y}^1)\}$ and $\mathcal{A}_2 \triangleq \{(\mathbf{u}^2,\mathbf{y}^2)\}$, meaning that $\mathcal{A}_1$ is a specified set of pairs of segmented time functions $\mathbf{u}^1$ and $\mathbf{y}^1$, and likewise for $\mathcal{A}_2$. (Strictly speaking, $\mathcal{A}_1$ and $\mathcal{A}_2$ are families of such sets of pairs, with t_0 and t playing the role of parameters; see *1.4.17*.) Then $\mathcal{A}_2\mathcal{A}_1$ is defined by $\mathcal{A}_2\mathcal{A}_1 = \{(\mathbf{u}^1,\mathbf{y}^2)\}$, where $\mathbf{u}^1$ and $\mathbf{y}^2$ are time functions such that there exists $\mathbf{y}^1$ $(= \mathbf{u}^2)$ with the property $(\mathbf{u}^1,\mathbf{y}^1) \in \mathcal{A}_1$ and $(\mathbf{u}^2,\mathbf{y}^2) \in \mathcal{A}_2$. It is of interest to note that the associative property of the product of $\mathcal{A}_1$ and $\mathcal{A}_2$ is an immediate consequence of this definition. It should also be noted that this way of defining the product of $\mathcal{A}_2$ and $\mathcal{A}_1$ coincides with the conventional way of defining the product of two relations.[1]

Interconnection of N objects

A basic question that arises when we interconnect N objects $\mathcal{A}_1, \ldots, \mathcal{A}_N$ in an arbitrary fashion—and which does not manifest itself in the case of a tandem combination of $\mathcal{A}_1, \ldots, \mathcal{A}_N$—is whether the interconnection is *determinate*, i.e., whether the input-output-state relation of the interconnection is determined by the input-output-state relations of its constituents.

The notion of determinateness will be discussed in greater detail in Chap. *2*, Sec. *6*. For our present purposes, it will suffice to remark that, essentially, an interconnection $\mathcal{A}$ is determinate if, given the input to $\mathcal{A}$ and the initial states of the components $\mathcal{A}_1, \ldots, \mathcal{A}_N$, one can determine, at least in principle, the outputs of $\mathcal{A}_1, \ldots, \mathcal{A}_N$. As will be shown in Chap. *2* (*2.6.10*), if $\mathcal{A}$ is determinate, then the composite state vector $\mathbf{s}(t) = (\mathbf{s}^1(t), \ldots, \mathbf{s}^N(t))$, where $\mathbf{s}^i(t) \triangleq$ state vector of $\mathcal{A}_i$ at time t, qualifies as the state of $\mathcal{A}$ at time t. In the special case of a tandem combination of $\mathcal{A}_1$ and $\mathcal{A}_2$, this implies that the state of $\mathcal{A}_2\mathcal{A}_1$ is $(\mathbf{s}^1(t),\mathbf{s}^2(t))$—a result which we have already used in our discussion of the tandem combination in the preceding subsection.

As in the case of a tandem combination, a basic problem that arises in the analysis of an interconnection $\mathcal{A}$ is that of the characterization and determination of Σ, the state space of $\mathcal{A}$. We approach this

[1] See, for example, Ref. *11*, p. 41.

problem in a similar way, i.e., by regarding $\mathcal{A}$ as a limiting form of an *initially free* interconnection $\mathcal{A}(t_0)$ which is derived from $\mathcal{A}$ by interposing switches between $\mathcal{A}_1, \ldots, \mathcal{A}_N$ which remain open until t_0, with $t_0 \to -\infty$ (Figs. *1.10.1* and *1.10.2*). It is understood, of course, that the input to $\mathcal{A}$ would be specified for, say, $t \geq t_1$, with the input to $\mathcal{A}(t_0)$ between t_0 and t_1 influencing only the state of $\mathcal{A}(t_0)$ at time t_1.

Again, as in the case of a tandem combination, the state space of $\mathcal{A}$ can be characterized in a formal way by the relation (*1.10.21*)

26 $$\Sigma = \lim_{t_0 \to -\infty} \Sigma(t_0,t)$$

where $\Sigma(t_0,t)$ denotes the state space of $\mathcal{A}(t_0)$ at time t, that is,

$$\Sigma(t_0,t) = \{\mathbf{s}(t) | \mathbf{s}(t_0) \in \Sigma_1 \times \cdots \times \Sigma_N, \mathbf{u} \in R[\mathbf{u}]\}$$

where $\mathbf{u}$ denotes the input to $\mathcal{A}$ and the components of $\mathbf{s}(t)$ are related to those of $\mathbf{s}(t_0)(= \boldsymbol{\alpha})$ by *9*. The determination of Σ from *26* in the general case of an arbitrary interconnection presents many complex and as yet unsolved problems. We shall not concern ourselves with the discussion of these problems, since in the case of systems that are of primary interest to us in this text, namely, linear differential systems, both $\Sigma(t_0,t)$ and Σ can generally be determined in a direct fashion by using a technique sketched in Chap. *2* (*2.6.8*).

Since the relation between a *free interconnection* of $\mathcal{A}_1, \ldots, \mathcal{A}_N$ on the one hand and a *constrained interconnection* on the other is identical with that of a free and constrained tandem combination (*1.10.10* and *1.10.23*), we shall not define these terms in a formal way. Instead, we shall concern ourselves with defining other terms which enter into the description of an interconnection of N objects, with the understanding that by an "interconnection" we mean a "constrained interconnection," which in turn can be regarded as a limiting form of a free interconnection.

27 **Definition** Let $\mathcal{A}$ be a collection $\{\mathcal{A}_i\}$ of N not necessarily oriented objects with respective terminal vectors $\mathbf{v}^i = (v_1^i, \ldots, v_{n_i}^i)$ (*1.4.1 et seq.*). The joining (connecting) of a terminal of $\mathcal{A}_i$ labeled, say, v_μ^i to a terminal of $\mathcal{A}_j$ labeled, say, v_ν^j signifies that $v_\mu^i(t) = v_\nu^j(t)$ for all t in T. The variable v_μ^i (and v_ν^j) is said to be a *shared* or *common* terminal variable. If v_μ^i is an input variable and v_ν^j is an output variable or vice versa, then v_μ^i (and v_ν^j) is said to be an *input-output variable.*

The joining of v_μ^i and v_ν^j is subject to the following compatibility conditions.

(I) v_μ^i and v_ν^j are not both output variables.

(II) If v_μ^i is an output variable and v_ν^j is an input variable, then the range of v_μ^i is contained in that of v_ν^j.

(III) If both v^i_μ and v^j_ν are input variables, then the range of the shared variable is the intersection of the ranges of v^i_μ and v^j_ν.

28 **Definition** The collection $\{\mathcal{A}_i\}$ is said to be an *interconnected set*, or, more simply, an *interconnection*, of the $\mathcal{A}_i$ if every object in the collection shares at least one terminal variable with one or more other objects in this collection. If some or possibly none of the $\mathcal{A}_i$ share terminal variables, the collection is said to be *partially interconnected.* Relations of the form $v^i_\mu = v^j_\nu$ or their equivalents (e.g., Kirchhoff voltage and current law relations in electrical networks) constitute the *interconnection relations* for the $\mathcal{A}_i$.

29 *Examples* In Fig. *1.10.1* is shown an interconnected set of three multipoles $\mathcal{A}_1$, $\mathcal{A}_2$, $\mathcal{A}_3$. Here the interconnection relations are $y^1_1 = u^2_1$, $y^2_1 = u^3_1$, $y^3_3 = u^1_2$, $y^3_1 = u^2_2$. In Fig. *1.10.5a* is shown a partially interconnected set of three 2-poles $\mathcal{A}$, $\mathcal{B}$, $\mathcal{C}$; in Fig. *1.10.5b* is shown the direct product of two 2-poles; and in Fig. *1.10.5c* is shown a degenerate case of a single interconnected object. To subsume this case under the definition given above, the terminals y_1, u_1 can be regarded as being connected by a fictitious unitor (*1.9.8*). ◁

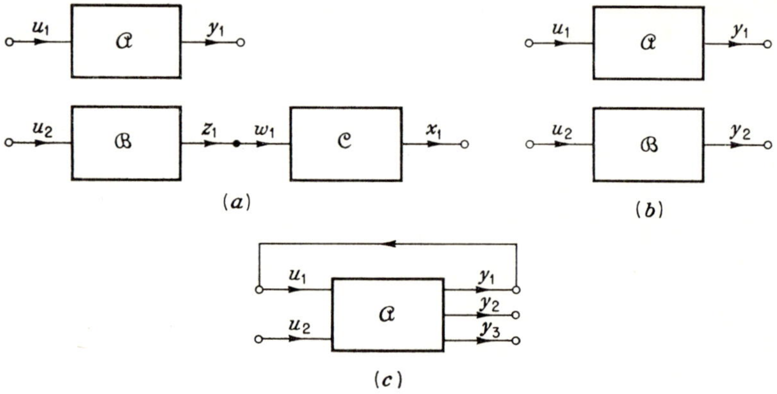

Fig. 1.10.5 (*a*) Partial interconnection; (*b*) direct product; (*c*) self-interconnection.

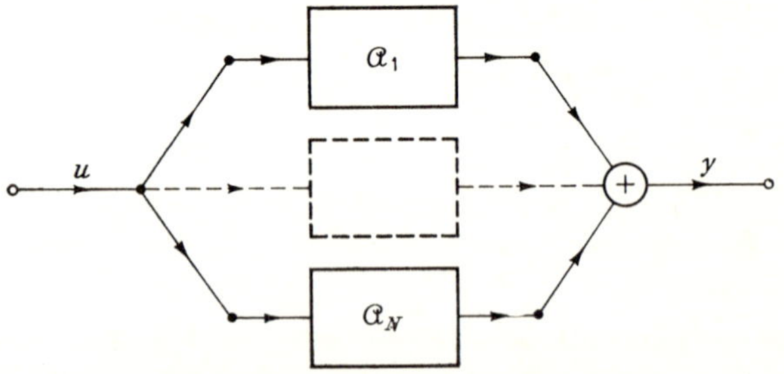

Fig. 1.10.6 Parallel combination.

There are several simple types of interconnection that are encountered quite frequently in system analysis. One is the tandem combination that we have already discussed. Another is the *parallel combination* that is illustrated in Fig. *1.10.6*. The parallel combination of $\mathcal{A}_1, \ldots, \mathcal{A}_N$ is denoted by $\mathcal{A}_1 + \cdots + \mathcal{A}_N$, with the object $\mathcal{A} = \mathcal{A}_1 + \cdots + \mathcal{A}_N$ referred to as the *sum* of $\mathcal{A}_1, \ldots, \mathcal{A}_N$. The input-output-state relation of $\mathcal{A}$ is expressed by

30 $$y = \sum_{i=1}^{N} \bar{\mathbf{A}}_i(\boldsymbol{\alpha}^i;\mathbf{u}^i)$$

with the state space of $\mathcal{A}$ being a subset Σ of the product space $\Sigma_1 \times \cdots \times \Sigma_N$.

31 *Exercise* Give a characterization of Σ as a limiting form of the state space of the free sum of $\mathcal{A}_1, \ldots, \mathcal{A}_N$. ◁

By combining series and parallel combinations of 2-poles, one can construct more complex structures of the form shown in Fig. *1.10.7*.

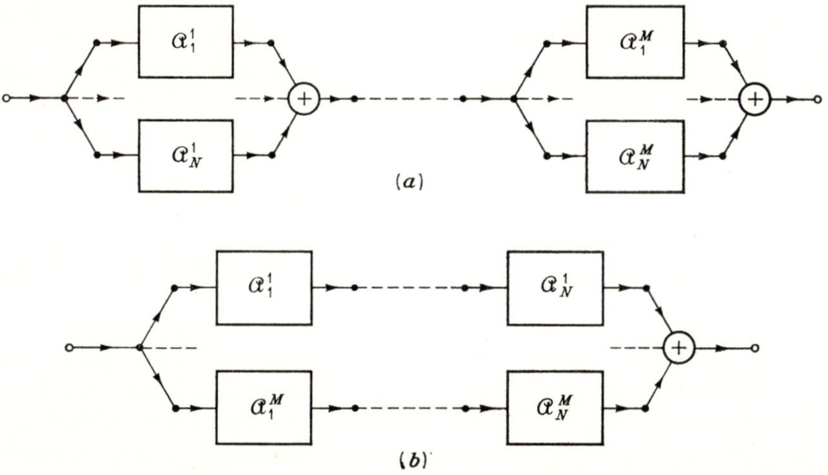

Fig. 1.10.7 (*a*) Series-parallel combination; (*b*) parallel-series combination.

The structure on top is called the *series-parallel* combination, while that below is called the *parallel-series* combination. In symbols, the series-parallel combination is represented by the expression

32 $$(\mathcal{A}_1^M + \cdots + \mathcal{A}_N^M)(\mathcal{A}_1^{M-1} + \cdots + \mathcal{A}_N^{M-1}) \cdots (\mathcal{A}_1^1 + \cdots + \mathcal{A}_N^1)$$

whereas the parallel-series combination is represented by

33 $$\mathcal{A}_N^1 \cdots \mathcal{A}_1^1 + \mathcal{A}_N^2 \cdots \mathcal{A}_1^2 + \cdots + \mathcal{A}_N^M \cdots \mathcal{A}_1^M$$

where the $\mathfrak{A}_i^\mu$ ($\mu = 1, \ldots, M, i = 1, \ldots, N$) denote the component 2-poles of these structures. (For an illustration of *33* see *3.8.11*.)

In dealing with interconnections of objects we shall frequently use in a tacit manner the following two basic principles.

34 **Associative principle** If $\mathfrak{A}_1, \ldots, \mathfrak{A}_N$ are components of an interconnected set $\mathfrak{A}$, then any subset of $\mathfrak{A}_1, \ldots, \mathfrak{A}_N$ may be regarded as a component of $\mathfrak{A}$. For example, in Fig. *1.10.1* the combination of $\mathfrak{A}_2$ and $\mathfrak{A}_3$ may be regarded as a single object with input u_1^2 and outputs y_2^3 and y_3^3.

35 **Substitution principle** If an oriented component $\mathfrak{A}_i$ of $\mathfrak{A}$ is replaced by an equivalent component (in the sense of Definition *2.5.8*), then the resulting interconnection is equivalent to $\mathfrak{A}$. ◁

These statements are *principles* in the sense that they are of general validity and that in specific contexts they can be formulated as provable theorems. We shall not attempt to do this in the present text, except for particular cases involving for the most part linear systems.

36 *Comment* In the foregoing discussion we have stressed the interconnection of oriented rather than nonoriented objects for two reasons. First, in later chapters we shall be concerned in the main with oriented objects. Second, it can be shown that under mildly restrictive conditions which are satisfied in the case of linear systems, an interconnection of nonoriented objects which is characterized by a set of terminal relations can be replaced by an interconnection of oriented objects characterized by input-output relations which are identical with the terminal relations of the nonoriented interconnection.[1] For our purposes it will suffice to illustrate this point by a simple

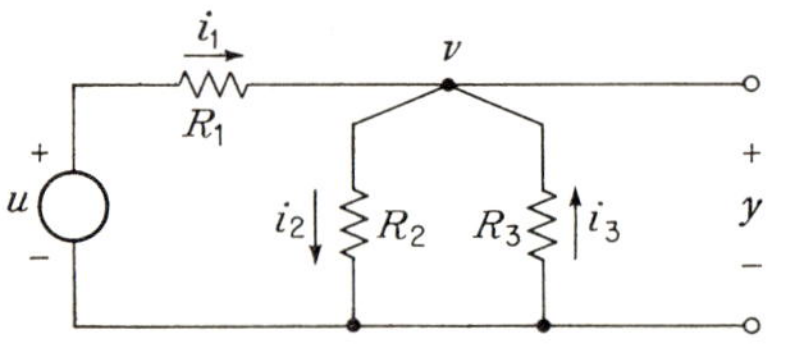

Fig. 1.10.8 Network of Example *1.10.37*.

37 *Example* Consider the nonoriented network of Fig. *1.10.8*, which is an interconnection of nonoriented resistors R_1, R_2, and R_3. Note that in the case of an electrical network a special convention is used to represent the constraints imposed by interconnections. Specifically, the joining of R_1, R_2, and R_3 at a node v (Fig. *1.10.8*) signifies that

38 $$i_1 - i_2 + i_3 = 0$$

which is a statement of Kirchhoff's current law applied to the currents

[1] This is basic to the signal-flow-graph analysis of electrical networks. See Ref. *9*.

i_1, i_2, and i_3 flowing through R_1, R_2, and R_3, respectively.[1] In effect, this equation constitutes an interconnection relation for this network (see *28*).

The terminal relations for R_1, R_2, and R_3 are

39 R_1: $u - y = R_1 i_1$

R_2: $y = R_2 i_2$

R_3: $y = -R_3 i_3$

Thus, the network as a whole is characterized by the terminal relations *38* and *39*.

Now consider the interconnection of the oriented scalors shown in Fig. *1.10.9*, in which the number inside each block represents its scale

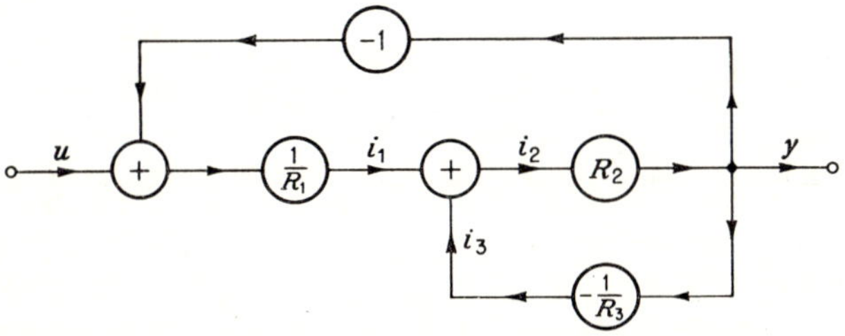

Fig. 1.10.9 Block-diagram representation of the network of Fig. *1.10.8*.

factor (*1.9.8*). Here we have an interconnection of oriented elements, with the input-output and interconnection relations reading

$$i_1 = \frac{1}{R_1}(u - y) \qquad i_3 = -\frac{1}{R_3}y$$

$$y = R_2 i_2 \qquad i_2 = i_1 + i_3$$

Since these equations are identical with the terminal relations of the original network, it follows that the network of Fig. *1.10.8* is equivalent to the interconnection of oriented scalors and adders of Fig. *1.10.9*. This is the point we wanted to illustrate.

Flow graphs and signal-state graphs

A diagrammatic representation of the interconnection of two or more objects, such as shown in Fig. *1.10.1*, is essentially a statement of the orientation of each object in the interconnected set together with an indication of which terminal variables are shared by two or more objects. This information can be conveyed in a number of alternative ways, two of which are discussed very briefly in the sequel.

[1] We assume, of course, that the reader is familiar with elementary electric circuit analysis. A careful discussion of the fundamentals of circuit analysis can be found in Refs. *3*, *4*, and *5*, among others.

First, one can employ the notation and terminology of signal-flow graphs, which will be discussed in greater detail in Chap. *9*, Sec. *15*. In this case, each terminal variable is associated with a node of an oriented graph, and an input-output relation of the form

$$y_1 = f_1(u_1, \ldots, u_n)$$

is represented as a set of oriented branches leading from nodes $u_1, \ldots,$ u_n to node y (Fig. *1.9.3a*).

As an illustration, on applying this convention to the interconnected set of Fig. *1.10.1*, we obtain the signal-flow graph shown in Fig. *1.10.10*.

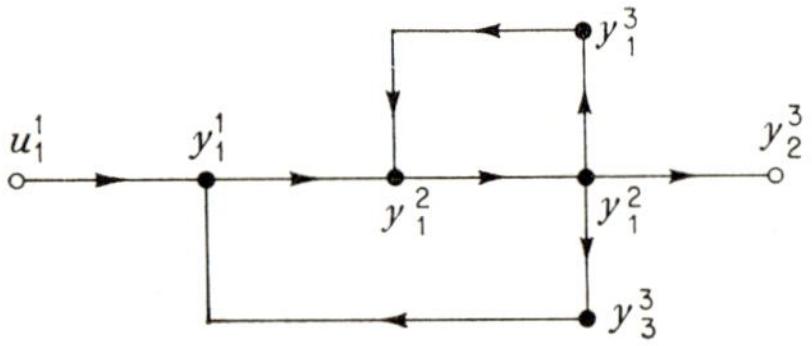

Fig. 1.10.10 Signal-flow graph of the system of Fig. *1.10.1*.

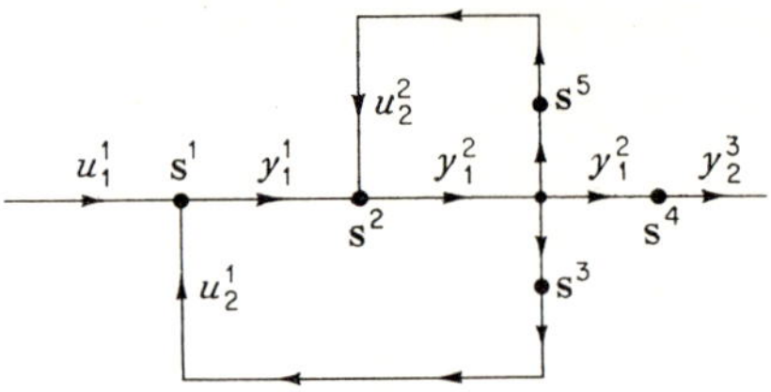

Fig. 1.10.11 Signal-state graph of the system of Fig. *1.10.1*.

Note that it is simpler in appearance than the block diagram of Fig. *1.10.1*. However, this simplicity is attained at a cost of some loss of information, since in order to make the signal-flow-graph representation applicable, it is necessary to decompose each object into component objects each having a single output terminal (see *1.9.3*). As a result, in the signal-flow-graph representation of an interconnected set of objects the individual objects lose their identity. For example, Fig. *1.10.10* does not indicate whether the nodes labeled y_1^3, y_2^3, y_3^3, and y_1^2 represent a single object with one input and three outputs (i.e., the object $\mathfrak{A}_3$ in Fig. *1.10.1*) or whether they represent three distinct 2-poles.

Another and, from our viewpoint, more important shortcoming of the conventional signal-flow-graph representation is that it makes no provision for associating a state with an object. As was pointed out previously (see *1.9.3 et seq.*), this can be done by the use of the signal-state-graph representation. Applying this mode of representation to the interconnected set of Fig. *1.10.1*, we obtain the signal-state graph shown in Fig. *1.10.11*. Here $\mathbf{s}^1$ represents the state vector associated with $\mathfrak{A}_1$, $\mathbf{s}^2$ is the state vector associated with $\mathfrak{A}^2$, and $\mathbf{s}^3$, $\mathbf{s}^4$, and $\mathbf{s}^5$ are the state vectors associated with the 2-poles into which $\mathfrak{A}_3$ is decomposed (*1.9.3*). Note that $\mathbf{s}^3$, $\mathbf{s}^4$, and $\mathbf{s}^5$ can be identified with $\mathbf{s}$, the state of $\mathfrak{A}_3$. However, if each 2-pole into which $\mathfrak{A}_3$ is decomposed is regarded as a separate object, it may be possible that vectors with smaller number of components than **s** may qualify as state vectors for the objects in question. It is for this reason that we use the different symbols $\mathbf{s}^1$, $\mathbf{s}^2$, and $\mathbf{s}^3$ to denote the state vectors of these 2-poles even though **s** qualifies as a state vector for all of them.

We shall not dwell any further on the details of various graphical ways of representing an interconnection of a set of oriented objects. The brief discussion presented above will suffice for our immediate purposes.

System (formal definition)

After this drawn-out introduction, we are finally in a position to define the notion of a system in fairly concrete terms.

40 **Definition** An *abstract system*, or, simply, a *system*, $\mathcal{S}$ is a partially interconnected set (*1.10.28*) of abstract objects $\mathcal{A}_1$, $\mathcal{A}_2$, $\mathcal{A}_3$, . . . , termed the *components* of $\mathcal{S}$. The components of $\mathcal{S}$ may be oriented or nonoriented; they may be finite or infinite in number; and each of them may be associated with a finite or infinite number of terminal variables. ◁

In the sequel we shall assume, unless otherwise noted, that $\mathcal{S}$ has a finite number of components each of which is an oriented object associated with a finite number of terminal variables.

Under the above definition, a system may be regarded as a single object, and conversely any single object may be regarded as a system. If a system comprising the objects $\mathcal{A}_1, \ldots, \mathcal{A}_N$ is regarded as a single object, then the set of terminal variables of $\mathcal{S}$ is the set-theoretic union of the terminal variables of its components. If the components of $\mathcal{S}$ are oriented, then so is $\mathcal{S}$, with the input variables of $\mathcal{S}$ being those input variables of its components which are not input-output variables (*1.10.27*). The aggregate of the shared terminal variables as well as the nonshared output variables of the components of $\mathcal{S}$ constitutes the set of output variables of $\mathcal{S}$. Any subset of the output variables of $\mathcal{S}$ may be disregarded and thus categorized as suppressed output variables of $\mathcal{S}$ (see Sec. *4*).

41 *Example* For the system (interconnected set) shown in Fig. *1.10.1* the terminal variables of $\mathcal{S}$ are u_1^1, u_2^1, y_1^1, u_2^2, y_1^2, y_2^3. (Note that $y_1^1 = u_1^2$, $u_2^2 = y_1^3$, $u_2^1 = y_3^3$.) The variable u_1^1 is the input to $\mathcal{S}$. The output variables are y_1^1, y_1^2, y_1^3, y_2^3, y_3^3 and of these all but y_2^3 are suppressed, leaving y_2^3 as the output of $\mathcal{S}$.

42 *Example* For the system shown in Fig. *1.10.5a* the terminal variables are u_1, y_1, u_2, z_1, x_1. Of these, z_1 is shared. y_1, z_1, and x_1 are the output variables of the system. The variable z_1 is suppressed.

43 *Example* For the nonoriented system shown in Fig. *1.10.8* the terminal variables are u, i_1, i_2, i_3, y. If the system is oriented by designating, say, u as the input and y as the output, then the output variables are i_1, i_2, i_3, and y, and of these i_1, i_2, and i_3 are suppressed. ◁

A question of basic importance in system analysis is the following. Suppose that $\mathcal{S}$ comprises a set of oriented objects $\mathcal{A}_1, \ldots, \mathcal{A}_N$ each of which is completely characterized by an input-output-state relation and all of which are interconnected in a specified way. Is $\mathcal{S}$—regarded as a single object—completely characterized by the aggregate of the input-output-state relations for its components and the interconnection relations? It is easy to construct examples which show that this is not the case in general. However, the answer to this question is in the affirmative for a wide category of systems which subsumes those studied in the book. We shall consider this question in greater detail in Chap. *2* in connection with the discussion of *determinate* systems.

This closes our introduction to the language, notation, and basic concepts of system theory. In the succeeding chapters, we shall examine in greater detail the ramifications and interrelations of the concepts introduced in this chapter, and we shall apply them to the study of a special and yet basic class of systems, namely, the class of linear systems.

REFERENCES AND SUGGESTED READING

1 Apostol, T. M.: "Mathematical Analysis," Addison-Wesley Publishing Company, Inc., Reading, Mass., 1957.

2 Kelley, J. L.: "General Topology," D. Van Nostrand Company, Inc., Princeton, N.J., 1955.

3 Van Valkenburg, M. E.: "Network Analysis," Prentice-Hall, Inc., Englewood Cliffs, N.J., 1955.

4 Seshu, S., and N. Balabanian: "Linear Network Analysis," John Wiley & Sons, Inc., New York, 1959.

5 Friedland, B., O. Wing, and R. Ash: "Principles of Linear Networks," McGraw-Hill Book Company, Inc., New York, 1961.

6 Moore, E. F.: Gedanken Experiments on Sequential Machines, in "Automata Studies," Princeton University Press, Princeton, N.J., 1956.

7 Bellman, R. E., and R. Kalaba: A Note on Interrupted Stochastic Control Processes, *Inf. and Cont.*, vol. 4, pp. 346–349, 1961.

8 Ginsburg, S.: On the Length of the Smallest Uniform Experiment Which Distinguishes the Terminal States of a Machine, *J. Assoc. Comput. Mach.*, vol. 5, pp. 266–280, 1958.

9 Mason, S. J., and H. J. Zimmermann: "Electronic Circuits, Signals and Systems," John Wiley & Sons, Inc., New York, 1960.

10 Bellman, R. E.: "Dynamic Programming," Princeton University Press, Princeton, N.J., 1957.

11 Halmos, P. R.: "Naive Set Theory," D. Van Nostrand Company, Inc., Princeton, N.J., 1960.

Concepts and properties associated with state and state equations

2

1 *Introduction*

In the preceding chapter no attempt was made to interrelate and draw consequences from the various concepts which were introduced there. This we shall begin to do in the present chapter.

Although our main concern in this book is with linear systems, the discussion of input-output-state relations in the following sections is couched in general terms, without any specialization, except through examples, to linear systems. The purpose of this approach is to put the reader in a position from which he can view linear system theory in a broader perspective and thereby acquire a better understanding of the roles played by such general concepts as state equivalence and system equivalence in the special context of linear systems.

Before proceeding to discuss the main topic of this chapter, it will be helpful to recapitulate some of the terminology and notation introduced in Chap. *1*.

Recapitulation of terminology and notation

The symbol $\mathcal{A}$ represents an abstract oriented object with input $\mathbf{u}$, output $\mathbf{y}$, and state $\mathbf{s}$. For each fixed t, the variables $\mathbf{u}(t)$, $\mathbf{y}(t)$, and $\mathbf{s}(t)$ range over the input space $R[\mathbf{u}(t)]$, the output space $R[\mathbf{y}(t)]$, and state space Σ, respectively. These spaces are assumed to be independent of time. In most of the applications we shall be concerned with, $R[\mathbf{u}(t)]$, $R[\mathbf{y}(t)]$, and Σ will be finite-dimensional linear vector spaces. The input segment space of $\mathcal{A}$ is a space $R[\mathbf{u}]$ to which the input segments $\mathbf{u} \triangleq \mathbf{u}_{(t_0,t]}$ are constrained to belong. The same statement applies to the output segment space $R[\mathbf{y}]$ (see Remark *1.4.3* and Definition *1.4.17*).

A generic element of Σ is denoted by $\boldsymbol{\alpha}$ and is called a state of $\mathcal{A}$. When dealing with two or more abstract objects, it may be expedient to use different symbols, say, $\boldsymbol{\alpha}$ to denote a state of $\mathcal{A}$ and $\boldsymbol{\beta}$ to denote a state of $\mathcal{B}$. In some cases $\boldsymbol{\alpha}$ will represent a particular fixed state of $\mathcal{A}$. In other cases $\boldsymbol{\alpha}$ may play the role of a variable ranging over Σ. Whether it is one or the other will usually be clear from the context. If necessary to avoid ambiguity, fixed values of $\boldsymbol{\alpha}$ will be denoted by symbols such as $\boldsymbol{\alpha}_i$ and $\boldsymbol{\alpha}_j$.

The symbols $\mathbf{u}_{[t_0,t_1)}$ and $\mathbf{u}_{(t_0,t_1]}$ represent segments of a time function $\mathbf{u}$ over the semiclosed intervals $[t_0,t_1)$ and $(t_0,t_1]$, respectively. More specifically, $\mathbf{u}_{(t_0,t_1]}$ is the set $\{(t,u(t))|t_0 < t \leq t_1\}$, and likewise for $\mathbf{u}_{[t_0,t_1)}$. The segment $\mathbf{u}_{(t_0,t]}$ will frequently be written in the abbreviated form $\mathbf{u}$ and will be referred to as the input segment or simply input. The same statement applies to $\mathbf{y}$, with $\mathbf{y}_{(t_0,t]}$ representing the output or the response segment of $\mathcal{A}$. The time t_0 represents an arbitrary initial time in T, the range of t. It should be understood that when we say "for all $\mathbf{u}$" or, more specifically, "for all $\mathbf{u}$ in the input segment space of $\mathcal{A}$," we vary not only the values of $\mathbf{u}$ over the observation interval $(t_0,t_1]$ but also the end points of the observation interval. When we say $\mathbf{u}$ is equal to $\mathbf{y}$ (or $\mathbf{u}$ is identical with $\mathbf{y}$) and write $\mathbf{u} = \mathbf{y}$ or

$$\mathbf{u}_{(t_0,t]} = \mathbf{y}_{(t_0,t]}$$

we mean that $\mathbf{u}(\tau) = \mathbf{y}(\tau)$ for all τ in the interval $(t_0,t]$. In order to reduce the cumbersomeness of notation, quantifiers such as $\forall\mathbf{u}$ (for all $\mathbf{u}$), $\forall t$ (for all t), $\forall t_0$ (for all t_0), etc., will frequently be omitted from equations when their presence is clearly implied by the context.

An oriented object $\mathcal{A}$ is characterized by its input-output relation

1 $$\mathcal{A}(\mathbf{u},\mathbf{y}) = 0$$

or equivalently

2 $$\mathbf{y} = \mathcal{A}(\mathbf{u})$$

with the understanding that *2* does not imply that $\mathbf{y}$ is a single-valued function of $\mathbf{u}$. A pair of time functions $(\mathbf{u}_{(t_0,t]},\mathbf{y}_{(t_0,t]})$, or, for short, $(\mathbf{u},\mathbf{y})$, which respectively belong to the input and output segment spaces of $\mathcal{A}$ constitutes an input-output pair for $\mathcal{A}$ if $\mathbf{u}$ and $\mathbf{y}$ satisfy the input-output relation *1* (or *2*). An input-output relation is in effect a means of defining the set of all input-output pairs which belong to $\mathcal{A}$ (see *1.4.17* and *1.4.26*). Essentially, $\mathcal{A}$ is a relation, written as $\mathcal{A} = \{(\mathbf{u},\mathbf{y})\}$, with domain $R[\mathbf{u}]$ and range $R[\mathbf{y}]$ (see *1.4.29*).

An abstract object $\mathcal{A}$ is completely characterized by its input-output-state relation. This relation will be written in two equivalent forms: (I)

3 $$\mathbf{y}(t) = \mathbf{A}(\mathbf{s}(t_0);\mathbf{u}_{(t_0,t]})$$

or, for short,

4 $$\mathbf{y}(t) = \mathbf{A}(\mathbf{s}(t_0);\mathbf{u})$$

and (II)

5 $$\mathbf{y}_{(t_0,t]} = \bar{\mathbf{A}}(\mathbf{s}(t_0);\mathbf{u}_{(t_0,t]})$$

or, for short,

6 $$\mathbf{y} = \bar{\mathbf{A}}(\mathbf{s}(t_0);\mathbf{u})$$

Equation *3* gives the value of the output at time t as a function of $\mathbf{s}(t_0)$, the state of $\mathcal{A}$ at time t_0, and the values of the input between t_0 and t. Equation *5* expresses the relation between the response segment $\mathbf{y}$, the input segment $\mathbf{u}$, and the state of $\mathcal{A}$ at time t_0. The overbar in $\bar{\mathbf{A}}$ serves to emphasize that $\bar{\mathbf{A}}(\mathbf{s}(t_0);\mathbf{u})$ is a segment of a time function rather than its value at a single point. When an equation of the form *3* or *5* is referred to as an input-output-state relation, it is implied that it satisfies the mutual- and self-consistency conditions (see *1.6.2 et seq.*).

An experiment on $\mathcal{A}$ consists in applying to $\mathcal{A}$ an input $\mathbf{u}_{(t_0,t_1]}$ and observing the resulting response segment $\mathbf{y}_{(t_0,t_1]}$. (It should be emphasized that when we write $\mathbf{u}_{(t_0,t_1]}$, we do not imply that the input vanishes outside the interval $(t_0,t_1]$. All we mean is that the input is observed over the interval $(t_0,t_1]$. The same comment applies to $\mathbf{y}$.) The state of $\mathcal{A}$ at time t_0 is the initial or the starting state of $\mathcal{A}$. (Equivalently, the initial state may be taken to be $\mathbf{s}(t_0-)$, in which case the input segment is understood to be $\mathbf{u}_{[t_0,t)}$; see *1.5.5*.) The state of $\mathcal{A}$ at time t_1 is the terminal state of $\mathcal{A}$. When we wish to place in evidence that the initial state of $\mathcal{A}$ is a point $\boldsymbol{\alpha}$ in Σ, the input-output-state relation *6* is written as

7 $$\mathbf{y} = \bar{\mathbf{A}}(\boldsymbol{\alpha};\mathbf{u})$$

The same thing applies to the other forms of the input-output-state relation. If the initial state is $\boldsymbol{\alpha}$ and the terminal state is $\boldsymbol{\alpha}^*$, we shall say that $\boldsymbol{\alpha}$ is taken into $\boldsymbol{\alpha}^*$ by $\mathbf{u}$. We shall also say that $(\mathbf{u},\mathbf{y})$ is an input-output pair associated with or with respect to $\boldsymbol{\alpha}$.

When no confusion can result, the symbol $\mathbf{uu}'$ will be used to denote a segment consisting of a segment $\mathbf{u}$ followed by a segment $\mathbf{u}'$ (see Fig. *1.6.1*). Correspondingly, the response of $\mathcal{A}$ to such a segment will be written as $\mathbf{yy}'$, where

8 $$\mathbf{yy}' = \bar{\mathbf{A}}(\boldsymbol{\alpha};\mathbf{uu}')$$

with $\boldsymbol{\alpha}$ being the state of $\mathcal{A}$ at the beginning of $\mathbf{u}$.

The symbol $\mathbf{s}(t)$ represents the state of $\mathcal{A}$ at time t. The state at time t is related to the state at time t_0 by an equation of the form

9 $$\mathbf{s}(t) = \mathbf{s}(\mathbf{s}(t_0);\mathbf{u}_{(t_0,t]}) \qquad t > t_0$$

where $\mathbf{s}$ is a function defined on the product space $\Sigma \times R[\mathbf{u}]$. This equation is referred to as the state equation of $\mathcal{A}$ (in explicit form). (See *1.8.2 et seq.*) When to this equation is adjoined the limiting equation resulting from letting t_0 approach t from the left in the input-output-state relation *3*, the resulting pair of equations constitutes the state equations of $\mathfrak{A}$ (in explicit form). Usually, the state equations in question read

10
$$\begin{aligned} \mathbf{s}(t) &= \mathbf{s}(\mathbf{s}(t_0);\mathbf{u}_{(t_0,t]}) \\ \mathbf{y}(t) &= \mathbf{g}(\mathbf{s}(t),\mathbf{u}(t),t) \end{aligned}$$

where $\mathbf{g}$ is a function defined on the product space $\Sigma \times R[\mathbf{u}(t)] \times T$.

In stating assertions, theorems, definitions, etc., we shall frequently be using the standard mathematical symbol $\Rightarrow$ to denote implication. Thus, if $\{a\}$ and $\{b\}$ are two statements, then $\{a\} \Rightarrow \{b\}$, or, simply, $a \Rightarrow b$, should be read "a implies b," in the sense that if a is true, then b is true.[1] Similarly, $a \Leftrightarrow b$ reads "a implies and is implied by b."

11 *Example* Let x be a real number. Then $\{x > 2\} \Rightarrow \{x > 1\}$ means "if (it is true that) $x > 2$, then (it is also true that) $x > 1$."

12 *Example* Let x, y, and z be real numbers. Then $\{x = y \text{ and } y = z\} \Rightarrow \{x = z\}$ means "if x is equal to y and y is equal to z, then x is equal to z." Here, for short, the words enclosed in parentheses in Example *11* are omitted.

13 *Example* Let x and y be time functions. Then

$$\{x \geq y\} \Leftrightarrow \{\forall t(x(t) \geq y(t))\}$$

means that "$x \geq y$" implies and is implied by "$x(t) \geq y(t)$ for all t," or, equivalently, "$x \geq y$ if and only if $x(t) \geq y(t)$ for all t." $\triangleleft$

We are now ready to take advantage of the compactness afforded by the above notation in defining several basic concepts relating to the notion of state.

By far the most important concept among the concepts in question is that of *equivalent states.* At this point, we shall merely define the notion of state equivalence and make limited use of it in clarifying the concept of the state of a system at time t. A more detailed discussion of the notion of equivalence in the context of linear systems will be presented in Chaps. *3* and *4*.

2 *State equivalence*

Let $\mathfrak{A}$ be a system with state space Σ. Suppose that $\boldsymbol{\alpha}_i$ and $\boldsymbol{\alpha}_j$ are two states of $\mathfrak{A}$, that is, $\boldsymbol{\alpha}_i \in \Sigma$ and $\boldsymbol{\alpha}_j \in \Sigma$. Consider an experiment on $\mathfrak{A}$ under which an input $\mathbf{u}$ ($\triangleq \mathbf{u}_{(t_0,t_1]}$) is applied to $\mathfrak{A}$ starting at time t_0 and ending at time t_1, with $\mathfrak{A}$ initially in state $\boldsymbol{\alpha}_i$. Denote the segment of

[1] For a more complete discussion of the significance of implication see, for example, Refs. *1* and *2*.

the corresponding output over the observation interval $(t_0,t_1]$ by $\bar{\mathbf{A}}(\boldsymbol{\alpha}_i;\mathbf{u})$.

Consider now the same experiment but with the difference that the initial state is $\boldsymbol{\alpha}_j$ rather than $\boldsymbol{\alpha}_i$. It is conceivable that for the particular input $\mathbf{u}$ applied to $\mathcal{A}$ the segments $\bar{\mathbf{A}}(\boldsymbol{\alpha}_i;\mathbf{u})$ and $\bar{\mathbf{A}}(\boldsymbol{\alpha}_j;\mathbf{u})$ might be identical. It is also conceivable that this may be true not only for the $\mathbf{u}$ in question but for all inputs in the input segment space of $\mathcal{A}$. If this is the case, then $\boldsymbol{\alpha}_i$ and $\boldsymbol{\alpha}_j$ are *equivalent states*,[1] written as $\boldsymbol{\alpha}_i \simeq \boldsymbol{\alpha}_j$, in the sense that $\mathcal{A}$ starting in state $\boldsymbol{\alpha}_i$ exhibits exactly the same behavior as $\mathcal{A}$ starting in state $\boldsymbol{\alpha}_j$. More formally, we can express this idea as a

1 **Definition** $\boldsymbol{\alpha}_i$ is *equivalent* to $\boldsymbol{\alpha}_j$ if and only if for all inputs $\mathbf{u}$ the response segment of $\mathcal{A}$ starting in state $\boldsymbol{\alpha}_i$ is identical with the response segment of $\mathcal{A}$ starting in state $\boldsymbol{\alpha}_j$.

In symbols,

2 $$\{\boldsymbol{\alpha}_i \simeq \boldsymbol{\alpha}_j\} \Leftrightarrow \{\bar{\mathbf{A}}(\boldsymbol{\alpha}_i;\mathbf{u}) = \bar{\mathbf{A}}(\boldsymbol{\alpha}_j;\mathbf{u})\} \qquad \forall \mathbf{u}$$

where the notation $\bar{\mathbf{A}}(\boldsymbol{\alpha}_i;\mathbf{u}) = \bar{\mathbf{A}}(\boldsymbol{\alpha}_j;\mathbf{u})$ signifies that, over the observation interval on which $\mathbf{u}$ is defined, the response segment of $\mathcal{A}$ to $\mathbf{u}$ starting in state $\boldsymbol{\alpha}_i$ is identical with the response segment of $\mathcal{A}$ to $\mathbf{u}$ starting in state $\boldsymbol{\alpha}_j$. The universal quantifier $\forall \mathbf{u}$ signifies that this is true for *all* $\mathbf{u}$ in the input segment space of $\mathcal{A}$, that is, for all $\mathbf{u}$ in $R[\mathbf{u}]$, with t_0 and t_1 varying over T. $\triangleleft$

An immediate consequence of this definition is the *transitivity of state equivalence*. That is, if $\boldsymbol{\alpha}_i \simeq \boldsymbol{\alpha}_j$ and $\boldsymbol{\alpha}_j \simeq \boldsymbol{\alpha}_k$, then $\boldsymbol{\alpha}_i \simeq \boldsymbol{\alpha}_k$. This follows at once from the transitivity of equality of segments of time functions. Specifically, if $\mathbf{u}_{(t_0,t]} = \mathbf{v}_{(t_0,t]}$ and $\mathbf{v}_{(t_0,t]} = \mathbf{w}_{(t_0,t]}$, then $\mathbf{u}_{(t_0,t]} = \mathbf{w}_{(t_0,t]}$. Thus, if $\bar{\mathbf{A}}(\boldsymbol{\alpha}_i;\mathbf{u}) = \bar{\mathbf{A}}(\boldsymbol{\alpha}_j;\mathbf{u})$ for all $\mathbf{u}$ and $\bar{\mathbf{A}}(\boldsymbol{\alpha}_j;\mathbf{u}) = \bar{\mathbf{A}}(\boldsymbol{\alpha}_k;\mathbf{u})$ for all $\mathbf{u}$, then $\bar{\mathbf{A}}(\boldsymbol{\alpha}_i;\mathbf{u}) = \bar{\mathbf{A}}(\boldsymbol{\alpha}_k;\mathbf{u})$ for all $\mathbf{u}$. Q.E.D.

On negating the definition of equivalent states we obtain the implication: If $\boldsymbol{\alpha}_i$ and $\boldsymbol{\alpha}_j$ are *nonequivalent*, then there exists an input $\mathbf{u}$ such that the response of $\mathcal{A}$ to $\mathbf{u}$ starting in state $\boldsymbol{\alpha}_i$ is not identical with the response of $\mathcal{A}$ to $\mathbf{u}$ starting in state $\boldsymbol{\alpha}_j$. In symbols,

3 $$\{\boldsymbol{\alpha}_i \not\simeq \boldsymbol{\alpha}_j\} \Leftrightarrow \{\exists \mathbf{u}[\bar{\mathbf{A}}(\boldsymbol{\alpha}_i;\mathbf{u}) \neq \bar{\mathbf{A}}(\boldsymbol{\alpha}_j;\mathbf{u})]\}$$

where the expression on the right reads, "there exists a $\mathbf{u}$ such that the response segment $\bar{\mathbf{A}}(\boldsymbol{\alpha}_i;\mathbf{u})$ is not equal to the response segment $\bar{\mathbf{A}}(\boldsymbol{\alpha}_j;\mathbf{u})$." An input $\mathbf{u}$ having this property will be referred to as a *distinguishing input* for $\boldsymbol{\alpha}_i$ and $\boldsymbol{\alpha}_j$, and it will be denoted by the symbol $\mathbf{u}(\boldsymbol{\alpha}_i,\boldsymbol{\alpha}_j)$. Because of its suggestiveness, we shall frequently employ the adjective *distinguishable* to describe two states that are nonequivalent. When $\boldsymbol{\alpha}_i$ and $\boldsymbol{\alpha}_j$ are vectors, $\boldsymbol{\alpha}_i$ and $\boldsymbol{\alpha}_j$ will be said to be *distinct* if $\boldsymbol{\alpha}_i \neq \boldsymbol{\alpha}_j$ in the

[1] The notion of equivalence in the context of finite-state systems was introduced by E. F. Moore, and independently by D. Huffman. See Refs. *3* and *4*.

usual sense of equality of vectors [for example, (2,5) = (2,5) and (2,5) ≠ (4,5)]. Note that distinctness does not imply distinguishability.

4 *Example* Consider the object described in Example *1.7.19*. In this case the state space is the set of numbers $\{0,1,2, \ldots ,11\}$; the input space is the set of numbers $\{1,2, \ldots ,6\}$; the output space is the set of numbers $\{0,1\}$; t ranges over $\{0,1,2, \ldots\}$; and the input-output-state relation is

5 $$y(t) = [s(t_0) + u(t_0) + u(t_0 + 1) + \cdots + u(t - 1)] \bmod 2 \qquad t > t_0$$

where $t_0 \triangleq$ initial time, $u(t) \triangleq$ input at time t, $y(t) \triangleq$ output at time t, and r mod 2 denotes the least nonnegative residue of r which results when multiples of 2 are subtracted from or added to r.

Now, it is clear that for any two integers a and b, we can write

6 $$(a + b) \bmod 2 = (a \bmod 2 + b) \bmod 2$$

Consequently, *5* may be rewritten as

$$y(t) = [s(t_0) \bmod 2 + u(t_0) + \cdots + u(t - 1)] \bmod 2 \qquad t > t_0$$

The form of this relation indicates that if α_i and α_j are both even or both odd, for example, $\alpha_i = 2$ and $\alpha_j = 4$, or $\alpha_i = 3$ and $\alpha_j = 7$, then α_i and α_j are equivalent; for $\alpha_i \bmod 2 = \alpha_j \bmod 2$ if α_i and α_j are both even or both odd. Thus, the states 0, 2, 4, 6, 8, and 10 are equivalent to one another and so are the states 1, 3, 5, 7, 9, 11. It should be noted that no even state can be equivalent to an odd state. For example, 3 is nonequivalent to 4 because the input $u(t_0) = 0$ gives the output 1 when $\mathcal{A}$ is initially in state 3 and the output 0 when $\mathcal{A}$ is initially in state 4.

7 *Example* Consider the simple electrical network shown in Fig. *2.2.1a*.

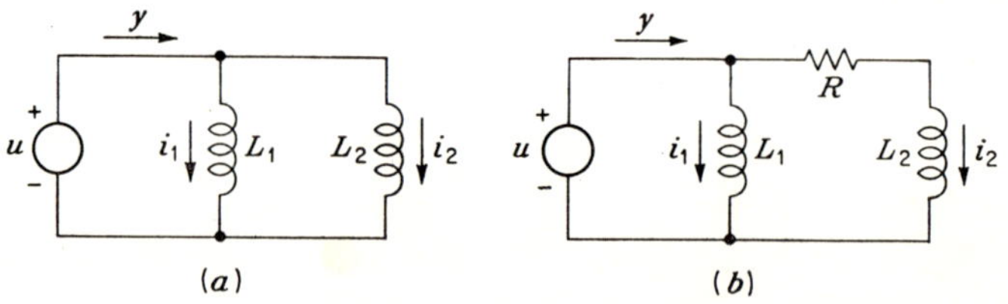

Fig. 2.2.1 (*a*) System with equivalent states; (*b*) system in reduced form.

Here the voltage u is identified with the input, the current y with the output, and the currents $i_1(t)$ and $i_2(t)$ through L_1 and L_2 with the components of the state vector $\mathbf{s}(t)$ (see *1.7.28*). The input-output-state relation—as can be readily verified directly—reads

8 $$y(t) = i_1(t_0) + i_2(t_0) + \left(\frac{1}{L_1} + \frac{1}{L_2}\right) \int_{t_0}^{t} u(\xi)\, d\xi$$

where the first two terms can be represented as the scalar product of the vector (1,1) and the initial state vector $\mathbf{s}(t_0) = (i_1(t_0), i_2(t_0))$.

Now from inspection of *8* it is clear that the only characteristic of $\mathbf{s}(t_0)$ that matters in so far as $y(t)$ is concerned is the sum of the components of $\mathbf{s}(t_0)$. Thus, any two states $\boldsymbol{\alpha}_i$ and $\boldsymbol{\alpha}_j$ such that the sums of their respective components are equal must necessarily be equivalent. For example, $\boldsymbol{\alpha}_i = (5,-2)$ and $\boldsymbol{\alpha}_j = (1,2)$ are equivalent states of the network in question. State equivalences of this type will be discussed in greater detail in Chap. *4*.

9 *Example* It is instructive also to consider a system which has no equivalent states. A simple example of such a system is shown in Fig. *2.2.1b*. This system, call it $\mathcal{A}$, differs from that of Fig. *2.2.1a* only in the addition of a resistor R between L_1 and L_2. As before, the state vector is $\mathbf{s}(t) = (i_1(t), i_2(t))$. For convenience, $R = L_1 = L_2 = 1$.

By using the techniques described in Chap. *4*, the input-output-state relation of $\mathcal{A}$ is readily found to be

10 $$y(t) = i_1(t_0) + i_2(t_0)e^{-(t-t_0)} + \int_{t_0}^{t} (1 + e^{-(t-\xi)})u(\xi)\, d\xi$$

The state space of $\mathcal{A}$ is $\mathcal{R}^2$; that is, Σ is the set of all ordered 2-tuples of real numbers.

Now from inspection of *10* it is clear that in order that a state of $\mathcal{A}$, say (ρ_1,ρ_2), be equivalent to a state (γ_1,γ_2) (where $\rho_1,\gamma_1,\rho_2,\gamma_2$ are real numbers representing the values of i_1 and i_2, respectively), it is necessary and sufficient that

$$\alpha_1 + \alpha_2 e^{-(t-t_0)} = \gamma_1 + \gamma_2 e^{-(t-t_0)}$$

for all t_0 and all $t \geq t_0$. Clearly, this is not possible unless $\alpha_1 = \gamma_1$ and $\alpha_2 = \gamma_2$. Consequently, no two distinct states of $\mathcal{A}$ can be equivalent.

As we shall see later, the system of Example *9* is but a special instance of a *reduced system*. More specifically:

11 **Definition** A system $\mathcal{A}$ is *reduced* or is in *reduced form* if there are no distinct states in its state space which are equivalent to one another. In other words, $\mathcal{A}$ is *reduced* if all of its states are distinguishable.[1] $\lhd$

Consider now the following basic question. Suppose that one is given an input-output-state relation for $\mathcal{A}$ in the form (*2.1.7*)

12 $$\mathbf{y} = \bar{\mathbf{A}}(\boldsymbol{\alpha};\mathbf{u})$$

where $\boldsymbol{\alpha}$ denotes the state of $\mathcal{A}$ at time t_0 (that is, $\mathbf{s}(t_0) = \boldsymbol{\alpha}$), with $\boldsymbol{\alpha}$

[1] In the case of linear systems there is a close connection between the notions of a "reduced" system and a "completely observable" system. The latter notion will be discussed in Chap. *11*.

ranging over the state space Σ. How can one determine from *12* the set of all states in Σ which are equivalent to a particular state $\boldsymbol{\alpha}_0$?

Suppose that an experimenter who does not know that $\mathcal{a}$ is initially in state $\boldsymbol{\alpha}_0$ applies to $\mathcal{a}$ a particular input $\mathbf{u}^1$ and observes the corresponding output $\mathbf{y}^1$. Since he has the input-output-state relation *12*, he can, at least in principle, determine all possible initial states $\boldsymbol{\alpha}$ with respect to which $(\mathbf{u}^1,\mathbf{y}^1)$ is an input-output pair (*2.1.7 et seq.*). Let this set of states be denoted by $Q(\boldsymbol{\alpha}_0;\mathbf{u}^1)$. In symbols,

13 $$Q(\boldsymbol{\alpha}_0;\mathbf{u}^1) = \{\boldsymbol{\alpha}|\bar{\mathbf{A}}(\boldsymbol{\alpha}_0;\mathbf{u}^1) = \bar{\mathbf{A}}(\boldsymbol{\alpha};\mathbf{u}^1)\}$$

Next, the experimenter takes another copy of $\mathcal{a}$ (in the same initial state $\boldsymbol{\alpha}_0$), applies an input $\mathbf{u}^2$, and observes the output $\mathbf{y}^2$. Let the set of states with respect to which $(\mathbf{u}^2,\mathbf{y}^2)$ constitutes an input-output pair be denoted by $Q(\boldsymbol{\alpha}_0;\mathbf{u}^2)$. Then, at the conclusion of the second experiment, the experimenter can assert that $\boldsymbol{\alpha}_0$ must lie in the intersection of $Q(\boldsymbol{\alpha}_0;\mathbf{u}^1)$ and $Q(\boldsymbol{\alpha}_0;\mathbf{u}^2)$.

By repeating this experiment (infinitely many times, if necessary) the experimenter arrives at the conclusion that $\boldsymbol{\alpha}_0$ must lie in the intersection of the family of sets $Q(\boldsymbol{\alpha}_0;\mathbf{u}^1)$, $Q(\boldsymbol{\alpha}_0;\mathbf{u}^2)$, $Q(\boldsymbol{\alpha}_0;\mathbf{u}^3)$, ... [Note that the intersection is nonempty since $\boldsymbol{\alpha}_0 \in Q(\boldsymbol{\alpha}_0;\mathbf{u}^i)$ for all i.] On denoting this intersection by $Q(\boldsymbol{\alpha}_0)$ and dropping the superscript on $\mathbf{u}$, we have in symbols

14 $$Q(\boldsymbol{\alpha}_0) = \bigcap_{\mathbf{u}} Q(\boldsymbol{\alpha}_0;\mathbf{u}) = \bigcap_{\mathbf{u}} \{\boldsymbol{\alpha}|\bar{\mathbf{A}}(\boldsymbol{\alpha}_0;\mathbf{u}) = \bar{\mathbf{A}}(\boldsymbol{\alpha};\mathbf{u})\}$$

where $\bigcap_{\mathbf{u}}$ denotes the intersection taken over all $\mathbf{u}$ in the input segment space of $\mathcal{a}$. Thus, the experimenter's conclusion after performing all these experiments would be that the initial state belongs to $Q(\boldsymbol{\alpha}_0)$, and no further experimentation can give him additional information about the initial state of $\mathcal{a}$.

From the way in which state equivalence is defined, it is evident that the set $Q(\boldsymbol{\alpha}_0)$ defined by *14* is the *set of all states in* Σ *which are equivalent to* $\boldsymbol{\alpha}_0$. Essentially, this is an immediate consequence of the following relation between the operations $\bigcap_{\mathbf{u}}$ and $\forall\mathbf{u}$. (To place this relation in a clearer perspective, we formulate it in symbols different from those used above.)

15 **Relation between $\bigcap$ and $\forall$** The relation in question may be stated as follows. Let R be a relation, i.e., a set of ordered pairs, $R \triangleq \{(v,w)\}$, with domain D. [The domain of R is the set of all v such that $(v,w) \in R$ for some w.] Define the sets $Q(v)$ and Q by

16 $$Q(v) \triangleq \{w|(v,w) \in R\}$$

and

17 $$Q \triangleq \bigcap_{v \in D} Q(v)$$

Then, Q can be expressed in the form

18 $$Q = \{w | \forall v[(v,w) \in R]\} \qquad v \in D$$

The equivalence between *17* and *18* follows at once from the fact that a point belongs to the intersection $\bigcap_v Q(v)$ if and only if it belongs to every set in the family $\{Q(v)\}$, $v \in D$. ◁

To apply this equivalence to the case under consideration, we identify R with the relation $\bar{\mathbf{A}}(\boldsymbol{\alpha}_0;\mathbf{u}) = \bar{\mathbf{A}}(\boldsymbol{\alpha};\mathbf{u})$. Then, from the definition

$$\text{set of all states equivalent to } \boldsymbol{\alpha}_0 = \{\boldsymbol{\alpha} | \forall \mathbf{u}[\bar{\mathbf{A}}(\boldsymbol{\alpha}_0;\mathbf{u}) = \bar{\mathbf{A}}(\boldsymbol{\alpha};\mathbf{u})]\}$$

we can infer at once the

19 **Assertion** The set of all states in Σ equivalent to $\boldsymbol{\alpha}_0$ is given by

20 $$Q(\boldsymbol{\alpha}_0) = \bigcap_{\mathbf{u}} \{\boldsymbol{\alpha} | \bar{\mathbf{A}}(\boldsymbol{\alpha}_0;\mathbf{u}) = \bar{\mathbf{A}}(\boldsymbol{\alpha};\mathbf{u})\} \qquad \triangleleft$$

One immediate application that we can make of *15* is to prove statement *1.7.1* concerning the equivalence of states in the set Q_t. Specifically, it will be recalled that the set Q_t was defined as the intersection (see *1.6.38*)

21 $$Q_t(\boldsymbol{\alpha}_0;\mathbf{u}) = \bigcap_{\mathbf{u}'} Q(\boldsymbol{\alpha}_0;\mathbf{u},\mathbf{u}')$$

where

22 $$Q(\boldsymbol{\alpha}_0;\mathbf{u},\mathbf{u}') = \{\boldsymbol{\alpha} | \mathbf{y}' = \bar{\mathbf{A}}(\boldsymbol{\alpha};\mathbf{u}')\}$$

and where $\mathbf{y}'$ is defined by the relation

23 $$\mathbf{y}\mathbf{y}' = \bar{\mathbf{A}}(\boldsymbol{\alpha}_0;\mathbf{u}\mathbf{u}')$$

in which $\mathbf{u}\mathbf{u}'$ denotes a segment consisting of a segment $\mathbf{u}$ followed by a segment $\mathbf{u}'$ and $\mathbf{y}\mathbf{y}'$ is the corresponding response segment of $\mathcal{a}$, with $\mathcal{a}$ initially in state $\boldsymbol{\alpha}_0$.

Let $\boldsymbol{\alpha}'$ be any state in $Q_t(\boldsymbol{\alpha}_0;\mathbf{u})$. Then by *21* and *22* we can express $\mathbf{y}'$ as $\mathbf{y}' = \bar{\mathbf{A}}(\boldsymbol{\alpha}';\mathbf{u}')$, and hence *21* can be written as

24 $$Q_t(\boldsymbol{\alpha}_0;\mathbf{u}) = \bigcap_{\mathbf{u}'} \{\boldsymbol{\alpha} | \bar{\mathbf{A}}(\boldsymbol{\alpha}';\mathbf{u}') = \bar{\mathbf{A}}(\boldsymbol{\alpha};\mathbf{u}')\}$$

On applying *15* to *24*, we obtain

$$Q_t(\boldsymbol{\alpha}_0;\mathbf{u}) = \{\boldsymbol{\alpha} | \forall \mathbf{u}'[\bar{\mathbf{A}}(\boldsymbol{\alpha}';\mathbf{u}') = \bar{\mathbf{A}}(\boldsymbol{\alpha};\mathbf{u}')]\}$$

which, in words, states that $Q_t(\boldsymbol{\alpha}_0;\mathbf{u})$ is the set of all states equivalent to α'. Consequently, all states in $Q_t(\alpha_0;u)$ are equivalent to one another. Q.E.D.

It will be helpful at this point to summarize our discussion by restating in a more precise form the tentative definition of the state of $\mathcal{A}$ at time t which was given in *1.7.1*.

25 **Definition** Let $\mathcal{A}$ be characterized by the input-output-state relation

$$\mathbf{y}(t) = \mathbf{A}(\boldsymbol{\alpha}_0;\mathbf{u}_{(t_0,t]}) \qquad \boldsymbol{\alpha}_0 \in \Sigma$$

or equivalently

26 $$\mathbf{y} = \bar{\mathbf{A}}(\boldsymbol{\alpha}_0;\mathbf{u})$$

where $\boldsymbol{\alpha}_0$ denotes the state of $\mathcal{A}$ at time t_0.

Then the *state of* $\mathcal{A}$ *at time* t, $t > t_0$, is denoted by $\mathbf{s}(t)$ and is defined to be any element of the set

27 $$Q_t(\boldsymbol{\alpha}_0;\mathbf{u}) = \{\boldsymbol{\alpha}|\forall\mathbf{u}'[\bar{\mathbf{A}}(\boldsymbol{\alpha}_0;\mathbf{u})\bar{\mathbf{A}}(\boldsymbol{\alpha};\mathbf{u}') = \bar{\mathbf{A}}(\boldsymbol{\alpha}_0;\mathbf{uu}')]\}$$

where $\mathbf{u} \triangleq \mathbf{u}_{(t_0,t]}$; $\mathbf{u}' \triangleq \mathbf{u}_{(t,t_1]}$; $\mathbf{uu}' = \mathbf{u}_{(t_0,t_1]}$; $\bar{\mathbf{A}}(\boldsymbol{\alpha}_0;\mathbf{u}) = \mathbf{y}_{(t_0,t]} \triangleq \mathbf{y}$; $\bar{\mathbf{A}}(\boldsymbol{\alpha};\mathbf{u}') = \mathbf{y}_{(t,t_1]} \triangleq \mathbf{y}'$; $\bar{\mathbf{A}}(\boldsymbol{\alpha}_0;\mathbf{u})\bar{\mathbf{A}}(\boldsymbol{\alpha};\mathbf{u}') = \mathbf{yy}'$; and $\boldsymbol{\alpha} \in \Sigma$, with the understanding that $\mathbf{uu}'$ denotes a segment of the input comprised of $\mathbf{u}$ followed by $\mathbf{u}'$, and likewise for $\mathbf{yy}'$ and $\bar{\mathbf{A}}(\boldsymbol{\alpha}_0;\mathbf{u})\bar{\mathbf{A}}(\boldsymbol{\alpha};\mathbf{u}')$. [Note, in particular, that $\bar{\mathbf{A}}(\boldsymbol{\alpha}_0;\mathbf{u})\bar{\mathbf{A}}(\boldsymbol{\alpha};\mathbf{u}')$ denotes the response segment $\bar{\mathbf{A}}(\boldsymbol{\alpha}_0;\mathbf{u})$ *followed* by the response segment $\bar{\mathbf{A}}(\boldsymbol{\alpha};\mathbf{u}')$ and should not be interpreted as the product of $\bar{\mathbf{A}}(\boldsymbol{\alpha}_0;\mathbf{u})$ and $\bar{\mathbf{A}}(\boldsymbol{\alpha};\mathbf{u}')$.]

28 *Comment* Despite its different appearance, the set $Q_t(\boldsymbol{\alpha}_0;\mathbf{u})$ defined by *27* is the same as the set $Q_t(\boldsymbol{\alpha}_0;\mathbf{u})$ defined by *21*. The reason is this. In the first place, we have already shown that $Q_t(\boldsymbol{\alpha}_0;\mathbf{u})$ can be expressed by *24*, in which $\mathbf{y}'$ is defined by *23*. Now the statement: for all $\mathbf{u}'$

29 $$\mathbf{y}' = \bar{\mathbf{A}}(\boldsymbol{\alpha};\mathbf{u}')$$

and

30 $$\mathbf{yy}' = \bar{\mathbf{A}}(\boldsymbol{\alpha}_0;\mathbf{uu}')$$

can be combined into a single statement by substituting *29* into *30*, with the result reading

31 $$\forall\mathbf{u}'[\mathbf{y}\bar{\mathbf{A}}(\boldsymbol{\alpha};\mathbf{u}') = \bar{\mathbf{A}}(\boldsymbol{\alpha}_0;\mathbf{uu}')]$$

But

$$\mathbf{y} = \bar{\mathbf{A}}(\boldsymbol{\alpha}_0;\mathbf{u})$$

and hence *31* can be expressed as

$$\forall\mathbf{u}'[\bar{\mathbf{A}}(\boldsymbol{\alpha}_0;\mathbf{u})\bar{\mathbf{A}}(\boldsymbol{\alpha};\mathbf{u}) = \bar{\mathbf{A}}(\boldsymbol{\alpha}_0;\mathbf{uu}')]$$

which establishes the equivalence of *27* and *21*. Q.E.D.

As was pointed out in Chap. 1 (*1.8.1 et seq.*), Definition *25* not only implies that $\mathbf{s}(t)$ is a function of $\boldsymbol{\alpha}_0$ and $\mathbf{u}_{(t_0,t]}$ but also provides, at least in principle, a means of deducing the functional dependence of $\mathbf{s}(t)$ on $\mathbf{s}(t_0)$ and $\mathbf{u}_{(t_0,t]}$ from the input-output-state relation

32 $$\mathbf{y}(t) = \bar{\mathbf{A}}(\mathbf{s}(t_0);\mathbf{u}_{(t_0,t]})$$

where $\boldsymbol{\alpha}_0 \triangleq \mathbf{s}(t_0) \triangleq$ initial state of $\mathcal{A}$. In other words, the state equation (*1.8.2*)

33 $$\mathbf{s}(t) = \mathbf{s}(\mathbf{s}(t_0);\mathbf{u}_{(t_0,t]})$$

which is the expression for $\mathbf{s}(t)$ as a function of $\mathbf{s}(t_0)$ and $\mathbf{u}_{(t_0,t]}$, is derivable from the input-output-state relation *32*. We shall establish some of the important properties of this equation and the state of $\mathcal{A}$ at time t in the following section.

3 *Basic properties of state and state equations*

A basic property of $\mathbf{s}(t)$ which is an immediate consequence of its definition is the

1 **Separation property of $\mathbf{s}(t)$** By the definition of $\mathbf{s}(t)$ (*1.7.1*), $\mathbf{s}(t)$ is an element of the set $Q_t(\boldsymbol{\alpha}_0;\mathbf{u})$, which is the set of all $\boldsymbol{\alpha}$ satisfying the relation (*2.2.27*)

2 $$\forall \mathbf{u}'[\bar{\mathbf{A}}(\boldsymbol{\alpha}_0;\mathbf{u})\bar{\mathbf{A}}(\boldsymbol{\alpha};\mathbf{u}') = \bar{\mathbf{A}}(\boldsymbol{\alpha}_0;\mathbf{u}\mathbf{u}')]$$

On replacing $\boldsymbol{\alpha}_0$ in *2* by $\mathbf{s}(t_0)$ and $\boldsymbol{\alpha}$ by $\mathbf{s}(t)$, we obtain the identity

3 $$\bar{\mathbf{A}}(\mathbf{s}(t_0);\mathbf{u}\mathbf{u}') = \bar{\mathbf{A}}(\mathbf{s}(t_0);\mathbf{u})\bar{\mathbf{A}}(\mathbf{s}(t);\mathbf{u}') \qquad \forall \mathbf{s}(t_0)\forall \mathbf{u}\forall \mathbf{u}'$$

which, expressed in words, states: for any input consisting of a segment $\mathbf{u}$ followed by a segment $\mathbf{u}'$, the response of $\mathcal{A}$ to $\mathbf{u}\mathbf{u}'$ starting in state $\mathbf{s}(t_0)$ consists of the response segment $\bar{\mathbf{A}}(\mathbf{s}(t_0);\mathbf{u})$ followed by the response segment $\bar{\mathbf{A}}(\mathbf{s}(t);\mathbf{u}')$, where $\mathbf{s}(t)$ is the state of $\mathcal{A}$ at time t (see Fig. *2.3.1*). This basic property of $\mathbf{s}(t)$ will be referred to as the *response separation property*, since $\mathbf{s}(t)$ in a sense "separates" the response segment $\bar{\mathbf{A}}(\mathbf{s}(t_0);\mathbf{u})$ from $\bar{\mathbf{A}}(\mathbf{s}(t);\mathbf{u}')$. ◁

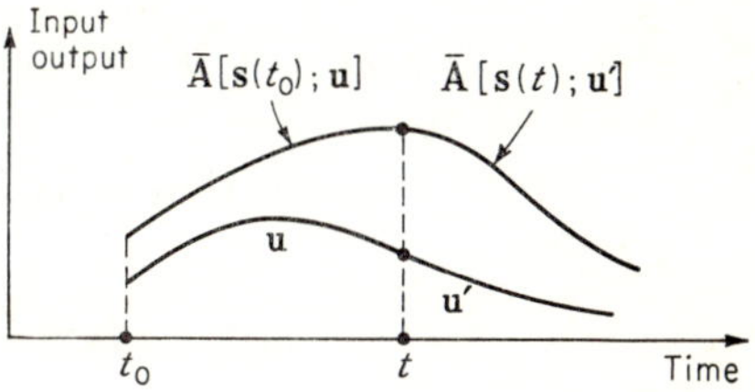

Fig. 2.3.1 Response separation property.

The response separation property of $\mathbf{s}(t)$ can be expressed more compactly by replacing $\mathbf{s}(t_0)$ in *3* by $\boldsymbol{\alpha}_0$ and regarding $\mathbf{s}(t)$ as the terminal state $\boldsymbol{\alpha}_0^*$ of $\mathfrak{A}$ into which the initial state $\boldsymbol{\alpha}_0$ is taken by $\mathbf{u}$ (see *2.1.7 et seq.*). In this form, *3* reads

4 $$\bar{\mathbf{A}}(\boldsymbol{\alpha}_0;\mathbf{uu}') = \bar{\mathbf{A}}(\boldsymbol{\alpha}_0;\mathbf{u})\bar{\mathbf{A}}(\boldsymbol{\alpha}_0^*;\mathbf{u}')$$

where it is understood that *4* holds for all $\boldsymbol{\alpha}_0 \in \Sigma$, all $\mathbf{u}$, and all $\mathbf{u}'$.

Since in most cases the input-output-state relation of $\mathfrak{A}$ has the form of an expression for $\mathbf{y}(t)$ as a function of $\mathbf{s}(t_0)$ and $\mathbf{u}_{(t_0,t]}$ rather than an expression for the response segment $\mathbf{y}_{(t_0,t]}$ as a function of $\mathbf{s}(t_0)$ and $\mathbf{u}_{(t_0,t]}$, it is desirable to cast *4* into a form that involves $\mathbf{A}(\boldsymbol{\alpha}_0;\mathbf{u})$ rather than $\bar{\mathbf{A}}(\boldsymbol{\alpha}_0,\mathbf{u})$. For this purpose, let τ be a time intermediate between t_0 and t. Then it is evident that the identity [for all $\mathbf{s}(t_0)$, $\mathbf{u}$, and τ]

5 $$\mathbf{A}(\mathbf{s}(t_0);\mathbf{u}_{(t_0,t]}) = \mathbf{A}(\mathbf{s}(\tau);\mathbf{u}_{(\tau,t]}) \qquad t_0 < \tau \leq t$$

where $\mathbf{A}(\mathbf{s}(\tau);\mathbf{u}_{(\tau,t]})$ denotes the response of $\mathfrak{A}$ at time t to $\mathbf{u}_{(\tau,t]}$, with $\mathfrak{A}$ initially in state $\mathbf{s}(\tau)$, expresses the separation property *4*. In the sequel, we shall be using for the most part *5* rather than *4* as a statement of the response separation property.

6 *Example* Consider the system of Example *1.7.19* and, for concreteness, let $t_0 = 0$, $t = 3$, $t_1 = 7$, $u_{[t_0,t_1)} = 3\ 1\ 2\ 2\ 1\ 5\ 4$ and $s(0) = 3$.[1] Then by inspection of *1.7.21*, we have

$$\bar{A}(3;\ 3\ 1\ 2\ 2\ 1\ 5\ 4) = 0\ 1\ 1\ 1\ 0\ 1\ 1$$

Now the state of $\mathfrak{A}$ at time $t = 3$ is given by

$$x(3) = (3 + 3 + 1 + 2) \bmod 12 = 9$$

and hence

$$\bar{A}(3;\ 3\ 1\ 2) = 0\ 1\ 1$$
$$\bar{A}(9;\ 2\ 1\ 5\ 4) = 1\ 0\ 1\ 1$$

We have the verification

$$\bar{A}(3;\ 3\ 1\ 2\ 2\ 1\ 5\ 4) = \bar{A}(3;\ 3\ 1\ 2)\bar{A}(9;\ 2\ 1\ 5\ 4)$$

Here the initial state *3* plays the role of $\boldsymbol{\alpha}_0$ in *4*; $\mathbf{u}$ is 3 1 2, $\mathbf{u}'$ is 2 1 5 4, and $\boldsymbol{\alpha}_0^*$ is 9.

7 *Example* Suppose that $\mathfrak{A}$ is characterized by the input-output-state relation

8 $$y(t) = s(t_0)e^{-(t-t_0)} + \int_{t_0}^{t} e^{-(t-\xi)}u(\xi)\,d\xi$$

where $s(t_0)$ and $u(t)$ range over the real line, and u contains no delta

[1] Note that, strictly speaking, $u_{[t_0,t_1)}$ is the set of pairs (0,3), (1,1), (2,2), (3,2), (4,1), (5,5), (6,4) (see *1.3.3*). Note also that the corresponding response segment is observed over $(t_0,t_1]$.

functions. Then we can readily verify that with $s(t)$ given by

9 $$s(t) = s(t_0)e^{-(t-t_0)} + \int_{t_0}^{t} e^{-(t-\xi)}u(\xi)\,d\xi$$

the input-output-state relation *8* has the response separation property *5*. Specifically, on letting $t_0 \leq \tau \leq t$, we can write *8* as

$$y(t) = s(t_0)e^{-(t-t_0)} + \int_{t_0}^{\tau} e^{-(t-\xi)}u(\xi)\,d\xi + \int_{\tau}^{t} e^{-(t-\xi)}u(\xi)\,d\xi$$

Making use of *9*, we have

10 $$s(\tau) = s(t_0)e^{-(\tau-t_0)} + \int_{t_0}^{\tau} e^{-(t-\xi)}u(\xi)\,d\xi$$

and hence the expression for $y(t)$ can be written as

11 $$y(t) = s(\tau)e^{-(t-\tau)} + \int_{\tau}^{t} e^{-(t-\xi)}u(\xi)\,d\xi$$

which is of the same form as *8*, with $s(t_0)$ replaced by $s(\tau)$. Q.E.D.

There are a number of important connections between the separation property and the self-consistency conditions discussed in Chap. *1* (see *1.6.2*). For convenience in reference, we now state the relevant facts in the form of short assertions and theorems.

12 **Assertion** If a relation of the form

13 $$\mathbf{y}(t) = \mathbf{A}(\boldsymbol{\alpha};\mathbf{u}_{(t_0,t]}) \qquad \boldsymbol{\alpha} \in \Sigma,\ \mathbf{u} \in R[\mathbf{u}]$$

in which $\mathbf{A}$ is a function defined on $\Sigma \times R[\mathbf{u}]$, has the response separation property (for all $\boldsymbol{\alpha}$ and $\mathbf{u}$)

14 $$\mathbf{A}(\boldsymbol{\alpha};\mathbf{u}_{(t_0,t]}) = \mathbf{A}(\boldsymbol{\alpha}^*;\mathbf{u}_{(\tau,t]}) \qquad t_0 < \tau \leq t$$

in which $\boldsymbol{\alpha}^* \in \Sigma$ and $\boldsymbol{\alpha}^*$ depends only on $\boldsymbol{\alpha}$ and $\mathbf{u}_{(t_0,\tau]}$, then *13* satisfies the three self-consistency conditions and $\boldsymbol{\alpha}^*$ qualifies as the state of $\mathcal{A}$ at time τ, with $\boldsymbol{\alpha}$ being the state of $\mathcal{A}$ at time t_0. Also, *13* qualifies as an input-output-state relation for $\mathcal{A}$. (See *1.6.2.*) (Note that since we do not start with a characterization of $\mathcal{A}$ in terms of an input-output relation, there is no need to consider the mutual-consistency condition.)

Proof The first self-consistency condition is satisfied because $\mathbf{y}(t)$ is uniquely determined through *13* by $\boldsymbol{\alpha}$ and $\mathbf{u}_{(t_0,t]}$. As for the second self-consistency condition, the separation property *14* without the qualification "$\boldsymbol{\alpha}^*$ depends only on $\boldsymbol{\alpha}$ and $\mathbf{u}_{(t_0,\tau]}$" is in effect a statement of it (see *1.6.15*). With this qualification added, every set $Q(\boldsymbol{\alpha};\mathbf{u},\mathbf{u}')$ (*1.6.29*) will contain $\boldsymbol{\alpha}^*$ and hence the intersection $Q_\tau(\boldsymbol{\alpha};\mathbf{u})$ will be nonempty by virtue of containing $\boldsymbol{\alpha}^*$. This shows that all three self-consistency conditions are satisfied and hence by Definition *1.6.2* $\boldsymbol{\alpha}^*$ qualifies as the state of $\mathcal{A}$ at time τ and *14* qualifies as its input-output-state relation. Q.E.D.

15 *Comment* The importance of Assertion *12* stems from the fact that it reduces the verification of self-consistency conditions for a given relation of the form *13* to demonstrating that *13* has the response separation property *14*, with $\boldsymbol{\alpha}^*$ dependent only on $\boldsymbol{\alpha}$ and $\mathbf{u}_{(t_0,\tau]}$. Actually, we have already used this technique in an implicit fashion in a number of examples, among them *1.7.10*, *1.7.12*, and *1.7.19*.

16 **Theorem** If

17 $$\mathbf{y}(t) = \mathbf{A}(\mathbf{s}(t_0);\mathbf{u}_{(t_0,t]})$$

is an input-output-state relation for $\mathcal{A}$, then the state equation induced by *17* (see *1.8.2 et seq.*),

18 $$\mathbf{s}(t) = \mathbf{s}(\mathbf{s}(t_0);\mathbf{u}_{(t_0,t]})$$

has the *state separation property* expressed by

19 $$\mathbf{s}(\mathbf{s}(t_0);\mathbf{u}_{(t_0,t]}) \simeq \mathbf{s}(\mathbf{s}(\tau);\mathbf{u}_{(\tau,t]}) \qquad t_0 < \tau \leq t$$

for all $\mathbf{s}(t_0)$ and $\mathbf{u}$, with $\mathbf{s}(\tau)$ given by *18* for $\mathbf{u}_{(t_0,\tau]}$.

Proof We shall establish this result by contradiction. Specifically, with reference to Fig. *2.3.2*, assume that

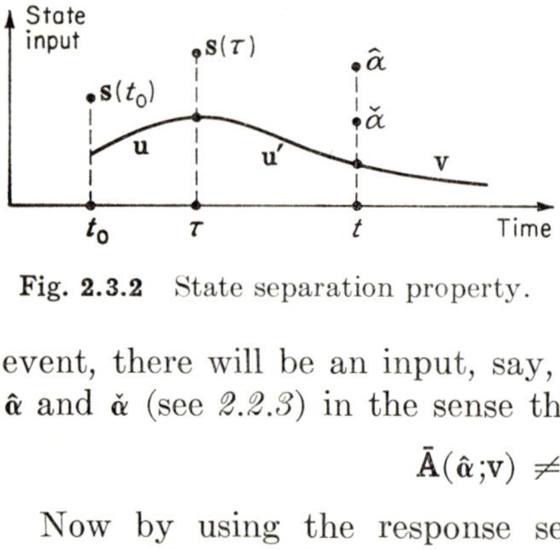

Fig. 2.3.2 State separation property.

20 $$\mathbf{s}(\mathbf{s}(t_0);\mathbf{u}_{(t_0,t]}) = \hat{\boldsymbol{\alpha}}$$

21 $$\mathbf{s}(\mathbf{s}(\tau);\mathbf{u}_{(\tau,t]}) = \check{\boldsymbol{\alpha}}$$

and $\hat{\boldsymbol{\alpha}} \not\simeq \check{\boldsymbol{\alpha}}$, that is, the state at time t yielded by *20* is not equivalent to the state at time t yielded by *21*. In that event, there will be an input, say, $\mathbf{v}_{(t,t_1]}$, that will distinguish between $\hat{\boldsymbol{\alpha}}$ and $\check{\boldsymbol{\alpha}}$ (see *2.2.3*) in the sense that

22 $$\bar{\mathbf{A}}(\hat{\boldsymbol{\alpha}};\mathbf{v}) \neq \bar{\mathbf{A}}(\check{\boldsymbol{\alpha}};\mathbf{v})$$

Now by using the response separation property *4* the response segment of $\mathcal{A}$ to $\mathbf{uu'v}$ starting in state $\mathbf{s}(t_0)$ can be split up in two ways:

23 $$\bar{\mathbf{A}}(\mathbf{s}(t_0);\mathbf{uu'v}) = \bar{\mathbf{A}}(\mathbf{s}(t_0);\mathbf{u})\bar{\mathbf{A}}(\mathbf{s}(\tau);\mathbf{u'v})$$

24 $$\bar{\mathbf{A}}(\mathbf{s}(t_0);\mathbf{uu'v}) = \bar{\mathbf{A}}(\mathbf{s}(t_0);\mathbf{uu'})\bar{\mathbf{A}}(\hat{\boldsymbol{\alpha}};\mathbf{v})$$

Applying the separation property again, we have

25 $$\bar{\mathbf{A}}(\mathbf{s}(\tau);\mathbf{u'v}) = \bar{\mathbf{A}}(\mathbf{s}(\tau);\mathbf{u'})\bar{\mathbf{A}}(\check{\boldsymbol{\alpha}};\mathbf{v})$$

26 $$\mathbf{A}(\mathbf{s}(t_0);\mathbf{uu'}) = \bar{\mathbf{A}}(\mathbf{s}(t_0);\mathbf{u})\bar{\mathbf{A}}(\mathbf{s}(\tau);\mathbf{u'})$$

On substituting *25* and *26* into *23* and *24*, respectively, and noting *22*, we arrive at the conclusion

$$\bar{\mathbf{A}}(\mathbf{s}(t_0);\mathbf{uu'v}) \neq \bar{\mathbf{A}}(\mathbf{s}(t_0);\mathbf{uu'v})$$

which violates the first self-consistency condition and hence is a contradiction. Consequently, $\hat{\alpha} \simeq \check{\alpha}$ and the theorem is proved.

27 *Comment* The state separation property of the state equation *18* is one of its key characteristics. We have already encountered special cases of it in examples and will encounter it in a more general way in Chaps. *3*, *4*, and *5*. To cite just one illustration, in Example *7* the state equation reads

$$s(t) = s(t_0)e^{-(t-t_0)} + \int_{t_0}^{t} e^{-(t-\xi)}u(\xi)\,d\xi \qquad t \geq t_0$$

and the state separation property *19* assumes the form

$$s(t) = s(\tau)e^{-(t-\tau)} + \int_{\tau}^{t} e^{-(t-\xi)}u(\xi)\,d\xi$$

28 **Theorem** If $\mathcal{A}$ is characterized by equations of the form

29 $$\mathbf{s}(t) = \mathbf{s}(\mathbf{s}(t_0);\mathbf{u}_{(t_0,t]}) \qquad t > t_0$$

30 $$\mathbf{y}(t) = \mathbf{g}(\mathbf{s}(t),\mathbf{u}(t),t)$$

where $\mathbf{s}(t_0) \in \Sigma$, $\mathbf{s}(t) \in \Sigma$, $\mathbf{s}$ is a function on $\Sigma \times R[\mathbf{u}]$, and $\mathbf{g}$ is a function on $\Sigma \times R[\mathbf{u}(t)] \times T$, and if *29* has the state separation property *19*, then $\mathbf{s}(t)$ qualifies as the state of $\mathcal{A}$ at time t and *29* and *30* constitute the state equations of $\mathcal{A}$ (see *1.8.5 et seq.*).

Proof On substituting *29* into *30* we obtain the relation

31 $$\mathbf{y}(t) = \mathbf{g}(\mathbf{s}(\mathbf{s}(t_0);\mathbf{u}_{(t_0,t]}),\mathbf{u}(t),t)$$

which is of the form

32 $$\mathbf{y}(t) = \mathbf{A}(\mathbf{s}(t_0);\mathbf{u}_{(t_0,t]})$$

Since *29* has the state separation property, we can write

33 $$\mathbf{s}(\mathbf{s}(t);\mathbf{u}_{(t_0,t]}) \simeq \mathbf{s}(\mathbf{s}(\tau);\mathbf{u}_{(\tau,t]})$$

and on substituting *33* into *31*, we have

34 $$\mathbf{y}(t) = \mathbf{g}(\mathbf{s}(\mathbf{s}(\tau);\mathbf{u}_{(\tau,t]}),\mathbf{u}(t),t)$$

Now *34* can be obtained from *31* by replacing t_0 with τ. Hence

$$\mathbf{y}(t) = \mathbf{A}(\mathbf{s}(\tau);\mathbf{u}_{(\tau,t]})$$

and

$$\mathbf{A}(\mathbf{s}(t_0);\mathbf{u}_{(t_0,t]}) = \mathbf{A}(\mathbf{s}(\tau);\mathbf{u}_{(\tau,t]}) \qquad t_0 < \tau \leq t$$

which establishes that $\mathcal{A}$ has the response separation property *5*. Furthermore, since $\mathbf{s}(\tau)$ is determined uniquely in terms of $\mathbf{s}(t_0)$ and $\mathbf{u}_{(t_0,\tau]}$ by *29*, it follows by Assertion *12* that *32* qualifies as the input-output-state relation for $\mathcal{A}$, with $\mathbf{s}(t)$ representing the state of $\mathcal{A}$ at time t and *29* and *30* constituting the state equations of $\mathcal{A}$. Q.E.D.

35 *Comment* This theorem adds insight into the key role of the state separation property of the state equation and its close relation with the self-consistency conditions introduced in Chap. 1. Together with its corollary (Corollary *36*, below), Theorem *28* provides an indirect but effective way of verifying that the self-consistency conditions are satisfied. For this reason, it will find many applications in later chapters, particularly in Chap. *4*.

It should be observed that the proof given above remains valid when the expression for $\mathbf{y}(t)$ (*30*) involves not only $\mathbf{u}(t)$ but also one or more derivatives of $\mathbf{u}$ at time t. In fact, the theorem holds true under even weaker assumptions on the form of $\mathbf{y}(t)$. However, the form assumed here will suffice for most of our purposes.

36 **Corollary** If $\mathcal{A}$ is characterized by differential state equations of the *canonical* form

37 $$\dot{\mathbf{s}}(t) = \mathbf{f}(\mathbf{s}(t),\mathbf{u}(t),t)$$

38 $$\mathbf{y}(t) = \mathbf{g}(\mathbf{s}(t),\mathbf{u}(t),t)$$

where $\mathbf{f}$ and $\mathbf{g}$ are functions on $\Sigma \times R[\mathbf{u}(t)] \times T$ such that *37* has a unique solution for $\mathbf{s}(t)$, then $\mathbf{s}(t)$ qualifies as the state of $\mathcal{A}$ at time t.

Proof. On integrating *37* between t and t_0, we have

39 $$\mathbf{s}(t) = \mathbf{s}(t_0) + \int_{t_0}^{t} \mathbf{f}(\mathbf{s}(\xi),\mathbf{u}(\xi),\xi)\, d\xi \qquad t > t_0$$

which may be regarded as an implicit form of the state equation

$$\mathbf{s}(t) = \mathbf{s}(\mathbf{s}(t_0);\mathbf{u}_{(t_0,t]})$$

Note that, by hypothesis, $\mathbf{s}(t)$ is determined uniquely by $\mathbf{s}(t_0)$ and $\mathbf{u}_{(t_0,t]}$.

Now it is trivial to verify that *39* has the separation property. Specifically, for $t_0 < \tau \leq t$ we can write *39* in the form

$$\mathbf{s}(t) = \mathbf{s}(t_0) + \int_{t_0}^{\tau} \mathbf{f}(\mathbf{s}(\xi),\mathbf{u}(\xi),\xi)\, d\xi + \int_{\tau}^{t} \mathbf{f}(\mathbf{s}(\xi),\mathbf{u}(\xi),\xi)\, d\xi$$

or

$$\mathbf{s}(t) = \mathbf{s}(\tau) + \int_{\tau}^{t} \mathbf{f}(\mathbf{s}(\xi),\mathbf{u}(\xi),\xi)\, d\xi$$

since

$$\mathbf{s}(\tau) = \mathbf{s}(t_0) + \int_{t_0}^{\tau} \mathbf{f}(\mathbf{s}(\xi),\mathbf{u}(\xi),\xi)\, d\xi$$

by *39*. The conclusion then follows from Theorem *28*.

40 **Remark** It is a well-known result in the theory of ordinary differential equations that a sufficient condition for *37* to have a unique solution (for a fixed $\mathbf{u}$) satisfying a given initial condition $\mathbf{s}(t_0) = \boldsymbol{\alpha}$ is the Lipschitz condition,[1] namely, that there exists a finite constant L

[1] See, for instance, Refs. *5* and *6*.

depending on $\mathbf{u}$ such that for all $\mathbf{s}(t)$ and $\mathbf{s}'(t) \in \Sigma$

41 $$\|\mathbf{f}(\mathbf{s}(t),\mathbf{u}(t)) - \mathbf{f}(\mathbf{s}'(t),\mathbf{u}(t))\| \leq L\|\mathbf{s}(t) - \mathbf{s}'(t)\| \qquad t \in T$$

where $\| \quad \|$ denotes the euclidean norm (see Appendix *C.16.5*). ◁

In later chapters, Corollary *36* will be applied only to linear systems. In the case of such systems, *36* and *41* lead to the following

42 **Conclusion** If $\mathcal{A}$ is characterized by linear differential state equations of the form

43 $$\begin{aligned} \dot{\mathbf{s}}(t) &= \mathbf{A}(t)\mathbf{s}(t) + \mathbf{B}(t)\mathbf{u}(t) \\ \mathbf{y}(t) &= \mathbf{C}(t)\mathbf{s}(t) + \mathbf{D}(t)\mathbf{u}(t) \end{aligned}$$

where $\mathbf{A}(t), \ldots, \mathbf{D}(t)$ are (possibly) time-dependent continuous matrices, then *41* is satisfied with

44 $$L = \sum_{i,j} \sup_{t \in T} |a_{ij}(t)|$$

where $a_{ij}(t)$ denotes the ijth element of $\mathbf{A}(t)$, and consequently $\mathbf{s}(t)$ qualifies as the state of $\mathcal{A}$ at time t (see also *6.2.2*).

45 *Comment* State equations of the form *43* will be referred to as the *canonical state equations* of $\mathcal{A}$. As we shall see in Chap. *4*, this is not the most general form that the state equations of a linear system can assume. ◁

As was pointed out in Chap. *1* (*1.6.44*), it may be advantageous in certain cases to relabel the points of Σ in a way that may simplify the form of the state equations of $\mathcal{A}$ or place in evidence some of their special features. In this connection, the following assertion is of relevance.

46 **Assertion** Let $\mathcal{A}$ be a system characterized by an input-output-state relation

47 $$\mathbf{y}(t) = \mathbf{A}(\boldsymbol{\alpha};\mathbf{u}) \qquad \boldsymbol{\alpha} \triangleq \mathbf{s}(t_0),\ \mathbf{u} \triangleq \mathbf{u}_{(t_0,t]}$$

which induces the state equations (see *1.8.7* and *1.8.8*)

48 $$\begin{aligned} \mathbf{s}(t) &= \mathbf{s}(\boldsymbol{\alpha};\mathbf{u}) \\ \mathbf{y}(t) &= \mathbf{g}(\mathbf{s}(t),\mathbf{u}(t),t) \end{aligned}$$

Let $\boldsymbol{\mu}$ be a one-one mapping of the state space Σ onto itself which takes $\check{\boldsymbol{\alpha}}$ into $\boldsymbol{\alpha}$. That is,

$$\boldsymbol{\mu}(\check{\boldsymbol{\alpha}}) = \boldsymbol{\alpha} \qquad \text{and} \qquad \boldsymbol{\mu}^{-1}(\boldsymbol{\alpha}) = \check{\boldsymbol{\alpha}}$$

where $\boldsymbol{\mu}^{-1}$ denotes the mapping inverse to $\boldsymbol{\mu}$. Then the relation

49 $$\mathbf{y}(t) = \check{\mathbf{A}}(\check{\boldsymbol{\alpha}};\mathbf{u}) \triangleq \mathbf{A}(\boldsymbol{\mu}(\check{\boldsymbol{\alpha}});\mathbf{u})$$

is also an input-output-state relation for $\mathcal{A}$, with the corresponding state equations reading

50
$$\begin{aligned}\check{\mathbf{s}}(t) &= \check{\mathbf{s}}(\check{\boldsymbol{\alpha}};\mathbf{u}) \triangleq \boldsymbol{\mu}^{-1}(\mathbf{s}(\boldsymbol{\mu}(\check{\boldsymbol{\alpha}});\mathbf{u}))\\ \mathbf{y}(t) &= \check{\mathbf{g}}(\check{\mathbf{s}}(t),\mathbf{u}(t),t) \triangleq \mathbf{g}(\boldsymbol{\mu}(\check{\mathbf{s}}(t)),\mathbf{u}(t),t)\end{aligned}$$

Proof It is obvious that all that is involved here is a relabeling of the elements of Σ via the mapping $\boldsymbol{\alpha} = \boldsymbol{\mu}(\check{\boldsymbol{\alpha}})$. Clearly, such relabeling does not affect the satisfying of self-consistency conditions; and hence if *47* is an input-output-state relation for $\mathcal{A}$, so is *49*. The same statement applies to the state equations *50* (see also *2.5.14*).

4 *Equivalent states of two or more systems*

In our discussion of the notion of equivalent states in Sec. *2* we have restricted our attention to the states of a single system $\mathcal{A}$. More generally, one can speak of equivalent states belonging to two or more different systems. Thus, consider two systems $\mathcal{A}$ and $\mathcal{B}$ having identical input function spaces and let $\boldsymbol{\alpha}$ be a state of $\mathcal{A}$ and $\boldsymbol{\beta}$ be a state of $\mathcal{B}$. Furthermore, let $\mathbf{y}$ denote the response segment of $\mathcal{A}$ and $\mathbf{w}$ that of $\mathcal{B}$ when the starting states are respectively $\boldsymbol{\alpha}$ and $\boldsymbol{\beta}$ and the common input to $\mathcal{A}$ and $\mathcal{B}$ is $\mathbf{u}$. Thus

1
$$\mathbf{y} = \bar{\mathbf{A}}(\boldsymbol{\alpha};\mathbf{u})$$

and

2
$$\mathbf{w} = \bar{\mathbf{B}}(\boldsymbol{\beta};\mathbf{u})$$

As a natural extension of the notion of equivalent states belonging to a single system $\mathcal{A}$, two states such as $\boldsymbol{\alpha}$ and $\boldsymbol{\beta}$ which belong to two systems $\mathcal{A}$ and $\mathcal{B}$ are defined to be *equivalent* if and only if for all inputs $\mathbf{u}$ in the input segment space of $\mathcal{A}$ and $\mathcal{B}$ the response of $\mathcal{A}$ starting in state $\boldsymbol{\alpha}$ is the same as the response of $\mathcal{B}$ starting in state $\boldsymbol{\beta}$.

In symbols, we have the

3 **Definition** $\{\boldsymbol{\alpha} \simeq \boldsymbol{\beta}\} \Leftrightarrow \{\bar{\mathbf{A}}(\boldsymbol{\alpha};\mathbf{u}) = \bar{\mathbf{B}}(\boldsymbol{\beta};\mathbf{u})\} \qquad \forall \mathbf{u}$

4 *Comment* We are tacitly restricting ourselves to a concept of state equivalence in which the time at which $\mathbf{u}$ is applied to $\mathcal{A}$ and $\mathcal{B}$ is immaterial. To be more general, we would have to make the qualification that $\boldsymbol{\alpha}$ and $\boldsymbol{\beta}$ are equivalent for $t_0 \geq \rho$, if $\bar{\mathbf{A}}(\boldsymbol{\alpha};\mathbf{u}) = \bar{\mathbf{B}}(\boldsymbol{\beta};\mathbf{u})$ for all $\mathbf{u}$ starting at $t_0 \geq \rho$. In this sense, when we say that $\boldsymbol{\alpha}$ and $\boldsymbol{\beta}$ are equivalent, we mean that they are equivalent for all t_0. Although we shall have no occasion to use the more general concept of equivalence

in this book, it does sometimes play a significant role in the case of time-varying systems.

5 *Example* A simple illustration of equivalent states belonging to different systems is provided by the two circuits shown in Fig. *2.4.1*.

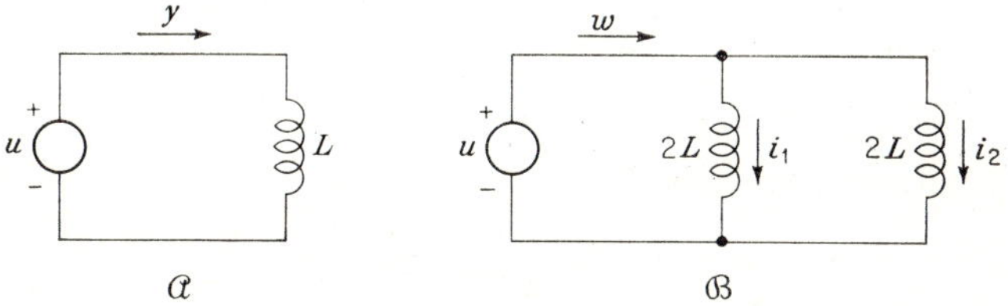

Fig. 2.4.1 Equivalent systems, $\mathcal{A} \equiv \mathcal{B}$.

Here u is the applied voltage and y and w represent the currents flowing through the terminals. The state of $\mathcal{A}$ at time t may be identified with the current through L; that is, $s(t) = y(t)$. Similarly, the state of $\mathcal{B}$ at time t is a 2-vector $\mathbf{s}(t) = (i_1(t), i_2(t))$ whose components are the currents flowing through the two inductors (see *1.7.28*). The input-output-state relations for $\mathcal{A}$ and $\mathcal{B}$ are:

6 $$\mathcal{A}: \qquad y(t) = y(t_0) + \frac{1}{L}\int_{t_0}^{t} u(\xi)\, d\xi \qquad s(t_0) = y(t_0)$$

7 $$\mathcal{B}: \qquad w(t) = i_1(t_0) + i_2(t_0) + \frac{1}{L}\int_{t_0}^{t} u(\xi)\, d\xi \qquad \mathbf{s}(t_0) = (i_1(t_0), i_2(t_0))$$

Let $y(t_0) = \alpha = 5$, say. Then $\boldsymbol{\beta} = (4,1)$ is equivalent to α. In fact, every state of $\mathcal{B}$ such that the sum of its two components is equal to 5 is equivalent to the state $\alpha = 5$.

8 *Example* Consider two systems $\mathcal{A}$ and $\mathcal{B}$ with respective state vectors (2-vectors) $\mathbf{s}(t) = (i_1(t), i_2(t))$ and $\boldsymbol{\sigma}(t) = (j_1(t), j_2(t))$ and input-output-state relations

9 $$\mathcal{A}: \qquad y(t) = i_1(t_0) + 2i_2(t_0) + \int_{t_0}^{t} u(\xi)\, d\xi$$

10 $$\mathcal{B}: \qquad w(t) = 2j_1(t_0) + 3j_2(t_0) + \int_{t_0}^{t} u(\xi)\, d\xi$$

From inspection of these relations it is clear that the state (0,0) of $\mathcal{A}$ is equivalent to the state (0,0) of $\mathcal{B}$. More generally, any two states of $\mathcal{A}$ and $\mathcal{B}$ for which the equality

11 $$i_1(t) + 2i_2(t) = 2j_1(t) + 3j_2(t)$$

holds are equivalent. For example, $(1,3) \simeq (2,1)$.

A pair of equivalent states belonging to two different systems (or, as a special case, to the same system) have an important property which is stated in the following

12 **Assertion** Let α and β be two equivalent states belonging to $\mathfrak{A}$ and $\mathfrak{B}$, respectively. Let α^* and β^* denote the respective states into which α and β are taken by an input $\mathbf{u}$, with $\mathfrak{A}$ and $\mathfrak{B}$ starting in α and β respectively. Then, for every $\mathbf{u}$ in the input segment space of $\mathfrak{A}$ and $\mathfrak{B}$, $\alpha^* \simeq \beta^*$. In short, if α and β are equivalent, so are the states into which they are taken by any input in the input segment space of $\mathfrak{A}$ and $\mathfrak{B}$.

Proof Suppose that $\alpha^* \not\simeq \beta^*$. Then by *2.2.3* there exists at least one input, say, $\mathbf{v}$, such that

13 $$\bar{\mathbf{A}}(\alpha^*;\mathbf{v}) \neq \bar{\mathbf{B}}(\beta^*;\mathbf{v})$$

By the response separation property (*2.3.4*), we can write

14 $$\bar{\mathbf{A}}(\alpha;\mathbf{uv}) = \bar{\mathbf{A}}(\alpha;\mathbf{u})\bar{\mathbf{A}}(\alpha^*;\mathbf{v})$$

and

15 $$\bar{\mathbf{B}}(\beta;\mathbf{uv}) = \bar{\mathbf{B}}(\beta;\mathbf{u})\bar{\mathbf{B}}(\beta^*;\mathbf{v})$$

and since α and β are equivalent states, we have

16 $$\bar{\mathbf{A}}(\alpha;\mathbf{uv}) = \bar{\mathbf{B}}(\beta;\mathbf{uv})$$

Clearly, *14* to *16* are inconsistent with *13*. Consequently, $\alpha^* \simeq \beta^*$ by contradiction. (Note that we are implicitly using the first part of the assumption of uniform reachability stated in *17*.)

It will be convenient for later purposes to reword Assertion *12* in terms of the notion of reachable states, which is defined below. The notion of reachability in the context of linear systems will be discussed in greater detail in Chap. *11*.

17 **Definition** A state α^* of $\mathfrak{A}$ is *reachable from a state* α of $\mathfrak{A}$ if and only if there exists an input segment $\mathbf{u}$ (of finite length) in the input segment space of $\mathfrak{A}$ such that α is taken into α^* by $\mathbf{u}$. Unless stated to the contrary, it will be assumed that the states of $\mathfrak{A}$ have the following uniform reachability property: If there exists an input segment $\mathbf{u}$ which takes a state α into a state α^*, then for each t_1 there is a t_0 and an input segment $\mathbf{v}_{(t_0,t_1]}$ which takes α [$= \mathsf{s}(t_0)$] into α^* [$= \mathsf{s}(t_1)$]; and for each t_0 there is a t_1 and an input segment $\mathbf{w}_{(t_0,t_1]}$ which takes α into α^*.

To illustrate, in Example *1.7.19* every state is reachable from every other state in the state space. Following the terminology used by E. F. Moore in his work on sequential machines (see Ref. *3*), systems having this property will be said to be *strongly connected*.[1]

[1] In the context of linear systems, "strong connectedness" is closely related to so-called "complete controllability." A detailed discussion of controllability, observability, and related concepts is presented in Chap. *11*.

In Example *5*, in the case of system $\mathfrak{A}$ every state is reachable from the state 0 (which corresponds to zero current through L). On the other hand, in the case of system $\mathfrak{B}$ the state $(-1,1)$ is not reachable from the state (0,0), because on application of an input voltage $\mathbf{u}_{(t_0,t_1]}$ to $\mathfrak{B}$ starting in (0,0), the resulting terminal state is the 2-vector

$$\mathbf{s}(t_1) = \left(\frac{1}{2L}\int_{t_0}^{t} u(t)\,dt, \frac{1}{2L}\int_{t_0}^{t} u(t)\,dt\right)$$

in which the first and second components are always equal to one another. Consequently, there does not exist a $\mathbf{u}_{(t_0,t_1]}$ such that $\mathbf{s}(t_1) = (-1,1)$.

An obvious and yet important property of reachable states is contained in the following

18 **Assertion** Let Σ_α denote the set of all states in the state space of $\mathfrak{A}$ which are reachable from $\boldsymbol{\alpha}$. Then Σ_α is a *closed* set in the sense that if $\boldsymbol{\alpha}_0$ is any initial state in Σ_α, then every state which is reachable from $\boldsymbol{\alpha}_0$ also belongs to Σ_α. (Note that this does not necessarily imply that Σ_α is a closed set in the usual set-theoretic sense.) $\lhd$

For convenient reference, we state in the form of an assertion a reworded version of *12* in terms of the notion of reachable states.

19 **Assertion** If $\boldsymbol{\alpha}^*$ and $\boldsymbol{\beta}^*$ are reachable from two equivalent states $\boldsymbol{\alpha}$ and $\boldsymbol{\beta}$ by the same input, then $\boldsymbol{\alpha}^*$ and $\boldsymbol{\beta}^*$ are equivalent states. $\lhd$

The notion of reachability also helps to clarify a question concerning the transition from a free to a constrained interconnection (see Chap. *1*, *Sec. 10*). This question is discussed in the following

20 **Remark** In terms of reachable states, the state space $\Sigma(t_0,t)$ of a free interconnection of $\mathfrak{A}_1, \ldots, \mathfrak{A}_N$ at time t (see *1.10.26*) may be characterized as follows: $\Sigma(t_0,t)$ is the set of all states in the product space $\Sigma_1 \times \cdots \times \Sigma_N$ which are reachable (at time t), with the initial states (at time t_0) ranging over $\Sigma_1 \times \cdots \times \Sigma_N$. $\lhd$

When we pass by a limiting process from $\Sigma(t_0,t)$ to Σ, which is the state space of the constrained interconnection of $\mathfrak{A}_1, \ldots, \mathfrak{A}_N$ (see *1.10.21*), the following question arises with respect to the validity of *19* in the limit as $t_0 \to -\infty$. Suppose that $\mathbf{u}$ is defined over $(-\infty,\infty)$ and that $\mathbf{u}_{(t_0,t]}$ takes $\boldsymbol{\alpha}$ and $\boldsymbol{\beta}$ (at time t_0) into $\boldsymbol{\alpha}^*(t)$ and $\boldsymbol{\beta}^*(t)$, respectively, at time t. Now if $\boldsymbol{\alpha} \simeq \boldsymbol{\beta}$ and $\boldsymbol{\alpha}^*(t)$ and $\boldsymbol{\beta}^*(t)$ converge to $\boldsymbol{\alpha}^*$ and $\boldsymbol{\beta}^*$, respectively, as $t_0 \to -\infty$, can we assert that the limit states $\boldsymbol{\alpha}^*$ and $\boldsymbol{\beta}^*$ are equivalent?

To discuss the question in more concrete terms, let us assume for simplicity that the state spaces of $\mathfrak{A}$ and $\mathfrak{B}$ are finite-dimensional euclidean spaces, with the norm (length) of an n-vector $\mathbf{s}(t)$ defined by (see *C.16.5 et seq.*)

21
$$\|\mathbf{s}(t)\| = \sqrt{|s_1(t)|^2 + \cdots + |s_n(t)|^2}$$

[Note that *21* is also defined for $\mathbf{s}(t)$ with complex-valued components.] Then, the statement "$\mathbf{s}(t)$ converges to $\boldsymbol{\alpha}^*$ as $t \to \infty$" means that $\|\mathbf{s}(t) - \boldsymbol{\alpha}^*\| \to 0$ as $t \to \infty$; that is, given any $\varepsilon > 0$, there exists a $T(\varepsilon)$ such that $\|\mathbf{s}(t) - \boldsymbol{\alpha}^*\| \leq \varepsilon$ for all $t \geq T(\varepsilon)$.

It is easy to show that a sufficient condition for the answer to the above question to be in the affirmative is that $\mathcal{A}$ and $\mathcal{B}$ are "well behaved" in the sense specified below for $\mathcal{A}$.

Let $\mathcal{A}$ be characterized by an input-output-state relation

22
$$\mathbf{y}(t) = \mathbf{A}(\boldsymbol{\alpha};\mathbf{u}_{(t_0,t]})$$

and let $\|\ \ \|$ be a suitably defined norm (not necessarily a euclidean norm) on the space of time functions $\mathbf{y}$ over (t_0, ∞). Then $\mathcal{A}$ is *continuous* at $\boldsymbol{\alpha}$ (for a fixed $\mathbf{u}$) if, roughly speaking, "small" changes in $\boldsymbol{\alpha}$ produce "small" changes in $\mathbf{y}$, or, more precisely, if given any $\varepsilon > 0$, there exists a $\delta(\varepsilon)$ such that

23
$$\|\boldsymbol{\alpha}' - \boldsymbol{\alpha}\| \leq \delta(\varepsilon) \Rightarrow \|\mathbf{y} - \mathbf{y}'\| \leq \varepsilon \qquad \forall \boldsymbol{\alpha}'$$

where $\mathbf{y}'(t) = \mathbf{A}(\boldsymbol{\alpha}';\mathbf{u}_{(t_0,t]})$. Note that this definition is closely related to that of *stability* of $\mathcal{A}$ (see *7.2.2*).

If both $\mathcal{A}$ and $\mathcal{B}$ have this property for all inputs, then we can make the following

24 **Assertion** Let $\boldsymbol{\alpha}^*(t)$ and $\boldsymbol{\beta}^*(t)$ be the respective states of $\mathcal{A}$ and $\mathcal{B}$ into which $\boldsymbol{\alpha}$ and $\boldsymbol{\beta}$ are taken by an input $\mathbf{u}_{(t_0,t]}$. Let $\boldsymbol{\alpha}^*(t) \simeq \boldsymbol{\beta}^*(t)$ for all t_0 and let $\boldsymbol{\alpha}^*(t)$ and $\boldsymbol{\beta}^*(t)$ converge to $\boldsymbol{\alpha}^*$ and $\boldsymbol{\beta}^*$, respectively, as $t_0 \to -\infty$. Then the limit states $\boldsymbol{\alpha}^*$ and $\boldsymbol{\beta}^*$ are equivalent if $\mathcal{A}$ and $\mathcal{B}$ are continuous at $\boldsymbol{\alpha}^*$ and $\boldsymbol{\beta}^*$, respectively, for all $\mathbf{u}$ starting at t.

Proof The proof is a simple exercise in the use of ε's and δ's. Specifically, on using the subscripts a and b to identify the ε's, δ's, and outputs associated with $\mathcal{A}$ and $\mathcal{B}$, respectively, we have

25
$$t_0 \leq T_a[\delta_a(\varepsilon)] \Rightarrow \|\boldsymbol{\alpha}(t) - \boldsymbol{\alpha}^*\| \leq \delta_a(\varepsilon) \Rightarrow \|\mathbf{y} - \mathbf{y}_a^*\| \leq \varepsilon$$
$$t_0 \leq T_b[\delta_b(\varepsilon)] \Rightarrow \|\boldsymbol{\beta}(t) - \boldsymbol{\beta}^*\| \leq \delta_b(\varepsilon) \Rightarrow \|\mathbf{y} - \mathbf{y}_b^*\| \leq \varepsilon$$

where T_a has the same significance as in *21 et seq.* and $\mathbf{y}_a^* = \bar{\mathbf{A}}(\boldsymbol{\alpha}^*;\mathbf{u})$. From *25* it follows at once by using the triangle inequality (see Appendix *C.16.4*)

$$t_0 \leq \min \{T_a[\delta_a(\varepsilon)], T_b[\delta_b(\varepsilon)]\} \Rightarrow \|\mathbf{y}_a^* - \mathbf{y}_b^*\| \leq 2\varepsilon$$

which implies that $\mathbf{y}_a^* = \mathbf{y}_b^*$. Q.E.D.

This concludes for the present our discussion of the notion of equivalent states. We turn next to another notion that plays a central role in system theory—the notion of *system equivalence.*

5 *System equivalence and related concepts*

In Secs. *2* and *4* we defined the notion of state equivalence and established some of its basic properties. In this section, we shall examine a related and very important concept, namely, the concept of *system equivalence.*

As the name implies, two systems $\mathcal{A}$ and $\mathcal{B}$ are *equivalent* if, roughly speaking, they exhibit the same behavior when subjected to the same input. To lend greater precision to this idea, it is convenient to define equivalence in terms of the ability of an experimenter to distinguish between $\mathcal{A}$ and $\mathcal{B}$ on the basis of applying to a black box which is known to contain $\mathcal{A}$ or $\mathcal{B}$ an input of his choice and observing the response. If, no matter what input is applied, the experimenter cannot determine whether the black box contains $\mathcal{A}$ or $\mathcal{B}$, then $\mathcal{A}$ and $\mathcal{B}$ are *equivalent.* Otherwise, they are *nonequivalent*, or *distinguishable.*

It is clear that in order to define equivalence in more concrete terms, one must consider a variety of different situations. For example, the experimenter may have just one copy of the black box at his disposal or he may have as many as he wishes. He may know the initial state of the black box or he may not. The range of inputs which he may be permitted to apply may be limited, etc. To cover these and other situations, we need more than one concept of equivalence. In the sequel, we shall define two basic kinds of equivalence and establish some connections between them. We begin with the notion of

Weak equivalence (equivalence under a single experiment)

Let $\mathcal{A}$ and $\mathcal{B}$ be two oriented systems characterized by their input-output relations, or input-output-state relations, or as families of their respective spaces of input-output pairs (see *1.4.17 et seq.*). We assume that $\mathcal{A}$ and $\mathcal{B}$ have the same input segment space $R[\mathbf{u}]$, $\mathbf{u} \triangleq \mathbf{u}_{(t_0,t_1]}$, and we denote by W the union of their output segment spaces.

Underlying the definition given below is the assumption that the experimenter is presented with a single black box which contains either $\mathcal{A}$ or $\mathcal{B}$ and that he can choose any input $\mathbf{u}$ in $R[\mathbf{u}]$. The response of the black box to $\mathbf{u}$ is denoted by $\mathbf{w}$. The experimenter is to decide on whether the black box contains $\mathcal{A}$ or $\mathcal{B}$ from the knowledge of the pair $(\mathbf{u},\mathbf{w})$ and nothing else.

1 **Definition** $\mathcal{A}$ is said to *contain* $\mathcal{B}$, written as $\mathcal{A} \supset \mathcal{B}$, if every input-output pair for $\mathcal{B}$ is also an input-output pair for $\mathcal{A}$. Geometrically, this means that the set of points (input-output pairs) in the product space $R[\mathbf{u}] \times W$ which represents $\mathcal{B}$ is a subset of the set of points (input-output pairs) which represents $\mathcal{A}$ (see *1.4.17* and Fig. *2.5.1*). (Note that this must hold for all observation intervals $(t_0,t_1]$.)

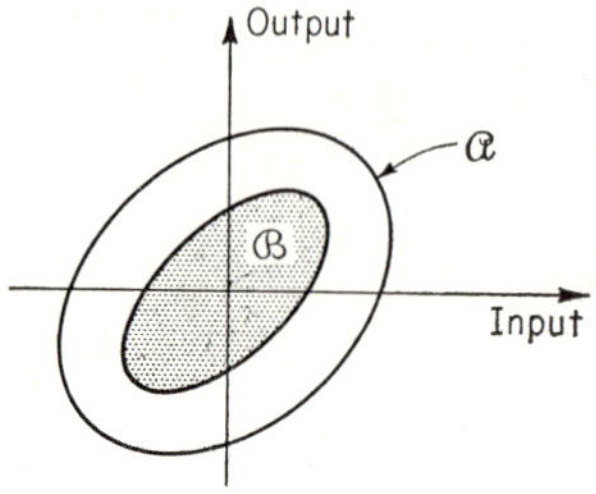

Fig. 2.5.1 Geometrical representation of "$\mathcal{A}$ contains $\mathcal{B}$."

2 *Example* Suppose that $\mathcal{A}$ and $\mathcal{B}$ are characterized by the input-output relations

3
$$\mathcal{A}: \qquad \frac{dy}{dt} = \frac{d^2u}{dt^2} \qquad T = [0, \infty)$$
$$\mathcal{B}: \qquad y = \frac{du}{dt}$$

where u is a real-valued time function. Clearly, every input-output pair for $\mathcal{B}$ is also an input-output pair for $\mathcal{A}$. For example, $(t,1), t > 0$, is an input-output pair for $\mathcal{B}$ as well as $\mathcal{A}$. On the other hand, $(t^2 + 2t, 2t + 1)$, $t > 0$, is an input-output pair for $\mathcal{A}$ but not for $\mathcal{B}$. Thus, $\mathcal{B}$ is contained in $\mathcal{A}$. [Note that this is a consequence of the fact that the input-output relation of $\mathcal{A}$ is the result of differentiating the input-output relation of $\mathcal{B}$. The same thing applies, more generally, when we act on both sides of an input-output relation with a differential operator (see *3.8.11*).]

4 **Definition** If $\mathcal{A}$ contains $\mathcal{B}$ and $\mathcal{B}$ contains $\mathcal{A}$, then $\mathcal{A}$ and $\mathcal{B}$ are *weakly equivalent* or *equivalent under a single experiment.* This means that every input-output pair for $\mathcal{A}$ is also an input-output pair for $\mathcal{B}$, and vice versa. Weak equivalence is denoted by $\mathcal{A} = \mathcal{B}$.

In terms of input-output-state relations rather than input-output relations, the definition of weak equivalence assumes the following form.

5 **Definition** $\mathcal{A}$ and $\mathcal{B}$ are *weakly equivalent* or *equivalent under a single experiment* if and only if to every input $\mathbf{u}$ and initial state $\boldsymbol{\alpha}$ in $\mathcal{A}$ there is a state $\boldsymbol{\beta}$ (depending on $\mathbf{u}$ and $\boldsymbol{\alpha}$) in $\mathcal{B}$ such that the response of $\mathcal{A}$ to $\mathbf{u}$ starting in state $\boldsymbol{\alpha}$ is identical with the response of $\mathcal{B}$ to $\mathbf{u}$ starting in state $\boldsymbol{\beta}$, and vice versa.

In symbols

6
$$\{\mathcal{A} = \mathcal{B}\} \Leftrightarrow \{\forall \boldsymbol{\alpha} \forall \mathbf{u} \exists \boldsymbol{\beta} [\bar{\mathbf{A}}(\boldsymbol{\alpha};\mathbf{u}) = \bar{\mathbf{B}}(\boldsymbol{\beta};\mathbf{u})]\}$$
and
$$\{\forall \boldsymbol{\beta} \forall \mathbf{u} \exists \boldsymbol{\alpha} [\bar{\mathbf{A}}(\boldsymbol{\alpha};\mathbf{u}) = \bar{\mathbf{B}}(\boldsymbol{\beta};\mathbf{u})]\}$$

Here $\exists$ denotes "there exists" and the order of the quantifiers in

$\forall\alpha\forall\mathbf{u}\exists\beta$ indicates that β depends on α and $\mathbf{u}$. Likewise, $\forall\beta\forall\mathbf{u}\exists\alpha$ indicates that α depends on β and $\mathbf{u}$.

7 *Comment* Definition *5* is equivalent to Definition *4*. This is an immediate consequence of the mutual-consistency condition—a condition which must be satisfied by an input-output-state relation (see *1.6.5*). Under this condition, to every input-output pair $(\mathbf{u},\mathbf{y})$ for $\mathcal{A}$ there corresponds at least one state α in the state space of $\mathcal{A}$ with respect to which $(\mathbf{u},\mathbf{y})$ satisfies the input-output-state relation, and conversely, every pair $(\mathbf{u},\mathbf{y})$ with $\mathbf{u} \in R[\mathbf{u}]$ and $\mathbf{y} \in R[\mathbf{y}]$ satisfying this relation is an input-output pair for $\mathcal{A}$. ◁

The justification for calling weak equivalence "equivalence under a single experiment" is provided by the following observation. If $\mathcal{A}$ and $\mathcal{B}$ have the property expressed by *6*, then no matter whether $\mathcal{A}$ or $\mathcal{B}$ is in the black box, no matter what the initial state is, and no matter what input $\mathbf{u}$ the experimenter chooses, the response $\mathbf{w}$ will always be attributable to both $\mathcal{A}$ and $\mathcal{B}$ and hence the experimenter will be unable to determine whether the black box contains $\mathcal{A}$ or $\mathcal{B}$ from the knowledge of the pair $(\mathbf{u},\mathbf{w})$. In this sense, $\mathcal{A}$ and $\mathcal{B}$ are indistinguishable by any single experiment and consequently are equivalent under a single experiment.

If $\mathcal{A}$ and $\mathcal{B}$ are indistinguishable by any single experiment, it does not follow that they could not be distinguished if the experimenter had at his disposal more than one copy of the black box, each copy containing, say, $\mathcal{A}$ in an initial state α. In that case, the experimenter could perform different experiments on the different copies of the black box and thereby base his conclusion not on just one input-output pair $(\mathbf{u},\mathbf{w})$—as in a single experiment—but on as many different input-output pairs as there are copies of the black box at his disposal. As was defined before (Chap. *1*, Sec. *8*), if the experimenter has available as many black boxes as he desires, all containing the same system in the same initial state (not known to the experimenter), then the experiment in question is said to be a *multiple experiment*. If even under such an experiment the experimenter is unable to determine whether the copies of the black box contain $\mathcal{A}$ or $\mathcal{B}$, then $\mathcal{A}$ and $\mathcal{B}$ are said to be *indistinguishable under a multiple experiment*. Clearly, indistinguishability under a multiple experiment is a stronger (i.e., more restrictive) property than indistinguishability under a single experiment.

The concept of indistinguishability under a multiple experiment underlies the

8 **Definition: system equivalence** $\mathcal{A}$ *is equivalent* to $\mathcal{B}$, written $\mathcal{A} \equiv \mathcal{B}$, if and only if to every state α in the state space of $\mathcal{A}$ there corresponds at least one equivalent state β in the state space of $\mathcal{B}$, and vice versa. ◁

To cast this definition into the form of a logical formula, observe that the statement "to every state $\boldsymbol{\alpha}$ in $\mathcal{A}$ there is an equivalent state $\boldsymbol{\beta}$ in $\mathcal{B}$" is expressed by

$$\forall \boldsymbol{\alpha} \exists \boldsymbol{\beta} \forall \mathbf{u}[\bar{\mathbf{A}}(\boldsymbol{\alpha};\mathbf{u}) = \bar{\mathbf{B}}(\boldsymbol{\beta};\mathbf{u})]$$

Making use of this observation, we have

9 $$\{\mathcal{A} \equiv \mathcal{B}\} \Leftrightarrow \{\forall \boldsymbol{\alpha} \exists \boldsymbol{\beta} \forall \mathbf{u}[\bar{\mathbf{A}}(\boldsymbol{\alpha};\mathbf{u}) = \bar{\mathbf{B}}(\boldsymbol{\beta};\mathbf{u})]\}$$

and $$\{\forall \boldsymbol{\beta} \exists \boldsymbol{\alpha} \forall \mathbf{u}[\bar{\mathbf{A}}(\boldsymbol{\alpha};\mathbf{u}) = \bar{\mathbf{B}}(\boldsymbol{\beta};\mathbf{u})]\}$$

Note that the only difference between this definition and that of weak equivalence (Definition *6*) lies in the inversion of the order of quantifiers. Specifically, in *6* the order in the first factor in the right-hand member is $\forall \boldsymbol{\alpha} \forall \mathbf{u} \exists \boldsymbol{\beta}$ (which indicates that $\boldsymbol{\beta}$ depends on $\boldsymbol{\alpha}$ and $\mathbf{u}$), whereas in *9* it is $\forall \boldsymbol{\alpha} \exists \boldsymbol{\beta} \forall \mathbf{u}$, which signifies that $\boldsymbol{\beta}$ depends only on $\boldsymbol{\alpha}$. It is this independence of $\boldsymbol{\beta}$ of $\mathbf{u}$ that is the crux of the difference between *6* and *9*, and it is this property that makes $\boldsymbol{\beta}$ equivalent to $\boldsymbol{\alpha}$.

It remains to be shown that equivalence in the sense of Definition *8* implies and is implied by indistinguishability under a multiple experiment.

It is obvious that if *8* is satisfied, then $\mathcal{A}$ and $\mathcal{B}$ are indistinguishable under a multiple experiment. For, suppose that the copies of the black box contain, say, $\mathcal{A}$ in state $\boldsymbol{\alpha}$ and that $\boldsymbol{\beta}$ is an equivalent state in $\mathcal{B}$. Then no matter what input the experimenter applies to each copy of the black box, the observed response will be identical to what $\mathcal{B}$ would produce starting in state $\boldsymbol{\beta}$. Thus, the experimenter will be unable to distinguish between $\mathcal{A}$ and $\mathcal{B}$. The same thing is true more generally when to each state in the state space of $\mathcal{A}$ there corresponds an equivalent state in the state space of $\mathcal{B}$, and vice versa.

To show the converse requires more elaborate arguments which will be given here only for the simple case where $\mathcal{A}$ and $\mathcal{B}$ are finite-state systems, i.e., the state spaces of $\mathcal{A}$ and $\mathcal{B}$ contain a finite number of states.[1]

Specifically, suppose the right-hand side of *9* is false. Then we will have established the converse once we have shown that $\mathcal{A}$ and $\mathcal{B}$ are distinguishable under a multiple experiment.

To show that $\mathcal{A}$ and $\mathcal{B}$ are distinguishable if the right-hand side of *9* is false, we note that in this case the negation[2] of the right-hand side is true, i.e., either (I) $\exists \boldsymbol{\alpha} \forall \boldsymbol{\beta} \exists \mathbf{u}[\bar{\mathbf{A}}(\boldsymbol{\alpha};\mathbf{u}) \neq \bar{\mathbf{B}}(\boldsymbol{\beta};\mathbf{u})]$ is true or (II)

[1] A proof for continuous-state systems which makes use of Zorn's lemma was suggested by A. Chang, a student at the University of California.

[2] To negate a logical expression containing $\forall$, $\exists$, "and," "or," $\Rightarrow$, and propositions $x, y, \ldots, z$, replace $\forall$ by $\exists$, $\exists$ by $\forall$, "and" by "or," "or" by "and," $\Rightarrow$ by $\not\Leftarrow$, and every proposition by its negation. See Ref. *2* for a more detailed discussion.

$\exists\beta\forall\alpha\exists\mathbf{u}[\bar{\mathbf{A}}(\alpha;\mathbf{u}) \neq \bar{\mathbf{B}}(\beta;\mathbf{u})]$ is true. (I) means that there is a state α in $\mathcal{A}$ such that for every state β in $\mathcal{B}$ there is an input $\mathbf{u}$ which can distinguish between α and β. The same interpretation applies to (II).

Assume that $\mathcal{A}$ has n states $\alpha_1, \ldots, \alpha_n$ and $\mathcal{B}$ has m states $\beta_1, \ldots, \beta_m$ and that (I) is true for, say, α_1. Then, the experimenter may be able to distinguish between $\mathcal{A}$ and $\mathcal{B}$ if the black boxes contain $\mathcal{A}$ in a state α_1. Specifically, assume that the experimenter is presented with k black boxes [$k \geq \max(m,n)$] numbered $1, \ldots, k$, each of which contains $\mathcal{A}$ in state α_1, with the experimenter being unaware of what object is in the black boxes. In this case, the experimenter will be able to distinguish between $\mathcal{A}$ and $\mathcal{B}$ by applying to copy i, $i = 1, \ldots, m$, an input $\mathbf{u}^i$ which distinguishes between α_1 and β_i [i.e., an input $\mathbf{u}^i$ such that $\bar{\mathbf{A}}(\alpha_1;\mathbf{u}^i) \neq \bar{\mathbf{B}}(\beta_i;\mathbf{u}^i)$]. For the observed input-output pairs $(\mathbf{u}^1,\mathbf{y}^1), \ldots, (\mathbf{u}^m,\mathbf{y}^m)$ will collectively lead the experimenter to the conclusion that the black boxes cannot contain $\mathcal{B}$ in any of its states $\beta_1, \ldots, \beta_m$. Hence the black boxes must contain $\mathcal{A}$.

10 *Comment* Note that the above reasoning is based on the tacit assumption that the definition of "distinguishability under a multiple experiment" is obtained from that of "indistinguishability under a multiple experiment" by the process of negation (see footnote on page 92). In this sense, $\mathcal{A}$ and $\mathcal{B}$ are *distinguishable under a multiple experiment* if there exists an initial state of $\mathcal{A}$ (or $\mathcal{B}$) such that if the black boxes contain $\mathcal{A}$ (or $\mathcal{B}$) in this state, then there exists a multiple experiment which will make the experimenter conclude that the black boxes contain $\mathcal{A}$ (or $\mathcal{B}$). A stronger definition of distinguishability would be one under which the experimenter can distinguish between $\mathcal{A}$ and $\mathcal{B}$ by a multiple experiment *regardless* of the initial state of the black box. It is easy to show that, in some cases, this stronger definition is implied by the weaker definition used above. One such case is that of so-called strongly connected systems (see *2.4.17 et seq*).

11 **Remark** It is obvious from the way in which system equivalence is defined by *8* that if $\mathcal{A}$ and $\mathcal{B}$ are equivalent, then they are also weakly equivalent. Although the converse is not true in general, it is true for systems that are of principal interest to us, namely, systems whose input-output relations have the form of linear differential equations. This basic property of linear systems will be established in Chap. *3*, Sec. *9*.

Before we proceed to illustrate the notion of system equivalence by an example, it will be helpful to state two basic properties of system equivalence. The first of these is the property of

12 **Transitivity** If $\mathcal{A}$ is equivalent to $\mathcal{B}$ and $\mathcal{B}$ is equivalent to $\mathcal{C}$, then $\mathcal{A}$ is equivalent to $\mathcal{C}$. In symbols,

$$\{\mathcal{A} \equiv \mathcal{B} \text{ and } \mathcal{B} \equiv \mathcal{C}\} \Rightarrow \{\mathcal{A} \equiv \mathcal{C}\}$$

This property is an immediate consequence of the transitivity of state equivalence (see *2.2.1 et seq.*).

13 *Question* Does transitivity apply to weak equivalence? ◁

The second property is

14 **Equivalence under one-one mapping** Consider a system $\mathcal{A}$ which is characterized by an input-output-state relation

$$\mathbf{y}(t) = \mathbf{A}(\boldsymbol{\alpha};\mathbf{u}) \qquad \boldsymbol{\alpha} \in \Sigma$$

Let $\boldsymbol{\mu}$ be a one-one mapping of Σ onto itself, with the image of $\boldsymbol{\alpha}$ under $\boldsymbol{\mu}$ being $\check{\boldsymbol{\alpha}}$, that is,

$$\check{\boldsymbol{\alpha}} = \boldsymbol{\mu}(\boldsymbol{\alpha}) \qquad \text{and} \qquad \boldsymbol{\alpha} = \boldsymbol{\mu}^{-1}(\check{\boldsymbol{\alpha}})$$

Then the system $\check{\mathcal{A}}$ characterized by the input-output-state relation

$$\begin{aligned} \mathbf{y} &= \mathbf{A}(\boldsymbol{\mu}^{-1}(\check{\boldsymbol{\alpha}});\mathbf{u}) \\ &\triangleq \check{\mathbf{A}}(\check{\boldsymbol{\alpha}};\mathbf{u}) \end{aligned}$$

is equivalent to $\mathcal{A}$. That this is true follows at once from the fact that a one-one mapping on Σ constitutes merely a relabeling of its points (see *2.3.46*).

We shall illustrate the notion of system equivalence by a simple example which is constructed by the use of *14*.

15 *Example* Consider two systems: (I) $\mathcal{A}$, characterized by the input-output-state relation

16 $$y(t) = (2e^{-(t-t_0)} - e^{-2(t-t_0)})\alpha_1 + (e^{-(t-t_0)} - e^{-2(t-t_0)})\alpha_2 + \int_{t_0}^{t} (e^{-(t-\xi)} - e^{-2(t-\xi)})u(\xi)\,d\xi \qquad t > t_0$$

where α_1 and α_2 are the components of $\boldsymbol{\alpha}$ and $\Sigma = \mathcal{R}^2$, and (II) $\mathcal{B}$, characterized by the input-output-state relation

17 $$y(t) = (3e^{-(t-t_0)} - 5e^{-2(t-t_0)})\beta_1 + (3e^{-(t-t_0)} - 2e^{-2(t-t_0)})\beta_2 + \int_{t_0}^{t} (e^{-(t-\xi)} - e^{-2(t-\xi)})u(\xi)\,d\xi \qquad t > t_0$$

where β_1 and β_2 are the components of $\boldsymbol{\beta}$.

Inspection of *16* and *17* shows that to an arbitrary $\boldsymbol{\alpha}$ there corresponds an equivalent $\boldsymbol{\beta}$ if and only if the following relation holds for all $t > t_0$:

18 $$(2e^{-(t-t_0)} - e^{-2(t-t_0)})\alpha_1 + (e^{-(t-t_0)} - e^{-2(t-t)_0})\alpha_2 = (3e^{-(t-t_0)} - 5e^{-2(t-t_0)})\beta_1 + (3e^{-(t-t_0)} - 2e^{-2(t-t_0)})\beta_2$$

On equating the coefficients of like exponentials in *18*, we have

$$\alpha_1 = -2\beta_1 + \beta_2 \qquad \alpha_2 = 7\beta_1 + \beta_2$$

Clearly, this system of equations can be solved for β_1 and β_2 in terms of α_1 and α_2, and vice versa. Consequently, for every state $\boldsymbol{\alpha}$ in $\mathcal{A}$ there is one equivalent state β in $\mathcal{B}$, and vice versa. This establishes that $\mathcal{A} \equiv \mathcal{B}$. ◁

In later chapters we shall frequently be concerned with the problem of representing a given system $\mathcal{A}$ as an interconnection of components of specified types, e.g., integrators, differentiators, scalors, and adders. This motivates the following

19 **Definition** Let $\Gamma_1, \ldots, \Gamma_k$ be a set of classes of systems (for example, $\Gamma_1 \triangleq$ class of integrators with real or complex time constants, $\Gamma_2 \triangleq$ class of scalors). Then an interconnection of components $\mathcal{A}_1, \ldots, \mathcal{A}_N$ each of which belongs to one of the Γ_i, $i = 1, \ldots, k$, is a *realization of $\mathcal{A}$ in terms of* $\Gamma_1, \ldots, \Gamma_k$ if the interconnection in question, call it $\mathcal{B}$, is equivalent to $\mathcal{A}$. A realization of $\mathcal{A}$ need not be physically realizable. ◁

As we noted in the last section of Chap. *1*, a basic property which a given interconnection may or may not possess is that of *determinateness*. If an interconnection $\mathcal{A}$ is determinate, then a basic question is: How can one associate a state vector with $\mathcal{A}$ and what is the corresponding state space of $\mathcal{A}$? This and related questions are discussed on a general level in the following section.

6 *The state of an interconnection of systems*

Consider a system $\mathcal{A}$ which consists of an interconnection (*1.10.28*) of oriented components $\mathcal{A}_1, \ldots, \mathcal{A}_N$ (Fig. *2.6.1*) each of which is characterized by an input-output-state relation

1 $$\mathbf{y}^i(t) = \mathbf{A}_i(\mathbf{s}^i(t_0);\mathbf{u}^i_{(t_0,t]}) \qquad i = 1, \ldots, N$$

where $\mathbf{s}_i(t) \in \Sigma_i \triangleq$ state space of $\mathcal{A}_i$. Note that we use superscripts

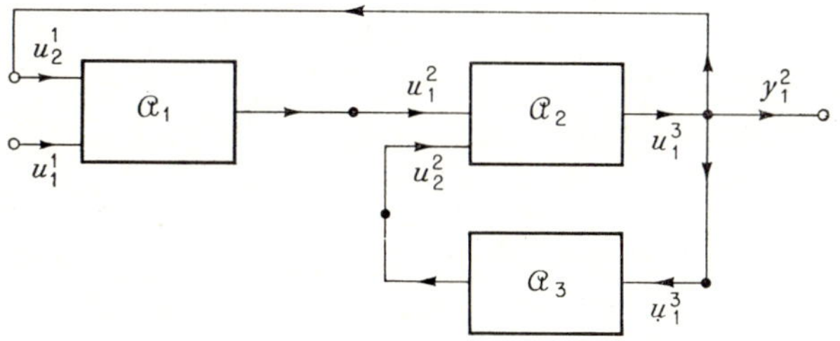

Fig. 2.6.1 An interconnection of $\mathcal{A}_1$, $\mathcal{A}_2$, $\mathcal{A}_3$.

to identify the input, output, and state vectors associated with $\mathcal{A}_i$, and we shall use subscripts to label the components of these vectors, for example, $u_j^i \triangleq j$th component of $\mathbf{u}^i$.

In accordance with *1.10.40*, the output of $\mathcal{A}$ is defined to be the composite vector

2 $$\mathbf{y} \triangleq (\mathbf{y}^1, \ldots, \mathbf{y}^N)$$

with the understanding that one or more of the components of $\mathbf{y}$ may be suppressed. As for the input to $\mathcal{A}$, it comprises all input variables of the $\mathcal{A}_i$ which are not input-output variables (*1.10.27*), and it is denoted by $\mathbf{u}$. [For example, in Fig. *2.6.1* the input to $\mathcal{A}$ is u_1^1 and the output vector is $(u_2^1, u_1^2, u_1^3, u_2^2)$.]

Let $\mathcal{A}(t_0)$ denote an initially free interconnection (*1.10.26*) which is derived from $\mathcal{A}$ by severing all connections between the $\mathcal{A}_i$ until $t = t_0$. Furthermore, let $\boldsymbol{\alpha}_0^1, \ldots, \boldsymbol{\alpha}_0^N$ denote the initial states of $\mathcal{A}_1, \ldots, \mathcal{A}_N$, with the composite initial state vector $\boldsymbol{\alpha}_0 = (\boldsymbol{\alpha}_0^1, \ldots, \boldsymbol{\alpha}_0^N)$ ranging over the product space $\Sigma_1 \times \cdots \times \Sigma_N$.

A basic property which $\mathcal{A}$ may or may not possess is that of *determinateness.* It is convenient to define this notion in terms of $\mathcal{A}(t_0)$ as follows.

3 **Definition** An interconnection $\mathcal{A}$ is *determinate* if, for all t_0, t, $\mathbf{u}$, and $\boldsymbol{\alpha}_0$, the output of $\mathcal{A}(t_0)$, $\mathbf{y}(t)$, is determined uniquely by $\mathbf{u}_{(t_0,t]}$ and $\boldsymbol{\alpha}_0$ through the input-output-state relations *1*, $i = 1, \ldots, N$ [and, of course, the interconnection relations (*1.10.28*)]. If $\mathcal{A}$ is not determinate, it is said to be *indeterminate.*

4 *Example* In the case of the system of Fig. *2.6.1*, $\mathcal{A}$ is determinate if $u_2^1(t)$, $u_1^2(t)$, $u_1^3(t)$, and $u_2^2(t)$ are determined uniquely by $\boldsymbol{\alpha}_0^1$, $\boldsymbol{\alpha}_0^2$, $\boldsymbol{\alpha}_0^3$, and $u^1_{1(t_0,t]}$. As will be shown in Chap. 3 (*3.9.12*), any linear time-invariant differential system is determinate. However, this is not generally true of nonlinear systems. For example, in the case of system $\mathcal{B}$ of Fig. *2.8.2*, if $L = -1$, $R = 2(i)^{-\frac{1}{2}}$, $t_0 = 0$, and $u(t) = 0$ for $t \geq 0$, then the differential equation

5 $$-\frac{di}{dt} + 2\sqrt{i} = 0$$

has at least two distinct solutions (I) $i(t) = 0$ for $t \geq 0$, and (II) $i(t) = t^2$ for $t \geq 0$, corresponding to the same initial state $i(0) = 0$. This is explained by the fact that *5* does not imply that if $i(0) = 0$, then all derivatives of i with respect to t are zero at $t = 0$. Note that *5* violates the Lipschitz condition (*2.3.40*).

Now let $\mathcal{A}$ be a determinate system and let $\mathcal{A}(t_0)$ be the corresponding free interconnection of the components of $\mathcal{A}$. Then given $\boldsymbol{\alpha}_0$ and $\mathbf{u}_{(t_0,t]}$ we can find $\mathbf{y}_{(t_0,t]}$ and hence the inputs $\mathbf{u}^i_{(t_0,t]}$ to the $\mathcal{A}_i$, $i = 1, \ldots, N$. From these and the $\boldsymbol{\alpha}_o^i$ we can determine through the state equations of the $\mathcal{A}_i$ the state $\mathbf{s}^i(t)$ of $\mathcal{A}_i$ at time t. The states of the $\mathcal{A}_i$ define the composite state vector

$$\mathbf{s}(t) = (\mathbf{s}^1(t), \ldots, \mathbf{s}^N(t))$$

which ranges over the space $\Sigma(t_0,t)$ (see *2.4.20*), where $\Sigma(t_0,t)$ is the set of all states in the product space $\Sigma_1 \times \cdots \times \Sigma_N$ which are reachable at time t from initial states in $\Sigma_1 \times \cdots \times \Sigma_N$ at time t_0. The state space of $\mathcal{A}$, then, is formally expressed by the limit, if it exists,

6 $$\Sigma = \lim_{t_0 \to -\infty} \Sigma(t_0,t)$$

where Σ is a subset of $\Sigma_1 \times \cdots \times \Sigma_N$. (See *1.10.21*.)

7 **Remark** Through the use of Σ, we can avoid introducing the free interconnection $\mathcal{A}(t_0)$ into the definition of a determinate system. Specifically, we can say that $\mathcal{A}$ is *determinate* if for every initial state $\boldsymbol{\alpha}_0$ in Σ and every input $\mathbf{u}$, the output $\mathbf{y}$ is uniquely determined by $\mathbf{u}$ and $\boldsymbol{\alpha}_0$. In the sequel, we shall be using for the most part this definition of determinateness rather than Definition *3*.

8 **Remark** Note that if $\mathcal{A}$ is a time-invariant system (see *1.7.6* and *3.2.36*), then clearly $\Sigma(t_0,t)$ has the inclusion property: $t_2 \geq t_1 \Rightarrow \Sigma(t_0,t_2) \subset \Sigma(t_0,t_1)$. (Why?) In this case, it may be more appropriate to define Σ by the relation

9 $$\Sigma = \bigcap_{t_0} \Sigma(t_0,t) \qquad t_0 < t$$

which differs from *6* in that every point of the space Σ defined by *9* belongs to all the $\Sigma(t_0,t)$, whereas some points of the space Σ defined by *7* may not belong to any of the $\Sigma(t_0,t)$ for finite t_0 and t.

In general, the determination of Σ—whether defined by *6* or *9*—is by no means a trivial problem. However, our primary concern in this text is with linear differential systems, and in the case of such systems Σ can be determined quite easily by virtue of the fact (which will be established in Chap. *4*) that $\mathbf{s}(t)$ is defined by a finite number of the derivatives of the input and output at time t. To illustrate, suppose that $\mathcal{A}$ comprises three components $\mathcal{A}_1$, $\mathcal{A}_2$, $\mathcal{A}_3$ (plus an adder) whose states are related to their respective inputs and outputs by the expressions

$\mathcal{A}_1$: $\quad s^1 = y^1$

$\mathcal{A}_2$: $\quad s^2 = y^2$

$\mathcal{A}_3$: $\quad s^3 = u^3$

with the interconnection relation being $u^3 = y^1 + y^2$ and

$$\Sigma_1 = \Sigma_2 = \Sigma_3 = \mathcal{R}^1$$

Then Σ is a linear subspace of $\mathcal{R}^3$ defined by the relation $s^1 + s^2 = s^3$.

It should be noted that in this example the states of $\mathcal{A}_1$, $\mathcal{A}_2$, and $\mathcal{A}_3$ do not involve any derivatives of their inputs and outputs. As will be shown later, in the case of linear differential systems it is always possible to represent a system as a combination of integrators, differentiators, scalors, and adders, with the states of the integrators and differentiators at time t being, respectively, their outputs and inputs at time t (see Chap. *4*, Sec. *6*). In terms of such a representation, the expressions for the states of components have the above form and, in this sense, the simple example considered here is illustrative of the general case. However, if the system is not represented in this form and the states of the $\mathcal{A}_i$ do involve the derivatives of their respective inputs and outputs, then the constraints on the $\mathbf{s}^i$ induced by the interconnection relations will in general involve linear combinations of the $\mathbf{s}^i$ and their derivatives, inputs to the $\mathcal{A}_i$ and their derivatives, and outputs of the $\mathcal{A}_i$ and their derivatives. Of these constraints only those that are holonomic in the $\mathbf{s}^i$ (that is, involve only the $\mathbf{s}^i$ and not their derivatives) are of relevance to the determination of Σ. For example, if

$$s_1^1 = s_1^2 \qquad s_2^1 = \dot{s}_1^2 + 2s_1^2 - 3u_1^1 \qquad s_1^3 = y_1^3$$

then Σ is a three-dimensional subspace of $\mathcal{R}^4$ defined by $s_1^1 = s_1^2$. (Note that $s_j^i \triangleq j$th component of the state of $\mathcal{A}_i$.)

Determinateness theorem

We have not yet shown that the vector $\mathbf{s}(t)$ defined by *5* qualifies as the state vector of $\mathcal{A}$. This is demonstrated in the

10 **Theorem** Let $\mathcal{A}$ be a determinate interconnection of $\mathcal{A}_1, \ldots, \mathcal{A}_N$, with the input and output of $\mathcal{A}$ denoted by $\mathbf{u}$ and $\mathbf{y}$, respectively, and let $\mathbf{s}(t) = (\mathbf{s}^1(t_0), \ldots, \mathbf{s}^N(t_0))$ be an initial state vector in Σ, where Σ is a subset of $\Sigma_1 \times \cdots \times \Sigma_N$ formally defined by *6*. Furthermore, let $\mathbf{s}^i(t)$ be the state of $\mathcal{A}_i$ at time t, under the assumption that the input to $\mathcal{A}$ is $\mathbf{u}_{(t_0,t]}$ and the initial states of $\mathcal{A}_1, \ldots, \mathcal{A}_N$ are $\mathbf{s}^1(t_0), \ldots, \mathbf{s}^N(t_0)$, respectively. Then the composite vector

11
$$\mathbf{s}(t) = (\mathbf{s}^1(t), \ldots, \mathbf{s}^N(t))$$

qualifies as the state of $\mathcal{A}$ at time t, with Σ being the state space of $\mathcal{A}$. In plain words, this means that if $\mathcal{A}$ is a determinate system, then the state of $\mathcal{A}$ at time t is the aggregate of the states of its components at time t.

Proof. To say that $\mathfrak{a}$ is determinate is equivalent to saying that $\mathbf{s}(t)$ satisfies the first self-consistency condition (*1.6.11*). Thus, it will suffice to show that the input-output-state relation of $\mathfrak{a}$ has the response separation property, for then it will follow from Assertion *2.3.12* that $\mathbf{s}(t)$ qualifies as the state of $\mathfrak{a}$ at time t. ◁

In order to avoid obscuring the essential simplicity of the argument by a multiplicity of symbols and equations, we shall establish the desired property by considering a concrete simple case such as illustrated in Fig. *2.6.6*. It will then be easy to see how the response separation property can be established in the same way for an arbitrary determinate interconnection.

The input-output-state relations for the system in question read

12 $$u^2(t) = A_1[\mathbf{s}^1(t_0);u^1_{1(t_0,t]},u^1_{2(t_0,t]}]$$

13 $$u^1_2(t) = A_2[\mathbf{s}^2(t_0);u^2_{(t_0,t]}]$$

By hypothesis, these equations can be solved uniquely for $u^2(t)$ and $u^1_2(t)$ in terms of $u^1_{1(t_0,t]}$ and $[\mathbf{s}^1(t_0),\mathbf{s}^2(t_0)]$, yielding expressions of the form

14 $$u^2(t) = A^1[\mathbf{s}^1(t_0),\mathbf{s}^2(t_0);u^1_{1(t_0,t]}]$$

15 $$u^1_2(t) = A^2[\mathbf{s}^1(t_0),\mathbf{s}^2(t_0);u^1_{1(t_0,t]}]$$

Now since *12* and *13* are input-output-state relations, they have the response separation property (*2.3.5*) expressed by the relations

16 $$u^2(t) = A_1[\mathbf{s}^1(\tau);u^1_{1(\tau,t]},u^1_{2(\tau,t]}]$$

17 $$u^1_2(t) = A_2[\mathbf{s}^2(\tau);u^2_{(\tau,t]}]$$

where τ is an arbitrary intermediate time between t_0 and t and $\mathbf{s}^1(\tau)$ and $\mathbf{s}^2(\tau)$ denote the states of $\mathfrak{a}_1$ and $\mathfrak{a}_2$, respectively, at time τ.

The fact that *12* and *13* can be solved uniquely for $u^2(t)$ and $u^1_2(t)$ in the form *14* and *15* implies that *16* and *17* can similarly be solved for $u^2(t)$ and $u^1_2(t)$, yielding expressions which differ from *14* and *15* only in that t_0 is replaced by τ. Thus

18 $$u^2(t) = A^1[\mathbf{s}^1(\tau),\mathbf{s}^2(\tau);u^1_{1(\tau,t]}]$$

19 $$u^1_2(t) = A^2[\mathbf{s}^1(\tau),\mathbf{s}^2(\tau);u^1_{1(\tau,t]}]$$

From comparison of these equations with *14* and *15* it follows that $\mathfrak{a}$ has the response separation property and hence, by Assertion *2.3.12*, the composite state vector $\mathbf{s}(t)$ defined by *11* qualifies as the state vector of $\mathfrak{a}$.

Special types of determinate systems

There are several types of system which are determinate by virtue of the structure of interconnections between their components or a special nature of their input-output-state relationships. In the latter category are linear differential systems, which will be defined formally in Chap. *3*. In the former category are the two types defined below.

20 **Definition** A system $\mathcal{A}$ is *cycle-free* if its signal-state graph (Chap. *1*, Sec. *9*), or, equivalently, signal-flow graph, has no cycles (loops, closed paths).[1] ◁

21 *Note* For our present purposes, it will be convenient to represent the state associated with each node in a signal-state graph as an additional branch variable which enters the node in question. This convention is illustrated in Fig. *2.6.2*, which depicts system $\mathcal{A}$ of Fig. *2.6.1*. In effect, this mode of representation amounts to nothing more than reducing the rectangular blocks in the diagrammatic representation of Fig. *1.9.1* to small circles (nodes) which serve to represent the components of $\mathcal{A}$. In this way, the structure of the interconnections of $\mathcal{A}$ (that is, the graph of $\mathcal{A}$) is placed in a clearer perspective.

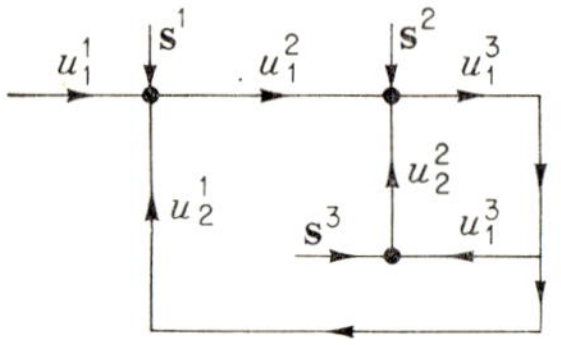

Fig. 2.6.2 Signal-state graph for the system of Fig. 2.6.1.

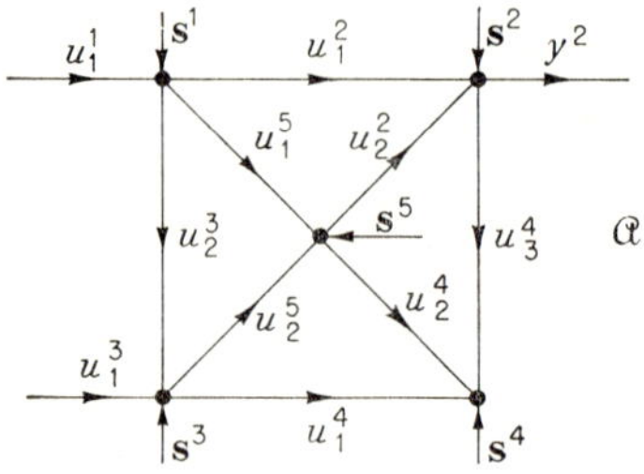

Fig. 2.6.3 A cycle-free system.

22 *Example* In Fig. *2.6.2* is illustrated the signal-state graph of a system (system $\mathcal{A}$, Fig. *2.6.1*) which has loops. In Fig. *2.6.3* is shown a cycle-free system. (Note that the labeling of branches is determined by their roles as inputs rather than outputs. For example, the branch connecting $\mathcal{A}_1$ to $\mathcal{A}_3$ represents the output of $\mathcal{A}_1$ and an input to $\mathcal{A}_3$. It is labeled u_2^3 because it is an input to $\mathcal{A}_3$; the other input to $\mathcal{A}_3$ is labeled u_1^3. Also, all the branches leaving a node are understood to represent the same output variable. For example, the branches u_1^2 and u_1^5 leaving $\mathcal{A}_1$ both represent the output of $\mathcal{A}_1$ and hence $u_1^2 = u_1^5$. Similarly, $u_2^5 = u_1^4$.)

Cycle-free systems have an intuitively obvious property which is stated in the following

[1] We assume that the reader is familiar with the elementary aspects of signal-flow graphs. An exposition of signal-flow graphs and related topics is presented in Chap. *9*, Sec. *15*.

23 **Theorem** Any cycle-free system $\mathcal{A}$ is determinate.

Proof Consider the signal-state graph of a cycle-free system $\mathcal{A}$ such as shown in Fig. *2.6.3*. Here u_1^1 and u_1^3 are the inputs to $\mathcal{A}$ and y^2 is the output. The other branch variables such as u_1^2 and u_2^2 represent the suppressed output variables. The state variables $\mathbf{s}^1, \ldots, \mathbf{s}^5$ are treated in the same way as the inputs to $\mathcal{A}$. That is, in so far as the graph-theoretic properties of the signal-state graph are concerned, all branches entering a node and not connected to other nodes are regarded as external inputs to the node, regardless of whether they are associated with input or state variables. For example, in Fig. *2.6.3*, u_1^1 and $\mathbf{s}^1$ are external inputs to node 1, $\mathbf{s}^2$ is an external input to node 2, and u_1^3 and $\mathbf{s}^3$ are external inputs to node 3. (The number of a node can be identified from the superscripts of input variables associated with it.) A node which has external inputs will be referred to as an *input node*, and all other nodes will be called *internal nodes*. [Note that internal nodes are associated with input-output variables (*1.10.27*).]

To begin with, we make the assumption that $\mathcal{A}$ is *well formed* in the sense that every internal node of $\mathcal{A}$ can be reached (via one or more oriented paths) from an input node. This ensures that there are no components of $\mathcal{A}$ which have no input applied to them.

The property that $\mathbf{s}^i$ and $\mathbf{u}^i$ determine the output of $\mathcal{A}_i$, $i = 1, \ldots, N$, translates itself into the following property of branch variables: The values of all the branches entering a node determine the values of all the branches leaving that node. (As was pointed out previously, the variables associated with the branches leaving a node have the same value, namely, the output of the component which is represented by the node in question. However, this is not relevant to the arguments which will be employed below.)

Using the terminology defined above, the statement of Theorem *23* can be restated as follows.

24 **Restatement of Theorem** *23* Let $\mathcal{A}$ be a well-formed cycle-free oriented graph in which each branch is associated with a variable, with the defining property that, for each node i, the values of branches leaving i are determined uniquely by the values of branches entering i. Furthermore, $\mathcal{A}$ contains one or more nodes (input nodes) such that at least one of the branches entering each such node is not connected to any other node (i.e., represents an external input). Then, the values of all such branches (external inputs) determine the values of all branches of $\mathcal{A}$. $\triangleleft$

We establish first the following property of $\mathcal{A}$.

25 **Property** $\mathcal{A}$ has at least one input node such that all branches entering it represent external inputs. For example, in Fig. *2.6.3* node 1 is an input node with this property.

For, suppose that this is false. Then all input nodes have the property that at least one branch entering each such node leaves an internal node. But, since $\mathcal{A}$ is well formed, every internal node is reachable from at least one input node. Consequently, every input node is reachable from at least one input node.

Now let 1, 2, . . . , k, say, designate input nodes. If 1 is reachable from 1, then $\mathcal{A}$ has a cycle—which is a contradiction. Suppose that 1 is reachable from 2. Then if 2 is reachable from 1, we have a cycle—which is a contradiction. Suppose 2 is reachable from 3. Repeating this argument, we finally exhaust all nodes in the set 1, . . . , k and are forced to conclude that there must be a cycle containing one or more input nodes. Since this is a contradiction, we have shown that $\mathcal{A}$ has Property *25*.

Now we can delete from the graph every input node having this property, together with all the branches entering it, without affecting the values of other branches in the graph. (This is illustrated in Fig. *2.6.4*.) At this stage, we know the values of the branches which leave the deleted nodes and hence can regard them as external inputs with respect to the graph that remains after the removal of the input nodes in question. The remaining graph, call it $\mathcal{A}'$, is, like $\mathcal{A}$, cycle-free and well formed. On applying previously used arguments to $\mathcal{A}'$ and then repeating the process, we establish, after not more than $N - 1$ steps ($N \triangleq$ number of nodes in $\mathcal{A}$) that the values of all branches in $\mathcal{A}$ are determined by the values of its external inputs. Q.E.D.

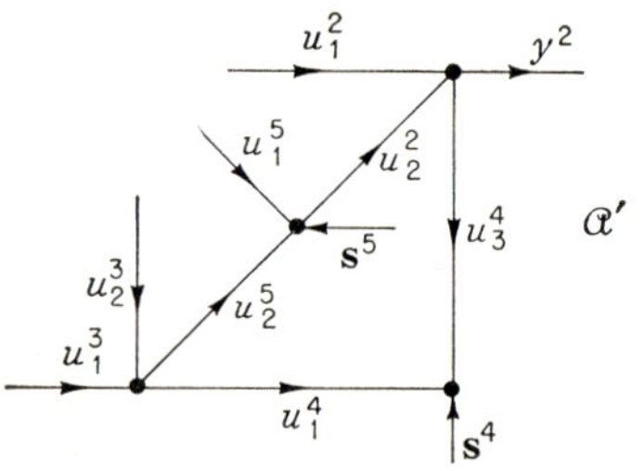

Fig. 2.6.4 Result of removing a node from the system of Fig. 2.6.3.

We turn next to the consideration of another class of determinate systems which, unlike the class discussed above, need not be cycle-free. As a preliminary, we introduce a notion which is relevant to the characterization of determinate systems.

26 **Definition** A system $\mathcal{A}$ is said to have *state-determined output* if for each t its output at time t is uniquely determined by its state at time t. This implies that the state equations of such a system have the form

27 $$\mathbf{s}(t) = \mathbf{s}(\mathbf{s}(t_0);\mathbf{u}_{(t_0,t]})$$

28 $$\mathbf{y}(t) = \mathbf{g}(\mathbf{s}(t),t)$$

where $\mathbf{g}$ is a function defined on $\Sigma \times T$.

29 *Example* A very simple example of a system with a state-determined output is an integrator (*1.9.10*), which is characterized by the input-output-state relation

30 $$y(t) = y(t_0) + \int_{t_0}^{t} u(\xi)\, d\xi \qquad t > t_0$$

or more precisely

31 $$y(t) = y(t_0-) + \int_{t_0}^{t} u(\xi)\, d\xi \qquad t \geq t_0$$

As was shown earlier (*1.7.10*), $y(t)$ qualifies as $s(t)$, the state of the integrator at time t. Thus, $s(t)$ determines $y(t)$.

More generally, as will be shown in Chap. 4 (see *4.5.19 et seq.*), any system whose input-output relation is of the form

32 $$a_n \frac{d^n y}{dt^n} + \cdots + a_1 \frac{dy}{dt} + a_0 y = b_m \frac{d^m u}{dt^m} + \cdots + b_0 u \qquad a_n \neq 0$$

has state-determined output provided $m < n$.

A simple and yet important example of a discrete-time system with state-determined output is a unit-delay element defined by the input-output relation (*1.7.25*)

33 $$y(t) = u(t-1) \qquad t \geq t_0 + 1,\ T = \{0,1,2, \cdots\}$$

As in the case of the integrator (which is in effect a continuous-time analog of the delay element), the state of the unit-delay element at time t is $y(t)$. Thus, the unit-delay element has state-determined output.

A notion which is closely related to the notion of a system with state-determined output is that of a system with state-determined input. Although this notion is not of direct relevance to the question of system determinateness, we define it for future reference.

34 **Definition** A system $\mathcal{A}$ is said to have *state-determined input* if for each t its input at time t is uniquely determined by its state at time t.

35 *Example* A very simple illustration of such a system is provided by the differentiator, i.e., a system characterized by the input-output relation

$$y = \frac{du}{dt}$$

As was pointed out previously (*1.7.10*), the state of a differentiator at time t is $u(t)$. Thus the input to the differentiator at time t is determined by $s(t)$. The same thing is true of any system whose input-output relation is of the form *32* with y and u interchanged.

Now consider a system $\mathcal{A}$ which is an interconnection of $\mathcal{A}_1, \ldots, \mathcal{A}_N$. We assume that the signal-state graph of $\mathcal{A}$ contains one or more cycles (closed paths) and that each $\mathcal{A}_i$ is characterized by differential state equations of the canonical form (see *2.3.36*)

36 $$\dot{\mathbf{s}}^i(t) = \mathbf{f}_i(\mathbf{s}^i(t),\mathbf{u}^i(t),t) \qquad i = 1, \ldots, N$$

37 $$\mathbf{y}^i(t) = \mathbf{g}_i(\mathbf{s}^i(t),\mathbf{u}^i(t),t)$$

We assume further that the $\mathbf{f}_i$ satisfy the usual sufficient conditions (*2.3.40*) which ensure the existence and uniqueness of solutions of *36* for all $\mathbf{u}^i$ in $R[\mathbf{u}^i]$.

Suppose that every cycle in $\mathcal{A}$ contains at least one node which represents a component with state-determined output. For concreteness, let $\mathcal{A}_2$ in Fig. *2.6.2* be such a component. Then, if at time t_0 we "split" the node in question in the manner illustrated in Fig. *2.6.5*, we shall know not only $\mathbf{s}^2(t_0)$ but also $\mathbf{y}^2(t_0)$, which is the output of $\mathcal{A}_2$ and an input to other nodes.

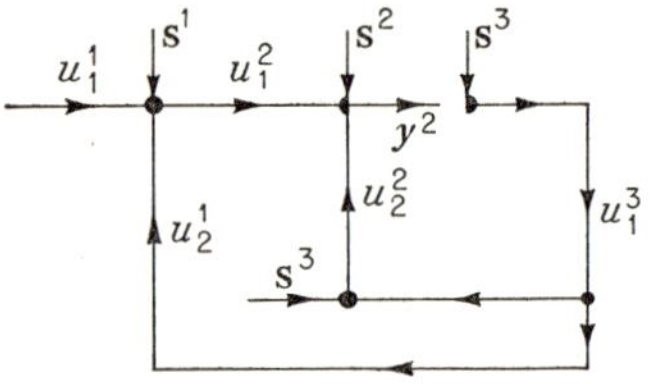

Fig. 2.6.5 Effect of node splitting.

By breaking all the cycles of $\mathcal{A}$ in this fashion, we transform the original system into a cycle-free system in which the states of all components and the values of all inputs are known at time t_0. (Note that in Fig. *2.6.5* the splitting of $\mathcal{A}_2$ breaks both cycles in $\mathcal{A}$.) On substituting these values into the state equations *36* and *37* we obtain $\dot{\mathbf{s}}^i(t_0)$ and $\mathbf{y}^i(t_0)$. Then, by using $\dot{\mathbf{s}}^i(t_0)$ and $\mathbf{s}^i(t_0)$, we can compute via *36* and *37* the value of $\mathbf{s}^i(t)$ at $t_0 + dt$, which in turn yields $\mathbf{u}^i(t_0 + dt)$ and which in combination with the known inputs at time $t_0 + dt$ makes it possible to recompute the $\mathbf{y}^i(t_0)$ and repeat the process. This heuristic argument suggests the following

38 **Theorem** Let $\mathcal{A}$ be an interconnection of $\mathcal{A}_1, \ldots, \mathcal{A}_N$ containing one or more cycles. If each $\mathcal{A}_i$ is characterized by canonical state equations of the form *36* and *37* and if each cycle in $\mathcal{A}$ contains at least one component which has state-determined output (*26*), then $\mathcal{A}$ can be characterized by canonical state equations of the form *36* and *37*.

Proof The proof is quite straightforward and is left as an exercise for the reader. To indicate how it proceeds, it will suffice to consider a simple system such as shown in Fig. *2.6.6*, in which $\mathcal{A}_2$ has state-

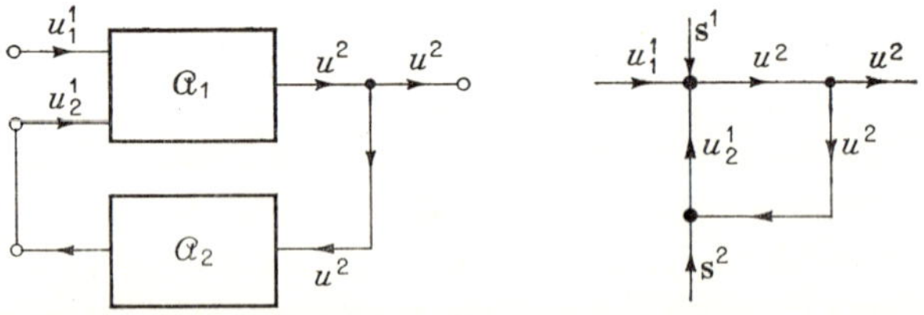

Fig. 2.6.6 Illustration for Theorem 2.6.38.

determined output, so that we have

39 $\mathcal{A}_1$:
$$\dot{\mathbf{s}}^1(t) = \mathbf{f}_1(\mathbf{s}^1(t),u_1^1(t),u_2^1(t),t)$$
$$u^2(t) = g_1(\mathbf{s}^1(t),u_1^1(t),u_2^1(t),t)$$

40 $\mathcal{A}_2$:
$$\dot{\mathbf{s}}^2(t) = \mathbf{f}_2(\mathbf{s}^2(t),u^2(t),t)$$
$$u_2^1(t) = g_2(\mathbf{s}^2(t),t)$$

Upon elimination of $u_2^1(t)$, these equations yield a pair of differential equations of the form

$$\dot{\mathbf{s}}^1(t) = \mathbf{f}_1^*(\mathbf{s}^1(t),\mathbf{s}^2(t),u_1^1(t),t)$$
$$\dot{\mathbf{s}}^2(t) = \mathbf{f}_2^*(\mathbf{s}^1(t),\mathbf{s}^2(t),u_1^1(t),t)$$

in which $\mathbf{f}_1^*$ and $\mathbf{f}_2^*$ are defined by $\mathbf{f}_1$, g_1, $\mathbf{f}_2$, and g_2, and $\mathbf{s}(t) \triangleq (\mathbf{s}^1(t),\mathbf{s}^2(t))$ plays the role of the state of $\mathcal{A}$. Thus, the state equations of $\mathcal{A}$ can be put into the canonical form

41
$$\dot{\mathbf{s}}(t) = \mathbf{f}(\mathbf{s}(t),u_1^1(t),t)$$
$$u^2(t) = g(\mathbf{s}(t),u_1^1(t),t)$$

which is what we wanted to demonstrate.

42 *Comment* Note that Theorem *38* does not assert that $\mathcal{A}$ is a determinate system. However, it follows from this theorem that $\mathcal{A}$ is determinate if the **f** and **g** functions in the canonical state equations which characterize $\mathcal{A}$ are such that the differential equation *36* has a unique solution. In particular, in the case of linear time-varying systems, it follows from Theorem *38* and Conclusion *2.3.42* that $\mathcal{A}$ is determinate if the matrices characterizing each $\mathcal{A}_i$ are continuous functions of time and every cycle in $\mathcal{A}$ has at least one component with state-determined output. More generally, in most physical systems of the feedback type every loop (cycle) contains a component which introduces some delay and which can be approximated by a delay element of the type defined in *1.7.25*. Such delay elements have state-determined output, and it is largely for this reason that most physical systems are determinate.

The question of determinateness of linear systems will be examined in greater detail in Chap. *3*, Sec. *9*. For the present, we return to the discussion of system equivalence.

7 *Further properties of system equivalence*

Equivalence of interconnections

Consider two collections of pairwise equivalent systems $\mathcal{A}_1, \ldots, \mathcal{A}_N$ and $\mathcal{B}_1, \ldots, \mathcal{B}_N$, with $\mathcal{A}_i \equiv \mathcal{B}_i$, $i = 1, \ldots, N$. Let $\mathcal{A}$ be a determinate interconnection of the $\mathcal{A}_i$ and let $\mathcal{B}$ be the same interconnection of the $\mathcal{B}_i$. On intuitive grounds we expect that, in general, $\mathcal{A} \equiv \mathcal{B}$. Indeed, we can readily deduce the equivalence of $\mathcal{A}$ and $\mathcal{B}$ in a formal way by a repeated application of the substitution principle (*1.10.35*).

There are two special cases of this property which are of frequent use in deducing the equivalence of two systems from equivalences of other systems. These properties follow from the

1 **Assertion** If $\mathcal{A} \equiv \mathcal{B}$ and $\mathcal{C} \equiv \mathcal{D}$, then (I) the free product (*1.10.10*) of $\mathcal{A}$ and $\mathcal{C}$ is equivalent to the free product of $\mathcal{B}$ and $\mathcal{D}$ and (II) the free sum (*1.10.30 et seq.*) of $\mathcal{A}$ and $\mathcal{C}$ is equivalent to the free sum of $\mathcal{B}$ and $\mathcal{D}$.

Proof It is easy to prove (I) directly, without relying on the substitution principle. The proof of (II) is left as an exercise for the reader. Let $\boldsymbol{\alpha}$, $\boldsymbol{\beta}$, $\boldsymbol{\gamma}$, $\boldsymbol{\delta}$ denote the states of $\mathcal{A}$, $\mathcal{B}$, $\mathcal{C}$, $\mathcal{D}$ at time t_0, with $\boldsymbol{\alpha}$, $\boldsymbol{\beta}$, $\boldsymbol{\gamma}$, $\boldsymbol{\delta}$ ranging over the respective state spaces Σ_a, Σ_b, Σ_c, Σ_d. Furthermore, let $\boldsymbol{\alpha}(t)$, $\boldsymbol{\beta}(t)$, $\boldsymbol{\gamma}(t)$, and $\boldsymbol{\delta}(t)$ be the states into which $\boldsymbol{\alpha}$, $\boldsymbol{\beta}$, $\boldsymbol{\gamma}$, and $\boldsymbol{\delta}$ are taken by $\mathbf{u}_{(t_0,t]}$. By *2.6.23* and *2.6.10* the composite states $[\boldsymbol{\alpha}(t),\boldsymbol{\gamma}(t)]$ and $[\boldsymbol{\beta}(t),\boldsymbol{\delta}(t)]$ qualify as the states of the free products $\mathcal{A}\mathcal{C}$ and $\mathcal{B}\mathcal{D}$, respectively, at time t. The state spaces of these products at time t are denoted by $\Sigma_{ac}(t_0,t)$ and $\Sigma_{bd}(t_0,t)$, and are defined by *1.10.16*.

Since $\mathcal{A} \equiv \mathcal{B}$, to every initial state $\boldsymbol{\alpha}$ of $\mathcal{A}$ there is an equivalent state, say, $\boldsymbol{\beta}$, of $\mathcal{B}$, and vice versa. Let $(\boldsymbol{\alpha},\boldsymbol{\beta})$ be such a pair of equivalent states and let $(\boldsymbol{\gamma},\boldsymbol{\delta})$ be an analogous pair for $\mathcal{C}$ and $\mathcal{D}$. Then it is obvious that the pair $(\boldsymbol{\alpha},\boldsymbol{\gamma})$, regarded as an initial state of the free product $\mathcal{A}\mathcal{C}$, is equivalent to the pair $(\boldsymbol{\beta},\boldsymbol{\delta})$, which plays the role of the initial state of $\mathcal{B}\mathcal{D}$. Now by hypothesis $[\boldsymbol{\alpha}(t),\boldsymbol{\gamma}(t)]$ and $[\boldsymbol{\beta}(t),\boldsymbol{\delta}(t)]$ are reachable, respectively, from $(\boldsymbol{\alpha},\boldsymbol{\gamma})$ and $(\boldsymbol{\beta},\boldsymbol{\delta})$ via $\mathbf{u}_{(t_0,t]}$. Consequently, by Assertion *2.4.19*, $[\boldsymbol{\alpha}(t),\boldsymbol{\gamma}(t)]$ and $[\boldsymbol{\beta}(t),\boldsymbol{\delta}(t)]$ are equivalent states. Furthermore, from the definitions of $\Sigma_{ac}(t_0,t)$ and $\Sigma_{bd}(t_0,t)$ it follows that the statement "to every state $(\boldsymbol{\alpha},\boldsymbol{\gamma})$ in the product space $\Sigma_a \times \Sigma_c$ there is an equivalent state $(\boldsymbol{\beta},\boldsymbol{\delta})$ in the product space $\Sigma_b \times \Sigma_d$" implies "to every state $[\boldsymbol{\alpha}(t),\boldsymbol{\gamma}(t)]$ in $\Sigma_{ac}(t_0,t)$ there is an equivalent state $[\boldsymbol{\beta}(t),\boldsymbol{\delta}(t)]$ in $\Sigma_{bd}(t_0,t)$." Now since the product of $\mathcal{A}$ and $\mathcal{C}$ is initially free, $(\boldsymbol{\alpha},\boldsymbol{\gamma})$ can be chosen arbitrarily in $\Sigma_a \times \Sigma_b$. From this and the preceding implication it follows that $\mathcal{A}\mathcal{C} \equiv \mathcal{B}\mathcal{D}$, with $\mathcal{A}\mathcal{C}$ and $\mathcal{B}\mathcal{D}$ interpreted as free products.

Having established Assertion *1*, we can deduce from it and Assertion *2.4.24* a corresponding property for constrained products (*1.10.23*). This property is stated in the

2 **Corollary** If $\mathcal{A}$, $\mathcal{B}$, $\mathcal{C}$, $\mathcal{D}$ are well behaved in the sense of *2.4.24* and if $\mathcal{A} \equiv \mathcal{B}$ and $\mathcal{C} \equiv \mathcal{D}$, then the constrained product $\mathcal{A}\mathcal{C}$ is equivalent to the constrained product $\mathcal{B}\mathcal{D}$.

Conditional equivalence

Consider two systems $\mathcal{A}$ and $\mathcal{B}$ such that there is a state $\boldsymbol{\alpha}_0$ in the state space of $\mathcal{A}$ which is equivalent to a state $\boldsymbol{\beta}_0$ in the state space of $\mathcal{B}$. Let Σ_{α_0} and Σ_{β_0} denote the closed sets of states (see *2.4.18*) which are reachable from $\boldsymbol{\alpha}_0$ and $\boldsymbol{\beta}_0$, respectively. Then it is obvious that Σ_{α_0} and Σ_{β_0} have the property expressed by the

3 **Assertion** To every state $\boldsymbol{\alpha}$ in Σ_{α_0} there is at least one equivalent state $\boldsymbol{\beta}$ in Σ_{β_0}, and vice versa.

Proof Let $\boldsymbol{\alpha}$ be an arbitrary state in Σ_{α_0} and let $\mathbf{u}$ be an input which takes $\boldsymbol{\alpha}_0$ into $\boldsymbol{\alpha}$. (Such an input exists because all the states in Σ_{α_0} are reachable from $\boldsymbol{\alpha}_0$.) Then by *2.4.19* the state $\boldsymbol{\beta}$ into which $\boldsymbol{\beta}_0$ is taken by $\mathbf{u}$ is equivalent to $\boldsymbol{\alpha}$, and likewise for an arbitrary state in Σ_{β_0}.

4 **Corollary** If $\boldsymbol{\alpha}_0 \simeq \boldsymbol{\beta}_0$ and all states in the state spaces of $\mathcal{A}$ and $\mathcal{B}$ are reachable from $\boldsymbol{\alpha}_0$ and $\boldsymbol{\beta}_0$, respectively, then $\mathcal{A} \equiv \mathcal{B}$. (A special instance of this is *2.9.5*.)

Based on *3*, we make the following

5 **Definition** $\mathcal{A}$ and $\mathcal{B}$ are *conditionally equivalent*, written as $\mathcal{A} \stackrel{*}{=} \mathcal{B}$, if there are closed subsets Σ_{α_0} and Σ_{β_0} in the state spaces of $\mathcal{A}$ and $\mathcal{B}$, respectively, such that to every state in Σ_{α_0} there is an equivalent state in Σ_{β_0}, and vice versa. ◁

In effect, if $\mathcal{A}$ and $\mathcal{B}$ are conditionally equivalent, then they behave like equivalent systems so long as their initial states are restricted to the respective subsets Σ_{α_0} and Σ_{β_0}. This implies that, under this restriction, the various properties of system equivalence such as those expressed by *1* also apply to conditional equivalence. In particular,

6 $$\{\mathcal{A} \stackrel{*}{=} \mathcal{B} \text{ and } \mathcal{C} \stackrel{*}{=} \mathcal{D}\} \Rightarrow \{\mathcal{A}\mathcal{C} \stackrel{*}{=} \mathcal{B}\mathcal{D}\}$$

under the same conditions as stated in *2*. We shall use this relation later on in our discussion of zero-state equivalence and inverse systems.

8 *Zero state, ground state, and equilibrium state*

The heading of this section lists three states which by virtue of their special properties play important roles in system analysis. A given system $\mathcal{A}$ may or may not have these states and, if it has them, they may or may not be coincident. Before defining the states in question, it will be convenient to have on record the following simple

1 **Definition** A time function which has the value 0 (in scalar or vector sense) for $t \geq t_0$ is called a *null function* and is denoted by $\mathbf{0}$, or, if necessary, by $\mathbf{0}(t_0)$. If an input $\mathbf{u}$ is zero for $t \geq t_0$, then it is said to be a *zero input.* $\triangleleft$

We begin with the definition of

Zero state

2 **Definition** A state $\boldsymbol{\theta}$ of $\mathcal{A}$ is a *zero state* if, for all t_0, the response of $\mathcal{A}$ to the zero input, with $\mathcal{A}$ starting in $\boldsymbol{\theta}$, is a null function. In symbols

3 $$\{\boldsymbol{\theta} \text{ is a } \textit{zero state} \text{ of } \mathcal{A}\} \Leftrightarrow \{A(\boldsymbol{\theta};\mathbf{0}) = \mathbf{0}\} \qquad \forall t_0 \qquad \triangleleft$$

Stated informally, if $\mathcal{A}$ is initially in its zero state and no input is applied to $\mathcal{A}$ (that is, $\mathbf{u} = \mathbf{0}$), then there is no output (or, equivalently, the output is a null function). It should be noted that $\mathcal{A}$ may have no zero state, or it may have one or more nonequivalent zero states.

4 *Example* The system discussed in Example *2.2.4* has six zero states: 0, 2, 4, 6, 8, 10. However, all these states are equivalent to one another.

5 *Example* System $\mathcal{A}$ in Example *2.4.5* has a unique zero state, namely $\theta = 0$ (which corresponds to the condition where no current flows through L). In system $\mathcal{B}$, (0,0) is a zero state, as are all states of the form $(\rho, -\rho)$, where ρ is any real number. Note that all such states are equivalent to (0,0), and hence the zero state (0,0) is unique (to within equivalent states).

6 *Example* The following example demonstrates that a system can have two or more nonequivalent zero states. Consider the network $\mathcal{A}$, shown in Fig. *2.8.1*, in which the input is a voltage u, the output is a

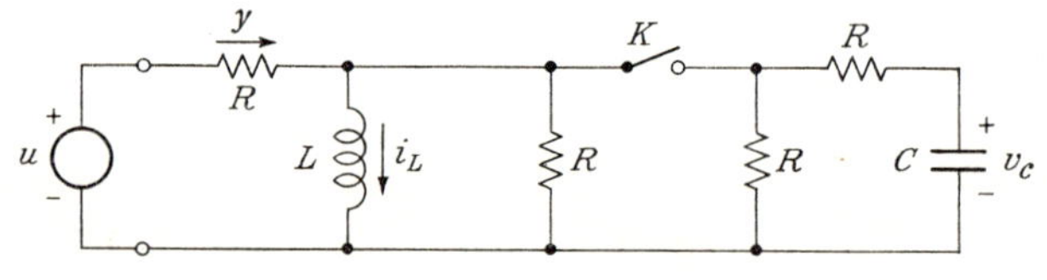

Fig. 2.8.1 A system with nonunique zero state.

current y, and K is a switch whose position is characterized by a binary variable σ, namely, $\sigma = 0$ if K is open and $\sigma = 1$ if K is closed. It is easy to verify that the state of $\mathcal{A}$ at time t can be identified with the 3-vector $\mathbf{s}(t) = (i_L(t), v_C(t), \sigma(t))$, where $i_L(t)$ is the current through L, $v_C(t)$ is the voltage across C, and $\sigma(t)$ is the value of σ at time t. Now, (0,0,0) is a zero state of $\mathcal{A}$, as is the state (0,0,1). Obviously, these two states are nonequivalent.

Ground state

A notion which frequently coincides in meaning with that of the zero state but is not coextensive with it is that of the *ground state*. Specifically, consider a system $\mathcal{A}$ in initial state $\mathbf{s}(t_0) = \boldsymbol{\alpha}$ to which a zero input is applied at t_0. Let us fix our attention on $\mathbf{s}(t)$, the state of $\mathcal{A}$ at time t, $t > t_0$, and examine what may happen to $\mathbf{s}(t)$ as $t \to \infty$.

Clearly, there are two possibilities: (I) $\mathbf{s}(t)$ may converge to a fixed state, say $\boldsymbol{\gamma}$, as $t \to \infty$; (II) $\mathbf{s}(t)$ may not converge as $t \to \infty$ (see *2.4.20 et seq.*).

Let us restrict our attention to (I). Here again we have two possibilities: (*a*) the limiting state $\boldsymbol{\gamma}$ depends on the initial state $\boldsymbol{\alpha}$; (*b*) $\boldsymbol{\gamma}$ is independent of $\boldsymbol{\alpha}$.

If (I) and (*b*) hold, the limiting terminal state $\boldsymbol{\gamma}$ is called the *ground state* of $\mathcal{A}$. Roughly, the ground state is the state into which $\mathcal{A}$ settles eventually when no input is applied. More formally, we have the

7 **Definition** The *ground state* of $\mathcal{A}$, if it exists, is the limiting terminal state of $\mathcal{A}$ when the zero input is applied to $\mathcal{A}$, provided the limiting state $\boldsymbol{\gamma}$ is the same for all initial states. In symbols,

8 $$\{\boldsymbol{\gamma} \text{ is the } \textit{ground state} \text{ of } \mathcal{A}\} \Leftrightarrow \{\boldsymbol{\gamma} = \lim_{t\to\infty} \mathbf{s}(\boldsymbol{\alpha};\mathbf{0}),\ \boldsymbol{\gamma} \text{ independent of } \boldsymbol{\alpha}\}$$

where $\mathbf{s}(\mathbf{s}(t_0);\mathbf{u}_{(t_0,t]})$ denotes the state of $\mathcal{A}$ at time t (*2.2.33*). Note that $\boldsymbol{\gamma}$, if it exists, is unique by virtue of the uniqueness of the limit. ◁

The following examples serve to illustrate the notion of the ground state and the differences between it and the zero state of a system.

9 *Example* Consider two networks $\mathcal{A}$ and $\mathcal{B}$ shown in Fig. *2.8.2*. In both cases, the state at time t can be identified with the current $i(t)$

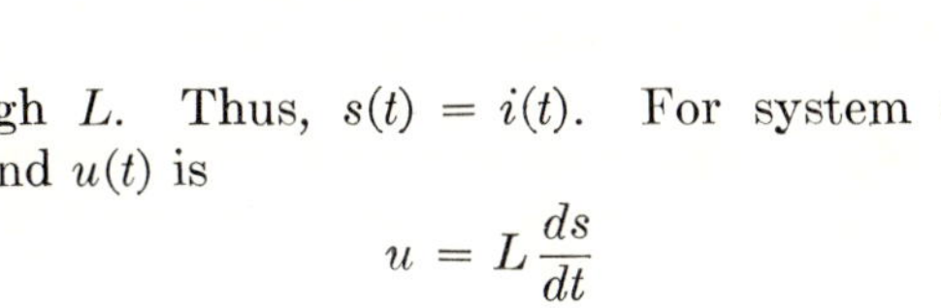

Fig. 2.8.2 $\mathcal{A}$ has zero state but not ground state; $\mathcal{B}$ has both zero state and ground state.

flowing through L. Thus, $s(t) = i(t)$. For system $\mathcal{A}$, the relation between $s(t)$ and $u(t)$ is

$$u = L\frac{ds}{dt}$$

Integrating this equation between t_0 and t, we obtain the state equation

10 $$s(t) = s(t_0) + \frac{1}{L}\int_{t_0}^{t} u(\xi)\, d\xi$$

which for the input $u(t) \equiv 0$ reduces to

11 $$s(t) = s(t_0) = \text{initial current through } L,\ t > t_0$$

Correspondingly, for system $\mathcal{B}$ we have

$$u = Rs + L\frac{ds}{dt}$$

which upon integration yields

12 $$s(t) = e^{-(R/L)(t-t_0)}s(t_0) + \frac{1}{L}\int_{t_0}^{t} e^{-(R/L)(t-\xi)}u(\xi)\,d\xi$$

and which for the zero input reduces to

13 $$s(t) = e^{-(R/L)(t-t_0)}s(t_0)$$

Inspection of *11* and *13* shows that $\mathcal{A}$ does not have a gound state—even though $s(t)$ converges to $s(t_0)$ as $t \to \infty$ [in a trivial sense, of course, since $s(t)$ is a constant] because the limiting state depends on the initial state. On the other hand, if $R/L > 0$, then $\mathcal{B}$ does have a ground state, namely $\gamma = 0$, since $s(t) \to 0$ as $t \to \infty$ for all values of $s(t_0)$. Note that both $\mathcal{A}$ and $\mathcal{B}$ have a zero state, $\theta = 0$. Thus, $\mathcal{A}$ has a zero state but not a ground state, whereas $\mathcal{B}$ has both a zero state θ and a ground state γ, with $\theta = \gamma = 0$.

14 *Example* The system shown in Fig. *2.8.1* (see Example *6*), with $R > 0$, $L > 0$, $C > 0$, does not have a ground state. The reason is that all initial states of the form $(a,b,0)$, where a and b are real numbers, tend to the limiting state $(0,0,0)$ as $t \to \infty$. On the other hand, all initial states of the form $(a,b,1)$ tend to $(0,0,1)$. Since $(0,0,0) \not\approx (0,0,1)$, the limiting state depends on the initial state and hence is not a ground state. ◁

The notion of the ground state is closely related to the notion of asymptotic stability in the large. This connection will be discussed in greater detail in Chap. *7*.

Equilibrium state

Another notion which frequently coincides in meaning with the notions of the ground state and the zero state is that of the *equilibrium state*. Roughly, $\mathbf{n}$ is an *equilibrium state* of $\mathcal{A}$ if it does not change when no input is applied to $\mathcal{A}$. More formally, we have the

15 **Definition** $\mathbf{n}$ is an *equilibrium state* of $\mathcal{A}$ if on application to $\mathcal{A}$ of the zero input, with $\mathcal{A}$ initially (at t_0) in state $\mathbf{n}$, $\mathbf{s}(t)$ remains equal to $\mathbf{n}$ for all $t \geq t_0$. In symbols,

$$\{\mathbf{n} \text{ is an equilibrium state of } \mathcal{A}\} \Leftrightarrow \{\mathbf{s}(\mathbf{n};\mathbf{0}) = \mathbf{n}, t \geq t_0\}$$

where $\mathbf{s}(\mathbf{n};\mathbf{0})$ denotes the state of $\mathcal{A}$ at time t when the state of $\mathcal{A}$ at time t_0 is $\mathbf{n}$ and the zero input is applied starting at t_0.

16 *Example* In system $\mathcal{A}$, Example *9*, 0 is the equilibrium state as well as the zero state but not the ground state. In system $\mathcal{B}$, Example *9*, 0 is the equilibrium state, the zero state, and the ground state. In the system of Example *6*, (0,0,0) and (0,0,1) are equilibrium states. (This shows that an equilibrium state need not be unique.)

17 *Exercise* Construct an example of a system which has an equilibrium state which is neither a zero state nor a ground state.

Zero-state, zero-input, and steady-state response

In later chapters we shall frequently be dealing with various special types of responses of $\mathcal{A}$. The more important of these are defined below.

18 **Definition** The *zero-state response* of $\mathcal{A}$ to $\mathbf{u}$ ($\mathbf{u} \triangleq \mathbf{u}_{(t_0,t]}$) is the response of $\mathcal{A}$ to $\mathbf{u}$ with $\mathcal{A}$ initially in its zero state $\boldsymbol{\theta}$. The zero-state response at time t is denoted by $\mathbf{A}(\boldsymbol{\theta};\mathbf{u})$ or, for short, $\mathbf{A}(\mathbf{u})$. Correspondingly, for response segments we shall use the notation $\bar{\mathbf{A}}(\boldsymbol{\theta};\mathbf{u})$ or $\bar{\mathbf{A}}(\mathbf{u})$.

19 **Definition** The *ground-state response* of $\mathcal{A}$ to $\mathbf{u}$ has the same meaning as the zero-state response, except that the initial state is $\boldsymbol{\gamma}$ (ground state) rather than $\boldsymbol{\theta}$. The ground-state response at time t is denoted by $\mathbf{A}(\boldsymbol{\gamma};\mathbf{u})$. ◁

An important type of response which can be conveniently defined in terms of the ground-state response is the *steady-state response.* Specifically,

20 **Definition** The *steady-state response* of $\mathcal{A}$ to an input $\mathbf{u}_{(t_0,t]}$ is denoted by $\mathbf{A}_\infty(\mathbf{u})$ and is defined by the limit, if it exists, of the ground-state response of $\mathcal{A}$ to $\mathbf{u}$ as $t_0 \to -\infty$.[1] In symbols,

21
$$\mathbf{A}_\infty(\mathbf{u}) = \lim_{t_0 \to -\infty} \mathbf{A}(\boldsymbol{\gamma};\mathbf{u}_{(t_0,t]})$$

22 *Example* Consider the system $\mathcal{B}$ of Example *9*, with $R = L = 1$, and assume that $u(t) = \sin t$, $-\infty < t < \infty$. Since $\boldsymbol{\gamma} = 0$, the steady-state response is given by

$$\begin{aligned} y(t) &= \lim_{t_0 \to -\infty} \int_{t_0}^{t} e^{-(t-\xi)} \sin \xi \, d\xi \\ &= \tfrac{1}{2} (\sin t - \cos t) \end{aligned}$$

Note that the same result obtains for any initial state $\boldsymbol{\alpha}$.

[1] Usually $\mathcal{A}$ and $\mathbf{u}$ are such that, in the right-hand member of *21*, $\boldsymbol{\gamma}$ can be replaced by an arbitrary initial state $\boldsymbol{\alpha}$ without affecting the limiting value of the response as $t_0 \to -\infty$.

23 **Definition** The response of $\mathcal{A}$ to a zero input, with $\mathcal{A}$ initially in state $s(t_0) = \boldsymbol{\alpha}$, is the *zero-input response* of $\mathcal{A}$ *starting in state* $\boldsymbol{\alpha}$. The zero-input response at time t is denoted by $\mathbf{A}(\boldsymbol{\alpha};\mathbf{0})$, with the corresponding response segment denoted by $\bar{\mathbf{A}}(\boldsymbol{\alpha};\mathbf{0})$. [If it is necessary to place in evidence the observation interval, the notation $\bar{\mathbf{A}}(\boldsymbol{\alpha};\mathbf{0}_{(t_0,t]})$ may be employed.]

24 *Example* The zero-input response of the system $\mathcal{B}$ of Example *9*, with $R = L = 1$, is expressed by

$$\mathbf{A}(\alpha;\mathbf{0}) = \alpha e^{-(t-t_0)} \qquad t > t_0$$

9 *Zero-state and zero-input equivalence*

There are two special types of system equivalence which will be encountered frequently in later chapters. One is the *zero-state equivalence* and the other is *zero-input equivalence.* These terms are defined below.

1 **Definition** $\mathcal{A}$ and $\mathcal{B}$ are *zero-state equivalent,* written $\mathcal{A} \doteq \mathcal{B}$, if to every zero state of $\mathcal{A}$ there corresponds an equivalent zero state of $\mathcal{B}$, and vice versa. ◁

In most of the cases with which we shall deal, $\mathcal{A}$ and $\mathcal{B}$ will have unique zero states (to within equivalent states) which will be denoted by $\boldsymbol{\theta}_a$ and $\boldsymbol{\theta}_b$, respectively. Then

$$\{\mathcal{A} \doteq \mathcal{B}\} \Leftrightarrow \{\boldsymbol{\theta}_a \simeq \boldsymbol{\theta}_b\}$$

2 **Definition** $\mathcal{A}$ and $\mathcal{B}$ are *zero-input equivalent* if to every state $\boldsymbol{\alpha}$ in $\mathcal{A}$ there corresponds a state $\boldsymbol{\beta}$ in $\mathcal{B}$, and vice versa, such that

$$\bar{\mathbf{A}}(\boldsymbol{\alpha};\mathbf{0}) = \bar{\mathbf{B}}(\boldsymbol{\beta};\mathbf{0}) \qquad \forall t_0 \forall t$$

That is, the zero-input responses of $\mathcal{A}$ and $\mathcal{B}$ starting in states $\boldsymbol{\alpha}$ and $\boldsymbol{\beta}$, respectively, are identical over all observation intervals $(t_0,t]$. ◁

The nature of these definitions suggests that two systems may be zero-state or zero-input equivalent without being equivalent in the more general sense of Definition *2.5.8*. Since this point will be discussed in detail in Chap. *3*, we confine ourselves at this juncture to a simple

3 *Example* Consider the two networks shown in Fig. *2.9.1*. Here the state space of $\mathcal{A}$ consists of a single element denoted by (1) (see *1.8.17*), while the state of $\mathcal{B}$ at time t is a 2-vector $\mathbf{s}(t) = (s_1(t),s_2(t))$ (see

1.7.28) whose components are $s_1(t)$ [$\triangleq i_L(t)$] and $s_2(t)$ [$\triangleq v_C(t)$], where $i_L(t)$ and $v_C(t)$ are respectively the current through the inductor and the voltage across the capacitor. The ground state (as well as the zero state of $\mathcal{A}$) is (1), while that of $\mathcal{B}$ is (0,0).

Now $\mathcal{B}$ is readily recognized to be a constant-resistance network (with constant input impedance of 1 ohm). This means that if $\mathcal{B}$ is initially (at t_0) in its zero state, then for an arbitrary input u the relation $y(t) = u(t)$ holds for all $t \geq t_0$. This in turn implies that $\mathcal{A}$ and $\mathcal{B}$ are zero-state equivalent.

On the other hand, $\mathcal{A}$ and $\mathcal{B}$ are not equivalent. To demonstrate this, it is sufficient to exhibit a state of $\mathcal{B}$ and an input u such that the

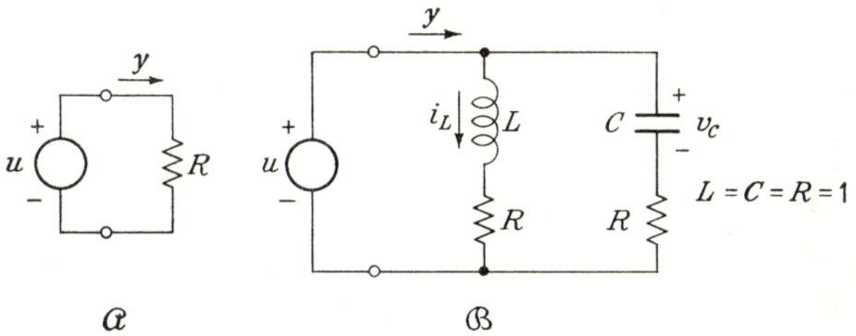

Fig. 2.9.1 $\mathcal{A}$ and $\mathcal{B}$ are zero-state equivalent although $\mathcal{A}$ and $\mathcal{B}$ are not equivalent.

response of $\mathcal{B}$ to u starting in this state is not identical with the response of $\mathcal{A}$ to u starting in state (1). Take u to be the zero input and let the initial state of $\mathcal{B}$ be (0,−1). By using elementary circuit analysis techniques, the corresponding output current is readily found to be given by

$$y(t) = e^{-(t-t_0)} \qquad t \geq t_0$$

whereas in the case of $\mathcal{A}$ the corresponding output current is 0. This proves that $\mathcal{A}$ and $\mathcal{B}$ are not equivalent.

4 *Comment* The above example illustrates a general property of linear time-invariant systems which will be stated in Chap. *3*, Sec. *6*. Although we have not yet given formal definitions of some of the terms in which this property is formulated, it will be helpful to have it stated at this point, without taking the time and space to define formally the terms with which the reader is certain to have prior familiarity.

Briefly, let $\mathcal{A}$ and $\mathcal{B}$ denote linear time-invariant systems with transfer functions $H_a(s)$ and $H_b(s)$, respectively, where $s \triangleq$ complex frequency. Then $\mathcal{A}$ and $\mathcal{B}$ are zero-state equivalent if and only if $H_a(s) = H_b(s)$ for all s. For example, in Example *3* the transfer function of $\mathcal{A}$ is 1 and that of $\mathcal{B}$ is s/s, which is also 1. Thus, $\mathcal{A}$ and $\mathcal{B}$ are zero-state equivalent. It is important to note, however, that if $\mathcal{A}$ and $\mathcal{B}$ have identical transfer functions, it does not necessarily follow that $\mathcal{A}$ and $\mathcal{B}$ are equivalent in the sense of Definition *2.5.8*.

5 **Remark** Although zero-state equivalence of $\mathcal{A}$ and $\mathcal{B}$ does not imply equivalence of $\mathcal{A}$ and $\mathcal{B}$, it does imply more than just equivalence between a state of $\mathcal{A}$ and a state of $\mathcal{B}$. Specifically, it implies that $\mathcal{A}$ and $\mathcal{B}$ are conditionally equivalent in the sense of Definition *2.7.5*. Furthermore, by Corollary *2.7.4*, if $\mathcal{A} \doteq \mathcal{B}$ and all the states in the state spaces of $\mathcal{A}$ and $\mathcal{B}$ are reachable from $\boldsymbol{\theta}_a$ and $\boldsymbol{\theta}_b$, respectively, then $\mathcal{A} \equiv \mathcal{B}$.

6 *Example* The systems $\mathcal{A}$ and $\mathcal{B}$ of Example *2.5.15* have the property that all states in $\Sigma_a = \Sigma_b = \mathcal{R}^2$ are reachable from the zero state $\boldsymbol{\theta}_a = \boldsymbol{\theta}_b = (0,0)$. (This is established in a general way in Chapter *11*.) Note that in this case the verification of the equivalence $\mathcal{A} \equiv \mathcal{B}$ involves in effect the demonstration of zero-state equivalence of $\mathcal{A}$ and $\mathcal{B}$, which is expressed by *2.5.18*.

The concepts of zero-state as well as zero-input equivalence will find extensive use in later chapters. For the present, we turn to the notions of inverse and converse systems. These notions are discussed in general terms in the following section, leaving specific applications involving linear systems to Chap. *4*.

10 *Inverse and converse systems*

The concepts of inverse and converse systems are closely related to one another, with the latter being somewhat more general. We begin with the

1 **Definition** $\mathcal{A}$ and $\mathcal{B}$ are *converse systems* if every input-output pair $(\mathbf{u},\mathbf{y})$ for $\mathcal{A}$ has the property that $(\mathbf{y},\mathbf{u})$ (with $\mathbf{y}$ as input and $\mathbf{u}$ as output) is an input-output pair for $\mathcal{B}$, and vice versa.

2 *Example* Let $\mathcal{A}$ (an integrator) and $\mathcal{B}$ (a differentiator) be characterized by the input-output relations

3 $\mathcal{A}$: $$\frac{dy}{dt} = u$$

4 $\mathcal{B}$: $$y = \frac{du}{dt}$$

Clearly, $\mathcal{A}$ and $\mathcal{B}$ are converse systems. For example, $(2t,t^2)$, $t \geq t_0$, is an input-output pair for $\mathcal{A}$ and $(t^2,2t)$, $t \geq t_0$, is an input-output pair for $\mathcal{B}$; more generally, for any v in the input function space of $\mathcal{A}$, $(\dot{v},v)$ is an input-output pair for $\mathcal{A}$ and $(v,\dot{v})$ is an input-output pair for $\mathcal{B}$. Note that if $\mathcal{A}$ and $\mathcal{B}$ are converse systems, then the input function space of $\mathcal{A}$ is the output function space of $\mathcal{B}$ and, conversely, the input function space of $\mathcal{B}$ is the output function space of $\mathcal{A}$.

5 *Comment* In effect, converse systems can be regarded as opposite orientations of an object. Thus, with reference to Example *2*, let $\mathcal{C}$ be a nonoriented object characterized by the terminal relation

$$\frac{dv_2}{dt} = v_1$$

Then $\mathcal{A}$ is the result of orienting $\mathcal{C}$ by designating v_1 as the input and v_2 as the output, while $\mathcal{B}$ results from the opposite designation of v_2 as the input and v_1 as the output.

Another significant point which comes to light in Example *2* is that the state of $\mathcal{A}$ at time t has the same meaning as the state of $\mathcal{B}$ at time t, viz., in both cases $s(t)$ is $v_2(t)$. This is a special instance of a general property of linear differential systems which will be established in Chap. *4* (*4.4.27*), to the effect that the state of such systems is independent of the orientation. ◁

Looked upon from another point of view, $\mathcal{A}$ and $\mathcal{B}$ are converse systems if, roughly speaking, $\mathcal{A}$ is capable of "undoing" whatever $\mathcal{B}$ is capable of doing to its input, and vice versa. This idea is embodied in a more specific form in the concept of inverse systems which is defined below.

6 **Definition** Let $\mathcal{A}$ and $\mathcal{B}$ be characterized by input-output-state relations of the form

7 $\mathcal{A}$: $$\mathbf{y} = \bar{\mathbf{A}}(\boldsymbol{\alpha};\mathbf{u}) \qquad \boldsymbol{\alpha} \in \Sigma_a$$

8 $\mathcal{B}$: $$\mathbf{w} = \bar{\mathbf{B}}(\boldsymbol{\beta};\mathbf{v}) \qquad \boldsymbol{\beta} \in \Sigma_b$$

where $\mathbf{u}$ and $\mathbf{v}$ denote input segments to $\mathcal{A}$ and $\mathcal{B}$, respectively, and $\mathbf{y}$ and $\mathbf{w}$ represent the corresponding output segments. (The input function space of $\mathcal{A}$ is assumed to be the output function space of $\mathcal{B}$, and conversely.) Then $\mathcal{A}$ and $\mathcal{B}$ are *inverse systems* (or $\mathcal{A}$ *is inverse to* $\mathcal{B}$ or $\mathcal{A}$ *is inverse of* $\mathcal{B}$) if and only if to every state $\boldsymbol{\alpha}$ of $\mathcal{A}$ there is a state $\boldsymbol{\beta}_\alpha$ of $\mathcal{B}$ such that

9 $$\bar{\mathbf{B}}(\boldsymbol{\beta}_\alpha;\bar{\mathbf{A}}(\boldsymbol{\alpha};\mathbf{u})) = \mathbf{u} \qquad \forall \mathbf{u}$$

and conversely to every state $\boldsymbol{\beta}$ of $\mathcal{B}$ there corresponds a state $\boldsymbol{\alpha}_\beta$ of $\mathcal{A}$ such that

10 $$\bar{\mathbf{A}}(\boldsymbol{\alpha}_\beta;\bar{\mathbf{B}}(\boldsymbol{\beta};\mathbf{v})) = \mathbf{v} \qquad \forall \mathbf{v}$$

If $\mathcal{A}$ is inverse to $\mathcal{B}$, it is denoted by $\mathcal{B}^{-1}$, and conversely, $\mathcal{B}$ is denoted by $\mathcal{A}^{-1}$. Correspondingly, $\boldsymbol{\beta}_\alpha$ is denoted by $\boldsymbol{\alpha}^{-1}$. $\mathcal{A}$ will be said to be *invertible* if it has an inverse.

Note that it follows from this definition that if *9* and *10* are satisfied with the states $\boldsymbol{\alpha}$, $\boldsymbol{\beta}_\alpha$ and $\boldsymbol{\beta}$, $\boldsymbol{\alpha}_\beta$, respectively, then they are also satisfied with the states $\boldsymbol{\alpha}_\beta$, $\boldsymbol{\beta}$ and $\boldsymbol{\beta}_\alpha$, $\boldsymbol{\alpha}$.

11 *Example* Consider two systems $\mathcal{A}$ and $\mathcal{B}$ which are characterized by the input-output relations

12 $\mathcal{A}$: $$\frac{dy}{dt} + y = u$$

13 $\mathcal{B}$: $$w = \frac{dv}{dt} + v$$

Note that in the case of $\mathcal{A}$, the scalar $y(t)$, representing the output at time t, qualifies as the state of $\mathcal{A}$ at time t, that is, $s(t) = y(t)$. The state space of $\mathcal{A}$ is the real line $\Sigma_a = \mathcal{R}^1 = (-\infty, \infty)$.

On the other hand, in the case of $\mathcal{B}$, it is the input at time t, $v(t)$, that qualifies as $s(t)$ (see *1.7.10*). This is consistent with Comment *5*.

The input-output-state relations for $\mathcal{A}$ and $\mathcal{B}$ read (see *1.9.10* and *1.7.10*)

14 $\mathcal{A}$: $$y(t) = y(t_0-)e^{-(t-t_0)} + \int_{t_0}^{t} e^{-(t-\xi)}u(\xi)\,d\xi \qquad t \geq t_0$$

15 $\mathcal{B}$: $$w(t) = \frac{dv}{dt} + v - v(t_0-)\delta(t-t_0) \qquad t \geq t_0$$

Denote $y(t_0-)$ and $v(t_0-)$ by α and β, respectively. Then it is easy to verify (see Appendix *A*) that

16 $$B(\beta;A(\alpha;u)) = u(t) + (\alpha - \beta)\delta(t - t_0)$$

and

17 $$A(\alpha;B(\beta;u)) = v(t) + \alpha - \beta$$

Thus, *9* and *10* are satisfied with $\alpha = \beta$. This establishes that $\mathcal{A}$ and $\mathcal{B}$ are inverse systems. More generally, this same result can be established for any pair of systems which are obtained through opposite orientation of a linear differential system (see *4.4.44*). Thus, the systems $\mathcal{A}$ and $\mathcal{B}$ of Example *11* illustrate merely a special instance of this general property. [Note that *16* and *17* express the outputs of the free products (*1.10.10*) $\mathcal{B}\mathcal{A}$ and $\mathcal{A}\mathcal{B}$, respectively.]

What is the relation between the notions of converse and inverse systems? In the case of Example *11*, $\mathcal{A}$ and $\mathcal{B}$ are both converse and inverse systems. More generally, it is evident from *6* that if $\mathcal{A}$ and $\mathcal{B}$ are inverse systems, then they are also converse systems. The reverse implication, however, is not necessarily true. Specifically, from inspection of Definition *1* it is clear that to say that $\mathcal{A}$ and $\mathcal{B}$ are converse systems is the same as saying that to every input $\mathbf{u}$ and every state $\boldsymbol{\alpha}$ of $\mathcal{A}$ there corresponds a state $\boldsymbol{\beta}_{\alpha,\mathbf{u}}$ of $\mathcal{B}$ such that *9* holds, and conversely. Thus, the crux of the difference between Definition *1* and Definition *6* is that in the former $\boldsymbol{\beta}$ may depend on both $\boldsymbol{\alpha}$ and $\mathbf{u}$, whereas in the latter $\boldsymbol{\beta}$ depends only on $\boldsymbol{\alpha}$. (The same statement applies to the dependence of $\boldsymbol{\alpha}$ on $\boldsymbol{\beta}$ and $\mathbf{v}$.)

The difference between the case where $\boldsymbol{\beta}$ depends on $\boldsymbol{\alpha}$ alone and the case where $\boldsymbol{\beta}$ depends on both $\boldsymbol{\alpha}$ and $\mathbf{u}$ is essentially the difference between weak equivalence and equivalence (see *2.5.4* and *2.5.8*). As will be established in Chap. *3*, Sec. *9*, linear differential systems have the basic property that if $\mathcal{A}$ and $\mathcal{B}$ are weakly equivalent, then $\mathcal{A}$ and $\mathcal{B}$ are equivalent. Thus if $\mathcal{A}$ and $\mathcal{B}$ are linear differential systems, then the statement "$\mathcal{A}$ and $\mathcal{B}$ are inverse systems" implies and *is implied by* the statement "$\mathcal{A}$ and $\mathcal{B}$ are converse systems."

What can be said about inverse systems which are connected in tandem, as in Fig. *2.10.1*, or in reverse order? Here $\mathcal{A}^{-1}$ denotes the inverse of $\mathcal{A}$ and $\boldsymbol{\alpha}^{-1}$ represents the state $\boldsymbol{\beta}_{\alpha}$ of Definition *6*.

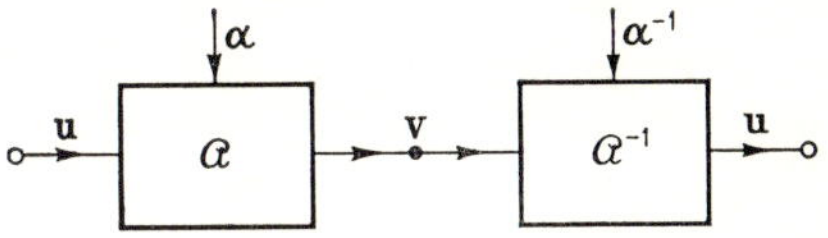

Fig. 2.10.1 Illustration of the notion of inverse.

If we restrict the initial states of the product $\mathcal{A}\mathcal{A}^{-1}$ and $\mathcal{A}^{-1}\mathcal{A}$ to states of the form $(\boldsymbol{\alpha},\boldsymbol{\alpha}^{-1})$ (note that such states form a closed set in the product space $\Sigma_a \times \Sigma_b$, where Σ_b denotes the state space of $\mathcal{A}^{-1}$), then Definition *6* implies that the systems $\mathcal{A}\mathcal{A}^{-1}$ and $\mathcal{A}^{-1}\mathcal{A}$ are equivalent to a unitor (*1.9.8*), that is, $\mathcal{A}\mathcal{A}^{-1} \equiv \mathcal{A}^{-1}\mathcal{A} \equiv I$. Conversely, if $\mathcal{A}\mathcal{A}^{-1} \equiv \mathcal{A}^{-1}\mathcal{A} \equiv I$, then it is clear that $\mathcal{A}$ and $\mathcal{A}^{-1}$ satisfy the conditions of Definition *6*. Thus, we can make the

18 **Assertion** $\{\mathcal{A}$ and $\mathcal{A}^{-1}$ are inverse systems$\} \Leftrightarrow \{\mathcal{A}\mathcal{A}^{-1} \overset{*}{\equiv} \mathcal{A}^{-1}\mathcal{A} \overset{*}{\equiv} I\}$ with the understanding that the initial states of $\mathcal{A}\mathcal{A}^{-1}$ and $\mathcal{A}^{-1}\mathcal{A}$ are restricted to states of the form $(\boldsymbol{\alpha},\boldsymbol{\alpha}^{-1})$, where $\boldsymbol{\alpha}$ is an arbitrary state of $\mathcal{A}$ and $\boldsymbol{\alpha}^{-1}$ is its correspondent in $\mathcal{A}^{-1}$ (see *6*). We use the symbol $\overset{*}{\equiv}$ rather than $\equiv$ in *18* to stress that the equivalence in question is conditional in the sense of *2.7.5*.

19 *Example* In Example *11*, α^{-1} is a scalar numerically equal to α. Thus, for any initial state of the form (α,α), *16* and *17* give the following:

$$\bar{A}^{-1}(\alpha^{-1};\bar{A}(\alpha;u)) = u$$
$$\bar{A}(\alpha;\bar{A}^{-1}(\alpha^{-1};v)) = v$$

which is consistent with *18*. ◁

Assertion *18* provides a partial answer to the question of what happens when $\mathcal{A}$ and $\mathcal{A}^{-1}$ are connected in tandem, in the sense that it shows that if $\mathcal{A}$ and $\mathcal{A}^{-1}$ are started in the respective states α and α^{-1}, then $\mathcal{A}\mathcal{A}^{-1}$ and $\mathcal{A}^{-1}\mathcal{A}$ are conditionally equivalent to a unitor. However, this question remains: Under what conditions is the product $\mathcal{A}\mathcal{A}^{-1}$ (or $\mathcal{A}^{-1}\mathcal{A}$) equivalent to I unconditionally?

It is easy to construct examples of systems which have this property in one direction only, i.e., are such that either $\mathcal{A}\mathcal{A}^{-1} \equiv I$ or $\mathcal{A}^{-1}\mathcal{A} \equiv I$ but not both simultaneously. This suggests the following

20 **Definition** $\mathcal{A}_r^{-1}$ is a *right-constrained* inverse of $\mathcal{A}$ if $\mathcal{A}_r^{-1}$ is inverse to $\mathcal{A}$ and, in addition,

21 $$\mathcal{A}\mathcal{A}_r^{-1} \equiv I$$

Similarly, $\mathcal{A}_l^{-1}$ is a *left-constrained* inverse of $\mathcal{A}$ if $\mathcal{A}_l^{-1}$ is inverse to $\mathcal{A}$ and, in addition,

22 $$\mathcal{A}_l^{-1}\mathcal{A} \equiv I$$

The adjective "constrained" serves to indicate that $\mathcal{A}_r^{-1}$, say, and $\mathcal{A}$ have the property that, when $\mathcal{A}_r^{-1}$ and $\mathcal{A}$ are connected in tandem, the constraint imposed by the interconnection (see *2.6.8*) automatically forces the states of $\mathcal{A}_r^{-1}$ and $\mathcal{A}$ to be the corresponding states α^{-1} and α. The same statement applies to $\mathcal{A}_l^{-1}$. ◁

Examples of right- and left-constrained inverses will be given in Chap. *4*, Sec. *4*. At this point, it will suffice to point out that a differentiator is a left-constrained inverse of an integrator but not vice versa. This is a special instance of a more general result which is stated in Chap. *4* (*4.4.44*).

Further properties of inverse systems

From the defining relations *9* and *10* it is a simple matter to deduce several basic properties of $\mathcal{A}^{-1}$. For convenient reference, we state them below in the form of assertions.

23 **Assertion** If $\mathcal{A}^{-1}$ exists, then it is unique (to within equivalent systems). *Proof* Let $\mathcal{A}_1^{-1}$ and $\mathcal{A}_2^{-1}$ be inverse to $\mathcal{A}$. Assume that an input $\mathbf{u}$ is applied to $\mathcal{A}_1^{-1}$ starting in state $\boldsymbol{\alpha}_1^{-1}$ and that the corresponding response is $\mathbf{v}$. By *9*, there exists a state, say $\boldsymbol{\alpha}_1$, of $\mathcal{A}$ such that the response of $\mathcal{A}$ to $\mathbf{v}$ starting in state $\boldsymbol{\alpha}_1$ is $\mathbf{u}$. Now by *10*, there exists a state, say $\boldsymbol{\alpha}_2^{-1}$, of $\mathcal{A}_2^{-1}$ such that the response of $\mathcal{A}_2^{-1}$ to $\mathbf{u}$ starting in state $\boldsymbol{\alpha}_2^{-1}$ is $\mathbf{v}$. Since this is true for all $\mathbf{u}$, it follows that to every state of $\mathcal{A}_1^{-1}$ there is an equivalent state of $\mathcal{A}_2^{-1}$. Similar argument shows that to every state of $\mathcal{A}_2^{-1}$ there is an equivalent state of $\mathcal{A}_1^{-1}$. Consequently, $\mathcal{A}_1^{-1}$ and $\mathcal{A}_2^{-1}$ are equivalent in the sense of Definition *2.5.8*.

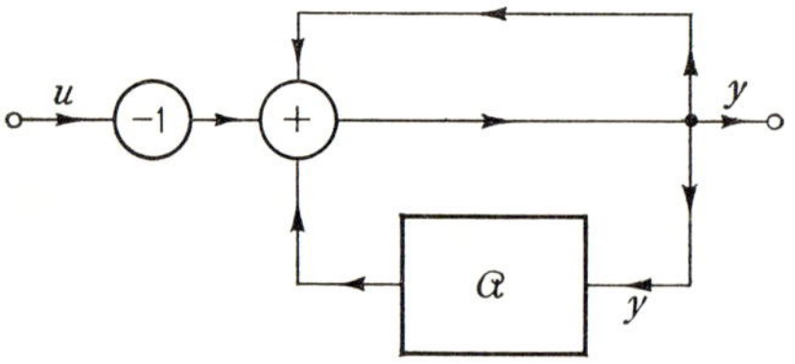

Fig. 2.10.2 Realization of the inverse of $\mathcal{A}$.

24 *Comment* It is of interest to note that the inverse of $\mathcal{A}$, if it exists, can be synthesized in the form of an interconnection of $\mathcal{A}$, an adder, and a scalor with scale factor -1, as shown in Fig. *2.10.2*. We leave it to the reader to verify that the system in question is equivalent to $\mathcal{A}^{-1}$.

25 **Assertion** If $\mathcal{A}$ is invertible and the response of $\mathcal{A}$ to an input $\mathbf{u}_1$ starting in any state $\boldsymbol{\alpha}$ is identical with the response of $\mathcal{A}$ to an input $\mathbf{u}_2$ starting in $\boldsymbol{\alpha}$, then $\mathbf{u}_1$ must be identical with $\mathbf{u}_2$. In symbols,

26
$$\{\bar{\mathbf{A}}(\boldsymbol{\alpha};\mathbf{u}_1) = \bar{\mathbf{A}}(\boldsymbol{\alpha};\mathbf{u}_2)\} \Rightarrow \{\mathbf{u}_1 = \mathbf{u}_2\} \qquad \forall\boldsymbol{\alpha}$$

for any $\mathbf{u}_1$ and $\mathbf{u}_2$ in the input segment space of $\mathcal{A}$.

This means that for each $\boldsymbol{\alpha}$ in the state space of $\mathcal{A}$, $\bar{\mathbf{A}}(\boldsymbol{\alpha};\mathbf{u})$ defines a one-one correspondence between the input segment space and the output segment space of $\mathcal{A}$.

Proof By Definition *6*, there exists a state $\boldsymbol{\alpha}^{-1}$ in $\mathcal{A}^{-1}$ such that

$$\bar{\mathbf{A}}^{-1}(\boldsymbol{\alpha}^{-1};\bar{\mathbf{A}}(\boldsymbol{\alpha};\mathbf{u}_1)) = \mathbf{u}_1 \qquad \text{and} \qquad \bar{\mathbf{A}}^{-1}(\boldsymbol{\alpha}^{-1};\bar{\mathbf{A}}(\boldsymbol{\alpha};\mathbf{u}_2)) = \mathbf{u}_2$$

and hence $\mathbf{u}_1 = \mathbf{u}_2$ by virtue of *26*.

27 **Assertion** If $\mathcal{A}$ has a unique zero state $\boldsymbol{\theta}$ and $\mathcal{A}^{-1}$ exists, then $\mathcal{A}^{-1}$ also has a unique zero state.

Proof From the definition of $\mathcal{A}^{-1}$, it follows that to the zero state $\boldsymbol{\theta}$ in $\mathcal{A}$ there corresponds a state, call it $\boldsymbol{\theta}^{-1}$, in $\mathcal{A}^{-1}$ such that

28
$$\bar{\mathbf{A}}^{-1}(\boldsymbol{\theta}^{-1};\bar{\mathbf{A}}(\boldsymbol{\theta};\mathbf{0})) = \mathbf{0}$$

Therefore $\boldsymbol{\theta}^{-1}$ is a zero state of $\mathcal{A}^{-1}$.

Next, suppose that $\mathcal{A}^{-1}$ has nonequivalent zero states, say $\boldsymbol{\theta}_1^{-1}$ and $\boldsymbol{\theta}_2^{-1}$. Then there exists a distinguishing input (see *2.2.3*), say, $\mathbf{v}$, such that

29
$$\mathbf{u}_1 = \bar{\mathbf{A}}^{-1}(\boldsymbol{\theta}_1^{-1};\mathbf{v}) \neq \bar{\mathbf{A}}^{-1}(\boldsymbol{\theta}_2^{-1};\mathbf{v}) \triangleq \mathbf{u}_2$$

But, by the definition of $\mathcal{A}^{-1}$ we must have

$$\bar{\mathbf{A}}(\boldsymbol{\theta};\mathbf{u}_1) = \bar{\mathbf{A}}(\boldsymbol{\theta};\mathbf{u}_2) = \mathbf{v}$$

which by Assertion *25* implies that $\mathbf{u}_1 = \mathbf{u}_2$. This contradicts *29* and hence proves the assertion.

30 **Assertion** If $\mathcal{A}^{-1}$ and $\mathcal{B}^{-1}$ exist, then

$$\{\mathcal{A} \equiv \mathcal{B}\} \Rightarrow \{\mathcal{A}^{-1} \equiv \mathcal{B}^{-1}\}$$

Proof Let $\mathbf{u}$ be the response of $\mathcal{A}^{-1}$ to $\mathbf{y}$ starting in state $\boldsymbol{\alpha}^{-1}$. By *10*, there exists a state $\boldsymbol{\alpha}$ of $\mathcal{A}$ such that the response of $\mathcal{A}$ to $\mathbf{u}$ starting in state $\boldsymbol{\alpha}$ is $\mathbf{y}$. Since $\mathcal{A} \equiv \mathcal{B}$, there exists a state $\boldsymbol{\beta}$ of $\mathcal{B}$ such that the response of $\mathcal{B}$ to $\mathbf{u}$ starting in state $\boldsymbol{\beta}$ is $\mathbf{y}$. Furthermore, by *9* there exists a state $\boldsymbol{\beta}^{-1}$ of $\mathcal{B}^{-1}$ such that the response of $\mathcal{B}^{-1}$ to $\mathbf{y}$ starting in state $\boldsymbol{\beta}^{-1}$ is $\mathbf{u}$. Consequently, to every state $\boldsymbol{\alpha}^{-1}$ of $\mathcal{A}$ there corresponds an equivalent state $\boldsymbol{\beta}^{-1}$ of $\mathcal{B}^{-1}$. By a similar argument, to every state $\boldsymbol{\beta}^{-1}$ of $\mathcal{B}^{-1}$ there corresponds an equivalent state $\boldsymbol{\alpha}^{-1}$ of $\mathcal{A}^{-1}$, and hence by Definition *2.5.8* $\mathcal{A}^{-1}$ and $\mathcal{B}^{-1}$ are equivalent systems.

31 **Assertion** If $\mathcal{A}^{-1}$ and $\mathcal{B}^{-1}$ exist and the product $\mathcal{A}\mathcal{B}$ is defined (see *1.10.23*), then

32 $$(\mathcal{A}\mathcal{B})^{-1} \equiv \mathcal{B}^{-1}\mathcal{A}^{-1}$$

Proof Exercise for the reader. (*Caution:* Note that the state space of $\mathcal{A}\mathcal{B}$ is not, in general, the product of the state spaces of $\mathcal{A}$ and $\mathcal{B}$.)

33 *Exercise* Construct a nontrivial example of a system which is its own inverse.

34 *Exercise* Show that $(\mathcal{A}^{-1})^{-1} \equiv \mathcal{A}$. ◁

In this as well as the preceding sections in this chapter we have presented and discussed in general terms such basic concepts as state equivalence, system equivalence, and inverse systems. Although we have illustrated the meaning of these notions with examples, their generality and abstractness may have made it somewhat difficult for the reader to concretize them in his own mind in terms of objects, devices, and concepts with which he is familiar. This is what should be expected at the present stage of our exposition. It is through the frequent use of the concepts in question in later chapters that the reader will eventually gain a clear understanding of the motivation behind them and the ways in which they can be applied to the analysis of linear systems.

REFERENCES AND SUGGESTED READING

1 Church, A.: "Introduction to Mathematical Logic," Princeton University Press, Princeton, N.J., 1956.

2 Kleene, S. C.: "Introduction to Metamathematics," D. Van Nostrand Company, Inc., Princeton, N.J., 1952.

3 Moore, E. F.: Gedanken Experiments on Sequential Machines, in "Automata Studies," Princeton University Press, Princeton, N.J., 1956.

4 Huffman, D.: The Synthesis of Sequential Switching Circuits, *J. Franklin Inst.*, vol. 257, pp. 161–190, 1954.

5 Birkhoff, G., and G. Rota: "Ordinary Differential Equations," Ginn and Company, Boston, 1962.

6 Coddington, E. A., and N. Levinson: "Theory of Ordinary Differential Equations," McGraw-Hill Book Company, Inc., New York, 1955.

7 Bellman, R.: "Adaptive Control Processes," Princeton University Press, Princeton, N.J., 1961.

8 Nemytskii, V. V., and V. V. Stepanov: "Qualitative Theory of Differential Equations," Princeton University Press, Princeton, N.J., 1960.

9 Birkhoff, G. D.: Dynamical Systems, *Amer. Math. Soc. Colloq. Publ.*, vol. 9, 1927.

10 Kalman, R. E.: On the General Theory of Control Systems, *Proc. First Intl. Cong. Automatic Cont., Moscow*, 1960, vol. 1, pp. 481–493, 1961, Butterworth & Co. (Publishers), Ltd., London.

Linearity and time invariance 3

1 Introduction

Like everything else, systems can be classified in a variety of ways according to their different properties. However, the more important modes of classification in system theory are dichotomies in the sense that they involve but two categories, say C and C', with C comprising all systems which have a specified property P and C' being the complement of C relative to the class of all systems. For example, in the case of the classification into linear and nonlinear systems, P is the property of *linearity* (which will be defined later in the chapter), C is the class of all *linear* systems, and C' is the class of all systems which are not linear, that is, are *nonlinear*.[1]

In the sequel, we shall focus our attention on two dichotomous modes of classification which play a particularly important role in system theory. These are summarized in the following table. As a

P	C	C'
Linearity	Linear systems	Nonlinear systems
Time invariance	Time-invariant systems	Time-varying (time-variant) systems

preliminary to defining the notions of linearity and time invariance, we begin with a brief

[1] In the literature the term "nonlinear" is sometimes used in such a broad sense that the class of nonlinear systems becomes virtually identical with the class of all systems. In this sense, the class of linear systems is merely a subclass of the class of nonlinear systems.

Recapitulation of terminology and notation

Symbols such as $\mathcal{A}$, $\mathcal{B}$, and $\mathcal{C}$ denote oriented systems. When dealing with a single system $\mathcal{A}$, its input, output, state, and state space are denoted by $\mathbf{u}$, $\mathbf{y}$, $\boldsymbol{\alpha}$, and Σ, respectively.[1] The symbol $\mathbf{s}(t)$ denotes the state of $\mathcal{A}$ at time t, and $\mathcal{A}$ is assumed to be characterized by an input-output-state relation

$$\mathbf{y}(t) = \mathbf{A}(\boldsymbol{\alpha};\mathbf{u})$$

which expresses the output at time t as a function of the initial state $\mathbf{s}(t_0) = \boldsymbol{\alpha}$ and the input segment $\mathbf{u} \triangleq \mathbf{u}_{(t_0,t]}$. In the above equation, $\boldsymbol{\alpha}$ ranges over the state space Σ and $\mathbf{u}$ ranges over the input segment space $R[\mathbf{u}]$ (see Remark *1.4.3* and Definitions *1.4.17* and *1.4.26*). When the left member of the input-output-state relation is the response segment $\mathbf{y} \triangleq \mathbf{y}_{(t_0,t]}$ rather than $\mathbf{y}(t)$, the input-output-state relation is written as

$$\mathbf{y} = \bar{\mathbf{A}}(\boldsymbol{\alpha};\mathbf{u})$$

It is understood that $\mathbf{u}$ and $\mathbf{y}$ are defined on the same observation interval and that $\boldsymbol{\alpha}$ is the state of $\mathcal{A}$ at the beginning of $\mathbf{u}$. If $\mathbf{u}$ vanishes for all $t \geq t_0$, it is said to be a zero input and is denoted by $\mathbf{0}$. The same thing applies to $\mathbf{y}$ and, more generally, to any time function $\mathbf{f}$.

If $\bar{\mathbf{A}}(\boldsymbol{\alpha};\mathbf{0}) = \mathbf{0}$, then $\boldsymbol{\alpha}$ is said to be a zero state of $\mathcal{A}$. Unless stated to the contrary, it is assumed that $\mathcal{A}$ has a unique zero state which is denoted by $\boldsymbol{\theta}$. The response segment of $\mathcal{A}$ to $\mathbf{u}$ when $\mathcal{A}$ is initially in its zero state is referred to as the zero-state response of $\mathcal{A}$ and is denoted by $\bar{\mathbf{A}}(\mathbf{u})$. Thus

$$\bar{\mathbf{A}}(\mathbf{u}) \triangleq \bar{\mathbf{A}}(\boldsymbol{\theta};\mathbf{u})$$

Note that the symbol $\mathbf{A}(\mathbf{u})$ represents $\mathbf{y}(t)$, the zero-state response of $\mathcal{A}$ at time t, whereas $\bar{\mathbf{A}}(\mathbf{u})$ denotes the zero-state response segment $\mathbf{y}_{(t_0,t]}$.

If $\boldsymbol{\alpha}$ is a state of $\mathcal{A}$ and $\boldsymbol{\beta}$ is a state of $\mathcal{B}$, then $\boldsymbol{\alpha}$ and $\boldsymbol{\beta}$ are equivalent states, written $\boldsymbol{\alpha} \simeq \boldsymbol{\beta}$, if and only if $\bar{\mathbf{A}}(\boldsymbol{\alpha};\mathbf{u}) = \bar{\mathbf{A}}(\boldsymbol{\beta};\mathbf{u})$ for all $\mathbf{u}$. $\mathcal{A}$ and $\mathcal{B}$ are equivalent systems, written $\mathcal{A} \equiv \mathcal{B}$, if to every state in the state space of $\mathcal{A}$ there is an equivalent state in the state space of $\mathcal{B}$ and vice versa. (It is understood that $\mathcal{A}$ and $\mathcal{B}$ have the same input function space.) $\mathcal{A}$ and $\mathcal{B}$ are said to be zero-state equivalent, written as $\mathcal{A} \doteq \mathcal{B}$, if the zero state of $\mathcal{A}$ is equivalent to the zero state of $\mathcal{B}$. $\mathcal{A}$ and $\mathcal{B}$ are zero-input equivalent if for every state $\boldsymbol{\alpha}$ of $\mathcal{A}$ there is a state $\boldsymbol{\beta}$ of $\mathcal{B}$, and vice versa, such that the zero-input response of $\mathcal{A}$ starting

[1] At a later point in this chapter where we begin to employ the Laplace transformation, the state of $\mathcal{A}$ will be denoted by $\mathbf{x}$ in order to avoid confusion with the complex variable s.

in state α is identical (for $t \geq t_0$) with the zero-input response of $\mathcal{B}$ starting in state β.

Unless stated to the contrary, the variable t is assumed to range over the real line $(-\infty, \infty)$. The initial time is denoted by t_0 or, when necessary, by t_0-. The input $\mathbf{u}(t)$ and output $\mathbf{y}(t)$ are usually, but not always, assumed to be vector-valued and real. When not defined in explicit terms, the input segment space $R[\mathbf{u}]$ is generally understood to be the space of time functions on $(t_0,t]$ which are finitely differentiable in the distribution sense (see *1.3.7*). Statements such as "holds for all $\mathbf{u}$" should be interpreted to mean "holds for all $\mathbf{u}$ in $R[\mathbf{u}]$, with both t_0 and t ranging over the real line." It is generally understood that the states of $\mathcal{A}$ have the following uniform reachability property: If there exists an input segment $\mathbf{u}$ which takes a state α into a state α^*, then for each t_1 there is a t_0 and an input segment $\mathbf{v}_{(t_0,t_1]}$ which takes α $(= \mathbf{s}(t_0))$ into α^* $(= \mathbf{s}(t_1))$; and for each t_0 there is a t_1 and an input segment $\mathbf{w}_{(t_0,t_1]}$ which takes α into α^*.

2 *Time invariance*

For our purposes, it will be convenient to begin our discussion with the concept of *time invariance*, which was briefly mentioned in *1.7.6* and which we shall need in our exposition of linearity and related properties in subsequent sections.

Roughly speaking, a system $\mathcal{A}$ is *time-invariant* if its characteristics do not change with time. For example, a network of fixed resistors, capacitors, and inductors is a time-invariant system. So is a system characterized by the input-output relation

1 $$y(t) = u(t) + u^2(t)$$

On the other hand, the system characterized by the input-output relation

$$y(t) = tu(t) + u^2(t)$$

is not time-invariant (i.e., is time-varying). Any system which undergoes aging (slow variation in characteristics) is, strictly speaking, a time-varying system. In fact, almost all physical systems are time-varying. Thus, the notion of a time-invariant system is essentially an idealization of a physical system whose characteristics change very slowly in relation to variations in the input.

We proceed to cast the rough definition given above into a more precise form. As a preliminary, we digress briefly to consider the notion of a translation operator.

Translation operators

2 **Definition** A *translation operator*[1] $\mathbf{T}_\delta$ is a system characterized by the input-output relation

3 $$\bar{\mathbf{T}}_\delta(\mathbf{u}_{(t_0,t]}) = \mathbf{u}_{(t_0+\delta,t+\delta]} \qquad \forall\delta\forall\mathbf{u}$$

Essentially, the action of $\mathbf{T}_\delta$ on its input $\mathbf{u}$ consists in translating $\mathbf{u}$ by a fixed amount δ along the time axis (see Fig. *3.2.1*). We write symbolically

4 $$\mathbf{y}(t) = \mathbf{T}_\delta(\mathbf{u})$$

or

5 $$\mathbf{y} = \bar{\mathbf{T}}_\delta(\mathbf{u})$$

where $\bar{\mathbf{T}}_\delta(\mathbf{u})$ denotes the response segment $\mathbf{y}_{(t_0,t]}$ and $\mathbf{u}$ is the input segment $\mathbf{u}_{(t_0,t]}$. Note that a positive shift δ corresponds to a *delay*

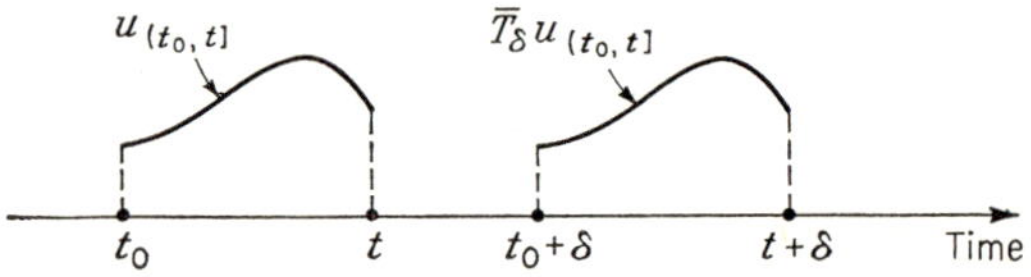

Fig. 3.2.1 $\bar{T}_\delta u_{(t_0,t]}$ denotes the translate of the segment $u_{(t_0,t]}$ by amount $\delta > 0$.

of δ units of time, whereas a negative shift δ corresponds to an *advance* of $-\delta$ units. In particular, a translation operator $\mathbf{T}_\delta$ in which δ is positive and u is scalar-valued, corresponds to a delayor (*1.7.25*) starting in its zero state. For $\delta = 0$, $\mathbf{T}_\delta$ reduces to a unitor, that is, $\mathbf{T}_0 = \mathbf{I}$ (see *1.9.8*).

From the definition of $\mathbf{T}_\delta$, it follows at once that $\mathbf{T}_\delta$ has the properties expressed by the following relations.

6 $$\bar{\mathbf{T}}_\delta(k\mathbf{u}) = k\bar{\mathbf{T}}_\delta(\mathbf{u}) \qquad \forall\delta\forall\mathbf{u} \text{ and all real constants } k$$

where $\mathbf{u}$ denotes a time function in the input function space $R[\mathbf{u}]$. This equation expresses the property of *homogeneity* defined in *3.3.5*.

7 $$\bar{\mathbf{T}}_\delta(\mathbf{u}_1 + \mathbf{u}_2) = \bar{\mathbf{T}}_\delta(\mathbf{u}_1) + \bar{\mathbf{T}}_\delta(\mathbf{u}_2) \qquad \forall\delta \text{ and all } \mathbf{u}_1,\mathbf{u}_2 \text{ in } R[\mathbf{u}]$$

where $\mathbf{u}_1$ and $\mathbf{u}_2$ denote time functions in the input function space

[1] Here and elsewhere in this book the term "operator" is usually, but not always, employed to describe a system whose only possible initial state is its zero state. We write $\mathbf{T}_\delta$ when we wish to emphasize that $\mathbf{T}_\delta$ is an operator, and, correspondingly, we write T_δ when we do not make the assumption that the initial state is restricted to be the zero state.

$R[\mathbf{u}]$. Equation *7* expresses the property of *additivity*, which is defined in *3.3.11*.

8 $$\mathbf{T}_{\delta_1}\mathbf{T}_{\delta_2} \doteq \mathbf{T}_{\delta_2}\mathbf{T}_{\delta_1} \doteq \mathbf{T}_{\delta_1+\delta_2} \qquad \forall\delta_1\forall\delta_2$$

9 $$\mathbf{T}_\delta\mathbf{T}_{-\delta} \doteq \mathbf{T}_0 = \mathbf{I} \qquad \forall\delta$$

These relations express the commutativity of translation operators. In mathematical terms, *8* and *9* indicate that the $\mathbf{T}_\delta, -\infty < \delta < \infty$, form a commutative group under multiplication, with $\mathbf{T}_0$ (the unitor) playing the role of the unit element.

10 *Comment* Note that in stating the above properties of translation operators we are tacitly assuming that the input function space $R[\mathbf{u}]$ has the following three characteristics: (I) if $\mathbf{u}$ belongs to $R[\mathbf{u}]$, so does every translate of $\mathbf{u}$; (II) if $\mathbf{u}$ belongs to $R[\mathbf{u}]$, then so does $k\mathbf{u}$, where k is an arbitrary real constant; and (III) if $\mathbf{u}_1$ and $\mathbf{u}_2$ belong to $R[\mathbf{u}]$, then so does their sum $\mathbf{u}_1 + \mathbf{u}_2$. The last two assumptions imply that $R[\mathbf{u}]$ is a linear vector space (see Appendix *C.2.1*). ◁

This concludes our brief digression on translation operators.

Zero-state and zero-input time invariance

As a preliminary to formulating a general definition of time invariance, it will be helpful to consider two particular forms of it, namely, *zero-state time invariance* and *zero-input time invariance.*

As these terms imply, zero-state time invariance is a form of time invariance exhibited when the system is started in its zero state. Correspondingly, zero-input time invariance is the form of time invariance associated with the zero-input response of the system.

11 *Comment* It is customary to call a system "time-invariant" if it is zero-state time-invariant in the sense defined below. The conventional definition of time invariance is not adequate for our purposes because it makes no provision for an initial state other than the zero state. Furthermore, it is easy to construct examples of time-varying systems (see Examples *33* and *38*) which are zero-state time-invariant and hence would be misclassified as time-invariant systems under the conventional definition of time invariance. ◁

The definition of zero-state time invariance is worded as follows.

12 **Definition** A system $\mathfrak{A}$ is *zero-state time-invariant* if, for all inputs $\mathbf{u}$ and all time shifts δ, the zero-state response of $\mathfrak{A}$ to $\mathbf{u} \triangleq \mathbf{u}_{(t_0,t]}$ is a translate (by amount $-\delta$ along the t axis) of its zero-state response to the translated input $\bar{\mathbf{T}}_\delta(\mathbf{u})$.

In symbols,

13 $$\{\mathfrak{A} \text{ is } \textit{zero-state time invariant}\} \Leftrightarrow \{\bar{\mathbf{A}}(\bar{\mathbf{T}}_\delta(\mathbf{u})) = \bar{\mathbf{T}}_\delta(\bar{\mathbf{A}}(\mathbf{u}))\} \qquad \forall\delta\forall\mathbf{u}$$

In other words, $\mathcal{A}$ is zero-state time-invariant if and only if the shape of the zero-state response of $\mathcal{A}$ to any input $\mathbf{u}$ depends only on the shape of $\mathbf{u}$ and not on the time of application of $\mathbf{u}$ (that is, t_0). ◁

Definition *12* can be put into a somewhat more compact form by noting that the zero-state equivalence

$$\mathcal{A}T_\delta \doteq T_\delta\mathcal{A}$$

implies and is implied by

$$\bar{\mathbf{A}}(\bar{\mathbf{T}}_\delta(\mathbf{u})) = \bar{\mathbf{T}}_\delta(\bar{\mathbf{A}}(\mathbf{u})) \qquad \forall \mathbf{u}$$

Thus, *13* may be replaced by

14 $$\{\mathcal{A} \text{ is } \textit{zero-state time-invariant}\} \Leftrightarrow \{\mathcal{A}T_\delta \doteq T_\delta\mathcal{A}\} \qquad \forall \delta$$

the right-hand side of which signifies that $\mathcal{A}$ commutes with all translation operators $\mathbf{T}_\delta$. Note that it does not matter whether one writes $\mathbf{T}_\delta$ or T_δ in equations in which zero-state equivalence is involved.

Actually, in order to conclude that $\mathcal{A}$ is zero-state time-invariant, it suffices to verify that *14* holds for all δ in any finite closed interval. This follows from the

15 **Assertion** If the equivalence $\mathcal{A}T_\delta \doteq T_\delta\mathcal{A}$ holds for all real δ in some finite closed interval $[a,b]$, then it holds for all real values of δ.

Proof Let δ_1 and δ_2 be arbitrary time shifts in the interval $[a,b]$. Then, on postmultiplying both sides of the equivalence[1]

$$\mathcal{A}T_{\delta_1} \doteq T_{\delta_1}\mathcal{A}$$

by T_{δ_2} and making use of the equivalence $\mathcal{A}T_{\delta_2} \doteq T_{\delta_2}\mathcal{A}$, we obtain (see *8*)

16 $$\mathcal{A}T_{\delta_1+\delta_2} \doteq T_{\delta_1+\delta_2}\mathcal{A}$$

which shows that if $\mathcal{A}$ commutes with T_{δ_1} and T_{δ_2}, then it also commutes with $T_{\delta_1+\delta_2}$.

Next, starting with

17 $$\mathcal{A}T_{\delta_2} \doteq T_{\delta_2}\mathcal{A}$$

and premultiplying both sides by $T_{-\delta_2}$, we get

18 $$T_{-\delta_2}\mathcal{A}T_{\delta_2} \doteq \mathcal{A}$$

which yields

$$T_{\delta_1-\delta_2}\mathcal{A} \doteq \mathcal{A}T_{\delta_1-\delta_2}$$

[1] Note that pre- and postmultiplication is justified by *2.7.6*. Note also that we are making tacit use of the associative and substitution principles *1.10.34* and *1.10.35*.

upon postmultiplication of both sides of *18* by $T_{-\delta_2}$, premultiplication of the resulting expression by T_{δ_1}, and making use of *17*. Consequently, if $\mathfrak{A}$ commutes with T_{δ_1} and T_{δ_2}, $\delta_1, \delta_2 \in [a,b]$, then it also commutes with $T_{\delta_1 \pm \delta_2}$. This implies that $\mathfrak{A}$ commutes with all T_δ in which $|\delta| \leq |b - a|$, which by virtue of *16* implies that $\mathfrak{A}$ commutes with all T_δ. ◁

Since there will be numerous examples of zero-state time-invariant systems in later chapters, we shall confine ourselves at this point to two simple illustrations.

19 *Example* Consider the system of Example *2.3.7*, which is characterized by the input-output-state relation

$$y(t) = e^{-(t-t_0)}y(t_0-) + \int_{t_0}^{t} e^{-(t-\xi)}u(\xi)\, d\xi \qquad t \geq t_0$$

in which $y(t_0-)$ represents the state at time t_0-. The zero-state response of this system is given by

20 $$y(t) = \int_{t_0}^{t} e^{-(t-\xi)}u(\xi)\, d\xi \qquad t \geq t_0$$

Now, if the input is shifted by δ sec in time, the value of the output at time $t + \delta$ becomes

21 $$\int_{t_0+\delta}^{t+\delta} e^{-(t-\xi+\delta)}u(\xi - \delta)\, d\xi$$

which is identical with the expression obtained by changing the variable of integration in *20* from ξ to $\xi - \delta$. Hence, the value of the zero-state response at time $t + \delta$ for the translated input is numerically equal to the value of the zero-state response at time t for the original input. Consequently, *13* is satisfied and the system in question is zero-state time-invariant.

22 *Example* By changing the time scale in a nonuniform fashion in accordance with the relation $t \to \log t$, the zero-state time-invariant system of Example *19* is transformed into a time-varying system characterized by the input-output-state relation

23 $$y(t) = \frac{t_0}{t}\, y(t_0-) + \frac{1}{t}\int_{t_0}^{t} u(\xi)\, d\xi \qquad t \geq t_0 > 0$$

This system is a special case of systems of the Euler-Cauchy type which are touched upon in *3.8.29*. ◁

A basic property of zero-state time-invariant systems is expressed by the following

24 **Theorem** Zero-state time-invariant systems form a class closed under system addition, multiplication, and inversion. More specifically, if

$\mathcal{A}$ and $\mathcal{B}$ are zero-state time-invariant, then so are $\mathcal{A} + \mathcal{B}$, $\mathcal{A}\mathcal{B}$, and, if it exists, $\mathcal{A}^{-1}$.

Proof By hypothesis,

25 $$\mathcal{A}T_\delta \doteq T_\delta\mathcal{A}$$

26 $$\mathcal{B}T_\delta \doteq T_\delta\mathcal{B}$$

for all δ. Adding *25* and *26*, we have

27 $$(\mathcal{A} + \mathcal{B})T_\delta \doteq T_\delta\mathcal{A} + T_\delta\mathcal{B} \qquad \forall\delta$$

and, on making use of *7*, *27* becomes

$$(\mathcal{A} + \mathcal{B})T_\delta \doteq T_\delta(\mathcal{A} + \mathcal{B}) \qquad \forall\delta$$

which by *14* implies that $\mathcal{A} + \mathcal{B}$ is zero-state time-invariant.

Next, replacing $\mathcal{B}T_\delta$ in $\mathcal{A}\mathcal{B}T_\delta$ by $T_\delta\mathcal{B}$ and then replacing $\mathcal{A}T_\delta$ by $T_\delta\mathcal{A}$, we obtain the equivalence

$$\mathcal{A}\mathcal{B}T_\delta \doteq T_\delta\mathcal{A}\mathcal{B} \qquad \forall\delta$$

which demonstrates that $\mathcal{A}\mathcal{B}$ is zero-state time-invariant.

Finally, assume that $\mathcal{A}$ has an inverse $\mathcal{A}^{-1}$. Then on premultiplying and postmultiplying both sides of the equivalence

28 $$\mathcal{A}^{-1}\mathcal{A} \doteq \mathrm{I}$$

by T_δ, we obtain

$$T_\delta\mathcal{A}^{-1}\mathcal{A} \doteq \mathcal{A}^{-1}\mathcal{A}T_\delta \doteq T_\delta \qquad \forall\delta$$

which, upon making use of *25*, becomes

29 $$T_\delta\mathcal{A}^{-1}\mathcal{A} \doteq \mathcal{A}^{-1}T_\delta\mathcal{A} \qquad \forall\delta$$

Postmultiplying both sides of *29* by $\mathcal{A}^{-1}$, we have

$$T_\delta\mathcal{A}^{-1} \doteq \mathcal{A}^{-1}T_\delta \qquad \forall\delta$$

which, by *14*, implies that $\mathcal{A}^{-1}$ is zero-state time-invariant.

Zero-input time invariance

We turn next to the notion of zero-input time invariance, which can be defined in a manner quite analogous to that of Definition *12*. Specifically,

30 **Definition** A system $\mathcal{A}$ is *zero-input time-invariant* if and only if for all initial states $\boldsymbol{\alpha}$ in the state space Σ, all initial times t_0, and all time shifts δ the zero-input response of $\mathcal{A}$ starting in state $\boldsymbol{\alpha}$ at time $t_0 - \delta$ is identical (to within a translation by amount δ along the time axis) with the zero-input response of $\mathcal{A}$ starting in state $\boldsymbol{\alpha}$ at time t_0.

In symbols,

31 $\{\mathcal{A} \text{ is } \textit{zero-input time-invariant}\} \Leftrightarrow \{\bar{\mathbf{A}}(\mathbf{s}(t_0 - \delta);\mathbf{0}) = \bar{\mathbf{T}}_\delta \bar{\mathbf{A}}(\mathbf{s}(t_0);\mathbf{0})\} \quad \forall \delta \forall \mathbf{s}(t_0)$

where $\bar{\mathbf{A}}(\mathbf{s}(t_0);\mathbf{0})$ denotes the zero-input response of $\mathcal{A}$ starting in state $\mathbf{s}(t_0) = \boldsymbol{\alpha}$ at time t_0, and likewise for $\bar{\mathbf{A}}(\mathbf{s}(t_0 - \delta);\mathbf{0})$, with $\mathbf{s}(t_0 - \delta) = \boldsymbol{\alpha}$.

32 *Comment* Our motive for introducing the notion of zero-input time invariance is the following. We shall see at a later point in this section that, in general, a system $\mathcal{A}$ may be zero-state time-invariant without being time-invariant in the more strict sense of Definition *36*. Furthermore, a system may be both zero-state and zero-input time-invariant without being time-invariant in the sense of Definition *36*. This is not possible, however, in the case of linear systems. Thus, if a linear system is both zero-input and zero-state time-invariant, then it is necessarily time-invariant. It is for this reason that the notion of zero-input time invariance is of significance in the case of linear systems. ◁

In general, a system may be zero-state time-invariant without being zero-input time-invariant, and vice versa. This possibility is illustrated by the following simple examples.

33 *Example* Let the input-output-state relation of $\mathcal{A}$ be of the form

$$y(t) = ts_1(t_0-) + e^{-(t-t_0)}s_2(t_0-) + \int_{t_0}^{t} e^{-(t-\xi)}u(\xi)\,d\xi \qquad t \geq t_0$$

where $\mathbf{s}(t_0-) = (s_1(t_0-), s_2(t_0-))$ represents the initial state vector of $\mathcal{A}$. Then $\mathcal{A}$ is zero-state time-invariant but not zero-input time-invariant. [Note that $\mathcal{A}$ can be realized as a parallel combination of an integrator $1/(p+1)$ and a source (see *1.8.22*) which has the form of a grounded integrator $1/p$ in tandem with a time-varying scalor t.]

34 *Example* Consider a system $\mathcal{A}$ whose input-output relation is

$$\frac{dy}{dt} + y(t) = tu(t)$$

This system is zero-input but not zero-state time-invariant.

Time invariance

We are now ready to formulate a more general definition of time invariance which subsumes zero-state and zero-input time invariance as special cases. Specifically, suppose that the initial state of $\mathcal{A}$ is a state $\boldsymbol{\alpha}$ not necessarily equivalent to $\boldsymbol{\theta}$. Then, by analogy with the definition of zero-state time invariance, $\mathcal{A}$ will be said to be *time-invariant with respect to the initial state* $\boldsymbol{\alpha}$ if and only if

35 $$\bar{\mathbf{T}}_\delta(\bar{\mathbf{A}}(\boldsymbol{\alpha};\mathbf{u})) = \bar{\mathbf{A}}(\boldsymbol{\alpha};\bar{\mathbf{T}}_\delta(\mathbf{u})) \qquad \forall \delta \forall \mathbf{u}$$

that is, if the response segment of $\mathcal{A}$ to $\bar{\mathbf{T}}_\delta(\mathbf{u})$ ($\mathbf{u}$ shifted by δ units of time), with $\mathcal{A}$ initially in state $\boldsymbol{\alpha}$, is identical with the shifted response $\bar{\mathbf{T}}_\delta(\bar{\mathbf{A}}(\boldsymbol{\alpha};\mathbf{u}))$ of $\mathcal{A}$ to $\mathbf{u}$, with $\mathcal{A}$ initially in state $\boldsymbol{\alpha}$.

If this property holds for *all* states in the state space of $\mathcal{A}$, then $\mathcal{A}$ will be said to be *time-invariant*. More formally,

36 **Definition** $\mathcal{A}$ is *time-invariant* if and only if for all starting states $\boldsymbol{\alpha}$, all inputs $\mathbf{u}$, and all shifts δ

$$\bar{\mathbf{T}}_\delta(\bar{\mathbf{A}}(\boldsymbol{\alpha};\mathbf{u})) = \bar{\mathbf{A}}(\boldsymbol{\alpha};\bar{\mathbf{T}}_\delta(\mathbf{u}))$$

or, more compactly,

37 $$\{\mathcal{A} \text{ is } \textit{time-invariant}\} \Leftrightarrow \{\mathcal{A}\mathbf{T}_\delta \equiv \mathbf{T}_\delta\mathcal{A}\} \qquad \forall\delta$$

where the relation $\mathcal{A}\mathbf{T}_\delta \equiv \mathbf{T}_\delta\mathcal{A}$ should be interpreted as system equivalence, with $\mathbf{T}_\delta$—but not necessarily $\mathcal{A}$—starting in its zero state, and with $\mathcal{A}\mathbf{T}_\delta$ and $\mathbf{T}_\delta\mathcal{A}$ interpreted as free products (see *1.10.10*). ◁

Clearly, if $\mathcal{A}$ has a zero state and is time-invariant, then it is also zero-state time-invariant. The converse, however, is not true, as Example *33* and the less trivial example given below demonstrate.

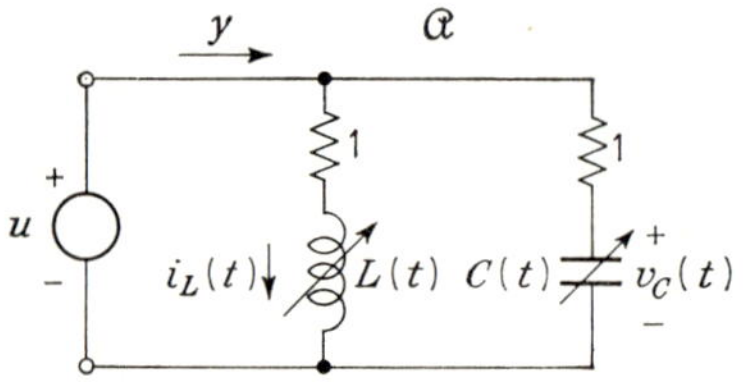

Fig. 3.2.2 $\mathcal{A}$ is a time-varying network which is zero-state time-invariant.

38 *Example* Consider the network $\mathcal{A}$ shown in Fig. *3.2.2*, in which u is the input, y is the output, and $L(t)$ and $C(t)$ are functions of time such that $L(t) = C(t) > 0$ for all t. The state of $\mathcal{A}$ at time t can be taken to be the 2-vector (see *1.7.28*)

$$\mathbf{s}(t) = (i_L(t), v_C(t))$$

where i_L is the current through L and v_C is the voltage across C. Using the well-known expression (see Ref. *1*)

$$y(t) = \exp\left[-\int_{t_0}^{t} a(\xi)\,d\xi\right] y(t_0) + \int_{t_0}^{t} d\xi\, u(\xi) \exp\left[-\int_{\xi}^{t} a(\eta)\,d\eta\right] \qquad t \geq t_0$$

for the solution of the first-order differential equation

$$\frac{dy}{dt} + a(t)y = u(t)$$

the expression for the input-output-state relation of $\mathcal{A}$ is readily found to be

39 $$y(t) = \frac{C(t)}{C(t_0)} \exp\left[-\int_{t_0}^{t} \frac{d\xi}{C(\xi)}\right] [i_L(t_0) - v_C(t_0)] + u(t)$$

Clearly, if $\mathcal{A}$ is started in its zero state, $\boldsymbol{\theta} = (0,0)$, or in any state in which the two components are equal [which are the states reachable (*2.4.17*) from the zero state], then $y(t) = u(t)$. This implies that $\mathcal{A}$ is zero-state time-invariant. On the other hand, if the starting state is such that $i_L(t_0) \neq v_C(t_0)$, then the first term in *39* does not vanish and $\mathcal{A}$ is not time-invariant with respect to the starting state in question. Hence $\mathcal{A}$ is not time-invariant. ◁

An important connection between time invariance and zero-state time invariance is provided by the following

40 **Assertion** If $\mathcal{A}$ is zero-state time-invariant and zero-state equivalent to a time-invariant system, then $\mathcal{A}$ is time-invariant with respect to every state which is reachable from the zero state.

Proof Let $\mathcal{B}$ be a time-invariant system which is zero-state equivalent to $\mathcal{A}$. Consider an arbitrary input $\mathbf{u}$ and assume that the zero state of $\mathcal{A}$ is taken by $\mathbf{u}$ into a state $\boldsymbol{\alpha}$ in the state space of $\mathcal{A}$ and, correspondingly, the zero state of $\mathcal{B}$ is taken into a state $\boldsymbol{\beta}$ in the state space of $\mathcal{B}$. By Assertion *2.4.12*, $\boldsymbol{\alpha}$ and $\boldsymbol{\beta}$ are equivalent states. Consequently, for any input $\mathbf{v}$ applied to $\mathcal{A}$ and $\mathcal{B}$ starting in $\boldsymbol{\alpha}$ and $\boldsymbol{\beta}$, respectively, we have

41 $$\bar{\mathbf{A}}(\boldsymbol{\alpha};\mathbf{v}) = \bar{\mathbf{B}}(\boldsymbol{\beta};\mathbf{v})$$

Now since $\mathcal{B}$ is a time-invariant system, we can write (by *35*)

$$\bar{\mathbf{T}}_\delta(\bar{\mathbf{B}}(\boldsymbol{\beta};\mathbf{v})) = \bar{\mathbf{B}}(\boldsymbol{\beta};\bar{\mathbf{T}}_\delta(\mathbf{v})) \qquad \forall\delta\forall\mathbf{v}$$

which by virtue of *41* implies (see *2.4.17*)

$$\bar{\mathbf{T}}_\delta(\bar{\mathbf{A}}(\boldsymbol{\alpha};\mathbf{v})) = \bar{\mathbf{A}}(\boldsymbol{\alpha};\bar{\mathbf{T}}_\delta(\mathbf{v})) \qquad \forall\delta\forall\mathbf{v}$$

and thus shows that $\mathcal{A}$ is time-invariant with respect to $\boldsymbol{\alpha}$. Clearly, the same is true of any state which is reachable from the zero state of $\mathcal{A}$. Q.E.D.

The various types of time invariance which we have defined in the foregoing discussion all involve the initial state of $\mathcal{A}$. It is possible, however, to define still another, weaker, kind of time invariance which does not have this characteristic. Specifically, we have the

42 **Definition** Let $(\mathbf{u}_{(t_0,t]},\mathbf{y}_{(t_0,t]})$ [or $(\mathbf{u},\mathbf{y})$ for short] be an input-output pair for $\mathcal{A}$ (see *1.4.17*), with $(\mathbf{u},\mathbf{y})$ ranging over the space $R_{(t_0,t]}$ of input-output pairs. Then $\mathcal{A}$ is *weakly time-invariant* if and only if for all δ and all $(\mathbf{u},\mathbf{y})$ in $R_{(t_0,t]}$

43 $$\{(\mathbf{u},\mathbf{y}) \text{ is an input-output pair for } \mathcal{A}\} \Leftrightarrow \{(\bar{\mathbf{T}}_\delta(\mathbf{u}),\bar{\mathbf{T}}_\delta(\mathbf{y})) \text{ is an input-output pair for } \mathcal{A}\}$$

That is, if $(\mathbf{u},\mathbf{y})$ is an input-output pair for $\mathcal{A}$, then so is every translate of $(\mathbf{u},\mathbf{y})$. ◁

In general, if $\mathfrak{A}$ is weakly time-invariant, then it is not necessarily time-invariant in the sense of Definition *36*. In the case of linear differential systems, however, weak time invariance implies time invariance. As will be seen in Sec. *9*, this is a consequence of the fact that in the case of linear systems weak equivalence implies equivalence (see *3.9.1*).

This concludes for the present our discussion of time invariance. We turn next to another basic concept, namely, *linearity*.

3 *Basic aspects of linearity: additivity and homogeneity*

Although our main concern in this text is with systems characterized by differential equations, our discussion of the notion of linearity and related concepts will be conducted on a considerably more general level. Though this makes it initially somewhat more difficult for the reader to relate the various assertions, theorems, etc., to those properties of linear systems with which he has prior familiarity, it does have the longer-run advantage of giving one a deeper insight into the basic properties which are common to all linear systems and not just systems governed by differential equations.

The concept of linearity

The term *linearity* connotes a relation of proportionality between the cause (input) and the effect (output). Actually, linearity implies more than proportionality—it implies the superposability of causes and their respective effects.

How can this vague statement be formulated in a more precise form? It is customary to formulate it precisely via the notions of *additivity* and *homogeneity*, under the tacit assumption that the initial state is the zero state of the system. Such an assumption has two drawbacks. In the first place, it does not provide a basis for deciding whether a given system $\mathfrak{A}$ is linear or nonlinear when the initial state of $\mathfrak{A}$ is not its zero state. Second, under the conventional definition of linearity a nonlinear system $\mathfrak{A}$ may be misclassified as a linear system if $\mathfrak{A}$ happens to behave like a linear system when its initial state is its zero state. Three such cases are cited in Examples *3.4.11*, *3.4.12*, and *3.4.24*.

In this chapter we shall formulate a definition of linearity (Definition *3.4.25*) which is not predicated on the assumption that the initial state is the zero state. However, instead of stating this definition at the outset, we shall first lay a groundwork for it by establishing the basic properties of homogeneity, additivity, and two special forms of linearity, namely zero-state linearity and zero-input linearity.

In defining these notions we shall be making the following assumptions about the system under consideration (which will be denoted by $\mathcal{A}$).

1 **Assumption** $\mathcal{A}$ has a unique zero state $\boldsymbol{\theta}$.

2 **Assumption** The input and output function spaces of $\mathcal{A}$ are *linear* function spaces. This implies that if $\mathbf{u}_1$ and $\mathbf{u}_2$ are two arbitrary time functions in $R[\mathbf{u}]$, the input function space of $\mathcal{A}$, then the time function $\mathbf{u} = a\mathbf{u}_1 + b\mathbf{u}_2$, where a and b are arbitrary constants, is likewise an element of $R[\mathbf{u}]$. The same statement applies to the output function space of $\mathcal{A}$. Beyond the assumption that the input and output function spaces of $\mathcal{A}$ are linear, we shall generally not make any restrictive assumptions on their elements other than that they are finitely differentiable in the distribution sense (see *1.3.7* and *1.4.29*).

3 **Assumption** The state space of $\mathcal{A}$ is a linear vector space. This implies that if $\boldsymbol{\alpha}_1$ and $\boldsymbol{\alpha}_2$ are arbitrary states in Σ, then any linear combination of them, $a\boldsymbol{\alpha}_1 + b\boldsymbol{\alpha}_2$, is likewise a state in Σ.

4 *Comment* The reason for Assumption *2* is the following. In the sequel, we shall frequently be comparing the response of $\mathcal{A}$ to an input $\mathbf{u}$ with the response of $\mathcal{A}$ to $k\mathbf{u}$, where k is an arbitrary constant, and also the responses of $\mathcal{A}$ to $\mathbf{u}_1$ and $\mathbf{u}_2$ with the response to the sum $\mathbf{u}_1 + \mathbf{u}_2$. Clearly, in order that both $k\mathbf{u}$ and $\mathbf{u}_1 + \mathbf{u}_2$ be admissible inputs for all $\mathbf{u}_1,\mathbf{u}_2$ in $R[\mathbf{u}]$ and all k, it is necessary that $R[\mathbf{u}]$ be a linear function space in the sense stated above. ◁

We begin with a discussion of the notion of

Homogeneity

Essentially, *homogeneity* denotes a relation of proportionality between the output and input. Stated more precisely:

5 **Definition** A system $\mathcal{A}$ is *homogeneous* if and only if for all inputs $\mathbf{u}$ in the input segment space of $\mathcal{A}$ and for all real constants k, the zero-state response of $\mathcal{A}$ to the input $k\mathbf{u}$ is k times the zero-state response of $\mathcal{A}$ to $\mathbf{u}$.

In symbols,

6 $$\{\mathcal{A} \text{ is } \textit{homogeneous}\} \Leftrightarrow \{\bar{\mathbf{A}}(k\mathbf{u}) = k\bar{\mathbf{A}}(\mathbf{u})\} \qquad \forall k \forall \mathbf{u}$$

where $\bar{\mathbf{A}}(\mathbf{u})$ denotes the zero-state response of $\mathcal{A}$ to $\mathbf{u}$ (see *2.8.18*).

Equivalently, let $\mathcal{K}$ denote a scalor with scale factor k (see *1.9.8*). Then $k\mathbf{u}$ can be regarded as the response of $\mathcal{K}$ to $\mathbf{u}$, and $\bar{\mathbf{A}}(k\mathbf{u})$ as the response of $\mathcal{A}\mathcal{K}$ (tandem combination of $\mathcal{A}$ and $\mathcal{K}$, with $\mathcal{K}$ preceding $\mathcal{A}$) to $\mathbf{u}$. In terms of these symbols, the statement "$\bar{\mathbf{A}}(k\mathbf{u}) = k\bar{\mathbf{A}}(\mathbf{u})\forall\mathbf{u}$" has the same meaning as the statement "$\mathcal{A}\mathcal{K}$ is zero-state equivalent to $\mathcal{K}\mathcal{A}$." ◁

This implies that homogeneity can also be defined as follows:

7 **Definition** A system $\mathcal{A}$ is *homogeneous* if for all real scalors $\mathcal{K}$, $\mathcal{A}\mathcal{K}$ is zero-state equivalent of $\mathcal{K}\mathcal{A}$. ◁

Actually, in order to check on whether or not the zero-state equivalence $\mathcal{A}\mathcal{K} \doteq \mathcal{K}\mathcal{A}$ holds for all real values of k, it is sufficient to verify that this equivalence holds for all values of k in any finite interval $[-a,a]$. The justification for this is provided by the following

8 **Assertion** If $\mathcal{A}\mathcal{K} \doteq \mathcal{K}\mathcal{A}$ for all scalors with k in the range $-a \leq k \leq a$, then $\mathcal{A}\mathcal{K} \doteq \mathcal{K}\mathcal{A}$ for all real k.
Proof Let $k \in [-a,a]$. Write k in the form $k = p/r$, where $p \in [-a,a]$, $p \neq 0$, and r is such that $p/r \in [-a,a]$. Then

$$\bar{\mathbf{A}}\left(\frac{p}{r}\mathbf{u}\right) = \frac{p}{r}\bar{\mathbf{A}}(\mathbf{u})$$

and on making use of $\bar{\mathbf{A}}(p\mathbf{u}) = p\bar{\mathbf{A}}(\mathbf{u})$, we have

$$p\bar{\mathbf{A}}\left(\frac{1}{r}\mathbf{u}\right) = \frac{p}{r}\bar{\mathbf{A}}(\mathbf{u})$$

or

$$\bar{\mathbf{A}}\left(\frac{1}{r}\mathbf{u}\right) = \frac{1}{r}\bar{\mathbf{A}}(\mathbf{u})$$

This shows that if $\mathcal{A}\mathcal{K} \doteq \mathcal{K}\mathcal{A}$ holds for all k in the interval $[-a,a]$, then it also holds for all k of the form $1/r$, where r is the ratio of any two numbers in $[-a,a]$. Since the range of $1/r$ is the entire real line, the assertion is proved. ◁

The following two simple examples will suffice for the present to illustrate the notion of homogeneity.

9 *Example* Consider a system $\mathcal{A}$ whose zero-state response is defined by the relation

$$y(t) = \int_{t_0}^{t} e^{-(t-\xi)}u(\xi)\,d\xi \qquad t \geq t_0$$

where u is a real-valued time function. Clearly, $\mathcal{A}$ is a homogeneous system, since if the response of $\mathcal{A}$ to u is y, then the response of $\mathcal{A}$ to ku is ky for all k and u.

10 *Example* A squarer is a memoryless system (see *1.8.17*) characterized by the input-output relation

$$y(t) = [u(t)]^2$$

where u is a real-valued time function. If y is the response of the squarer to u, then its response to ku is k^2y. Consequently, a squarer is not a homogeneous system. ◁

We turn next to the notion of

Additivity

Broadly speaking, additivity denotes the superposability of causes and effects. Thus, if a "cause" $\mathbf{u}_1$ produces the "effect" $\mathbf{y}_1$ and if a "cause" $\mathbf{u}_2$ produces the "effect" $\mathbf{y}_2$, then the superposability of causes and effects implies that the "cause" $\mathbf{u}_1 + \mathbf{u}_2$ would produce the "effect" $\mathbf{y}_1 + \mathbf{y}_2$.

This somewhat vague concept is formulated more precisely in the following

11 **Definition** A system $\mathcal{A}$ is *additive* if and only if for any pair of time functions $\mathbf{u}_1$ and $\mathbf{u}_2$ in the input segment space of $\mathcal{A}$, the zero-state response of $\mathcal{A}$ to the sum $\mathbf{u}_1 + \mathbf{u}_2$ is the sum of the zero-state responses of $\mathcal{A}$ to $\mathbf{u}_1$ and $\mathbf{u}_2$.

In symbols,

12 $$\{\mathcal{A} \text{ is } \textit{additive}\} \Leftrightarrow \{\bar{\mathbf{A}}(\mathbf{u}_1 + \mathbf{u}_2) = \bar{\mathbf{A}}(\mathbf{u}_1) + \bar{\mathbf{A}}(\mathbf{u}_2)\} \qquad \forall \mathbf{u}_1 \forall \mathbf{u}_2 \quad \triangleleft$$

Let us now examine some of the implications of additivity. First, suppose that $\mathcal{A}$ is additive. Then, it is trivial to show that $\mathcal{A}$ is also *finitely additive*, in the sense that the zero-state response of $\mathcal{A}$ to any finite sum of time functions belonging to the input segment space of $\mathcal{A}$ is the sum of the respective responses. In symbols (omitting the quantifiers $\forall \mathbf{u}_1 \cdots \forall \mathbf{u}_n$ for simplicity), we have

13 $$\{\bar{\mathbf{A}}(\mathbf{u}_1 + \mathbf{u}_2) = \bar{\mathbf{A}}(\mathbf{u}_1) + \bar{\mathbf{A}}(\mathbf{u}_2)\} \Rightarrow \{\bar{\mathbf{A}}(\mathbf{u}_1 + \cdots + \mathbf{u}_n) = \bar{\mathbf{A}}(\mathbf{u}_1) + \cdots + \bar{\mathbf{A}}(\mathbf{u}_n)\}$$

For, any finite sum such as $\mathbf{v} = \mathbf{u}_1 + \cdots + \mathbf{u}_n$ can be written as $\mathbf{v} = (\mathbf{u}_1 + \cdots + \mathbf{u}_{n-1}) + \mathbf{u}_n$. Then, on applying *12* to $\mathbf{v}$, we can write

14 $$\bar{\mathbf{A}}(\mathbf{u}_1 + \cdots + \mathbf{u}_n) = \bar{\mathbf{A}}(\mathbf{u}_1 + \cdots + \mathbf{u}_{n-1}) + \bar{\mathbf{A}}(\mathbf{u}_n)$$

On repeating this process for $\mathbf{u}_1 + \cdots + \mathbf{u}_{n-1}, \mathbf{u}_1 + \cdots + \mathbf{u}_{n-2}, \ldots,$ we eventually arrive at the identity

15 $$\bar{\mathbf{A}}(\mathbf{u}_1 + \cdots + \mathbf{u}_n) = \bar{\mathbf{A}}(\mathbf{u}_1) + \cdots + \bar{\mathbf{A}}(\mathbf{u}_n) \qquad \text{Q.E.D.}$$

An important implication of additivity concerns the relation between additivity and homogeneity. This implication is contained in the

16 **Assertion** If $\mathcal{A}$ is additive, then the relation

17 $$\mathcal{A}\mathcal{K} \doteq \mathcal{K}\mathcal{A}$$

where $\mathcal{K}$ is a scalor with scale factor k, holds for all rational values of k.

Comment The significance of this property of additive systems is this: If a system is additive, then *16* implies that, for all practical purposes, it is also homogeneous, since (I) any real constant k can be approximated as closely as desired by a rational number and (II) in any physical system a small difference between k and a rational approximation to k, say, $\hat{k}$, will result in a correspondingly small difference between the responses to $k\mathbf{u}$ and $\hat{k}\mathbf{u}$.

Proof Let $\mathcal{A}$ be an additive system. Set $\mathbf{u}_1 = \mathbf{u}_2 = \cdots = \mathbf{u}_n = \mathbf{u}$ in *15*. This leads to the consequence

18 $$\{\mathcal{A} \text{ is additive}\} \Rightarrow \{\bar{\mathbf{A}}(n\mathbf{u}) = n\bar{\mathbf{A}}(\mathbf{u})\}$$

for all $\mathbf{u} \in R[\mathbf{u}]$ and all finite, nonnegative, integral values of n. Now

$$\bar{\mathbf{A}}(n\mathbf{u} - n\mathbf{u}) = \bar{\mathbf{A}}(\mathbf{0}) = \mathbf{0}$$

since the zero-state response of $\mathcal{A}$ to the zero input is $\mathbf{0}$.

On the other hand,

$$\begin{aligned}\bar{\mathbf{A}}(n\mathbf{u} - n\mathbf{u}) &= \bar{\mathbf{A}}(n\mathbf{u}) + \bar{\mathbf{A}}(-n\mathbf{u}) \\ &= n\bar{\mathbf{A}}(\mathbf{u}) + \bar{\mathbf{A}}(-n\mathbf{u})\end{aligned}$$

by additivity and *18*. Consequently,

19 $$\{\mathcal{A} \text{ is additive}\} \Rightarrow \{\bar{\mathbf{A}}(n\mathbf{u}) = n\bar{\mathbf{A}}(\mathbf{u})\}$$

for all $\mathbf{u}$ and all finite, integral values of n.

Now let

$$\mathbf{v} \triangleq \frac{n}{m}\mathbf{u}$$

where m and n are finite integers. (Note that $\mathbf{v} \in R[\mathbf{u}]$ because $R[\mathbf{u}]$ is a linear function space.) On replacing $\mathbf{u}$ by $(m/n)\mathbf{v}$ in *19*, we get

$$\{\mathcal{A} \text{ is additive}\} \Rightarrow \left\{\bar{\mathbf{A}}(m\mathbf{v}) = n\bar{\mathbf{A}}\left(\frac{m}{n}\mathbf{v}\right)\right\} \qquad \forall\mathbf{v}$$

or equivalently

$$\{\mathcal{A} \text{ is additive}\} \Rightarrow \left\{\bar{\mathbf{A}}\left(\frac{m}{n}\mathbf{v}\right) = \frac{m}{n}\bar{\mathbf{A}}(\mathbf{v})\right\} \qquad \forall\mathbf{v}$$

which shows that *16* holds for all rational values of k. Q.E.D.

20 **Remark** Since additivity "almost" implies homogeneity (in the sense of *16*), it is natural to inquire if the converse is also true in some approximate sense. That it is not is demonstrated by the following simple counterexamples. However, the artificiality of these examples indicates that it is very unlikely that a physical system picked at random would exhibit homogeneity but not additivity.

21 *Example* Let $\mathcal{A}$ be a system which in response to a time function u produces a train of pulses of the form shown in Fig. *3.3.1*. The pulses occur at the points where $u(t) = 0$, and the amplitude of the pulse occurring at, say, $t_i[u(t_i) = 0]$ is equal to $\dot{u}(t_i)$, that is, to the slope of

u at t_i. The input function space of $\mathcal{A}$ is assumed to be the class of all differentiable functions.

Clearly, $\mathcal{A}$ is a homogeneous system, since if u is multiplied by any real constant k, the output of $\mathcal{A}$ is multiplied by the same constant. On the other hand, $\mathcal{A}$ is not an additive system, as is demonstrated by the simple case where $u_1(t) \equiv t$ and $u_2(t) \equiv t^2 - 1$.

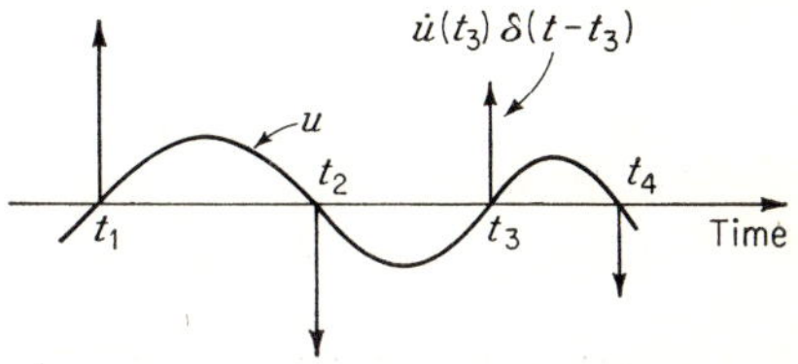

Fig. 3.3.1 Example of a system which is homogeneous but not additive.

22 *Example* Consider a memoryless system $\mathcal{A}$ whose zero-state response to an input u is given by

$$y(t) = \sqrt[3]{[u(t)]^3 + [\dot{u}(t)]^3}$$

This is a homogeneous system, since the response of $\mathcal{A}$ to ku is k times the response of $\mathcal{A}$ to u. On the other hand, $\mathcal{A}$ is clearly not additive. ◁

The following simple example exhibits a system which has the additivity property.

23 *Example* Consider the system characterized by the input-output-state relation

$$y(t) = e^{-(t-t_0)}\alpha + \int_{t_0}^{t} e^{-(t-\xi)}u(\xi)\,d\xi \qquad t \geq t_0$$

in which α is the state at time t_0-, with α ranging over $\mathcal{R}^1$ (the real line). Here the zero state is $\theta = 0$, and the zero-state response is expressed by

$$A(u) = \int_{t_0}^{t} e^{-(t-\xi)}u(\xi)\,d\xi \qquad t \geq t_0$$

Now, if $u = u_1 + u_2$, then $y = y_1 + y_2$, where

$$y_i(t) = \int_{t_0}^{t} e^{-(t-\xi)}u_i(\xi)\,d\xi \qquad i = 1, 2, t \geq t_0$$

Consequently, the system in question is additive.

This concludes our discussion of homogeneity and additivity. We shall employ these concepts in Sec. *4* to define two special cases of linearity, namely, zero-state linearity and zero-input linearity, with a view to using these cases as steppingstones to a more general definition of linearity in which no special assumptions are made about either the input or the initial state.

4 *Zero-state and zero-input linearity*

The notions of zero-state and zero-input linearity bear the same relation to one another as do the notions of zero-state and zero-input time invariance. For this reason, there are in what follows many points of parallelism with the material in Sec. *2*.

We begin with a discussion of

Zero-state linearity

In the engineering as well as the mathematical literature it is customary to equate linearity with additivity and homogeneity. In other words, under the conventional definition of linearity, a system $\mathcal{A}$ is linear if it is both additive and homogeneous.

The trouble with this definition is that one can construct examples of systems (Examples *11*, *12*, and *24*) which are linear under the conventional definition and yet exhibit clearly nonlinear behavior when started in a state other than the zero state. Thus, the conventional definition of linearity is in effect a description of a restricted type of linearity which may be appropriately called *zero-state linearity*. More formally,

1 **Definition** A system $\mathcal{A}$ is *zero-state linear* if and only if it is both additive and homogeneous. ◁

For our purposes, it will be convenient to put this definition into a somewhat different but equivalent form which is given below.

2 **Definition** Let $\mathbf{u}$ and $\mathbf{v}$ be arbitrary time functions in the input segment space of $\mathcal{A}$ and let k be an arbitrary real constant. Then $\mathcal{A}$ is *zero-state linear* if and only if the zero-state response of $\mathcal{A}$ to $k(\mathbf{u} - \mathbf{v})$ is k times the zero-state response of $\mathcal{A}$ to $\mathbf{u}$ minus k times the zero-state response of $\mathcal{A}$ to $\mathbf{v}$.

In symbols,

3 $\{\mathcal{A} \text{ is } \textit{zero-state linear}\} \Leftrightarrow \{\bar{\mathbf{A}}(k(\mathbf{u} - \mathbf{v})) = k\bar{\mathbf{A}}(\mathbf{u}) - k\bar{\mathbf{A}}(\mathbf{v})\}$ for all $\mathbf{u},\mathbf{v}$

in $R[\mathbf{u}]$ and all real k. [Note that $\bar{\mathbf{A}}(\mathbf{u})$ denotes the response segment of $\mathcal{A}$ to $\mathbf{u}$, with $\mathcal{A}$ initially in its zero state.] ◁

To demonstrate the equivalence of Definitions *1* and *2*, first set $\mathbf{v} = 0$ in *3*. It follows that $\mathcal{A}$ is homogeneous. Next, set $k = 1$ and $\mathbf{v} = -\mathbf{v}'$, where $\mathbf{v}' \in R[\mathbf{u}]$. It follows that $\mathcal{A}$ is additive. Conversely, if $\mathcal{A}$ is additive and homogeneous, it follows at once that

$$\bar{\mathbf{A}}(k(\mathbf{u} - \mathbf{v})) = k\bar{\mathbf{A}}(\mathbf{u}) - k\bar{\mathbf{A}}(\mathbf{v}) \quad \text{Q.E.D.}$$

4 *Example* The system of Example *2.2.9* is zero-state linear. So is the system of Example *2.3.7*.

Linearity with respect to an initial state

As an extension of the notion of zero-state linearity we shall say that system $\mathcal{A}$ is *linear with respect to an initial state* $\boldsymbol{\alpha}$ if *3* holds when the initial state is $\boldsymbol{\alpha}$. More specifically:

5 **Definition** A system $\mathcal{A}$ is *linear with respect to an initial state* $\boldsymbol{\alpha}$ if and only if the relation

6 $$k\bar{\mathbf{A}}(\boldsymbol{\alpha};\mathbf{u}) - k\bar{\mathbf{A}}(\boldsymbol{\alpha};\mathbf{v}) = \bar{\mathbf{A}}(\boldsymbol{\theta};k(\mathbf{u} - \mathbf{v}))$$

where $\bar{\mathbf{A}}(\boldsymbol{\alpha};\mathbf{u})$ denotes the response of $\mathcal{A}$ to $\mathbf{u}$ starting in state $\boldsymbol{\alpha}$ and $\bar{\mathbf{A}}(\boldsymbol{\theta};\mathbf{u})$ is the zero-state response of $\mathcal{A}$ to $\mathbf{u}$, holds for all real constants k and all $\mathbf{u},\mathbf{v}$ in the input segment space of $\mathcal{A}$. (Note that *6* reduces to *3* when $\boldsymbol{\alpha} = \boldsymbol{\theta}$.) ◁

Our motive in introducing this notion is the following. It is not difficult to construct examples of systems that are zero-state linear (i.e., linear in the conventional sense of the term) and yet are not linear with respect to all possible initial states. Although such systems behave like linear systems if started in the zero state, they are not "completely" linear, since there exist starting states with respect to which their behavior is nonlinear. Thus, Definition *5* serves the purpose of providing a basis for differentiating between (I) systems that are linear with respect to all possible starting states and (II) systems that are zero-state linear without being linear with respect to all possible starting states. Systems of the latter type are illustrated by Examples *11* and *12*.

A connection between zero-state linearity and linearity with respect to $\boldsymbol{\alpha}$ is provided by the following

7 **Assertion** If $\mathcal{A}$ is zero-state linear, then it is also linear with respect to all initial states which are reachable from the zero state (*2.4.17*).

Proof We have to show that if (I) $\mathcal{A}$ is zero-state linear and (II) $\boldsymbol{\alpha}$ is any state which is reachable from $\boldsymbol{\theta}$, then

$$k\bar{\mathbf{A}}(\boldsymbol{\alpha};\mathbf{u}) - k\bar{\mathbf{A}}(\boldsymbol{\alpha};\mathbf{v}) = \bar{\mathbf{A}}(\boldsymbol{\theta};k(\mathbf{u} - \mathbf{v})) \qquad \forall k \forall \mathbf{u} \forall \mathbf{v}$$

Suppose that $\boldsymbol{\alpha}$ is reachable from $\boldsymbol{\theta}$ by an input $\mathbf{w}$. Then, by the response separation property (*2.3.4*), we can write

8 $$\bar{\mathbf{A}}(\boldsymbol{\theta};\mathbf{wu}) = \bar{\mathbf{A}}(\boldsymbol{\theta};\mathbf{w}) + \bar{\mathbf{A}}(\boldsymbol{\alpha};\mathbf{u})$$

9 $$\bar{\mathbf{A}}(\boldsymbol{\theta};\mathbf{wv}) = \bar{\mathbf{A}}(\boldsymbol{\theta};\mathbf{w}) + \bar{\mathbf{A}}(\boldsymbol{\alpha};\mathbf{v})$$

where $\mathbf{wu}$ represents the segment resulting from adjoining $\mathbf{u}$ to $\mathbf{w}$ (with $\mathbf{u}$ following $\mathbf{w}$). [Note that $\bar{\mathbf{A}}(\boldsymbol{\theta};\mathbf{w})$ and $\bar{\mathbf{A}}(\boldsymbol{\alpha};\mathbf{u})$ are nonoverlapping segments.]

On subtracting *9* from *8* and multiplying both sides by k, we obtain

$$k\bar{\mathbf{A}}(\boldsymbol{\alpha};\mathbf{u}) - k\bar{\mathbf{A}}(\boldsymbol{\alpha};\mathbf{v}) = k\bar{\mathbf{A}}(\boldsymbol{\theta};\mathbf{wu}) - k\bar{\mathbf{A}}(\boldsymbol{\theta};\mathbf{wv})$$

Now, since $\mathcal{A}$ is zero-state linear, we have

$$\begin{aligned} k\bar{\mathbf{A}}(\boldsymbol{\theta};\mathbf{wu}) - k\bar{\mathbf{A}}(\boldsymbol{\theta};\mathbf{wv}) &= \bar{\mathbf{A}}(\boldsymbol{\theta};k(\mathbf{wu} - \mathbf{wv})) \\ &= \bar{\mathbf{A}}(\boldsymbol{\theta};k(\mathbf{u} - \mathbf{v})) \end{aligned}$$

and hence

$$k\bar{\mathbf{A}}(\boldsymbol{\alpha};\mathbf{u}) - k\bar{\mathbf{A}}(\boldsymbol{\alpha};\mathbf{v}) = \bar{\mathbf{A}}(\boldsymbol{\theta};k(\mathbf{u} - \mathbf{v}))$$

for all **u**,**v** in the input segment space of $\mathcal{A}$ and all real constants k.

10 **Corollary** If $\mathcal{A}$ is zero-state linear and if all the states of $\mathcal{A}$ are reachable from its zero state, then $\mathcal{A}$ is linear with respect to all possible initial states. ◁

The significance of Assertion *7* and its corollary is this. If a system is zero-state linear, then its nonlinearity can manifest itself only if the starting state is not the zero state or any state which is reachable from the zero state. Thus, if there are no states which are not reachable from the zero state, then zero-state linearity implies linearity with respect to all possible initial states.

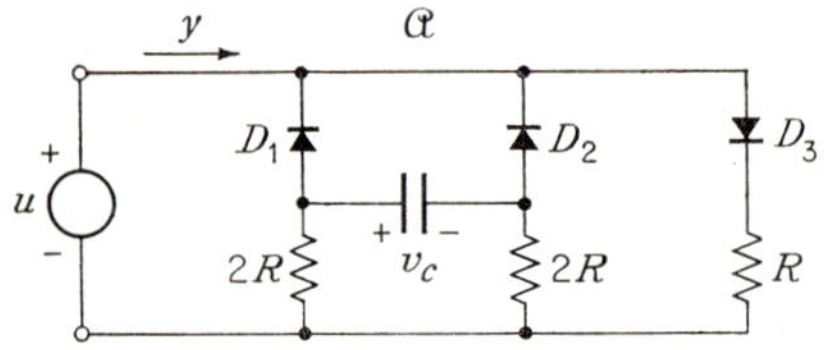

Fig. 3.4.1 Example of a nonlinear network which is zero-state linear.

11 *Example*[1] Consider the system $\mathcal{A}$ shown in Fig. *3.4.1*. Here D_1, D_2, and D_3 are ideal diodes,[2] the voltage u is the input, and the current y is the output. The voltage v_C across the capacitor C qualifies as the state of the system, and the zero state corresponds to $v_C = 0$.

If $v_C = 0$ initially (at t_0-), then the currents through the branches containing D_1 and D_2 will be equal for $t \geq t_0$ and consequently v_C will remain equal to zero for $t \geq t_0$. By virtue of this, C can be removed from the network without affecting the current y. Thus, the system of Fig. *3.4.1* is zero-state equivalent to the networks $\mathcal{A}'$,$\mathcal{A}''$ shown in Fig. *3.4.2*. These networks in turn are equivalent to a constant resistance R, which is a memoryless linear system. This implies that $\mathcal{A}$ is zero-state linear.

[1] This simple example was suggested by D. Cargille, a student at the University of California.

[2] An ideal diode is a memoryless system characterized by the input-output relation $v(t) = 0$ for $i(t) \geq 0$ and $v(t) = \infty$ for $i(t) < 0$, where $i(t)$ is the current through the diode in the direction of the arrowhead and $v(t)$ is the voltage across the diode.

On the other hand, $\mathcal{A}$ is not linear with respect to all initial states in the sense of Definition *5*, since if it is started in any state α other than the zero state (that is, $v_C = \alpha \neq 0$ at t_0-), then the relation between u and y is not linear for the starting state α. Note also that the state $v_C = \alpha$, $\alpha \neq 0$, is not reachable from the zero state $v_C = 0$.

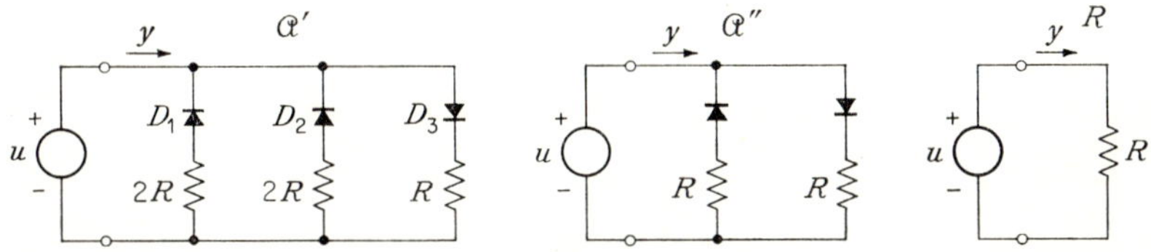

Fig. 3.4.2 $\mathcal{A}'$ is equivalent to $\mathcal{A}''$, which in turn is equivalent to R.

12 *Example* Consider the network $\mathcal{A}$, shown in Fig. *3.4.3*, in which R is a nonlinear resistance and G is a nonlinear conductance such that $R = G$ for equal values of their respective arguments and L is numerically equal to C. Essentially, this is merely a concrete instance of two networks $\mathfrak{N}$ and $\mathfrak{N}'$ (within the dashed-line rectangles) which are dual to one another in the sense that the voltages in $\mathfrak{N}$ are numerically equal to the corresponding currents in $\mathfrak{N}'$, and vice versa. Note that $\mathfrak{N}$ and $\mathfrak{N}'$ are converse to one another in the sense of Definition *2.10.1*. This implies that, if the dependence of R on the current passing through it is such that $\mathfrak{N}^{-1}$ (inverse of $\mathfrak{N}$) exists, then $\mathfrak{N}'$ is equivalent to $\mathfrak{N}^{-1}$.

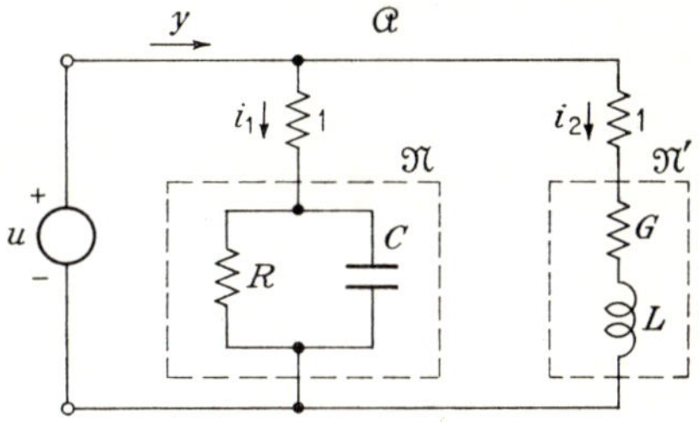

Fig. 3.4.3 Example of a nonlinear network which is zero-state linear.

Assuming that this is the case, if $\mathfrak{N}$ is initially in its zero state (zero voltage across C), then the current through $\mathfrak{N}$ at time t can be expressed symbolically as

13 $$i_1 = (I + N)^{-1}(u)$$

where $N(i_1)$ represents the zero-state response (voltage across $\mathfrak{N}$) to current i_1 and I is the identity operator. On the other hand, the current through $\mathfrak{N}'$, with $\mathfrak{N}'$ starting in its zero state (zero current through L), is given by

14 $$i_2 = (I + N^{-1})^{-1}(u)$$

Thus, the zero-state response of $\mathcal{A}$ to u is expressed by

15 $$y = [(I + N)^{-1} + (I + N^{-1})^{-1}](u)$$

Now for any $\mathfrak{N}$—regardless of whether $\mathfrak{N}$ is linear or nonlinear—which has a unique zero state and an inverse $\mathfrak{N}^{-1}$, we can write

16 $$I + N^{-1} \doteq (I + N)N^{-1}$$

and hence

$$(I + N^{-1})^{-1} \doteq [(I + N)N^{-1}]^{-1}$$
$$\doteq N(I + N)^{-1}$$

by virtue of *16*. Consequently, *15* can be rewritten as

$$\begin{aligned} y &= [(I + N)^{-1} + N(I + N)^{-1}](u) \\ &= (I + N)(1 + N)^{-1}(u) \\ &= I(u) \\ &= u \end{aligned}$$

This shows that any network of the form shown in Fig. *3.4.3*, in which $\mathfrak{N}$ and $\mathfrak{N}'$ are dual networks, is zero-state equivalent to a unit resistor and hence is zero-state linear. On the other hand, $\mathfrak{A}$ is not linear in the sense of Definition *5* for the starting state (1,0), where the first and second components respectively represent the voltage across C and the current through L. [This is easily verified by calculating y at $t = t_0$ and showing that $y(t_0)$ is no equal to $u(t_0)$.] Note that in consequence of the duality of $\mathfrak{N}$ and $\mathfrak{N}'$ only those states of $\mathfrak{A}$ in which the first and second components are numerically equal are reachable from the zero state.

Zero-input linearity

The above examples demonstrate that, in general, zero-state linearity does not imply linearity with respect to all possible initial states. Furthermore, it is not difficult to construct examples of systems (see Example *24*) which have the stronger property of linearity (in the sense of Definition *5*) with respect to all possible starting states and yet are clearly nonlinear. This indicates that the class of linear systems is narrower than the class of systems which are linear with respect to all possible starting states.

The additional property that we must require of a system $\mathfrak{A}$ in order to justify our calling $\mathfrak{A}$ "a linear system" is the property of *zero-input linearity* defined below.

17 **Definition** Let $\mathfrak{A}$ be a system whose state space Σ is a linear vector space. Then $\mathfrak{A}$ is *zero-input linear* if its zero-input response is a homogeneous and additive function of the initial state. That is,

18 $$\bar{\mathbf{A}}(k\boldsymbol{\alpha};\mathbf{0}) = k\bar{\mathbf{A}}(\boldsymbol{\alpha};\mathbf{0})$$

for all $\boldsymbol{\alpha}$ in Σ and all real constants k; and

19
$$\bar{\mathbf{A}}(\boldsymbol{\alpha}' + \boldsymbol{\alpha}''; \mathbf{0}) = \bar{\mathbf{A}}(\boldsymbol{\alpha}';\mathbf{0}) + \bar{\mathbf{A}}(\boldsymbol{\alpha}'';\mathbf{0})$$

for all $\boldsymbol{\alpha}'$, $\boldsymbol{\alpha}''$ in Σ, where $\bar{\mathbf{A}}(\boldsymbol{\alpha};\mathbf{0})$ denotes the zero-input response (segment) of $\mathcal{A}$ starting in state $\boldsymbol{\alpha}$. (Note that *18* and *19* can be combined into the single relation

20
$$\bar{\mathbf{A}}(k(\boldsymbol{\alpha}' - \boldsymbol{\alpha}'');\mathbf{0}) = k\bar{\mathbf{A}}(\boldsymbol{\alpha}';\mathbf{0}) - k\bar{\mathbf{A}}(\boldsymbol{\alpha}'';\mathbf{0}) \qquad \forall\boldsymbol{\alpha}'\forall\boldsymbol{\alpha}''\forall k$$

in the manner of Definition *2*.)

21 *Comment* Since one-one mapping of the state space onto itself does not affect the character of a system (see *2.3.46 et seq.*) it is sufficient to require that the zero-input response be a homogeneous and additive function of not necessarily $\boldsymbol{\alpha}$ itself, but of some variable $\hat{\boldsymbol{\alpha}}$ which is in one-one correspondence with $\boldsymbol{\alpha}$. However, the more restrictive wording of Definition *17* is adequate for our purposes. ◁

A system may be zero-input linear without being zero-state linear, and vice versa. This possibility is illustrated by the following examples.

22 *Example* The system characterized by the input-output-state relation

23
$$y(t) = e^{-(t-t_0)}\alpha + \int_{t_0}^{t} e^{-(t-\xi)}u^2(\xi)\, d\xi \qquad t \geq t_0$$

where α represents the initial state of the system, is zero-input linear but not zero-state linear. (Here $\Sigma = \mathcal{R}^1 \triangleq$ real line.)

24 *Example* The system characterized by the input-output-state relation

$$y(t) = (e^{-(t-t_0)}\alpha_1 + e^{-2(t-t_0)}\alpha_2)^2 + \int_{t_0}^{t} (e^{-(t-\xi)} + e^{-2(t-\xi)})u(\xi)\, d\xi \qquad t \geq t_0$$

where $\boldsymbol{\alpha} = (\alpha_1,\alpha_2)$ is the initial state vector, is zero-state linear but not zero-input linear. (Here $\Sigma = \mathcal{R}^2 \triangleq$ space of pairs of real numbers.)

Definition of linearity

We are now ready to formulate a general definition of linearity which subsumes both zero-state linearity and zero-input linearity. It reads:

25 **Definition** A system $\mathcal{A}$ is *linear* if and only if:
(I) $\mathcal{A}$ is linear with respect to all possible initial states; that is

26
$$k\bar{\mathbf{A}}(\boldsymbol{\alpha};\mathbf{u}) - k\bar{\mathbf{A}}(\boldsymbol{\alpha};\mathbf{v}) = \bar{\mathbf{A}}(\boldsymbol{\theta};k(\mathbf{u} - \mathbf{v}))$$

for all $\mathbf{u},\mathbf{v}$ in $R[\mathbf{u}]$ and all real constants k (see Definition *5*); and
(II) $\mathcal{A}$ is zero-input linear; that is

27
$$k\bar{\mathbf{A}}(\boldsymbol{\alpha}';\mathbf{0}) - k\bar{\mathbf{A}}(\boldsymbol{\alpha}'';\mathbf{0}) = \bar{\mathbf{A}}(k(\boldsymbol{\alpha}' - \boldsymbol{\alpha}'');\mathbf{0})$$

for all $\boldsymbol{\alpha}',\boldsymbol{\alpha}''$ in Σ (see Definition *17*). It is understood that Σ, $R[\mathbf{u}]$, and $R[\mathbf{y}]$ are linear spaces (see *3.3.2*). ◁

This definition constitutes the starting point of our analysis of linear systems. We proceed to study some of the immediate implications of Definition *25* in the following section.

5 *Linearity and some of its implications*

The decomposition property

There are several basic properties of linear systems which can easily be deduced from Definition *3.4.25*. Specifically, on setting $k = 1$ and $\mathbf{v} = \mathbf{0}$ in *3.4.26*, we obtain the identity

1 $$\bar{\mathbf{A}}(\boldsymbol{\alpha};\mathbf{u}) = \bar{\mathbf{A}}(\boldsymbol{\alpha};\mathbf{0}) + \bar{\mathbf{A}}(\boldsymbol{\theta};\mathbf{u}) \qquad \forall\boldsymbol{\alpha}\forall\mathbf{u}$$

which, in words, means that the response of $\mathcal{A}$ to $\mathbf{u}$ starting in state $\boldsymbol{\alpha}$ is identical with the zero-input response of $\mathcal{A}$ starting in state $\boldsymbol{\alpha}$ plus the zero-state response of $\mathcal{A}$ to $\mathbf{u}$.

Any, not necessarily linear, system whose input-output-state relation admits of this representation will be said to have the *decomposition property*. More formally,

2 **Definition** A system $\mathcal{A}$ has the *decomposition property* if its response to any input $\mathbf{u}$, with $\mathcal{A}$ initially in any state $\boldsymbol{\alpha}$, can be expressed in the form *1*, that is, as the sum of the zero-input response $\bar{\mathbf{A}}(\boldsymbol{\alpha};\mathbf{0})$ and the zero-state response $\bar{\mathbf{A}}(\boldsymbol{\theta};\mathbf{u})$. ◁

Equation *1* shows that every linear system has the decomposition property. The converse, however, is not true, as is demonstrated by Examples *3.4.22* and *3.4.24*.

An alternative definition of linearity

If a system $\mathcal{A}$ is linear, then in consequence of Definition *3.4.25* it has the decomposition property, and in addition, it is zero-state and zero-input linear. Conversely, it is easy to establish that if $\mathcal{A}$ has these three properties, then it is linear in the sense of Definition *3.4.25*. Specifically, suppose that $\mathcal{A}$ has the decomposition property. Then, on writing

3 $$k\bar{\mathbf{A}}(\boldsymbol{\alpha};\mathbf{u}) = k\bar{\mathbf{A}}(\boldsymbol{\alpha};\mathbf{0}) + k\bar{\mathbf{A}}(\boldsymbol{\theta};\mathbf{u}) \qquad \forall\boldsymbol{\alpha}\forall\mathbf{u}\forall k$$

4 $$k\bar{\mathbf{A}}(\boldsymbol{\alpha};\mathbf{v}) = k\bar{\mathbf{A}}(\boldsymbol{\alpha};\mathbf{0}) + k\bar{\mathbf{A}}(\boldsymbol{\theta};\mathbf{v}) \qquad \forall\boldsymbol{\alpha}\forall\mathbf{v}\forall k$$

and subtracting *4* from *3*, we have

5 $$k\bar{\mathbf{A}}(\boldsymbol{\alpha};\mathbf{u}) - k\bar{\mathbf{A}}(\boldsymbol{\alpha};\mathbf{v}) = k\bar{\mathbf{A}}(\boldsymbol{\theta};\mathbf{u}) - k\bar{\mathbf{A}}(\boldsymbol{\theta};\mathbf{v}) \qquad \forall\boldsymbol{\alpha}\forall\mathbf{u}\forall\mathbf{v}\forall k$$

If, in addition, $\mathcal{A}$ is zero-state linear, then *5* can be replaced by

$$k\bar{\mathbf{A}}(\boldsymbol{\alpha};\mathbf{u}) - k\bar{\mathbf{A}}(\boldsymbol{\alpha};\mathbf{v}) = \bar{\mathbf{A}}(\boldsymbol{\theta};k(\mathbf{u}-\mathbf{v})) \qquad \forall\boldsymbol{\alpha}\forall\mathbf{u}\forall\mathbf{v}\forall k$$

This relation shows that if $\mathcal{A}$ has the decomposition property and is zero-state linear, then $\mathcal{A}$ is linear with respect to all possible initial states $\boldsymbol{\alpha}$. Consequently, if $\mathcal{A}$ has the decomposition property and is both zero-state and zero-input linear, then $\mathcal{A}$ qualifies as a linear system under Definition *3.4.25*.

This conclusion permits us to replace Definition *3.4.25* by the equivalent

6 **Definition** A system $\mathcal{A}$ is *linear* if and only if it has the following three properties:

Decomposition property (*2*)

7 $$\bar{\mathbf{A}}(\boldsymbol{\alpha};\mathbf{u}) = \bar{\mathbf{A}}(\boldsymbol{\alpha};\mathbf{0}) + \bar{\mathbf{A}}(\boldsymbol{\theta};\mathbf{u}) \qquad \forall\boldsymbol{\alpha}\forall\mathbf{u}$$

Zero-state linearity (*3.4.1*)

8 $$\bar{\mathbf{A}}(\boldsymbol{\theta};k(\mathbf{u}-\mathbf{v})) = k\bar{\mathbf{A}}(\boldsymbol{\theta};\mathbf{u}) - k\bar{\mathbf{A}}(\boldsymbol{\theta};\mathbf{v}) \qquad \forall\mathbf{u}\forall\mathbf{v}\forall k$$

Zero-input linearity (*3.4.17*)

9 $$\bar{\mathbf{A}}(k(\boldsymbol{\alpha}'-\boldsymbol{\alpha}'');\mathbf{0}) = k\bar{\mathbf{A}}(\boldsymbol{\alpha}';\mathbf{0}) - k\bar{\mathbf{A}}(\boldsymbol{\alpha}'';\mathbf{0}) \qquad \forall\boldsymbol{\alpha}'\forall\boldsymbol{\alpha}''\forall k \qquad \triangleleft$$

In what follows, we shall be using for the most part Definition *6*, rather than the earlier Definition *3.4.25*, as the definition of linearity. We shall forego illustrating Definition *6* at this point, since there will be numerous examples of linear systems in the sequel.

A fact concerning equivalent states and equivalent systems

Before proceeding to study various special features of linear systems, it will be helpful to establish a simple fact concerning state equivalence which will be of use later in this section and which is characteristic of systems having the decomposition property. Expressed in the form of an assertion, it reads

10 **Assertion** If $\mathcal{A}$ has the decomposition property, then two states of $\mathcal{A}$, say $\boldsymbol{\alpha}'$ and $\boldsymbol{\alpha}''$, are equivalent (*2.2.1*) if and only if the zero-input response of $\mathcal{A}$ starting in state $\boldsymbol{\alpha}'$ is identical with the zero-input response of $\mathcal{A}$ starting in state $\boldsymbol{\alpha}''$. In symbols,

11 $$\{\boldsymbol{\alpha}' \simeq \boldsymbol{\alpha}''\} \Leftrightarrow \{\bar{\mathbf{A}}(\boldsymbol{\alpha}';\mathbf{0}) = \bar{\mathbf{A}}(\boldsymbol{\alpha}'';\mathbf{0})\}$$

Comment An immediate consequence of Assertion *10* is that if $\mathcal{A}$ and $\mathcal{B}$ have the decomposition property, then $\mathcal{A} \equiv \mathcal{B}$ if $\mathcal{A}$ and $\mathcal{B}$ are both zero-input and zero-state equivalent (see *2.9.1* and *2.9.2*).

Proof We have

$$\bar{\mathbf{A}}(\boldsymbol{\alpha}';\mathbf{u}) = \bar{\mathbf{A}}(\boldsymbol{\alpha}';\mathbf{0}) + \bar{\mathbf{A}}(\boldsymbol{\theta};\mathbf{u})$$
$$\bar{\mathbf{A}}(\boldsymbol{\alpha}'';\mathbf{u}) = \bar{\mathbf{A}}(\boldsymbol{\alpha}'';\mathbf{0}) + \bar{\mathbf{A}}(\boldsymbol{\theta};\mathbf{u})$$

from which it follows at once that $\bar{\mathbf{A}}(\boldsymbol{\alpha}';\mathbf{u}) = \bar{\mathbf{A}}(\boldsymbol{\alpha}'';\mathbf{u})$ for all $\mathbf{u}$ if and only if $\bar{\mathbf{A}}(\boldsymbol{\alpha}';\mathbf{0}) = \bar{\mathbf{A}}(\boldsymbol{\alpha}'';\mathbf{0})$. Note that *11* necessitates that the zero state $\boldsymbol{\theta}$ be unique (to within equivalent states).

12 *Example* As a simple illustration of *10*, consider the network of Example *2.2.7*, for which the input-output-state relation reads

$$y(t) = i_1(t_0-) + i_2(t_0-) + \left(\frac{1}{L_1} + \frac{1}{L_2}\right) \int_{t_0}^{t} u(\xi)\, d\xi \qquad t \geq t_0$$

Here $\boldsymbol{\alpha}$ is a 2-vector with components $\alpha_1 \triangleq i_1(t_0-)$ and $\alpha_2 \triangleq i_2(t_0-)$, and the zero-input response is given by

$$\bar{\mathbf{A}}(\boldsymbol{\alpha};\mathbf{0}) = \alpha_1 + \alpha_2 \qquad t \geq t_0$$

Thus, two states $\boldsymbol{\alpha}' \triangleq (\alpha_1',\alpha_2')$ and $\boldsymbol{\alpha}'' \triangleq (\alpha_1'',\alpha_2'')$ are equivalent if and only if $\alpha_1' + \alpha_2' = \alpha_1'' + \alpha_2''$. Note that any state of the form $\boldsymbol{\alpha} = (\alpha_1, -\alpha_1)$ is equivalent to the zero state $\boldsymbol{\theta} = (0,0)$. ◁

The following corollary of Assertion *10* follows at once from *10* and the zero-input linearity of $\mathcal{A}$.

13 **Corollary** If $\boldsymbol{\alpha}'$ and $\boldsymbol{\alpha}''$ are equivalent states of a linear system $\mathcal{A}$, then every linear combination $\boldsymbol{\alpha} = a\boldsymbol{\alpha}' + b\boldsymbol{\alpha}''$ is a state equivalent to $\boldsymbol{\alpha}'$ (and $\boldsymbol{\alpha}''$). This implies that the sets of equivalent states of $\mathcal{A}$ constitute *linear varieties* in Σ. (A linear variety is a translate of a subspace of Σ.)

Special features of linear systems

The importance of the concept of linearity stems mainly from two special features of linear systems. The first of these is the relative ease with which the response of a linear system $\mathcal{A}$ to a given input $\mathbf{u}$ can be determined. Specifically, by virtue of the decomposition property, the response can be expressed in the form

$$\mathbf{A}(\boldsymbol{\alpha};\mathbf{u}) = \mathbf{A}(\boldsymbol{\alpha};\mathbf{0}) + \mathbf{A}(\boldsymbol{\theta};\mathbf{u})$$

which separates the effect of $\mathbf{u}$ from that of the "initial excitation" represented by $\boldsymbol{\alpha}$. Furthermore, by virtue of the zero-state linearity of $\mathcal{A}$, the problem of determining the zero-state response $\mathbf{A}(\boldsymbol{\theta};\mathbf{u})$, or, for short, $\mathbf{A}(\mathbf{u})$, can be reduced to that of resolving $\mathbf{u}$ into simpler components $\boldsymbol{\phi}_1, \boldsymbol{\phi}_2, \ldots$ and determining the zero-state response to each component separately. More specifically, suppose that $\mathbf{u}$ is resolved into a set of component time functions $\boldsymbol{\phi}_1, \boldsymbol{\phi}_2, \ldots$ by expressing $\mathbf{u}$ as an infinite series

14 $$\mathbf{u} = a_1\boldsymbol{\phi}_1 + a_2\boldsymbol{\phi}_2 + \cdots$$

where the coefficients a_λ, $\lambda = 1, 2, \ldots$, are constants representing the "weights" of the component time functions $\boldsymbol{\phi}_1, \boldsymbol{\phi}_2, \ldots$ in **u**. Now, if $\mathcal{A}$ is countably additive, then the zero-state response of $\mathcal{A}$ to **u** can be expressed as an infinite series

15 $$\mathbf{A}(\mathbf{u}) = a_1\mathbf{A}(\boldsymbol{\phi}_1) + a_2\mathbf{A}(\boldsymbol{\phi}_2) + \cdots$$

where $\mathbf{A}(\boldsymbol{\phi}_\lambda)$ represents the zero-state response of $\mathcal{A}$ to the component $\boldsymbol{\phi}_\lambda$, $\lambda = 1, 2, \ldots$. Thus, if the $\boldsymbol{\phi}_\lambda$ are chosen in such a way that the determination of the zero-state response of $\mathcal{A}$ to $\boldsymbol{\phi}_\lambda$, $\lambda = 1, 2, \ldots$, is a significantly simpler problem than the direct calculation of $\mathbf{A}(\mathbf{u})$, then it may be advantageous to determine $\mathbf{A}(\mathbf{u})$ indirectly by (I) resolving **u** into the $\boldsymbol{\phi}_\lambda$, (II) calculating $\mathbf{A}(\boldsymbol{\phi}_\lambda)$, $\lambda = 1, 2, \ldots$, and (III) obtaining $\mathbf{A}(\mathbf{u})$ by summing the series represented by *15*.

This basic procedure appears in various guises in many of the methods used to analyze the behavior of linear systems. In general, the set of component time functions $\{\boldsymbol{\phi}_\lambda\}$ is a continuum rather than a countable set—as is assumed in the above discussion—and the summations in *14* and *15* are integrals in which λ plays the role of the variable of integration. The basic idea, however, remains the same.

16 *Example* In the case of a complex Fourier series, the component time functions are of the form $e^{j\lambda\omega_0 t}$, where λ ranges over all integers and ω_0 is a real constant (fundamental frequency).

17 *Example* In the case of the Fourier integral representation, the component time functions form a continuum $\{e^{j\omega t}\}$, in which ω ranges over all real numbers.

18 *Example* The Laplace transform technique for the analysis of linear systems (see Appendix *B*) is based essentially on the resolution of u and y into component time functions of the form e^{st}, where s is a complex variable ranging over a line parallel to the axis of imaginaries in the s plane. Note that both the Laplace transform and the Fourier integral techniques are particularly effective in dealing with systems in which the determination of the response to $e^{j\omega t}$ or e^{st} is a simple problem. It is for this reason that the Laplace and Fourier transform techniques are used so widely in the analysis of linear time-invariant systems.

19 *Example* A particularly simple set of component time functions is the set of delta functions $\{\delta(t - \lambda)\}$, in which λ ranges over the real line. In this case, the resolution of a scalar time function u into the $\delta(t - \lambda)$ is trivially simple (see Appendix *A*), but the problem of finding the zero-state response to $\delta(t - \lambda)$ is frequently not much easier

than that of finding the zero-state response to u. Nevertheless, there is generally a distinct advantage, at least for purposes of representation of input-output relations, in resolving the input time function into the delta functions.

The closure theorem

The second special feature of linear systems finds its expression in the following

20 **Theorem** The class of linear systems is closed under the operations of system addition, multiplication, and inversion. Stated more explicitly, if $\mathcal{A}$ and $\mathcal{B}$ are linear systems, then so are $\mathcal{A} + \mathcal{B}$ and $\mathcal{A}\mathcal{B}$. Furthermore, if $\mathcal{A}$ has an inverse $\mathcal{A}^{-1}$, then $\mathcal{A}^{-1}$ is a linear system. (In effect, Theorem *20* states that any interconnection of linear systems is a linear system.)

Proof To prove this theorem, we have to establish that $\mathcal{A} + \mathcal{B}$, $\mathcal{A}\mathcal{B}$, and $\mathcal{A}^{-1}$ have the decomposition property and are both zero-state and zero-input linear. We begin with establishing this for $\mathcal{A} + \mathcal{B}$.

Let $\boldsymbol{\alpha}$ and $\boldsymbol{\beta}$ denote the states of $\mathcal{A}$ and $\mathcal{B}$, respectively. Then by Theorem *2.6.10* the state of $\mathcal{A} + \mathcal{B} \triangleq \mathcal{C}$ can be taken to be $\boldsymbol{\gamma} \triangleq (\boldsymbol{\alpha},\boldsymbol{\beta})$ (see Theorem *2.6.23*), with the zero state of $\mathcal{C}$ being $(\boldsymbol{\theta},\boldsymbol{\theta})$. (*Exercise:* Show that the zero state of $\mathcal{C}$ is unique.)

In terms of the input-output-state relations of $\mathcal{A}$ and $\mathcal{B}$, the input-output-state relation of $\mathcal{C}$ reads

21 $$\bar{\mathbf{C}}(\boldsymbol{\gamma};\mathbf{u}) = \bar{\mathbf{A}}(\boldsymbol{\alpha};\mathbf{0}) + \bar{\mathbf{B}}(\boldsymbol{\beta};\mathbf{0}) + \bar{\mathbf{A}}(\boldsymbol{\theta};\mathbf{u}) + \bar{\mathbf{B}}(\boldsymbol{\theta};\mathbf{u})$$

It is evident from inspection of *21* that $\mathcal{C}$ has the decomposition property and is both zero-state and zero-input linear. Consequently, $\mathcal{C}$ is linear.

Next, consider $\mathcal{C} \triangleq \mathcal{A}\mathcal{B}$. In this case, recalling the definition of the product of two systems (*1.10.23*) and making use of the linearity of $\mathcal{A}$ and $\mathcal{B}$, the expression for the input-output-state relation of $\mathcal{C}$ becomes

22 $$\bar{\mathbf{C}}(\boldsymbol{\gamma};\mathbf{u}) = \bar{\mathbf{A}}(\boldsymbol{\alpha};\mathbf{0}) + \bar{\mathbf{A}}(\boldsymbol{\theta};\bar{\mathbf{B}}(\boldsymbol{\beta};\mathbf{0})) + \bar{\mathbf{A}}(\boldsymbol{\theta};\bar{\mathbf{B}}(\boldsymbol{\theta};\mathbf{u}))$$

where the second term represents the zero-state response of $\mathcal{A}$ to the zero-input response of $\mathcal{B}$ and the third term represents the zero-state response of $\mathcal{A}$ to the zero-state response of $\mathcal{B}$ to $\mathbf{u}$.

Now, on setting $\mathbf{u} = \mathbf{0}$ in *22*, the expression for the zero-input response of $\mathcal{C}$ is found to be

23 $$\bar{\mathbf{C}}(\boldsymbol{\gamma};\mathbf{0}) = \bar{\mathbf{A}}(\boldsymbol{\alpha};\mathbf{0}) + \bar{\mathbf{A}}(\boldsymbol{\theta};\bar{\mathbf{B}}(\boldsymbol{\beta};\mathbf{0}))$$

To verify that $\mathcal{C}$ is zero-input linear, we note that since $\boldsymbol{\gamma} \triangleq (\boldsymbol{\alpha},\boldsymbol{\beta})$, $k\boldsymbol{\gamma} = (k\boldsymbol{\alpha},k\boldsymbol{\beta})$ for all real k. Likewise, if $\boldsymbol{\gamma}' \triangleq (\boldsymbol{\alpha}',\boldsymbol{\beta}')$ and $\boldsymbol{\gamma}'' \triangleq (\boldsymbol{\alpha}'',\boldsymbol{\beta}'')$, then $\gamma' - \gamma'' = (\alpha' - \alpha'', \beta' - \beta'')$. Thus,

24 $$\begin{aligned}\bar{\mathbf{C}}(k(\gamma' - \gamma'');\mathbf{0}) &= \bar{\mathbf{A}}(k(\alpha' - \alpha'');\mathbf{0}) + \bar{\mathbf{A}}(\theta;\bar{\mathbf{B}}(k(\beta' - \beta'');\mathbf{0}))\\ &= k\bar{\mathbf{A}}(\alpha';\mathbf{0}) - k\bar{\mathbf{A}}(\alpha'';\mathbf{0}) + k\bar{\mathbf{A}}(\theta;\bar{\mathbf{B}}(\beta';\mathbf{0}))\\ &\qquad - k\bar{\mathbf{A}}(\theta;\bar{\mathbf{B}}(\beta'';\mathbf{0}))\\ &= k\bar{\mathbf{C}}(\gamma';\mathbf{0}) - k\bar{\mathbf{C}}(\gamma'';\mathbf{0}) \qquad \forall\gamma'\ \forall\gamma''\ \forall k\end{aligned}$$

by virtue of the linearity of $\mathcal{A}$ and $\mathcal{B}$. Consequently, $\mathcal{C}$ is zero-input linear.

In a similar fashion, the zero-state response of $\mathcal{C}$ is found by setting $\gamma = \theta_c \triangleq (\theta,\theta)$ = zero state of $\mathcal{C}$ in *22*. This yields

25 $$\bar{\mathbf{C}}(\theta_c;\mathbf{u}) = \bar{\mathbf{A}}(\theta;\bar{\mathbf{B}}(\theta;\mathbf{u}))$$

Again, for any $\mathbf{u}',\mathbf{u}''$ in $R[\mathbf{u}]$ and any real k, we can write

26 $$\begin{aligned}\bar{\mathbf{C}}(\theta_c;k(\mathbf{u}' - \mathbf{u}'')) &= k\bar{\mathbf{A}}(\theta;\mathbf{B}(\theta;\mathbf{u}')) - k\bar{\mathbf{A}}(\theta;\mathbf{B}(\theta;\mathbf{u}''))\\ &= k\bar{\mathbf{C}}(\theta_c;\mathbf{u}') - k\bar{\mathbf{C}}(\theta_c;\mathbf{u}'')\end{aligned}$$

which shows that $\mathcal{C}$ is zero-state linear. Furthermore, on comparing the expressions for $\bar{\mathbf{C}}(\gamma;\mathbf{0})$ (*23*) and $\bar{\mathbf{C}}(\theta_c;\mathbf{u})$ (*25*) with that for $\bar{\mathbf{C}}(\gamma;\mathbf{u})$ (*22*), we can write

$$\bar{\mathbf{C}}(\gamma;\mathbf{u}) = \bar{\mathbf{C}}(\gamma;\mathbf{0}) + \bar{\mathbf{C}}(\theta_c;\mathbf{u})$$

which implies that $\mathcal{C}$ has the decomposition property. This conclusion, together with the just established zero-state and zero-input linearity of $\mathcal{C}$, implies that $\mathcal{C}$ is linear.

Finally, consider $\mathcal{A}^{-1}$. We shall show first that $\mathcal{A}^{-1}$ is zero-state linear. For this purpose, define

27 $$\mathbf{v}' \triangleq \bar{\mathbf{A}}(\theta;\mathbf{u}') \qquad \mathbf{v}'' \triangleq \bar{\mathbf{A}}(\theta;\mathbf{u}'')$$

where $\mathbf{u}'$ and $\mathbf{u}''$ are arbitrary inputs in $R[\mathbf{u}]$, and let the input to $\mathcal{A}$ be $k(\mathbf{u}' - \mathbf{u}'')$ with $\mathcal{A}$ in its zero state. Then the input to $\mathcal{A}^{-1}$ is $k(\mathbf{v}' - \mathbf{v}'')$, and since $\mathcal{A}^{-1}$ is inverse to $\mathcal{A}$, its zero-state response to $k(\mathbf{v}' - \mathbf{v}'')$ must be $k(\mathbf{u}' - \mathbf{u}'')$. Thus,

$$\bar{\mathbf{A}}^{-1}(\theta;k(\mathbf{v}' - \mathbf{v}'')) = k\mathbf{u}' - k\mathbf{u}'' \qquad \forall\mathbf{u}'\forall\mathbf{u}''$$

where $\bar{\mathbf{A}}^{-1}(\theta;\mathbf{v})$ denotes the zero-state response of $\mathcal{A}^{-1}$ to $\mathbf{v}$, and since

28 $$\bar{\mathbf{A}}^{-1}(\theta;\mathbf{v}') = \mathbf{u}' \qquad \bar{\mathbf{A}}^{-1}(\theta;\mathbf{v}'') = \mathbf{u}''$$

we have

29 $$\bar{\mathbf{A}}^{-1}(\theta;k(\mathbf{v}' - \mathbf{v}'')) = k\bar{\mathbf{A}}^{-1}(\theta;\mathbf{v}') - k\bar{\mathbf{A}}^{-1}(\theta;\mathbf{v}'') \qquad \forall\mathbf{v}'\forall\mathbf{v}''\forall k$$

where $\mathbf{v}'$ and $\mathbf{v}''$ range over the output segment space of $\mathcal{A}$ [(which is assumed to coincide with the input segment space of $\mathcal{A}^{-1}$ (see *2.10.6*)]. This establishes that $\mathcal{A}^{-1}$ is zero-state linear.

Next, suppose that $\mathcal{A}$ is initially in state $\boldsymbol{\alpha}$ and $\mathcal{A}^{-1}$ is in a corresponding state $\boldsymbol{\alpha}^{-1}$ (see *2.10.6*). Under these conditions, let the input to $\mathcal{A}^{-1}$ be zero and let the zero-input response of $\mathcal{A}^{-1}$, which is $\bar{\mathbf{A}}^{-1}(\boldsymbol{\alpha}^{-1};\mathbf{0})$, be applied to $\mathcal{A}$. Since the output of $\mathcal{A}$ must be zero, we have

$$\bar{\mathbf{A}}(\boldsymbol{\alpha};\bar{\mathbf{A}}^{-1}(\boldsymbol{\alpha}^{-1};\mathbf{0})) = \mathbf{0}$$

which can be written as

30 $$\bar{\mathbf{A}}(\boldsymbol{\alpha};\mathbf{0}) = -\bar{\mathbf{A}}(\boldsymbol{\theta};\bar{\mathbf{A}}^{-1}(\boldsymbol{\alpha}^{-1};\mathbf{0}))$$

by virtue of the decomposition property of $\mathcal{A}$.

Under the same conditions, let the input to $\mathcal{A}$ be $\mathbf{u}$ and let the output of $\mathcal{A}$,

31 $$\mathbf{w} \triangleq \bar{\mathbf{A}}(\boldsymbol{\alpha};\mathbf{u}) = \bar{\mathbf{A}}(\boldsymbol{\alpha};\mathbf{0}) + \bar{\mathbf{A}}(\boldsymbol{\theta};\mathbf{u})$$

be applied to $\mathcal{A}^{-1}$ in state $\boldsymbol{\alpha}^{-1}$. In this case, the output of $\mathcal{A}^{-1}$ must be $\mathbf{u}$ and hence

32 $$\bar{\mathbf{A}}^{-1}(\boldsymbol{\alpha}^{-1};\bar{\mathbf{A}}(\boldsymbol{\alpha};\mathbf{0}) + \bar{\mathbf{A}}(\boldsymbol{\theta};\mathbf{u})) = \mathbf{u}$$

Now let $\mathbf{v}$ be the zero-state response of $\mathcal{A}$ to $\mathbf{u}$. Then

$$\mathbf{v} = \bar{\mathbf{A}}(\boldsymbol{\theta};\mathbf{u}) \qquad \text{and} \qquad \bar{\mathbf{A}}^{-1}(\boldsymbol{\theta};\mathbf{v}) = \mathbf{u}$$

since $\mathbf{v}$ applied to $\mathcal{A}^{-1}$ (in zero state) should yield $\mathbf{u}$. Thus, *32* may be rewritten as

33 $$\bar{\mathbf{A}}^{-1}(\boldsymbol{\alpha}^{-1};\, \bar{\mathbf{A}}(\boldsymbol{\alpha};\mathbf{0}) + \mathbf{v}) = \bar{\mathbf{A}}^{-1}(\boldsymbol{\theta};\mathbf{v})$$

or, in view of *31*,

$$\bar{\mathbf{A}}^{-1}(\boldsymbol{\alpha}^{-1};\mathbf{w}) = \bar{\mathbf{A}}^{-1}(\boldsymbol{\theta};\, \mathbf{w} - \bar{\mathbf{A}}(\boldsymbol{\alpha};\mathbf{0}))$$

which in turn may be written as

34 $$\bar{\mathbf{A}}^{-1}(\boldsymbol{\alpha}^{-1};\mathbf{w}) = \bar{\mathbf{A}}^{-1}(\boldsymbol{\theta};\mathbf{w}) - \bar{\mathbf{A}}^{-1}(\boldsymbol{\theta};\bar{\mathbf{A}}(\boldsymbol{\alpha};\mathbf{0}))$$

by virtue of the zero-state linearity of $\mathcal{A}^{-1}$.

On substituting $\bar{\mathbf{A}}(\boldsymbol{\alpha};\mathbf{0})$ as given by *30* into *34* and making use of the identity

$$\bar{\mathbf{A}}^{-1}(\boldsymbol{\theta};\bar{\mathbf{A}}(\boldsymbol{\theta};\bar{\mathbf{A}}^{-1}(\boldsymbol{\alpha}^{-1};\mathbf{0}))) = \bar{\mathbf{A}}^{-1}(\boldsymbol{\alpha}^{-1};\mathbf{0})$$

we obtain

35 $$\bar{\mathbf{A}}^{-1}(\boldsymbol{\alpha}^{-1};\mathbf{w}) = \bar{\mathbf{A}}^{-1}(\boldsymbol{\alpha}^{-1};\mathbf{0}) + \bar{\mathbf{A}}^{-1}(\boldsymbol{\theta};\mathbf{w})$$

which establishes that $\mathcal{A}^{-1}$ has the decomposition property.

It remains to be shown that $\mathcal{A}^{-1}$ is zero-input linear. To this end, let $\mathcal{A}$ be initially in state $\boldsymbol{\alpha}$ and let the input to $\mathcal{A}$ be zero. Then, on applying the output of $\mathcal{A}$, which is $\bar{\mathbf{A}}(\boldsymbol{\alpha};\mathbf{0})$, to $\mathcal{A}^{-1}$ (in state $\boldsymbol{\alpha}^{-1}$)

we should get zero output; i.e.,

36 $$\bar{\mathbf{A}}^{-1}(\boldsymbol{\alpha}^{-1};\bar{\mathbf{A}}(\boldsymbol{\alpha};\mathbf{0})) = \mathbf{0}$$

Since $\mathcal{A}^{-1}$ has the decomposition property, *36* may be written as

$$\bar{\mathbf{A}}^{-1}(\boldsymbol{\alpha}^{-1};\mathbf{0}) = -\bar{\mathbf{A}}^{-1}(\boldsymbol{\theta};\mathbf{A}(\boldsymbol{\alpha};\mathbf{0}))$$

From this relation and the zero-state linearity of $\mathcal{A}$ and $\mathcal{A}^{-1}$ it follows at once that $\mathcal{A}^{-1}$ is zero-input linear. This completes our proof of Theorem *20*. ◁

An incidental and yet interesting identity which we used in the course of proving Theorem *20* is given by *30*. In view of the importance of the property expressed by *30*, we state it as an

37 **Assertion** Let $\mathcal{A}$ be a (not necessarily linear) system having the decomposition property and let $\mathcal{A}$ have an inverse $\mathcal{A}^{-1}$. Then, given any initial state $\boldsymbol{\alpha}$ of $\mathcal{A}$, one can always find an input $\mathbf{u}$ such that the zero-state response of $\mathcal{A}$ to $\mathbf{u}$ is equal to the negative of the zero-input response of $\mathcal{A}$ starting in state $\boldsymbol{\alpha}$. Specifically, $\mathbf{u}$ is given by

38 $$\mathbf{u} = \bar{\mathbf{A}}^{-1}(\boldsymbol{\alpha}^{-1};\mathbf{0})$$

where $\boldsymbol{\alpha}^{-1}$ is the state of $\mathcal{A}^{-1}$ corresponding to the state $\boldsymbol{\alpha}$ of $\mathcal{A}$.

In plain words, this means that the initial excitation of $\mathcal{A}$, which is defined by $\boldsymbol{\alpha}$, can be traded for an input $\mathbf{u}$ given by *38*.

39 *Example* As a simple illustration, let $\mathcal{A}$ be a system characterized by the input-output relation

40 $$\frac{dy}{dt} + y = u$$

The input-output state relation of this system reads (see *2.10.14*)

41 $$y(t) = y(t_0-)e^{-(t-t_0)} + \int_{t_0}^{t} e^{-(t-\xi)}u(\xi)\,d\xi \qquad t \geq t_0$$

with $y(t_0-) = \alpha$ representing the state at time t_0-.

Correspondingly, the input-output-state relation of the inverse system is given by (see *2.10.15*)

$$y(t) = -u(t_0-)\delta(t - t_0) + \frac{dy}{dt} + u$$

and if the state of $\mathcal{A}$ is α, then the corresponding state of $\mathcal{A}^{-1}$ is $-u(t_0) = -y(t_0-) = -\alpha$.

Now the zero-input response of $\mathcal{A}$ starting in state α is

$$A(\alpha;0) = \alpha e^{-(t-t_0)} \qquad t \geq t_0$$

and by Assertion *37* the negative of this response can be obtained by applying to $\mathcal{A}$ the input

42 $$A^{-1}(\alpha^{-1};0) = -\alpha\delta(t - t_0)$$

with $\mathcal{A}$ starting in zero state. This is easily verified by substituting *42* into *41*, with $y(t_0-)$ set equal to zero.

6 *Further implications of linearity*

Representation of input-output-state relations

In this section we shall focus our attention on the properties of input-output-state relations of linear systems.

We begin by noting that in consequence of the decomposition property (*3.5.2*), the input-output-state relation of any[1] system $\mathcal{A}$ admits of the representation

1 $$\mathbf{y}(t) \triangleq \mathbf{A}(\boldsymbol{\alpha};\mathbf{u}) = \mathbf{A}(\boldsymbol{\alpha};\mathbf{0}) + \mathbf{A}(\mathbf{u}) \qquad \boldsymbol{\alpha} \in \Sigma,\ \mathbf{u} \in R[\mathbf{u}]$$

where $\mathbf{u} = \mathbf{u}_{(t_0,t]}$ is the input segment, with $\mathbf{u}$ ranging over a linear input segment space $R[\mathbf{u}]$; $\boldsymbol{\alpha}$ is the initial state (at $t = t_0-$) (see Remark *1.5.5*) of $\mathcal{A}$, with $\boldsymbol{\alpha}$ ranging over a linear state space Σ; $\mathbf{A}(\boldsymbol{\alpha};\mathbf{0})$ denotes the zero-input response of $\mathcal{A}$ at time t, with $\mathcal{A}$ starting in state $\boldsymbol{\alpha}$ at time t_0-; and $\mathbf{A}(\mathbf{u})$ is the zero-state response of $\mathcal{A}$ at time t to $\mathbf{u} \triangleq \mathbf{u}_{(t_0,t]}$. When there is need to place in evidence the fact that $\boldsymbol{\alpha}$ is the state of $\mathcal{A}$ at time t_0-, *1* will be written in the form

2 $$\mathbf{y}(t) = \mathbf{A}(\mathbf{x}(t_0-);\mathbf{0}) + \mathbf{A}(\mathbf{u})$$

where $\mathbf{x}(t_0-)$ denotes the state of $\mathcal{A}$ at time t_0-. [In this and subsequent sections we denote the state vector by $\mathbf{x}$ in place of $\mathbf{s}$ in order to avoid confusion with the Laplace transform variable s. Furthermore, we shall frequently be forced to depart from our notation for time functions in order to avoid conflict with the customary notations for delta functions and Laplace transforms (see Remark *1.3.3*).]

To simplify our analysis, we shall make the convenient but inessential assumption that $\mathbf{u}$ and $\mathbf{y}$ are scalar-valued. The results obtained under this assumption can readily be extended to the general case where $\mathbf{u}$ and $\mathbf{y}$ are vector-valued. The general case will be treated in Chaps. *5* and *6*.

Fixing our attention on *1*, we note that the second term in the right-hand member represents the result of acting on u with a linear operator

[1] Henceforth it will be understood that $\mathcal{A}$ is a linear system unless explicitly noted to the contrary.

defined on the input segment space $R[u]$. From the basic theorems of Fréchet and Riesz concerning the representation of linear functionals and the extensions of these theorems to distributions,[1] it follows that the term $A(u)$ admits of a formal representation

3
$$A(u) = \int_{t_0}^{t} h(t,\xi)u(\xi)\, d\xi$$

where the kernel $h(t,\xi)$ may contain delta functions of various orders, with the understanding that if the integrand in *3* contains delta functions at t_0 or t, then the limits of integration in *3* should be t_0- and $t+$. To express *3* in a rigorous form requires the use of the language of the theory of distributions. For our purposes, *3* will be sufficient, even though it is not well defined by mathematical standards.

Impulse response

The kernel $h(t,\xi)$ of *3* has a simple interpretation. Specifically, on applying a unit impulse (delta function) $\delta(t - \xi)$ to $\mathcal{A}$, with $\mathcal{A}$ starting in its zero state, and changing the variable of integration in *3* to ξ' in order to avoid confusion with ξ, we have

4
$$\begin{aligned} \text{zero-state response of } \mathcal{A} \text{ to } \delta(t - \xi) &= \int_{t_0}^{t} h(t,\xi')\delta(\xi' - \xi)\, d\xi' \\ &= h(t,\xi) \qquad \text{for } t \geq \xi \geq t_0 \\ &= 0 \qquad \text{for } \xi > t \geq t_0 \end{aligned}$$

by the sifting property of delta functions (*A.2.10*) and the definition of zero state. Consequently, we can make the

5 **Assertion** The zero-state response of $\mathcal{A}$ to an input u is given by

6
$$A(u) = \int_{t_0}^{t} h(t,\xi)u(\xi)\, d\xi \qquad t_0 \leq \xi \leq t$$

where $h(t,\xi)$ is the zero-state response of $\mathcal{A}$ at time t to a unit impulse $\delta(t - \xi)$ applied at time ξ. The zero-state response $h(t,\xi)$ is called the *impulse response* of $\mathcal{A}$. Note that by *4* $h(t,\xi)$ vanishes for $t < \xi$.[2] ◁

A detailed discussion of the properties of impulse response will be presented in Chap. *8*. Here we shall restrict ourselves to giving two simple examples and establishing a few elementary properties which will be needed in this chapter.

7 *Example* The zero-state response of the system of Example *3.3.23* is

[1] For an exposition of these theorems see Ref. *2*. For a discussion of distributions see Appendix *A*.

[2] It should be remembered that we have assumed throughout that $\mathcal{A}$ is a nonanticipative system (*1.8.13*). Without this assumption, $h(t,\xi)$ need not vanish for $t < \xi$.

given by

$$A(u) = \int_{t_0}^{t} e^{-(t-\xi)}u(\xi)\,d\xi$$

Consequently, the impulse response of this system reads

$$h(t,\xi) = e^{-(t-\xi)} \qquad t \geq \xi$$
$$h(t,\xi) = 0 \qquad t < \xi$$

or more compactly,

$$h(t,\xi) = 1(t-\xi)e^{-(t-\xi)}$$

where the unit step function $1(t-\xi)$ serves to truncate $h(t,\xi)$ to the left of $t = \xi$.

8 *Example* The zero-state response of the system of Example *3.2.22* is given by

$$A(u) = \frac{1}{t}\int_{t_0}^{t} u(\xi)\,d\xi \qquad t \geq t_0 > 0$$

In this case

$$h(t,\xi) = 1(t-\xi)\frac{1}{t} \qquad \xi > 0$$

Zero-state time invariance and equivalence

We observe that the system of Example *7* is time-invariant and that its impulse response depends only on the difference of t and ξ. This is a special case of a more general property which is now stated in the form of an

9 **Assertion** A system $\mathcal{A}$ is zero-state time-invariant (*3.2.12*) if and only if its impulse response $h(t,\xi)$ is of the form $h(t-\xi)$.

Proof To prove this assertion, it is expedient to represent $h(t,\xi)$ in the form

10 $$h(t,\xi) = w(\tau,t)$$

where $\tau \triangleq t-\xi$, and then demonstrate that $w(\tau,t)$ is independent of t. (The variable τ as defined here is sometimes referred to as the "age variable." Essentially, the age variable is reversed time with the "present" serving as the reference point.)

Now by definition, $h(t,\xi)$, or equivalently, $w(\tau,t)$, is the zero-state response of $\mathcal{A}$ to $\delta(t-\xi)$. Consequently, the zero-state response of $\mathcal{A}$ to $\delta(t-(\xi+\lambda))$, where λ is an arbitrary shift in ξ, must be $h(t,\xi+\lambda)$, or, equivalently, $w(\tau-\lambda, t)$. On the other hand, if $\mathcal{A}$ is zero-state time-invariant, then by Definition *3.2.12*, the zero-state response of $\mathcal{A}$ to $\delta(t-(\xi+\lambda))$ must be identical with $h(t-\lambda, \xi)$ or, equivalently, $w(\tau-\lambda, t-\lambda)$. This implies that

$$w(\tau-\lambda, t-\lambda) = w(\tau-\lambda, t) \qquad \forall t \forall \tau \forall \lambda$$

which in turn implies that $w(\tau,t)$ does not depend on its second argument. Conversely, if $h(t,\xi)$ is of the form $h(t - \xi)$, then by using *6* it is trivial to demonstrate that $\mathcal{A}$ is zero-state time-invariant (see *3.2.19*). ◁

With respect to zero-state equivalence of $\mathcal{A}$ and $\mathcal{B}$, we have the following obvious consequence of the definition of h.

11 **Assertion** Let $\mathcal{A}$ and $\mathcal{B}$ be associated with impulse responses $h_{\mathcal{A}}$ and $h_{\mathcal{B}}$, respectively. Then $\mathcal{A}$ and $\mathcal{B}$ are zero-state equivalent (*2.9.1*) if and only if $h_{\mathcal{A}} = h_{\mathcal{B}}$.

Impulse response of sum and product

Consider two linear systems $\mathcal{A}$ and $\mathcal{B}$ and let $\mathcal{S}$ and $\mathcal{P}$ be their sum $\mathcal{A} + \mathcal{B}$ and product $\mathcal{A}\mathcal{B}$, respectively (see *1.10.23* and *1.10.30*). It is natural to raise the question: How are the impulse responses of $\mathcal{S} \triangleq \mathcal{A} + \mathcal{B}$ and $\mathcal{P} \triangleq \mathcal{A}\mathcal{B}$ related to those of $\mathcal{A}$ and $\mathcal{B}$?

Before answering this question it should be noted that, by Theorem *3.5.20*, both $\mathcal{A} + \mathcal{B}$ and $\mathcal{A}\mathcal{B}$ are linear systems and as such have unique zero states. In what follows, the zero state of $\mathcal{A}$ and $\mathcal{B}$ will be denoted by $\boldsymbol{\theta}'$, while that of $\mathcal{A} + \mathcal{B}$ and $\mathcal{A}\mathcal{B}$ will be denoted by $\boldsymbol{\theta}''$, with the understanding that $\boldsymbol{\theta}'' = (\boldsymbol{\theta}',\boldsymbol{\theta}')$.

First consider $\mathcal{A} + \mathcal{B}$. By the definition of $\mathcal{A} + \mathcal{B}$, we have

12 $$\bar{\mathbf{S}}(\boldsymbol{\theta}'';u) = \bar{\mathbf{A}}(\boldsymbol{\theta}';u) + \bar{\mathbf{B}}(\boldsymbol{\theta}';u) \qquad \forall u$$

and hence by the definition of the impulse response (*3.6.5*)

13 $$h_{\mathcal{S}}(t,\xi) = h_{\mathcal{A}}(t,\xi) + h_{\mathcal{B}}(t,\xi) \qquad \forall t \forall \xi$$

where $h_{\mathcal{S}}(t,\xi)$ denotes the impulse response of $\mathcal{S}$ and likewise for $\mathcal{A}$ and $\mathcal{B}$.

Next consider $\mathcal{A}\mathcal{B}$. In this case, we have

$$\bar{\mathbf{P}}(\boldsymbol{\theta}'';u) = \bar{\mathbf{A}}(\boldsymbol{\theta}';\bar{\mathbf{B}}(\boldsymbol{\theta}';u)) \qquad \forall u$$

and upon making use of the expression for the zero-state response of $\mathcal{A}$:

$$A(\boldsymbol{\theta}';v) = \int_{t_0}^{t} h_{\mathcal{A}}(t,\lambda)v(\lambda)\,d\lambda \qquad t \geq t_0$$

where $$v = B(\boldsymbol{\theta}'',u) \qquad u = \delta(t - \xi)$$

we find the desired relation

14 $$h_{\mathcal{P}}(t,\xi) = \int_{\xi}^{t} h_{\mathcal{A}}(t,\lambda)h_{\mathcal{B}}(\lambda,\xi)\,d\lambda$$

For zero-state time-invariant systems, this relation reduces to

15 $$h_{\mathcal{P}}(t) = \int_{0}^{t} h_{\mathcal{A}}(t - \lambda)h_{\mathcal{B}}(\lambda)\,d\lambda$$

which means that $h_{\mathcal{P}}$ is given by the convolution of $h_{\mathcal{A}}$ and $h_{\mathcal{B}}$ (see Appendix *B.3.8*).

In summary, we can make the

16 **Assertion** Let $h_{\mathcal{A}}$ and $h_{\mathcal{B}}$ be the impulse responses of $\mathcal{A}$ and $\mathcal{B}$, respectively. Then the impulse responses of $\mathcal{S} \triangleq \mathcal{A} + \mathcal{B}$ and $\mathcal{P} \triangleq \mathcal{A}\mathcal{B}$ are given by *13* and *14*.

Transfer function

It is frequently convenient in dealing with linear systems—particularly of the time-invariant type—to characterize the zero-state response in terms of the *transfer function* rather than the impulse response of the system. A detailed discussion of the notion of the transfer function and its properties will be presented in Chap. *9*. At this point, we shall merely give its definition, illustrate it by a simple example, and cite a few properties that will be needed in the interim.

17 **Definition** Let $\mathcal{A}$ be a time-invariant system with impulse response $h(t - \xi)$ and let $h(t)$ be Laplace transformable (see Appendix *B*). Then the *transfer function* of $\mathcal{A}$ is denoted by $H(s)$ and is defined to be the Laplace transform of $h(t)$. In symbols,

18
$$\begin{aligned} H(s) &\triangleq \mathcal{L}\{h(t)\} \\ &= \int_{0-}^{\infty} h(t)e^{-st}\,dt \end{aligned}$$

or equivalently

$$H(s) = \int_{-\infty}^{\infty} h(t - \xi)e^{-s(t-\xi)}\,d\xi$$

with the understanding that $h(t) = 0$ for $t < \xi$.

More generally, if $\mathcal{A}$ is a time-varying system with impulse response $h(t,\xi)$, then its transfer function is denoted by $H(s,t)$ and is defined by the relation

19
$$H(s,t) = \int_{-\infty}^{\infty} h(t,\xi)e^{-s(t-\xi)}\,d\xi$$

20 *Comment* It is customary to define the transfer function in two ways: (I) as the Laplace transform of the impulse response, which is what we have done above, and (II) in terms of the steady-state response of $\mathcal{A}$ to an input of the form $u(t) = e^{st}$. Since the latter definition is valid only under certain restrictive conditions involving the stability of $\mathcal{A}$, its formulation is deferred until Chap. *9*.

21 *Example* Let the impulse response of $\mathcal{A}$ be expressed by

$$h(t) = 1(t)(e^{-t} + e^{-2t})$$

Then the transfer function of $\mathcal{A}$ is given by

$$H(s) = \mathcal{L}\,\{h(t)\} = \frac{2s + 3}{(s + 1)(s + 2)}$$

22 **Remark** As the reader readily recognizes, this is merely a special case of the transfer function of a linear time-invariant system which is characterized by a differential equation. We shall be able to give a general expression for the transfer function of such a system once we have established the identity of its zero state, which will be done in Chap. *4*. At this point, it will suffice to state without proof that the transfer function of a system characterized by the input-output relation

$$a_n \frac{d^n y}{dt^n} + \cdots + a_0 y = b_m \frac{d^m u}{dt^m} + \cdots + b_0 u$$

where the a's and b's are constants, is given by the ratio

$$H(s) = \frac{b_m s^m + \cdots + b_0}{a_n s^n + \cdots + a_0} \qquad \triangleleft$$

A basic property of the transfer function is expressed by the following

23 **Assertion** Let $\mathcal{A}$ be a time-invariant system with transfer function $H(s)$ and let y be its zero-state response to u, where u is assumed to start at $t_0 = 0$. Then the Laplace transforms of y and u (assuming that they exist) are related to one another by

24
$$Y(s) = H(s)U(s)$$

where $Y(s)$ and $U(s)$ are the Laplace transforms of y and u, respectively.
Proof By Assertion *5*, we have

$$y(t) = \int_0^t h(t - \xi)u(\xi)\, d\xi$$

and on applying the Laplace transformation to both sides of this equation we derive

$$Y(s) = H(s)U(s)$$

by the convolution property of the Laplace transforms (see Appendix *B.3.11*).

25 **Assertion** Let $\mathcal{A}$ and $\mathcal{B}$ be time-invariant systems with transfer functions $H_{\mathcal{A}}(s)$ and $H_{\mathcal{B}}(s)$, respectively. Then the transfer functions of the sum $\mathcal{S} = \mathcal{A} + \mathcal{B}$ and the product $\mathcal{P} = \mathcal{A}\mathcal{B}$ are given by

$$H_{\mathcal{S}}(s) = H_{\mathcal{A}}(s) + H_{\mathcal{B}}(s)$$
$$H_{\mathcal{P}}(s) = H_{\mathcal{A}}(s)H_{\mathcal{B}}(s)$$

We omit the proof because it is trivial.

26 **Remark** Note that the transfer function of the product of time-invariant systems is independent of the order in which the systems are connected. This implies that if $\mathcal{A}$ and $\mathcal{B}$ are time-invariant, then they are zero-state commutative in the sense that $\mathcal{A}\mathcal{B}$ is zero-state equivalent to $\mathcal{B}\mathcal{A}$. As we shall see in Chap. *4*, Sec. *8*, it is not always true that $\mathcal{A}\mathcal{B}$ is equivalent to $\mathcal{B}\mathcal{A}$ if $\mathcal{A}$ and $\mathcal{B}$ are time-invariant.

Proper, strictly proper, and improper systems

An important characteristic of a system $\mathcal{A}$ is whether or not its zero-state response contains derivatives of the input, and if so, of what order. It is convenient to define this characteristic in terms of the impulse response of $\mathcal{A}$. This motivates the following

27 **Definition** A system $\mathcal{A}$ is *strictly proper* if $h(t,\xi)$ does not contain any delta functions. $\mathcal{A}$ is *proper* if $h(t,\xi)$ contains no delta functions of order 1 or higher. That is, $h(t,\xi)$ may contain $\delta(t-\xi)$ but not $\delta^{(1)}(t-\xi)$, $\delta^{(2)}(t-\xi)$, . . . , where $\delta^{(n)}(t)$ denotes a delta function of nth order (see Appendix *A*). $\mathcal{A}$ is *improper* if it is not proper.

28 *Example* The impulse response of the system $\mathcal{A}$ characterized by the input-output relation

$$\frac{dy}{dt} + y = u(t)$$

is (see Example *7*)

$$h(t-\xi) = 1(t-\xi)e^{-(t-\xi)}$$

Clearly, $\mathcal{A}$ is a strictly proper system. On the other hand, the impulse response of the system characterized by

$$y = \frac{du}{dt} + u$$

is

$$h(t-\xi) = \delta(t-\xi) + \delta^{(1)}(t-\xi)$$

Consequently, this system is improper.

29 *Comment* In a more general way, it will be shown in Chap. *4* that if a system is characterized by an input-output relation which has the form of a differential equation

$$a_n\frac{d^ny}{dt^n} + \cdots + a_0y = b_m\frac{d^mu}{dt^m} + \cdots + b_0u \qquad a_n \neq 0,\ b_m \neq 0$$

in which the a's and b's are constants, then $\mathcal{A}$ is strictly proper if $m \leq n-1$, $\mathcal{A}$ is proper if $m \leq n$, and $\mathcal{A}$ is improper if $m > n$. The same result holds for systems in which the a's and b's depend on time (see *A.7.13*). ◁

This concludes for the present our brief discussion of the notions of the impulse response and the transfer function—both of which have to do with the zero-state response of $\mathcal{A}$. We turn next to the properties of the zero-input response of $\mathcal{A}$.

Properties of zero-input response

We recall that the input-output-state relation *1* comprises two terms—the zero-input response and the zero-state response of $\mathcal{A}$. The zero-input response (at time t) is expressed by $\mathbf{A}(\mathbf{x}(t_0-);\mathbf{0})$ where $\mathbf{x}(t_0-)$ denotes the initial state of $\mathcal{A}$ at time t_0-, with $\mathbf{x}(t_0-)$ ranging over the state space Σ. When needed, the zero-input response is written as $\mathbf{A}(\boldsymbol{\alpha};\mathbf{0})$ to indicate that the initial state is $\mathbf{x}(t_0-) = \boldsymbol{\alpha}$, $\boldsymbol{\alpha} \in \Sigma$.

In what follows, we shall concern ourselves exclusively with systems whose state space is finite-dimensional. Such systems will be said to be of *finite order*. Furthermore, since the components of a state vector $\mathbf{x}(t)$ are in general complex numbers, we shall identify Σ with the space of m-tuples of complex numbers, denoted by $\mathcal{C}^m$. For convenient reference, we formalize this as an

30 **Assumption** The state space Σ is the space $\mathcal{C}^m$ of ordered m-tuples of complex numbers. $\lhd$

By the definition of zero-input linearity (Definition *3.4.17*) the (scalar-valued) zero-input response $A(\mathbf{x}(t_0-);0)$ is, for fixed t, a linear functional[1] on $\mathcal{C}^m$, which is a linear vector space of dimension m. Now any linear functional f on a finite-dimensional linear vector space such as Σ admits of the representation (see Appendix C)

$$f(\boldsymbol{\alpha}) = \gamma_1\alpha_2 + \cdots + \gamma_m\alpha_m \qquad \boldsymbol{\alpha} \in \mathcal{C}^m$$

where the γ_i, $i = 1, \ldots, m$, are scalar constants and

$$\boldsymbol{\alpha} = (\alpha_1, \ldots, \alpha_m)$$

Consequently, the zero-input response $A(\mathbf{x}(t_0-);0)$ must necessarily be of the form

31 $$A(\mathbf{x}(t_0-);0) = \phi_1(t,t_0)x_1(t_0-) + \cdots + \phi_m(t,t_0)x_m(t_0-)$$

where $\mathbf{x}(t_0-) \triangleq (x_1(t_0-), \ldots, x_m(t_0-))$ and the time functions $\phi_i(t,t_0)$, $i = 1, \ldots, m$, play the role of constants for fixed t and t_0.

The functions $\phi_1(t,t_0), \ldots, \phi_m(t,t_0)$ in *31* have a simple interpretation. Specifically, let $\boldsymbol{\alpha}^i$ be the ith coordinate vector $(0,0, \ldots ,1,0, \ldots ,0)$ in $\mathcal{C}^m$. Then from *31* we have

32 $$\phi_i(t,t_0) = A(\boldsymbol{\alpha}^i;0) \qquad i = 1, \ldots, m$$

That is, $\phi_i(t,t_0)$ is the zero-input response of $\mathcal{A}$ starting in the state $(0, \ldots ,1, \ldots ,0)$ (with 1 in the ith place) at time t_0-.

Before proceeding further, it will be helpful to summarize our conclusions up to this point in the form of an

[1] A functional is a scalar-valued function.

33 **Assertion** Let $\mathcal{A}$ be a system of finite order with a state space Σ of dimension m. Then if the input and output of $\mathcal{A}$ are scalar-valued, the input-output-state relation of $\mathcal{A}$ admits of the representation

34 $$y(t) = \langle \boldsymbol{\phi}(t,t_0),\mathbf{x}(t_0-)\rangle + \int_{t_0}^{t} h(t,\xi)u(\xi)\,d\xi \qquad t \geq t_0$$

where $\boldsymbol{\phi}(t,t_0) \triangleq (\phi_1(t,t_0), \ldots, \phi_m(t,t_0))$
$\phi_i(t,t_0) \triangleq$ zero-input response of $\mathcal{A}$ starting in the state $(0, \ldots, 1, \ldots, 0)$ at time t_0-
$\mathbf{x}(t_0-) \triangleq$ initial state of $\mathcal{A}$ at time t_0-, $\mathbf{x}(t_0-) \in \Sigma$
$h(t,\xi) \triangleq$ impulse response of $\mathcal{A} \triangleq$ zero-state response of $\mathcal{A}$ to $\delta(t-\xi)$
$\langle \mathbf{a},\mathbf{b}\rangle \triangleq$ scalar product of vectors **a** and **b**[1] $\triangleleft$

We have established (Assertion *9*) that $\mathcal{A}$ is zero-state time-invariant if and only if the impulse response $h(t,\xi)$ is of the form $h(t-\xi)$. In exactly the same way, it can readily be shown that $\mathcal{A}$ is zero-input time-invariant (*3.2.30*) if and only if $\boldsymbol{\phi}(t,t_0)$ is of the form $\boldsymbol{\phi}(t-t_0)$. Now by virtue of the decomposition property, $\mathcal{A}$ is time-invariant if and only if it is both zero-state and zero-input time-invariant. Combining these observations, we have the

35 **Assertion** A system $\mathcal{A}$ is time-invariant if and only if its input-output-state relation admits of the representation

36 $$y(t) = \langle \boldsymbol{\phi}(t-t_0),\mathbf{x}(t_0-)\rangle + \int_{t_0}^{t} h(t-\xi)u(\xi)\,d\xi \qquad t \geq t_0 \quad \triangleleft$$

In the following subsection we shall focus our attention on the properties of the zero-input response functions $\phi_i(t,t_0)$, $i = 1, \ldots, m$. In order to simplify our analysis, we shall assume that $\mathcal{A}$ is a time-invariant system. Then at appropriate points we shall indicate how the results obtained under the assumption of time invariance can be modified for or extended to time-varying systems.

Basis functions

Since the zero-input response functions of a time-invariant system have the form $\phi_1(t-t_0), \ldots, \phi_m(t-t_0)$, there is no loss in generality in assuming that $t_0 = 0$—which is what we shall do in the sequel.

The time functions $\phi_1(t), \ldots, \phi_m(t)$, or $\phi_1, \ldots, \phi_m$ for short, may or may not be linearly independent (see Appendix *C*). If they are not, there must be a subset of n functions, say, $\phi_1, \ldots, \phi_n$ for convenience in numbering, such that (I) $\phi_1, \ldots, \phi_n$ are linearly

[1] Strictly speaking, when $\boldsymbol{\phi}$ is complex the scalar product should be written $\langle \boldsymbol{\phi}^*,\mathbf{x}\rangle$ (see *C.5.1*), where $\boldsymbol{\phi}^*$ denotes the complex conjugate of $\boldsymbol{\phi}$. For simplicity of notation we shall omit the complex conjugate symbol in this and the following chapter since $\boldsymbol{\phi}(t_0,t)$ will usually be real-valued.

independent, (II) every time function ϕ_i, $i = 1, \ldots, m$, is expressible as a linear combination of $\phi_1, \ldots, \phi_n$. Such time functions constitute a set of *basis functions* or, more simply, a *basis set* or a *basis* for $\mathcal{A}$.

Suppose that the expressions for $\phi_{n+1}, \ldots, \phi_m$ in terms of $\phi_1, \ldots, \phi_n$ read

$$\phi_{n+1} = a_{n+1,1}\phi_1 + \cdots + a_{n+1,n}\phi_n$$
$$\cdots\cdots\cdots\cdots\cdots\cdots\cdots\cdots$$
$$\phi_m = a_{m1}\phi_1 + \cdots + a_{mn}\phi_n$$

where the a's are constant coefficients. On substituting these expressions into the expression for the zero-input response of $\mathcal{A}$,

$$A(\mathbf{x}(0-);0) = \phi_1(t)x_1 + \cdots + \phi_m(t)x_m$$

where for simplicity we write x_i for $x_i(0-)$, $i = 1, \ldots, m$, we obtain after minor rearrangement of terms

37 $$A(\mathbf{x}'(0-);0) = \phi_1(t)x_1' + \cdots + \phi_n(t)x_n'$$

where the vector $\mathbf{x}' = (x_1', \ldots, x_n')$ is related to the vector

$$\mathbf{x} = (x_1, \ldots, x_m)$$

by

38 $$\mathbf{x}' = \mathbf{\Gamma}\mathbf{x}$$

with $\mathbf{\Gamma}$ expressed by

$$\mathbf{\Gamma} = \begin{bmatrix} 1 & 0 & \cdots & 0 & a_{n+1,1} & \cdots & a_{m,1} \\ \cdots & \cdots & \cdots & \cdots & \cdots & \cdots & \cdots \\ 0 & 0 & \cdots & 1 & a_{n+1,n} & \cdots & a_{m,n} \end{bmatrix}$$

Since this matrix is of rank n, it follows that if $\mathbf{x}$ ranges over $\mathcal{C}^m$ then $\mathbf{x}'$ ranges over $\mathcal{C}^n$. More specifically, the fact that *36* is an input-output-state relation (see *1.6.2*) implies, in conjunction with *38*, that every zero-input response is representable as *37* and, conversely, for every $\mathbf{x}'$ in $\mathcal{C}^n$ *37* represents a zero-input response of $\mathcal{A}$. Thus, the state space of $\mathcal{A}$ may be taken to be $\mathcal{C}^n$ instead of $\mathcal{C}^m$. [A more detailed discussion of the connection between $\mathbf{x}(0-)$, $\mathbf{x}'(0-)$, as well as the connection between different basis functions for $\mathcal{A}$, is presented in Chap. *4*, Sec. *2*. In that section, the state vectors $\mathbf{x}$ and $\hat{\mathbf{x}}$ have the same significance as $\mathbf{x}'$ and $\mathbf{x}$, respectively, in the foregoing discussion.]

The observations made above lead us to the following more complete definition of basis functions.

39 **Definition** A set of (linearly independent) zero-input responses $\phi_1, \ldots, \phi_n$ constitutes a set of *basis functions* for a system $\mathcal{A}$ with state space $\Sigma = \mathcal{C}^m$, $m \geq n$, ($\mathcal{C}^m \triangleq$ space of m-tuples of complex numbers), if and only if every zero-input response of $\mathcal{A}$ is representable in the form

40 $$A(\mathbf{x}(0-);0) = \phi_1(t)\alpha_1 + \cdots + \phi_n(t)\alpha_n \qquad t \geq 0$$

where $\mathbf{x}(0-) \epsilon \mathcal{C}^m$ and $\boldsymbol{\alpha} = (\alpha_1, \ldots, \alpha_n)$ ranges over $\mathcal{C}^n$, and, conversely, for every $\boldsymbol{\alpha}$ in $\mathcal{C}^n$ *40* expresses a zero-input response of $\mathcal{A}$. [Note that it is immaterial whether the right-hand member of *40* is expressed in terms of $\mathbf{x}'(0-)$ or $\boldsymbol{\alpha}$, since $\boldsymbol{\alpha}$ plays the role of a generic value of $\mathbf{x}'(0-)$.]

The number n is defined to be the *order* of $\mathcal{A}$. In effect, the order of $\mathcal{A}$ is the dimension of $\mathcal{C}^n$, which is the "smallest" state space that can be associated with $\mathcal{A}$. This point is discussed more fully in Chap. *4*, Sec. *2*.

41 **Remark** Definition *39* applies to time-varying systems also, provided *40* is replaced by

$$A(\mathbf{x}(t_0-);0) = \phi_1(t,t_0)\alpha_1 + \cdots + \phi_n(t,t_0)\alpha_n \qquad t \geq t_0$$

with the understanding that, in order to qualify as a set of basis functions for $\mathcal{A}$, the zero-input responses $\phi_1(t,t_0), \ldots, \phi_n(t,t_0)$ must be linearly independent for *all* values of t_0.

42 *Note* If the state space Σ is $\mathcal{R}^m$ (space of m-tuples of real numbers) rather than $\mathcal{C}^m$, then $\mathcal{C}^m$ and $\mathcal{C}^n$ in *39* are replaced by $\mathcal{R}^m$ and $\mathcal{R}^n$ respectively.

43 *Example* Consider a system $\mathcal{A}$ which is characterized by the input-output-state relation ($t_0 = 0$)

$$y(t) = e^{-t}x_1(0-) + e^{-2t}x_2(0-) + (2e^{-t} + e^{-2t})x_3(0-) + \int_{t_0}^{t} [e^{-(t-\xi)} + e^{-2(t-\xi)}]u(\xi)\,d\xi \qquad t \geq t_0$$

with $\mathbf{x}(0-)$ ranging over $\mathcal{R}^3$.

In this case, the zero-input responses are expressed by $\phi_1(t) = e^{-t}$, $\phi_2(t) = e^{-2t}$, and $\phi_3(t) = 2e^{-t} + e^{-2t}$, $t \geq 0$, and $\phi_1(t)$ and $\phi_2(t)$ qualify as basis functions for $\mathcal{A}$. In terms of these, the zero-input response reads

44 $$A(\mathbf{x}(0-);0) = e^{-t}x_1'(0-) + e^{-2t}x_2'(0-) \qquad t \geq 0$$

where $x_1'(0-) = x_1(0-) + 2x_3(0-)$ and $x_2'(0-) = x_2(0-) + x_3(0-)$. Thus, every zero-input response of $\mathcal{A}$ is of the form *44* with $\mathbf{x}'(0-) \epsilon \mathcal{R}^2$ and, conversely, for every $\mathbf{x}'(0-) \epsilon \mathcal{R}^2$ *44* represents a zero-input response of $\mathcal{A}$. Note that $\mathcal{A}$ is a system of order 2.

45 **Remark** The basis functions $\phi_1(t), \ldots, \phi_n(t)$ are defined by *39* only for nonnegative values of t. As we shall see in the next section, it is

frequently convenient to extend the range of definition of the basis functions to the entire time axis. This can easily be done in a "natural" fashion when the basis functions are linear combinations of terms of the form $r_j(t)e^{\lambda_j t}$, $j = 1, 2, \ldots, n$, where the $r_j(t)$ are polynomials in t and the λ_j are constants. In this case—which is typical of systems characterized by constant-coefficient differential equations—the value of $\phi_i(t)$ for negative t is defined to be the value resulting from substituting the negative value in question in each term $r_j(t)e^{\lambda_j t}$ in the representation of $\phi_i(t)$. Equivalently, let $\Phi_i(s)$ denote the Laplace transform of $\phi_i(t)$. Then, $\phi_i(-t)$ is defined to be the negative of the inverse Laplace transform of $\Phi_i(-s)$ evaluated at $|t|$. For example, if $\phi_i(t) = e^{-t}$ for $t \geq 0$, then $\Phi_i(s) = 1/(s+1)$ and the value of $\phi_i(t)$ for negative t is given by $-\mathcal{L}^{-1}\{1/(1-s)\}$, which is $1(t)e^t$ evaluated at $|t|$. Thus, for $t \leq 0$, the value of $\phi_i(t)$ is defined to be $e^{|t|}$. Obviously, this is equivalent to saying that the value of $\phi_i(t)$ for negative t is e^{-t}.

In general, it will be clear from the context whether the basis functions are defined only for $t \geq 0$ or for all t. When we wish to emphasize that it is defined for all t, the basis function will be referred to as an *extended basis function*, whereas a basis function defined for only $t \geq 0$ will be said to be *one-sided*. The adjective *two-sided* will be reserved for time functions which are defined for all t, but not necessarily through an extension of their definition for $t \geq 0$.

Systems in reduced form

So far we have not made any assumptions concerning the form of the basis functions of $\mathcal{A}$. This does not mean, however, that they can be chosen in an arbitrary fashion. Indeed, as will be shown in Sec. 7, the fact that *1* is an input-output-state relation places severe limitations on the class of functions which can serve as basis functions for $\mathcal{A}$.

As a preliminary to determining the nature of these limitations, we shall establish several equivalence properties which will be needed at later points.

To begin with, we recall that a system $\mathcal{A}$ is in *reduced form* (*2.2.11*) if no distinct states in the state space of $\mathcal{A}$ are equivalent. It is easy to verify that $\mathcal{A}$ is in reduced form if its input-output-state relation is expressed in terms of a set of basis functions of $\mathcal{A}$. More specifically:

46 **Assertion** If the input-output-state relation of $\mathcal{A}$ has the form

$$y(t) = \langle \boldsymbol{\phi}(t - t_0), \mathbf{x}(t_0-) \rangle + \int_{t_0}^{t} h(t - \xi)u(\xi)\, d\xi$$

where $\mathbf{x}(t_0-) \in \mathcal{C}^n$, $\boldsymbol{\phi}(t) = (\phi_1(t), \ldots, \phi_n(t))$, and $\phi_1(t), \ldots, \phi_n(t)$ are basis functions for $\mathcal{A}$, then $\mathcal{A}$ is in reduced form.

Proof We have to show that if $\boldsymbol{\alpha}'$ and $\boldsymbol{\alpha}''$ are distinct states (that is, $\boldsymbol{\alpha}' \neq \boldsymbol{\alpha}''$), then they cannot be equivalent. For, if $\boldsymbol{\alpha}'$ is equivalent to

$\boldsymbol{\alpha}''$, then by *3.5.10*

47 $$\langle \boldsymbol{\phi}(t - t_0), \boldsymbol{\alpha}' \rangle = \langle \boldsymbol{\phi}(t - t_0), \boldsymbol{\alpha}'' \rangle \qquad t \geq t_0$$

or
$$\langle \boldsymbol{\phi}(t - t_0), (\boldsymbol{\alpha}' - \boldsymbol{\alpha}'') \rangle = 0$$

which implies $\boldsymbol{\alpha}' = \boldsymbol{\alpha}''$ by virtue of the linear independence of $\phi_1(t - t_0)$, $\ldots$, $\phi_n(t - t_0)$.

48 *Note* In proving Assertion *46* we have not made any use of the time invariance of $\mathcal{A}$. Consequently, Assertion *46* is valid for time-varying systems (see Remark *41*). ◁

In addition to the property expressed by Assertion *46*, we shall need another result which concerns the connection between the basis functions of two equivalent systems. Expressed in the form of an assertion, it reads

49 **Assertion** Let $\mathcal{A}$ and $\mathcal{B}$ be zero-input equivalent systems of order n and let $\phi_1, \ldots, \phi_n$ and $\varphi_1, \ldots, \varphi_n$ be basis functions for $\mathcal{A}$ and $\mathcal{B}$, respectively. Then every basis function of $\mathcal{A}$ is expressible as a linear combination of basis functions of $\mathcal{B}$, and vice versa.

Proof By the definition of zero-input equivalence (*2.9.2*), to every state $\boldsymbol{\alpha}$ of $\mathcal{A}$ there corresponds a state $\boldsymbol{\beta}$ of $\mathcal{B}$, and vice versa, such that

50 $$\bar{A}(\boldsymbol{\alpha};0) = \bar{B}(\boldsymbol{\beta};0)$$

of which *3.5.10* is a special case.

Expressed in terms of zero-input responses, *50* implies that to every $\boldsymbol{\alpha}$ in $\mathcal{C}^n$ there is a $\boldsymbol{\beta}$ in $\mathcal{C}^n$, and vice versa, such that

51 $$\langle \boldsymbol{\phi}(t), \boldsymbol{\alpha} \rangle = \langle \boldsymbol{\varphi}(t), \boldsymbol{\beta} \rangle \qquad t \geq 0$$

From this and the linear independence of the basis functions $\phi_1, \ldots, \phi_n$ and $\varphi_1, \ldots, \varphi_n$ it follows at once that every ϕ_i, $i = 1, \ldots, n$, is a linear combination of $\varphi_1, \ldots, \varphi_n$, and vice versa. We omit the details at this point since they are given in a more general setting in *4.2.4*.

52 **Corollary** If $\phi_1, \ldots, \phi_n$ and $\varphi_1, \ldots, \varphi_n$ are two different sets of basis functions for $\mathcal{A}$, then every basis function in the first set is expressible as a linear combination of basis functions in the second set and vice versa. In symbols, this means that the row vectors $\boldsymbol{\phi} = (\phi_1, \ldots, \phi_n)$ and $\boldsymbol{\varphi} = (\varphi_1, \ldots, \varphi_n)$ are related to one another by

$$\boldsymbol{\phi} = \boldsymbol{\varphi}\boldsymbol{\Gamma}$$

where $\boldsymbol{\Gamma}$ is a nonsingular matrix.

7 *Basis functions and state equations*

As was pointed out in Sec. *6*, the fact that the relation

1 $$y(t) = \langle \boldsymbol{\phi}(t - t_0), \mathbf{x}(t_0) \rangle + \int_{t_0}^{t} h(t - \xi) u(\xi)\, d\xi$$

is an input-output-state relation for $\mathcal{A}$ and as such satisfies the self-consistency conditions *1.6.11*, *1.6.15*, and *1.6.35*, or, equivalently, has the separation properties *2.3.4* and *2.3.19*, constrains the choice of the basis functions $\phi_1, \ldots, \phi_n$ and induces a relation between them and the impulse response $h(t - \xi)$. In this section, we shall examine the nature of these constraints and, by their use, deduce the state equations for $\mathcal{A}$ in both explicit and differential forms. We shall not, however, concern ourselves at present with the solution of these equations. We shall do this in Chap. *5*.

In addition to making use of the fact that *1* is an input-output-state relation, we shall make use of the following assumptions about $\boldsymbol{\phi}$, h, and u.

I $\boldsymbol{\phi}$ is an n-vector whose components $\phi_1, \ldots, \phi_n$ are basis functions for $\mathcal{A}$ (see *3.6.39*). Note that this implies that $\mathcal{A}$ is in reduced form (*3.6.46*).

II The basis functions h,u and their derivatives of all finite orders have at most a finite number of discontinuities and delta functions over any finite observation interval.

Derived properties of basis functions

We turn our attention first to those properties of basis functions that follow from the response separation property *2.3.4* in the case of zero input.

Specifically, for each τ in $(t_0,t]$, the zero-input response of $\mathcal{A}$ at time t starting in $\mathbf{x}(t_0)$ must be identical with the zero-input response of $\mathcal{A}$ at time t starting in state $\mathbf{x}(\tau)$. Thus, the basis functions must satisfy the identity

$$\langle \boldsymbol{\phi}(t - t_0), \mathbf{x}(t_0) \rangle = \langle \boldsymbol{\phi}(t - \tau), \mathbf{x}(\tau) \rangle \qquad \forall t_0 \forall t,\ t > t_0$$

for all τ in $(t_0,t]$.

For $t_0 = 0$ and $\mathbf{x}(t_0) = (1,0, \ldots, 0)$, this identity reduces to

2 $$\phi_1(t) = \sum_{i=1}^{n} x_i(\tau)\phi_i(t - \tau) \qquad 0 < \tau \le t$$

which implies that every basis function can be expressed as a linear combination of delayed basis functions $\phi_1(t - \tau), \ldots, \phi_n(t - \tau)$.

We shall refer to this property as the *translation property* of basis functions.

An important consequence of the translation property is contained in the following

3 **Assertion** Every basis function of $\mathcal{A}$ is infinitely differentiable (i.e., has derivatives of every finite order) for $t > 0$.

Proof Suppose that $\phi_1(t)$ has a jump at, say, $t = a$. Then by virtue of *2* at least one of the functions $\phi_1(t - \tau), \ldots, \phi_n(t - \tau)$ must have a jump at $t = a$, or, equivalently, the vector $\boldsymbol{\phi}(t)$ must have a jump at $a - \tau$. Since τ can take any value in $(0,t]$, this implies that $\boldsymbol{\phi}(t)$ will have an uncountable number of jumps in the interval $(0,t]$, which contradicts assumption (II) above. Consequently, $\phi_1(t)$ cannot have any jumps for $t > 0$, and the same statement applies to $\phi_2(t), \ldots, \phi_n(t)$. However, if $\phi_1(t)$ is a single delta function at $t = 0$, then $\mathbf{x}(\tau) = 0$ and the contradiction does not arise. Nor does it arise if $\phi_1(t)$ is a linear combination of delta functions at $t = 0$ and functions which are infinitely differentiable for $t > 0$.

Similar reasoning rules out the presence of discontinuities in the derivatives of $\phi_1(t), \ldots, \phi_n(t)$ for $t > 0$. Furthermore, on successive differentiation of both sides of *2* and repetition of the arguments used above, we conclude that $\phi_1(t)$ is infinitely differentiable for $t > 0$, with the possibility that $\phi_1(t)$ may contain delta functions at $t = 0$. The same thing holds for $\phi_2(t), \ldots, \phi_n(t)$. ◁

For the present we shall assume that there are no delta functions among the basis functions of $\mathcal{A}$. [In effect, this means that we are assuming that $\mathcal{A}$ is a proper system (*3.6.27*).] This assumption will be removed at a later point in this section.

Relation between $\mathbf{x}(t_0)$ and $\mathbf{x}(\tau)$

Having established that the basis functions are infinitely differentiable, we can deduce further relations from *2* by differentiating both sides of *2* $n - 1$ times with respect to t. This yields, together with *2*, the following n relations:

4 $$\langle \boldsymbol{\phi}^{(\lambda-1)}(t - t_0),\mathbf{x}(t_0)\rangle = \langle \boldsymbol{\phi}^{(\lambda-1)}(t - \tau),\mathbf{x}(\tau)\rangle \qquad \lambda = 1, \ldots, n$$

which hold for all t_0, all $t \geq t_0$, and all τ intermediate between t_0 and t.

5 *Comment* It makes but little difference in our analysis whether the basis functions $\phi_1, \ldots, \phi_n$ are assumed to be one-sided or extended (*3.6.45*). If they are one-sided, then $\phi_i^{(\lambda)}(t)$, $i = 1, \ldots, n$, $\lambda = 0, \ldots, n - 1$, is defined for all $t > 0$ but not at $t = 0$, and hence τ must be restricted to the open interval (t_0,t). On the other hand, if the $\phi_i(t)$ are extended, then $\phi_i^{(\lambda)}(0)$ is defined. Thus, strictly speaking we must write $\phi_i^{(\lambda)}(0+)$ for one-sided basis functions and we may write $\phi_i^{(\lambda)}(0)$ for extended basis

functions. This implies that, when we write $\phi_i^{(\lambda)}(0)$, we mean that either the basis functions are extended or else the basis functions are one-sided and that we have a tacit understanding that $\phi_i^{(\lambda)}(0)$ should be interpreted as $\phi_i^{(\lambda)}(0+)$, $i = 1, \ldots, n$, $\lambda = 0, \ldots, n-1$. In particular, when relations such as *4* are stated to hold for $t \geq t_0$ and $t_0 \leq \tau \leq t$ (rather than $t > t_0$ and $t_0 < \tau < t$) it is understood that either the basis functions in question are extended or, if they are not, $\phi_i^{(\lambda)}(0) \triangleq \phi_i^{(\lambda)}(0+)$, $\mathbf{h}(0) \triangleq \mathbf{h}(0+)$, and likewise for other symbols which will appear presently in our analysis. It should be noted that, since at this stage of our analysis we have not yet defined how the basis functions may be extended (this is done in *21*), the latter interpretation is more appropriate. ◁

The n relations *4* can be expressed more compactly in the matrix form

6 $$\mathbf{\Phi}(t - t_0)\mathbf{x}(t_0) = \mathbf{\Phi}(t - \tau)\mathbf{x}(\tau) \qquad \forall t_0, \forall t \geq t_0, t_0 \leq \tau \leq t$$

where $\mathbf{\Phi}$ is an $n \times n$ matrix whose rows are the vectors $\boldsymbol{\phi}(t), \ldots, \boldsymbol{\phi}^{(n-1)}(t)$. That is,

7 $$\mathbf{\Phi}(t) = \begin{bmatrix} \phi_1(t) & \cdots & \phi_n(t) \\ \phi_1^{(1)}(t) & \cdots & \phi_n^{(1)}(t) \\ \cdots & \cdots & \cdots \\ \phi_1^{(n-1)}(t) & \cdots & \phi_n^{(n-1)}(t) \end{bmatrix}$$

with the understanding that $\mathbf{\Phi}(t)$ is defined only for $t > 0$ if the basis functions are one-sided, in which case $\mathbf{\Phi}(0)$ should be interpreted as $\mathbf{\Phi}(0+)$.

On setting $\tau = t$, *6* yields

8 $$\mathbf{\Phi}(0)\mathbf{x}(t) = \mathbf{\Phi}(t - t_0)\mathbf{x}(t_0) \qquad t \geq t_0$$

which determines $\mathbf{x}(t)$ uniquely—and not just to within equivalent states—provided $\mathbf{\Phi}(0)$ is nonsingular. Note that since $\mathfrak{A}$ is in reduced form, any two states $\mathbf{x}'$ and $\mathbf{x}''$ which are distinct are necessarily nonequivalent (see *3.6.46*).

9 **Remark** If $\mathbf{\Phi}(0)$ is nonsingular, then by Corollary *3.6.52* we can always "normalize" the basis functions $\phi_1(t), \ldots, \phi_n(t)$ in such a way that $\mathbf{\Phi}(0) = \mathbf{I}$. Specifically, if $\mathbf{\Phi}(0) \neq \mathbf{I}$ and $\mathbf{\Phi}(0)$ is nonsingular, then we can construct a normalized set of basis functions $\varphi_1(t), \ldots, \varphi_n(t)$ from $\phi_1(t), \ldots, \phi_n(t)$ by the linear transformation

10 $$\boldsymbol{\varphi} = \boldsymbol{\phi}\mathbf{\Phi}^{-1}(0)$$

where $\boldsymbol{\varphi}$ and $\boldsymbol{\phi}$ are the row vectors $(\varphi_1, \ldots, \varphi_n)$ and $(\phi_1, \ldots, \phi_n)$, respectively.

State transition matrix and its properties

For convenience, we shall assume hereafter that the basis functions $\phi_1(t), \ldots, \phi_n(t)$ are normalized, so that $\mathbf{\Phi}(0) = \mathbf{I}$. Under this assumption, *8* becomes

11 $$\mathbf{x}(t) = \mathbf{\Phi}(t - t_0)\mathbf{x}(t_0) \qquad \forall t_0, \forall t \geq t_0$$

In this equation, $\mathbf{\Phi}(t - t_0)$ plays the role of an operator which takes the initial state $\mathbf{x}(t_0)$ into the state $\mathbf{x}(t)$. For this reason, $\mathbf{\Phi}(t)$ is called the *state transition matrix*. We shall derive only a few of its basic properties at this point, with a fuller discussion in Chaps. *5* and *6*.

12 **Property**[1] If the basis functions are linearly independent over every subinterval of $[0, \infty)$, then $\mathbf{\Phi}(t)$ is nonsingular for $t \geq 0$.

Proof The determinant of $\mathbf{\Phi}(t)$ is the Wronskian,[2] $W(t)$, of the basis functions $\phi_1(t), \ldots, \phi_n(t)$. A basic property of the Wronskian (see Ref. *5*, p. 48) is that if $W(t) = 0$ for $t \geq 0$, then there exists a subinterval of $[0, \infty)$ over which the time functions $\phi_1, \ldots, \phi_n$ are linearly dependent. Since by hypothesis the basis functions are linearly independent over every subinterval of $[0, \infty)$, there must exist a positive value of t, say, t_1, such that $W(t_1) \neq 0$ or, equivalently, $\mathbf{\Phi}(t_1)$ is nonsingular. Now on letting $t - t_0 = t_1$ in *6* and noting that *6* holds for all $\mathbf{x}(t_0) \in \mathcal{C}^n$ and all $\tau \in [t_0,t_1]$, it follows that $\mathbf{\Phi}(t)$ is nonsingular for all $t \in [0,t_1]$ and, in particular, $\mathbf{\Phi}(0)$ is nonsingular. Then by *9* we can assume that $\mathbf{\Phi}(0) = \mathbf{I}$ and correspondingly express the state at time t as $\mathbf{x}(t) = \mathbf{\Phi}(t - t_0)\mathbf{x}(t_0)$.

On invoking the state separation property (*2.3.16*) of this state equation it follows that if $\mathbf{\Phi}(t)$ is nonsingular for $t \in [0,t_1]$ then it is nonsingular for all $t \geq 0$. Specifically, let $0 \leq \tau - t_0 \leq t_1$ and $0 \leq t - \tau \leq t_1$. Then if $\mathbf{x}(t_0)$ ranges over $\mathcal{C}^n$, so will $\mathbf{x}(\tau) = \mathbf{\Phi}(\tau - t_0)\mathbf{x}(t_0)$ since $\mathbf{\Phi}(\tau - t_0)$ is nonsingular. By the same argument

$$\mathbf{x}(t) = \mathbf{\Phi}(t - \tau)\mathbf{x}(\tau)$$

ranges over $\mathcal{C}^n$ if $\mathbf{x}(\tau)$ ranges over $\mathcal{C}^n$. Consequently, $\mathbf{x}(t)$ will range over $\mathcal{C}^n$ if $\mathbf{x}(t_0)$ ranges over $\mathcal{C}^n$. This implies that $\mathbf{\Phi}(t - t_0)$ is nonsingular for all $t - t_0$ in the interval $[0,2t_1]$. On repeating this argument, we deduce that $\mathbf{\Phi}(t)$ is nonsingular for all t in the intervals $[0,4t_1]$, $[0,8t_1]$, ... In this way, we can demonstrate that $\mathbf{\Phi}(t)$ is nonsingular for all $t \geq 0$ if it is nonsingular over $[0,t_1]$.

In the sequel we shall assume, except where otherwise noted, that the basis functions $\phi_1, \ldots, \phi_n$ are linearly independent not only over $[0, \infty)$ but also over every subinterval of this interval. As shown above, under this assumption $\mathbf{\Phi}(t)$ is nonsingular for $t \geq 0$.

[1] A. C. Chang has shown that the nonsingularity of $\mathbf{\Phi}(t)$ can be deduced from the translation property (*2 et seq.*).

[2] For a discussion of the Wronskian see Ref. *1* or Ref. *5*.

13 **Property** $\mathbf{\Phi}(t)$ has the multiplicative (semigroup) property

14
$$\mathbf{\Phi}(t+\xi) = \mathbf{\Phi}(t)\mathbf{\Phi}(\xi) \qquad \forall \xi \geq 0, \forall t \geq 0$$

This follows at once from *11* by writing $\mathbf{x}(t+\xi)$ in two ways:

$$\mathbf{x}(t+\xi) = \mathbf{\Phi}(t+\xi)\mathbf{x}(0)$$

and

$$\begin{aligned}\mathbf{x}(t+\xi) &= \mathbf{\Phi}(t)\mathbf{x}(\xi)\\ &= \mathbf{\Phi}(t)\mathbf{\Phi}(\xi)\mathbf{x}(0)\end{aligned}$$

and asserting that the relation $\mathbf{\Phi}(t+\xi)\mathbf{x}(0) = \mathbf{\Phi}(t)\mathbf{\Phi}(\xi)\mathbf{x}(0)$ holds for all $\mathbf{x}(0)$.

15 **Property** $\mathbf{\Phi}(t)$ has the commutative property

$$\mathbf{\Phi}(t)\mathbf{\Phi}(\xi) = \mathbf{\Phi}(\xi)\mathbf{\Phi}(t) \qquad \forall t \geq 0, \forall \xi \geq 0$$

This follows at once from *14*.

16 **Property** $\mathbf{\Phi}(t)$ is the solution of the first-order matrix differential equation

17
$$\dot{\mathbf{\Phi}}(t) = \dot{\mathbf{\Phi}}(0)\mathbf{\Phi}(t) \qquad \mathbf{\Phi}(0) = \mathbf{I},\ t \geq 0$$

This equation follows from *14* [with the order of $\mathbf{\Phi}(t)$ and $\mathbf{\Phi}(\xi)$ interchanged] by differentiating both sides first with respect to t and then with respect to ξ. Thus

18
$$\frac{\partial}{\partial t}\mathbf{\Phi}(t+\xi) = \mathbf{\Phi}(\xi)\dot{\mathbf{\Phi}}(t)$$

$$\frac{\partial}{\partial \xi}\mathbf{\Phi}(t+\xi) = \dot{\mathbf{\Phi}}(\xi)\mathbf{\Phi}(t)$$

and since

$$\frac{\partial}{\partial t}\mathbf{\Phi}(t+\xi) = \frac{\partial}{\partial \xi}\mathbf{\Phi}(t+\xi)$$

we have

19
$$\mathbf{\Phi}(\xi)\dot{\mathbf{\Phi}}(t) = \dot{\mathbf{\Phi}}(\xi)\mathbf{\Phi}(t)$$

which becomes *17* upon setting $\xi = 0$. Note that *15* and *18* imply that $\mathbf{\Phi}(t)$ has the commutative property

20
$$\mathbf{\Phi}(t)\dot{\mathbf{\Phi}}(\xi) = \dot{\mathbf{\Phi}}(\xi)\mathbf{\Phi}(t) \qquad \forall t \geq 0, \forall \xi \geq 0$$

The extended state transition matrix

If the basis functions $\phi_1, \ldots, \phi_n$ are one-sided (*3.6.45*), then the matric differential equation *17* [in which $\dot{\mathbf{\Phi}}(0)$ should be interpreted as $\dot{\mathbf{\Phi}}(0+)$] provides a convenient means of extending the definition of $\mathbf{\Phi}(t)$ to negative values of t. Thus we have the

21 **Definition** The *(extended) state transition matrix* $\mathbf{\Phi}(t)$ is defined for all t as the solution of the differential equation

22 $$\dot{\mathbf{\Phi}}(t) = \dot{\mathbf{\Phi}}(0)\mathbf{\Phi}(t) \qquad \mathbf{\Phi}(0) = \mathbf{I}$$

Correspondingly, the extended basis functions $\phi_1(t), \ldots, \phi_n(t)$ are defined for all t by the elements of the first row of $\mathbf{\Phi}(t)$.

23 *Comment* Note that the adjective "extended" is put in parentheses to indicate that it is used only when there is a need to emphasize that $\mathbf{\Phi}(t)$ is defined for all t rather than just for $t > 0$. Note also that $\dot{\mathbf{\Phi}}(0)$ in *22* is in effect $\dot{\mathbf{\Phi}}(t)\Big]_{t=0+}$, where $\mathbf{\Phi}(t)$ is defined by *7* in terms of one-sided basis functions. ◁

There is an important relation between $\mathbf{\Phi}^{-1}(t)$ and $\mathbf{\Phi}(-t)$ that can easily be deduced from this definition. Specifically, on differentiating both sides of the identity

$$\mathbf{\Phi}^{-1}(t)\mathbf{\Phi}(t) = \mathbf{I}$$

with respect to t, we find that $\mathbf{\Phi}^{-1}(t)$ is the solution of the differential equation

24 $$\dot{\mathbf{\Phi}}^{-1}(t) = -\dot{\mathbf{\Phi}}(0)\,\mathbf{\Phi}^{-1}(t) \qquad \mathbf{\Phi}^{-1}(0) = \mathbf{I}$$

On the other hand, *22* implies that

$$\dot{\mathbf{\Phi}}(-t) = -\dot{\mathbf{\Phi}}(0)\mathbf{\Phi}(-t) \qquad \mathbf{\Phi}(0) = \mathbf{I}$$

and therefore

25 $$\mathbf{\Phi}^{-1}(t) = \mathbf{\Phi}(-t) \qquad \forall t$$

This relation can be used as an alternative means of extending the definition of $\mathbf{\Phi}(t)$ to negative t.

26 **Remark** By starting with *21* as the definition of $\mathbf{\Phi}(t)$, it is easy to show (see Chap. *5*, Sec. *2*) that Properties *12*, *13*, *15*, and *20* hold for *all* values of t_0, t, τ, and ξ. Henceforth it will be understood, unless stated to the contrary, that the state transition matrix $\mathbf{\Phi}(t)$ and the basis functions $\phi_1(t), \ldots, \phi_n(t)$ are defined for all t by *21*, and that, in consequence, Properties *12*, *13*, *15*, and *20* hold for all t.

State equations

In the preceding discussion we have obtained several basic properties of the state transition matrix $\mathbf{\Phi}(t)$ by invoking the separation and uniqueness properties of the zero-input response of $\mathcal{A}$. By proceeding in a similar fashion in the general case of a nonzero input, we shall obtain in the sequel the state equations of $\mathcal{A}$ and a relation between the impulse response and the state transition matrix.

Specifically, in the general case the response separation property

2.3.4 implies that the input-output-state relation *1* must satisfy the identity

$$\langle \boldsymbol{\phi}(t - t_0), \mathbf{x}(t_0) \rangle + \int_{t_0}^{t} h(t - \xi) u(\xi)\, d\xi$$
$$= \langle \boldsymbol{\phi}(t - \tau), \mathbf{x}(\tau) \rangle + \int_{\tau}^{t} h(t - \xi) u(\xi)\, d\xi$$

for all t_0, all $t \geq t_0$, and all τ, $t_0 \leq \tau \leq t$. On setting $t_0 = 0$, $u(t) = \delta(t)$, and $\mathbf{x}(t_0) = \mathbf{0}$, this identity yields

27 $$h(t) = \langle \boldsymbol{\phi}(t - \tau), \mathbf{x}(\tau) \rangle \qquad t \geq \tau,\ \tau \geq 0$$

which shows that the impulse response is expressible as a linear combination of basis functions. By *3*, this implies that, in general, the impulse response is a linear combination of delta functions and functions that are infinitely differentiable for $t > 0$. More particularly, if there are no delta functions among the basis functions of $\mathcal{A}$, then $h(t)$ is infinitely differentiable for $t > 0$. This is the case to which we shall restrict our attention in the sequel, deferring until a later point (see *45 et seq.*) the discussion of the general case.

Returning to the case where u is not a delta function and differentiating the identity with respect to τ, we have

28 $$h(t - \tau) u(\tau) = \frac{d}{d\tau} (\langle \boldsymbol{\phi}(t - \tau), \mathbf{x}(\tau) \rangle)$$

or more explicitly

$$h(t - \tau) u(\tau) = - \langle \dot{\boldsymbol{\phi}}(t - \tau), \mathbf{x}(\tau) \rangle + \langle \boldsymbol{\phi}(t - \tau), \dot{\mathbf{x}}(\tau) \rangle$$

On substituting $h(t - \tau)u(\tau)$ as given by this equation into the expression for $y(t)$,

$$y(t) = \langle \boldsymbol{\phi}(t), \mathbf{x}(0) \rangle + \int_0^t h(t - \tau) u(\tau)\, d\tau$$

we obtain after cancellation of the term $\langle \boldsymbol{\phi}(t), \mathbf{x}(0) \rangle$ the expression for the output at time t as a function of the state at time t. It reads

29 $$y(t) = \langle \boldsymbol{\phi}(0), \mathbf{x}(t) \rangle$$

(Note that the same result can be obtained by setting $t = t_0$ in *1*.)

Next, on differentiating both sides of *28* $n - 1$ times with respect to t, we obtain the following n relations:

$$h^{(\lambda-1)}(t - \tau) u(\tau) = \frac{d}{d\tau} (\langle \boldsymbol{\phi}^{(\lambda-1)}(t - \tau), \mathbf{x}(\tau) \rangle) \qquad \lambda = 1, \ldots, n$$

which may be expressed more compactly in the matrix form

30 $$\mathbf{h}(t - \tau) u(\tau) = \frac{d}{d\tau} [\boldsymbol{\Phi}(t - \tau) \mathbf{x}(\tau)] \qquad t \geq \tau$$

where $\boldsymbol{\Phi}(t)$ is the state transition matrix defined by *7* or, equivalently, *21*, and $\mathbf{h}(t)$ is a column n vector whose λth element is $h^{(\lambda-1)}(t)$; that is,

31 $$\mathbf{h}(t) = \begin{bmatrix} h(t) \\ h^{(1)}(t) \\ \cdot \\ \cdot \\ \cdot \\ h^{(n-1)}(t) \end{bmatrix} \qquad t \geq 0$$

with the understanding that $\mathbf{h}(0) \triangleq \mathbf{h}(0+)$. (Why? See Assertion *40* and the definition of extended impulse response.)

On integrating both sides of *30* with respect to τ between the limits t_0 and t, and making use of the fact that $\boldsymbol{\Phi}(0) = \mathbf{I}$ (see *9*), we obtain the relation

32 $$\mathbf{x}(t) = \boldsymbol{\Phi}(t - t_0)\mathbf{x}(t_0) + \int_{t_0}^{t} \mathbf{h}(t - \xi)u(\xi)\, d\xi \qquad t \geq t_0$$

which constitutes a state equation for $\mathcal{A}$ in explicit form (*1.8.8 et seq.*). Observe that it reduces to *11* when u is a zero input.

By differentiating both sides of *1* $n - 1$ times with respect to t and comparing the result with *32* (taking note of *7* and *31*), we find that the state of $\mathcal{A}$ at time t is expressed in terms of the output and its derivatives by the simple relation

33 $$\mathbf{x}(t) = (y(t), \ldots, y^{(n-1)}(t))$$

However, a more careful analysis shows that *33* is valid only if $h^{(n-1)}(t)$, interpreted as the $(n - 1)$st derivative of $h(t)$ in the distribution sense, has no delta functions at $t = 0$. As we shall see in Chap. *4*, Sec. *4*, this corresponds to the case where $\mathcal{A}$ is characterized by an input-output relation of the form

$$(a_n p^n + \cdots + a_0)y = u \qquad a_n \neq 0$$

In the more general case where $\mathcal{A}$ is characterized by an input-output relation of the form

$$(a_n p^n + \cdots + a_0)y = (b_m p^m + \cdots + b_0)u$$

where $b_m \neq 0$, for some $m \geq 1$, the vector $[y(t), \ldots, y^{(n-1)}(t)]$ *does not* qualify as the state of $\mathcal{A}$ at time t (see Chap. *4*, Sec. *5*).

State impulse response

The state equation *32* can be used to obtain a simple interpretation for $\mathbf{h}(t)$. Specifically, setting $t_0 = 0$, $\mathbf{x}(t_0) = \mathbf{0}$, and $u(t) = \delta(t)$ in *32* yields

34 $$\begin{aligned} \mathbf{h}(t) &= \mathbf{x}(t)\Big]_{\mathbf{x}(0-)=0,\, u(t)=\delta(t)} && \forall t \geq 0 \\ \mathbf{h}(t) &= 0 && \text{for } t < 0 \end{aligned}$$

which means that $\mathbf{h}(t)$ is the state in which $\mathcal{a}$ is at time t given that $\mathcal{a}$ is in its zero state at time $0-$ and $u(t) = \delta(t)$. In view of this interpretation, $\mathbf{h}$ will be referred to as the *state impulse response* of $\mathcal{a}$. Note that $h(t)$, the impulse response of $\mathcal{a}$, is merely the first component of the vector $\mathbf{h}(t)$ (see *31*).

State equation in differential form

Equation *32* expresses the state of $\mathcal{a}$ at time t as a function of the initial state t_0 and the input segment $u_{(t_0,t]}$. To obtain a corresponding state equation for $\mathcal{a}$ in the differential form, it is sufficient to differentiate both sides of the identity *28* $n - 1$ times with respect to t. This yields n relations

35 $$h^{(\lambda-1)}(t - \tau)u(\tau) = -\langle \dot{\boldsymbol{\phi}}^{(\lambda-1)}(t - \tau),\mathbf{x}(\tau)\rangle + \langle \boldsymbol{\phi}^{(\lambda-1)}(t - \tau),\dot{\mathbf{x}}(\tau)\rangle \qquad \lambda = 1, \ldots, n$$

which, upon setting $\tau = t$ and noting that $\boldsymbol{\Phi}(0) = \mathbf{I}$, becomes (in matrix form)

36 $$\dot{\mathbf{x}}(t) = \dot{\boldsymbol{\Phi}}(0)\mathbf{x}(t) + \mathbf{h}(0)u(t)$$

where $\mathbf{h}$ is defined by *31*, with the understanding that $\mathbf{h}(0)$ should be interpreted as $\mathbf{h}(0+)$. Equation *36* is the desired state equation for $\mathcal{a}$ in the differential form.

37 *Comment* Strictly speaking, we should set $\tau = t-$ rather than $\tau = t$ in *35*, since $h^{(\lambda-1)}(t)$ may not be defined for $t = 0$ (see Example *43*). This yields

$$\dot{\mathbf{x}}(t-) = \dot{\boldsymbol{\Phi}}(0+)\mathbf{x}(t-) + \mathbf{h}(0+)u(t-) \qquad \forall t$$

which, upon the replacement of $t-$ with t, becomes *36*, with $\mathbf{h}(0+)$ and $\dot{\boldsymbol{\Phi}}(0+)$ in place of $\mathbf{h}(0)$ and $\dot{\boldsymbol{\Phi}}(0)$. Since $\dot{\boldsymbol{\Phi}}(0+) = \dot{\boldsymbol{\Phi}}(0)$ (by *21*), we can express the state equation in the form *36*, with the understanding that $\mathbf{h}(0)$ stands for $\mathbf{h}(0+)$. [We use $\mathbf{h}(0+)$ rather than $\mathbf{h}(0)$ because $\mathbf{h}(t)$ may contain delta functions at $t = 0$.]

Before proceeding further, it will be helpful to summarize the results obtained so far in the form of an

38 **Assertion** Let $\mathcal{a}$ be a strictly proper system (*3.6.27*) of order n characterized by the input-output-state relation

$$y(t) = \langle \boldsymbol{\phi}(t - t_0),\mathbf{x}(t_0)\rangle + \int_{t_0}^{t} h(t - \xi)u(\xi)\, d\xi \qquad t \geq t_0$$

Then the state equations of $\mathcal{a}$ in the differential form read

39 $$\dot{\mathbf{x}}(t) = \dot{\boldsymbol{\Phi}}(0)\mathbf{x}(t) + \mathbf{h}(0+)u(t) \qquad \forall t$$
$$y(t) = \langle \boldsymbol{\phi}(0),\mathbf{x}(t)\rangle$$

and the state at time t is given by

$$\mathbf{x}(t) = \mathbf{\Phi}(t - t_0)\mathbf{x}(t_0) + \int_{t_0}^{t} \mathbf{h}(t - \xi)u(\xi)\, d\xi \qquad t \geq t_0$$

where $\mathbf{h}(t)$ is the state impulse response given by *31* and *34*, and $\mathbf{\Phi}(t)$ is the state transition matrix defined by *7* for $t \geq 0$ in terms of one-sided basis functions (*3.6.45*), with the extended state transition matrix defined for all t as the solution of the differential equation *22*,

$$\dot{\mathbf{\Phi}}(t) = \dot{\mathbf{\Phi}}(0)\mathbf{\Phi}(t) \qquad \mathbf{\Phi}(0) = \mathbf{I} \qquad \forall t$$

The extended state transition matrix $\mathbf{\Phi}(t)$ has the properties expressed by *12*, *13*, *15*, *20*, and *25* for all t_0, t, τ, and ξ.

Connection between the state transition matrix and the impulse response

Based on the interpretation of the state impulse response expressed by *34*, it is a simple matter to deduce an expression for $\mathbf{h}(t)$ in terms of the state transition matrix $\mathbf{\Phi}(t)$. The connection between $\mathbf{h}(t)$ and $\mathbf{\Phi}(t)$ is stated in the

40 **Assertion** The state impulse response $\mathbf{h}(t)$ is related to the state transition matrix $\mathbf{\Phi}(t)$ by the equation

41 $$\mathbf{h}(t) = 1(t)\mathbf{\Phi}(t)\mathbf{h}(0+)$$

where $1(t)$ is the unit step function.

Proof The state of $\mathfrak{A}$ at time t is given by *32* or, equivalently, by the solution of the state equation

$$\dot{\mathbf{x}}(t) = \dot{\mathbf{\Phi}}(0)\mathbf{x}(t) + \mathbf{h}(0+)u(t)$$

with the state $\mathbf{x}(0-)$ serving as the prescribed initial condition at time $0-$.

From this and *34* it follows that $\mathbf{h}(t)$ is the solution of the differential equation

42 $$\dot{\mathbf{h}}(t) = \dot{\mathbf{\Phi}}(0)\mathbf{h}(t) + \mathbf{h}(0+)\delta(t)$$

subject to the initial condition $\mathbf{h}(0-) = \mathbf{0}$. Thus, to prove the assertion, it is sufficient to show that $\mathbf{h}(t)$ as expressed by *41* is the solution in question.

On substituting *41* in *42* and noting that as a result of the discontinuity at $t = 0$, the derivative of $\mathbf{h}(t)$ [in the distribution sense (*3.8.2 et seq.*)] is given by

$$\frac{d}{dt}\left(1(t)\mathbf{\Phi}(t)\mathbf{h}(0+)\right) = \mathbf{\Phi}(0)\mathbf{h}(0+)\delta(t) + 1(t)\dot{\mathbf{\Phi}}(t)\mathbf{h}(0+)$$

we have

$$\boldsymbol{\Phi}(0)\mathbf{h}(0+)\delta(t) + 1(t)\dot{\boldsymbol{\Phi}}(t)\mathbf{h}(0+) = \dot{\boldsymbol{\Phi}}(0)\boldsymbol{\Phi}(t)\mathbf{h}(0+) + \mathbf{h}(0+)\delta(t)$$

which is an identity since $\boldsymbol{\Phi}(0) = \mathbf{I}$ and $\dot{\boldsymbol{\Phi}}(0)\boldsymbol{\Phi}(t) = \dot{\boldsymbol{\Phi}}(t)$ (by *22*). Consequently, $1(t)\boldsymbol{\Phi}(t)\mathbf{h}(0+)$ is the solution of *42* satisfying the initial condition $\mathbf{h}(0-) = \mathbf{0}$. Q.E.D.

43 *Example* Consider the system characterized by the input-output-state relation

$$y(t) = e^{-t}x_1(0-) + e^{-2t}x_2(0-) + \int_{t_0}^{t} (e^{-(t-\xi)} + e^{-2(t-\xi)})u(\xi)\,d\xi \qquad t \geq t_0$$

In this case, $\phi_1(t) = e^{-t}$, $\phi_2(t) = e^{-2t}$, $\dot{\phi}_1(t) = -e^{-t}$, $\dot{\phi}_2(t) = -2e^{-2t}$, and

$$\boldsymbol{\Phi}(t) = \begin{bmatrix} e^{-t} & e^{-2t} \\ -e^{-t} & -2e^{-2t} \end{bmatrix} \qquad \boldsymbol{\Phi}(0) = \begin{bmatrix} 1 & 1 \\ -1 & -2 \end{bmatrix}$$

Since the basis functions are not normalized, we first form the normalized state transition matrix $\boldsymbol{\Phi}'(t)$ (*9*)

$$\begin{aligned} \boldsymbol{\Phi}'(t) &= \boldsymbol{\Phi}(t)\boldsymbol{\Phi}^{-1}(0) \\ &= \begin{bmatrix} 2e^{-t} - e^{-2t} & e^{-t} - e^{-2t} \\ -2e^{-t} + 2e^{-2t} & -e^{-t} + 2e^{-2t} \end{bmatrix} \end{aligned}$$

Now

$$\begin{aligned} h(t) &= 1(t)(e^{-t} + e^{-2t}) \\ h^{(1)}(t) &= 1(t)(-e^{-t} - 2e^{-2t}) \end{aligned}$$

and

$$\mathbf{h}(0+) = \begin{bmatrix} 2 \\ -3 \end{bmatrix}$$

It is easy to check that $\mathbf{h}(t) = 1(t)\boldsymbol{\Phi}'(t)\mathbf{h}(0+)$.

The state equations in differential form are expressed by

$$\begin{aligned} \dot{\mathbf{x}}(t) &= \begin{bmatrix} 0 & 1 \\ -2 & -3 \end{bmatrix} \mathbf{x}(t) + \begin{bmatrix} 2 \\ -3 \end{bmatrix} u(t) \\ y(t) &= \langle (1,0), \mathbf{x}(t) \rangle \end{aligned}$$

where $\mathbf{x}(t)$ is a column 2-vector.

44 *Comment* The relation *41* between the state transition matrix and the state impulse response provides a natural way of extending the definition of $\mathbf{h}(t)$ to negative t. Specifically, the *extended state impulse response* will be understood to be given by the relation $\mathbf{h}(t) = \boldsymbol{\Phi}(t)\mathbf{h}(0+)$ for all t.

Improper systems

Throughout the preceding analysis we have made the assumption that there are no delta functions among the basis functions of $\mathcal{A}$. This assumption rules out improper systems (*3.6.27*), for, as we shall see in

Chap. *4*, Sec. *5*, the basis functions of such systems include delta functions of various orders.

In what follows we shall merely sketch an extension of the results obtained so far to the case of improper systems, deferring a more detailed discussion until Chap. *4*.

By virtue of *3.7.3* and *3.7.27*, the input-output-state relation of an improper system of order r must necessarily comprise two kinds of terms: (I) infinitely differentiable basis functions and that part of the impulse response which is a linear combination of them and (II) delta functions and the remaining part of the impulse response which comprises delta functions of various orders. More specifically, $y(t)$ can be partitioned into two parts (with the subscripts p and δ identifying the proper and delta-function terms, respectively):

45 $$y(t) = y_p(t) + y_\delta(t)$$

of the form

46 $$y_p(t) = \phi_1(t - t_0)x_1(t_0) + \cdots + \phi_n(t - t_0)x_n(t_0) + \int_{t_0}^{t} h_p(t - \xi)u(\xi)\, d\xi$$

and

47 $$y_\delta(t) = \delta(t - t_0)x_{n+1}(t_0-) + \cdots + \delta^{(r-n-1)}(t - t_0)x_r(t_0-) + \int_{t_0+}^{t} h_\delta(t - \xi)u(\xi)\, d\xi$$

where $\phi_1, \ldots, \phi_n$ are infinitely differentiable basis functions, h_p is a linear combination of $\phi_1, \ldots, \phi_n$, and h_δ is a linear combination of delta functions of orders up to and including $r - n$.

This partitioning amounts essentially to representing $\mathcal{A}$ as a sum (parallel combination) of two systems $\hat{\mathcal{A}}$ and $\mathfrak{D}$ characterized by *46* and *47*, respectively, with $\hat{\mathcal{A}}$ being a strictly proper system (*3.6.27*) and $\mathfrak{D}$ being a system of the differential operator type (*4.4.1*). Note that the state of $\mathcal{A}$ has r components, the first n of which are associated with $\hat{\mathcal{A}}$ and the rest with $\mathfrak{D}$. The corresponding state equations are given in Chap. *4*, Sec. *5*, and we shall forego reproducing them at this point.

48 *Note* All of our results concerning the state transition matrix, state equations, etc., have been obtained under the simplifying assumption that $\mathcal{A}$ is a time-invariant system. Analogous results for time-varying systems will be established in Chap. *6*, Sec. *4*.

Validity of input-output-state relations for reversed time

In writing the input-output-state relation *1* and the state equation *32* which was derived from it, we make the usual qualification that $t \geq t_0$. Actually, it is easy to show that these equations hold for *all t* provided the state transition matrix, the state impulse response, and the impulse response which occur in these equations are assumed to be extended rather than one-sided (see *21* and *44*). This implies that we can specify the state of $\mathcal{A}$ at time t_0 and determine via these equations the state and the output of $\mathcal{A}$ at time t *before* t_0 in terms of $\mathbf{x}(t_0)$ and the values of the input over the interval $[t,t_0]$. In this case, time is, in a sense, reversed, and the state $\mathbf{x}(t_0)$ plays the role of the *terminal* (see *1.7.5 et seq.*) rather than the initial state of the system. [When we wish to distinguish between this case and the usual one where $\mathbf{x}(t_0)$ is the initial state, we shall refer to a state equation as a *backward* or *forward* state equation according as $t < t_0$ or $t \geq t_0$.]

Stated more formally, we have the

49 **Theorem** The input-output-state relation

50
$$y(t) = \langle \boldsymbol{\phi}(t - t_0), \mathbf{x}(t_0) \rangle + \int_{t_0}^{t} h(t - \xi) u(\xi) \, d\xi$$

and the state equations

51
$$\mathbf{x}(t) = \boldsymbol{\Phi}(t - t_0)\mathbf{x}(t_0) + \int_{t_0}^{t} \mathbf{h}(t - \xi) u(\xi) \, d\xi$$
$$y(t) = \langle \boldsymbol{\phi}(0), \mathbf{x}(t) \rangle$$

hold for all t and t_0 provided the basis functions $\phi_1, \ldots, \phi_n$, the state transition matrix $\boldsymbol{\Phi}(t)$, the impulse response $h(t)$, and the state impulse response $\mathbf{h}(t)$ are understood to be extended rather than one-sided. [The extended state transition matrix is defined for negative t by the relation

$$\boldsymbol{\Phi}(-t) = \boldsymbol{\Phi}^{-1}(t)$$

or, equivalently, as the solution of the differential equation *22*. The extended state impulse response is given in terms of the extended $\boldsymbol{\Phi}(t)$ by

$$\mathbf{h}(t) = \boldsymbol{\Phi}(t)\mathbf{h}(0)$$

Correspondingly, the extended basis functions and the impulse response are given by the elements of the first row of the extended $\boldsymbol{\Phi}(t)$ and $\mathbf{h}(t)$, respectively.]

Proof Since the state equations *51* are equivalent to *50*, and since $y(t)$ is a linear combination of the components of $\mathbf{x}(t)$, it will suffice to demonstrate that the theorem holds true for $\mathbf{x}(t)$.

To this end, let $t \geq t_0$ and let τ be an intermediate time between t_0 and t, $t_0 \leq \tau \leq t$. Then, the theorem is proved if we can demonstrate that the value of $\mathbf{x}(\tau)$ computed by using the forward state equation with $\mathcal{A}$ starting in state $\mathbf{x}(t_0)$ is identical with that yielded by the backward equation with $\mathbf{x}(t)$ playing the role of the terminal state.

On replacing $\mathbf{h}(t)$ by $\mathbf{\Phi}(t)\mathbf{h}(0)$, the forward state equation yields

52 $$\mathbf{x}(\tau) = \mathbf{\Phi}(\tau - t_0)\mathbf{x}(t_0) + \int_{t_0}^{\tau} \mathbf{\Phi}(\tau - \xi)\mathbf{h}(0)u(\xi)\,d\xi \qquad \tau \geq t_0$$

53 $$\mathbf{x}(t) = \mathbf{\Phi}(t - t_0)\mathbf{x}(t_0) + \int_{t_0}^{t} \mathbf{\Phi}(t - \xi)\mathbf{h}(0)u(\xi)\,d\xi \qquad t \geq t_0$$

On the other hand, the backward state equation yields the following expression for the state at time τ:

54 $$\mathbf{x}'(\tau) = \mathbf{\Phi}(\tau - t)\mathbf{x}(t) + \int_{t}^{\tau} \mathbf{\Phi}(\tau - \xi)\mathbf{h}(0)u(\xi)\,d\xi$$

so that we have to show that $\mathbf{x}'(\tau) = \mathbf{x}(\tau)$.

On substituting $\mathbf{x}(t)$ as given by *53* into *54*, we have

$$\mathbf{x}'(\tau) = \mathbf{\Phi}(\tau - t)\mathbf{\Phi}(t - t_0)\mathbf{x}(t_0) + \int_{t_0}^{t} \mathbf{\Phi}(\tau - t)\mathbf{\Phi}(t - \xi)\mathbf{h}(0)u(\xi)\,d\xi + \int_{t}^{\tau} \mathbf{\Phi}(\tau - \xi)\mathbf{h}(0)u(\xi)\,d\xi$$

and since by *14* and *36*

$$\mathbf{\Phi}(\tau - t)\mathbf{\Phi}(t - t_0) = \mathbf{\Phi}(\tau - t_0)$$
$$\mathbf{\Phi}(\tau - t)\mathbf{\Phi}(t - \xi) = \mathbf{\Phi}(\tau - \xi)$$

we have

$$\mathbf{x}'(\tau) = \mathbf{\Phi}(\tau - t_0)\mathbf{x}(t_0) + \int_{t_0}^{\tau} \mathbf{\Phi}(\tau - \xi)\mathbf{h}(0)u(\xi)\,d\xi$$

which is identical with $\mathbf{x}(\tau)$. Q.E.D.

An immediate consequence of Theorem *49* is the

55 **Corollary** If $\mathcal{A}$ is characterized by an input-output-state relation of the form *50*—which implies that $\mathcal{A}$ is in reduced form (see *3.6.46*)—then $\mathcal{A}$ is initial state determinable (*1.8.12 et seq.*) in the following sense. Given $u_{(t_0,t]}$ and $y_{(t_0,t]}$, one can uniquely determine the initial state $\mathbf{x}(t_0)$. ◁

Specifically, the knowledge of $u_{(t_0,t]}$ and $y_{(t_0,t]}$ yields the values of $u(t)$, $y(t)$, and their derivatives at $t-$. These in turn determine by *4.5.18* the state of $\mathcal{A}$ at time $t-$. Then, using *51*, we obtain

$$\mathbf{x}(t_0) = \mathbf{\Phi}(t_0 - t)\left[\mathbf{x}(t-) + \int_{t}^{t_0} \mathbf{h}(t - \xi)u(\xi)\,d\xi\right]$$

which shows that the initial state is uniquely determined by the input-output pair $(u_{(t_0,t]}, y_{(t_0,t]})$.

56 *Comment* From the results obtained in this section we can draw a significant conclusion, namely, that any linear system of finite order, subject to assumptions (I) and (II), can be characterized by state equations in the differential form. As we shall see in Chap. 4, this in turn implies that any system of finite order can be characterized by

an input-output relation having the form of a linear differential equation. Thus, by restricting ourselves to systems of finite order, we have in effect restricted ourselves to systems whose behavior is governed by linear differential equations. Such systems are the main concern of this text.

8 *Differential and discrete-time systems*

Like the class of all systems, the class of linear systems can be categorized in many different ways. In what follows, we shall focus our discussion on two principal types of linear systems, namely, *differential systems* and *discrete-time systems,* with the differential systems receiving the bulk of our attention. Actually, our main concern in this section will be with matters pertaining to terminology and notation, with more detailed analyses of the systems in question to follow in later chapters.

Linear differential systems

We begin our discussion with linear differential systems, which, as the name implies, are linear systems whose behavior is governed by differential equations. More specifically, by a *linear differential system* we mean a system whose input and output are related by one or more ordinary (rather than partial) differential equations. In the engineering literature, such systems are commonly referred to as *linear lumped-parameter systems.*

In essence, the importance of linear differential systems stems from three factors. First, most of the linear systems encountered in the physical world are of this type or can be approximated by systems of this type. Second, linear differential systems have been investigated more extensively than any other category of linear systems. And third, the study of linear differential systems serves as a convenient introduction, if not a prerequisite, to that of most other types of linear systems. For example, the properties of linear discrete-time systems parallel very closely those of linear differential systems and in many cases can be found or established by very similar techniques.

Before giving a general definition of a linear differential system, it will be helpful to observe that by a linear time-invariant differential system $\mathcal{A}$ one usually means a system characterized by a differential equation of the form

1 $$a_n \frac{d^n y}{dt^n} + \cdots + a_0 y = b_m \frac{d^m u}{dt^m} + \cdots + b_0 u$$

in which u is a scalar input time function, y is a scalar output time function, and the a_i, $i = 0, \ldots, n$, and b_j, $j = 0, \ldots, m$, are constant coefficients.

More generally, in the case of a linear differential system which is not necessarily time-invariant, the coefficients a_i and b_j may depend on time. Furthermore, u and y may be related to one another by a system of linear differential equations rather than by a single differential equation as in *1*. Also, the input u and output y may be vectors rather than scalars, in which case the a_i and b_j are matrices, possibly time-dependent.

Now as was pointed out in Chap. *1*, Sec. *4*, an input-output relation such as *1* is merely a way of defining the class of all input-output pairs which can be associated with $\mathcal{A}$. Thus, $(u_{(t_0,t]}, y_{(t_0,t]})$, or simply (u,y), is an input-output pair for $\mathcal{A}$ if and only if (u,y) satisfies the differential equation *1*. This naturally leads to the question: What do we mean—in precise terms—when we say that u and y satisfy *1*?

To answer this question, we must examine more carefully the meaning of the derivatives such as dy/dt and d^2u/dt^2 at the point at which the observation interval begins, i.e., at $t = t_0$. As a first step in this direction, let us employ the usual symbol p to denote the operation of differentiation. That is, $p\,u = du/dt$, with the understanding that $p\,u$ represents the derivative of u in the distribution sense (see Appendix *A*), which means, informally, that the derivative of a unit step function $1(t)$ is a delta function $\delta(t)$ and, more generally,

2
$$\begin{aligned} p\,1(t) &= \delta(t) \\ p\,\delta(t) &= \delta^{(1)}(t) \\ &\cdots\cdots\cdots \\ p\,\delta^{(n)}(t) &= \delta^{(n+1)}(t) \end{aligned}$$

where $\delta^{(n)}(t)$ denotes a delta function of nth order occurring at $t = 0$. This implies that, if u has a jump at, say, $t = 0$, then $p\,u$ will have a delta function at $t = 0$ expressed by $[u(0+) - u(0-)]\delta(t)$. We call $p\,u$ the "derivative in the distribution sense" for two main reasons. First, to differentiate between $p\,u$ and the derivative of u in the conventional sense—which does not exist when u has jumps or delta functions—and second, to indicate that $p\,u$ can be defined in a rigorous fashion by using the concepts and techniques of the theory of distributions. Although we shall not use the theory of distributions as such in this text, a brief self-contained exposition of it is presented in Appendix *A*.

Let us consider now what happens when p acts on a time function u which is not defined for $t < t_0$, for example, an input u which is defined for $t \geq t_0$ but not $t < t_0$. Then $p\,u$ is not uniquely defined by u. However, $p\,u$ is determined uniquely by u for $t \geq t_0$ if one has the value of u at t_0-, since the delta function in $p\,u$ at t_0 is given by $(u(t_0+) - u(t_0-))\delta(t - t_0)$ and the rest of $p\,u$ is defined by the values of u for $t \geq t_0$. These observations motivate the following

3 **Definition** If u is a time function defined for $t \geq t_0$ and u is differentiable in the distribution sense, then

4
$$p\,u(t) \triangleq -u(t_0-)\delta(t - t_0) + p\,1(t - t_0)u(t)$$

and the $p^n\,u(t)$, $n = 2, 3, \ldots$, are defined by

5
$$p^2\,u(t) = -u(t_0-)\delta^{(1)}(t - t_0) - u^{(1)}(t_0-)\delta(t - t_0) + p^2\,1(t - t_0)u(t)$$

6
$$p^3\,u(t) = -u(t_0-)\delta^{(2)}(t - t_0) - u^{(1)}(t_0-)\delta^{(1)}(t - t_0) - u^{(2)}(t_0-)\delta(t - t_0) + p^3\,1(t - t_0)u(t)$$

. . . .

Note To be consistent with our notation for time functions, we should write $p\,u$ rather than $p\,u(t)$ in *4*. Our lack of consistency in *4* is explained in Remark *1.3.3*. Note also that if $\mathbf{u}$ is a vector-valued time function, then $p\,\mathbf{u}$ is defined as above, with u replaced throughout by $\mathbf{u}$.

7 *Example* Let $u(t) = 3e^{-t}$ for $t \geq 0$ and $u(0-) = 1$, $u^{(1)}(0-) = 2$. Then,

$$\begin{aligned} p\,u(t) &= -\delta(t) + p\,1(t)3e^{-t} \\ &= -\delta(t) + 3\delta(t) - 3e^{-t} \\ &= 2\delta(t) - 3e^{-t} \qquad t \geq 0 \end{aligned}$$

Similarly,

$$p^2\,u(t) = 2\delta^{(1)}(t) - 5\delta(t) + 3e^{-t} \qquad t \geq 0$$

8 *Comment* As we shall see in Chap. *4*, Sec. *4*, p represents a differentiator $\mathfrak{D}$ which is characterized by the input-output-state relation *4*. Thus, the terms $-u(t_0-)\delta(t - t_0)$ and $p\,1(t - t_0)u(t)$ in *4* respectively represent the zero-input response and the zero-state response of $\mathfrak{D}$, with $u(t_0-)$ playing the role of the initial state of $\mathfrak{D}$.

9 *Comment* If u represents a segment $u_{(t_0,t_1]}$, then it will be understood that $p\,u$ should be interpreted as in *4*, with the delta function

$$(u(t_1+) - u(t_1-))\delta(t - t_1)$$

not included in the interval of observation $(t_0,t_1]$.

10 **Remark** In view of the fact that the symbol $p\,u$ is frequently used in many different and not always well-defined ways, it is sometimes necessary to indicate whether $p\,u$ should be understood in the sense of *4* or in the sense of *4* but with $u(t_0-)$ set equal to zero. When it is necessary to differentiate between these two cases, $p\,u$ will be referred to as an *indefinite derivative* when $p\,u$ is defined by *4* and as a *definite derivative* when $u(t_0-)$ in *4* is taken to be zero. In the sequel,

$p\,u$ and, more generally, $p^k\,u$, $k = 1, 2, \ldots$, will be understood to be indefinite derivatives unless stated otherwise. (Note that when $p\,u$ is a definite derivative, p represents a differentiator in its zero state. Note also that the notions of definite and indefinite derivatives are analogs of the notions of definite and indefinite integrals.)

Differential operators

We are now in a position to clarify what is meant by a differential equation such as *1*. For this purpose, we make the

11 **Definition** An *indefinite differential operator*, or, simply, a *differential operator of order n*, $n = 0, 1, 2, \ldots$, is a polynomial in p of the form

12 $$L(p) = a_n p^n + \cdots + a_0 \qquad a_n \neq 0$$

where p^k, $k = 0, 1, 2, \ldots$, is defined by *3* and the a_i, $i = 0, \ldots, n$, are scalar coefficients. If p in *12* is a definite derivative, then $L(p)$ is referred to as a *definite* differential operator. If the coefficients in *12* are matrices, $L(p)$ is said to be a *matric* differential operator. If $a_0, \ldots, a_n$ are time-dependent, they are written as $a_0(t), \ldots, a_n(t)$.

Note that in view of Comment *8*, a differential operator may be regarded as a parallel-series combination (*1.10.33*) of differentiators and scalors.

13 *Example* Let $L(p) = p^2 + 2p + 3$ and let $u(t) = e^{-t}$ for $t \geq 0$. Then

$$L(p)\,u = -u(0-)\delta^{(1)}(t) - (u^{(1)}(0-) + 2u(0-))\delta(t) + (p^2 + 2p + 3)1(t)e^{-t} \qquad t \geq 0$$

and

$$(p^2 + 2p + 3)1(t)e^{-t} = \delta^{(1)}(t) + \delta(t) + 2e^{-t} \qquad t \geq 0$$

14 *Comment* There is a lack of consistency in both the engineering and mathematical literature of ordinary differential equations with regard to the numbering of the coefficients in a differential operator. In many texts a differential operator is written as

15 $$L(p) = a_0 p^n + \cdots + a_n$$

rather than as in *12*, and in some both *12* and *15* are used. In what follows, we shall be somewhat inconsistent too, employing *12* in some chapters and *15* in others, depending on the purposes of the analysis.

Differential equations

Turning to the differential equation *1*, we make the following

16 **Definition** A pair (u,y), with u and y starting at t_0, *satisfies* the differential equation

17 $$a_n \frac{d^n y}{dt^n} + \cdots + a_0 y = b_m \frac{d^m u}{dt^m} + \cdots + b_0 u$$

if and only if

18 $$(a_n p^n + \cdots + a_0)y = (b_m p^m + \cdots + b_0)u \qquad \text{for all } t \geq t_0$$

where

19 $$\begin{aligned} L &\triangleq L(p) \triangleq a_n p^n + \cdots + a_0 \qquad a_n \neq 0 \\ M &\triangleq M(p) \triangleq b_m p^m + \cdots + b_0 \qquad b_m \neq 0 \end{aligned}$$

are indefinite differential operators in the sense of Definition *11*, with the understanding that the delta functions in Ly and Mu at t_0 are included in the interval $t \geq t_0$.

20 *Example* Suppose that

21 $$(p^2 + 3p + 2)y = (p + 3)u$$

Then a pair (u,y), with u and y starting at, say, $t_0 = 0$, satisfies *21* if there exist values of $y(0-)$, $y^{(1)}(0-)$, and $u(0-)$ such that the relation

$$\begin{aligned} -y(0-)\delta^{(1)}(t) - [3y(0-) + y^{(1)}(0-)]\delta(t) + (p^2 + 3p + 2)1(t)y(t) \\ = -u(0-)\delta(t) + (p + 3)1(t)u(t) \qquad t \geq 0 \end{aligned}$$

is satisfied by u and y, with the delta functions included in the interval $t \geq 0$. For example, the pair $(e^{-3t}, 2e^{-t} - 2e^{-2t})$, $t \geq 0$, satisfies *21* with $y(0-) = 0$, $y^{(1)}(0-) = 2$, $u(0-) = 1$. ◁

Note that when we say that a system $\mathcal{A}$ is *characterized* by an input-output relation of the form *18*, we mean that $\mathcal{A}$ admits as input-output pairs only those time functions (u,y) (with u and y defined over an observation interval starting at t_0) which satisfy *17* in the sense of Definition *16*.

22 **Remark** In the sequel, the larger of the two numbers m,n in *19* will be referred to as the *order* of the differential equation *17*. Thus

23 $$\text{order of } \{Ly = Mu\} \triangleq \max(n,m)$$

As we shall see in Chap. *4*, the order of the differential equation $Ly = Mu$ which characterizes $\mathcal{A}$ is identical with the order of $\mathcal{A}$ in the sense of Definition *3.6.39*.

24 **Remark** Suppose that $\mathcal{A}$ is characterized by *17* as well as by its input-output-state relation

25 $$y(t) = \langle \boldsymbol{\phi}(t - t_0), \mathbf{x}(t_0-) \rangle + \int_{t_0}^{t} h(t,\xi)u(\xi)\, d\xi$$

Now by the mutual-consistency condition, every input-output pair which satisfies *17* also satisfies *25*, and vice versa. This implies that the input-output-state relation *25* constitutes what is called in the theory of differential equations a *general solution* of *17*.

Definition of a linear differential system

With these definitions as a background, we are in a position to give a precise meaning to the notion of a linear differential system.

Informally, $\mathfrak{A}$ is a *linear differential system* if $\mathfrak{A}$ can be represented as an interconnection of adders, scalors, and differentiators. This informal definition underlies the more general formal

26 **Definition** A system $\mathfrak{A}$ with vector input $\mathbf{u}$ and vector output $\mathbf{y}$ is a *linear differential* system if and only if the input-output relation of $\mathfrak{A}$ admits of the representation

27
$$\mathbf{L}\mathbf{y} + \mathbf{K}\tilde{\mathbf{y}} = \mathbf{M}\mathbf{u}$$

where $\mathbf{L}$, $\mathbf{M}$, and $\mathbf{K}$ are, in general, matric differential operators (not necessarily with constant coefficients) and the components of $\tilde{\mathbf{y}}$ are scalar time functions playing the role of suppressed output variables of $\mathfrak{A}$ (*1.10.40*). Furthermore, if $\tilde{\mathbf{y}}$ has r components, $r = 0, 1, 2, 3, \ldots$, then $\mathbf{K}$ must be a $k \times r$ matrix, $k > r$. If the matric differential operators $\mathbf{L}$, $\mathbf{M}$, and $\mathbf{K}$ have constant matrices as coefficients of powers of p, then $\mathfrak{A}$ is a time-invariant linear differential system in the sense of Definitions *3.2.36* and *3.4.25*. (This will follow from the results established in Chap. *4*.)

28 *Example* A system characterized by the input-output relation

$$\frac{d^2y}{dt^2} + 3\frac{dy}{dt} + 3y = \frac{du}{dt} + 2u$$

is a linear differential time-invariant system. Here

$$L = p^2 + 3p + 3$$
$$M = p + 2$$
$$K = 0$$

29 *Example* A system characterized by the input-output relation

30
$$a_n t^n \frac{d^n y}{dt^n} + \cdots + a_1 t \frac{dy}{dt} + a_0 y = b_m t^m \frac{d^m y}{dt^m} + \cdots + b_1 t u + b_0 u$$

in which the a's and b's are constant coefficients, is a linear differential time-varying system. Here

$$L = a_n t^n p^n + \cdots + a_1 t p + a_0$$
$$M = b_m t^m p^m + \cdots + b_1 t p + b_0$$
$$K = 0$$

31 *Comment* Systems of this type are sometimes referred to as *Euler-Cauchy* systems. It is easy to verify that if in the differential equation characterizing a linear time-invariant system the variable t is replaced by $\log t'$, the resulting system is of the Euler-Cauchy type. In other words, Euler-Cauchy systems can be derived from linear differential time-invariant systems by the change of time scale $t \to \log t$ (see Example *3.2.22* and Ref. *3*).

32 *Example* The system characterized by the system of equations

33
$$\begin{aligned}(3p^2 - 2p + 1)y_1 + 2p^2 y_2 &= -u_1 + 3u_2 \\ (p^2 - p)y_1 + (p^2 + p + 1)y_2 &= 2u_1 + u_2\end{aligned}$$

in which $\mathbf{u} = (u_1,u_2)$ and $\mathbf{y} = (y_1,y_2)$ are the input and output vectors, respectively, is a linear differential time-invariant system. On rewriting *33* in the matrix form

$$\begin{bmatrix} 3 & 2 \\ 1 & 1 \end{bmatrix} \frac{d^2\mathbf{y}}{dt^2} + \begin{bmatrix} -2 & 0 \\ -1 & 1 \end{bmatrix} \frac{d\mathbf{y}}{dt} + \begin{bmatrix} 1 & 0 \\ 0 & 1 \end{bmatrix} \mathbf{y} = \begin{bmatrix} -1 & 3 \\ 2 & 1 \end{bmatrix} \mathbf{u}$$

it is readily seen that L and M are given by

$$\mathbf{L} = \begin{bmatrix} 3p^2 - 2p + 1 & 2p^2 \\ p^2 - p & p^2 + p + 1 \end{bmatrix}$$
$$\mathbf{M} = \begin{bmatrix} -1 & 3 \\ 2 & 1 \end{bmatrix}$$

In all of the above examples the operator $\mathbf{K}$ is zero. The following example illustrates a system in which this is not the case.

34 *Example* Consider the network $\mathfrak{A}$ shown in Fig. *3.8.1*, in which u is the input current, y is the output voltage, and the node voltages $\tilde{y}_1$ and $\tilde{y}_2$ are the suppressed variables. Using nodal analysis (see, for example, Ref. *6*), the equations relating u,y and $\tilde{y}_1,\tilde{y}_2$ are found to be

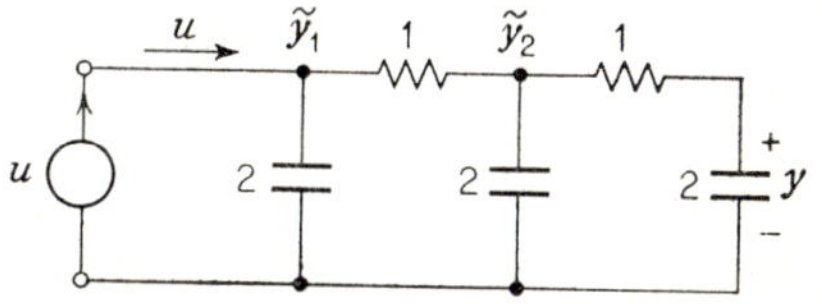

Fig. 3.8.1 Network of Example *3.8.34*.

$$\begin{aligned} 0 &= u - (2p + 1)\tilde{y}_1 + \tilde{y}_2 \\ y &= -\tilde{y}_1 + (2 + 2p)\tilde{y}_2 \\ -(2p + 1)y &= -\tilde{y}_2 \end{aligned}$$

These equations are of the general form

$$\mathbf{Ly} + \mathbf{K\tilde{y}} = \mathbf{Mu}$$

in which $\tilde{\mathbf{y}} = (\tilde{y}_1,\tilde{y}_2)$,

$$\mathbf{L} = \begin{bmatrix} 0 \\ 1 \\ -(2p + 1) \end{bmatrix} \qquad \mathbf{M} = \begin{bmatrix} 1 \\ 0 \\ 0 \end{bmatrix} \qquad \text{and}$$

$$\mathbf{K} = \begin{bmatrix} (2p + 1) & -1 \\ 1 & -(2 + 2p) \\ 0 & 1 \end{bmatrix}$$

Note that **K** is a 3 × 2 matric differential operator—as it should be in order to satisfy the condition stated in Definition *26*.

35 *Comment* In Chap. *4*, Sec. *7*, it will be shown that if $\mathcal{A}$ is characterized by an input-output relation of the form *27*, then there exists an equivalent system $\mathcal{B}$ which is characterized by a single differential equation which does not involve suppressed output variables. The techniques by which a system of differential equations of the form *27* can be reduced to an equivalent single differential equation are discussed in Chaps. *4* and *6*. This remark implies that the class of linear systems characterized by *27* is no more general than that characterized by *27* with $\mathbf{K} = \mathbf{0}$.

This concludes our discussion of the terminology and notation pertaining to linear differential systems. We turn next to a closely analogous class of systems, namely, linear discrete-time systems. Because of this analogy, our discussion of linear discrete-time systems will parallel closely that of linear differential systems and will be very brief.

Discrete-time systems

As the name implies, a characteristic feature of *discrete-time systems*, or, as they are frequently called, *sampled-data systems*, is the discreteness of time. Thus, in a discrete-time system the input u and output y are sequences of the form $u_0u_1u_2 \cdots y_0y_1y_2 \cdots$, where u_λ denotes the value of u at time t_λ, with λ ranging over integers. The u_λ are called the *samples* of u, with the samples being *equispaced* if $t_\lambda - t_{\lambda-1} \triangleq T_0$ is a constant independent of λ. T_0 is called the *sampling interval*. In what follows, we shall restrict our attention to systems in which the samples of u and y are equispaced, and for convenience we shall assume that $T_0 = 1$. Under this assumption, t ranges over the integers , . . . , $-1, 0, 1, \ldots$ and the value of u at time t reads u_t.

Central to the representation and characterization of discrete-time systems is the *unit delay operator* D, which is defined below. (For simplicity we assume that $t_0 = 0$ and that u_t ranges over scalars.)

36 **Definition** Let u denote a sequence $u_0u_1u_2 \cdots$ defined for $t = 0, 1, 2, \ldots$. Then Du is a sequence y defined by

$$y_t = u_{t-1} \qquad t = 0, 1, 2, \ldots$$

and $D^k u$, $k = 1, 2, \ldots$, is defined recursively by

37 $$D^k u = D(D^{k-1}u)$$

with $D^0 = 1$. Thus

$$D^1u = u_{-1}u_0u_1u_2 \cdots$$
$$D^2u = u_{-2}u_{-1}u_0u_1 \cdots$$
$$D^3u = u_{-3}u_{-2}u_{-1}u_0u_1 \cdots$$

Note that Du is not defined for all $t \geq 0$ (with t ranging over the integers) if u_{-1} is not specified. Thus D is an *indefinite* operator in the same sense as p (Remark *10*). D will be said to be *definite* if u_{-1} in Du is set equal to zero. Unless stated to the contrary, D will be understood to be indefinite.

38 *Comment* In effect, D represents a unit delayor (*1.7.25*) whose input function space is the space of sequences $u_0 u_1 u_2 \cdots$, with u_{t-1} playing the role of the state of D at time t. Thus, a definite D is a unit delayor which is initially in its zero state. Equivalently, a definite D is a translation operator (*3.2.2*) with delay $\delta = 1$.

39 *Example* Let $u = 3\ 1\ 2\ 2\ 0\ 1$. Then $Du = u_{-1}\ 3\ 1\ 2\ 2\ 0\ 1$, and if D is definite, $Du = 0\ 3\ 1\ 2\ 2\ 0\ 1$.

By analogy with *11* and *26*, we make the following definitions.

40 **Definition** An *indefinite delay operator*, or, simply, a *delay operator of order* n, $n = 0, 1, 2, \ldots$, is a polynomial in D of the form

41
$$L(D) = a_n D^n + \cdots + a_0 \qquad a_n \neq 0$$

where D^k, $k = 0, 1, 2, \ldots$, is defined by *36* and the a_i are scalar coefficients. If the coefficients in *41* are matrices, $L(D)$ is said to be a *matric* delay operator. If $a_0, \ldots, a_n$ are time-dependent, they are written as $a_0(t), \ldots, a_n(t)$. Essentially, $L(D)$ represents a parallel-series combination of delayors and scalors. [Note that $L(D)\mathbf{u}$ for vector-valued $\mathbf{u}$ is defined in exactly the same way as for scalar-valued u.]

42 **Definition** A system $\mathcal{A}$ with vector-valued $\mathbf{u}_t$ and $\mathbf{y}_t$ and t ranging over integers is a *linear discrete-time system* if and only if the input-output relation of $\mathcal{A}$ admits of the representation

43
$$\mathbf{Ly} + \mathbf{K\tilde{y}} = \mathbf{Mu}$$

where $\mathbf{L}$, $\mathbf{M}$, and $\mathbf{K}$ are, in general, matric delay operators (not necessarily with constant coefficients), and the components of $\mathbf{\tilde{y}}$ are suppressed output variables (*1.10.40*). As in Definition *26*, if $\mathbf{\tilde{y}}$ has r components, $r = 0, 1, 2, \ldots$, then $\mathbf{K}$ must be a $k \times r$ matrix, $k > r$. If the operators $\mathbf{L}$, $\mathbf{M}$, and $\mathbf{K}$ have constant matrices as coefficients of powers of D, then $\mathcal{A}$ is a time-invariant discrete-time system.

44 *Comment* In effect, a linear discrete-time system is an interconnection of scalors, adders, and delayors. As in the case of linear differential systems, if a linear discrete-time system is characterized by an input-output relation of the form *43* with $\mathbf{K} \neq \mathbf{0}$, then there exists an equivalent system $\mathcal{B}$ which is characterized by an input-output relation of the form *43* but with $\mathbf{K} = \mathbf{0}$. Thus, there is no loss of generality in assuming that a linear discrete-time system is characterized by an input-output relation of the form $\mathbf{Ly} = \mathbf{Mu}$.

45 *Example* Let $\mathcal{A}$ be characterized by the input-output relation

46 $$(D^2 + D + 2)y = (D + 1)u$$

Then $\mathcal{A}$ is a linear time-invariant discrete-time system with

$$L(D) = D^2 + D + 2$$

$M(D) = D + 1$, and $\mathbf{K} = \mathbf{0}$. Note that a pair (u,y) (with u and y starting at 0) satisfies *46* if and only if for some y_{-2}, y_{-1}, and u_{-1} the u_t and y_t for $t \geq 0$ satisfy the relations

$$\begin{aligned} y_{-2} + y_{-1} + 2y_0 &= u_{-1} + u_0 \\ y_{-1} + y_0 + 2y_1 &= u_0 + u_1 \\ y_0 + y_1 + 2y_2 &= u_1 + u_2 \\ &\cdots\cdots \\ y_{t-2} + y_{t-1} + 2y_t &= u_{t-1} + u_t \\ &\cdots\cdots \end{aligned}$$

These relations can be expressed more compactly by the use of z transforms, a brief exposition of which is presented in Chap. *10*. ◁

This concludes our brief discussion of the notation and basic terminology for linear discrete-time systems. We do not dwell further on systems of this type because our principal concern in the following chapters will be with differential systems.

9 *Two basic properties of linear systems*

In this final section of Chap. *3* we shall establish two basic properties of linear systems of finite order—one having to do with system equivalence and the other with determinateness (*2.6.3*). We begin with the

Connection between weak equivalence and equivalence

As was pointed out in Chap. *2* (*2.5.10*), weak equivalence does not, in general, imply equivalence. Thus, if $\mathcal{A}$ and $\mathcal{B}$ are two systems which are weakly equivalent [in the sense that every input-output pair (u,y) for $\mathcal{A}$ is also an input-output pair for $\mathcal{B}$, and vice versa (see *2.5.4*)], then it is not necessarily true that $\mathcal{A} \equiv \mathcal{B}$.

In this regard, linear systems have an important property which finds its expression in the following

1 **Theorem** Let $\mathcal{A}$ and $\mathcal{B}$ be linear differential systems of finite order. If $\mathcal{A}$ and $\mathcal{B}$ are weakly equivalent, then $\mathcal{A}$ and $\mathcal{B}$ are equivalent in the sense of Definition *2.5.8*.

Proof We shall prove this theorem under the simplifying but not essential assumptions that $\mathcal{A}$ and $\mathcal{B}$ are time-invariant and that the input and output are scalar-valued.

Let $\mathcal{A}$ and $\mathcal{B}$ be characterized by input-output-state relations of the form *3.7.1*,

2 $$\mathcal{A}: \qquad y(t) = \langle \boldsymbol{\phi}(t), \boldsymbol{\alpha} \rangle + \int_0^t h_a(t-\xi)u(\xi)\, d\xi$$

3 $$\mathcal{B}: \qquad y(t) = \langle \hat{\boldsymbol{\phi}}(t), \boldsymbol{\beta} \rangle + \int_0^t h_b(t-\xi)u(\xi)\, d\xi$$

where $t_0 = 0$, the basis functions are linearly independent over every subinterval of $[0, \infty)$ (see *3.7.12*), and h_a and h_b represent the impulse responses of $\mathcal{A}$ and $\mathcal{B}$, respectively. The state vectors $\boldsymbol{\alpha}$ and $\boldsymbol{\beta}$ are assumed to range over $\mathcal{C}^n$ (space of n-tuples of complex numbers).

First, let us consider input-output pairs of the form $(0,y)$, where y represents a zero-input response. By Assertion *3.6.49*, if every zero-input response of $\mathcal{A}$ is also a zero-input response of $\mathcal{B}$ and vice versa, then every basis function of $\mathcal{A}$ is expressible as a linear combination of basis functions of $\mathcal{B}$ and vice versa. This implies that, if $\mathcal{A}$ and $\mathcal{B}$ are weakly equivalent, then we can use the same set of basis functions, say, $\boldsymbol{\phi}(t)$, in both *2* and *3*, with the result that *3* can be replaced by a relation of the form

4 $$\mathcal{B}: \qquad y(t) = \langle \boldsymbol{\phi}(t), \boldsymbol{\beta} \rangle + \int_0^t h_b(t-\xi)u(\xi)\, d\xi$$

where $\boldsymbol{\beta}$ (which is not the same $\boldsymbol{\beta}$ as in *3*) ranges over $\mathcal{C}^n$.

The theorem is proved if we can show that $h_a(t) = h_b(t)$ for all t; for then $\mathcal{A}$ and $\mathcal{B}$ are both zero-input and zero-state equivalent, and hence are equivalent. To this end, we shall demonstrate that the statement "$\mathcal{A}$ and $\mathcal{B}$ are weakly equivalent" implies "$\mathcal{A}$ and $\mathcal{B}$ are zero-state equivalent."

Specifically, let (u,y) be an input-output pair for $\mathcal{A}$, with $\mathcal{A}$ initially in its zero state, and let

5 $$\begin{aligned} u(t) &= 0 && \text{for } 0 \le t < T \\ u(t) &= \delta(t-T) && \text{for } t \ge T \end{aligned}$$

Then the response of $\mathcal{A}$ to u is of the form

6 $$\begin{aligned} y(t) &= 0 && \text{for } 0 \le t < T \\ y(t) &= h_a(t-T) && \text{for } t \ge T \end{aligned}$$

By hypothesis, there must be a state in $\mathcal{B}$, say, $\boldsymbol{\beta}$, such that the response of $\mathcal{B}$ to u, with $\mathcal{B}$ initially in state $\boldsymbol{\beta}$, is identical with y. Using *2* and *4* to *6*, this condition yields

7 $$\langle \boldsymbol{\phi}(t), \boldsymbol{\beta} \rangle = 0 \qquad \text{for } 0 \le t < T$$

8 $$\langle \boldsymbol{\phi}(t), \boldsymbol{\beta} \rangle + h_b(t-T) = h_a(t-T) \qquad \text{for } t \ge T$$

Now *7* implies that $\boldsymbol{\beta} = \mathbf{0}$, since the basis functions are assumed to be linearly independent over every subinterval of $[0, \infty)$. [For, by differentiating *7* $n - 1$ times, we get $\boldsymbol{\Phi}(t)\boldsymbol{\beta} = \mathbf{0}$ [$\boldsymbol{\Phi}(t) \triangleq$ state transition matrix (*3.7.7*)], which implies that $\boldsymbol{\beta} = \mathbf{0}$.] Therefore, on substituting $\boldsymbol{\beta} = \mathbf{0}$ in *8*, we have

9 $$h_a(t - T) = h_b(t - T) \qquad t \geq T$$

and since T is arbitrary, $h_a(t) = h_b(t)$ for all $t \geq 0$, which proves the theorem. ◁

We leave it to the reader to prove appropriately restricted forms of the following corollaries of Theorem *1*.

10 **Corollary** If $\mathcal{A}$ is weakly time-invariant (*3.2.42*), then $\mathcal{A}$ is time-invariant.

11 **Corollary** If $\mathcal{A}$ and $\mathcal{B}$ are converse systems (*2.10.1*), then $\mathcal{A}$ and $\mathcal{B}$ are inverse to one another.

Determinateness of linear systems

We turn next to another basic property of linear systems, namely the *determinateness* of linear systems (*2.6.3* and *2.6.7*).

Essentially, linear systems are determinate in the sense that if $\mathcal{A}$ is an interconnection of linear systems $\mathcal{A}_1, \ldots, \mathcal{A}_N$, with the state vector of $\mathcal{A}_i$ at time t, $\mathbf{x}^i(t)$, ranging over Σ_i, then the composite vector $\mathbf{x}(t) = (\mathbf{x}^1(t), \ldots, \mathbf{x}^N(t))$ which takes values in a subset Σ of the product space $\Sigma_1 \times \cdots \times \Sigma_N$ (see *2.6.6 et seq.*) qualifies as a state vector of $\mathcal{A}$. In practical terms, this means that the problem of associating a state vector $\mathbf{x}$ with $\mathcal{A}$ can be reduced to a number of simpler problems, each involving the association of a state vector $\mathbf{x}^i$ with a component $\mathcal{A}_i$ of $\mathcal{A}$. This approach to the problem of associating a state vector with a differential system will be discussed more fully in Chap. *4*, Sec. *6*.

We shall not attempt to establish the determinateness property of linear systems in its most general form. For our purposes, it will suffice to establish the following restricted version of it.

12 **Theorem** Let $\mathcal{A}$ be an interconnection of $\mathcal{A}_1, \ldots, \mathcal{A}_N$ and adders, where $\mathcal{A}_i$, $i = 1, \ldots, N$, is a linear time-invariant differential system of finite order with scalar input u^i and scalar output y^i. Then $\mathcal{A}$ is a determinate system in the sense of Remark *2.6.7*.

Proof Let $\boldsymbol{\alpha}^i$ denote the initial state of $\mathcal{A}_i$, $i = 1, \ldots, N$. Then by the decomposition property (*3.5.2*), the output of $\mathcal{A}_i$ can be expressed as

13 $$y^i = A_i(\boldsymbol{\alpha}^i;0) + A_i(u^i)$$

where $A_i(\boldsymbol{\alpha}^i;0)$ denotes the zero-input response of $\mathcal{A}_i$ starting in state $\boldsymbol{\alpha}^i$ and $A_i(u^i)$ is the zero-state response of $\mathcal{A}_i$ to u^i. Furthermore, by

3.6.23, the Laplace transform of $A_i(u^i)$ is given by $H_i(s)U^i(s)$, where $H_i(s)$ is the transfer function of $\mathcal{A}_i$ and $U^i(s)$ is the Laplace transform of u^i (t_0 is assumed to be 0). Thus, each $\mathcal{A}_i$ can be replaced by $\mathcal{A}_i$ in its zero state plus a source $A_i(\boldsymbol{\alpha}^i;0)$ which is connected to the output of $\mathcal{A}_i$ as shown in Fig. *3.9.1*. For convenience, the source is denoted by $A_i(s)$, where $A_i(s)$ stands for the Laplace transform of $A_i(\boldsymbol{\alpha}^i;0)$.

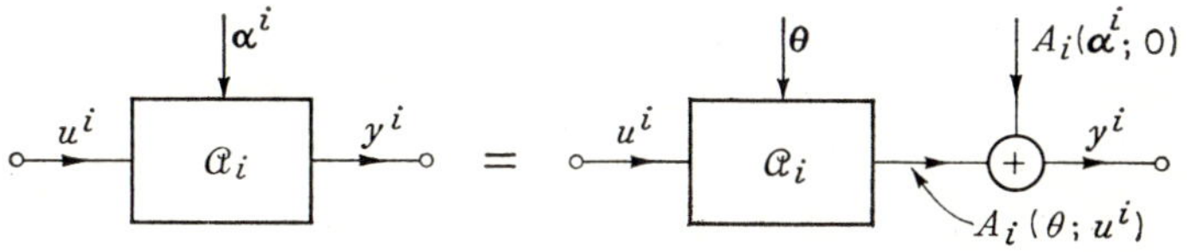

Fig. 3.9.1 Trading initial state for addition to output.

On performing the replacement for each $\mathcal{A}_i$ we obtain a signal-flow graph with branch gains $H_1(s), \ldots, H_N(s)$, a simple example of which is shown in Fig. *3.9.2*. Here $N = 4$ with the four branches labeled $H_1(s), \ldots, H_4(s)$ corresponding to $\mathcal{A}_1, \ldots, \mathcal{A}_4$, respectively. The input to $\mathcal{A}_3$, say, is the signal $V_3(s)$, which is associated with the node (node 3) from which the branch labeled $H_3(s)$ emanates. The external

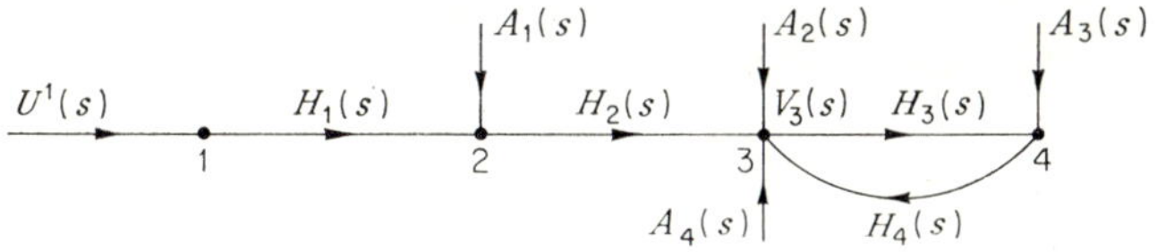

Fig. 3.9.2 System resulting from replacing each component by its equivalent in zero state.

input is $U^1(s)$; the internal inputs are $V_2(s)$, $V_3(s)$, and $V_4(s)$; and $A_1(s), \ldots, A_4(s)$ play the roles of fictitious external inputs.

More generally, if there are n nodes and m external inputs $U^1(s), \ldots, U^m(s)$, then by using elementary properties of signal-flow graphs (see Chap. *9*, Sec. *15*), we can express the V's in terms of U's by relations of the form

14

$$
\begin{aligned}
V_1(s) &= G_{11}(s)W_1(s) + \cdots + G_{1n}(s)W_n(s) \\
&\cdots\cdots\cdots\cdots\cdots\cdots\cdots\cdots \\
V_n(s) &= G_{n1}(s)W_1(s) + \cdots + G_{nn}(s)W_n(s)
\end{aligned}
$$

where $V_\lambda(s) \triangleq$ signal associated with node λ and $W_\lambda(s)$ is the sum of all external inputs applied at node λ. For example, in Fig. *3.9.2*

$$W_1(s) = U^1(s) \qquad W_2(s) = A_1(s) \qquad W_3(s) = A_2(s) + A_4(s) \qquad \text{and} \qquad W_4(s) = A_3(s)$$

In these equations, $G_{\mu\lambda}(s)$ represents the transmission function from node λ to node μ, which is given by Mason's formula (see *9.15.14*)

15 $$G_{\mu\lambda}(s) = \frac{\sum_k G_{\mu\lambda}^k(s)\,\Delta_{\mu\lambda}^k(s)}{\Delta(s)}$$

where the summation is over all forward paths from node λ to node μ, $G_{\mu\lambda}^k(s)$, $\mu,\lambda = 1, \ldots, n$, denotes the gain of kth forward path (i.e., the product of the gains of all branches constituting the forward path in question), $\Delta(s)$ is the determinant of the graph, and $\Delta_{\mu\lambda}^k(s)$ is the determinant of the graph resulting from the original graph when all branches of the kth forward path and all branches having a node in common with it are deleted.

Since the $\mathcal{A}_i$ are linear time-invariant systems, the $H_i(s)$ are rational functions of s. Furthermore, the numerator and denominator of $G_{\mu\lambda}(s)$ are rational functions of $H_1(s), \ldots, H_N(s)$, and hence the numerator and denominator of $G_{\mu\lambda}(s)$ are likewise rational functions of s. Consequently, $G_{\mu\lambda}(s)$ is a rational function of s. This implies that, if we exclude the degenerate case where the numerator and denominator of $G_{\mu\lambda}(s)$ vanish identically for all s, the $V_i(s)$ can be expressed via *14* and *15* in terms of the inputs $U^1(s), \ldots, U^m(s)$ and the initial states $\boldsymbol{\alpha}^1, \ldots, \boldsymbol{\alpha}^N$. Consequently, $\mathcal{A}$ is a determinate system provided there is at least one value of s for which the numerator and denominator of $G_{\mu\lambda}(s)$, $\lambda,\mu = 1, \ldots, n$, do not vanish simultaneously. This minor qualification can be disregarded for all practical purposes.

16 *Comment* The proof given above involves two main points. First, by the use of the decomposition property, $\mathcal{A}$ is reduced to an interconnection of component systems $\mathcal{A}_1, \ldots, \mathcal{A}_N$, which are in their zero state, plus "fictitious" sources which are determined by the initial states of $\mathcal{A}_1, \ldots, \mathcal{A}_N$ and which are connected to the outputs of $\mathcal{A}_1, \ldots, \mathcal{A}_N$. Needless to say, this reduction is applicable to any system whose components have the decomposition property, and not just a linear system.

The second step involves showing that the reduced system is, so to speak, *zero-state determinate* in the sense that given the external as well as the fictitious inputs, one can determine the input to each of the components of $\mathcal{A}$ if they are initially in their zero state. In the case of linear time-invariant systems, this is readily established by the use of Mason's formula.

In the absence of an analogous formula for linear time-varying systems, one can regard the statement of Theorem *12* as a rule of procedure which may fail in special cases. Thus, if by associating a state vector $\mathbf{x}$ with $\mathcal{A}$ by the use of this rule we succeed in deriving the state equations (or input-output-state relations) for $\mathcal{A}$ in terms of the state equations (or input-output-state relations) of its components $\mathcal{A}_1, \ldots, \mathcal{A}_N$, then $\mathbf{x}$ qualifies as a state vector for $\mathcal{A}$, and we have solved our problem. We cannot guarantee in advance—as we can in the case of time-invariant differential systems of finite order—that the application of the rule in question will be successful. However, we can assert that a linear time-varying system is determinate if it satisfies the conditions of Comment *2.6.42*. Actually, most linear time-varying systems satisfy those conditions or can be made to satisfy them through a rearrangement of components. ◁

This concludes our discussion of general properties of linear systems. In the next chapter we shall focus our attention on various techniques for associating a state vector $\mathbf{x}$ with a given time-invariant system $\mathcal{A}$ and finding the corresponding state equations of $\mathcal{A}$.

REFERENCES AND SUGGESTED READING

1 Ince, E. L.: "Ordinary Differential Equations," Dover Publications, Inc., New York, 1926.
2 Lévy, P.: "Problèmes Concrets d'Analyse Fonctionelle," Gauthier-Villars, Paris, 1951.
3 Aseltine, J. A.: "Transform Method in Linear System Analysis," McGraw-Hill Book Company, Inc., New York, 1958.
4 Yosida, K.: "Lectures on Differential and Integral Equations," Interscience Publishers, Inc., New York, 1960.
5 Hurewicz, W.: "Lectures on Ordinary Differential Equations," John Wiley & Sons, Inc., New York, 1958.
6 Van Valkenburg, M. E.: "Network Analysis," Prentice-Hall, Inc., Englewood Cliffs, N.J., 1955.

State vectors and state equations of time-invariant differential systems

4

1 *Introduction*

Our main objective in this chapter is to develop a systematic approach to the problem of associating a state vector and state equations with a linear system characterized by a differential equation or a system of differential equations with constant coefficients. In more concrete terms, let $\mathcal{A}$ be a differential system (*3.8.26*) defined by an input-output relation of the form

1 $$\mathbf{Ly} + \mathbf{K\tilde{y}} = \mathbf{Mu}$$

where $\mathbf{L}$, $\mathbf{K}$, and $\mathbf{M}$ are constant-coefficient matric differential operators, $\mathbf{u}$ is the input vector, $\mathbf{y}$ is the output vector, and $\mathbf{\tilde{y}}$ is the suppressed output vector (*1.10.40*). The questions of principal interest to us are these: How can one associate with *1* a state vector $\mathbf{x}$ which satisfies the mutual- and self-consistency conditions *1.6.5*, *1.6.11*, *1.6.15*, and *1.6.35*? How can one determine the input-output-state relation and state equations of $\mathcal{A}$? How can one synthesize $\mathcal{A}$ in the form of an interconnection of adders, scalors, integrators, and differentiators? What are the forms which the state equations of $\mathcal{A}$ assume in various special cases? Is there a canonical form for the state equations of a system of finite order? How can one eliminate the suppressed output variables from *1* and thereby determine a system $\mathcal{B}$ which is equivalent to $\mathcal{A}$ and which is characterized by an input-output relation in which $\mathbf{K} = \mathbf{0}$?

In the course of answering these questions we shall solve several subsidiary problems and deduce a number of properties of input-output-state relations which bear on the main problem at hand, namely, the association of a state vector with $\mathcal{A}$. Essentially, our approach to this problem will be the following: First, by the use of the Laplace transformation or other techniques, we find an expression for the general solution (see *3.8.24*) of the differential equation in terms of which $\mathcal{A}$ is characterized. Second, we identify the constants or combinations of constants appearing in the general solution with the components of the state vector at time $0-$ and thereby obtain an expression for $\mathbf{x}(0-)$ in terms of the values at $0-$ of $\mathbf{u}$, $\mathbf{y}$, $\tilde{\mathbf{y}}$, and a finite number of their derivatives.[1] In this way the general solution in question becomes a "tentative" input-output-state relation for $\mathcal{A}$. [The adjective "tentative" is used here to emphasize that at this stage of the process it remains to be shown that the tentative input-output-state relation satisfies the self-consistency conditions or, equivalently, has the response separation property (see *2.3.12*).] Then, by replacing $0-$ in the expression for $\mathbf{x}(0-)$ by t, we obtain an expression for $\mathbf{x}(t)$ in terms of $\mathbf{u}(t)$, $\mathbf{y}(t)$, $\tilde{\mathbf{y}}(t)$, and their derivatives. Finally, on forming $\dot{\mathbf{x}}(t)$ through the differentiation of $\mathbf{x}(t)$ and eliminating $\mathbf{y}(t)$, $\tilde{\mathbf{y}}(t)$, and their derivatives from the equations defining $\mathbf{x}(t)$ and $\dot{\mathbf{x}}(t)$, we obtain the state equation of $\mathcal{A}$. Once these equations have been obtained, we invoke Corollary *2.3.36 et seq.* and thereby establish that $\mathbf{x}$ qualifies as a state vector for $\mathcal{A}$ and that the tentative input-output-state relation is indeed an input-output-state relation for $\mathcal{A}$.

This approach will serve as a basis for another, less cumbersome, technique which accomplishes the association of a state vector with $\mathcal{A}$ through the process of realizing $\mathcal{A}$ in the form of an interconnection of adders, scalors, integrators, and differentiators. In this method, we shall make use of the determinateness of linear systems (Theorem *3.9.12*) and the equivalence between $\mathcal{A}$ and its realization (see *2.5.19*), which allow us to identify the components of $\mathbf{x}$ with the state variables of the integrators and differentiators in the realization of $\mathcal{A}$. The so-called partial-fraction technique for associating a state vector with $\mathcal{A}$ is a special case of this approach.

In brief, these are the basic features of the approaches we shall employ in this chapter to associate a state vector with a system characterized by a differential equation. Before we can develop these approaches in greater detail, it will be necessary for us to establish several properties of input-output-state relations of linear time-invariant systems which bear on the question of uniqueness of $\mathbf{x}$ and the connection between different modes of associating a state vector $\mathbf{x}$ with $\mathcal{A}$.

2 *Properties of input-output-state relations*

1 *Note* In this and the following sections the initial state will be assumed to be defined at time t_0- rather than t_0. Correspondingly, the observation interval will be taken to be $[t_0,t)$ unless otherwise specified. ◁

[1] The term "derivative" is used here in its wider sense to denote a derivative of any order k, $k = 1, 2, 3, \ldots$.

In order to avoid obscuring the main lines of the argument with cumbersome notation, we shall assume for simplicity that the input u and output y of $\mathcal{A}$ are scalar-valued time functions. This assumption will not play an essential role in our discussion, and it is a simple matter to extend the results obtained under this assumption to the general case in which u and y are vector-valued.

We begin with a recapitulation of certain basic facts which were established in Chap. *3*. Specifically, it was shown in Chap. *3*, Sec. *6*, that the input-output-state relation of a time-invariant linear system $\mathcal{A}$ of order n (with scalar input u and scalar output y) admits of the representation (for $t_0 = 0$)

2

$$y(t) = \langle \boldsymbol{\phi}(t), \mathbf{x}(0-) \rangle + \int_0^t h(t - \xi) u(\xi)\, d\xi \qquad t \geq 0$$

where (I) $\mathbf{x}(0-)$ is the initial state of $\mathcal{A}$ at time $0-$.

(II) $\mathbf{x}(0-)$ ranges over $\mathcal{C}^n$, the space of n-tuples of complex numbers.

(III) $\boldsymbol{\phi}(t)$ is a basis vector whose components $\phi_1(t), \ldots, \phi_n(t)$ represent the zero-input responses of $\mathcal{A}$ starting in the initial states $(1,0, \ldots ,0)$, $(0,1,0, \ldots ,0)$, $\ldots$, $(0, \ldots ,0,1)$, respectively.

(IV) $\phi_1, \ldots, \phi_n$ are linearly independent time functions and have a nonvanishing Wronskian.

(V) $\langle \boldsymbol{\phi}(t), \mathbf{x}(0-) \rangle$ is the scalar product of $\boldsymbol{\phi}(t)$ and $\mathbf{x}(0-)$.

(VI) $h(t)$ is the impulse response of $\mathcal{A}$, that is, the zero-state response of $\mathcal{A}$ to the unit impulse $\delta(t)$.

(VII) $h(t)$ is the inverse Laplace transform of $H(s)$, the transfer function of $\mathcal{A}$.

Let us disregard for the moment *2* and assume that $\mathcal{A}$ is associated with a state vector $\hat{\mathbf{x}}$, with $\hat{\mathbf{x}}$ ranging over a linear vector space $\mathcal{C}^m$ of dimension m. By Assertion *3.6.33*, the corresponding input-output-state relation for $\mathcal{A}$ must necessarily be of the form

3

$$y(t) = \langle \hat{\boldsymbol{\phi}}(t), \hat{\mathbf{x}}(0-) \rangle + \int_0^t \hat{h}(t - \xi) u(\xi)\, d\xi \qquad t \geq 0$$

where the various symbols have the same meaning as their counterparts in *2*(I) to *2*(VII), except that the components of $\hat{\boldsymbol{\phi}}(t)$ need not be linearly independent time functions.

Comparing *3* with *2*, it is clear that $\hat{h}(t)$ and $h(t)$ must be equal for all t, since $\hat{h}$ and h represent the same time function, namely, the zero-state response of $\mathcal{A}$ to $\delta(t)$. The question is: What is the relationship, if any, between $\hat{\boldsymbol{\phi}}$, $\boldsymbol{\phi}$, $\mathbf{x}(0-)$, and $\hat{\mathbf{x}}(0-)$? This question is answered in the following

4 **Assertion** Let $\mathcal{A}$ be a linear system of order n which is associated with (I) a state vector $\mathbf{x}$ ranging over a state space $\mathcal{C}^n$ of dimension n, with the corresponding input-output-state relation having the form *2*, and (II) a state vector $\hat{\mathbf{x}}$ ranging over a state space $\mathcal{C}^m$ of dimension m, with the corresponding input-output-state relation reading

5 $$y(t) = \langle \hat{\boldsymbol{\phi}}(t), \hat{\mathbf{x}}(0-) \rangle + \int_0^t h(t - \xi)u(\xi)\, d\xi \qquad t \geq 0$$

Then $m \geq n$ and $\mathbf{x}(0-)$ is related to $\hat{\mathbf{x}}(0-)$ by

6 $$\mathbf{x}(0-) = \mathbf{\Gamma}\hat{\mathbf{x}}(0-)$$

where $\mathbf{\Gamma}$ is an $n \times m$ matrix of rank n. In particular, if $m = n$, then $\mathbf{x}(0-)$ and $\hat{\mathbf{x}}(0-)$ range over the same state space and the linear correspondence between $\mathbf{x}(0-)$ and $\hat{\mathbf{x}}(0-)$ is one-one.

Proof Since *2* and *5* are input-output-state relations for $\mathfrak{a}$, they satisfy the mutual-consistency condition *1.6.5*, meaning that every input-output pair for $\mathfrak{a}$ satisfies both *2* and *5* and, conversely, every pair (u,y) satisfying either *2* or *5* is an input-output pair for $\mathfrak{a}$. This in turn implies that every input-output pair (u,y) satisfying *2* also satisfies *5*, and vice versa. In particular, for input-output pairs of the form $(0,y)$ (which represent zero-input responses of $\mathfrak{a}$) this implies that to every vector $\mathbf{x}(0-)$ in $\mathcal{C}^n$ there corresponds a vector $\hat{\mathbf{x}}(0-)$ in $\mathcal{C}^m$, and vice versa, such that

7 $$\langle \boldsymbol{\phi}(t), \mathbf{x}(0-) \rangle = \langle \hat{\boldsymbol{\phi}}(t), \hat{\mathbf{x}}(0-) \rangle \qquad \text{for all } t \geq 0$$

Thus, let $\hat{\boldsymbol{\gamma}}^1 = (\hat{\gamma}_1^1, \ldots, \hat{\gamma}_m^1), \ldots, \hat{\boldsymbol{\gamma}}^n = (\hat{\gamma}_1^n, \ldots, \hat{\gamma}_m^n)$ be m-vectors in $\mathcal{C}^m$ corresponding to the unit n-vectors $(1, \ldots, 0)$, $(0,1, \ldots, 0)$, $\ldots$, $(0, \ldots, 1)$ in $\mathcal{C}^n$. (Note that $\boldsymbol{\gamma}^1, \ldots, \boldsymbol{\gamma}^n$ are not necessarily unique.) For these vectors *7* yields

$$\begin{aligned} \phi_1 &= \hat{\gamma}_1^1\hat{\phi}_1 + \cdots + \hat{\gamma}_m^1\hat{\phi}_m \\ &\cdots\cdots\cdots\cdots \\ \phi_n &= \hat{\gamma}_1^n\hat{\phi}_1 + \cdots + \hat{\gamma}_m^n\hat{\phi}_m \end{aligned}$$

or more compactly,

8 $$\boldsymbol{\phi} = \hat{\boldsymbol{\phi}}\hat{\mathbf{\Gamma}}$$

where $\hat{\mathbf{\Gamma}}$ is a constant matrix whose ijth element is $\hat{\gamma}_i^j$ and

$$\boldsymbol{\phi} = (\phi_1, \ldots, \phi_n) \qquad \hat{\boldsymbol{\phi}} = (\hat{\phi}_1, \ldots, \hat{\phi}_m)$$

are row vectors.

Since $\phi_1, \ldots, \phi_n$ are linearly independent time functions, *8* cannot be true unless at least n of the m time functions $\hat{\phi}_1, \ldots, \hat{\phi}_m$ are linearly independent. This implies that $m \geq n$ and hence that $\mathcal{C}^m$ is of dimension $m \geq n$.

In a similar fashion, let $\boldsymbol{\gamma}^1 = (\gamma_1^1, \ldots, \gamma_n^1), \ldots, \boldsymbol{\gamma}^m = (\gamma_1^m, \ldots, \gamma_n^m)$ be m n-vectors in $\mathcal{C}^n$ corresponding to the m unit vectors $(1, \ldots, 0)$,

(0, . . . ,1) in $\mathcal{C}^m$. In this case *7* yields

$$\hat{\phi}_1 = \gamma_1^1\phi_1 + \cdots + \gamma_n^1\phi_n$$
$$\cdots\cdots\cdots\cdots$$
$$\hat{\phi}_m = \gamma_1^m\phi_1 + \cdots + \gamma_n^m\phi_n$$

or more compactly,

9 $$\hat{\boldsymbol{\phi}} = \boldsymbol{\phi}\boldsymbol{\Gamma}$$

where $\boldsymbol{\Gamma}$ is a constant matrix whose *ij*th element is γ_i^j.

Since $\boldsymbol{\phi}$ is an n-vector, *9* implies that not more than n of the m time functions $\hat{\phi}_1, \ldots, \hat{\phi}_m$ can be linearly independent. On the other hand, we have already shown that not fewer than n of the m time functions $\hat{\phi}_1, \ldots, \hat{\phi}_m$ are linearly independent. Consequently, we can assert that *exactly n of the m time functions* $\hat{\phi}_1, \ldots, \hat{\phi}_m$ *are linearly independent.* This leads at once to the conclusion that the matrices $\hat{\boldsymbol{\Gamma}}$ and $\boldsymbol{\Gamma}$ in *8* and *9* are of rank n; for otherwise the number of linearly independent components in the left-hand members of *8* and *9* would be less than n. (Note that the rank of $\hat{\boldsymbol{\Gamma}}$ and $\boldsymbol{\Gamma}$ cannot exceed n because they are $m \times n$ and $n \times m$ matrices, respectively.)

Next, on substituting *9* in *7*, we have

10 $$\langle\boldsymbol{\phi},\mathbf{x}(0-)\rangle = \langle\boldsymbol{\phi}\boldsymbol{\Gamma},\hat{\mathbf{x}}(0-)\rangle \qquad t \geq 0$$

and since the components of $\boldsymbol{\phi}$ are linearly independent time functions, *10* implies

11 $$\mathbf{x}(0-) = \boldsymbol{\Gamma}\hat{\mathbf{x}}(0-) \qquad \text{for all } \mathbf{x}(0-) \text{ in } \mathcal{C}^n \text{ and all } \hat{\mathbf{x}}(0-) \text{ in } \mathcal{C}^m$$

which is what we wanted to demonstrate.

12 **Remark** Let $\mathcal{A}$ and $\hat{\mathcal{A}}$ be systems characterized by the input-output-state relations *2* and *5*, respectively. To say that $\mathbf{x}$ and $\hat{\mathbf{x}}$ are two different representations for the state vector of $\mathcal{A}$ is the same as saying that $\mathcal{A}$ and $\hat{\mathcal{A}}$ are equivalent systems (see *2.3.46*), with $\mathbf{x}$ being the state of $\mathcal{A}$ and $\hat{\mathbf{x}}$ the state of $\hat{\mathcal{A}}$. From this point of view, *7* may be regarded as expressing the condition of zero-input equivalence of $\mathcal{A}$ and $\hat{\mathcal{A}}$ (*2.9.2*). (Note that $\mathcal{A}$ and $\hat{\mathcal{A}}$ are zero-state equivalent by virtue of the relation $h = \hat{h}$.) Also, the condition $m \geq n$ means that the dimension of the state space of any system equivalent to $\mathcal{A}$ cannot be less than n. If $m > n$, then the correspondence between $\mathcal{C}^m$ and $\mathcal{C}^n$ is not one-one, which implies that some of the states in $\mathcal{C}^m$ are equivalent to one another and hence that $\hat{\mathcal{A}}$ is not in reduced form (see *2.2.11*). This is consistent with the result that $\mathbf{x}(0-) = \boldsymbol{\Gamma}\hat{\mathbf{x}}(0-)$ and that $\hat{\mathbf{x}}(0-)$ cannot be similarly expressed in terms of $\mathbf{x}(0-)$ when $m > n$.

Relation between x and x̂ when $m = n$

The case that is of particular interest to us is that where $\mathbf{x}(0-)$ and $\hat{\mathbf{x}}(0-)$ are vectors of the same dimension n. In this case, $\mathbf{\Gamma}$ and $\hat{\mathbf{\Gamma}}$ are $n \times n$ matrices of rank n. Furthermore, to the relation

13 $$\mathbf{x}(0-) = \mathbf{\Gamma}\hat{\mathbf{x}}(0-) \qquad \forall \mathbf{x} \forall \hat{\mathbf{x}}$$

which was derived from *7* and *9* by invoking the linear independence of $\phi_1, \ldots, \phi_n$, we can add the relation

14 $$\hat{\mathbf{x}}(0-) = \hat{\mathbf{\Gamma}}\mathbf{x}(0-) \qquad \forall \mathbf{x} \forall \hat{\mathbf{x}}$$

which follows in the same way from *7*, *8*, and the linear independence of $\hat{\phi}_1, \ldots, \hat{\phi}_n$. Equations *13* and *14* taken together imply that $\mathbf{\Gamma}$ is nonsingular and that $\hat{\mathbf{\Gamma}}$ and $\mathbf{\Gamma}$ are inverse to one another, i.e.,

15 $$\hat{\mathbf{\Gamma}} = \mathbf{\Gamma}^{-1}$$

Note that the nonsingularity of $\mathbf{\Gamma}$ follows also from the fact that $\mathbf{\Gamma}$ is of rank n. This result enables us to state the following corollary of Assertion *4*.

16 **Corollary** If a linear system $\mathcal{A}$ of order n is associated with a state vector $\mathbf{x}$ ranging over a linear vector space $\mathcal{C}^n$ of dimension n, then any other way of associating with $\mathcal{A}$ a state vector of the same dimension as $\mathbf{x}$ must necessarily lead to a vector $\hat{\mathbf{x}}$ such that

17 $$\mathbf{x}(0-) = \mathbf{\Gamma}\hat{\mathbf{x}}(0-)$$

where $\mathbf{\Gamma}$ is a constant nonsingular $n \times n$ matrix defined by *9*. ◁

We can strengthen this result by showing that if *17* holds at $t = 0-$, then it holds for all $t \geq 0$. This is done in the following

18 **Lemma** If the basis functions $\boldsymbol{\phi}$ and $\hat{\boldsymbol{\phi}}$ in *2* and *3* are related by

$$\hat{\boldsymbol{\phi}} = \boldsymbol{\phi}\mathbf{\Gamma}$$

and the initial state vectors are related by

$$\mathbf{x}(0-) = \mathbf{\Gamma}\hat{\mathbf{x}}(0-)$$

where $\mathbf{\Gamma}$ is a nonsingular $n \times n$ matrix, then for all inputs u in the input segment space of $\mathcal{A}$ the states $\mathbf{x}$ and $\hat{\mathbf{x}}$ at time t are related by

$$\mathbf{x}(t) = \mathbf{\Gamma}\hat{\mathbf{x}}(t) \qquad \forall t \geq 0$$

Proof It is easy to verify that when the basis functions are not normalized, the differential state equation satisfied by $\mathbf{x}$ (see *3.7.9*) assumes the form (see *3.7.36*)

19 $$\mathbf{\Phi}(0)\dot{\mathbf{x}}(t) = \dot{\mathbf{\Phi}}(0)\mathbf{x}(t) + \mathbf{h}(0)u(t)$$

where, as in *3.7.7*, $\mathbf{\Phi}(t)$ is expressed by

$$\mathbf{\Phi}(t) = \begin{bmatrix} \phi_1(t) & \cdots & \phi_n(t) \\ \cdots & \cdots & \cdots \\ \phi_1^{(n-1)}(t) & \cdots & \phi_n^{(n-1)}(t) \end{bmatrix} \qquad t \geq 0$$

and $\mathbf{h}(t)$ is an n-vector whose ith element is the $(i - 1)$st derivative of the impulse response $h(t)$; that is,

20 $$\mathbf{h}(t) = \begin{bmatrix} h(t) \\ \cdots \\ h^{(n-1)}(t) \end{bmatrix} \qquad t \geq 0$$

Since $\hat{\mathbf{x}}$, like $\mathbf{x}$, is a state vector for $\mathcal{A}$, it too must satisfy an equation of the same form as *19*:

21 $$\hat{\mathbf{\Phi}}(0)\dot{\hat{\mathbf{x}}}(t) = \dot{\hat{\mathbf{\Phi}}}(0)\hat{\mathbf{x}}(t) + \mathbf{h}(0)u(t)$$

[Note that $\hat{\mathbf{h}}(0) = \mathbf{h}(0)$ because $\hat{\mathbf{h}}(t) \equiv \mathbf{h}(t)$.]

Furthermore, since the relation $\hat{\boldsymbol{\phi}} = \boldsymbol{\phi}\mathbf{\Gamma}$ holds for all $t \geq 0$, we have [with the understanding that $\boldsymbol{\phi}^{(i-1)}(0) \triangleq \boldsymbol{\phi}^{(i-1)}(0+)$ and likewise for $\hat{\boldsymbol{\phi}}$]

$$\hat{\boldsymbol{\phi}}^{(i-1)} = \boldsymbol{\phi}^{(i-1)}\mathbf{\Gamma} \qquad t \geq 0,\ i = 1, \ldots, n$$

and therefore

22 $$\hat{\mathbf{\Phi}}(t) = \mathbf{\Phi}(t)\mathbf{\Gamma} \qquad t \geq 0$$

On replacing $\hat{\mathbf{\Phi}}(0)$ in *19* by $\mathbf{\Phi}(0)\mathbf{\Gamma}$, eliminating $\mathbf{h}(0)u(t)$ from *19* and *21*, and defining

$$\mathbf{z}(t) = \mathbf{x}(t) - \mathbf{\Gamma}\hat{\mathbf{x}}(t)$$

we have

$$\mathbf{\Phi}(0)\dot{\mathbf{z}}(t) = \dot{\mathbf{\Phi}}(0)\mathbf{z}$$

or

23 $$\dot{\mathbf{z}}(t) = \mathbf{\Phi}^{-1}(0)\dot{\mathbf{\Phi}}(0)\mathbf{z}$$

since $\mathbf{\Phi}(0)$ is a nonsingular matrix.

Now by hypothesis

$$\mathbf{z}(0-) = \mathbf{x}(0-) - \mathbf{\Gamma}\hat{\mathbf{x}}(0-) = \mathbf{0}$$

and since $\mathbf{z}(t)$ is the solution of *23* with $\mathbf{z}(0-) = \mathbf{0}$, it follows that[1] $\mathbf{z}(t) = \mathbf{0}$ for $t \geq 0$ and hence

24 $$\mathbf{x}(t) = \mathbf{\Gamma}\hat{\mathbf{x}}(t) \qquad \forall t \geq 0$$

which is what we set out to demonstrate.

[1] The solution of vector differential equations of first order is discussed in greater detail in Chap. *5*, Sec. *2*.

Relation between x and $\hat{\mathbf{x}}$

The foregoing analysis leads us to an important conclusion, namely, that the different ways of associating a state vector with $\mathfrak{A}$ amount merely to attaching different labels (connected by a one-one linear mapping) to the points of the state space $\mathcal{C}^n$ (see *2.3.46*). This conclusion is stated more precisely in the following

25 **Theorem** If a linear time-invariant system $\mathfrak{A}$ of order n is associated with a state vector $\mathbf{x}$ ranging over $\mathcal{C}^n$, then any other way of associating a state vector with $\mathfrak{A}$ of the same dimension as $\mathbf{x}$ must necessarily lead to a vector $\hat{\mathbf{x}}$ which is related to $\mathbf{x}$ by

$$\mathbf{x} = \mathbf{\Gamma}\hat{\mathbf{x}} \qquad \forall t$$

where $\mathbf{\Gamma}$ is a nonsingular constant $n \times n$ matrix defined by *9*.

From this it follows (see *2.3.46*) that if the state equations of $\mathfrak{A}$ corresponding to $\mathbf{x}$ are of the canonical form (*2.3.45*)

26
$$\dot{\mathbf{x}} = \mathbf{A}\mathbf{x} + \mathbf{b}u$$
$$y = \langle \mathbf{c},\mathbf{x} \rangle + d_0 u$$

where $\mathbf{A}$ is a constant matrix, $\mathbf{b}$ and $\mathbf{c}$ are constant vectors, and d_0 is a constant scalar, then the state equations of $\mathfrak{A}$ corresponding to $\hat{\mathbf{x}}$ are

27
$$\dot{\hat{\mathbf{x}}} = \hat{\mathbf{A}}\hat{\mathbf{x}} + \hat{\mathbf{b}}u$$
$$y = \langle \hat{\mathbf{c}},\hat{\mathbf{x}} \rangle + d_0 u$$

where

28
$$\hat{\mathbf{A}} = \mathbf{\Gamma}^{-1}\mathbf{A}\mathbf{\Gamma}$$
$$\hat{\mathbf{b}} = \mathbf{\Gamma}^{-1}\mathbf{b}$$
$$\hat{\mathbf{c}} = \mathbf{c}\mathbf{\Gamma}$$

29 *Comment* The form of the relation between $\hat{\mathbf{A}}$ and $\mathbf{A}$ in *28* suggests that by a suitable choice of the state vector $\hat{\mathbf{x}}$ it may be possible to make $\hat{\mathbf{A}}$ a diagonal matrix or, at least, put it into the Jordan canonical form (see *D.5.21*). As will be shown in Sec. *9*, the partial-fraction technique for associating a state vector with $\mathfrak{A}$ leads to state equations in which $\hat{\mathbf{A}}$ has the Jordan canonical form.

30 *Comment* Theorem *25* remains valid when $\hat{\mathbf{x}}$ ranges over a space $\mathcal{C}^m$ of dimension $m > n$. However, in this case $\mathbf{\Gamma}$ is not a square matrix and the correspondence between $\mathbf{x}$ and $\hat{\mathbf{x}}$ is not one-one (see Assertion *4*).

31 **Remark** In deducing Theorem *25* we have not used in any essential way the assumption that $\mathfrak{A}$ is a time-invariant system. Consequently, Theorem *25* holds true when $\mathfrak{A}$ is a time-varying system, provided the

following replacements of symbols are made in the statement of Theorem *25* as well as in the analysis which precedes it (see *3.6.33*).

(I)	$0 \rightarrow t_0$
(II)	$\boldsymbol{\phi}(t) \rightarrow \boldsymbol{\phi}(t,t_0)$
(III)	$\boldsymbol{\Phi}(t) \rightarrow \boldsymbol{\Phi}(t,t_0)$
(IV)	$h(t - \xi) \rightarrow h(t,\xi)$
(V)	$\mathbf{h}(t - \xi) \rightarrow \mathbf{h}(t,\xi)$
(VI)	$\mathbf{A} \rightarrow \mathbf{A}(t)$, $\mathbf{b} \rightarrow \mathbf{b}(t)$, $\mathbf{c} \rightarrow \mathbf{c}(t)$, $\mathbf{d}_0 \rightarrow \mathbf{d}_0(t)$

It is important to note that $\boldsymbol{\Phi}(t_0,t)$ must be nonsingular for all t_0 and all $t \geq t_0$.

Association of a state vector with $\mathcal{A}$

By exhibiting the connection between all possible ways in which a state vector can be associated with $\mathcal{A}$ and by showing that, in effect, the different ways of defining $\mathbf{x}$ involve merely different ways of expressing the zero-input response of $\mathcal{A}$ as a linear combination of some set of basis functions of $\mathcal{A}$, Theorem *25* provides a base for constructing systematic approaches to the problem of associating a state vector with any given system of finite order. One such approach will be developed in the following sections along the lines sketched in Sec. *1*. In more detailed terms, given a linear time-invariant system $\mathcal{A}$, we shall associate a state vector $\mathbf{x}$ with $\mathcal{A}$ through the following steps.

1. A general solution to the differential equation which characterizes $\mathcal{A}$ is found by the use of the Laplace transformation (or other techniques if $\mathcal{A}$ is not time-invariant).[1]

2. The constants appearing in the general solution are identified with the components of $\mathbf{x}(0-)$. This yields an expression for $\mathbf{x}(0-)$ [or $\mathbf{x}(t_0-)$ if $t_0 \neq 0$], in terms of the initial values of the input, output, and the suppressed output vectors, and, at the same time, identifies the basis functions $\phi_1, \ldots, \phi_n$ (which must be linearly independent).

3. With the components of $\mathbf{x}(0-)$ substituted for the constants in the general solution and $0-$ replaced by t_0-, the general solution becomes a tentative input-output-state relation—"tentative" in the sense that at this stage of the process it remains to be shown that it qualifies as an input-output-state relation under the mutual- and self-consistency conditions (*1.6.2*). (It should be emphasized that the function of the first three steps is merely that of leading to a likely candidate for the state vector of $\mathcal{A}$ rather than that of proving that the expression for $\mathbf{x}(t)$ in terms of u, y, $\tilde{\mathbf{y}}$, and their derivatives qualifies as a state vector for $\mathcal{A}$.)

[1] Actually, all that we need to determine $\mathbf{x}$ are the terms which represent the initial conditions. This point is elaborated upon in *4.5.58*.

4. In the relations between $\mathbf{x}(0-)$ and the initial values of u, y, the suppressed output vector $\tilde{\mathbf{y}}$, and their derivatives, the initial time $0-$ is replaced by t. This yields a tentative expression for the state of $\mathcal{A}$ at time t in terms of $u(t)$, $y(t)$, $\tilde{\mathbf{y}}(t)$, and the derivatives. Again, this expression for $\mathbf{x}(t)$ is "tentative" in the sense that $\mathbf{x}(t)$ cannot be claimed to represent the state of $\mathcal{A}$ at time t (for all $t \geq 0$) until it has been demonstrated directly or indirectly that $\mathbf{x}(t)$ qualifies as the state of $\mathcal{A}$ under Definition *1.6.2*.

5. The state equations of $\mathcal{A}$ are obtained by forming the expressions for $\dot{\mathbf{x}}(t)$ and $y(t)$ and eliminating all variables other than $\mathbf{x}(t)$ and $u(t)$ from these expressions. If the state equations have the canonical form (*2.3.45*)

32
$$\begin{aligned}\dot{\mathbf{x}} &= \mathbf{Ax} + \mathbf{Bu}\\ \mathbf{y} &= \mathbf{Cx} + \mathbf{Du}\end{aligned}$$

then $\mathbf{x}(t)$ qualifies as the state of the system at time t by virtue of Corollary *2.3.36 et seq.*, which imply that $\mathbf{x}$ satisfies the mutual- and self-consistency conditions. If the state equations do not have the canonical form [e.g., in the case of systems of the differential operator type (see *4.4.17*)], then it is necessary to show by direct means that $\mathbf{x}$ satisfies the mutual- and self-consistency conditions or, equivalently, that $\mathbf{x}$ has the state separation property *2.3.19*.

We now proceed to develop this approach in greater detail, deferring until Sec. *9* the description of an alternative and very effective technique which is based on the determinateness of linear systems. We begin with a simple class of systems, namely, systems of the *reciprocal differential operator type* which will be used in subsequent sections as a prototype for systems of more general form.

3 *Systems of the reciprocal differential operator type*

Consider a nonoriented object $\mathcal{B}$ with scalar-valued terminal variables v_1 and v_2 (*1.4.1*) which is characterized by a differential equation of the form (see *3.8.11*)

1
$$L(p)v_1 = v_2$$

where $L(p) = a_n p^n + \cdots + a_0$, $a_n \neq 0$, and the a_λ, $\lambda = 1, \ldots, n$, are real or complex constants. Such an object can be oriented in but two ways: (I) by letting $v_1 \triangleq y \triangleq$ output and $v_2 \triangleq u \triangleq$ input, and (II) by letting $v_1 \triangleq u \triangleq$ input and $v_2 \triangleq y =$ output. By the first mode of orientation we obtain what we call a *reciprocal differential operator system* $\mathfrak{R}$, which is characterized by the input-output relation

$$L(p)y = u \qquad L(p) = a_n p^n + \cdots + a_0$$

and which for convenience will be denoted by $1/L(p)$, while by the second mode we obtain a *differential operator system* $\mathfrak{D}$, denoted by $L(p)$, which is characterized by the input-output relation

$$y = L(p)u \qquad L(p) = a_n p^n + \cdots + a_0$$

In this section, we shall focus our attention on systems of the reciprocal differential operator type. Systems of the differential operator type and their relation to systems of the reciprocal differential operator type will be considered in Sec. *4*.

Input-output-state relations

As was pointed out in Sec. *2*, the first step toward associating a state vector with and finding the corresponding input-output-state relation for a system characterized by a differential equation consists in finding the general solution of this equation. Thus, our first task is to find a general solution of the differential equation

2 $$L(p)y = u \qquad L(p) = a_n p^n + \cdots + a_0 \qquad a_n \neq 0$$

which characterizes $\mathfrak{R}$. This can be done by drawing on the well-known facts in the theory of ordinary differential equations. For our purposes, however, it will be more convenient not to rely on the classical theory and instead derive the expression for the general solution directly through the use of the Laplace transformation.

Specifically, suppose that a pair of time functions $u_{(t_0,\infty)}, y_{(t_0,\infty)}$, or (u,y) for short, satisfies (see *3.8.16*) the differential equation *2*; that is, (u,y) is an input-output pair for $\mathfrak{R}$. Then, by virtue of the constancy of the coefficients of *2*, every pair of the form $(T_\delta u, T_\delta y)$, where T_δ is a translation operator (*3.2.2*) with arbitrary shift δ, will also satisfy *2*. With this in mind, we can—without any loss in generality—set $t_0 = 0$ in order to make it convenient to apply the Laplace transformation to the determination of the general solution of *2*.

Thus, let $u_{(0,\infty)}, y_{(0,\infty)}$, or (u,y) for short, be an input-output pair satisfying *2*. On applying the Laplace transformation to both sides of *2* and making use of the formula (see *B.3.19*)

3 $$\mathfrak{L}\{p^r f\} = s^r F(s) - s^{r-1} f(0-) - \cdots - f^{(r-1)}(0-) \qquad r = 1, 2, \ldots$$

where $F(s)$ is the Laplace transform of $f(t)$[1] and $f^{(\lambda)}(0-)$, $\lambda = 1, 2, \ldots,$

[1] Note that to be consistent with our notation for functions, we should say "F is the Laplace transform of f" rather than "$F(s)$ is the Laplace transform of $f(t)$." This and other departures from consistency in notation in this text are dictated either by expediency or by the desirability of avoiding conflict with standard notation (see Remark *1.3.3*).

is the initial value of the λth derivative of f at $t = 0-$, we obtain the relation

4 $$(a_n s^n + \cdots + a_0)Y(s) = U(s) + a_n y^{(n-1)}(0-) + (a_n s + a_{n-1})y^{(n-2)}(0-) + \cdots + (a_n s^{n-1} + \cdots + a_1)y(0-)$$

Upon replacing $a_n s^n + \cdots + a_0$ by $L(s)$, dividing both sides by $L(s)$, and rearranging terms, *4* becomes

5 $$Y(s) = \frac{U(s)}{L(s)} + \sum_{\lambda=1}^{n} \frac{(a_n s^{n-\lambda} + \cdots + a_\lambda)}{L(s)} y^{(\lambda-1)}(0-)$$

where $Y(s) = \mathcal{L}\{y(t)\}$ and $U(s) = \mathcal{L}\{u(t)\}$. At this point, we can assert that if (u,y) is an input-output pair satisfying *2*, then the pair of their respective Laplace transforms $(U(s),Y(s))$ satisfies *5*. Furthermore, since the correspondence between a time function and its Laplace transform is one-one, the inverse Laplace transforms of $U(s)$ and $Y(s)$ are $u(t)$ and $y(t)$, respectively. Consequently, on applying the inverse Laplace transformation to both sides of *5* and making use of the convolution property (*B.3.11*), we obtain

6 $$y(t) = \sum_{\lambda=1}^{n} y^{(\lambda-1)}(0-)\phi_\lambda(t) + \int_0^t h(t-\xi)u(\xi)\,d\xi \qquad t \geq 0$$

where

$$h(t) = \mathcal{L}^{-1}\left\{\frac{1}{L(s)}\right\} = \text{impulse response of } \mathcal{R} \text{ (see } \mathit{3.6.5}\text{)}$$

7 $$\frac{1}{L(s)} \triangleq H(s) = \text{transfer function of } \mathcal{R} \text{ (see } \mathit{3.6.17}\text{)}$$

$$\phi_\lambda(t) = \mathcal{L}^{-1}\left\{\frac{a_n s^{n-\lambda} + \cdots + a_\lambda}{L(s)}\right\} \qquad \lambda = 1, \ldots, n$$

$$= \text{zero-input response of } \mathcal{R} \text{ (see } \mathit{3.6.32}\text{)}$$

Thus, if (u,y) is an input-output pair satisfying *2*, then (u,y) also satisfies *6*. This implies, more specifically, that to every input-output pair (u,y) for *2* there corresponds a (not necessarily unique) n-tuple of complex numbers $(\alpha_1, \ldots, \alpha_n)$ in $\mathcal{C}^n$ ($\mathcal{C}^n \triangleq$ space of n-tuples of complex numbers) such that (u,y) satisfies *6* for $t \geq 0$ with

$$y^{(\lambda-1)}(0-) = \alpha_\lambda \qquad \lambda = 1, \ldots, n$$

Next, we must show that for every n-tuple $(\alpha_1, \ldots, \alpha_n)$ in $\mathcal{C}^n$, the expression for y given by *6* with $y^{(\lambda-1)}(0-)$ replaced by α_λ, $\lambda = 1, \ldots, n$, defines an input-output pair for *2*, that is, *6* is a solution of *2* for each $(\alpha_1, \ldots, \alpha_n)$ in $\mathcal{C}^n$. As a preliminary, we establish that the ϕ_λ and h in *6* have the properties asserted below.

8 **Assertion** The functions ϕ_λ, $\lambda = 1, \ldots, n$, are linearly independent solutions of the homogeneous differential equation

9 $$L(p)\phi_\lambda = 0 \qquad t \geq 0,\ \lambda = 1, \ldots, n$$

satisfying the initial conditions

10 $$\phi_\lambda^{(\lambda-1)}(0-) = 1$$

11 $$\phi_\lambda^{(\mu-1)}(0-) = 0 \qquad \text{for } \mu \neq \lambda,\ \mu,\lambda = 1, \ldots, n$$

Proof On applying the Laplace transformation to both sides of *9* and making use of *10* and *11*, we obtain the following expression for the Laplace transform of ϕ_λ:

12 $$\Phi_\lambda(s) = \mathcal{L}\{\phi_\lambda(t)\} = \frac{a_n s^{n-\lambda} + \cdots + a_\lambda}{L(s)}$$

Since *12* is identical with the expression for $\mathcal{L}\{\phi_\lambda(t)\}$ in *7*, it follows that ϕ_λ as defined by *7* is a solution of *9* satisfying the initial conditions *10* and *11*.

The linear independence of the ϕ_λ is an immediate consequence of the linear independence of their respective Laplace transforms. To establish the latter, suppose that the $\Phi_\lambda(s)$ are not linearly independent. Then there exist constants $\rho_1, \ldots, \rho_n$ not all zero such that

13 $$\rho_1\Phi_1(s) + \cdots + \rho_n\Phi_n(s) \equiv 0$$

Now on substituting the $\Phi_\lambda(s)$ into *13*, multiplying both sides by $L(s)$ [note that $L(s) \not\equiv 0$ by virtue of the assumption that $a_n \neq 0$], and setting the coefficients of powers of s equal to zero, we obtain the system of linear equations

$$\begin{aligned} \rho_1 a_n &= 0 \\ \rho_1 a_{n-1} + \rho_2 a_n &= 0 \\ \cdots\cdots\cdots& \\ \rho_1 a_1 + \cdots + \rho_n a_n &= 0 \end{aligned}$$

which implies $\rho_1 = \rho_2 = \cdots = \rho_n = 0$ (since $a_n \neq 0$). This contradicts the assumption that not all $\rho_1, \ldots, \rho_n$ are zero and thus establishes the linear independence of the $\Phi_\lambda(s)$. Q.E.D.

14 **Assertion** The function h is the solution of the differential equation

15 $$L(p)h(t) = \delta(t)$$

satisfying the initial conditions

16 $$h^{(\lambda-1)}(0-) = 0 \qquad \lambda = 1, \ldots, n$$

Proof By applying the Laplace transformation to both sides of *15* and making use of the initial conditions *16*, we obtain

$$\mathfrak{L}\{h(t)\} = \frac{1}{L(s)}$$

which agrees with *7*. This verifies that h is the impulse response of $\mathfrak{R}$ [zero-state response of $\mathfrak{R}$ to $\delta(t)$] and that the transfer function of $\mathfrak{R}$ is given by

$$H(s) = \mathfrak{L}\{h(t)\} = \frac{1}{L(s)} \qquad \triangleleft$$

We are now ready to verify by direct substitution that for every $(\alpha_1, \ldots, \alpha_n)$ in $\mathfrak{C}^n$ the expression

17 $$y(t) = \sum_{\lambda=1}^{n} \alpha_\lambda \phi_\lambda(t) + \int_0^t h(t-\xi)u(\xi)\,d\xi$$

satisfies *2*. Specifically, on substituting *17* into *2* and making use of *9*, *2* reduces to

$$L(p) \int_0^t h(t-\xi)u(\xi)\,d\xi = u(t) \qquad t \geq 0$$

We can readily verify that this is an identity by applying *B.3.24*. Thus, we have

$$\begin{aligned} L(p) \int_0^t h(t-\xi)u(\xi)\,d\xi &= \int_0^t [L(p)h(t-\xi)]u(\xi)\,d\xi \\ &= \int_0^t \delta(t-\xi)u(\xi)\,d\xi \\ &= u(t) \qquad t \geq 0 \end{aligned}$$

by virtue of *15* and the sifting property of delta functions (see *A.2.11*).

Expression for the general solution for arbitrary t_0

We have established that every input-output pair (u,y) (starting at $t_0 = 0$) which satisfies the input-output relation *2* also satisfies *6* and, conversely, every pair (u,y) (starting at $t_0 = 0$) which satisfies *6* also satisfies *2*. This implies that *6* is the general solution of *2*, or, equivalently, that *6* satisfies the mutual-consistency condition (*1.6.5*).

To obtain the form of the general solution for arbitrary t_0, let $(u_{(t_0,\infty)}, y_{(t_0,\infty)})$ be any input-output pair satisfying *2*. Then by virtue of the constancy of the coefficients in *2*, the shifted pair $(T_{-t_0}u_{(t_0,\infty)}, T_{-t_0}y_{(t_0,\infty)})$ will satisfy *2* (as well as *6*) since $T_{-t_0}u_{(t_0,\infty)}$ and $T_{-t_0}y_{(t_0,\infty)}$ start at $t = 0$. More specifically, let (see *3.2.2*)

$$\hat{u} \triangleq T_{-t_0}u_{(t_0,\infty)}$$
$$\hat{y} \triangleq T_{-t_0}y_{(t_0,\infty)}$$

to simplify the notation. Then

18 $$\hat{y}(t) = \sum_{\lambda=1}^{n} \hat{y}^{(\lambda-1)}(0-)\phi_\lambda(t) + \int_0^t h(t-\xi)\hat{u}(\xi)\,d\xi \qquad t \geq 0$$

and since

$$\hat{u}(t) = u(t_0 + t) \qquad \forall t$$
$$\hat{y}(t) = y(t_0 + t) \qquad \forall t$$

18 may be written as

$$y(t_0 + t) = \sum_{\lambda=1}^{n} y^{(\lambda-1)}(t_0-)\phi_\lambda(t) + \int_0^t h(t-\xi)u(t_0+\xi)\,d\xi$$

which upon the change of variables $t \to t - t_0$, $\xi \to \xi - t_0$, yields

19 $$y(t) = \sum_{\lambda=1}^{n} y^{(\lambda-1)}(t_0-)\phi_\lambda(t-t_0) + \int_{t_0}^t h(t-\xi)u(\xi)\,d\xi \qquad t \geq t_0$$

This is the desired expression for the general solution of *2* for arbitrary t_0.

Expression for the input-output-state relation

It will be helpful at this point to summarize the results of the foregoing analysis in the form of an

20 **Assertion** Let $\mathcal{R}$ be a system characterized by the input-output relation

21 $$L(p)y = u$$

in which

$$L(p) = a_n p^n + \cdots + a_0 \qquad a_n \neq 0$$

and the a_i, $i = 1, \ldots, n$, are constants, not necessarily real.

Then, the expression for the general solution of *21* is

22 $$y(t) = \sum_{\lambda=1}^{n} y^{(\lambda-1)}(t_0-)\phi_\lambda(t-t_0) + \int_{t_0}^t h(t-\xi)u(\xi)\,d\xi \qquad t \geq t_0$$

where

23 $$h(t) = \mathcal{L}^{-1}\left\{\frac{1}{L(s)}\right\} = \text{impulse response of } \mathcal{R}$$

$$H(s) = \frac{1}{L(s)} = \text{transfer function of } \mathcal{R}$$

$$\phi_\lambda(t) = \mathcal{L}^{-1}\left\{\frac{a_n s^{n-\lambda} + \cdots + a_\lambda}{L(s)}\right\} \qquad \lambda = 1, \ldots, n$$

The time functions $\phi_1, \ldots, \phi_n$ are linearly independent and satisfy the differential equation

$$L(p)\phi_\lambda(t) = 0 \qquad \lambda = 1, \ldots, n$$

under the initial conditions *10* and *11*. Thus they qualify as a set of (one-sided) basis functions for $\mathcal{R}$ (see *3.6.39*). ◁

This completes step 1. Turning to step 2, we have to identify the constants appearing in *21* with the components of $\mathbf{x}(t_0-)$, the state of $\mathcal{R}$ at time t_0-. This can be done in an infinite variety of ways, depending on the choice of the basis functions (see *4.2.25*) in *22*. A natural, though not necessarily the most advantageous, way (as will be seen later) is to identify the basis functions with the functions $\phi_1, \ldots, \phi_n$ in *23*. In this case, the components of $\mathbf{x}(t_0-)$, which are the coefficients of the basis functions, are given by

$$x_1(t_0-) = y(t_0-)$$
$$\cdots\cdots\cdots\cdots$$
$$x_n(t_0-) = y^{(n-1)}(t_0-)$$

or, more compactly,

24 $$\mathbf{x}(t_0-) = (y(t_0-), \ldots, y^{(n-1)}(t_0-))$$

Another natural way is to let the basis functions—or rather, their Laplace transforms—be

25 $$\frac{s^{n-1}}{L(s)}, \ldots, \frac{s}{L(s)}, \frac{1}{L(s)}$$

In this case, the components of $\mathbf{x}(t_0-)$ are given by

26 $$\begin{aligned} x_1(t_0-) &= a_n y(t_0-) \\ x_2(t_0-) &= a_n y^{(1)}(t_0-) + a_{n-1}y(t_0-) \\ &\cdots\cdots\cdots\cdots\cdots\cdots \\ x_n(t_0-) &= a_n y^{(n-1)}(t_0-) + \cdots + a_1 y(t_0-) \end{aligned}$$

In the sequel, we shall focus our attention on the case where the first component of the initial state vector is equated to the output of $\mathcal{R}$ at t_0-, the second component to the first derivative of the output, etc. This implies that the state vector at time t is expressed by

27 $$\mathbf{x}(t) = (y(t), \ldots, y^{(n-1)}(t))$$

For convenience in referring to this case, we shall call $\mathbf{x}(t)$, as given by *27*, the *normal state vector of* $\mathcal{R}$ and the corresponding state equations the *normal state equations for* $\mathcal{R}$ (compare with *3.7.33*). The state space over which $\mathbf{x}$ ranges is $\mathcal{C}^n$, the space of n-tuples of complex numbers.

On replacing the initial values $y^{(\lambda-1)}(t_0-)$ in *22* by their expressions in terms of the components of $\mathbf{x}(t_0-)$, the general solution *22* becomes

28 $$y(t) = \langle \boldsymbol{\phi}(t - t_0), \mathbf{x}(t_0-) \rangle + \int_{t_0}^{t} h(t - \xi)u(\xi)\, d\xi \qquad t \geq t_0$$

where h is the impulse response of $\mathcal{R}$, $\boldsymbol{\phi}(t)$ represents the vector

29 $$\boldsymbol{\phi}(t) = (\phi_1(t), \ldots, \phi_n(t))$$

whose components are the basis functions

30 $$\phi_\lambda(t) = \mathcal{L}^{-1}\left\{\frac{a_n s^{n-\lambda} + \cdots + a_\lambda}{L(s)}\right\}$$

and $\langle \boldsymbol{\phi}(t - t_0), \mathbf{x}(t_0-) \rangle$ denotes the scalar product of the basis vector $\boldsymbol{\phi}(t - t_0)$ and the initial state vector $\mathbf{x}(t_0-)$. Equation *28* constitutes a tentative input-output-state relation for $\mathcal{R}$, since it remains to be shown that $\mathbf{x}(t)$ as defined by *27* qualifies as a state vector for $\mathcal{R}$.

State equations

To derive the state equations of $\mathcal{R}$, we form the expression for $\dot{\mathbf{x}}(t)$ by differentiating both sides of *27*. This yields

31 $$\dot{\mathbf{x}}(t) = (y^{(1)}(t), y^{(2)}(t), \ldots, y^{(n)}(t))$$

On the other hand, the differential equation *21* gives

$$y^{(n)}(t) = \frac{1}{a_n}(u(t) - a_0 y(t) - \cdots - a_{n-1}y^{(n-1)}(t))$$

and hence

$$\dot{\mathbf{x}}(t) = \left(y^{(1)}(t), y^{(2)}(t), \ldots, \frac{1}{a_n}(u(t) - a_0 y(t) - \cdots - a_{n-1}y^{(n-1)}(t))\right)$$

This relation shows that the components of $\dot{\mathbf{x}}(t)$ are linear combinations of the components of $\mathbf{x}(t)$ and $u(t)$. To exhibit these relations in a matrix form, let $x_i(t)$ and $\dot{x}_i(t)$, $i = 1, \ldots, n$, denote the ith components of $\mathbf{x}(t)$ and $\dot{\mathbf{x}}(t)$, respectively. Then *31* implies

$$\begin{aligned} \dot{x}_1(t) &= x_2(t) \\ \dot{x}_2(t) &= x_3(t) \\ &\cdots\cdots \\ \dot{x}_n(t) &= -\frac{a_0}{a_n}x_1(t) - \cdots - \frac{a_{n-1}}{a_n}x_n(t) + \frac{1}{a_n}u(t) \end{aligned}$$

or in matrix form,

32 $$\begin{bmatrix} \dot{x}_1(t) \\ \cdots \\ \cdots \\ \dot{x}_n(t) \end{bmatrix} = \begin{bmatrix} 0 & 1 & 0 & \cdots & 0 \\ 0 & 0 & 1 & \cdots & 0 \\ \cdot & \cdot & \cdot & \cdots & \cdot \\ -\frac{a_0}{a_n} & \cdots & \cdot & \cdots & -\frac{a_{n-1}}{a_n} \end{bmatrix} \begin{bmatrix} x_1(t) \\ \cdots \\ \cdots \\ x_n(t) \end{bmatrix} + \begin{bmatrix} 0 \\ 0 \\ \cdots \\ \frac{1}{a_n} \end{bmatrix} u(t)$$

which is a state equation of the canonical form $\dot{\mathbf{x}} = \mathbf{A}\mathbf{x} + \mathbf{b}u$, where $\mathbf{A}$ and $\mathbf{b}$ are a constant $n \times n$ matrix and an n-vector, respectively.

We have thus established that if $\mathbf{x}(t)$ is defined as in *27*, with $\mathbf{x}(t)$ ranging over $\mathcal{C}^n$, then $\mathbf{x}(t)$ satisfies the state equation *32*. This result, in conjunction with the relation

$$y(t) = x_1(t)$$

or, in matrix form,

$$y(t) = [1 \quad 0 \quad \cdots \quad 0] \begin{bmatrix} x_1(t) \\ \cdots \\ x_n(t) \end{bmatrix}$$

permits us to conclude by *2.3.42* that $\mathbf{x}(t)$ and the input-output-state relation *28* satisfy the mutual- and self-consistency conditions *1.6.5*, *1.6.11*, *1.6.15*, and *1.6.35*.

We summarize this and previous conclusions in the form of an

33 **Assertion** Let $\mathcal{R}$ be a system of the reciprocal differential operator type characterized by the input-output relation

$$L(p)y = u \qquad L(p) = a_n p^n + \cdots + a_0 \qquad a_n \neq 0$$

Then the vector

$$\mathbf{x}(t) = (y(t), \ldots, y^{(n-1)}(t))$$

qualifies as a state vector of $\mathcal{R}$ at time t. With this choice of $\mathbf{x}(t)$ (called the *normal state vector*), the state equations of $\mathcal{R}$ read

34
$$\dot{\mathbf{x}}(t) = \mathbf{A}\mathbf{x}(t) + \mathbf{b}u(t)$$
$$y(t) = \langle \mathbf{c}, \mathbf{x}(t) \rangle$$

where

$$\mathbf{A} = \begin{bmatrix} 0 & 1 & 0 & \cdots & 0 \\ 0 & 0 & 1 & \cdots & 0 \\ \cdot & \cdot & \cdot & \cdots & \cdot \\ -\dfrac{a_0}{a_n} & \cdot & \cdot & \cdots & -\dfrac{a_{n-1}}{a_n} \end{bmatrix} \qquad \mathbf{b} = \begin{bmatrix} 0 \\ \cdots \\ \dfrac{1}{a_n} \end{bmatrix}$$

$$\mathbf{c} = [1 \quad 0 \quad \cdots \quad 0]$$

The input-output-state relation of $\mathcal{R}$ is given by

35
$$y(t) = \langle \boldsymbol{\phi}(t - t_0), \mathbf{x}(t_0-) \rangle + \int_{t_0}^{t} h(t - \xi)u(\xi)\, d\xi \qquad t \geq t_0$$

where, to recapitulate,

$$h(t) = \mathcal{L}^{-1}\left\{\frac{1}{L(s)}\right\}$$

$$\boldsymbol{\phi}(t) = (\phi_1(t), \ldots, \phi_n(t))$$

$$\phi_\lambda(t) = \mathcal{L}^{-1}\left\{\frac{a_n s^{n-\lambda} + \cdots + a_\lambda}{L(s)}\right\} \qquad \lambda = 1, \ldots, n$$

$$\mathbf{x}(t_0-) = (y(t_0-), \ldots, y^{(n-1)}(t_0-))$$

36 *Example* Suppose that $\mathcal{R}$ is characterized by

$$\frac{d^2y}{dt^2} + 3\frac{dy}{dt} + 2y = u$$

Then

$$L(s) = s^2 + 3s + 2 = (s+1)(s+2)$$

and

$$h(t) = \mathcal{L}^{-1}\left\{\frac{1}{(s+1)(s+2)}\right\} = 1(t)(e^{-t} - e^{-2t})$$

$$\phi_1(t) = \mathcal{L}^{-1}\left\{\frac{s+3}{(s+1)(s+2)}\right\} = 1(t)(2e^{-t} - e^{-2t})$$

$$\phi_2(t) = \mathcal{L}^{-1}\left\{\frac{1}{(s+1)(s+2)}\right\} = 1(t)(e^{-t} - e^{-2t})$$

where $1(t)$ denotes the unit step function.

The components of the state vector at time t are

$$x_1(t) = y(t)$$
$$x_2(t) = \dot{y}(t)$$

Correspondingly, the input-output-state relation is given by

$$y(t) = (2e^{-(t-t_0)} - e^{-2(t-t_0)})x_1(t_0-) + (e^{-(t-t_0)} - e^{-2(t-t_0)})x_2(t_0-) + \int_{t_0}^{t} [e^{-(t-\xi)} - e^{-2(t-\xi)}]u(\xi)\, d\xi \qquad t \geq t_0$$

and the matrices and vectors which appear in the normal state equations *34* read

$$\mathbf{A} = \begin{bmatrix} 0 & 1 \\ -2 & -3 \end{bmatrix} \qquad \mathbf{b} = \begin{bmatrix} 0 \\ 1 \end{bmatrix} \qquad \mathbf{c} = [1 \quad 0]$$

37 *Exercise* Determine the input-output-state relation and the state equations for the system of Example *36* for the case where the state is defined by *26*. Find the matrix $\mathbf{\Gamma}$ (*4.2.9*) which relates the state defined by *26* to the normal state vector.

Some properties of $\mathcal{R}$

The input-output-state relation *35* provides us with a complete characterization of $\mathcal{R}$. Having this characterization in hand, it is a simple matter to deduce the several basic properties of $\mathcal{R}$ which are listed below.

38 **Property** $\mathcal{R}$ is a system of order n in the sense of Definition *3.6.39*. This follows at once from the fact that the number of (linearly independent) basis functions in the input-output-state relation *35* is n.

39 **Property** No matter how the state vector of $\mathcal{R}$ is defined, the dimension of the state space of $\mathcal{R}$ cannot be lower than n. This follows from Assertion *4.2.4*.

40 **Property** $\mathcal{R}$ is a linear system in the sense of Definition *3.5.6*.

41 **Property** $\mathcal{R}$ is a time-invariant system in the sense of Definition *3.2.36*. This follows from Assertion *3.6.35*.

42 **Property** The normal state equations are in reduced form (see *3.6.46*), which means that there are no distinct equivalent states in the state space $\mathcal{C}^n$.

43 **Property** The basis functions $\phi_1, \ldots, \phi_n$ defined by *30* are the zero-input responses of $\mathcal{R}$ starting, respectively, in the initial (unit vector) states $(1,0, \ldots ,0)$, $(0,1, \ldots ,0)$, $\ldots$, $(0,0, \ldots ,1)$. This follows from Assertion *8* and the way in which $\mathbf{x}$ is defined in terms of the initial values of y and its derivatives.

44 **Remark** Most of the results and conclusions arrived at in the preceding discussion apply with but minor modifications to the case where $\mathcal{R}$ is a time-varying system. Specifically, if the coefficients in $L(p)$ are functions of time rather than constants, that is, $L(p)$ is of the form

45 $$L(p,t) = a_n(t)p^n + \cdots + a_0(t) \qquad a_n(t) \neq 0 \qquad \forall t$$

then the vector

46 $$\mathbf{x}(t) = (y(t), \ldots ,y^{(n-1)}(t))$$

qualifies as a normal state vector for $\mathcal{R}$, just as it does in the time-invariant case. The normal state equations for $\mathcal{R}$ read

47 $$\begin{aligned} \dot{\mathbf{x}}(t) &= \mathbf{A}(t)\mathbf{x}(t) + \mathbf{b}(t)u(t) \\ y(t) &= \langle \mathbf{c}(t),\mathbf{x}(t) \rangle \end{aligned}$$

where $\mathbf{A}(t)$, $\mathbf{b}(t)$, and $\mathbf{c}(t)$ have exactly the same form *32* as their counterparts in the time-invariant case, with the $a_i(t)$ taking the place of a_i, $i = 0, \ldots , n$.

Special case: an integrator

A special and yet very important form of a system of the reciprocal differential operator type is an *integrator*, a system which we have used for illustrative purposes on many previous occasions. Our earlier definition of the notion of an integrator was informal in nature (*1.9.10*). We are now in a position to define it more precisely and associate an integrator with a state vector and state equations.

In its wide sense—which is the sense in which the term will be used in the sequel—an integrator is essentially a reciprocal differential operator system of order 1. More specifically,

48 **Definition** An *integrator* is a system characterized by an input-output relation of the form

49 $$(p + a)y = u$$

where a is a constant, not necessarily real. The reciprocal of a is called the *time constant* of the integrator. A *pure integrator* is one in which $a = 0$. In this case *49* becomes

50 $$p\,y = u$$

(Since the term "integrator" is commonly used to denote a device defined by *50* rather than *49*, we shall on occasion use the term "*impure, or imperfect, integrator*" to make it clear that the device we have in mind is characterized by *49* rather than *50*.) An integrator characterized by an input-output relation of the form *49* is denoted by $1/(p + a)$. ◁

Since an integrator is a system of the reciprocal differential operator type of order 1, its normal state vector has just one component, $y(t)$, which is the output of the integrator at time t. Furthermore, by Theorem *4.2.25*, this is the only way (to within a constant factor) in which a one-dimensional state vector can be associated with an integrator. Thus, we can make the

51 **Assertion** The state of an integrator at time t is necessarily of the form

$$x(t) = ky(t)$$

where k is a constant. Unless indicated to the contrary, k will be understood to be unity. ◁

The input-output-state relation and the state equations for an integrator are special cases of *34* and *35*, with $a_1 = 1$ and $a_0 = a$. They read

52 $$y(t) = x(t_0-)e^{-a(t-t_0)} + \int_{t_0}^{t} e^{-a(t-\xi)}u(\xi)\,d\xi \qquad t \geq t_0$$

and

53 $$\dot{x} = -ax + u$$
$$y = x$$

This concludes for the present our discussion of systems of the reciprocal differential operator type. We shall return to them in Sec. *4* and again in Sec. *9*, where an alternative technique for associating a state vector with $\mathcal{R}$ will be described.

4 Systems of the differential operator type

In this section, we consider systems of the differential operator type denoted by $\mathfrak{D}$ and characterized by an input-output relation of the form

1 $$y(t) = M(p)u$$

where $M(p) = b_m p^m + \cdots + b_0$, $b_m \neq 0$, and $b_0, \ldots, b_m$ are constants, not necessarily real. For brevity, a system characterized by *1* will be denoted by $M(p)$.

Our interest in systems of this type stems from two reasons: First, as will be shown in Secs. *5* and *6*, any differential system can be realized as a combination of systems of the reciprocal differential and differential operator type and, second, the systems characterized by $L(p)y = u$ and $y = L(p)u$ are inverse to each other in the sense of Definition *2.10.6*. This will be established at a later point in the present section.

In Sec. *3* we found the input-output-state relation and state equations for $\mathfrak{R}$. In this section we shall do the same for $\mathfrak{D}$, omitting the details of derivations and arguments which are parallel to those of Sec. *3*. We begin with the derivation of the general solution of *1*.

General solution

By applying the Laplace transformation to both sides of *1*, we obtain the following expression for the Laplace transform of the general solution:

2 $$Y(s) = M(s)U(s) - \sum_{\lambda=1}^{m} (b_m s^{m-\lambda} + \cdots + b_\lambda)u^{(\lambda-1)}(0-)$$

where $M(s) = b_m s^m + \cdots + b_0$ and $u^{(\lambda)}(0-) \triangleq$ initial value of the λth derivative of the input at $t = 0-$. Consequently, the general solution itself (for $t_0 = 0$) may be written as

3 $$y(t) = \sum_{\lambda=1}^{m} \phi_\lambda(t)u^{(\lambda-1)}(0-) + \int_0^t h(t-\xi)u(\xi)\,d\xi \qquad t \geq 0$$

where

4 $$\phi_\lambda(t) = \mathcal{L}^{-1}\{b_m s^{m-\lambda} + \cdots + b_\lambda\} = b_m \delta^{(m-\lambda)}(t) + \cdots + b_\lambda \delta(t) \qquad \lambda = 1, \ldots, m$$

$$h(t) = \mathcal{L}^{-1}\{M(s)\} = b_m \delta^{(m)}(t) + \cdots + b_0 \delta(t)$$

with $\delta^{(\lambda)}(t)$ representing a delta function of λth order (see *A.2.5*). By using exactly the same arguments as in Sec. *3*, we can easily establish the following facts about *3*.

(I) The basis functions $\phi_\lambda(t)$ are linearly independent, which implies that $\mathfrak{D}$ is a differential system of order m.

(II) The $\phi_\lambda(t)$ represent zero-input responses of $\mathfrak{D}$, with $\phi_\lambda(t)$ given by

5 $$\phi_\lambda(t) = M(p)u \qquad \lambda = 1, \ldots, m$$

where $u = 0$ for $t \geq 0$, and

6 $$\begin{aligned} &u^{(\lambda-1)}(0-) = 1 \\ &u^{(\mu-1)}(0-) = 0 \qquad \text{for all } \mu \ (\mu = 1, \ldots, m) \text{ different from } \lambda \end{aligned}$$

(III) $h(t)$ is the impulse response of $\mathfrak{D}$; that is,

7 $$h(t) = M(p)u$$

where $u = \delta(t)$ and $u(0-) = \cdots = u^{(m-1)}(0-) = 0$.

We are now ready to associate a state vector $\mathbf{x} = (x_1, \ldots, x_m)$ with $\mathfrak{D}$. We do this in a manner analogous to that used in Sec. *3*; that is, we let

$$x_1(0-) \triangleq u(0-),\ x_2(0-) \triangleq u^{(1)}(0-), \ldots, x_m(0-) \triangleq u^{(m-1)}(0-)$$

and call such a state vector the *normal* state vector of $\mathfrak{D}$. Under this mode of identification of the components of $\mathbf{x}(0-)$ with the constants appearing in the general solution, the expression for the state at time t becomes

8 $$\mathbf{x}(t) = (u(t), \ldots, u^{(m-1)}(t))$$

with the state space Σ being $\mathcal{C}^m$, the space of m-tuples of complex numbers. [Note that this implies that $u(t)$ may be complex-valued.]

Input-output-state relation

If the normal state vector of $\mathfrak{D}$ is defined to be the vector $\mathbf{x}(t)$ given by *8*, then the corresponding normal input-output-state relation for $\mathfrak{D}$ (for arbitrary t_0) is obtained by substituting $x_\lambda(0-)$ for each $u^{(\lambda-1)}(0-)$ in the general solution *3* and replacing $0-$ with t_0-. The result reads

9 $$y(t) = \langle \boldsymbol{\phi}(t - t_0), \mathbf{x}(t_0-) \rangle + \int_{t_0}^{t} h(t - \xi)u(\xi)\, d\xi \qquad t \geq t_0$$

where

10 $$\begin{aligned} &\mathbf{x}(t) = (u(t), \ldots, u^{(m-1)}(t)) \\ &\boldsymbol{\phi}(t) \triangleq (\phi_1(t), \ldots, \phi_m(t)) \\ &\phi_\lambda(t) \text{ is given by } 4,\ \lambda = 1, \ldots, m \\ &h(t) \text{ is given by } 4 \end{aligned}$$

11 *Comment* A question which suggests itself at this point is this: what is the connection between the input-output relation

12
$$\begin{aligned} y(t) &= M(p)u \\ &= (b_m p^m + \cdots + b_0)u \\ &= b_m u^{(m)}(t) + \cdots + b_0 u(t) \end{aligned}$$

and the input-output-state relation *9*?

On writing the latter relation in full, we have

13
$$y(t) = -\sum_{\lambda=1}^{m} u^{(\lambda-1)}(t_0-)[b_m\delta^{(m-\lambda)}(t-t_0) + \cdots + b_\lambda\delta(t-t_0)] + \int_{t_0}^{t} [b_m\delta^{(m)}(t-\xi) + \cdots + b_0\delta(t-\xi)]u(\xi)\,d\xi$$

Now, a delta function $\delta^{(\lambda)}(t-\xi)$ of order λ has the property

14
$$\int_{t_0}^{t} \delta^{(\lambda)}(t-\xi)u(\xi)\,d\xi = p^\lambda 1(t-t_0)u(t)$$

where $1(t-t_0)$ is a unit step function which serves to truncate u to the left of the point $t = t_0$. Thus, *13* may be written as

15
$$y(t) = \sum_{\lambda=1}^{m} u^{(\lambda-1)}(t_0-)[b_m\delta^{(m-\lambda)}(t-t_0) + \cdots + b_\lambda\delta(t-t_0)] + M(p)1(t-t_0)u(t)$$

Now we recall (see *3.8.3 et seq.*) that when p acts on a time function u which may have a discontinuity at $t = t_0$, we have

16
$$\begin{aligned} p\,u &= -\,u(t_0-)\delta(t-t_0) + p\,1(t-t_0)u(t) \\ p^2\,u &= -u^{(1)}(t_0-)\delta(t-t_0) - u(t_0-)\delta^{(1)}(t-t_0) + p^2\,1(t-t_0)u(t) \\ &\cdots\cdots\cdots\cdots\cdots\cdots \\ p^m\,u &= -u^{(m-1)}(t_0-)\delta(t-t_0) - \cdots - u(t_0-)\delta^{(m-1)}(t-t_0) + p^m\,1(t-t_0)u(t) \end{aligned}$$

Furthermore, we note that *15* results when the right-hand members of *16* are substituted for $p\,u, \ldots, p^m\,u$, respectively, in the input-output relation *12*. This shows that the input-output-state relation *13* can be derived directly from the input-output relation *12* merely by replacing $p\,u$, $p^2\,u, \ldots, p^m\,u$ in *12* with the equivalent expressions *16*. In effect, then, the input-output-state relation *9* serves merely to place in evidence the effect of a discontinuity in u or its derivatives at $t = t_0$.

It will be helpful at this point to summarize our conclusions in the form of an

17 **Assertion** Let $\mathfrak{D}$ be a system of the differential operator type

18
$$y = M(p)u \qquad M(p) = b_m p^m + \cdots + b_0 \qquad b_m \neq 0$$

and let the normal state vector of $\mathfrak{D}$ at time t be defined by

19 $$\mathbf{x}(t) = (u(t), \ldots, u^{(m-1)}(t))$$

Then the input-output-state relation of $\mathfrak{D}$ reads

20 $$y(t) = \langle \boldsymbol{\phi}(t - t_0), \mathbf{x}(t_0-) \rangle + \int_{t_0}^{t} h(t - \xi)u(\xi)\, d\xi \qquad t \geq t_0$$

or, more explicitly,

21 $$y(t) = \langle \boldsymbol{\phi}(t - t_0), \mathbf{x}(t_0-) \rangle + M(p)[1(t - t_0)u(t)]$$

where, to recapitulate,

22 $$\boldsymbol{\phi}(t) = (\phi_1(t), \ldots, \phi_n(t))$$

23 $$\phi_\lambda(t) = \mathfrak{L}^{-1}\{b_m s^{m-\lambda} + \cdots + b_\lambda\} = b_m \delta^{(m-\lambda)}(t) + \cdots + b_\lambda \delta(t) \qquad \lambda = 1, \ldots, m$$

24 $$h(t) = \mathfrak{L}^{-1}\{M(s)\} = b_m \delta^{(m)}(t) + \cdots + b_0 \delta(t)$$

25 *Comment* It is understood that any delta functions which u may have at $t = t_0$ are included in the term $1(t - t_0)u(t)$ and are acted upon by $M(p)$ in the usual fashion (see *A.2.5*).

26 *Comment* Note that at this step of our analysis *9* must be regarded as a tentative input-output-state relation since we have not yet verified that $\mathbf{x}$ as defined by *19* qualifies as a state vector for $\mathfrak{D}$ under the self-consistency conditions. We shall establish this after deriving the state equations for $\mathfrak{D}$.

27 **Remark** It is of interest to observe that the normal state vector associated with $\mathfrak{D}$ has exactly the same intrinsic significance as the normal state vector associated with $\mathfrak{R}$. Specifically, if we write the input-output relations of $\mathfrak{R}$ and $\mathfrak{D}$ in the nonoriented form

$$L(p)v_1 = v_2 \qquad L(p) = a_n p^n + \cdots + a_0 \qquad a_0 \neq 0$$

then the state vector $\mathbf{x}$ is expressed in both cases by

28 $$\mathbf{x} = (v_1, \ldots, v_1^{(n-1)})$$

regardless of which orientation is used. This implies that the definition of the state vector in terms of the variables entering into the differential equations which characterize a system is independent of the manner in which the system is oriented, i.e., independent of whether a particular variable is labeled as input or output. The reason for this lies in the fact that, in the process of deriving the input-output-state relation from the general solution, we make no real use of the assumption that u is the input and y is the output. This is true not only of systems of the reciprocal differential and differential types considered above but, more generally, of any linear differential system. It should be noted, however, that, although the definition of $\mathbf{x}$ is independent of the orientation, the form of the state equations satisfied by $\mathbf{x}$ depends very strongly on the orientation of the system.

Setting up the state equations

The state equations of $\mathfrak{D}$ corresponding to the normal state vector

$$\mathbf{x}(t) = (u(t), \ldots, u^{(m-1)}(t))$$

can readily be set up by following the procedure of Sec. *3*. Specifically, on forming the expression for $\dot{\mathbf{x}}(t)$ and comparing it with that for $\mathbf{x}(t)$, we obtain

29
$$\begin{aligned} x_1 &= u \\ \dot{x}_1 &= x_2 \\ \dot{x}_2 &= x_3 \\ &\cdots\cdots \\ \dot{x}_{m-1} &= x_m \end{aligned}$$

Furthermore, from the input-output relation

30
$$y = b_0u + b_1u^{(1)} + \cdots + b_mu^{(m)}$$

and *29*, we deduce

31
$$y = b_0x_1 + \cdots + b_{m-1}x_m + b_m\dot{x}_m$$

Equations *29*, together with *31*, constitute the desired state equations of $\mathfrak{D}$. We shall refer to them as the *normal state equations* corresponding to the normal state vector *19*.

32 *Example* Suppose that $\mathfrak{D}$ is characterized by the input-output relation

$$y = \frac{d^2u}{dt^2} + 3\frac{du}{dt} + 2u$$

In this case, the normal state vector is expressed by

$$\mathbf{x} = (u, \dot{u})$$

and the normal state equations *29* and *31* become

$$\begin{aligned} x_1 &= u \\ \dot{x}_1 &= x_2 \\ y &= 2x_1 + 3x_2 + \dot{x}_2 \end{aligned}$$

The input-output-state relation for $\mathfrak{D}$ reads

$$\begin{aligned} y(t) = (\delta^{(1)}(t - t_0) + 3\delta(t - t_0))x_1(t_0-) + \delta(t - t_0)x_2(t_0-) \\ + (p^2 + 3p + 2)1(t - t_0)u(t) \quad \triangleleft \end{aligned}$$

We observe that the state equations *29* and *31* are not of the same form as the state equations of a system of the reciprocal differential operator type (*4.3.32*), nor are they of the canonical form

$$\dot{\mathbf{x}} = \mathbf{Ax} + \mathbf{Bu}$$
$$\mathbf{y} = \mathbf{Cx} + \mathbf{Du}$$

Consequently, we cannot use Corollary *2.3.36*, as in Sec. 3, to show that $\mathbf{x}$ qualifies as a state vector of $\mathfrak{D}$.

Although it is an easy matter to verify that $\mathbf{x}$ as defined by *19* has the state separation property *2.3.19*, it will be instructive to demonstrate directly that $\mathbf{x}$ satisfies the four consistency conditions *1.6.5*, *1.6.11*, *1.6.15*, and *1.6.35*. This is what we shall do in the sequel.

Direct verification that x qualifies as a state vector

We have to verify that *21* is an input-output-state relation for the system characterized by the input-output relation

33 $$y(t) = (b_m p^m + \cdots + b_0)u$$

To avoid being influenced by notation, let us express *21* in the form

34 $$y(t) = \langle \boldsymbol{\phi}(t - t_0), \boldsymbol{\alpha} \rangle + M(p)[1(t - t_0)u(t)] \qquad t \geq t_0$$

where $\boldsymbol{\alpha}$ is an m-vector, $\boldsymbol{\alpha} = (\alpha_1, \ldots, \alpha_m)$, ranging over $\mathcal{C}^m$. [In effect, $\boldsymbol{\alpha}$ represents a point (state) in the state space Σ of $\mathfrak{D}$.] We employ the symbol $\boldsymbol{\alpha}$ rather than $\mathbf{x}(t_0-)$ in *34* because at this stage of our argument we have not yet demonstrated that the components of $\boldsymbol{\alpha}$ are the values of u and its derivatives at t_0-.

Since *34* is a general solution for *33*, it follows that *34* satisfies the mutual-consistency condition (see *3.8.24*). Furthermore, since the response is uniquely determined by $\boldsymbol{\alpha}$ and $u_{(t_0,t]}$, it follows that *34* satisfies the first self-consistency condition also. Thus, it will suffice to show that *34* satisfies the third self-consistency condition, since this condition implies the second self-consistency condition (see *1.6.39*).

We can simplify the argument without losing its essential features by carrying out the verification for the case where $m = 1$. Actually, this is all that we shall need at a later point (Secs. *8* and *9*) to develop an effective general technique for associating a state vector with a differential system of any finite order.

For $m = 1$, *33* and *34* become respectively

35 $$y = p\,u$$

36 $$y(t) = \alpha\delta(t - t_0) + p\,1(t - t_0)u(t)$$

where for simplicity we have set $b_0 = 0$, $b_1 = 1$. We assume furthermore that $u(t)$ is real-valued, which implies that α ranges over the real line $(-\infty, \infty)$.

Suppose that, starting at $t = 0$ with $\alpha = 0$, we apply to $\mathfrak{D}$ an input of the form shown in Fig. *4.4.1a* and observe the corresponding response y. Then u and y constitute an input-output pair (u,y) for *35* starting at $t = 0$ with $\alpha = 0$.

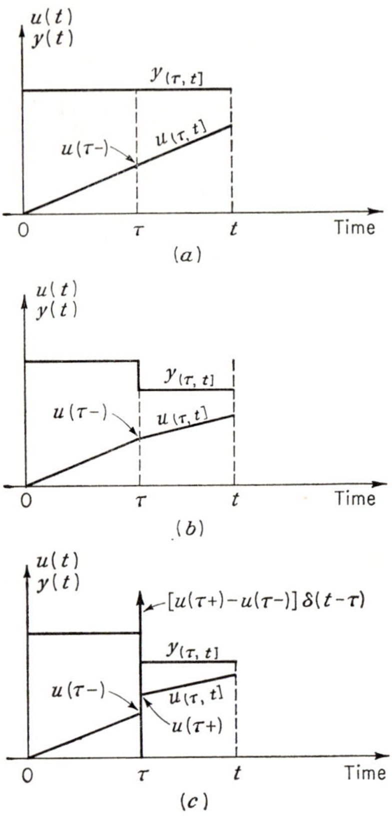

Fig. 4.4.1 Response of p. (*a*) u and $\dot{u}$ are continuous at τ; (*b*) u is continuous at τ; (*c*) u is discontinuous at τ.

Now let τ be a fixed intermediate time between 0 and t and consider the segments $u_{(\tau,t]}$ and $y_{(\tau,t]}$ of u and y, respectively, over the observation interval $(\tau,t]$. By the second self-consistency condition, there must be a value of α in $(-\infty,\infty)$ such that $(u_{(\tau,t]},y_{(\tau,t]})$ is an input-output pair starting at time τ. Indeed, we note that the term $p\ 1(t-\tau)u(t)$ in *36* contributes a delta function $u(\tau+)\delta(t-\tau)$ to the response at time τ. Since y does not contain a delta function at time τ, it follows that the term $u(\tau+)\delta(t-\tau)$ must be canceled by the term $\alpha\delta(t-\tau)$. This in turn implies that $\alpha = u(\tau+)$ or, equivalently, $\alpha = u(\tau-)$, since u is continuous at τ. This, then, is the value of α in *36* with respect to which $(u_{(\tau,t]},y_{(\tau,t]})$ is an input-output pair for *35*.

Next, assume that u is kept fixed over the interval $(0,\tau]$ and is varied over the interval $(\tau,t]$ in such a way as to preserve the continuity of u at τ. For example, a "new" u may have the appearance shown in Fig. *4.4.1b*.

Again, the same argument shows that the "new" pair $(u_{(\tau,t]},y_{(\tau,t]})$ is an input-output pair for *35* starting at τ with $\alpha = u(\tau-)$. More generally, we can draw the same conclusion for any "new" pair $(u_{(\tau,t]},y_{(\tau,t]})$ so long as u is continuous at τ.

Next, consider the case where u is not continuous at τ, as is illustrated in Fig. *4.4.1c*. In this case, the response y will contain a delta function at τ whose magnitude is equal to the value of the jump in u at τ. Specifically, the delta function in question is expressed by

$$(u(\tau+) - u(\tau-))\delta(t-\tau)$$

To account for this delta function in y, the value of α in *36* must be

$u(\tau-)$. Thus, we can conclude that even when we allow the "new" input u to have a discontinuity at τ, the value of α with respect to which $(u_{(\tau,t]},y_{(\tau,t]})$ is an input-output pair remains equal to $u(\tau-)$.

The same thing holds true when we allow u to have not only a discontinuity but also a delta function at τ. In sum, then, no matter what $u_{(\tau,t]}$ is, the value of α with respect to which $(u_{(\tau,t]},y_{(\tau,t]})$ is an input-output pair remains fixed and equal to $u(\tau-)$. This shows that *36* satisfies the third self-consistency condition, since $u(\tau-)$ belongs to all the sets $Q(\alpha;u_{(0,\tau]}u_{(\tau,t]})$ and hence to the intersection of these sets (see *1.6.35*). This in turn implies that $u(\tau-)$ is the state of $\mathfrak{D}$ (for $m = 1$) at time $\tau-$.

We have thus demonstrated in a direct fashion that the expression

$$y(t) = \alpha\delta(t - t_0) + p\,1(t - t_0)u(t)$$

in which α ranges over $\mathcal{R}^1 \triangleq (-\infty,\infty)$, qualifies as an input-output-state relation for the system characterized by the input-output relation

$$y = p\,u$$

Furthermore, we have shown that $u(t)$ qualifies as the state of the system at time t. Note that, by Theorem *4.2.25*, any other way of associating a state vector with the system in question cannot lead to anything other than $ku(t)$ [that is, a constant times $u(t)$] as the expression for the state of the system at time t.

More generally, we can demonstrate in the same fashion that *34*, with $\Sigma = \mathcal{C}^m$, satisfies the third self-consistency condition and that $\mathbf{x}(t) = (u(t), \ldots, u^{(m-1)}(t))$ qualifies as the state of $\mathfrak{D}$ at time t. It is much simpler, however, to arrive at the same conclusion by employing a technique described in Sec. *6*, which makes use of the determinateness theorem for linear systems (Theorem *3.9.12*) and the fact that the state of a differential operator system of order 1 can be identified with its input—which is what we have just demonstrated.

Special case: a differentiator

In Sec. *3* we defined an integrator as a reciprocal differential operator system of order 1. In a parallel manner, we can define a differentiator as a differential operator system of order 1. More precisely,

37 **Definition** A *differentiator* is a system characterized by an input-output relation of the form

38 $$y = (p + b)u$$

where b is a constant, not necessarily real. A *pure* differentiator is one in which $b = 0$. A differentiator which is characterized by an input-output relation of the form *38* is denoted by $(p + b)$. ◁

As in the case of an integrator, the normal state vector of a differentiator has just one component, $u(t)$, which is the input to the differentiator at time t. Again, we can make the

39 **Assertion** The state of a differentiator at time t is necessarily of the form

$$x(t) = ku(t)$$

where k is a constant. Unless indicated to the contrary, k will be assumed to be unity. ◁

The input-output-state relation and the state equations for a differentiator are special cases of *29* and *31*, with $b_1 = 1$ and $b_0 = b$. They read

40 $$y(t) = x(t_0-)\delta(t - t_0) + p\,1(t - t_0)u(t) + bu(t)$$

41 $$x = u$$
$$y = \dot{x} + bu$$

A connection between systems of differential and reciprocal differential type

Consider a system $\mathfrak{R}$ of the reciprocal differential operator type which is characterized by an input-output relation

42 $$L(p)y = u \qquad L(p) = a_n p^n + \cdots + a_0 \qquad a_n \neq 0$$

Correspondingly, let $\mathfrak{D}$ be a system of the differential operator type characterized by the input-output relation ($w \triangleq$ output, $v \triangleq$ input)

43 $$w = L(p)v \qquad L(p) = a_n p^n + \cdots + a_0 \qquad a_n \neq 0$$

Obviously, $\mathfrak{R}$ and $\mathfrak{D}$ are converse systems in the sense of Definition *2.10.1*, since if a pair of time functions (f,g) is an input-output pair for $\mathfrak{R}$, then (g,f) is an input-output pair for $\mathfrak{D}$, and vice versa.

By using the expressions for the input-output-state relations for $\mathfrak{R}$ and $\mathfrak{D}$, it is easy to show that $\mathfrak{R}$ and $\mathfrak{D}$ are not only converse but also inverse to one another (*2.10.6*). Furthermore, we can show that $\mathfrak{D}$ is a left- but not right-constrained inverse of $\mathfrak{R}$ (see *2.10.20*). More specifically, we can make the

44 **Assertion** Let $\mathfrak{R}$ be a system of the reciprocal differential operator type characterized by *42* and let $\mathfrak{D}$ be a system of the differential operator type characterized by *43*. Then $\mathfrak{R}$ and $\mathfrak{D}$ are inverse systems in the sense of Definition *2.10.6* and $\mathfrak{D}$ is a left- but not right-constrained inverse of $\mathfrak{R}$ in the sense of Definition *2.10.20*.

Proof Let $\bar{R}(\boldsymbol{\alpha};u)$ denote the response segment of $\mathfrak{R}$ to u (see *1.6.2*) starting in a state $\boldsymbol{\alpha} \in \mathfrak{C}^n$ ($\mathfrak{C}^n$ = state space of $\mathfrak{R}$ = state space of $\mathfrak{D}$) and let $\bar{D}(\boldsymbol{\beta};v)$ denote the response segment of $\mathfrak{D}$ to v starting in a state

$\boldsymbol{\beta}$ in $\mathcal{C}^n$. To demonstrate that $\mathcal{R}$ and $\mathcal{D}$ are inverse to one another, we have to show that (I) to every state $\boldsymbol{\alpha}$ in $\mathcal{C}^n$ there corresponds a state $\boldsymbol{\beta}(\boldsymbol{\alpha})$ in $\mathcal{C}^n$ such that if $\bar{R}(\boldsymbol{\alpha};u)$ is applied to $\mathcal{D}$ starting in $\boldsymbol{\beta}(\boldsymbol{\alpha})$, then the response of $\mathcal{D}$ is u, for all u in the input segment space of $\mathcal{R}$, and (II) to every state $\boldsymbol{\beta}$ in $\mathcal{C}^n$ there corresponds a state $\boldsymbol{\alpha}(\boldsymbol{\beta})$ in $\mathcal{C}^n$ such that if $\bar{D}(\boldsymbol{\beta};v)$ is applied to $\mathcal{R}$ starting in $\boldsymbol{\alpha}(\boldsymbol{\beta})$, then the response of $\mathcal{R}$ is v, for all v in the input segment space of $\mathcal{D}$.

Now the input-output-state relations of $\mathcal{R}$ and $\mathcal{D}$ in the Laplace transform form read (see *4.3.5* and *4.4.2*)

45 $$\mathcal{R}: \qquad Y(s) = \frac{U(s)}{L(s)} + \frac{\sum_{\lambda=1}^{n} (a_n s^{n-\lambda} + \cdots + a_\lambda) y^{(\lambda-1)}(0-)}{L(s)}$$

46 $$\mathcal{D}: \qquad W(s) = L(s)V(s) - \sum_{\lambda=1}^{n} (a_n s^{n-\lambda} + \cdots + a_\lambda) v^{(\lambda-1)}(0-)$$

where $W(s) \triangleq \mathcal{L}\{w(t)\}$ and $V(s) \triangleq \mathcal{L}\{v(t)\}$. If the initial state of $\mathcal{R}$ is $\boldsymbol{\alpha} = (\alpha_1, \ldots, \alpha_n)$, that is,

$$(y(0-), \ldots, y^{(n-1)}(0-)) \triangleq \boldsymbol{\alpha}$$

and the initial state of $\mathcal{D}$ is $\boldsymbol{\beta} = (\beta_1, \ldots, \beta_n)$, that is,

$$(v(0-), \ldots, v^{(n-1)}(0-)) \triangleq \boldsymbol{\beta}$$

then on setting $V(s) = Y(s)$ [which signifies that $\bar{R}(\boldsymbol{\alpha};u)$ is applied to $\mathcal{D}$] and substituting $Y(s)$ in place of $V(s)$ in *46*, we obtain the following expression for the Laplace transform of the response of $\mathcal{D}$:

47 $$W(s) = U(s) + \sum_{\lambda=1}^{n} (a_n s^{n-\lambda} + \cdots + a_\lambda)(\alpha_\lambda - \beta_\lambda)$$

Clearly, $W(s) \equiv U(s)$ if and only if $\alpha_\lambda = \beta_\lambda$, $\lambda = 1, \ldots, n$, that is, if $\boldsymbol{\alpha} = \boldsymbol{\beta}$. Furthermore, we obtain the same result on applying $\bar{D}(\boldsymbol{\beta};v)$ to $\mathcal{R}$ starting in state $\boldsymbol{\alpha}$.

This shows that (I) and (II) are satisfied when $\boldsymbol{\alpha} = \boldsymbol{\beta}$, that is, when both $\mathcal{R}$ and $\mathcal{D}$ are started at the same point in the state space $\mathcal{C}^n$. Consequently, $\mathcal{R}$ and $\mathcal{D}$ are inverse systems in the sense of Definition *2.10.6*.

Now suppose that $\mathcal{R}$ and $\mathcal{D}$ are connected in tandem as shown in Fig. *4.4.2a*. Then, since $\mathcal{R}$ has output-determined state and $\mathcal{D}$ has input-determined state (see *2.6.26* and *2.6.34*), the equality $y = v$ which is imposed by the connection implies the equality $\boldsymbol{\alpha} = \boldsymbol{\beta}$ of the states of $\mathcal{R}$ and $\mathcal{D}$. From this and *47* it follows that $\mathcal{D}$ is a left-constrained inverse of $\mathcal{R}$.

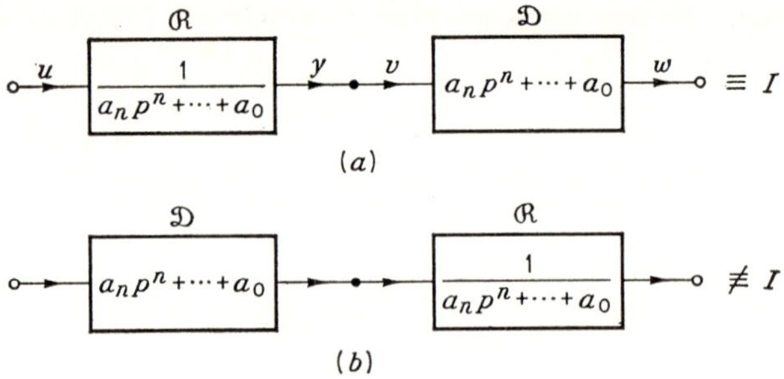

Fig. 4.4.2 Inverse systems. (*a*) $\mathfrak{D}\mathfrak{R} \equiv I$; (*b*) $\mathfrak{R}\mathfrak{D} \not\equiv I$. $\mathfrak{R}$ is a right-constrained inverse of $\mathfrak{D}$.

On the other hand, $\mathfrak{D}$ is not a right-constrained inverse of $\mathfrak{R}$ (Fig. *4.4.2b*), since in this case $\boldsymbol{\alpha}$ is not constrained to be equal to $\boldsymbol{\beta}$ by the tandem connection and $W(s) \not\equiv U(s)$ by *47*. (We disregard, of course, the trivial case where $n = 0$.)

48 **Remark** An important special case of Assertion *44* is one where $\mathfrak{R}$ is an integrator defined by

$$p\,y = u$$

and denoted by $1/p$, and $\mathfrak{D}$ is a differentiator defined by

$$w = p\,u$$

and denoted by p. By Assertion *44*, $1/p$ and p are inverse to one another. Furthermore, p is a left-constrained inverse of $1/p$, which in operational notation signifies that the constrained product $p \cdot \frac{1}{p}$ [which represents a tandem combination of $1/p$ and p, with p acting on the output of $1/p$ (see *1.10.23*)] is equivalent to the identity operator [i.e., a unitor (*1.9.8*)]. On the other hand, $\frac{1}{p} \cdot p$ *is not* equivalent to the identity operator, since p is not a right-constrained inverse of $1/p$. This observation clarifies the results obtained previously in Example *2.10.11*. (Specifically, the delta-function term in *2.10.16* vanishes when $\mathfrak{B}\mathfrak{A}$ is interpreted as a constrained product.) ◁

Another useful relation that can be drawn at once from *45* and *46* provides a rule for "trading" a time function added to the output of $\mathfrak{R}$ for one added to the input to $\mathfrak{R}$. The rule in question is contained in the

49 **Assertion** Let $\mathfrak{R}$ denote the system $1/L(p)$, with the initial state of $\mathfrak{R}$ at time t_0 being $\boldsymbol{\alpha} = (\alpha_1, \ldots, \alpha_n)$. Suppose that the response of $\mathfrak{R}$ to $\mathbf{u}$ starting in state $\boldsymbol{\alpha}$ is $\mathbf{y}$. Then to produce an output $\mathbf{y} + \mathbf{w}$, where $\mathbf{w}$ is a specified time function (defined for $t \geq t_0$), it is sufficient to (I) add to the input the time function

$$\mathbf{v} = L(p)\mathbf{w}$$

where $L(p)$ should be interpreted as an indefinite differential operator (*3.8.11*), and (II) subtract from the initial state of $\mathcal{R}$ the vector $(w(t_0-), \ldots, w^{(n-1)}(t_0-))$. ◁

We turn next to the consideration of a more general type of differential system which subsumes both the differential and reciprocal differential operator types of systems as special cases.

5 *State vectors and state equations for general differential systems*

By a general differential system (with scalar input u and scalar output y) we mean here a system characterized by an input-output relation of the form

1 $$\mathbf{L}(p)y + \mathbf{K}(p)\tilde{\mathbf{y}} = \mathbf{M}(p)u$$

where $\mathbf{L}(p)$ and $\mathbf{M}(p)$ are vector differential operators, $\mathbf{K}(p)$ is a matrix differential operator, and $\tilde{\mathbf{y}}$ is the suppressed output vector.

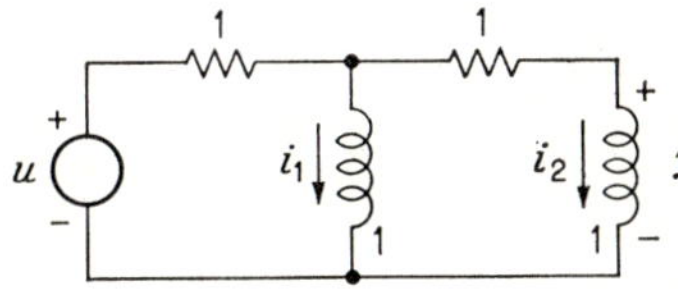

Fig. 4.5.1 Network of Example *4.5.2*.

2 *Example* Consider the network $\mathcal{a}$ shown in Fig. *4.5.1* in which u is the input voltage, y is the output voltage, and the currents i_1 and i_2 play the role of suppressed output variables. For simplicity, the element values are assumed to be unity.

The circuit equations for $\mathcal{R}$ read

$$\begin{aligned} u &= p\,i_1 + i_1 + i_2 \\ 0 &= p\,i_1 - p\,i_2 - i_2 \\ y &= p\,i_2 \end{aligned}$$

Upon rearranging these equations to put them into the form of *1*, we have

$$-\begin{bmatrix} 0 \\ 0 \\ 1 \end{bmatrix} y + \left[\begin{bmatrix} 1 & 0 \\ 1 & -1 \\ 0 & 1 \end{bmatrix} p + \begin{bmatrix} 1 & 1 \\ 0 & -1 \\ 0 & 0 \end{bmatrix}\right] \begin{bmatrix} i_1 \\ i_2 \end{bmatrix} = \begin{bmatrix} 1 \\ 0 \\ 0 \end{bmatrix} u$$

In this case,

$$\mathbf{L}(p) = -\begin{bmatrix} 0 \\ 0 \\ 1 \end{bmatrix} \qquad \mathbf{K}(p) = \begin{bmatrix} 1 & 0 \\ 1 & -1 \\ 0 & 1 \end{bmatrix} p + \begin{bmatrix} 1 & 1 \\ 0 & -1 \\ 0 & 0 \end{bmatrix} \qquad \mathbf{M}(p) = \begin{bmatrix} 1 \\ 0 \\ 0 \end{bmatrix}$$

3 *Example* Consider a system $\mathcal{R}$ characterized by an integrodifferential equation of the form

4 $$\left(p^2 + 2p + 3 + \frac{1}{p}\right) y = \left(p + \frac{2}{p}\right) u$$

where $1/p$ represents an integrator (*4.3.48*).

Let

5 $$\tilde{y}_1 = \frac{1}{p} y \qquad \tilde{y}_2 = \frac{1}{p} u$$

Then

$$p\,\tilde{y}_1 = y \qquad p\,\tilde{y}_2 = u$$

by the definition of an integrator. [This follows also from operating on both sides of *5* with p and making use of the fact that $p \cdot \frac{1}{p} = 1$ (see *4.4.48*).] Thus, *4* may be replaced by the system of equations

6 $$\begin{aligned}(p^2 + 2p + 3)y + \tilde{y}_1 &= p\,u + 2\tilde{y}_2 \\ p\,\tilde{y}_1 &= y \\ p\,\tilde{y}_2 &= u\end{aligned}$$

which is of the general form *1*, with

$$\mathbf{L}(p) = \begin{bmatrix}1\\0\\0\end{bmatrix} p^2 + \begin{bmatrix}2\\0\\0\end{bmatrix} p + \begin{bmatrix}3\\-1\\0\end{bmatrix}$$

$$\mathbf{K}(p) = \begin{bmatrix}0 & 0\\1 & 0\\0 & 1\end{bmatrix} p + \begin{bmatrix}1 & -2\\0 & 0\\0 & 0\end{bmatrix}$$

$$\mathbf{M}(p) = \begin{bmatrix}1\\0\\0\end{bmatrix} p + \begin{bmatrix}0\\0\\1\end{bmatrix}$$

In Sec. 7 we shall establish an important fact: that, given any general differential system $\mathcal{G}$, one can always find a system $\mathcal{Q}$ equivalent to $\mathcal{G}$ which is characterized by an input-output relation of the form

7 $$L(p)y = M(p)u$$

where

8 $$\begin{aligned}L(p) &= a_n p^n + \cdots + a_0 \qquad a_n \neq 0 \\ M(p) &= b_m p^m + \cdots + b_0 \qquad b_m \neq 0\end{aligned}$$

and the a's and b's are constant coefficients, not necessarily real. Thus, the class of systems defined by *1* is no more and no less general than that defined by *7*.

In view of this fact, we can focus our attention on differential systems characterized by *7*. This does not mean, however, that our approach to the problem of associating a state vector with a system $\mathcal{G}$ characterized by *1* will always involve—as a first step—the replacement of $\mathcal{G}$ with an equivalent system $\mathcal{A}$ characterized by *7*. We shall be able to associate a state vector with $\mathcal{G}$ in a direct fashion by the use of techniques developed in Secs. *6* to *8*.

State vectors and state equations

Let $\mathcal{A}$ denote a system characterized by an input-output relation of the form *7*. Our problem is to associate a state vector with $\mathcal{A}$ and find the corresponding state equations.

Before applying our standard approach (Sec. *2*), let us note that *7* may be replaced by two relations

9
$$\begin{aligned} L(p)y - v &= 0 \\ v &= M(p)u \end{aligned}$$

which is of the general form *1*, with v playing the role of a suppressed output variable. Consequently, $\mathcal{A}$ may be regarded as a tandem combination (Fig. *4.5.2*) of the differential operator system $M(p)$,

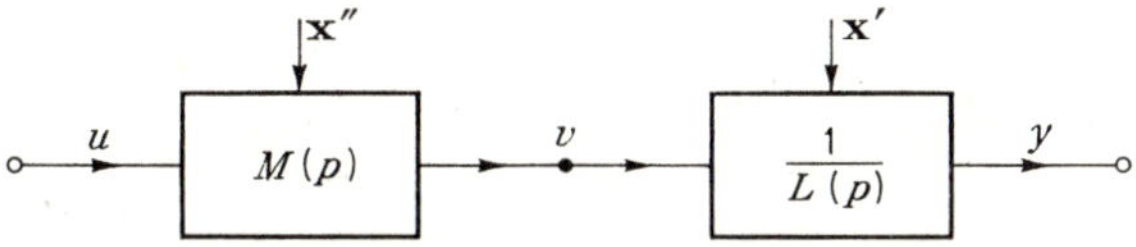

Fig. 4.5.2 A general system regarded as a tandem combination of $M(p)$ and $1/L(p)$.

followed by the reciprocal differential operator system $1/L(p)$ or, equivalently, as the product of $1/L(p)$ and $M(p)$. Accordingly, we can (and frequently shall) denote $\mathcal{A}$ by $\frac{1}{L(p)} \cdot M(p)$. [Note that this notation does not permit us, without additional justification, to treat $\frac{1}{L(p)} \cdot M(p)$ as if it were an ordinary ratio of two polynomials in p.]

We can make an immediate use of this observation to derive a way of associating a state vector with $\mathcal{A}$. Specifically, by the determinateness of linear systems (Theorem *3.9.12*), the vector

10
$$\mathbf{x} = (\mathbf{x}', \mathbf{x}'')$$

in which $\mathbf{x}'$ is the state vector for $1/L(p)$ and $\mathbf{x}''$ is the state vector for $M(p)$, qualifies as a state vector for $\mathcal{A}$. Thus, taking $\mathbf{x}'$ and $\mathbf{x}''$ to be the normal state vectors (*4.3.27* and *4.4.19*) of $1/L(p)$ and $M(p)$, respectively, we have

11 $$\mathbf{x}(t) = (y(t), \ldots, y^{(n-1)}(t); u(t), \ldots, u^{(m-1)}(t))$$

as the expression for the state of $\mathcal{A}$ at time t. Since $\mathbf{x}'$ ranges over $\mathcal{C}^n$ and $\mathbf{x}''$ ranges over $\mathcal{C}^m$, $\mathbf{x}$ ranges over $\mathcal{C}^{n+m}$. Thus, when the state vector of $\mathcal{A}$ is defined by *11*, the state space of $\mathcal{A}$ is $\mathcal{C}^{n+m}$, the space of all $(m + n)$-tuples of complex numbers.

The corresponding state equations of $\mathcal{A}$ can readily be set up by adjoining the state equations of $1/L(p)$ to those of $M(p)$. Specifically, we have for $1/L(p)$ (see *4.3.34*)

12 $$\begin{aligned} \dot{\mathbf{x}}' &= \mathbf{A}\mathbf{x}' + \mathbf{b}v \\ y &= \langle \mathbf{c},\mathbf{x}' \rangle \end{aligned}$$

where

$$\mathbf{A} = \begin{bmatrix} 0 & 1 & 0 & \cdots & 0 \\ 0 & 0 & 1 & \cdots & 0 \\ \cdot & \cdot & \cdot & \cdots & \cdot \\ -\frac{a_0}{a_n} & \cdot & \cdot & \cdots & -\frac{a_{n-1}}{a_n} \end{bmatrix} \qquad \mathbf{b} = \begin{bmatrix} 0 \\ 0 \\ \cdots \\ \frac{1}{a_n} \end{bmatrix} \qquad \mathbf{c} = [1 \quad 0 \quad \cdots \quad 0]$$

and for $M(p)$ (see *4.4.17*)

13 $$\begin{aligned} x_1'' &= u \\ \dot{x}_1'' &= x_2'' \\ &\cdots\cdots \\ \dot{x}_{m-1}'' &= x_m'' \\ v &= b_0 x_1'' + \cdots + b_{m-1} x_m'' + b_m \dot{x}_m'' \end{aligned}$$

On eliminating v from these equations, we obtain

14 $$\begin{aligned} \dot{\mathbf{x}}' &= \mathbf{A}\mathbf{x}' + \mathbf{b}(b_0 x_1'' + \cdots + b_{m-1} x_m'' + b_m \dot{x}_m'') \\ x_1'' &= u \\ \dot{x}_1'' &= x_2'' \\ &\cdots\cdots \\ \dot{x}_{m-1}'' &= x_m'' \\ y &= \langle \mathbf{c},\mathbf{x}' \rangle \end{aligned}$$

which are the desired state equations of $\mathcal{A}$.

A disadvantage of this way of associating a state vector with $\mathcal{A}$ is that it leads to a system which is not in reduced form (*3.6.46*). This manifests itself in the dimension of the state space $\mathcal{C}^{n+m}$ being equal to $n + m$ rather than max (m,n), which, as we shall see presently, is the order (*3.6.39*) of $\mathcal{A}$.

We can obtain a reduced form of $\mathcal{A}$ by employing the approach used in Secs. *3* and *4*, which starts with the derivation of the general solution of *7*. This is what we shall do in the sequel.

General solution

It will be convenient to consider first the case where $\mathcal{A}$ is a proper system in the sense of Definition *3.6.27*. This is equivalent to assuming that $\mathcal{A}$ is characterized by the input-output relation

15 $$L(p)y = M(p)u$$

where $L(p) = a_n p^n + \cdots + a_0$, $M(p) = b_n p^n + \cdots + b_0$, $a_n \neq 0$.

On applying the Laplace transformation to both sides of *15* we obtain after rearrangement of terms the following expression for the Laplace transform of the general solution (for $t_0 = 0$):

16 $$\begin{aligned} Y(s) = \frac{1}{L(s)} \{ & s^{n-1}[a_n y(0-) - b_n u(0-)] \\ & + s^{n-2}[a_n y^{(1)}(0-) - b_n u^{(1)}(0-) + a_{n-1} y(0-) - b_{n-1} u(0-)] \\ & + \cdots \cdots \\ & + [a_n y^{(n-1)}(0-) - b_n u^{(n-1)}(0-) \\ & \qquad + \cdots + a_1 y(0-) - b_1 u(0-)]\} \\ & + H(s)U(s) \end{aligned}$$

where

17 $$H(s) = \frac{M(s)}{L(s)} = \text{transfer function of } \mathcal{A} \text{ (see } 3.6.17)$$

Expression for the state vector

At this stage we are ready to associate a state vector $\mathbf{x}$ with $\mathcal{A}$ by identifying the components of $\mathbf{x}(0-)$ with the constants appearing in the general solution *16*. One way of doing this is to equate $x_1(0-)$ to the coefficient of s^{n-1}, $x_2(0-)$ to the coefficient of s^{n-2}, etc. In this way, we arrive at the following expression for $\mathbf{x}(t)$ in terms of $u(t)$, $y(t)$, and their derivatives. [Alternative expressions for $\mathbf{x}(t)$ will be given in Sec. *9*.]

18 $$\begin{aligned} x_1 &= a_n y - b_n u \\ x_2 &= a_n y^{(1)} - b_n u^{(1)} + a_{n-1} y - b_{n-1} u \\ &\cdots\cdots \\ x_n &= a_n y^{(n-1)} - b_n u^{(n-1)} + \cdots + a_1 y - b_1 u \end{aligned}$$

State equations

To show that $\mathbf{x}$ defined in this manner qualifies as a state vector for $\mathcal{A}$, it is sufficient to show (by Conclusion *2.3.42*) that $\dot{\mathbf{x}}$, $\mathbf{x}$, u, and y satisfy state equations of the canonical form

19
$$\dot{\mathbf{x}} = \mathbf{A}\mathbf{x} + \mathbf{B}\mathbf{u}$$
$$\mathbf{y} = \mathbf{C}\mathbf{x} + \mathbf{D}\mathbf{u}$$

This is easy to demonstrate. Specifically, on forming the expression for $\dot{\mathbf{x}}$ and comparing it with $\mathbf{x}$, we arrive at once at the following relations:

20
$$\begin{aligned} x_1 &= a_n y - b_n u \\ x_2 &= \dot{x}_1 + a_{n-1} y - b_{n-1} u \\ x_3 &= \dot{x}_2 + a_{n-2} y - b_{n-2} u \\ &\cdots\cdots\cdots\cdots \\ x_n &= \dot{x}_{n-1} + a_1 y - b_1 u \end{aligned}$$

In addition, the differential equation *15* yields

$$\dot{x}_n = -(a_0 y - b_0 u)$$

Solving these equations for $\dot{\mathbf{x}}$ and y in terms of $\mathbf{x}$ and u, we have

21
$$\begin{aligned} \dot{x}_1 &= \frac{1}{a_n}(-a_{n-1}x_1 + a_n x_2 + (a_n b_{n-1} - a_{n-1} b_n)u) \\ \dot{x}_2 &= \frac{1}{a_n}(-a_{n-2}x_1 + a_n x_3 + (a_n b_{n-2} - a_{n-2} b_n)u) \\ &\cdots\cdots\cdots\cdots\cdots\cdots \\ \dot{x}_{n-1} &= \frac{1}{a_n}(-a_1 x_1 + a_n x_n + (a_n b_1 - a_1 b_n)u) \\ \dot{x}_n &= \frac{1}{a_n}(-a_0 x_1 + (a_n b_0 - a_0 b_n)u) \end{aligned}$$

and
$$y = \frac{1}{a_n}(x_1 + b_n u)$$

These are the desired state equations of $\mathcal{A}$. We note that they have the canonical form *19*, with

22
$$\mathbf{A} = \frac{1}{a_n}\begin{bmatrix} -a_{n-1} & a_n & 0 & \cdots & 0 \\ -a_{n-2} & 0 & a_n & \cdots & 0 \\ \cdots & \cdots & \cdots & \cdots & \cdots \\ -a_1 & 0 & 0 & \cdots & a_n \\ -a_0 & 0 & 0 & \cdots & 0 \end{bmatrix} \qquad \mathbf{B} = \frac{1}{a_n}\begin{bmatrix} a_n b_{n-1} - a_{n-1} b_n \\ a_n b_{n-2} - a_{n-2} b_n \\ \cdots\cdots \\ \cdots\cdots \\ a_n b_0 - a_0 b_n \end{bmatrix}$$

$$\mathbf{C} = \begin{bmatrix} \frac{1}{a_n} & 0 & \cdots & 0 \end{bmatrix} \qquad \mathbf{D} = \begin{bmatrix} \frac{b_n}{a_n} \end{bmatrix}$$

23 *Comment* The expression for $\mathbf{x}$ and the state equations derived above correspond to the choice of basis functions defined by

$$\phi_\lambda(t) = \mathcal{L}^{-1}\left\{\frac{s^{n-\lambda}}{L(s)}\right\} \qquad \lambda = 1, \ldots, n$$

This choice is different from that which leads to the normal state vector in the case of systems of the reciprocal differential operator type, namely (see *4.3.30*)

$$\phi_\lambda(t) = \mathcal{L}^{-1}\left\{\frac{a_n s^{n-\lambda} + \cdots + a_\lambda}{L(s)}\right\} \qquad \lambda = 1, \ldots, n$$

With the latter choice, the **A** matrix in the state equations *19* is expressed by *4.3.32*; that is, it is identical with the **A** matrix of the reciprocal differential operator system $1/L(p)$. The corresponding realization of $\mathcal{A}$ is shown in Fig. *4.9.7*. We leave it as an exercise for the reader to set up the expression for the state vector and determine the state equations for this case (see *4.9.15 et seq.*).

Input-output-state relation

To obtain the expression for the input-output-state relation for $\mathcal{A}$, we invert $Y(s)$ as given by *16* and replace $0-$ by t_0-. The result reads

24 $$y(t) = \langle \boldsymbol{\phi}(t - t_0), \mathbf{x}(t_0-) \rangle + \int_{t_0}^{t} h(t - \xi) u(\xi)\, d\xi$$

where

25 $$\boldsymbol{\phi}(t) = (\phi_1(t), \ldots, \phi_n(t)) = \text{basis function vector}$$

$$\phi_\lambda(t) = \mathcal{L}^{-1}\left\{\frac{s^{n-\lambda}}{L(s)}\right\} \qquad \lambda = 1, \ldots, n$$

$$= \text{zero-input responses of } \mathcal{A}$$

$$h(t) = \mathcal{L}^{-1}\{H(s)\} = \text{impulse response of } \mathcal{A}$$

$$H(s) = \frac{M(s)}{L(s)} = \text{transfer function of } \mathcal{A}$$

$$\mathbf{x}(t) = (x_1(t), \ldots, x_n(t)) = \text{state vector of } \mathcal{A}$$

$$x_\lambda(t) = a_n y^{(\lambda-1)}(t) - b_n u^{(\lambda-1)}(t) + \cdots + a_{n-\lambda+1} y(t) - b_{n-\lambda+1} u(t) \qquad \lambda = 1, \ldots, n$$

From inspection of these relations we can draw the following conclusions.

26 **Conclusion** The zero-input responses $\phi_1(t), \ldots, \phi_n(t)$ are linearly independent and hence constitute a set of basis functions for $\mathcal{A}$. This follows at once from the linear independence of the Laplace transforms of $\phi_1(t), \ldots, \phi_n(t)$, which are, respectively, $\frac{s^{n-1}}{L(s)}, \ldots, \frac{1}{L(s)}$.

27 **Conclusion** $\mathcal{A}$ is a system of order n. This is a consequence of Conclusion *26* and Definition *3.6.39*.

28 **Conclusion** $\mathcal{A}$ is in reduced form; i.e., there are no distinct states in the state space of $\mathcal{A}$ which are equivalent to one another. This is a consequence of Conclusion *26* and Assertion *3.6.46*.

29 **Conclusion** $\mathcal{A}$ is a linear system in the sense of Definition *3.5.6* and a time-invariant system in the sense of Definition *3.2.36*. (The latter follows from *3.6.35*.)

30 **Conclusion** $\mathcal{A}$ is a proper system in the sense of Definition *3.6.27*. This follows from the fact that $h(t)$ is the inverse Laplace transform of a rational function $M(s)/L(s)$ in which the degree of the numerator does not exceed the degree of the denominator.

It will be helpful to summarize the main results of the foregoing analysis in the form of an

31 **Assertion** Let $\mathcal{A}$ be a proper differential system of order n which is characterized by an input-output relation of the form

$$L(p)y = M(p)u$$

$$L(p) = a_n p^n + \cdots + a_0 \qquad M(p) = b_n p^n + \cdots + b_n \qquad a_n \neq 0$$

Then the vector $\mathbf{x} = (x_1, \ldots, x_n)$ ranging over $\mathcal{C}^n$ and defined by

32 $$x_\lambda = a_n y^{(\lambda-1)} - b_n u^{(\lambda-1)} + \cdots + a_{n-\lambda+1} y - b_{n-\lambda+1} u \qquad \lambda = 1, \ldots, n$$

qualifies as a state vector for $\mathcal{A}$. Correspondingly, the input-output-state relation of $\mathcal{A}$ is given by *24* and the state equations of $\mathcal{A}$ are of the canonical form

$$\dot{\mathbf{x}} = \mathbf{A}\mathbf{x} + \mathbf{B}u$$

$$y = \mathbf{C}\mathbf{x} + \mathbf{D}u$$

where **A**, **B**, **C**, and **D** are given by *22*. ◁

As we shall see in Sec. *9*, the results derived above through the determination of the general solution of $\mathcal{A}$ can be obtained much more expeditiously by the use of the so-called realization technique. For this reason, we defer until Sec. *9* the discussion of alternative choices for the state vector **x** and the state equations to which they lead.

The case of improper $\mathcal{A}$

We turn next to the more general case where $\mathcal{A}$ is not necessarily a proper system. Specifically, let $\mathcal{A}$ be characterized by

33 $$L(p)y = M(p)u$$

$$L(p) = a_n p^n + \cdots + a_n \qquad M(p) = b_m p^m + \cdots + b_0$$

where m may be greater than n. For concreteness, we assume that $m > n$.

We can treat this case by using the standard approach employed in previous cases, i.e., by finding the general solution of *33* and identifying the constants appearing in this solution with the components of the initial state vector $\mathbf{x}(0-)$. It will be much simpler, however, to proceed indirectly, reducing the case in question to the case of a proper system through the use of a theorem which is proved below.

34 **Theorem** Let $\mathcal{A}$ be an improper system characterized by *33*, with $m > n$. By using long division, express $M(p)$ in the form

35 $$M(p) \equiv (d_k p^k + \cdots + d_1 p)L(p) + \hat{M}(p)$$

where $k = m - n$, $\hat{M}(p)$ is a polynomial of degree n, and the d's are constants. Then

$$\mathcal{A} \equiv \hat{\mathfrak{D}} + \hat{\mathcal{A}}$$

(which, in words, means that $\mathcal{A}$ is equivalent to a parallel combination of $\hat{\mathfrak{D}}$ and $\hat{\mathcal{A}}$), where $\hat{\mathfrak{D}}$ is a system of the differential operator type characterized by the input-output relation

$$w = (d_k p^k + \cdots + d_1 p)u \qquad w \triangleq \text{output of } \hat{\mathfrak{D}}$$

and $\hat{\mathcal{A}}$ is a proper system characterized by the input-output relation

$$L(p)v = \hat{M}(p)u \qquad v \triangleq \text{output of } \hat{\mathcal{A}}$$

The statement of this theorem is illustrated in Fig. *4.5.3*.

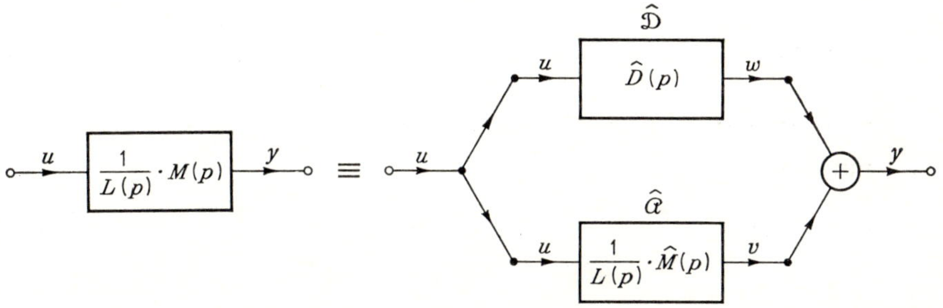

Fig. 4.5.3 Decomposition of an improper system into a proper system and a system of the differential operator type.

Proof It was shown in Chap. *3* (*3.5.10* and *3.6.49*) that two systems $\mathcal{A}$ and $\mathcal{B}$ are equivalent if they have the same basis functions and are zero-state equivalent. We shall show that this is true of $\mathcal{A}$ and $\hat{\mathfrak{D}} + \hat{\mathcal{A}}$.

Specifically, the transfer function of $\mathcal{A}$ is

36 $$H_{\mathcal{A}}(s) = \frac{M(s)}{L(s)}$$

On the other hand, the transfer function of $\hat{\mathfrak{D}} + \hat{\mathcal{A}}$ is (by *3.6.25*) the sum of transfer functions of $\hat{\mathfrak{D}}$ and $\hat{\alpha}$, that is,

$$\begin{aligned} H_{\hat{\mathfrak{D}}+\hat{\alpha}}(s) &= \frac{\hat{M}(s)}{L(s)} + d_k s^k + \cdots + d_1 s \\ &= \frac{\hat{M}(s) + L(s)(d_k s^k + \cdots + d_1 s)}{L(s)} \\ &= \frac{M(s)}{L(s)} \\ &= H_{\alpha}(s) \end{aligned}$$

by *35* and *36*. Thus, α and $\hat{\mathfrak{D}} + \hat{\alpha}$ have identical transfer functions and hence are zero-state equivalent (*3.6.11*).

Turning to the basis functions of α and $\hat{\mathfrak{D}} + \hat{\alpha}$, we recall from our earlier analyses of proper systems and systems of the differential operator type that the basis functions in question are given (in the Laplace transform form) by

$\hat{\alpha}$: $\quad \frac{s^{n-1}}{L(s)}, \frac{s^{n-2}}{L(s)}, \ldots, \frac{1}{L(s)} \quad$ (see *23*)

and

$\hat{\mathfrak{D}}$: $\quad 1, s, \ldots, s^{k-1} \quad$ (see *4.4.17*[1])

Hence the basis functions of $\hat{\mathfrak{D}} + \hat{\alpha}$ are

$\hat{\mathfrak{D}} + \hat{\alpha}$: $\quad \frac{s^{n-1}}{L(s)}, \ldots, \frac{s}{L(s)}, \frac{1}{L(s)}, 1, s, \ldots, s^{k-1}$

Note that these basis functions are linearly independent and are m in number, since $k = m - n$.

To find the basis functions of α, we apply the Laplace transformation to both sides of *33*, with u set equal to zero. The result, which is the expression for the Laplace transform of the zero-input response of α, is of the form

37 $$Y(s) = \frac{c_{m-1}s^{m-1} + \cdots + c_0}{L(s)}$$

where the c's are constants involving the initial values of u, y, and their derivatives. Writing *37* as

$$Y(s) = \beta_k s^{k-1} + \cdots + \beta_1 s + \beta_0 + \frac{\gamma_0}{L(s)} + \gamma_1 \frac{s}{L(s)} + \cdots + \gamma_{n-1} \frac{s^{n-1}}{L(s)}$$

where the β's and γ's are constants, shows that the Laplace transforms

[1] Note that the basis functions in *4.4.17* are linear combinations of $1, s, \ldots, s^{m-1}$.

of the basis functions of $\mathcal{A}$ are

38 $$\frac{s^{n-1}}{L(s)}, \ldots, \frac{s}{L(s)}, \frac{1}{L(s)}, 1, s, \ldots, s^{k-1}$$

and hence are identical with those of $\hat{\mathfrak{D}} + \hat{\mathcal{A}}$. This concludes the proof of Theorem *34*.

39 **Remark** In the course of proving Theorem *34* we have established that $\mathcal{A}$ has m basis functions if $m > n$. Previously we have shown (see *4.5.27*) that if $m \leq n$, then $\mathcal{A}$ has n basis functions. Consequently, the number of basis functions of $\mathcal{A}$ is given by max (m,n), regardless of which of these conditions holds. Thus, we can assert that the order (in the sense of Definition *3.6.39*) of a linear differential system characterized by an input-output relation of the form *33* is max (m,n). Note that this agrees with the definition of the order of a differential equation (*3.8.22*). ◁

Having established Theorem *34*, we can make an immediate application of it to the problem of associating a state vector and setting up the state equations of an improper system $\mathcal{A}$ characterized by an input-output relation of the form

40 $$L(p)y = M(p)u$$
$$L(p) = a_n p^n + \cdots + a_0 \qquad M(p) = b_m p^m + \cdots + b_0$$

Specifically, on replacing $\mathcal{A}$ by $\hat{\mathfrak{D}} + \hat{\mathcal{A}}$ in the manner illustrated in Fig. *4.5.3* and making use of the determinateness of linear systems (*3.9.12*), we reduce the problem of associating a state vector $\mathbf{x}$ with $\mathcal{A}$ to that of associating a state vector $\mathbf{x}'$ with $\hat{\mathcal{A}}$ and a state vector $\mathbf{x}''$ with $\hat{\mathfrak{D}}$. Then $\mathbf{x}$ is expressed as the composite vector

$$\mathbf{x} = (\mathbf{x}',\mathbf{x}'')$$

Since we know how to associate a state vector with a proper system such as $\hat{\mathcal{A}}$ as well as with a system of the differential operator type such as $\hat{\mathfrak{D}}$, the problem of associating a state vector with $\mathcal{A}$ may be regarded as solved. It is, however, actually more convenient not to associate a state vector with $\hat{\mathfrak{D}}$ and to use instead a hybrid type of representation for the output of $\mathcal{A}$ which is described below.

Hybrid state equations

Essentially, we do this: Let v denote the output of $\hat{\mathcal{A}}$ in Fig. *4.5.3* and let $\hat{D}(p)$ denote the differential operator $d_k p^k + \cdots + d_1 p$ which characterizes $\hat{\mathfrak{D}}$; that is,

41 $$\hat{D}(p) = d_k p^k + \cdots + d_1 p$$

Then the output of $\mathcal{A}$ can be written as

42 $$y = \hat{D}(p)u + v$$

Now $\hat{\mathcal{A}}$ is characterized by the input-output relation

43 $$L(p)v = \hat{M}(p)u$$

where

$$L(p) = a_n p^n + \cdots + a_0 \qquad \hat{M}(p) = \hat{b}_n p^n + \cdots + \hat{b}_0$$

with the $\hat{b}$'s determined by *35*. Let $\mathbf{x}$ denote the state vector of $\hat{\mathcal{A}}$ (rather than $\mathcal{A}$), defined in terms of v, u, and their derivatives via *32*, with b's in *32* replaced by $\hat{b}$'s. In terms of $\mathbf{x}$, the state equations of $\hat{\mathcal{A}}$ can be written as

44 $$\dot{\mathbf{x}} = \mathbf{A}\mathbf{x} + \mathbf{b}u$$
$$v = \langle \mathbf{c}, \mathbf{x} \rangle + d_0 u$$

where $\mathbf{A}$, $\mathbf{b}$, $\mathbf{c}$, and d_0 are given by *22*, with $d_0 = \hat{b}_n/a_n$.

On substituting v from *44* into *42*, we have

$$y = \hat{D}(p)u + \langle \mathbf{c}, \mathbf{x} \rangle + d_0 u$$

and on adjoining this relation to *44*, we obtain the equations

45 $$\dot{\mathbf{x}} = \mathbf{A}\mathbf{x} + \mathbf{b}u$$

46 $$y = \langle \mathbf{c}, \mathbf{x} \rangle + d_0 u + \hat{D}(p)u$$

which, strictly speaking, are not the state equations of $\mathcal{A}$ since $\mathbf{x}$ is the state vector of $\hat{\mathcal{A}}$ rather than $\mathcal{A}$. Nevertheless, since *45* and *46* serve the same purpose as the usual state equations, we shall refer to *45* and *46* as the *hybrid state equations* of $\mathcal{A}$, with the adjective "hybrid" indicating that $\mathbf{x}$ is the state vector of a part of $\mathcal{A}$, namely, $\hat{\mathcal{A}}$, and that *45* and *46* are in effect a mixture of state equations for $\hat{\mathcal{A}}$ and the input-output relation ($w = \hat{D}(p)u$) for $\hat{\mathfrak{D}}$.

When no confusion is likely to result, we shall drop the adjective "hybrid" and refer to *45* and *46* simply as the *state equations* of $\mathcal{A}$.

47 **Remark** Note that *46* can be expressed in a somewhat neater form by absorbing the term $d_0 u$ in $\hat{D}(p)$. (This is equivalent to transferring the scalar d_0 from $\hat{\mathcal{A}}$ to $\hat{\mathfrak{D}}$.) Thus, if we define

48 $$D(p) = d_k p^k + \cdots + d_1 p + d_0$$
$$= \hat{D}(p) + d_0$$

then the hybrid state equations become

49 $$\dot{\mathbf{x}} = \mathbf{A}\mathbf{x} + \mathbf{b}u$$
$$y = \langle \mathbf{c}, \mathbf{x} \rangle + D(p)u$$

This is the form in which the state equations of an improper system will usually be written in the sequel. ◁

It will be helpful at this point to summarize our conclusions in the form of an

50 **Assertion** Let $\mathcal{A}$ be an improper differential system characterized by the input-output relation

$$L(p)y = M(p)u$$

$$L(p) = a_n p^n + \cdots + a_0 \quad M(p) = b_m p^m + \cdots + b_0 \quad m - n \triangleq k > 0$$

By using long division write $M(p)$ in the form

$$M(p) = D(p)L(p) + \hat{M}(p)$$

where $D(p)$ is a polynomial in p of degree k and $\hat{M}(p)$ is a polynomial of degree $n - 1$. Let $\mathcal{D}$ be a system of the differential operator type characterized by the input-output relation

$$w = D(p)u$$

and let $\hat{\mathcal{A}}$ be the proper system characterized by the input-output relation

$$L(p)v = \hat{M}(p)u$$

Then $\mathcal{A}$ is equivalent to the sum (i.e., parallel combination) of $\mathcal{D}$ and $\hat{\mathcal{A}}$ (Fig. *4.5.3*) and is a system of order m.

Let $\mathbf{x}$ be a state vector associated with $\hat{\mathcal{A}}$ (for example, via *32*) and let the corresponding canonical state equations of $\hat{\mathcal{A}}$ be

$$\dot{\mathbf{x}} = \mathbf{A}\mathbf{x} + \mathbf{b}u$$
$$y = \langle \mathbf{c},\mathbf{x} \rangle$$

where, if $\mathbf{x}$ is defined by *32*, $\mathbf{A}$, $\mathbf{b}$, and $\mathbf{c}$ are given by *22*. Then the relations between the input to $\mathcal{A}$, the output of $\mathcal{A}$, and the state vector of $\hat{\mathcal{A}}$ are given by

51
$$\dot{\mathbf{x}} = \mathbf{A}\mathbf{x} + \mathbf{b}u$$
$$y = \langle \mathbf{c},\mathbf{x} \rangle + D(p)u$$

where $D(p)$ is a differential operator of the form

52
$$D(p) = d_k p^k + \cdots + d_1 p + d_0 \qquad k = m - n$$

These relations constitute the hybrid state equations or simply state equations of $\mathcal{A}$.

53 *Comment* Note that in Assertion *50* we state that $\hat{M}(p)$ is a polynomial of degree $n - 1$, whereas in Theorem *34* $\hat{M}(p)$ is stated to be of degree n. This difference is accounted for by our transferring the scalar d_0 from

$\hat{\mathfrak{A}}$ to $\hat{\mathfrak{D}}$ (see Remark *47*). This transfer does not affect the expressions for $\mathbf{A}$, $\mathbf{b}$, and the coefficients in $\hat{D}(p)$ (apart from d_0).

54 *Example* Consider an improper system $\mathfrak{A}$ which is characterized by the input-output relation

55 $$(p^2 + 3p + 2)y = (p^3 + 4p^2 + 6p + 5)u$$

In this case, we have

$$p^3 + 4p^2 + 6p + 5 = (p^2 + 3p + 2)(p + 1) + p + 3$$

and hence $\mathfrak{A}$ is equivalent to the parallel combination of $\hat{\mathfrak{A}}$ and $\hat{\mathfrak{D}}$ which are characterized by the input-output relations

$$\hat{\mathfrak{A}}: \qquad (p^2 + 3p + 2)v = (p + 3)u$$
$$\hat{\mathfrak{D}}: \qquad w = (p + 1)u$$

Using *32*, the expression for the state vector of $\hat{\mathfrak{A}}$ in terms of u, v, and their derivatives is found to be

56 $$\mathbf{x} = (v, \dot{v} + 3v - u)$$

and the corresponding state equations are given by (see *21*)

$$\begin{aligned} \dot{x}_1 &= -3x_1 + x_2 + u \\ \dot{x}_2 &= -2x_1 + 3u \\ v &= x_1 \end{aligned}$$

In terms of these equations, the hybrid state equations of $\mathfrak{A}$ read

57 $$\begin{aligned} \dot{x}_1 &= -3x_1 + x_2 + u \\ \dot{x}_2 &= -2x_1 + 3u \\ y &= x_1 + (p + 1)u \end{aligned}$$

58 **Remark** Some (but not all) of the expressions for the state vector and the state equations derived above carry over with minor modifications to the case where $\mathfrak{A}$ is a time-varying system characterized by an input-output relation of the form

$$L(p,t)y = M(p,t)u$$

where

59 $$\begin{aligned} L(p,t) &= a_n(t)\, p^n + \cdots + a_0(t) \\ M(p,t) &= b_n(t)\, p^n + \cdots + b_0(t) \end{aligned}$$

with $a_n(t) \neq 0$ for all t.

The time-varying case will be treated in detail in Chap. *6*. Here we shall merely sketch a direct technique for associating a state vector with $\mathfrak{A}$ which does not require the determination of the general solution, as does the approach used in the foregoing analysis.

The basic idea of the method is this. By replacing p and its powers in the differential equation by the expressions given by *3.8.3*, we place in evidence the initial conditions needed to determine the response. Then we group these in a suitable manner to form the components of an initial state vector $\mathbf{x}(t_0-)$ and thereafter use the usual procedure to determine the corresponding state equations. To illustrate, suppose that $n = 2$ and that the a's and b's are continuous time functions. Then, on using *3.8.3* and transferring the terms involving the initial values to the right-hand side, we have

60 $$L(p,t)1(t - t_0)y(t) = M(p,t)1(t - t_0)u(t) + \delta^{(1)}(t - t_0)[a_2(t)y(t_0-) - b_2(t)u(t_0-)] + \delta(t - t_0)[a_2(t)\dot{y}(t_0-) - b_2(t)\dot{u}(t_0-) + a_1(t)y(t_0-) - b_1(t)u(t_0-)]$$

Now by *A.5.17*

61 $$\delta^{(1)}(t - t_0)a_2(t) = \delta^{(1)}(t - t_0)a_2(t_0) - \delta(t - t_0)\dot{a}_2(t_0)$$

and likewise for $b_2(t)$. Consequently, *60* can be rewritten as

62 $$L(p,t)1(t - t_0)y(t) = M(p,t)1(t - t_0)u(t) + \delta^{(1)}(t - t_0)[a_2(t_0)y(t_0-) - b_2(t_0)u(t_0-)] + \delta(t - t_0)[a_2(t_0)\dot{y}(t_0-) - \dot{a}_2(t_0)y(t_0-) + a_1(t_0)y(t_0-) - b_2(t_0)\dot{u}(t_0-) + \dot{b}_2(t_0)u(t_0-) - b_1(t_0)u(t_0-)]$$

which suggests that $\mathbf{x}(t)$ be defined by

63 $$\begin{aligned} x_1(t) &= a_2(t)y(t) - b_2(t)u(t) \\ x_2(t) &= a_2(t)\dot{y}(t) - \dot{a}_2(t)y(t) + a_1(t)y(t) - b_2(t)\dot{u}(t) + \dot{b}_2(t)u(t) - b_1(t)u(t) \end{aligned}$$

Because of the presence of the derivatives of the coefficients, the expressions for the components of $\mathbf{x}(t)$ rapidly gain in complexity with increase in n. However, as will be shown in Chap. *6* (*6.4.12*), it is possible to define the state vector in such a way that the corresponding state equations have the same form as they have in the time-invariant case (see *4.9.15*).

64 *Note* If $\mathfrak{A}$ is improper, then, as in the time-invariant case, $\mathfrak{A}$ can be represented as a parallel combination of a proper system and a system of the differential operator type $\hat{\mathfrak{D}}$. However, $\hat{M}(p)$ and $\hat{D}(p)$ cannot be found by the use of long division—as they can be in the time-invariant case.

This concludes for the present our discussion of systems characterized by a single differential equation. We turn next to the analysis of systems represented by an interconnection of adders, scalors, integrators, and differentiators.

6 *State vectors and state equations for an interconnection of adders, scalors, integrators, and differentiators*

Systems having the form of an interconnection of adders, scalors, integrators, and differentiators are of interest to us for two related reasons. First, as we shall see in Sec. *9*, given any linear system $\mathcal{A}$ characterized by a single differential equation $L(p)y = M(p)u$ or by its state equations, one can always construct an interconnection of adders, scalors, integrators, and differentiators, call it $\mathcal{B}$, such that $\mathcal{B}$ and $\mathcal{A}$ are equivalent systems. Second, on the basis of this equivalence, we can develop an effective and intuitively appealing technique for associating a state vector with $\mathcal{A}$ which consists essentially in realizing $\mathcal{A}$ in the form of an interconnection of adders, scalors, integrators, and differentiators and then associating a state vector with and finding the state equations for the latter system. This technique will be described in detail in Sec. *9*.

Associating a state vector

Consider an interconnection of scalors, adders, integrators, and differentiators, an example of which is shown in Fig. *4.6.1*.

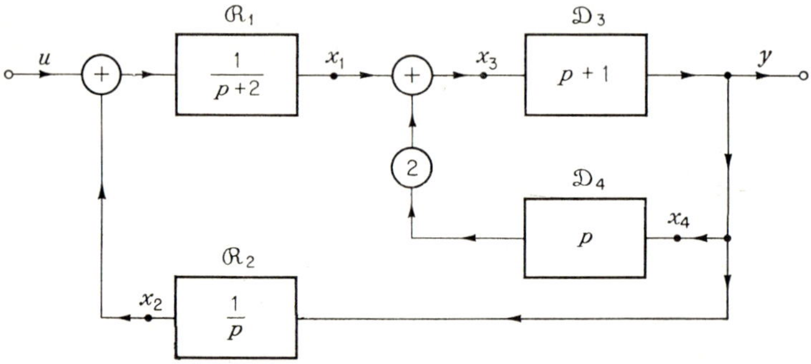

Fig. 4.6.1 A system comprising integrators, differentiators, adders, and scalors.

By Theorem *3.9.12*, such a system, call it $\mathcal{B}$, is determinate. Consequently, its state $\mathbf{x}$ can be identified with a vector $\mathbf{x} = (x_1, \ldots, x_n)$ whose components are the state vectors of those elements of $\mathcal{B}$ which are not memoryless, i.e., the integrators and differentiators. Now by Assertions *4.3.51* and *4.4.39*, the state of an integrator at time t is its output or, more generally, a constant times its output at t, while the state of a differentiator at time t is its input or, more generally, a constant times its input at t. From this it follows that a vector $\mathbf{x}$ whose components are the outputs of the integrators and the inputs to the differentiators qualifies as a state vector for $\mathcal{B}$.

1 **Remark** If $\mathcal{B}$ contains integrators and differentiators connected in tandem as illustrated in Fig. *4.6.2*, then, in order to reduce the dimension of $\mathbf{x}$, it is expedient to let the output of the integrator serve in the dual role of the state of both the integrator and the differentiator. For example, in Fig. *4.6.2* the component x_1 of $\mathbf{x}$ is the state of the integrator as well as that of the differentiator. ◁

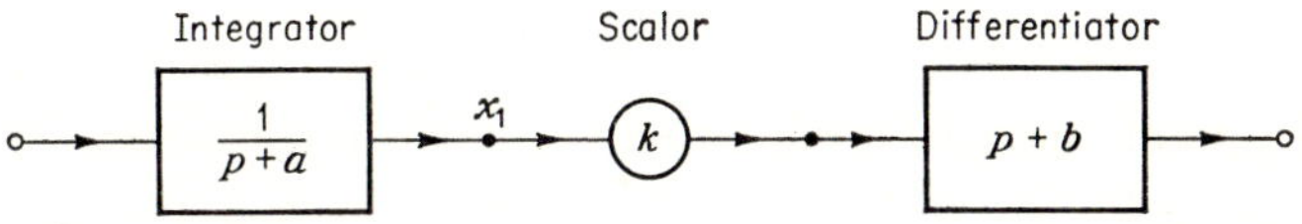

Fig. 4.6.2 Reduction of dimension of $\mathbf{x}$ by choosing x_1 to be the state of both $1/(p+a)$ and $p+b$.

The above observations lead us to the following simple rule for associating a state vector with $\mathcal{B}$.

2 **Rule** Let $\mathcal{B}$ be a system comprising scalors, adders, integrators, and differentiators. To associate a state vector $\mathbf{x} = (x_1, \ldots, x_n)$ with $\mathcal{B}$, assign a component of $\mathbf{x}$ to (I) the output of each integrator and (II) to the input of each differentiator which is not connected to the output of an integrator through a scalor.

3 *Note* For convenience in numbering, we assign the first l components of $\mathbf{x}$ ($l \triangleq$ number of integrators in $\mathcal{B}$) to the outputs of the integrators in $\mathcal{B}$ and the remaining $n - l$ components $x_{l+1}, \ldots, x_n$ to the inputs of those differentiators which are not connected to an integrator in the manner illustrated in Fig. *4.6.2*.

4 *Example* By applying this rule to the system of Fig. *4.6.1* we obtain $\mathbf{x} = (x_1,x_2,x_3,x_4)$, where $x_1 \triangleq$ output of integrator $\mathcal{R}_1$, $x_2 \triangleq$ output of $\mathcal{R}_2$, $x_3 \triangleq$ input to differentiator $\mathfrak{D}_3$, $x_4 \triangleq$ input to $\mathfrak{D}_4$.

5 *Example* For the system of Fig. *4.6.3* (which differs from that of Fig. *4.6.1* in the interchange of $\mathcal{R}_2$ and $\mathfrak{D}_4$), we have $\mathbf{x} = (x_1,x_2,x_3,x_4)$, with the x_i, $i = 1, \ldots, n$, having the same significance as in Example *4*.

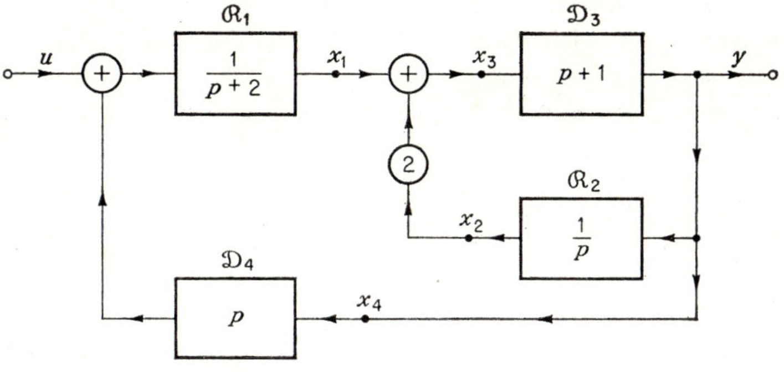

Fig. 4.6.3 System of Example *4.6.5*.

6 *Example* For the system of Fig. *4.6.4*, we have $\mathbf{x} = (x_1,x_2,x_3)$.

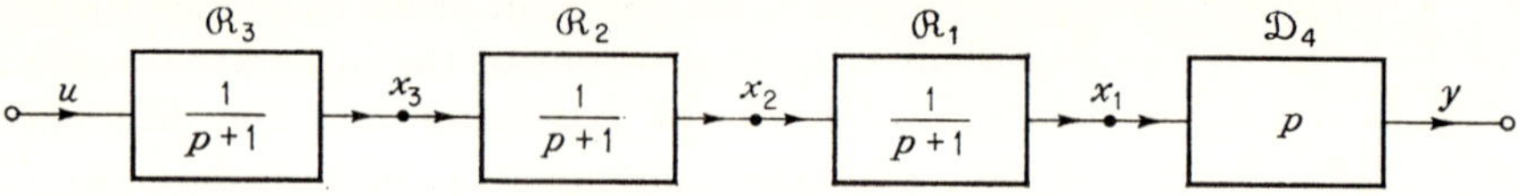

Fig. 4.6.4 System of Example *4.6.6*.

State equations

As a preliminary to formulating a rule for setting up the state equations for $\mathcal{B}$ we make the following obvious

7 *Observation* Let $\mathcal{R}_i$ be an integrator in $\mathcal{B}$ characterized by the input-output relation

$$(p + \gamma_i)y = u$$

where γ_i is a constant, not necessarily real. Then if the output of $\mathcal{R}_i$ is identified with x_i, the ith component of $\mathbf{x}$, we have

8
$$\begin{aligned} \text{output of } \mathcal{R}_i &= x_i \\ \text{input to } \mathcal{R}_i &= (p + \gamma_i)x_i \\ &= \dot{x}_i + \gamma_i x_i \end{aligned}$$

Similarly, let $\mathfrak{D}_j$ denote a differentiator defined by

$$y = (p + \gamma_j)u$$

Then if the input to $\mathfrak{D}_j$ is identified with x_j, we have

9
$$\begin{aligned} \text{input to } \mathfrak{D}_j &= x_j \\ \text{output of } \mathfrak{D}_j &= (p + \gamma_j)x_j \\ &= \dot{x}_j + \gamma_j x_j \end{aligned}$$

We are now ready to formulate a simple procedure for setting up the state equations.

10 **Procedure** Suppose that $\mathcal{B}$ is an interconnection of scalors, adders, integrators, and differentiators with which a state vector $\mathbf{x}$ is associated in accordance with Rule *2*. Clearly, the input and output of $\mathcal{B}$ remain undisturbed when any integrator, say $\mathcal{R}_i$, in $\mathcal{B}$ is removed and its place is taken by an input x_i applied to the terminal to which the output of $\mathcal{R}_i$ is connected, with $(p + \gamma_i)x_i$ playing the role of a suppressed output at the terminal to which the input of $\mathcal{R}_i$ is connected. Similarly, we can remove any differentiator, say $\mathfrak{D}_j$, and replace it by an input $(p + \gamma_j)x_j$ applied to the terminal to which the output of $\mathfrak{D}_j$ is connected, with x_j playing the role of a suppressed output at the terminal to which the input of $\mathfrak{D}_j$ is connected.

On removing all the integrators and the differentiators in $\mathcal{B}$ in this fashion we are left with a system $\hat{\mathcal{B}}$ which contains only adders and scalors. This is illustrated in Figs. *4.6.5* and *4.6.6*, which show the systems remaining after the removal of integrators and differentiators from the systems of Figs. *4.6.1* and *4.6.4*, respectively.

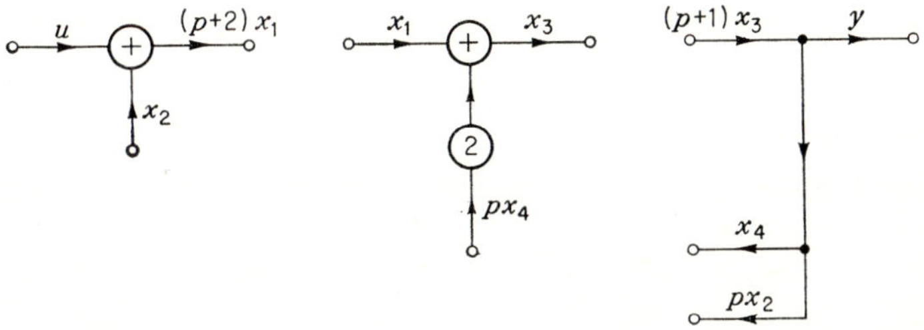

Fig. 4.6.5 System remaining after removal of integrators and differentiators from system of Fig. *4.6.1*.

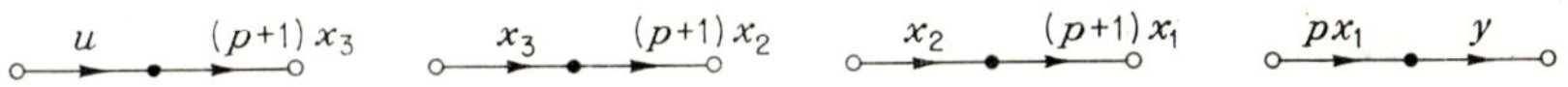

Fig. 4.6.6 System remaining after removal of integrators and differentiators from system of Fig. *4.6.4*.

In accordance with Note *3*, let $x_1, \ldots, x_l$ denote the outputs of the integrators in $\mathcal{B}$ and let $x_{l+1}, \ldots, x_n$ denote the inputs to the differentiators in $\mathcal{B}$, excluding those whose input terminals are connected directly or through scalors to the output terminals of an integrator (see Remark *1*). In terms of these variables, the inputs to $\hat{\mathcal{B}}$ are u, $x_1, \ldots, x_l$, $(p + \gamma_{l+1})x_{l+1}, \ldots, (p + \gamma_n)x_n$ and the outputs of $\hat{\mathcal{B}}$ are y, $(p + \gamma_1)x_1, \ldots, (p + \gamma_l)x_l$, $x_{l+1}, \ldots, x_n$. This is represented diagrammatically in Fig. *4.6.7*.

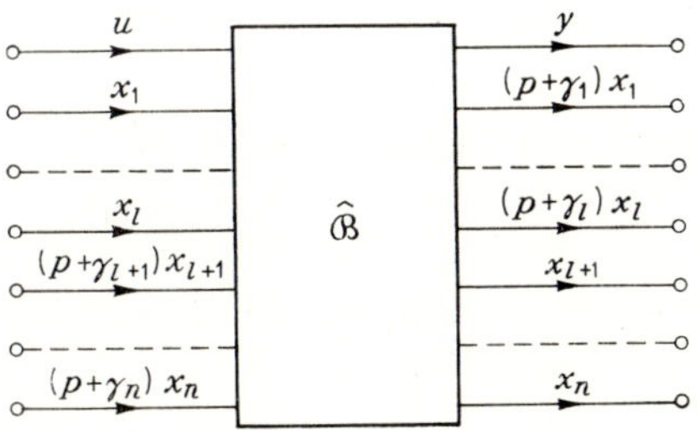

Fig. 4.6.7 $\hat{\mathcal{B}}$ and its input and output variables.

By this construction, we obtain from $\mathcal{B}$ a memoryless linear system $\hat{\mathcal{B}}$ such that the relations between the inputs and outputs of $\hat{\mathcal{B}}$ are exactly the same as the relations between u, y, and the components of $\mathbf{x}$ and their derivatives in $\mathcal{B}$. Having obtained $\hat{\mathcal{B}}$, the state equations of $\mathcal{B}$ can then be set up by inspection. For example, from Fig. *4.6.5* we obtain in this manner

$$(p + 2)x_1 = x_2 + u$$
$$x_3 = 2p\, x_4 + x_1$$
$$p\, x_2 = x_4$$
$$x_4 = (p + 1)x_3$$
$$y = (p + 1)x_3$$

Upon minor rearrangement, these equations read

11
$$\begin{aligned} \dot{x}_1 &= -2x_1 + x_2 + u \\ \dot{x}_2 &= x_4 \\ \dot{x}_4 &= \tfrac{1}{2}(x_3 - x_1) \\ \dot{x}_3 &= -x_3 + x_4 \\ y &= x_4 \end{aligned}$$

or more compactly,

$$\dot{\mathbf{x}} = \mathbf{A}\mathbf{x} + \mathbf{b}u$$
$$y = \langle \mathbf{c},\mathbf{x} \rangle$$

where

$$\mathbf{A} = \begin{bmatrix} -2 & 1 & 0 & 0 \\ 0 & 0 & 0 & 1 \\ 0 & 0 & -1 & 1 \\ -0.5 & 0 & 0.5 & 0 \end{bmatrix} \qquad \mathbf{b} = \begin{bmatrix} 1 \\ 0 \\ 0 \\ 0 \end{bmatrix} \qquad \mathbf{c} = [0 \quad 0 \quad 0 \quad 1]$$

Equations *11* are the desired state equations for the system of Fig. *4.6.1*.

In a similar fashion, for the system of Fig. *4.6.4* we derive from inspection of Fig. *4.6.6*

$$(p + 1)x_3 = u$$
$$(p + 1)x_2 = x_3$$
$$(p + 1)x_1 = x_2$$
$$y = p\, x_1$$

or

$$\begin{aligned} \dot{x}_1 &= -x_1 + x_2 \\ \dot{x}_2 &= -x_2 + x_3 \\ \dot{x}_3 &= -x_3 + u \\ y &= x_2 - x_1 \end{aligned}$$

Again, these equations are of the form *11*, with

$$\mathbf{A} = \begin{bmatrix} -1 & 1 & 0 \\ 0 & -1 & 1 \\ 0 & 0 & -1 \end{bmatrix} \qquad \mathbf{b} = \begin{bmatrix} 0 \\ 0 \\ 1 \end{bmatrix} \qquad \mathbf{c} = [-1 \quad 1 \quad 0]$$

General form of state equations

To find the general form of state equations, we note that $\hat{\mathcal{B}}$ is a memoryless linear system (see *1.8.17*) and hence the input variables $u, x_1, \ldots, x_l, (\dot{x}_{l+1} + \gamma_{l+1}x_{l+1}), \ldots, (\dot{x}_n + \gamma_n x_n)$ and the output variables $y, (\dot{x}_1 + \gamma_1 x_1), \ldots, (\dot{x}_l + \gamma_l x_l), x_{l+1}, \ldots, x_n$ must be related to one another by linear algebraic equations. For compactness, let $\mathbf{x}_{\mathcal{R}}$ and $\mathbf{x}_{\mathcal{D}}$ denote the vectors

12 $$\mathbf{x}_{\mathcal{R}} = (x_1, \ldots, x_l)$$

and

13 $$\mathbf{x}_{\mathcal{D}} = (x_{l+1}, \ldots, x_n)$$

respectively, in terms of which the state vector $\mathbf{x}$ can be represented in partitioned form as

$$\mathbf{x} = (\mathbf{x}_{\mathcal{R}}, \mathbf{x}_{\mathcal{D}})$$

Then, the relations between the variables in question can be written as

14 $$\begin{aligned} \dot{\mathbf{x}}_{\mathcal{R}} &= \mathbf{A}\mathbf{x}_{\mathcal{R}} + \mathbf{A}'\mathbf{x}_{\mathcal{D}} + \mathbf{A}''\dot{\mathbf{x}}_{\mathcal{D}} + \mathbf{b}u \\ \mathbf{x}_{\mathcal{D}} &= \mathbf{F}\mathbf{x}_{\mathcal{R}} + \mathbf{F}'\mathbf{x}_{\mathcal{D}} + \mathbf{F}''\dot{\mathbf{x}}_{\mathcal{D}} + \mathbf{g}u \\ y &= \langle \mathbf{c}, \mathbf{x}_{\mathcal{R}} \rangle + \langle \mathbf{c}', \mathbf{x}_{\mathcal{D}} \rangle + \langle \mathbf{c}'', \dot{\mathbf{x}}_{\mathcal{D}} \rangle + d_0 u \end{aligned}$$

where the coefficients on the right are constant matrices, vectors, and scalars. These equations express the general form of the state equations of an interconnection of adders, scalors, differentiators, and integrators.[1]

15 *Example* Consider the system $\mathcal{B}$ shown in Fig. *4.6.8*. Here x_1 is the output of $\mathcal{R}_1$ and $(p + 1)x_1$ is the corresponding input; x_2 is the output of $\mathcal{R}_2$, as well as the input to $\mathcal{D}_3$, and $p\, x_2$ is the corresponding input to $\mathcal{R}_2$ and $(p + 2)x_2$ is the corresponding output of $\mathcal{D}_3$; x_3 is the input to $\mathcal{D}_4$ and $p\, x_3$ is the corresponding output. By inspection, we have

$$\begin{aligned} (p + 1)x_1 &= u \\ p\, x_2 &= x_1 + u \\ x_3 &= (p + 2)x_2 + u \\ y &= x_1 + 3x_2 + (p + 2)x_2 + p\, x_3 \end{aligned}$$

which after rearrangement read

$$\begin{aligned} \dot{x}_1 &= -x_1 + u \\ \dot{x}_2 &= x_1 + u \\ x_3 &= \dot{x}_2 + 2x_2 + u = x_1 + 2x_2 + 2u \\ y &= x_1 + 3x_2 + \dot{x}_2 + 2x_2 + \dot{x}_3 = 2x_1 + 3x_2 + \dot{x}_3 + u \end{aligned}$$

[1] It should be noted that when $\mathcal{B}$ contains one or more scalors with infinite gain (*1.9.8*), the inputs to these scalors are constrained to be zero (so long as their outputs are finite), which is equivalent to "grounding" the system at the terminals in question. Such constraints give rise to additional linear relations between the components of $\mathbf{x}$ which are not exhibited explicitly in *14*.

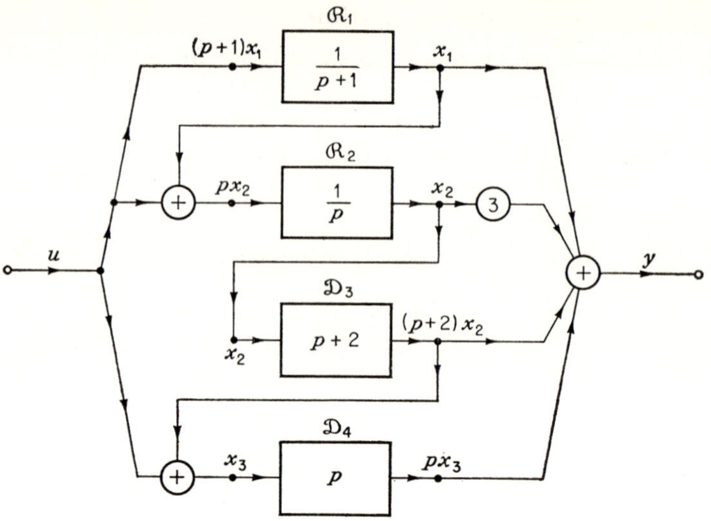

Fig. 4.6.8 System of Example *4.6.15*.

These are the desired state equations of $\mathcal{B}$. Note that they are of the general form *14*. ◁

As they stand, Eqs. *14* do not have the canonical form

16
$$\begin{aligned}\dot{\mathbf{x}} &= \mathbf{Ax} + \mathbf{b}u \\ y &= \langle \mathbf{c},\mathbf{x} \rangle + d_0 u\end{aligned}$$

nor are they in the hybrid form

17
$$\begin{aligned}\dot{\mathbf{x}} &= \mathbf{Ax} + \mathbf{b}u \\ y &= \langle \mathbf{c},\mathbf{x} \rangle + D(p)u\end{aligned}$$

Thus, the question arises: Can Eqs. *14* be put into the canonical form *16* or the hybrid form *17*, and, if so, by what means?

In the first place, it is evident that there are special cases in which Eqs. *14* have the canonical form *16*. For example, if $\mathcal{B}$ contains no differentiators, then $\mathbf{x} = \mathbf{x}_{\mathcal{R}}$ and *14* reduces to

18
$$\begin{aligned}\dot{\mathbf{x}} &= \mathbf{A}'\mathbf{x} + \mathbf{b}u \\ y &= \langle \mathbf{c},\mathbf{x} \rangle + d_0 u\end{aligned}$$

Furthermore, it is easy to construct examples of systems (e.g., the system of Fig. *4.6.1*) which contain both integrators and differentiators and yet lead to state equations of the form *16* (see *11*) either directly or after the elimination of one or more of the components of $\mathbf{x}_{\mathcal{R}}$ and $\mathbf{x}_{\mathcal{D}}$ from *14*. However, it is also easy to demonstrate that it is not always possible to reduce *14* to *16* merely by solving the state equations *14* for $\dot{\mathbf{x}}$ and y in terms of $\mathbf{x}$ and u. For example, if the state equations

14 read

19
$$\begin{aligned} \dot{x}_1 &= x_1 + \dot{x}_2 + u \\ x_2 &= 2x_1 - 2u \\ y &= x_1 \end{aligned}$$

then on eliminating x_2, we have

$$\begin{aligned} \dot{x}_1 &= -x_1 + 2\dot{u} - u \\ y &= x_1 \end{aligned}$$

which are not of the form *16*. Similarly, on eliminating x_1 we obtain

$$\begin{aligned} \dot{x}_2 &= -x_2 + 2\dot{u} - 4u \\ y &= \tfrac{1}{2}(x_2 + 2u) \end{aligned}$$

which likewise do not have the form in question.

In citing this example we tacitly assume that there exists an interconnection of adders, scalors, integrators, and differentiators such that equations *19* are its state equations. The assumption is justified by the following

20 **Assertion** Given any set of equations of the form *14*, one can always construct an interconnection of adders, scalors, integrators, and differentiators whose state equations are given by *14*. ◁

The manner in which such an interconnection can be constructed is easily understood from the

21 *Example* Consider the set of equations

22
$$\begin{aligned} \dot{x}_1 &= x_1 + x_2 + \dot{x}_2 + u \\ \dot{x}_2 &= 2x_1 + 3x_2 + x_3 \\ x_3 &= 2x_1 + \dot{x}_3 - \dot{x}_4 \\ x_4 &= 2x_1 + u \\ y &= x_1 + \dot{x}_3 + \dot{x}_4 + u \end{aligned}$$

By inspection of the structure of the system $\mathfrak{B}$ shown in Fig. *4.6.9*, it is easy to verify that the state equations of $\mathfrak{B}$ are given by *22*. The reason for this lies in the nature of the procedure which we use to construct $\mathfrak{B}$. Essentially, it amounts to the following simple rule. With each equation in *14* associate a pure integrator or a pure differentiator according as the left-hand member is a component of $\dot{\mathbf{x}}_{\mathfrak{R}}$ or $\mathbf{x}_{\mathfrak{D}}$. Label the input of the ith integrator $\dot{x}_i$ and its output x_i. Similarly, label the input of the jth differentiator x_j and its output $\dot{x}_j$. Let the input to each differentiator and integrator be the output of an adder, the inputs to which are weighted combinations of $x_1, \ldots, x_n$ and $\dot{x}_1, \ldots, \dot{x}_n$, with each combination representing a row on the

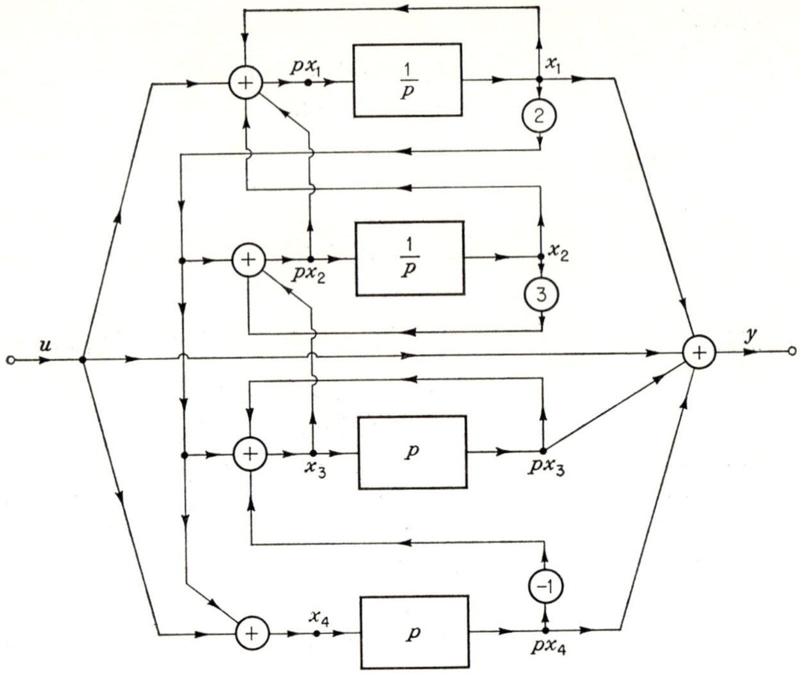

Fig. 4.6.9 System of Example *4.6.21*.

right-hand side of *14*. Finally, obtain y by forming the linear combination defined by the expression for y in *14*. ◁

In summary, Eqs. *14* cannot always be reduced to the canonical form merely by solving for $\dot{\mathbf{x}}$ and y in terms of $\mathbf{x}$ and u. However, as we shall see in Secs. *7* and *8*, it is always possible to replace $\mathcal{B}$ by an equivalent interconnection of adders, scalors, integrators, and differentiators whose state equations have the hybrid form (*4.5.45 et seq.*). Such a replacement can be interpreted as a linear change of variables $\mathbf{x} \rightarrow \mathbf{\Gamma x}'$ in which $\mathbf{x}$, $\mathbf{x}'$, and $\mathbf{\Gamma}$ play, respectively, the roles of $\mathbf{x}$, $\hat{\mathbf{x}}$, and $\mathbf{\Gamma}$ in *4.2.18*.

State equations for *RLC* networks

An *RLC* network, i.e., an electrical network comprising resistors, inductors, and capacitors, is a special case of an interconnection of adders, scalors, integrators and differentiators. Indeed, as will be shown in the sequel, any *RLC* network can be represented as an interconnection of adders, scalors, and integrators.[1]

More specifically, consider a network $\mathcal{a}$ such as is shown in Fig. *4.6.10*,

[1] We are tacitly assuming that the network in question is such that the corresponding interconnection does not contain scalors with infinite gain (see footnote on page 247). For more general treatment see Refs. *3* and *4*.

in which u ($\triangleq$ input voltage) is the input and y ($\triangleq$ output voltage) is the output. Note that $\mathcal{A}$ is an oriented system even though its components L_1, L_2, C_3, C_4, R_5, and R_6 are nonoriented.

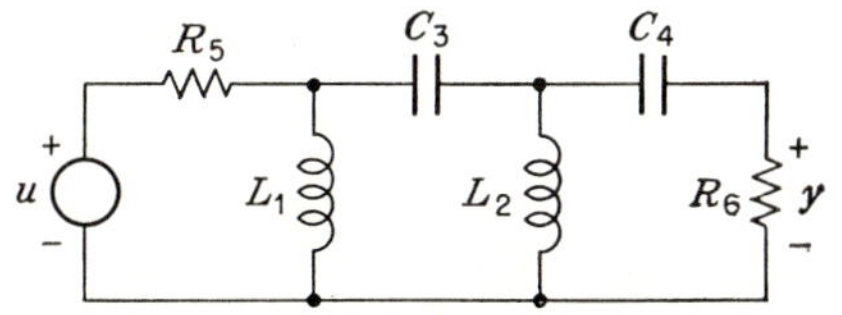

Fig. 4.6.10 An *RLC* network.

As was noted previously, the state equations *14* of an interconnection of scalors, adders, integrators, and differentiators assume the canonical form *18* when the interconnection in question does not contain differentiators. To achieve this condition in an electrical network comprising R's, L's, and C's, it is expedient to orient the L's and C's in such a way as to make them integrators. In concrete terms, this means that an inductor L characterized by

$$v = L\frac{di}{dt}$$

is oriented by identifying v with its input and i with its output. Similarly, a capacitor C defined by

$$i = C\frac{dv}{dt}$$

is oriented by making i its input and v its output. Correspondingly, the state of L at time t is taken to be $i(t)$ and that of C is taken to be $v(t)$.

Once $\mathcal{A}$ is oriented in this fashion, it becomes an interconnection of adders, scalors, and integrators to which we can apply Rule *2* and Procedure *10*. Rewording these in network terminology, we obtain the following rule for setting up the state equations of $\mathcal{A}$.

23 **Procedure** (I) Associate a state vector $\mathbf{x}$ with $\mathcal{A}$ by assigning components of $\mathbf{x}$ to the currents flowing through the inductors and the voltages appearing across the capacitors. (II) Replace each inductor L in $\mathcal{A}$ with a fictitious current source i_L, where $i_L \triangleq$ current flowing through L = state of L (Fig. *4.6.11a*). (III) Replace each capacitor C in $\mathcal{A}$ with a fictitious voltage source v_C, where $v_C \triangleq$ voltage across C = state of C (Fig. *4.6.11b*). (IV) As a result of these replacements, $\mathcal{A}$ is transformed into a network $\hat{\mathcal{A}}$ comprising only resistors, current sources, and voltage sources. (Observe that $\hat{\mathcal{A}}$ has the same significance as $\hat{\mathcal{B}}$ in *10*.) Find the output voltage appearing at the terminals of each fictitious current source i_L in $\mathcal{A}$ and equate it to $-Lpi_L$. Similarly, find the output current flowing through the terminals of each fictitious voltage source v_C in $\mathcal{A}$ and equate it to $-Cpv_C$. Finally, express the output y of $\mathcal{A}$ as a linear combination of all the voltage and

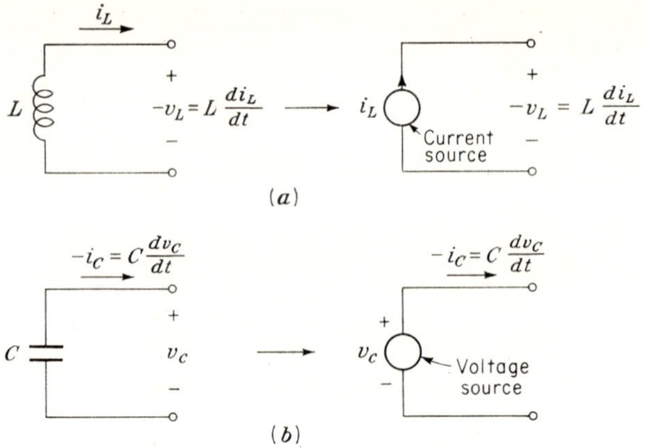

Fig. 4.6.11 (*a*) Replacement of an inductor by a current source, (*b*) replacement of a capacitor by a voltage source. (Note the polarities.)

current sources in $\mathcal{A}$ (i.e., including both the fictitious and external sources). The resulting equations are the state equations of $\mathcal{A}$.

24 *Example* Applying this procedure to the network of Fig. *4.6.10*, we associate with $\mathcal{A}$ the state vector $\mathbf{x} = (i_1, i_2, v_3, v_4)$. Then, on replacing L's and C's by the corresponding current and voltage sources, we obtain the structure shown in Fig. *4.6.12*. Finally, using elementary

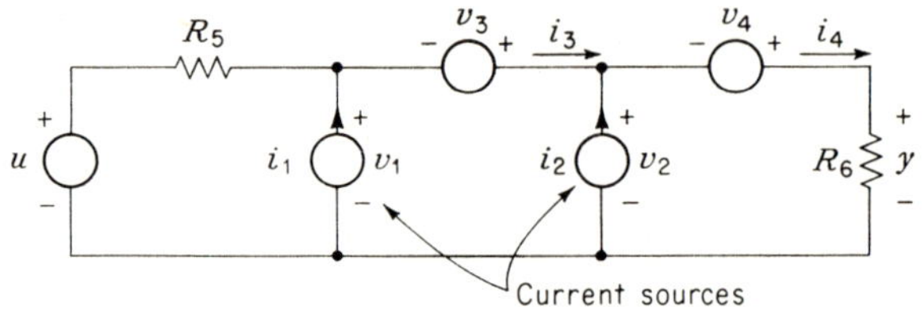

Fig. 4.6.12 The result of replacing inductors and capacitors in Fig. *4.6.10* by current and voltage sources.

circuit analysis techniques, we derive the following expressions for v_1, v_2, i_3, i_4, y:

$$-L_1 p i_1 = v_1 = \frac{1}{R_5 + R_6}(R_5R_6i_1 + R_5R_6i_2 - R_5v_3 - R_5v_4 + R_6u)$$

$$-L_2 p i_2 = v_2 = \frac{1}{R_5 + R_6}(R_5R_6i_1 + R_5R_6i_2 + R_6v_3 - R_5v_4 + R_6u)$$

$$-C_3 p v_3 = i_3 = \frac{1}{R_5 + R_6}(R_5i_1 - R_6i_2 + v_3 + v_4 + u)$$

$$-C_4 p v_4 = i_4 = \frac{1}{R_5 + R_6}(R_5i_1 + R_5i_2 + v_3 + v_4 + u)$$

$$y = \frac{1}{R_5 + R_6}(R_5R_6i_1 + R_5R_6i_2 + R_6v_3 + R_6v_4 + R_6u)$$

which when equated to $-L_1pi_1, \ldots, y$, respectively, yield the desired state equations for $\mathcal{A}$.

25 **Remark** The technique described above is applicable also to time-varying networks. However, it should be noted that when L and C depend on time, their terminal relations assume the following form:

26 L: $$v(t) = \frac{d}{dt}(L(t)i(t))$$

27 C: $$i(t) = \frac{d}{dt}(C(t)v(t))$$

28 *Exercise* Extend Procedure *23* to the case of networks containing mutual inductors, in addition to R's, L's, and C's. (A mutual inductor is a nonoriented 4-pole characterized by the terminal relations

29 $$\begin{aligned} v_1 &= L_{11}pi_1 + L_{12}pi_2 \\ v_2 &= L_{21}pi_1 + L_{22}pi_2 \end{aligned}$$

where $L_{11}, \ldots, L_{22}$ are constants.)

7 *Equivalence relations and properties of zero-input response*

Suppose that we are given a system $\mathcal{A}$ which is characterized by a differential equation or a system of differential equations, or is represented as an interconnection of systems of simpler types such as scalors, adders, integrators, and differentiators. Assume that our problem is to associate a state vector $\mathbf{x}$ with $\mathcal{A}$ and find the corresponding state equations for $\mathcal{A}$. Instead of attacking this problem directly, it is frequently expedient to replace $\mathcal{A}$ by an equivalent system, say $\mathcal{B}$, and then associate a state vector $\hat{\mathbf{x}}$ with, and find the corresponding state equations for, $\mathcal{B}$ rather than $\mathcal{A}$. By virtue of the equivalence of $\mathcal{A}$ and $\mathcal{B}$, the state vector $\hat{\mathbf{x}}$ will also qualify as a state vector for $\mathcal{A}$, and so will the state equations satisfied by it. Clearly, if $\mathcal{B}$ is easier to deal with than $\mathcal{A}$ or if it is of a type with which we know how to associate a state vector—whereas $\mathcal{A}$ is not—then there may be a distinct advantage in using this approach over trying to associate a state vector with $\mathcal{A}$ by direct means.

To cite an example, suppose that $\mathcal{A}$ is characterized by a differential equation of the form $Ly = Mu$. As an alternative to the approach described in Sec. *5*, one may try to find a realization for $\mathcal{A}$ in the form of an interconnection of adders, scalors, integrators, and differentiators and then associate a state vector with this realization by the use of the techniques of Sec. *6*. This is the basis for the so-called partial-fraction expansion technique which will be taken up in Sec. *9*.

For this as well as other reasons, it is important to have in hand a set of working rules for replacing a given system with one equivalent to it. The development of such rules is our main objective in this section.

A basic theorem

In Chap. *3* we established an important fact (*3.7.38*), namely, that a strictly proper time-invariant system $\mathfrak{A}$ of finite order n admits the representation

1
$$\dot{\mathbf{x}} = \mathbf{A}\mathbf{x} + \mathbf{b}u$$
$$y = \langle \mathbf{c},\mathbf{x} \rangle$$

where $\mathbf{A}$ is a constant $n \times n$ matrix, $\mathbf{b}$ and $\mathbf{c}$ are constant vectors, and $\mathbf{x}$ is a state vector ranging over $\mathcal{C}^n$ ($\triangleq$ space of n-tuples of complex numbers). In what follows, we shall establish another important fact, namely, that any system characterized by *1* or, more generally, by the hybrid state equations (*4.5.45 et seq.*)

2
$$\dot{\mathbf{x}} = \mathbf{A}\mathbf{x} + \mathbf{b}u$$
$$y = \langle \mathbf{c},\mathbf{x} \rangle + D(p)u$$

where $D(p)$ is a differential operator, is equivalent to a system characterized by a single differential equation of the form $L(p)y = M(p)u$. These two facts add up to the conclusion that, *given any strictly proper*[1] *time-invariant system* $\mathfrak{A}$ *of finite order, one can always find a differential system* $\mathfrak{B}$ *which is equivalent to* $\mathfrak{A}$ *and which is characterized by a single differential equation* $L(p)y = M(p)u$.

Our proof of the existence of such a $\mathfrak{B}$ will be constructive in nature, i.e., given any $\mathfrak{A}$, we shall be able to determine $L(p)$ and $M(p)$ such that the system $\frac{1}{L(p)} \cdot M(p)$ (see *4.5.9 et seq.*) is equivalent to $\mathfrak{A}$. As a preliminary, we shall derive the expression for the transfer function and the Laplace transform of the zero-input response of a system characterized by *1* and establish a connection between them.

Relation between the transfer function and zero-input response

Let $\mathfrak{A}$ be a system characterized by state equations of the form *1*, with $\mathbf{x}$ ranging over $\mathcal{C}^m$. To obtain the expression for the general solution

[1] The restriction to strictly proper systems is made here merely because *1* is proved in Chap. *3* only for such systems. Actually, as is pointed out there, it is not difficult to extend the result to improper systems.

of *1*, we apply the Laplace transformation to both sides of *1*, which yields

3 $$(s\mathbf{I} - \mathbf{A})\mathbf{X}(s) = \mathbf{x}(0-) + \mathbf{b}U(s)$$

4 $$Y(s) = \langle \mathbf{c}, \mathbf{X}(s) \rangle$$

where $\mathbf{I}$ is the $m \times m$ identity matrix, $U(s)$, $Y(s)$, and $\mathbf{X}(s)$ are the Laplace transforms of $u(t)$, $y(t)$, and $\mathbf{x}(t)$, respectively, and $\mathbf{x}(0-)$ is the initial state vector.

On eliminating $\mathbf{X}(s)$ from *3* and *4*, we obtain

5 $$Y(s) = \langle \mathbf{c}, (s\mathbf{I} - \mathbf{A})^{-1}\mathbf{x}(0-) \rangle + \langle \mathbf{c}, (s\mathbf{I} - \mathbf{A})^{-1}\mathbf{b}U(s) \rangle$$

where $(s\mathbf{I} - \mathbf{A})^{-1}$ denotes the inverse of the matrix $(s\mathbf{I} - \mathbf{A})$. This relation is in effect the desired expression for the general solution of *1* in the Laplace transform form. From *5*, the input-output-state relation for $\mathcal{A}$ can readily be derived by using the inverse Laplace transformation. For our present purposes, however, *5* will be sufficient. The input-output-state relation and the solution of state equations of the form *1* will be considered in detail in Chaps. *5* and *6*.

We note that the first term in *5* represents the Laplace transform of the zero-input response of $\mathcal{A}$. For convenience, we denote it by $Z(s;\mathbf{x}(0-))$:

6 $$Z(s;\mathbf{x}(0-)) = \langle \mathbf{c}, (s\mathbf{I} - \mathbf{A})^{-1}\mathbf{x}(0-) \rangle$$

The second term in *5* represents the Laplace transform of the zero-state response of $\mathcal{A}$ to u. Since the transfer function of $\mathcal{A}$ is the Laplace transform of the impulse response of $\mathcal{A}$—which in turn is the zero-state response of $\mathcal{A}$ to $\delta(t)$—and since $\mathcal{L}\{\delta(t)\} = 1$, it follows that the expression for the transfer function of $\mathcal{A}$ is[1]

7 $$H(s) = \langle \mathbf{c}, (s\mathbf{I} - \mathbf{A})^{-1}\mathbf{b} \rangle$$

We observe that $H(s)$ is related to $Z(s;\mathbf{x}(0-))$ in a very simple manner, namely,

8 $$H(s) = Z(s;\mathbf{b})$$

Thus, we can make the

9 **Assertion** Let $\mathcal{A}$ be a system characterized by the state equations

10 $$\dot{\mathbf{x}} = \mathbf{A}\mathbf{x} + \mathbf{b}u$$
$$y = \langle \mathbf{c}, \mathbf{x} \rangle$$

Then the transfer function of $\mathcal{A}$, $H(s)$, is the Laplace transform of the

[1] Our treatment of transfer functions in this section has the restricted objective of making it possible for us to formulate certain equivalence criteria for $\mathcal{A}$. A more general treatment of transfer functions and related questions is presented in Chap. *9*.

zero-input response of $\mathcal{A}$ starting in the state $\mathbf{x}(0-) = \mathbf{b}$. [Equivalently, the impulse response of $\mathcal{A}$, $h(t)$, is the zero-input response of $\mathcal{A}$ starting in $\mathbf{x}(0-) = \mathbf{b}$ (see also *45*).]

11 **Remark** Note that if $\mathcal{A}$ is a time-varying system in which $\mathbf{A}$ or $\mathbf{b}$ or $\mathbf{c}$ is time-dependent, it is still true that the *impulse response of* $\mathcal{A}$, $h(t,\xi)$ (see *3.6.5*), *is the zero-input response of* $\mathcal{A}$ *starting in* $\mathbf{x}(\xi-) = \mathbf{b}(\xi-)$. This is easily demonstrated without the use of the Laplace transformation by expressing $\dot{\mathbf{x}}$ as $-\mathbf{x}(\xi-)\delta(t-\xi) + p\,1(t-\xi)\mathbf{x}(t)$ (see *3.8.3 et seq.*) and noting that the impulse response $h(t,\xi)$ results when u is set equal to zero and $\mathbf{x}(\xi-) = \mathbf{b}(\xi-)$. ◁

We can draw useful results from Assertion *9* by examining $Z(s;\mathbf{x}(0-))$ and $H(s)$ in greater detail. Specifically, the matrix $(s\mathbf{I} - \mathbf{A})^{-1}$ has the form (before the cancellation of common factors in the numerator and denominator)

$$(s\mathbf{I} - \mathbf{A})^{-1} = \frac{\text{polynomial in } s \text{ of degree } m-1}{\det\,(s\mathbf{I} - \mathbf{A})}$$

where $\det\,(s\mathbf{I} - \mathbf{A})$ is $(-1)^m$ times the characteristic polynomial of $\mathbf{A}$ (see *C.18.1*). Consequently, $Z(s;\mathbf{x}(0-))$ and $H(s)$ must be of the form (before cancellation)

12 $$Z(s;\mathbf{x}(0-)) = \frac{r(s;\mathbf{x}(0-))}{d(s)}$$

13 $$H(s) = \frac{q(s)}{d(s)}$$

where $d(s) = \det\,(s\mathbf{I} - \mathbf{A})$, $q(s)$ is a polynomial of degree $m-1$, and $r(s;\mathbf{x}(0-))$ is a polynomial of degree $m-1$ whose coefficients are linear combinations of the components of $\mathbf{x}(0-)$. Furthermore, $r(s;\mathbf{b}) = q(s)$.

Let us reduce H and Z by canceling all common factors in $q(s)$ and $d(s)$, and $r(s;\mathbf{x}(0-))$ and $d(s)$, respectively, with the understanding that in Z we cancel only those factors which are common to $r(s;\mathbf{x}(0-))$ and $d(s)$ for all values of $\mathbf{x}(0-)$. In this way, Z and H are reduced to expressions of the form

14 $$Z(s;\mathbf{x}(0-)) = \frac{r'(s;\mathbf{x}(0-))}{d'(s)}$$

15 $$H(s) = \frac{q''(s)}{d''(s)}$$

where the primed and double-primed polynomials represent what is left of the original polynomials after cancellation.

16 **Remark** If the minimal polynomial of $\mathbf{A}$ (see *D.2.1*) is not identical with the characteristic polynomial of $\mathbf{A}$, then the ratio of the two will

be the greatest common divisor of $r(s;\mathbf{x}(0-))$ and $d(s)$ for all $\mathbf{x}(0-)$ (see *D.2.6*).

A basic lemma

The reduced form of $Z(s;\mathbf{x}(0-))$ has a basic property which is characteristic of time-invariant systems and which is frequently assumed to be true without proof. The statement and proof of this property are contained in the

17 **Lemma** Let $Z(s;\mathbf{x}(0-))$ be the Laplace transform of the zero-input response starting in state $\mathbf{x}(0-)$ of any linear time-invariant system $\mathcal{A}$. Suppose that after the cancellation of all factors common [for all $\mathbf{x}(0-)$] to the numerator and denominator of Z, the expression for $Z(s;\mathbf{x}(0-))$ in reduced form reads

18
$$Z(s;\mathbf{x}(0-)) = \frac{\alpha_1 s^{n-1} + \cdots + \alpha_{n-1}s + \alpha_n}{d'(s)} \triangleq Z'(s;\boldsymbol{\alpha})$$

where $d'(s)$ is a polynomial of degree n, the α_i, $i = 1, \ldots, n$, are coefficients dependent on $\mathbf{x}(0-)$, and the symbol $Z'(s;\boldsymbol{\alpha})$ is introduced to indicate that $Z(s;\mathbf{x}(0-))$ is in reduced form and that $Z(s;\mathbf{x}(0-))$ is expressed as a function of s and $\boldsymbol{\alpha} \triangleq (\alpha_1, \ldots, \alpha_n)$.

Then, for *every* $\boldsymbol{\alpha}$ in $\mathcal{C}^n$, $Z'(s;\boldsymbol{\alpha})$ is the Laplace transform of a zero-input response of $\mathcal{A}$. In other words, the α_i in *18* cannot be constrained to a subspace of $\mathcal{C}^n$ and, in particular, the numerator polynomial in *18* cannot have any of its powers missing for all values of $\boldsymbol{\alpha}$.

Proof The property in question is essentially a consequence of the time invariance of $\mathcal{A}$ and the superposability of zero-input responses (see *3.4.17*).

Specifically, the time invariance of $\mathcal{A}$ implies that, if z is a zero-input response of $\mathcal{A}$, then so is $\dot{z}$. To show this, write

19
$$\dot{\mathbf{x}} = \mathbf{A}\mathbf{x} \qquad u \equiv 0$$
$$z = \langle \mathbf{c},\mathbf{x} \rangle$$

which defines z to be a zero-input response $\mathcal{A}$. On differentiating both sides of *19* and denoting $\dot{\mathbf{x}}$ by $\mathbf{w}$, we have

$$\dot{\mathbf{w}} = \mathbf{A}\mathbf{w}$$
$$\dot{z} = \langle \mathbf{c},\mathbf{w} \rangle$$

which shows that $\dot{z}$ also satisfies *19* and hence, like z, is a zero-input response of $\mathcal{A}$. It is understood that, if z is interpreted as a one-sided response (see *3.6.45*), we do not include in $\dot{z}$ the delta function resulting from the differentiation of z at $t = 0$.

Consider now $Z'(s;\boldsymbol{\alpha})$. By the definition of $Z'(s;\boldsymbol{\alpha})$, *every* zero-input response of $\mathcal{A}$ can be obtained by assigning a particular value to $\boldsymbol{\alpha}$ in

$\mathcal{C}^n$. [For, $Z'(s;\boldsymbol{\alpha})$ is merely $Z(s;\mathbf{x}(0-))$, which in turn is the zero-input part of the Laplace transform of the general solution (see *4.5.16*).] What we want to show is the converse, namely, that for *every* $\boldsymbol{\alpha}$ in $\mathcal{C}^n$, $Z'(s;\boldsymbol{\alpha})$ is the Laplace transform of a zero-input response.

Suppose that this is not true, that is, $\boldsymbol{\alpha}$ is restricted to a subspace of $\mathcal{C}^n$, say $\mathcal{C}^k$, $k < n$. Then we can select a subset of components of $\boldsymbol{\alpha}$, say $\alpha_1, \ldots, \alpha_k$, such that all the components of $\boldsymbol{\alpha}$ can be represented as linear combinations of $\alpha_1, \ldots, \alpha_k$ and, furthermore, for *every* $\boldsymbol{\alpha} \triangleq (\alpha_1, \ldots, \alpha_k)$ in $\mathcal{C}^k$, $Z'(s;\boldsymbol{\alpha})$ is a zero-input response of $\mathfrak{A}$.

Written out in terms of $\alpha_1, \ldots, \alpha_k$, $Z'(s;\boldsymbol{\alpha})$ assumes the following form:

20 $$Z'(s;\boldsymbol{\alpha}) = \frac{\alpha_1 p_1(s) + \alpha_2 p_2(s) + \cdots + \alpha_k p_k(s)}{d'(s)}$$

where $p_1(s), \ldots, p_k(s)$ are linearly independent polynomials in s of degree not exceeding $n - 1$. [Note that if the $p_i(s)$ were not linearly independent, then we could reduce k still further—which by hypothesis cannot be done.] Since $Z'(s;\boldsymbol{\alpha})$ is a zero-input response (in the Laplace transform form) for every $\boldsymbol{\alpha}' = (\alpha_1, \ldots, \alpha_k)$ in $\mathcal{C}^k$, it follows that

$$\frac{p_1(s)}{d'(s)}, \ldots, \frac{p_k(s)}{d'(s)}$$

are also zero-input responses [corresponding to

$$\boldsymbol{\alpha}' = (1,0, \ldots, 0), \ldots, (0,0, \ldots, 1)$$

respectively].

Let $z_j(t)$, $j = 1, \ldots, k$, denote the zero-input response of which $p_j(s)/d'(s)$ is the Laplace transform. Since $p_j(s)$ is a polynomial of degree not exceeding $n - 1$ [where n is the degree of $d'(s)$], we can always write by long division

21 $$\frac{sp_j(s)}{d'(s)} = \gamma_j + \frac{q_j(s)}{d'(s)}$$

where $q_j(s)$, the remainder, is a polynomial of degree not exceeding $n - 1$ and γ_j is a constant which vanishes when the degree of $p_j(s)$ is lower than $n - 1$.

We can interpret *21* as follows. $q_j(s)/d'(s)$ is the Laplace transform of the derivative of $z_j(t)$, with the delta-function term $\gamma_j\delta(t)$ not included. In accordance with our understanding, then (see *19 et seq.*), $q_j(s)/d'(s)$ is the Laplace transform of $\dot{z}_j(t)$, $j = 1, \ldots, k$.

Since $\dot{z}_j(t)$ is a zero-input response of $\mathfrak{A}$ and since the Laplace transform of every zero-input response of $\mathfrak{A}$ is of the form *20*, we must have

22 $$\frac{q_j(s)}{d'(s)} = \frac{\alpha_1^j p_1(s) + \cdots + \alpha_k^j p_k(s)}{d'(s)} \qquad j = 1, \ldots, k$$

On replacing $q_j(s)$ in *22* by

$$q_j(s) = sp_j(s) - \gamma_j d'(s)$$

(which results from *21*), and clearing $d'(s)$, we obtain a set of relations between the $p_j(s)$ and $d'(s)$ which read

23
$$\begin{array}{l} (s - \alpha_1^1)p_1(s) - \alpha_2^1 p_2(s) - \cdots - \alpha_k^1 p_k(s) = \gamma_1 d'(s) \\ \cdots\cdots\cdots\cdots\cdots\cdots\cdots\cdots \\ -\alpha_1^k p_1(s) - \alpha_2^k p_2(s) - \cdots + (s - \alpha_k^k)p_k(s) = \gamma_k d'(s) \end{array}$$

We note that not all $\gamma_1, \ldots, \gamma_k$ can be zero in *23*, for this would be inconsistent with the linear independence of the $p_j(s)$ (see *20*). Thus, on solving these equations for the $p_j(s)$ by the Cramer rule, we obtain expressions of the form

24
$$p_j(s) = d'(s)\,\frac{r_j(s)}{\Delta(s)}$$

where

25
$$\Delta(s) = \begin{vmatrix} s - \alpha_1^1 & -\alpha_2^1 & \cdots & -\alpha_k^1 \\ \cdots & \cdots & \cdots & \cdots \\ -\alpha_1^k & -\alpha_2^k & \cdots & (s - \alpha_k^k) \end{vmatrix}$$

is a polynomial of degree not exceeding k and $r_j(s)$ is a polynomial of degree not exceeding $k - 1$.

Finally, on substituting *24* and *25* in *20*, and canceling $d'(s)$, we obtain the following expression for the Laplace transform of the zero-input response:

26
$$Z'(s;\boldsymbol{\alpha}) = \frac{(\alpha_1 r_1(s) + \cdots + \alpha_k r_k(s))}{\Delta(s)}$$

in which the degree of the numerator does not exceed $k - 1$ and hence is lower than $n - 1$ (since $k < n$) and the degree of the denominator does not exceed k and hence is lower than n. This implies that there is a common factor in the numerator and denominator of $Z'(s;\boldsymbol{\alpha})$ (as given by *20*) which upon cancellation leads to *26*. This contradicts the hypothesis that $Z'(s;\boldsymbol{\alpha})$ is in reduced form and thus proves the lemma.

An immediate consequence of Lemma *17* is the

27 **Corollary** The basis functions of $\mathcal{A}$ (see *4.5.23*) can be taken to be (in the Laplace transform form)

$$\frac{s^{n-1}}{d'(s)}, \ldots, \frac{s}{d'(s)}, \frac{1}{d'(s)}$$

where $d'(s)$ is the denominator of the reduced form of the Laplace transform of the zero-input response of $\mathcal{A}$ and n is the degree of $d'(s)$.

[This follows from the linear independence of the functions in question and the fact just established, namely, that for every $\boldsymbol{\alpha} \in \mathcal{C}^n$, $Z'(s;\boldsymbol{\alpha})$ is a zero-input response function (in the Laplace transform form).] [Note that we could not have made this assertion before proving Lemma *17*, for we could not rule out the possibility that some of the powers of s might be missing in the numerator of $Z'(s;\mathbf{x}(0-))$ (*18*) for all values of $\mathbf{x}(0-)$.] $\triangleleft$

In changing from these basis functions to another set, say $\Phi_1(s), \ldots, \Phi_n(s)$, it is frequently helpful to invoke the following

28 **Assertion** Let $Z'(s;\boldsymbol{\alpha})$ be the Laplace transform (in reduced form) of the zero-input response of $\mathcal{A}$ starting from a state $\boldsymbol{\alpha}$ in $\mathcal{C}^n$. If for every $\boldsymbol{\alpha} \in \mathcal{C}^n$ we can represent $Z'(s;\boldsymbol{\alpha})$ in the form

29
$$Z'(s;\boldsymbol{\alpha}) = \beta_1\Phi_1(s) + \cdots + \beta_n\Phi_n(s)$$

where $\Phi_1(s), \ldots, \Phi_n(s)$ are linearly independent functions of s, then $\Phi_1(s), \ldots, \Phi_n(s)$ qualify as a set of basis functions for $\mathcal{A}$ (in the Laplace transform form) and the vectors $\boldsymbol{\beta} = (\beta_1, \ldots, \beta_n)$ and $\boldsymbol{\alpha} = (\alpha_1, \ldots, \alpha_n)$ are in one-one correspondence with each other.

Proof By hypothesis, every zero-input response of $\mathcal{A}$ can be written (in the Laplace transform form)

30
$$Z'(s;\boldsymbol{\alpha}) = \alpha_1 \frac{s^{n-1}}{d'(s)} + \cdots + \alpha_n \frac{1}{d'(s)}$$

Consequently, to every n-tuple $(\beta_1, \ldots, \beta_n)$ which can result from the expansion of $Z'(s;\boldsymbol{\alpha})$ in the form *29* there will correspond an $\boldsymbol{\alpha}$ in $\mathcal{C}^n$ such that

31
$$\alpha_1 \frac{s^{n-1}}{d'(s)} + \cdots + \alpha_n \frac{1}{d'(s)} \equiv \beta_1\Phi_1(s) + \cdots + \beta_n\Phi_n(s)$$

Conversely, by Lemma *17*, for every $\boldsymbol{\alpha}$ in $\mathcal{C}^n$, *30* represents the Laplace transform of a zero-input response of $\mathcal{A}$; therefore, to every $\boldsymbol{\alpha}$ in $\mathcal{C}^n$ there corresponds a $\boldsymbol{\beta}$ in $\mathcal{C}^n$ such that *31* is satisfied. This implies that

32
$$\boldsymbol{\beta} = \boldsymbol{\alpha}\boldsymbol{\Gamma} \qquad \forall \boldsymbol{\alpha} \in \mathcal{C}^n$$

where $\boldsymbol{\Gamma}$ is an $n \times n$ constant matrix which associates a $\boldsymbol{\beta}$ to each $\boldsymbol{\alpha}$. (Note that this $\boldsymbol{\Gamma}$ has the same significance as in *4.2.25*.)

On denoting

33
$$\boldsymbol{\phi}(s) \triangleq (\Phi_1(s), \ldots, \Phi_n(s))$$
$$\hat{\boldsymbol{\phi}}(s) \triangleq \left(\frac{s^{n-1}}{d'(s)}, \ldots, \frac{1}{d'(s)}\right)$$

and substituting *32* in *31*, we have

$$\langle \boldsymbol{\alpha}, (\hat{\boldsymbol{\phi}}(s) - \boldsymbol{\Gamma}\boldsymbol{\phi}(s)) \rangle \equiv 0 \qquad \forall \boldsymbol{\alpha} \in \mathcal{C}^n$$

which implies that

34 $$\hat{\boldsymbol{\phi}}(s) - \boldsymbol{\Gamma}\Phi(s) \equiv 0$$

Now both $\Phi_1(s), \ldots, \Phi_n(s)$ and $s^{n-1}/d'(s), \ldots, 1/d'(s)$ are linearly independent sets of functions. Consequently, the matrix $\boldsymbol{\Gamma}$ in *32* must be of rank n. This implies that the correspondence between $\boldsymbol{\alpha}$ and $\boldsymbol{\beta}$ is one-one and hence that, for *every* $\boldsymbol{\beta}$ in $\mathcal{C}^n$, *29* represents the Laplace transform of a zero-input response of $\mathcal{A}$. Furthermore, from *34* and the nonsingularity of $\boldsymbol{\Gamma}$ it follows that every basis function $s^{\lambda-1}/d'(s)$ is a linear combination of $\Phi_1(s), \ldots, \Phi_n(s)$ and, conversely, every $\Phi_i(s)$ is a linear combination of $s^{n-1}/d'(s), \ldots, 1/d'(s)$. Consequently, $\Phi_1(s), \ldots, \Phi_n(s)$ qualify as a set of basis functions for $\mathcal{A}$. Q.E.D.

We can make an immediate application of this result to the partial-fraction expansion of $Z'(s;\boldsymbol{\alpha})$. Thus, suppose that

35 $$Z'(s;\boldsymbol{\alpha}) = \frac{\alpha_1 s^{n-1} + \cdots + \alpha_n}{(s-\lambda_1)^{\mu_1} \cdots (s-\lambda_p)^{\mu_p}} \qquad \mu_1 + \cdots + \mu_p = n$$

where the λ's denote the zeros of $d'(s)$ and the μ's are their multiplicities.

On expanding $Z'(s;\boldsymbol{\alpha})$ into partial fractions, we obtain an expression of the general form

36 $$\begin{aligned} Z'(s;\boldsymbol{\alpha}) = {} & \frac{\beta_1^1}{(s-\lambda_1)^{\mu_1}} + \frac{\beta_2^1}{(s-\lambda_1)^{\mu_1-1}} + \cdots + \frac{\beta_{\mu_1}^1}{(s-\lambda_1)} \\ & + \cdots\cdots\cdots\cdots \\ & + \frac{\beta_1^p}{(s-\lambda_p)^{\mu_p}} + \frac{\beta_2^p}{(s-\lambda_p)^{\mu_p-1}} + \cdots + \frac{\beta_{\mu_p}^p}{(s-\lambda_p)} \end{aligned}$$

Clearly, the n functions

37 $$\begin{array}{c} \dfrac{1}{(s-\lambda_1)^{\mu_1}}, \ldots, \dfrac{1}{(s-\lambda_1)} \\ \cdots\cdots\cdots\cdots \\ \dfrac{1}{(s-\lambda_p)^{\mu_p}}, \ldots, \dfrac{1}{(s-\lambda_p)} \end{array}$$

are linearly independent. Hence by Assertion *28* they qualify as a set of basis functions (in the Laplace transform form) for $\mathcal{A}$.

38 *Example* Suppose that $Z'(s;\boldsymbol{\alpha})$ is given by

$$Z'(s;\boldsymbol{\alpha}) = \frac{\alpha_1 s^4 + \alpha_2 s^3 + \alpha_3 s^2 + \alpha_4 s + \alpha_5}{(s+1)^2(s+2)^2(s+3)}$$

Then the functions

$$\frac{1}{(s+1)^2}, \frac{1}{(s+1)}, \frac{1}{(s+2)^2}, \frac{1}{(s+2)}, \frac{1}{(s+3)}$$

qualify as a set of basis functions for $\mathcal{A}$.

Relation between $Z(s;\mathbf{x}(0-))$ and $H(s)$

We are now prepared to resume our discussion of the relation between $Z(s;\mathbf{x}(0-))$ (*12*) and the transfer function $H(s)$ (*13*). Specifically, we note that if a factor, say $(s - a)$, is canceled from the numerator and denominator of $Z(s;\mathbf{x}(0-))$, then it will also be canceled from the numerator and denominator of $H(s)$, since (I) the numerator of $H(s)$ ($\triangleq q(s)$) is merely the numerator of $Z(s;\mathbf{x}(0-))$ ($\triangleq r(s;\mathbf{x}(0-))$) with $\mathbf{x}(0-)$ replaced by $\mathbf{b}$; and (II) no factor is canceled in Z unless it is a common factor for *all* values of $\mathbf{x}(0-)$. This observation leads to the following

39 **Lemma** If the transfer function $H(s)$ and the Laplace transform of the zero-input response $Z(s)$ are in reduced form, then the denominator of $H(s)$ divides the denominator of $Z(s)$. In symbols, this means that

40 $$d'(s) = \rho(s)d''(s)$$

where $\rho(s)$ is a polynomial in s. For reasons that will become clear later, we shall refer to $\rho(s)$ as the *restoring factor*. $\triangleleft$

We can use this lemma to derive a somewhat more explicit relation between the impulse response and the zero-input response than we did previously (*9*). Specifically, we had found that the impulse response, $h(t)$, may be identified with the zero-input response of $\mathcal{A}$ starting in state $\mathbf{b}$. We have also found (Lemma *17*) that, after reduction, the Laplace transform of the zero-input response admits the representation

41 $$Z'(s;\boldsymbol{\alpha}) = \alpha_1\Phi_1(s) + \cdots + \alpha_n\Phi_n(s)$$

where $\Phi_1(s), \ldots, \Phi_n(s)$ are a set of basis functions of $\mathcal{A}$ (in the Laplace transform form) and the vector $\boldsymbol{\alpha} = (\alpha_1, \ldots, \alpha_n)$ ranges over $\mathcal{C}^n$. [Correspondingly, the zero-input response $z(t)$ reads

42 $$z(t) = \alpha_1\phi_1(t) + \cdots + \alpha_n\phi_n(t) \qquad t \geq 0$$

where the basis functions $\phi_1(t), \ldots, \phi_n(t)$ are the inverse Laplace transforms of $\Phi_1(s), \ldots, \Phi_n(s)$, respectively.]

Now we can choose $\Phi_1(s), \ldots, \Phi_n(s)$ to be the functions given by *37*. Then, Lemma *39* implies that the partial-fraction expansion of the transfer function $H(s)$ must necessarily be of the form

43 $$H(s) = \gamma_1\Phi_1(s) + \cdots + \gamma_n\Phi_n(s)$$

where the γ_i, $i = 1, \ldots, n$, are constants such that $\gamma_i = 0$ if the factor corresponding to $\Phi_i(s)$ is present in $d'(s)$ but not in $d''(s)$ (see *39*). On the basis of this conclusion, we can make the

44 **Assertion** Let $\phi_1(t), \ldots, \phi_n(t)$ be the set of basis functions of a proper system $\mathcal{A}$ which are yielded by the partial-fraction expansion of $Z(s;\mathbf{x}(0-))$, the Laplace transform of the zero-input response of $\mathcal{A}$. Then the impulse response of $\mathcal{A}$ admits the representation

45 $$h(t) = \gamma_1\phi_1(t) + \cdots + \gamma_n\phi_n(t)$$

where the γ_i, $i = 1, \ldots, n$, are constants, with $\gamma_i = 0$ if $\Phi_i(s)$ is present in the partial-fraction expansion of $Z(s;\mathbf{x}(0-))$ but not in $H(s)$. [Note that the partial-fraction expansions of $Z(s;\mathbf{x}(0-))$ and $Z'(s;\boldsymbol{\alpha})$ are identical. The same statement applies to $H(s)$ and $H'(s)$.]

46 *Comment* To extend this assertion to an improper system, it is necessary to make a modification in *45*. Specifically, if $\mathcal{A}$ is an improper system, it can be decomposed into a strictly proper system $\hat{\mathcal{A}}$ and a system of the differential operator type $\hat{\mathfrak{D}}$ which is characterized by an input-output relation of the form (see *4.5.50*)

47 $$w = (d_0 + d_1 p + \cdots + d_k p^k)u$$

Now let $\phi_1(t), \ldots, \phi_n(t)$ be a set of basis functions for $\hat{\mathcal{A}}$. Then, using Assertion *44* and *47*, the impulse response of $\mathcal{A}$ can be expressed as

48 $$h(t) = \gamma_1\phi_1(t) + \cdots + \gamma_n\phi_n(t) + d_0\delta(t) + d_1\delta^{(1)}(t) + \cdots + d_k\delta^{(k)}(t)$$

By Assertion *28* and Assertion *44*, the functions $\phi_1(t), \ldots, \phi_n(t)$, $\delta(t), \ldots, \delta^{(k-1)}(t)$ constitute a set of basis functions for $\mathcal{A}$. In terms of these functions, *48* is an expression of the same general form as *45*, except for the last term $\delta^{(k)}(t)$—which is the derivative of the basis function $\delta^{(k-1)}(t)$. Consequently, we can conclude that when $\mathcal{A}$ is an improper system, its impulse response admits of the representation

49 $$h(t) = \gamma_1\phi_1(t) + \cdots + \gamma_n\phi_n(t) + d_0\phi_{n+1}(t) + \cdots + d_{k-1}\phi_{n+k}(t) + d_k\dot{\phi}_{n+k}(t)$$

where $\phi_\lambda(t) = \delta^{(\lambda-n-1)}(t)$, $\lambda = n + 1, \ldots, n + k$.

Determination of a system equivalent to $\mathcal{A}$

We are now ready to apply the results derived above to the following basic

50 *Problem* Let $\mathcal{A}$ be a system characterized by the state equations

51 $$\dot{\mathbf{x}} = \mathbf{A}\mathbf{x} + \mathbf{b}u$$
$$y = \langle \mathbf{c},\mathbf{x} \rangle$$

Find a system $\mathcal{B}$ characterized by a single differential equation of the form

$$L(p)y = M(p)u$$

such that $\mathcal{B}$ is equivalent to $\mathcal{A}$.

To solve this problem, we recall (*3.5.10*) that two linear systems $\mathcal{A}$ and $\mathcal{B}$ are equivalent if (I) $\mathcal{A}$ and $\mathcal{B}$ are zero-state equivalent—i.e., have the same transfer function—and (II) $\mathcal{A}$ and $\mathcal{B}$ are zero-input equivalent—i.e., have the same basis functions (or, more generally, are such that every basis function of $\mathcal{A}$ is expressible as a linear combination of basis functions of $\mathcal{B}$, and vice versa).

This fact, in conjunction with the results derived in the preceding subsection, leads us to a theorem which constitutes a solution to the problem before us.

52 **Theorem** Let $\mathcal{A}$ be a system characterized by *51* and let its transfer function $H_a(s)$ be (in reduced form)

53
$$H_a(s) = \frac{\tilde{M}(s)}{\tilde{L}(s)}$$

[Note that $\tilde{M}(s)$ and $\tilde{L}(s)$ correspond to $q''(s)$ and $d''(s)$ in *15*.] Furthermore, let the denominator of the Laplace transform of the zero-input response of $\mathcal{A}$ (in reduced form) be $\rho(s)\tilde{L}(s)$, where $\rho(s)$ is a polynomial in s (see *39*) and $\rho(s)\tilde{L}(s)$ is a polynomial of degree n in s. Then the system $\mathcal{B}$ characterized by the input-output relation

54
$$L(p)y = M(p)u$$

in which $L(p)$ and $M(p)$ are polynomials in p given by the products

55
$$L(p) = \rho(p)\tilde{L}(p)$$
$$M(p) = \rho(p)\tilde{M}(p)$$

is equivalent to $\mathcal{A}$.

Proof In the first place, $\mathcal{A}$ and $\mathcal{B}$ are zero-state equivalent, since, by construction, the transfer function of $\mathcal{B}$ is

$$\begin{aligned} H_b(s) &= \frac{M(s)}{L(s)} \\ &= \frac{\rho(s)\tilde{L}(s)}{\rho(s)\tilde{M}(s)} = \frac{\tilde{L}(s)}{\tilde{M}(s)} \\ &= H_a(s) \end{aligned}$$

In the second place, the basis functions of $\mathcal{B}$ can be taken to be (in the Laplace transform form) (see *4.5.26*)

$$\frac{s^{n-1}}{L(s)}, \ldots, \frac{s}{L(s)}, \frac{1}{L(s)}$$

But these are also the basis functions of $\mathcal{A}$ (see *27*): thus, $\mathcal{A}$ and $\mathcal{B}$ are both zero-input equivalent and zero-state equivalent, and hence are equivalent. Q.E.D.

56 *Comment* The importance of this theorem stems from the fact that it provides us with an effective technique for replacing a given differential system $\mathcal{A}$ with an equivalent system $\mathcal{B}$ which is characterized by a single differential equation.

The crux of the idea behind the technique in question is this. Given a system $\mathcal{A}$, it is generally a simple matter to find its transfer function $H_a(s)$. Now suppose that after determining $H_a(s)$ (in reduced form) as the ratio of two polynomials $\tilde{L}(s)$ and $\tilde{M}(s)$, we define a system $\tilde{\mathcal{B}}$ by the input-output relation

57 $$\tilde{L}(p)y = \tilde{M}(p)u$$

Clearly, the transfer function of $\tilde{\mathcal{B}}$ is identical with that of $\mathcal{A}$. However, this in itself does not allow us to conclude that $\tilde{\mathcal{B}}$ is equivalent to $\mathcal{A}$, since the basis functions of $\mathcal{A}$ and $\tilde{\mathcal{B}}$ may be different. Thus, our problem is to modify $\tilde{\mathcal{B}}$ in such a way as to leave its transfer function unchanged, while at the same time making its basis functions identical with those of $\mathcal{A}$.

We achieve this result by finding the "restoring factor" $\rho(s)$ which accounts for the canceled factors in $H_a(s)$. Once $\rho(s)$ is determined, we form the "restored" transfer function

$$\begin{aligned} H_b(s) &= \frac{\rho(s)}{\rho(s)} H_a(s) \\ &= \frac{\rho(s)\tilde{M}(s)}{\rho(s)\tilde{L}(s)} \end{aligned}$$

and define a corresponding differential system $\mathcal{B}$ by the input-output relation

$$L(p)y = M(p)u$$

where

58 $$\begin{aligned} L(p) &= \rho(p)\tilde{L}(p) \\ M(p) &= \rho(p)\tilde{M}(p) \end{aligned}$$

In this way, we obtain a system $\mathcal{B}$ which has both the same transfer function and the same basis functions as $\mathcal{A}$ and hence is equivalent to $\mathcal{A}$.

This explains the reason for calling $\rho(s)$ the restoring factor for $\mathcal{A}$. For the same reason, the technique provided by Theorem *52* for finding a system $\mathcal{B}$ which is characterized by a single differential equation and which is equivalent to a given system $\mathcal{A}$ will be termed the "restoration technique."

Although in the statement of Theorem *52* $\mathcal{A}$ is assumed to be a strictly proper system, both the theorem and the restoration technique which is based on it are applicable to any type of differential system.

The way in which the restoration technique is applied in various situations is illustrated by the following examples.

59 *Example* Consider a system $\mathcal{A}$ which is characterized by the state equations *51*, with

$$\mathbf{A} = \begin{bmatrix} -1 & 0 & 0 \\ 1 & -2 & 0 \\ 0 & 0 & -4 \end{bmatrix} \qquad \mathbf{b} = \begin{bmatrix} 1 \\ 0 \\ 0 \end{bmatrix} \qquad \mathbf{c} = [1 \quad -2 \quad 1]$$

In this case,

$$(s\mathbf{I} - \mathbf{A})^{-1} = \frac{1}{(s+1)(s+2)(s+4)} \begin{bmatrix} (s+2)(s+4) & 0 & 0 \\ s+4 & (s+1)(s+4) & 0 \\ 0 & 0 & (s+1)(s+2) \end{bmatrix}$$

and the Laplace transform of the zero-input response is given by

$$\begin{aligned} Z(s;\mathbf{x}(0-)) &= \langle \mathbf{c}, (s\mathbf{I} - \mathbf{A})^{-1}\mathbf{x}(0-)\rangle \\ &= \frac{s}{(s+1)(s+2)} x_1(0-) - \frac{2}{(s+2)} x_2(0-) \\ &\qquad + \frac{1}{(s+4)} x_3(0-) \\ &= \frac{\alpha_1 s^2 + \alpha_2 s + \alpha_3}{(s+1)(s+2)(s+4)} \\ &= Z'(s;\boldsymbol{\alpha}) \end{aligned}$$

Note that the numerator and the denominator of $Z(s;\mathbf{x}(0-))$ have no common factors for all $\mathbf{x}(0-)$ and hence $Z(s;\mathbf{x}(0-))$ is in reduced form.

Now the transfer function $H(s)$ is expressed by

$$\begin{aligned} H(s) &= Z(s;\mathbf{b}) \qquad \mathbf{b} = (1,0,0) \\ &= \frac{s}{(s+1)(s+2)} \end{aligned}$$

and hence the restoring factor $\rho(s)$ is given by

$$\rho(s) = (s+4)$$

Making use of *55*, the differential equation characterizing the equivalent system $\mathcal{B}$ is found to be

$$(p+1)(p+2)(p+4)y = (p+4)u$$

60 *Example* Consider an improper system $\mathcal{A}$ which is characterized by state equations of the form

$$\begin{aligned} x_1 &= u \\ \dot{x}_2 &= 0 \\ \dot{x}_3 &= x_2 - x_3 \\ y &= \dot{x}_1 + x_3 \end{aligned}$$

On applying the Laplace transformation to these equations and setting $u = 0$, we find the following expression for the Laplace transform of the zero-input response:

$$Z(s;\boldsymbol{\alpha}) = \frac{\alpha_1 s + \alpha_2}{s(s+1)}$$

where α_1 and α_2 are constants depending on $x_1(0-)$, $x_2(0-)$, and $x_3(0-)$.

On the other hand, on setting $x_1(0-) = x_2(0-) = x_3(0-) = 0$ and forming the ratio

$$H(s) = \frac{Y(s)}{U(s)}$$

where $Y(s)$ and $U(s)$ are the Laplace transforms of y and u, respectively, we have

$$H(s) = s$$

In this case, the restoring factor is given by

$$\rho(s) = s(s+1)$$

and hence the equivalent system $\mathcal{B}$ is characterized by the differential equation

$$p(p+1)y = p^2(p+1)u$$

61 *Example* Let $\mathcal{a}$ be represented by an interconnection of scalors, adders, integrators, and differentiators such as shown in Fig. *4.7.1.* The

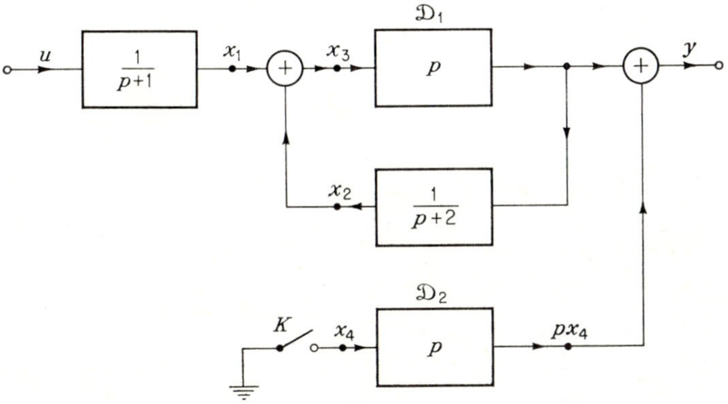

Fig. 4.7.1 System of Example *4.7.61.*

problem is to find an equivalent system $\mathcal{B}$ which is characterized by a single differential equation.

We associate a state vector $\mathbf{x} \triangleq (x_1,x_2,x_3,x_4)$ with $\mathcal{a}$ by the use of Rule *4.6.2.* The input to $\mathfrak{D}_2$ is assumed to be $x_4(0-)$ at $t = 0-$ and zero for $t \geq 0$. This is represented graphically by the switch K which "grounds" the input to $\mathfrak{D}_2$ starting at $t = 0$.

By using Procedure *4.6.10*, the state equations of $\mathcal{A}$ are readily found to be

$$\begin{aligned} \dot{x}_1 &= -x_1 + u \\ x_3 &= x_1 + x_2 \\ \dot{x}_2 &= -2x_2 + \dot{x}_3 \\ x_4 &= 0 \qquad t \geq 0 \\ y &= \dot{x}_4 + \dot{x}_3 \end{aligned}$$

On applying the Laplace transformation to these equations (with $u = 0$ for $t \geq 0$), and solving for the Laplace transform of the zero-input response, we have

$$Z(s;\mathbf{x}(0-)) = -\frac{x_1(0-)}{2(s+1)} + \tfrac{1}{2}x_1(0-) - x_3(0-) - x_4(0-)$$

On the other hand,

$$H(s) = \left.\frac{Y(s)}{U(s)}\right]_{\mathbf{x}(0-)=0} = \frac{s(s+2)}{2(s+1)}$$

In this case $\rho(s) = 1$ and hence $\mathcal{B}$ is characterized by

$$2(p+1)y = p(p+2)u$$

Note that $\mathcal{A}$ is an improper system.

Equivalence between integrodifferential and differential systems

The restoration technique can readily be used to establish the following

62 **Assertion** Let $\mathcal{A}$ be a system characterized by an input-output relation of the form

63 $$\left(a_n p^n + \cdots + a_0 + \frac{a_0'}{p} + \cdots + \frac{a_{n'}'}{p^{n'}}\right) y = \left(b_m p^m + \cdots + b_0 + \frac{b_0'}{p} + \cdots + \frac{b_{m'}'}{p^{m'}}\right) u$$

where $a_n, \ldots, a_0, a_0', \ldots, a_{n'}'$ and $b_m, \ldots, b_0, b_0', \ldots, b_{m'}'$ are constants. (Note that *63* is an integrodifferential equation.) Let r be the larger of the two integers n' and m', that is,

64 $$r \triangleq \max(n',m')$$

Then $\mathcal{A}$ is a system of order $r + \max(m,n)$ (see *3.6.39*), and $\mathcal{A}$ is equivalent to the system $\mathcal{B}$ characterized by the single differential equation

65 $$(a_n p^{n+r} + \cdots + a_{n'}' p^{r-n'})y = (b_m p^{m+r} + \cdots + b_{m'}' p^{r-m'})u$$

66 *Comment* In effect, this assertion legitimizes the commonly used procedure of acting on both sides of an integrodifferential equation with the operator p^r and replacing the product $p^r \cdot \frac{1}{p^\lambda}$ by $p^{r-\lambda}$. While it is true (by *4.4.48*) that $p^r \cdot \frac{1}{p^\lambda}$ is equivalent to $p^{r-\lambda}$ (in the sense of system equivalence), this in itself does not prove that $\mathcal{B}$ is equivalent to $\mathcal{A}$. The reason for this is that when we act on both sides of a differential or an integrodifferential equation with, say, p, the resulting equation characterizes a system which contains the original system (in the sense of Definition *2.5.1*) but is not necessarily equivalent to it.

Assertion *62* can be proved in several ways. A straightforward way is to apply the Laplace transformation to both sides of *63* and *65* and verify that the input-output-state relations of $\mathcal{A}$ and $\mathcal{B}$ (in the Laplace transform form) are identical. It is simpler, however, to use the restoration technique in the manner of Example *59*.

We leave the details of the proof as an exercise for the reader. ◁

In addition to the results concerning system equivalence which were established in the foregoing presentation, we shall need several more bearing on the equivalence between a system $\mathcal{A}$ and an interconnection of adders, scalors, integrators, and differentiators. These results are deduced in the following section.

8 *Further equivalence properties of time-invariant systems*

Let $\mathcal{A}$ be a system of order r characterized by an input-output relation of the form

1 $$L(p)y = M(p)u$$

where

2 $$\begin{aligned} L(p) &= a_n p^n + \cdots + a_0 \\ M(p) &= b_m p^m + \cdots + b_0 \\ r &= \max(m,n) \end{aligned}$$

Let $\mathcal{B}$ be a realization of $\mathcal{A}$ in the form of an interconnection of adders, scalors, integrators, and differentiators. Then, by the definition of a realization of $\mathcal{A}$, $\mathcal{A}$ and $\mathcal{B}$ are equivalent systems (see *2.5.19*).

To show that $\mathcal{B}$ is a realization of $\mathcal{A}$, we have to demonstrate that (I) $\mathcal{A}$ and $\mathcal{B}$ are zero-state equivalent and (II) $\mathcal{A}$ and $\mathcal{B}$ are zero-input equivalent. The verification of (I) is, in general, a simple matter, since to show that $\mathcal{A}$ and $\mathcal{B}$ are zero-state equivalent, it suffices to verify that $H_a(s)$, the transfer function of $\mathcal{A}$, is identical with $H_b(s)$, the transfer function of $\mathcal{B}$. The verification of (II), on the other hand, is a relatively cumbersome and not altogether straightforward problem. Thus, it is desirable to have criteria under which $\mathcal{B}$ can be asserted to be a realization of $\mathcal{A}$ without the necessity of demonstrating that $\mathcal{A}$ and $\mathcal{B}$ are zero-input equivalent.

One such criterion and several related results are established in the sequel. As a preliminary, we derive a simple bound on the number of integrators and differentiators in $\mathcal{B}$.

3 **Assertion** If $\mathcal{A}$ is of order r and $\mathcal{B}$ is a realization of $\mathcal{A}$ in the form of an interconnection of adders, scalors, integrators, and differentiators, then $\mathcal{B}$ must contain at least r integrators and differentiators.

Proof In Sec. *6* we have shown that the vector $\mathbf{x}$, whose components are the outputs of the integrators and the inputs to the differentiators in $\mathcal{B}$, qualifies as a state vector for $\mathcal{B}$. Thus, if m is the total number of integrators and differentiators in $\mathcal{B}$ and $\dim \Sigma$ is the dimension of the state space of $\mathcal{B}$, then

4 $$\dim \Sigma \leq m$$

since the dimension of the state space of $\mathcal{B}$ is bounded by the number of components in $\mathbf{x}$.

On the other hand, by Assertion *4.2.4* and Remark *4.2.12*, the dimension of the state space of any system equivalent to $\mathcal{A}$ cannot be lower than r, that is,

5 $$r \leq \dim \Sigma$$

Consequently,

6 $$m \geq r$$

which is what we wanted to establish.

A criterion of equivalence between $\mathcal{A}$ and $\mathcal{B}$

Consider now a system $\mathcal{A}$ characterized by an input-output relation of the form

7 $$L(p)y = M(p)u$$

where

$$L(p) = a_n p^n + \cdots + a_0 \qquad M(p) = b_m p^m + \cdots + b_0$$

with the order of $\mathcal{A}$ being $r \triangleq \max(m,n)$. Suppose that $\mathcal{B}$ is an interconnection of scalors, adders, integrators, and differentiators which satisfies the condition stated in Assertion *3* with the equality sign, that is, $\mathcal{B}$ has r integrators and differentiators. Furthermore, suppose that $\mathcal{B}$ is zero-state equivalent to $\mathcal{A}$, which implies that the transfer function of $\mathcal{B}$, $H_b(s)$, is identical with that of $\mathcal{A}$, or, more concretely,

$$H_b(s) = H_a(s) = \frac{M(s)}{L(s)}$$

Are these conditions sufficient to ensure that $\mathcal{B}$ is a realization of $\mathcal{A}$ or, in other words, that $\mathcal{B}$ is equivalent to $\mathcal{A}$? The answer to this question is in the negative, as the following simple counterexample demonstrates.

8 *Example* Let $\mathcal{A}$ be defined by

9 $$p(p+1)y = (p+1)u$$

and let $\mathcal{B}$ be a realization of the system characterized by the input-output relation

$$p^2 y = p u$$

Clearly, both $\mathcal{A}$ and $\mathcal{B}$ are of order 2 and $\mathcal{A}$ is zero-state equivalent to $\mathcal{B}$, since $H_a(s) = H_b(s) = 1/s$. A realization of $\mathcal{B}$ comprising two integrators is shown in Fig. *4.8.1*. (Note that the first integrator is

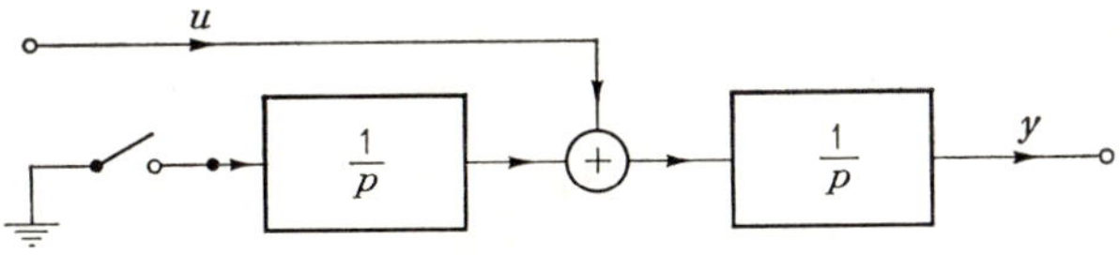

Fig. 4.8.1 System of Example *4.8.8*.

grounded through a switch, which means that its input is zero for $t \geq 0$. Such realizations will be treated in greater detail in Sec. *9*.)

Now $\mathcal{B}$ is not equivalent to $\mathcal{A}$ because it is not zero-input equivalent to $\mathcal{A}$, which in turn follows from the fact that the basis functions of $\mathcal{A}$ are $1/s(s+1)$ and $1/(s+1)$, whereas those of $\mathcal{B}$ are $1/s^2$ and $1/s$. Thus, the system of Fig. *4.8.1* is not equivalent to $\mathcal{A}$ even though it contains two integrators and is zero-state equivalent to $\mathcal{A}$.

We note that, in the case of the input-output relation *9*, $L(p)$ and $M(p)$ have a common factor $p+1$. Is it possible to construct a counterexample in which $L(p)$ and $M(p)$ do not have a factor in common? The answer to this question as well as a very useful criterion of equivalence between $\mathcal{A}$ and $\mathcal{B}$ is provided by the following

10 **Theorem** Let $\mathcal{A}$ be a system of order r characterized by an input-output relation of the form

$$L(p)y = M(p)u$$

in which the polynomials $L(p)$ and $M(p)$ do not have common factors (i.e., do not have one or more common zeros).

Let $\mathcal{B}$ be an interconnection of scalors, adders, integrators, and differentiators which is zero-state equivalent to $\mathcal{A}$ and which contains exactly r integrators and differentiators. Then $\mathcal{B}$ is a realization of $\mathcal{A}$; that is, $\mathcal{B}$ is equivalent to $\mathcal{A}$.

Proof It is easy to prove this theorem by contradiction. Specifically, suppose that (I) $\mathcal{B}$ is zero-state equivalent to $\mathcal{A}$, (II) $\mathcal{B}$ contains exactly r integrators and differentiators, (III) $L(p)$ and $M(p)$ have no common factors, and (IV) $\mathcal{B}$ is not equivalent to $\mathcal{A}$. We proceed to show that (I), . . . , (IV) are inconsistent.

By Theorem *4.7.52*, there exists a system, call it $\hat{\mathcal{B}}$, which is equivalent to $\mathcal{B}$ and which is characterized by a single differential equation of the form

11 $$\tilde{L}(p)y = \tilde{M}(p)u$$

where $\tilde{L}(p)$ and $\tilde{M}(p)$ are polynomials in p. Since $\mathcal{B}$ is zero-state equivalent to $\mathcal{A}$, so $\hat{\mathcal{B}}$ must be. Consequently, the operators $\tilde{L}(p)$ and $\tilde{M}(p)$ in *11* must necessarily be such that

12 $$\frac{\tilde{M}(s)}{\tilde{L}(s)} = \frac{M(s)}{L(s)}$$

in order to satisfy the requirement that the transfer function of $\mathcal{A}$ be identical with that of $\mathcal{B}$ and $\hat{\mathcal{B}}$, that is,

$$H_a(s) = \frac{M(s)}{L(s)}$$
$$= H_b(s) = H_{\hat{b}}(s) = \frac{\tilde{L}(s)}{\tilde{M}(s)}$$

Now since $M(s)$ and $L(s)$ have no factors in common, *11* implies that $\tilde{L}(p)$ and $\tilde{M}(p)$ must be of the form

13 $$\tilde{L}(p) = Q(p)L(p)$$
$$\tilde{M}(p) = Q(p)M(p)$$

where $Q(p)$ is a polynomial in p. Furthermore, by (IV), $\mathcal{A}$ is not equivalent to $\mathcal{B}$ and hence $\mathcal{A}$ is not equivalent to $\hat{\mathcal{B}}$. This implies that $Q(p)$ in *13* is a polynomial of degree not less than 1. [For, if $Q(p)$ were a constant, $\mathcal{A}$ and $\mathcal{B}$ would be characterized by the same input-output relation and hence would be equivalent.] This in turn implies that $\hat{r}$, the order of $\hat{\mathcal{B}}$ [which is equal to the highest power of p in $\tilde{L}(p)$ and $\tilde{M}(p)$], is greater than r, the order of $\mathcal{A}$.

Now $\mathcal{B}$ is a realization of $\hat{\mathcal{B}}$, since $\mathcal{B}$ is equivalent to $\hat{\mathcal{B}}$. By Assertion *3*, $\mathcal{B}$ must contain at least $\hat{r}$ integrators and differentiators. This contradicts the assumption that $\mathcal{B}$ contains exactly r integrators and differentiators and thus proves the theorem. ◁

We shall make extensive use of Theorem *10* in Sec. *9*, where it will play a key role in the development of a simple and yet effective technique for associating a state vector with a given differential system $\mathcal{A}$. For this reason, we limit its illustration at this point to one simple

14 *Example* Let $\mathcal{A}$ be a system of order 2 characterized by the input-output relation

$$(p + 1)(p + 2)y = p\,u$$

The system $\mathcal{B}$ shown in Fig. *4.8.2* comprises two integrators and is zero-state equivalent to $\mathcal{A}$. Therefore, by Theorem *10*, $\mathcal{B}$ is equivalent to $\mathcal{A}$.

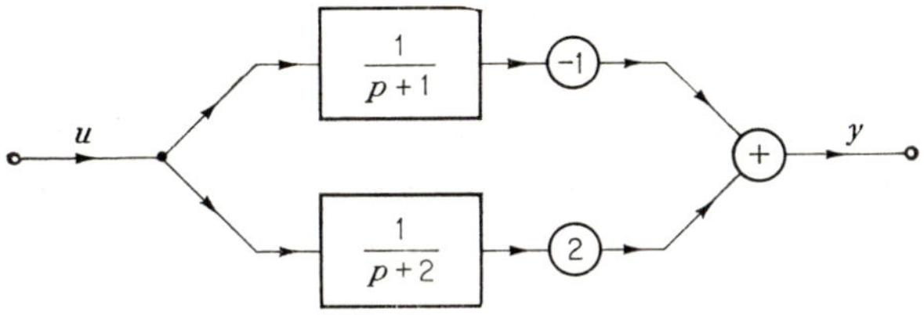

Fig. 4.8.2 System of Example *4.8.14*.

Commutativity

One of the most basic equivalence properties of linear time-invariant systems is the property of *commutativity*, which is expressed by the following

15 **Theorem** Let $\mathcal{A}_1$ and $\mathcal{A}_2$ be two systems characterized by $\frac{1}{L_1(p)} \cdot M_1(p)$ and $\frac{1}{L_2(p)} \cdot M_2(p)$, respectively, where L_1 and M_1 have no factors in common, and likewise for L_2 and M_2. [This notation signifies that $\mathcal{A}_1$ and $\mathcal{A}_2$ are characterized by input-output relations of the form $L_1(p)y = M_1(p)u$ and $L_2(p)y = M_2(p)u$, respectively (see *4.5.9 et seq.*).] Then $\mathcal{A}_1$ and $\mathcal{A}_2$ commute, i.e.,

$$\mathcal{A}_1\mathcal{A}_2 \equiv \mathcal{A}_2\mathcal{A}_1$$

(meaning that the tandem combination of $\mathcal{A}_1$ and $\mathcal{A}_2$, with $\mathcal{A}_1$ following $\mathcal{A}_2$, is equivalent to the tandem combination of $\mathcal{A}_1$ and $\mathcal{A}_2$ with $\mathcal{A}_1$ preceding $\mathcal{A}_2$) if and only if $M_1(p)$ has no factors in common with $L_2(p)$ and $M_2(p)$ has no factors in common with $L_1(p)$. ◁

We shall first establish three special cases of this theorem and then use them to prove the theorem itself.

Case 1 Let $\mathcal{A}_1$ and $\mathcal{A}_2$ be two differentiators represented by $(p + \gamma_1)$ and $(p + \gamma_2)$, respectively. Then $(p + \gamma_1) \cdot (p + \gamma_2)$ is equivalent to $(p + \gamma_2) \cdot (p + \gamma_1)$ for all γ_1, γ_2. [Here $(p + \gamma_1) \cdot (p + \gamma_2)$ and $(p + \gamma_2) \cdot (p + \gamma_1)$ represent the systems $\mathcal{A}_1\mathcal{A}_2$ and $\mathcal{A}_2\mathcal{A}_1$, respectively.]

Proof In Sec. *4* we have shown that the basis functions of a system of the differential operator type

$$y = (b_m p^m + \cdots + b_0)u \qquad b_m \neq 0$$

are given by 1, s, . . . , s^{m-1}. Consequently, the systems $\mathcal{A}_1\mathcal{A}_2$ and $\mathcal{A}_2\mathcal{A}_1$ have identical basis functions, namely, 1, s. Furthermore, $\mathcal{A}_1\mathcal{A}_2$ and $\mathcal{A}_2\mathcal{A}_1$ have the same transfer function

16 $$H_{12}(s) = H_{21}(s) = (s + \gamma_1)(s + \gamma_2)$$

From this it follows that $\mathcal{A}_1\mathcal{A}_2$ and $\mathcal{A}_2\mathcal{A}_1$ are zero-input as well as zero-state equivalent, and hence $\mathcal{A}_1\mathcal{A}_2 \equiv \mathcal{A}_2\mathcal{A}_1$. [Note that it is partly because of this commutative property of $(p + \gamma_1)$ and $(p + \gamma_2)$ that we usually write $(p + \gamma_1)(p + \gamma_2)$ instead of $(p + \gamma_1) \cdot (p + \gamma_2)$.]

Case 2 Let $\mathcal{A}_1$ and $\mathcal{A}_2$ be two integrators represented by $\dfrac{1}{p + \gamma_1}$ and $\dfrac{1}{p + \gamma_2}$, respectively. Then for all γ_1 and γ_2

17 $$\frac{1}{(p + \gamma_1)} \cdot \frac{1}{(p + \gamma_2)} \equiv \frac{1}{(p + \gamma_2)} \cdot \frac{1}{(p + \gamma_1)}$$

Note that, as in the case of differentiators, we usually write

$$\frac{1}{(p + \gamma_1)(p + \gamma_2)} \qquad \text{instead of} \qquad \frac{1}{(p + \gamma_1)} \cdot \frac{1}{(p + \gamma_2)}$$

Proof We repeat the arguments used in Case 1. Specifically, in Sec. *3* it was shown that the basis functions of a system of the reciprocal differential operator type

$$L(p)y = u$$

where

$$L(p) = a_n p^n + \cdots + a_0 \qquad a_n \neq 0$$

are given by

$$\frac{s^{n-1}}{L(s)}, \ldots, \frac{1}{L(s)}$$

Consequently, the systems $\mathcal{A}_1\mathcal{A}_2$ and $\mathcal{A}_2\mathcal{A}_1$ have identical basis functions, namely,

$$\frac{s}{s^2 + (\gamma_1 + \gamma_2)s + \gamma_1\gamma_2}, \frac{1}{s^2 + (\gamma_1 + \gamma_2)s + \gamma_1\gamma_2}$$

Furthermore, $\mathcal{A}_1\mathcal{A}_2$ and $\mathcal{A}_2\mathcal{A}_1$ have the same transfer function

$$H_{12}(s) = H_{21}(s) = \frac{1}{s^2 + (\gamma_1 + \gamma_2)s + \gamma_1\gamma_2}$$

Therefore, $\mathcal{A}_1\mathcal{A}_2$ and $\mathcal{A}_2\mathcal{A}_1$ are equivalent systems.

Case 3 Let $\mathcal{A}_1$ be a differentiator represented by $p + \gamma_1$ and let $\mathcal{A}_2$ be an integrator represented by $\dfrac{1}{p + \gamma_2}\cdot$ Then

18 $$(p + \gamma_1) \cdot \frac{1}{(p + \gamma_2)} \equiv \frac{1}{(p + \gamma_2)} \cdot (p + \gamma_1)$$

if and only if $\gamma_1 \neq \gamma_2$.

Proof To say that $\mathcal{A}_2\mathcal{A}_1$ is represented by $\dfrac{1}{(p + \gamma_2)} \cdot (p + \gamma_1)$ is equivalent to saying that $\mathcal{A}_2\mathcal{A}_1$ is characterized by the input-output relation

$$(p + \gamma_2)y = (p + \gamma_1)u$$

Such a system has a single basis function (see *4.5.26*) $\dfrac{1}{s + \gamma_2}\cdot$

Next, let us determine the basis functions of $(p + \gamma_1) \cdot \dfrac{1}{(p + \gamma_2)}$, that is, $\mathcal{A}_1\mathcal{A}_2$. If the output of $\mathcal{A}_2$ is denoted by v, then $\mathcal{A}_1\mathcal{A}_2$ is characterized by the equations

19 $$\begin{aligned} y &= (p + \gamma_1)v \\ (p + \gamma_2)v &= u \end{aligned}$$

in which v plays the role of a suppressed output variable (see *3.8.26*). On applying the Laplace transformation to both sides of *19* (with u set equal to zero for $t \geq 0$), the expression for the Laplace transform of the zero-input response is found to be

20 $$Z(s;v(0-)) = \frac{v(0-)(\gamma_1 - \gamma_2)}{s + \gamma_2}$$

Thus, $\mathcal{A}_1\mathcal{A}_2$ has just one basis function, $\dfrac{1}{s + \gamma_2}$, which is identical with that of $\mathcal{A}_2\mathcal{A}_1$, *provided* $\gamma_1 \neq \gamma_2$. ◁

A somewhat more illuminating (but also less convincing) way of deriving the same result is this: by the argument used in Case 2, $\dfrac{1}{p + \gamma_2}$ has just one basis function $\dfrac{1}{s + \gamma_2}$ or, as a function of time, $1(t)e^{-\gamma_2 t}$. Consequently, the basis function of $(p + \gamma_1) \cdot \dfrac{1}{(p + \gamma_2)}$ is given by the result of acting with the differential operator $p + \gamma_1$ on $1(t)e^{-\gamma_2 t}$. Deleting the delta function due to the discontinuity at $t = 0$ (see *4.7.19 et seq.*), the basis function in question is seen to be $1(t)e^{-\gamma_2 t}$ *provided* $\gamma_2 \neq \gamma_1$, for otherwise the time function resulting from acting with $p + \gamma_1$ on $1(t)e^{-\gamma_2 t}$ vanishes for $t > 0$.

We have shown so far that $\dfrac{1}{(p + \gamma_2)} \cdot (p + \gamma_1)$ and $(p + \gamma_1) \cdot \dfrac{1}{(p + \gamma_2)}$

have the same set of basis functions $\left(\text{namely, the function } \frac{1}{s+\gamma_2}\right)$ for all values of γ_1 and γ_2 excluding $\gamma_1 = \gamma_2$. Now the transfer functions of these two systems are identical for all values of γ_1 and γ_2, for

$$H_{12}(s) = \frac{s+\gamma_1}{s+\gamma_2} = H_{21}(s)$$

Consequently, $\mathcal{A}_1\mathcal{A}_2$ is equivalent to $\mathcal{A}_2\mathcal{A}_1$ if and only if $\gamma_1 \neq \gamma_2$, which is what we wanted to demonstrate. [Note that this result is in agreement with the conclusion reached in *4.4.44*, namely, that $(p + \gamma_1)$ is a left-constrained inverse of $\frac{1}{p+\gamma_1}$ but not vice versa.] ◁

Having established Cases 1, 2, and 3, it is a simple matter to reduce any given case to these special cases by repeated transposition of the first-order factors in $\mathcal{A}_1$ and $\mathcal{A}_2$. Specifically, suppose that $\mathcal{A}_1$ and $\mathcal{A}_2$ are represented by

$$\mathcal{A}_1: \quad \frac{1}{(p+1)(p+2)} \cdot (p+4)$$

$$\mathcal{A}_2: \quad \frac{1}{(p+3)} \cdot p$$

Then $\mathcal{A}_1\mathcal{A}_2$ is represented by

21 $$\mathcal{A}_1\mathcal{A}_2: \quad \frac{1}{(p+1)(p+2)} \cdot (p+4) \cdot \frac{1}{(p+3)} \cdot p$$

By making use of the substitution principle (*1.10.35*), we can replace *21* with the successive equivalent expressions

$$\frac{1}{(p+1)(p+2)} \cdot (p+4)p \cdot \frac{1}{(p+3)}$$

$$\frac{1}{(p+1)(p+2)} \cdot p(p+4) \cdot \frac{1}{(p+3)}$$

$$p \cdot \frac{1}{(p+1)(p+2)} \cdot (p+4) \cdot \frac{1}{(p+3)}$$

. . . .

$$\frac{1}{(p+3)} \cdot p \cdot \frac{1}{(p+1)(p+2)} \cdot (p+4)$$

with the last expression representing $\mathcal{A}_2\mathcal{A}_1$.

To prove the converse, suppose that the conditions of Theorem *15* are violated in that either $M_1(p)$ and $L_2(p)$ or $M_2(p)$ and $L_1(p)$ or both have factors in common. Then by successive transposition of factors which are not in common, we can put $\mathcal{A}_1\mathcal{A}_2$ and $\mathcal{A}_2\mathcal{A}_1$ into the forms $\mathcal{B}\mathcal{C}$ and $\mathcal{B}^*\mathcal{C}$, respectively, where $\mathcal{B}$, $\mathcal{B}^*$, and $\mathcal{C}$ are products of integrators and differentiators, with $\mathcal{B}$ and $\mathcal{B}^*$ containing the same factors but in reverse order. For example, if $\mathcal{A}_1$ and $\mathcal{A}_2$ are represented by

$$\mathcal{A}_1: \qquad \frac{1}{(p+1)(p+2)} \cdot p$$

$$\mathcal{A}_2: \qquad \frac{1}{p(p+3)}$$

then

22 $$\mathcal{A}_1\mathcal{A}_2 \equiv \mathcal{B}\mathcal{C} \equiv p \cdot \frac{1}{p} \frac{1}{(p+1)(p+2)(p+3)}$$

and

23 $$\mathcal{A}_2\mathcal{A}_1 \equiv \mathcal{B}^*\mathcal{C} \equiv \frac{1}{p} \cdot p \frac{1}{(p+1)(p+2)(p+3)}$$

where

24 $$\mathcal{C} \equiv \frac{1}{(p+1)(p+2)(p+3)}$$

$$\mathcal{B} \equiv p \cdot \frac{1}{p}$$

$$\mathcal{B}^* \equiv \frac{1}{p} \cdot p$$

25 **Remark** Note that $\mathcal{B}$ (or $\mathcal{B}^*$) can always be put into the form of a product $\mathcal{D}\mathcal{R}$, where $\mathcal{D}$ is a differential operator system and $\mathcal{R}$ is a reciprocal differential operator system. Thus, if $\mathcal{B} \equiv \mathcal{D}\mathcal{R}$, then $\mathcal{B}^* \equiv \mathcal{R}\mathcal{D}$, and vice versa. Furthermore, $\mathcal{D}$ and $\mathcal{R}$ are inverse to one another (see *4.4.44*) if the common factors in $L_1(p)$ and $M_2(p)$, and $L_2(p)$ and $M_1(p)$, are of multiplicity 1. ◁

The nonequivalence of $\mathcal{A}_1\mathcal{A}_2$ and $\mathcal{A}_2\mathcal{A}_1$ follows from the nonequivalence of $\mathcal{B}$ and $\mathcal{B}^*$. Specifically, if $\mathcal{B}$ is a product of factors of the form

26 $$(p+\gamma)^{\mu} \cdot \frac{1}{(p+\gamma)^{\lambda}}$$

then $\mathcal{B}^*$ is a product of the same factors in reverse order:

27 $$\frac{1}{(p+\gamma)^{\lambda}} \cdot (p+\gamma)^{\mu}$$

Now *26* and *27* are necessarily nonequivalent because the basis functions of *26* form a proper subset of the basis functions of *27*. [That is, every basis function of *26* is also a basis function of *27*, but there are basis functions of *27* which are not in the set of basis function of *26*. This is due to the fact that the operator $(p+\gamma)^{\mu}$ in *26* "annihilates" one or more of the basis functions of *27*. Note that this argument is similar to that used in Case 3.] This implies that the basis functions of either

$\mathcal{A}_1\mathcal{A}_2$ or $\mathcal{A}_2\mathcal{A}_1$—according as it is $\mathcal{B}$ or $\mathcal{B}^*$ that is of the form $\mathcal{R}\mathcal{D}$—form a proper subset of the basis functions of either $\mathcal{A}_2\mathcal{A}_1$ or $\mathcal{A}_1\mathcal{A}_2$, and this demonstrates that $\mathcal{A}_1\mathcal{A}_2$ and $\mathcal{A}_2\mathcal{A}_1$ are nonequivalent if $L_1(p)$ and $M_2(p)$ or $L_2(p)$ and $M_1(p)$ have factors in common. Q.E.D.

We are now ready to apply the equivalence properties established in this and the preceding section to the problem of associating a state vector with a given differential system. This forms the subject of the following section.

9 *Determination of state vector and state equations by the realization technique*

The various equivalence properties derived in the preceding sections and, in particular, Theorem *4.8.10*, provide us with a basis for an alternative approach to the problem of associating a state vector with a given differential system $\mathcal{A}$. For convenience, the approach in question will be referred to as the *realization technique.*

The essence of the realization technique is this: given a system $\mathcal{A}$ characterized by, say, an input-output relation of the form

$$L(p)y = M(p)u$$

where

$$L(p) = a_n p^n + \cdots + a_0$$
$$M(p) = b_m p^m + \cdots + b_0$$

we construct an equivalent system $\mathcal{B}$ which is a realization of $\mathcal{A}$ in the form of an interconnection of adders, scalors, integrators, and differentiators. Then we associate a state vector $\mathbf{x}$ with $\mathcal{B}$ by the use of Rule *4.6.2* and find the corresponding state equations for $\mathcal{A}$ by the techniques of Sec. *6*. Since $\mathcal{B}$ and $\mathcal{A}$ are equivalent systems, $\mathbf{x}$ qualifies as a state vector for $\mathcal{A}$ and the state equations of $\mathcal{B}$ may be regarded as being also the state equations of $\mathcal{A}$.

This approach is particularly effective when $L(p)$ and $M(p)$ have no factors in common, for in that case we can construct $\mathcal{B}$ simply by synthesizing the transfer function $H(s)$ through the usual techniques of circuit theory (which yield a system that is zero-state equivalent to $\mathcal{A}$) and then invoking Theorem *4.8.10* (assuming the number of integrators and differentiators in $\mathcal{B}$ is equal to the order of $\mathcal{A}$) to verify that $\mathcal{B}$ is equivalent to $\mathcal{A}$.

For purposes of comparison, we shall first apply the realization technique to the various types of systems which we have already studied in Secs. *3* to *5*. Then we shall develop a special version of the realization technique, namely, the so-called partial-fraction expansion technique, an important feature of which is that it yields state equations of the form *4.7.51* in which the matrix **A** is in the Jordan canonical form.

We begin with an analysis of

Systems of the reciprocal differential type

Let $\mathcal{A}$ be a system of the reciprocal differential type characterized by

1 $$(a_n p^n + \cdots + a_0)y = u \qquad a_n \neq 0$$

The transfer function of $\mathcal{A}$ is given by

2 $$H(s) = \frac{1}{a_n s^n + \cdots + a_0}$$

and it is easy to verify that the network $\mathcal{B}$ shown in Fig. *4.9.1* has the

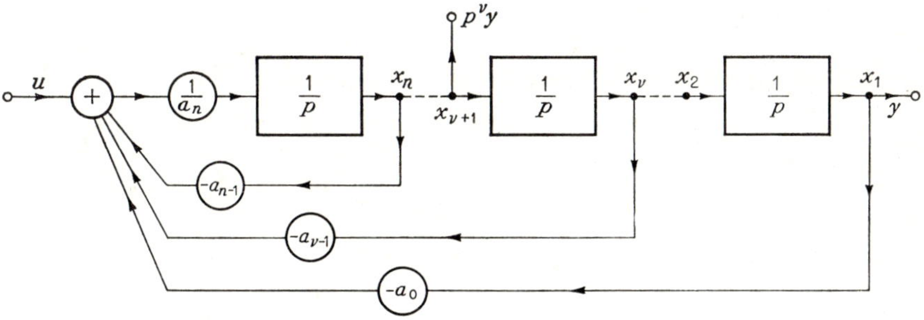

Fig. 4.9.1 Realization of $\dfrac{1}{a_n p^n + \cdots + a_0}$.

same transfer function and hence is zero-state equivalent to $\mathcal{A}$. Furthermore, $\mathcal{B}$ has exactly n integrators. Therefore, by Theorem *4.8.10*, $\mathcal{B}$ is a realization of $\mathcal{A}$.

To associate a state vector with $\mathcal{B}$, we use Rule *4.6.2*, with the results shown in Fig. *4.9.1*. The defining relations for $\mathbf{x}$ are

$$\begin{aligned} x_1 &= y \\ x_2 &= y^{(1)} \\ &\cdots\cdots \\ x_n &= y^{(n-1)} \end{aligned}$$

In this way, we arrive at what we previously called the normal state vector for $\mathcal{A}$ (see *4.3.27*). Needless to say, the corresponding state equations for $\mathcal{B}$ are identical with the normal state equations obtained previously (*4.3.33*).

As an illustration of what happens when we realize $\mathcal{A}$ in a different way, consider the following

3 *Example* Let $\mathcal{A}$ be characterized by the input-output relation

$$(p + 1)(p + 2)^2 y = u$$

In this case, it is trivial to verify that the system $\mathcal{B}$ shown in Fig. *4.9.2*

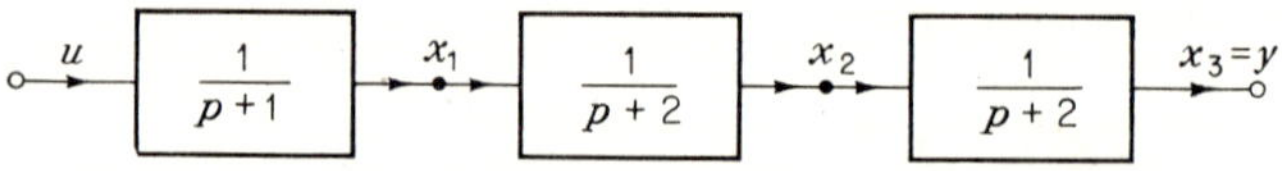

Fig. 4.9.2 Realization of the system of Example *4.9.3*.

constitutes a realization of $\mathcal{A}$. Choosing $\mathbf{x}$ as indicated, the state equations of $\mathcal{B}$ (and hence $\mathcal{A}$) are easily found by inspection of Fig. *4.9.2*:

$$\begin{aligned} \dot{x}_1 &= -x_1 + u \\ \dot{x}_2 &= -2x_2 + x_1 \\ \dot{x}_3 &= -2x_3 + x_2 \\ y &= x_3 \end{aligned}$$

We turn next to

Systems of the differential operator type

Let $\mathcal{A}$ be a system of the differential operator type characterized by the input-output relation

4 $$y = (b_m p^m + \cdots + b_0)u \qquad b_m \neq 0$$

A simple realization of $\mathcal{A}$ is shown in Fig. *4.9.3*.

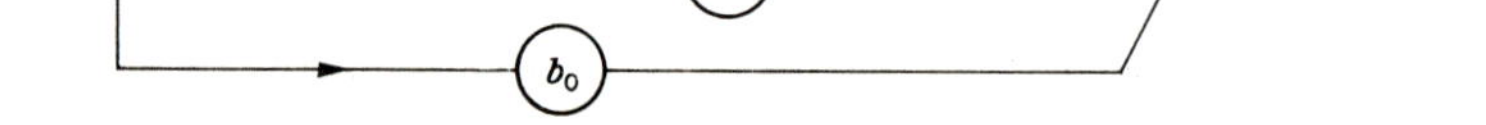

Fig. 4.9.3 Realization of $b_m p^m + \cdots + b_0$.

Associating a component of $\mathbf{x}$ with the input of each differentiator in $\mathcal{B}$, we have

5 $$\begin{aligned} x_1 &= u \\ x_2 &= u^{(1)} \\ &\cdots\cdots \\ x_m &= u^{(m-1)} \end{aligned}$$

which coincides with the definition of the normal state vector for $\mathcal{A}$ (see *4.4.8*). The corresponding state equations have the normal form *4.4.29*,

Systems of the general type

Let $\mathcal{A}$ be a system characterized by the input-output relation

6 $$(a_n p^n + \cdots + a_0)y = (b_n p^n + \cdots + b_0)u$$

and let us assume, to begin with, that $\mathcal{A}$ is a proper system of order n—which means that a_n, but not necessarily b_n, is different from zero.

Again, using Theorem *4.8.10* it is easy to verify that the system $\mathcal{B}$ shown in Fig. *4.9.4* constitutes a realization of $\mathcal{A}$. (Actually, this is true

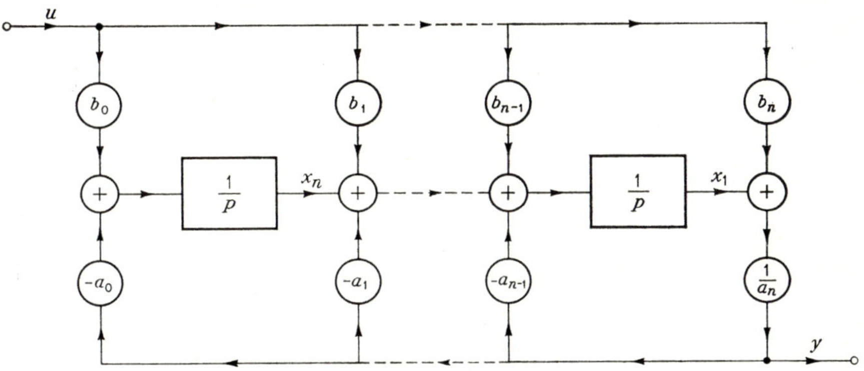

Fig. 4.9.4 Realization of $\dfrac{1}{(a_n p^n + \cdots + a_0)} \cdot (b_n p^n + \cdots + b_0)$, $a_n \neq 0$.

even when the left- and right-hand operators in *6* have one or more factors in common.)

Choosing $\mathbf{x}$ in accordance with Rule *4.6.2*, we obtain by inspection of Fig. *4.9.4* the following relations:

7 $$\begin{aligned} x_1 &= a_n y - b_n u \\ x_2 &= \dot{x}_1 + a_{n-1}y - b_{n-1}u \\ &\cdots\cdots\cdots\cdots\cdots \\ x_n &= \dot{x}_{n-1} + a_1 y - b_1 u \\ \dot{x}_n &= -a_0 y + b_0 u \end{aligned}$$

which are identical with the expressions found previously by the approach of Sec. *5* (see *4.5.18 et seq.*). Clearly, the technique used above is much simpler, since it does not require the determination of the general solution. However, a difficulty arises when $L(p)$ and $M(p)$ in *6* have factors in common, since in this event we cannot draw upon Theorem *4.8.10* to establish the equivalence between $\mathcal{A}$ and its realization $\mathcal{B}$. In such cases, we generally rely on a somewhat vague "continuity" argument to reason that if a structure such as $\mathcal{B}$ in Fig. *4.9.4* is equivalent to $\mathcal{A}$ for all values of a's and b's for which $L(p)$ and $M(p)$

have no factors in common, then $\mathcal{B}$ is equivalent to $\mathcal{A}$ for all values of a's and b's.

8 *Example* Consider a system $\mathcal{A}$ characterized by the input-output relation

$$p(p+1)(p+2)y = p(p^2+1)u$$

or equivalently

9 $$(p^3 + 3p^2 + 2p)y = (p^3 + p)u$$

In this case, system $\mathcal{B}$ of Fig. *4.9.4* reduces to the system shown in Fig. *4.9.5*. Note that p is a common factor in $L(p)$ and $M(p)$ and that

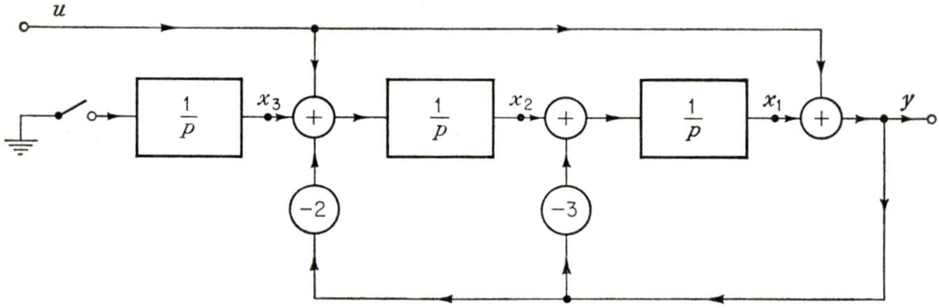

Fig. 4.9.5 System of Example *4.9.8*.

as a result of the absence of the coefficients a_0 and b_0 in *9*, the input to the first integrator in Fig. *4.9.5* is zero for $t \geq 0$. This is indicated in the diagram by a switch which "grounds" the first integrator for $t \geq 0$.

Alternative realizations

There are many alternative realizations of the system $\mathcal{A}$ characterized by *6*, of which the two structures described below are worthy of special attention because of the close relation between them and the structure of Fig. *4.9.1*, which constitutes a realization of a system of the reciprocal differential type which is characterized by *1*. [For convenient reference, we shall denote the latter system by $1/L(p)$ and $\mathcal{A}$ by $\dfrac{1}{L(p)} \cdot M(p)$.]

Specifically, by inspection of the structure of Fig. *4.9.1*, it is clear that if the output of $1/L(p)$ is y, then the input to the νth integrator (counting from right to left) is $p^\nu y$. From this it follows that we can obtain any linear combination of $y, \ldots, p^{n-1}y$ by "tapping" the inputs to the integrators and, more specifically, we can realize the system characterized by the operational expression

10 $$\mathcal{A}': \qquad (b_n p^n + \cdots + b_0) \cdot \frac{1}{(a_n p^n + \cdots + a_0)}$$

or, more simply, $M(p) \cdot \dfrac{1}{L(p)}$, by using the structure of Fig. *4.9.6*.

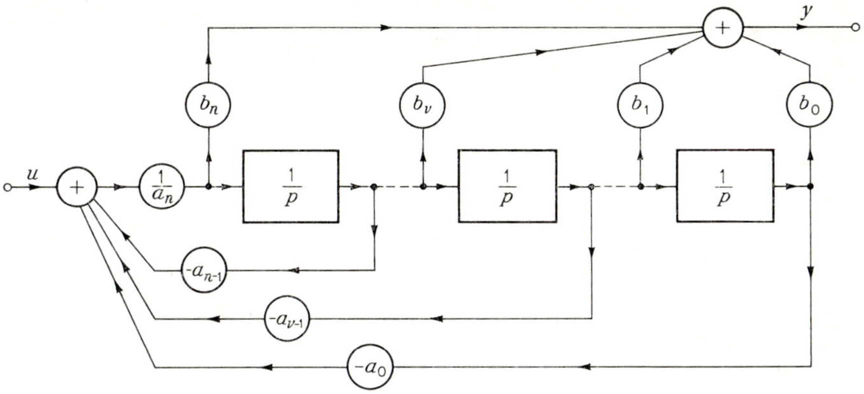

Fig. 4.9.6 Realization of $(b_n p^n + \cdots + b_0) \cdot \dfrac{1}{(a_n p^n + \cdots + a_0)}$.

It should be noted, however, that by virtue of Theorem *4.8.15*, $\mathfrak{A}'$ is *not* equivalent to $\mathfrak{A}$ $\left[\text{which is defined by } \frac{1}{L(p)} \cdot M(p) \text{ rather than } M(p) \cdot \frac{1}{L(p)}\right]$ unless $L(p)$ and $M(p)$ have no common factors. If this condition is satisfied, then the structure of Fig. *4.9.6* constitutes a realization of $\mathfrak{A}$, with the property (which is also possessed by the structure of Fig. *4.9.7*) that the corresponding **A** matrix in the state equations *4.7.51* has exactly the same form for this structure as it has for the realization of $1/L(p)$ in Fig. *4.9.1*.

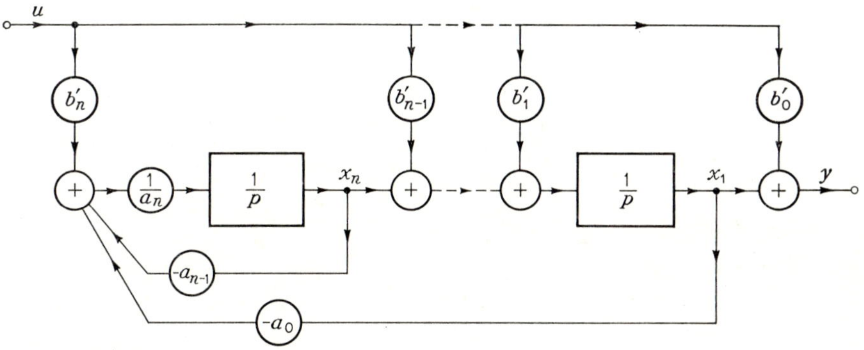

Fig. 4.9.7 Alternative realization of

$$\frac{1}{(a_n p^n + \cdots + b_0)} \cdot (b_n p^n + \cdots + b_0) \qquad a_n \neq 0$$

The second structure, which is shown in Fig. *4.9.7*, follows from the following observations.

(I) We can write *6* as $L(p)y = \tilde{u}$, where $\tilde{u} = M(p)u$. This implies that we can obtain the output of $\frac{1}{L(p)} \cdot M(p)$ by applying the input $\tilde{u} = M(p)u$ to the reciprocal differential system $1/L(p)$. Note that $M(p)$ should be interpreted as an indefinite differential operator in the sense of *3.8.11*.

(II) Suppose that we add to the output of $1/L(p)$ (which is in an initial state $\boldsymbol{\alpha}$) a time function w and raise the question: What is the time function v which if added to the input of $1/L(p)$ would produce the added output w? By Assertion *4.4.49*, the answer to this question is $v = L(p)w$, provided the initial state of $1/L(p)$ is changed from $\boldsymbol{\alpha}$ to $\boldsymbol{\alpha}'$, where

11 $$\boldsymbol{\alpha}' = \boldsymbol{\alpha} - [w(t_0-), \ldots, w^{(n-1)}(t_0-)]$$

(III) Suppose we take the structure of Fig. *4.9.1* and add a time function u_ν to the output of the νth integrator (counting from right to left), as shown in Fig. *4.9.8*. We observe that the structure to the left

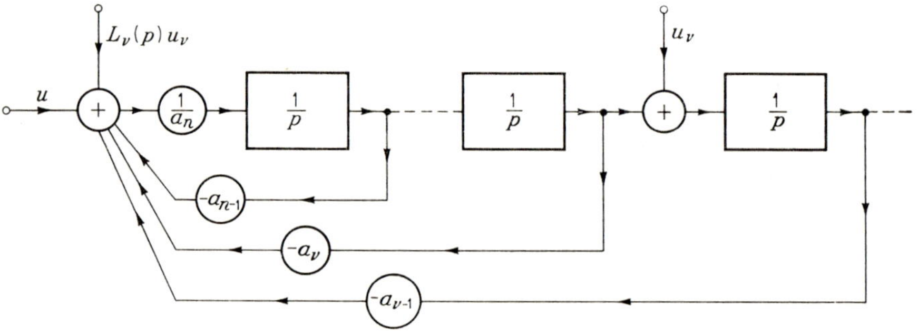

Fig. 4.9.8 Replacement of u_ν by an input applied at the input of $\frac{1}{(a_np^n + \cdots + a_0)}$.

of the point at which u_ν is injected is of the same form as that of Fig. *4.9.1*, except that the differential operator corresponding to it is

12 $$\begin{aligned} L_\nu(p) &= a_np^{n-\nu} + \cdots + a_\nu \qquad \nu = 1, \ldots, n-1 \\ L_0(p) &= L(p) \\ L_n(p) &= 1 \end{aligned}$$

with the last two relations signifying that u_0 and u_n are added to the output and input, respectively, of $1/L(p)$.

From this observation and (II) it follows that injecting u_ν at the input of the νth integrator is equivalent to adding the time function $L_\nu(p)u_\nu$ to the input of $1/L(p)$, as illustrated in Fig. *4.9.8*. It is understood, of course, that when we replace u_ν by $L_\nu(p)u_\nu$, we should also modify the states of the first $n - \nu$ integrators (from left to right) in Fig. *4.9.8* in accordance with *11*.

(IV) Finally, suppose that we make u_ν proportional to u, that is, $u_\nu = b'_\nu u$, $\nu = 0, \ldots, n$, where the b'_ν are as yet undetermined constants. By (III), this is equivalent to adding to the input of $1/L(p)$ the time function

$$b_0'L_0(p)u + b_1'L_1(p)u + \cdots + b_n'L_n(p)u$$

Thus, if we can choose the b_ν' in such a way as to make this time function identical with $\tilde{u}$ [see (I)], that is,

13 $$[b_0'L_0(p) + \cdots + b_n'L_n(p)]u = (b_np^n + \cdots + b_0)u$$

then the structure of Fig. *4.9.8* will have the same output as $\dfrac{1}{L(p)} \cdot M(p)$ for all u and hence will constitute a realization of $\mathcal{A}$.

It is easy to show that such b_ν', $\nu = 0, \ldots, n$, can always be found. Specifically, on substituting *12* into *13* and equating the coefficients of like terms, we obtain the following triangular system of equations:

14 $$\begin{aligned} b_n &= b_0'a_n \\ b_{n-1} &= b_0'a_{n-1} + b_1'a_n \\ &\cdots\cdots\cdots\cdots \\ b_0 &= b_0'a_0 + \cdots + b_{n-1}'a_{n-1} + b_n' \end{aligned}$$

which can always be solved for the b_ν' in terms of the b_ν and the a_ν, since $a_n \neq 0$.

On identifying the components of $\mathbf{x} = (x_1, \ldots, x_n)$ with the outputs of integrators in Fig. *4.9.7*, the state equations of $\mathcal{A}$ become

15 $$\begin{aligned} \dot{x}_1 &= x_2 + b_1'u \\ &\cdots\cdots\cdots \\ \dot{x}_{n-1} &= x_n + b_{n-1}'u \\ \dot{x}_n &= \frac{1}{a_n}(-a_0x_1 - \cdots - a_{n-1}x_n + b_n'u) \\ y &= x_1 + b_0'u \end{aligned}$$

with the components of the state vector expressed in terms of u, y, and their derivatives by

16 $$\begin{aligned} x_1 &= y - b_0'u \\ x_2 &= y^{(1)} - b_0'u^{(1)} - b_1'u \\ &\cdots\cdots\cdots\cdots \\ x_n &= y^{(n-1)} - b_0'u^{(n-1)} - \cdots - b_{n-1}'u \end{aligned}$$

For the time-varying case, equations analogous to *15* will be derived by a somewhat different technique in Chap. *6*, Sec. *4*.

Case of improper $\mathcal{A}$

The case of improper $\mathcal{A}$ can readily be reduced to the cases treated above by using the decomposition technique stated in Theorem *4.5.34*. In this way, $\mathcal{A}$ is realized as a parallel combination of a proper system

$\hat{\mathcal{A}}$ which comprises scalors, adders, and integrators, and a system $\mathcal{D}$ of the differential operator type whose realization comprises scalors, adders, and differentiators.

17 **Remark** It is possible to realize an improper $\mathcal{A}$ as a system $\mathcal{B}$ which does not contain any differentiators. One way of doing this is based on the observation (see *4.4.48*) that a differentiator represented by $(p + \gamma)$ is inverse to the integrator represented by $1/(p + \gamma)$ and hence can be realized in terms of the latter by the scheme of Comment *2.10.24* which is illustrated in Fig. *4.9.9.* (The triangular block represents an infinite gain amplifier.) Using the substitution described

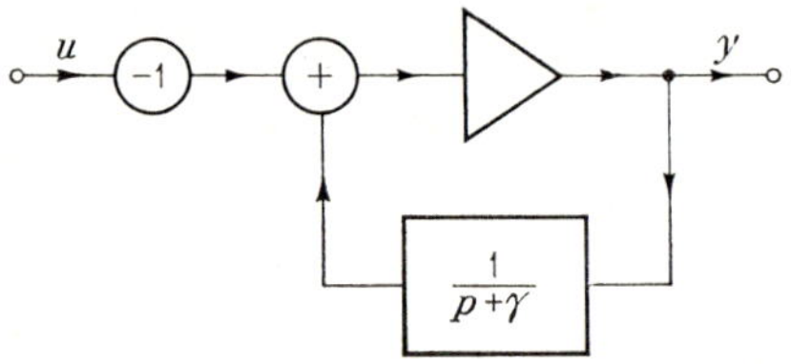

Fig. 4.9.9 Realization of $p + \gamma$. (The triangular block represents an amplifier with infinite gain.)

above in $\mathcal{D}$, $\mathcal{B}$ becomes a system comprising scalors, adders, integrators, and infinite gain amplifiers. It is more convenient, however, to realize $\mathcal{A}$ in the form shown in Fig. *4.9.10,* which employs a single infinite gain amplifier. By the use of Theorem *4.8.10,* it is easy to

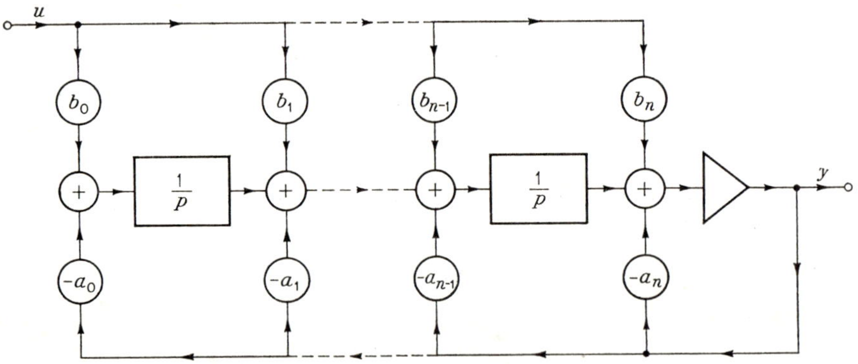

Fig. 4.9.10 Realization of a not necessarily proper system

$$\frac{1}{(a_n p^n + \cdot \cdot \cdot + a_0)} \cdot (b_n p^n + \cdot \cdot \cdot + b_0).$$

verify that this structure is equivalent to $\mathcal{A}$. Note that for $a_n = 0$, the realization of Fig. *4.9.4* reduces to that of Fig. *4.9.10.*

Partial-fraction expansion technique

The technique to be described below is merely a special case of the realization technique. Its most significant feature is that it leads to state equations in which the **A** matrix has a diagonal or, more generally, a Jordan canonical form (see *D.5.21*). Its main disadvantage is that it requires the factorization of $L(p)$.

Specifically, let $\mathcal{A}$ be a proper system characterized by an input-output relation

18 $$L(p)y = M(p)u$$

in which $L(p)$ has, or can be put into, the form

$$L(p) = (p - \lambda_1)^{\mu_1} \cdots (p - \lambda_q)^{\mu_q} \qquad \mu_1 + \cdots + \mu_q = n$$

in which $\lambda_1, \ldots, \lambda_q$ are the (distinct) zeros of $L(p)$ and $\mu_1, \ldots, \mu_q$ are their respective multiplicities. We assume that $M(p)$ and $L(p)$ have no factors in common.

If $\mathcal{A}$ is an improper system, then by Theorem *4.5.34* we can decompose it into a proper system $\hat{\mathcal{A}}$ and a system of the differential operator type $\hat{\mathfrak{D}}$. The procedure described below will then apply to $\hat{\mathcal{A}}$.

For simplicity, we consider first the case where $\mu_1 = \cdots = \mu_q = 1$, that is, all the zeros of $L(p)$ are simple.

Case of simple zeros

In this case, $\mathcal{A}$ is characterized by the input-output relation

$$(p - \lambda_1) \cdots (p - \lambda_n)y = M(p)u$$

where the λ_i, $i = 1, \ldots, n$, are distinct, the degree of $M(p)$ does not exceed n, and $(p - \lambda_i)$ is not a factor of $M(p)$, $i = 1, \ldots, n$.

The transfer function of $\mathcal{A}$ is given by

$$H(s) = \frac{M(s)}{(s - \lambda_1) \cdots (s - \lambda_n)}$$

and on expanding $H(s)$ in partial fractions, we have

$$H(s) = \frac{c_1}{s - \lambda_1} + \cdots + \frac{c_n}{s - \lambda_n} + d_0$$

where the c_i and d_0, $i = 1, \ldots, n$, are constants depending on $\lambda_1, \ldots, \lambda_n$ and the coefficients of $M(p)$, and where $d_0 = 0$ if $\mathcal{A}$ is a strictly proper system [i.e., if the degree of $M(p)$ does not exceed $n - 1$].

Now it is easy to verify by inspection that the system $\mathcal{B}$ shown in Fig. *4.9.11* has this transfer function and hence by Theorem *4.8.10* is a realization of $\mathcal{A}$.

On associating the components of **x** with the outputs of integrators in $\mathcal{B}$, the state equations of $\mathcal{B}$ are easily found by inspection. They read

$$\begin{aligned}\dot{x}_1 &= \lambda_1 x_1 + u \\ &\cdots\cdots\cdots \\ \dot{x}_n &= \lambda_n x_n + u \\ y &= c_1x_1 + \cdots + c_nx_n + d_0u\end{aligned}$$

For convenient reference, these results are summarized in the

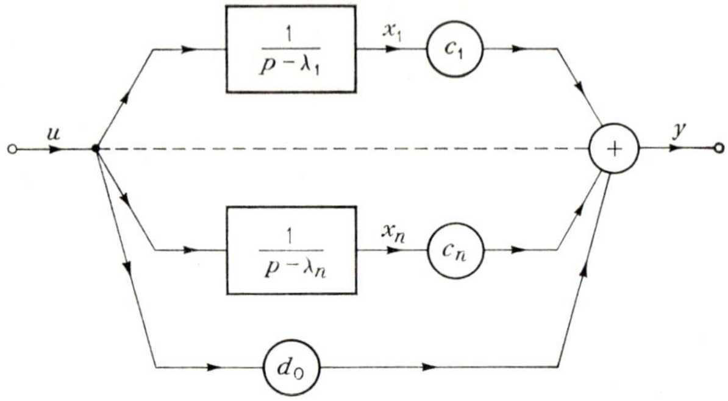

Fig. 4.9.11 Partial-fraction expansion in the case of simple poles.

19 **Assertion** Let $\mathcal{A}$ be a proper system characterized by the input-output relation

$$(p - \lambda_1) \cdots (p - \lambda_n)y = M(p)u$$

in which the λ_i, $i = 1, \ldots, n$, are distinct and no $(p - \lambda_i)$ is a factor of $M(p)$. Furthermore, let the partial-fraction expansion of the transfer function $H(s)$ be

20 $$H(s) = \frac{c_1}{s - \lambda_1} + \cdots + \frac{c_n}{s - \lambda_n} + d_0$$

Then the system $\mathcal{B}$ shown in Fig. *4.9.11* is a realization of $\mathcal{A}$ and the vector $\mathbf{x} = (x_1, \ldots, x_n)$ defined by

$$\dot{x}_i = \lambda_i x_i + u \qquad i = 1, \ldots, n$$

qualifies as a state vector for $\mathcal{A}$. The corresponding state equations of $\mathcal{A}$ read (in matrix form)

21 $$\begin{aligned}\dot{\mathbf{x}} &= \mathbf{A}\mathbf{x} + \mathbf{b}u \\ y &= \langle \mathbf{c},\mathbf{x} \rangle + d_0u\end{aligned}$$

where

22 $$\mathbf{A} = \begin{bmatrix} \lambda_1 & 0 & \cdots & 0 \\ 0 & \lambda_2 & \cdots & 0 \\ \cdot & \cdot & \cdots & \cdot \\ 0 & 0 & \cdots & \lambda_n \end{bmatrix} \qquad \mathbf{b} = \begin{bmatrix} 1 \\ 1 \\ \cdots \\ 1 \end{bmatrix} \qquad \mathbf{c} = [c_1c_2 \cdots c_n]$$

Note that under the assumption that the zeros of $L(p)$ are simple, the partial-fraction technique leads to state equations *21* in which the **A** matrix is in diagonal form.

Case of multiple zeros

For simplicity, we assume that $L(p)$ has just one multiple zero of order μ, in which case we write

$$L(p) = (p - \lambda_1)^\mu (p - \lambda_{\mu+1}) \cdots (p - \lambda_n)$$

[It is a trivial matter to extend the results obtained under this simplifying assumption to the case where $L(p)$ has more than one zero of multiplicity higher than 1.]

Proceeding as in the case of simple zeros, we expand the transfer function $H(s)$:

23 $$H(s) = \frac{M(s)}{(s - \lambda_1)^\mu (s - \lambda_{\mu+1}) \cdots (s - \lambda_n)}$$

into the partial fractions

24 $$H(s) = \frac{c_1}{(s - \lambda_1)^\mu} + \frac{c_2}{(s - \lambda_1)^{\mu-1}} + \cdots + \frac{c_\mu}{s - \lambda_1} + \frac{c_{\mu+1}}{s - \lambda_{\mu+1}} + \cdots + \frac{c_n}{s - \lambda_n} + d_0$$

The system $\mathcal{B}$ shown in Fig. *4.9.12* has this transfer function and hence, by Theorem *4.8.10*, is a realization of $\mathcal{A}$.

Associating the components of $\mathbf{x} = (x_1, \ldots, x_n)$ with the outputs of integrators in the manner indicated in Fig. *4.9.12*, the state equations

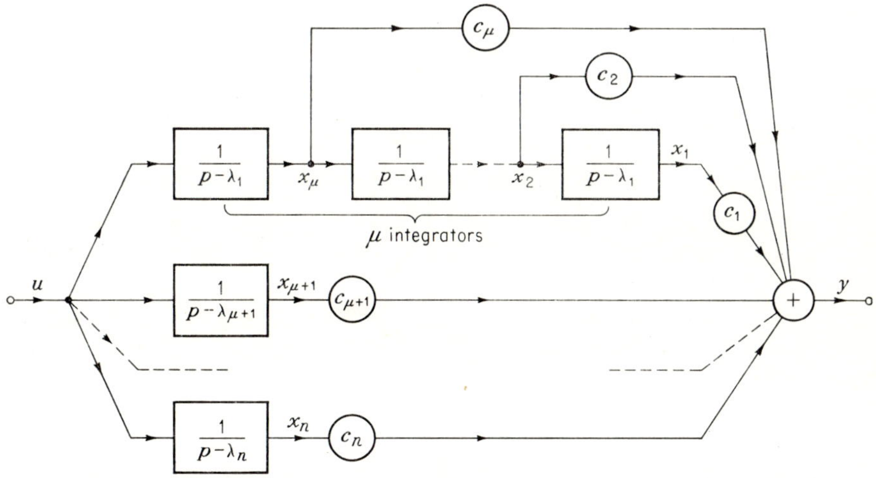

Fig. 4.9.12 Partial-fraction expansion in the case of multiple poles.

of $\mathfrak{A}$ become

25
$$
\begin{aligned}
\dot{x}_1 &= \lambda_1 x_1 + x_2 \\
&\cdots\cdots\cdots \\
\dot{x}_{\mu-1} &= \lambda_1 x_{\mu-1} + x_\mu \\
\dot{x}_\mu &= \lambda_1 x_\mu + u \\
&\cdots\cdots\cdots \\
\dot{x}_n &= \lambda_n x_n + u \\
y &= c_1 x_1 + \cdots + c_n x_n + d_0
\end{aligned}
$$

These results are summarized in the

26 **Assertion** Let $\mathfrak{A}$ be a proper system characterized by the input-output relation

27
$$(p - \lambda_1)^\mu (p - \lambda_{\mu+1}) \cdots (p - \lambda_n) y = M(p) u$$

in which $\lambda_1, \lambda_{\mu+1}, \ldots, \lambda_n$ are distinct and no $(p - \lambda_i)$, $i = 1, \mu + 1, \ldots, n$, is a factor of $M(p)$. Furthermore, let the partial-fraction expansion of the transfer function of $\mathfrak{A}$ be

28
$$H(s) = \frac{c_1}{(s - \lambda_1)^\mu} + \cdots + \frac{c_\mu}{s - \lambda_1} + \frac{c_{\mu+1}}{s - \lambda_{\mu+1}} + \cdots + \frac{c_n}{s - \lambda_n} + d_0$$

Then the system $\mathfrak{B}$ shown in Fig. *4.9.12* is a realization of $\mathfrak{A}$ and the vector $\mathbf{x} = (x_1, \ldots, x_n)$ defined as in Fig. *4.9.12* qualifies as a state vector for $\mathfrak{A}$. The corresponding state equations of $\mathfrak{A}$ read (in matrix form)

29
$$
\begin{aligned}
\dot{\mathbf{x}} &= \mathbf{A}\mathbf{x} + \mathbf{b}u \\
y &= \langle \mathbf{c}, \mathbf{x} \rangle + d_0 u
\end{aligned}
$$

where

$$
\mathbf{A} = \begin{bmatrix} \lambda_1 & 1 & 0 & \cdots & 0 \\ 0 & \lambda_1 & 1 & \cdots & 0 \\ 0 & 0 & \lambda_1 & \cdots & 0 \\ \cdot & \cdot & \cdot & \cdots & \cdot \\ 0 & 0 & 0 & \cdots & \lambda_n \end{bmatrix}
\qquad
\mathbf{b} = \begin{bmatrix} 0 \\ \cdot \\ \cdot \\ \cdot \\ 0 \\ 1 \\ \cdot \\ \cdot \\ \cdot \\ 1 \end{bmatrix}
\qquad
\mathbf{c} = [c_1 \cdots c_n]
$$

Note that the matrix $\mathbf{A}$ yielded by this procedure has the Jordan canonical form (*D.5.21*). The characteristic features of this form are placed more clearly in evidence when $L(p)$ has more than one multiple zero. For example, suppose that

$$L(p) = (p + 2)^4(p + 4)^2(p + 1)(p + 3)$$

Then **A** has the following form:

30
$$\mathbf{A} = \begin{bmatrix} -2 & 1 & 0 & 0 & 0 & 0 & 0 & 0 \\ 0 & -2 & 1 & 0 & 0 & 0 & 0 & 0 \\ 0 & 0 & -2 & 1 & 0 & 0 & 0 & 0 \\ 0 & 0 & 0 & -2 & 0 & 0 & 0 & 0 \\ 0 & 0 & 0 & 0 & -4 & 1 & 0 & 0 \\ 0 & 0 & 0 & 0 & 0 & -4 & 0 & 0 \\ 0 & 0 & 0 & 0 & 0 & 0 & -1 & 0 \\ 0 & 0 & 0 & 0 & 0 & 0 & 0 & -3 \end{bmatrix}$$

As was already stated, the main advantage of the partial-fraction expansion technique is that it leads to state equations in which the **A** matrix is in the Jordan canonical form or, when $L(p)$ has no multiple zeros, in the diagonal form. Its disadvantages are: (I) it requires the expansion of $1/L(p)$ into partial fractions—a process which is by no means trivial for high-order polynomials, (II) it leads to complex-valued state vectors when $\lambda_1, \ldots, \lambda_n$ are not all real, and (III) the relationship between the components of the state vector and the values of the output and its derivatives is considerably more complicated than in the case of the state vector defined by *7* or *16*.

31 *Example* Assume that $\mathcal{A}$ is characterized by the input-output relation

$$L(p)y = u$$

where

$$L(p) = (p + 1)^2(p^2 + 1)$$

and hence

$$H(s) = \frac{1}{(s + 1)^2(s^2 + 1)}$$

On expanding $H(s)$ into partial fractions, we have

$$H(s) = \frac{1}{2}\frac{1}{(s + 1)^2} + \frac{1}{2}\frac{1}{(s + 1)} - \frac{1}{4}\frac{1}{(s + j)} - \frac{1}{4}\frac{1}{(s - j)}$$

Correspondingly, the state equations read

$$\dot{\mathbf{x}} = \mathbf{A}\mathbf{x} + \mathbf{b}u$$
$$y = \langle \mathbf{c},\mathbf{x} \rangle$$

where

$$\mathbf{A} = \begin{bmatrix} -1 & 1 & 0 & 0 \\ 0 & -1 & 0 & 0 \\ 0 & 0 & -j & 0 \\ 0 & 0 & 0 & j \end{bmatrix} \qquad \mathbf{b} = \begin{bmatrix} 0 \\ 1 \\ 1 \\ 1 \end{bmatrix} \qquad \mathbf{c} = [\tfrac{1}{2}, \ \tfrac{1}{2}, \ -\tfrac{1}{4}, \ -\tfrac{1}{4}]$$

It is instructive to compare these matrices with their counterparts in the state equations *4.3.32*. On writing

$$L(p) = p^4 + 2p^3 + 2p^2 + 2p + 1$$

and making use of *4.3.32* we obtain for the latter

$$\mathbf{A} = \begin{bmatrix} 0 & 1 & 0 & 0 \\ 0 & 0 & 1 & 0 \\ 0 & 0 & 0 & 1 \\ -1 & -2 & -2 & -2 \end{bmatrix} \qquad \mathbf{b} = \begin{bmatrix} 0 \\ 0 \\ 0 \\ 1 \end{bmatrix} \qquad \mathbf{c} = [1, \quad 0, \quad 0, \quad 0]$$

This concludes our discussion of various techniques for associating a state vector with a differential system and determining the corresponding state equations of the system. We are now ready to proceed to study the behavior of differential systems in greater detail, based on the characterization of such systems by their state equations rather than input-output relations having the form of a single differential equation.

REFERENCES AND SUGGESTED READING

1 Laning, J. H., Jr., and R. H. Battin: "Random Processes in Automatic Control," McGraw-Hill Book Company, Inc., New York, 1956.

2 Lur'e, A. I.: Some Nonlinear Problems in the Theory of Automatic Control, *Gos. Izdat. Teh. Teo. Lit.*, Moscow, 1951.

3 Bashkow, T.: The *A* Matrix, New Network Description, *I.R.E. Trans. on Circuit Theory,* vol. CT-4, pp. 117–119, September, 1957.

4 Bryant, P. R.: The Explicit Form of Bashkow's *A* Matrix, *I.R.E. Trans. on Circuit Theory,* vol. CT-9, no. 3, pp. 303–306, 1962.

Linear time-invariant differential systems

5

1 *Introduction*

In the preceding chapters, we have introduced and discussed in general terms such basic concepts as state, linearity, time invariance, system equivalence, and differential systems. In what follows, we shall enter a new phase in our analysis. Specifically, we shall place greater stress on detail in our study of the relations between input, output, and state, with a view to developing a body of concrete analytical as well as computational techniques for the analysis and design of linear systems of various types.

The present chapter is concerned in the main with the development of methods for computing the response of time-invariant systems, with the discussion of time-varying systems deferred until the next chapter. In the first two sections we consider time-invariant systems described by their canonical state equations (*2.3.45*) $\dot{\mathbf{x}} = \mathbf{A}\mathbf{x} + \mathbf{B}\mathbf{u}$, $\mathbf{y} = \mathbf{C}\mathbf{x} + \mathbf{D}\mathbf{u}$. Since the main problem in determining the dependence of $\mathbf{y}$ on $\mathbf{u}$ lies in solving the differential equation $\dot{\mathbf{x}} = \mathbf{A}\mathbf{x} + \mathbf{B}\mathbf{u}$, we shall focus our attention primarily on this problem, treating first the zero-input (free-motion) case and then the forced-motion ($\mathbf{u} \neq 0$) case. In particular, in Sec. *3*, a Laplace transform technique is developed for computing $\exp(\mathbf{A}t)$—a matrix which plays a central role in relating the state at time t to an initial state $\mathbf{x}(t_0)$. In Secs. *4* and *5* we present the mode interpretation of the motion in state space; this mode interpretation is a generalization of the concept of "normal modes" usually encountered in the study of small vibrations about an equilibrium point of a conservative dynamical system. Our interpretation, however, is applicable to all linear time-invariant systems. In the final sections we focus our attention on the properties and methods of solution for systems represented by higher-order differential equations of the form

$$\sum_j L_{ij}(p)x_j(t) = F_i(p)u(t)$$

$i = 1, 2, \ldots, n$, because such systems occur frequently in practice.

Since our main concern in this and the following three chapters is with the solution of systems of differential equations, we shall find it expedient in some instances to modify our notations and terminology in order to bring them into closer conformity with those commonly used in the theory of ordinary differential equations. In most cases, these modifications will be minor and self-explanatory. For example, the initial state at time t_0 will be denoted by $\mathbf{x}_0$ rather than $\boldsymbol{\alpha}_0$ or $\boldsymbol{\alpha}$, and the state at time t (in the absence of input) will be expressed as $\mathbf{x}(t;\mathbf{x}_0,t_0)$. The observation interval will frequently be taken to be $[t_0,t]$ rather than $(t_0,t]$. The zero-input response (for the case where $\mathbf{y} = \mathbf{x}$) will be referred to as *free motion* to stress its physical significance, etc. As a rule, we shall not comment on minor changes in terminology and notations when their motivation is clear from the context.

2 *Linear time-invariant systems described by their state equations*

In this section we consider linear time-invariant systems represented by their state equations in differential form. As explained in the introduction, we can simplify our analysis by assuming that the output of the system is its state. Thus, our primary concern will be with computing the solutions of state equations and interpreting their properties.

Zero-input response (free motion)

Since the output is the state $\mathbf{x}$ of the system, and since the input is identically zero, the system is described by a vector differential equation

1 $$\dot{\mathbf{x}} = \mathbf{A}\mathbf{x} \qquad \mathbf{x}(t_0) = \mathbf{x}_0$$

where $\mathbf{A}$ is an $n \times n$ matrix with constant elements that are real or complex numbers, $\mathbf{x}(t)$ is an n-rowed column vector representing the state of the system at time t, and $\mathbf{x}_0$ is the initial state of the system. It is well known that the differential equation *1* has a unique solution (see *2.3.40 et seq.*). We shall denote by $\mathbf{x}(t;\mathbf{x}_0,t_0)$ the value at time t of the solution which reduces to $\mathbf{x}_0$ at t_0. As a first step giving an analytic expression for $\mathbf{x}(t;\mathbf{x}_0,0)$ we introduce the

2 **Lemma** Let $\mathbf{A}$ be a constant $n \times n$ matrix whose elements are real or complex numbers. Define

3 $$\exp(\mathbf{A}t) = \mathbf{I} + \mathbf{A}t + \cdots + \mathbf{A}^k \frac{t^k}{k!} + \cdots$$

Then the series converges absolutely for all finite t and uniformly in any finite interval.

Proof The assertion is equivalent to stating that the sequence of matrices $\mathbf{S}_N$, where

$$\mathbf{S}_N = \sum_{k=0}^{N} \mathbf{A}^k \frac{t^k}{k!}$$

converges absolutely and uniformly to a finite-sum matrix. Let $a_{ij}^{(k)}$ be the (i,j) element of $\mathbf{A}^k$, $k = 0, 1, 2, \ldots$; then the (i,j) element of $\mathbf{S}_N$ is $\sum_{k=0}^{N} a_{ij}^{(k)} \frac{t^k}{k!}$. We have to establish the absolute convergence of this series for all $i, j = 1, 2, \ldots, n$. In other words, we have to show that

4 $$\sum_{k=0}^{\infty} \left| a_{ij}^{(k)} \frac{t^k}{k!} \right| < \infty \qquad \forall t \text{ finite}, \forall i,j$$

Let $m = \max_i \left(\sum_{j=1}^{n} |a_{ij}| \right)$; m is a finite number which depends only on $\mathbf{A}$. Since $\mathbf{A}^{k+1} = \mathbf{A}\mathbf{A}^k$, we have

$$\max_{i,j} |a_{ij}^{(k+1)}| = \max_{i,j} \left| \sum_{l=1}^{n} a_{il} a_{lj}^{(k)} \right| \leq \max_{i,j} \left(\sum_{l=1}^{n} |a_{il}|\, |a_{lj}^{(k)}| \right)$$

$$\leq \left(\max_i \sum_{l=1}^{n} |a_{il}| \right) \left(\max_{l,j} |a_{lj}^{(k)}| \right)$$

Hence

$$\max_{i,j} |a_{ij}^{(k+1)}| \leq m \max_{i,j} |a_{ij}^{(k)}| \qquad k = 1, 2, \ldots$$

Since for $k = 0$, by the definition of m, $\max_{i,j} |a_{ij}| \leq m$, we get, by induction on k,

$$\max_{i,j} |a_{ij}^{(k)}| \leq m^k \qquad k = 1, 2, \ldots$$

Thus the series *4* is dominated by $\sum_{k=0}^{\infty} m^k \frac{|t|^k}{k!} = e^{m|t|}$; hence it converges absolutely for any finite t. Futhermore, by the Weierstrass M test it converges uniformly over any finite interval $[-T, +T]$.

5 *Exercise* Let

$$\mathbf{A} = \begin{bmatrix} \lambda_1 & 0 \\ 0 & \lambda_2 \end{bmatrix} \qquad \mathbf{B} = \begin{pmatrix} \lambda & 1 \\ 0 & \lambda \end{pmatrix} \qquad \mathbf{C} = \begin{bmatrix} \lambda & 1 & 0 & \cdot & \cdots & 0 \\ 0 & \lambda & 1 & 0 & \cdots & 0 \\ \cdot & \cdot & \cdot & \cdot & \cdot & \cdot \\ \cdot & \cdot & \cdot & \cdot & \lambda & 1 \\ 0 & \cdot & \cdot & \cdot & 0 & \lambda \end{bmatrix}$$

Use the defining power series *3* to show that

$$\exp(\mathbf{A}t) = \operatorname{diag}(e^{\lambda_1 t}, e^{\lambda_2 t}) \qquad \exp(\mathbf{B}t) = \begin{pmatrix} e^{\lambda t} & te^{\lambda t} \\ 0 & e^{\lambda t} \end{pmatrix}$$

$$\exp(\mathbf{C}t) = e^{\lambda t} \begin{bmatrix} 1 & t & \dfrac{t^2}{2!} & \cdots & \dfrac{t^{n-1}}{(n-1)!} \\ 0 & 1 & t & \cdots & \dfrac{t^{n-2}}{(n-2)!} \\ \cdot & \cdot & \cdot & \cdots & \cdot \\ 0 & 0 & 0 & \cdots & 1 \end{bmatrix}$$

6 *Exercise* Show that $\mathbf{AB} = \mathbf{BA} \Rightarrow [\exp(\mathbf{A}t)][\exp(\mathbf{B}t)] = [\exp(\mathbf{B}t)][\exp(\mathbf{A}t)]$, $\forall t$. Find an example to show that if $\mathbf{AB} \neq \mathbf{BA}$, the above conclusion need not hold.

7 **Lemma** Let $\mathbf{A}$ be an $n \times n$ constant matrix whose elements are real or complex numbers. Let $\mathbf{\Phi}(t)$ be the solution to the matrix differential equation

8 $$\dot{\mathbf{\Phi}}(t) = \mathbf{A\Phi}(t) \qquad \mathbf{\Phi}(0) = \mathbf{I}$$

Then

$$\mathbf{\Phi}(t) = \exp(\mathbf{A}t)$$

Proof Since $\mathbf{A}$ is constant, the solution is unique. The initial condition is satisfied by $\exp(\mathbf{A}t)$. Equation *3* is a power series in t which converges over any finite interval; therefore, it can be differentiated term by term. The resulting series is the derivative of $\exp(\mathbf{A}t)$, viz.:

$$\frac{d}{dt}\exp(\mathbf{A}t) = \mathbf{A} + \mathbf{A}^2 t + \mathbf{A}^3 \frac{t^2}{2!} + \cdots + \mathbf{A}^{k+1}\frac{t^k}{k!} + \cdots = \mathbf{A}\exp(\mathbf{A}t)$$

Therefore, $\exp(\mathbf{A}t)$ is a solution of *8*. ◁

Lemma *7* can also be proved easily by considering $\exp(\mathbf{A}t)$ as a function of a matrix (see *D.6.12*).

9 **Theorem** Let $\mathbf{A}$ be an $n \times n$ constant matrix whose elements are real or complex numbers. Let the system be described by its state equation

10 $$\dot{\mathbf{x}}(t) = \mathbf{Ax}(t) \qquad \mathbf{x}(t_0) = \mathbf{x}_0$$

where $\mathbf{x}_0$ is the prescribed state at time t_0. The resulting free motion is given by

11 $$\mathbf{x}(t;\mathbf{x}_0,t_0) = \exp[\mathbf{A}(t - t_0)]\mathbf{x}_0 \qquad -\infty < t < \infty$$

12 *Comment* The expression *11* gives, for all t in $(-\infty, \infty)$, the state at time t which is on the trajectory going through $\mathbf{x}_0$ at time t_0. A useful interpretation of *11* in the more general context of time-varying systems is the following. Let the matrix $\mathbf{\Phi}(t,t_0)$ be the solution of *8* such that $\mathbf{\Phi}(t_0,t_0) = \mathbf{I}$; that is, from *7*, $\mathbf{\Phi}(t,t_0) = \exp[\mathbf{A}(t - t_0)]$. Then by *11*,

$\mathbf{x}(t;\mathbf{x}_0,t_0)$ is the product of $\mathbf{\Phi}(t,t_0)$ and $\mathbf{x}_0$, that is, $\mathbf{\Phi}(t,t_0)$ is a linear transformation (see *C.9.1*) which maps the initial state $\mathbf{x}_0$ at t_0 into the state at time t. For this reason, $\mathbf{\Phi}(t,t_0)$ is called the *state transition matrix*. This point of view leads to another important fact: let $\bar{\mathbf{x}}(\mathbf{x}_0,t_0)$ be the trajectory going through $\mathbf{x}_0$ at time t_0 (that is, the vector-valued function defined for each t by *11*). The set of all solutions of *10* constitutes an n-dimensional linear vector space (see *C.2.1* and *C.4.11*); indeed, let $\mathbf{a}_1, \mathbf{a}_2, \ldots, \mathbf{a}_n$ constitute a basis for the state space and let $\mathbf{x}_0 = \sum_{i=1}^{n} \xi_i \mathbf{a}_i$ be the (unique) representation of $\mathbf{x}_0$ in that basis. Then

13 $$\mathbf{x}(t;\mathbf{x}_0,t_0) = \sum_{i=1}^{n} \xi_i \mathbf{x}(t;\mathbf{a}_i,t_0)$$

In other words, any solution of *10* can be uniquely represented by a suitable linear combination of $\mathbf{x}(t;\mathbf{a}_i,t_0)$, $i = 1, 2, \ldots, n$. It is relation *13* which is the basis for the mode interpretation which is developed in Sec. *4*.

Properties of exp $(\mathbf{A}t)$

14 **Assertion** The matrix exp $(\mathbf{A}t)$ is nonsingular for all finite t's. More precisely, let α be the trace of $\mathbf{A}$, that is, $\alpha = \sum_{i=1}^{n} a_{ii}$; then

15 $$\det[\exp(\mathbf{A}t)] = e^{\alpha t} \qquad -\infty < t < \infty$$

Proof Exercise for the reader (see *6.2.10*).

16 **Assertion** $\exp(\mathbf{A}t_1) \cdot \exp(\mathbf{A}t_2) = \exp[\mathbf{A}(t_1 + t_2)] \qquad \forall t_1, t_2$

Proof From *3*, we have

$$\exp(\mathbf{A}t_1) \cdot \exp(\mathbf{A}t_2) = \left(\sum_{k=0}^{\infty} \frac{\mathbf{A}^k t_1^k}{k!}\right)\left(\sum_{l=0}^{\infty} \frac{\mathbf{A}^l t_2^l}{l!}\right)$$
$$= \sum_{k,l=0}^{\infty} \mathbf{A}^{k+l} \frac{t_1^k t_2^l}{k!\,l!}$$

where we used the fact that the two power series are absolutely convergent. Now, reorder the summation by putting $k + l = n$; thus

$$\exp(\mathbf{A}t_1) \cdot \exp(\mathbf{A}t_2) = \sum_{n=0}^{\infty} \mathbf{A}^n \frac{1}{n!}\left[\sum_{l=0}^{n} \frac{n!}{(n-l)!\,l!} t_1^{n-l} t_2^l\right]$$
$$= \sum_{n=0}^{\infty} \mathbf{A}^n \frac{(t_1 + t_2)^n}{n!}$$

which is recognized to be exp $[\mathbf{A}(t_1 + t_2)]$ by Definition *3*.

17 *Comment* Since the R.H.S. of *16* is invariant under an interchange of t_1 and t_2, it follows that $\exp(\mathbf{A}t_1)$ and $\exp(\mathbf{A}t_2)$ commute for all t_1, t_2. Now let $t_2 = -t_1$; then together with *14* we get

18 $$[\exp(\mathbf{A}t)]^{-1} = \exp(-\mathbf{A}t)$$

Forced response

Let us now allow the input to differ from zero; we then have to consider the equation $\dot{\mathbf{x}}(t) = \mathbf{A}\mathbf{x}(t) + \mathbf{B}\mathbf{u}(t)$, where $\mathbf{B}$ is an $n \times r$ constant matrix and $\mathbf{u}(t)$ is an $r \times 1$ vector representing r scalar inputs. We shall simplify the notation and write $\mathbf{a}(t)$ for $\mathbf{B}\mathbf{u}(t)$.

19 **Theorem** Let the system be described by its equation in the state form

20 $$\dot{\mathbf{x}}(t) = \mathbf{A}\mathbf{x}(t) + \mathbf{a}(t)$$

where $\mathbf{A}$ is an $n \times n$ constant matrix and $\mathbf{a}(t)$ represents the effect of the input to the system. If the state at time t_0 is $\mathbf{x}_0$, then the state at time t is given by (see *6.2.22*)

21 $$\mathbf{x}(t;\mathbf{x}_0,t_0;\mathbf{a}_{(t_0,t]}) = \{\exp[\mathbf{A}(t - t_0)]\}\mathbf{x}_0 + \int_{t_0}^{t} \{\exp[\mathbf{A}(t-\tau)]\}\mathbf{a}(\tau)\,d\tau \qquad -\infty < t < \infty$$

or equivalently

22 $$\mathbf{x}(t;\mathbf{x}_0,t_0;\mathbf{a}_{(t_0,t]}) = \{\exp[\mathbf{A}(t - t_0)]\}\mathbf{x}_0 + \exp(\mathbf{A}t)\int_{t_0}^{t} [\exp(-\mathbf{A}\tau)]\mathbf{a}(\tau)\,d\tau \qquad -\infty < t < \infty$$

Proof The equivalence of *21* and *22* follows directly from *16* and *18*. To establish *22*, observe that for $t = t_0$ the R.H.S. reduces to $\mathbf{x}_0$, as it should. Next, differentiate the R.H.S. of *22*:

$$\mathbf{A}\{\exp[\mathbf{A}(t - t_0)]\}\mathbf{x}_0 + \mathbf{A}[\exp(\mathbf{A}t)]\int_{t_0}^{t} [\exp(-\mathbf{A}\tau)]\mathbf{a}(\tau)\,d\tau + [\exp(\mathbf{A}t)\exp(-\mathbf{A}t)]\mathbf{a}(t)$$

The last term reduces to $\mathbf{a}(t)$ by *18*. By comparison with *22*, we see that the expression above is of the form $\mathbf{A}\mathbf{x}(t;\mathbf{x}_0,t_0;\mathbf{a}_{(t_0,t]}) + \mathbf{a}(t)$. Therefore, the R.H.S. of *22* satisfies the differential equation *1* and the initial condition.

23 *Exercise* Consider the multiple-input linear time-invariant system described by

$$\dot{\mathbf{x}}(t) = \mathbf{A}\mathbf{x}(t) + \mathbf{B}\mathbf{u}(t)$$

where $\mathbf{A}$ is $n \times n$ and $\mathbf{B}$ is $n \times r$. Show that the control $\tilde{\mathbf{u}}_{(t_0,t_1]}$ will bring the system from the state $\mathbf{x}_0$ at time t_0 to the state $\mathbf{x}_1$ at time t_1 if and only if

$$\int_{t_0}^{t_1} [\exp(-\mathbf{A}\tau)]\mathbf{B}\tilde{\mathbf{u}}(\tau)\,d\tau = [\exp(-\mathbf{A}t_1)]\mathbf{x}_1 - [\exp(-\mathbf{A}t_0)]\mathbf{x}_0 \quad \triangleleft$$

We should like to give an interpretation of the second term of *21* and *22*. Let us introduce the notation

24 $$\mathbf{G}(t,\tau) = [\exp(\mathbf{A}t)] \cdot [\exp(-\mathbf{A}\tau)] \cdot 1(t-\tau) = 1(t-\tau)\exp[\mathbf{A}(t-\tau)]$$

$\mathbf{G}(t,\tau)$ is the Green function of the system *20*; because it is a matrix in the present case, we shall adopt the language of the physicist and refer to it as the *Green dyadic* of the system *20* (compare with *3.7.34 et seq.*). $\mathbf{G}(t,\tau)$ allows us to calculate the effect at time t of the input $\mathbf{a}$ acting in the interval $[\tau, \tau + \Delta\tau]$: for $\Delta\tau$ very small, this contribution is (see *21*)

$$\mathbf{G}(t,\tau)\mathbf{a}(\tau)\,d\tau$$

Another way of looking at this fact is to consider the case where $\mathbf{a}(t)$ is an impulse $\mathbf{c}$ occurring at time τ, that is, $\mathbf{a}(t) = \mathbf{c}\delta(t-\tau)$. The contribution to $\mathbf{x}(t)$ of this forcing function is given by the second term of *21*, namely

25 $$\mathbf{G}(t,\tau)\mathbf{c}$$

where t is the time at which the contribution is computed and τ is the time at which the impulse is applied.

The presence of the factor $1(t-\tau)$ in *24* is necessitated by the nonanticipative character of the system *20*; without this factor, an impulse $\mathbf{c}$ occurring at time τ would have some effect on $\mathbf{x}(t)$, according to *25*, even though τ might be later than t, that is, even though the input was applied later than the instant of observation.

With the notation *24* in mind, *21* may be rewritten for the case $\mathbf{x}_0 = \mathbf{0}$:

26 $$\mathbf{x}(t) = \int_{t_0}^{t} \mathbf{G}(t,\tau)\mathbf{a}(\tau)\,d\tau \qquad t \geq t_0$$

27 *Comment* Equation *24* can also be written in the form

$$\mathbf{G}(t,\tau) = [\exp(\mathbf{A}t)][\exp(\mathbf{A}\tau)^{-1}]1(t-\tau)$$

which is a form very close to the Green dyadic of time-varying systems (see *6.2.18*). The fact that the integral in *21* expresses the superposition property may be brought to light by an example borrowed from physics. Consider a uniform medium in which the electric charge density ρ is specified. The electrostatic potential ϕ at point $\mathbf{r}$ is given by

28 $$\phi(\mathbf{r}) = \int \frac{\rho(\mathbf{r}')}{4\pi\epsilon_0|\mathbf{r} - \mathbf{r}'|}\, d\mathbf{r}'$$

where the integral is carried over the whole space. There is a rough analogy between *26* and *28* in the sense that they both express the superposition property: $1/(4\pi\epsilon_0|\mathbf{r} - \mathbf{r}'|)$ is the potential at $\mathbf{r}$ due to a *unit* point change at $\mathbf{r}'$; $\mathbf{G}(t,\tau)\mathbf{c}$ gives the response at time t due to a "point" excitation $\mathbf{c}\delta(t - \tau)$.

3 *The computation of* exp (**A***t*)

It is not practicable to compute the matrix exp $(\mathbf{A}t)$ by the power series *5.2.3* except in very special cases or for small values of t. We can derive a closed-form expression for exp $(\mathbf{A}t)$ either as a special case of the study of the functions of a matrix or by a purely algebraic method based on the Laplace transform. The first method has the advantage of using the geometric insight resulting from the study of Appendixes *C* and *D*. We shall present it first. The second method, purely algebraic in its spirit, needs no special background, and we shall develop it to the extent of deriving all the important relationships obtained by the first method. From a practical computational point of view either method is effective.

exp (A*t*) as a particular case of a function of a matrix

The exponential function e^z is analytic everywhere in the finite complex plane; therefore, the fundamental formula for the function of a matrix is directly applicable (see *D.6.5*) and yields

1 $$\exp(\mathbf{A}t) = \sum_{k=1}^{\sigma} \sum_{l=0}^{m_k - 1} t^l e^{\lambda_k t} \mathbf{Z}_{kl}$$

where $\lambda_1, \lambda_2, \ldots, \lambda_\sigma$ are the distinct eigenvalues of $\mathbf{A}$ (hence $\lambda_i \neq \lambda_j$, $i,j = 1, 2, \ldots, \sigma$), m_k is the multiplicity of the eigenvalue λ_k as a zero of the minimal polynomial of $\mathbf{A}$ (see *18* below *et seq.* or *D.2.1*), and the matrices $\mathbf{Z}_{kl}$ have constant elements and depend exclusively on $\mathbf{A}$. The computation of the matrices $\mathbf{Z}_{kl}$ from $\mathbf{A}$ is presented in detail, together with an example, in Appendix *D*, Sec. *8*.

It should be stressed that if the minimal polynomial of $\mathbf{A}$ is not known, one may use for m_k the multiplicity of λ_k as a zero of the characteristic polynomial of $\mathbf{A}$, that is, the polynomial $\Delta(\lambda) = \det(\mathbf{A} - \lambda\mathbf{I})$. The only possible penalty for doing so is, in some cases, a certain amount of unnecessary computation, but the final result is the same (*D.8.27*). Whenever $\Delta(\lambda)$ has distinct zeros, then $m_k = 1$ for $k = 1, 2, \ldots, n$ and the minimal polynomial is identical with characteristic polynomial.

exp (A*t*) from the Laplace transform point of view

For very simple cases, say 2×2 matrices, the theoretical developments that follow are unnecessary. Consider *5.2.8*; take its Laplace transform. Clearly, $\exp(\mathbf{A}t)$, for $t \geq 0$, is the inverse Laplace transform of $(s\mathbf{I} - \mathbf{A})^{-1}$. This is easily computed to be of the form $\mathbf{B}(s)/d(s)$, where $\mathbf{B}(s)$ is a matrix whose elements are polynomials in s of degree $n - 1$ at most, and $d(s) = \det(s\mathbf{I} - \mathbf{A})$. By dividing all elements of $\mathbf{B}$ by the polynomial d, one obtains a matrix $\mathbf{R}$ whose elements are rational functions of s. [All elements have a common denominator $d(s)$.] Now perform a partial-fraction expansion on all elements of $\mathbf{R}$. The result is a sum of the form

$$\sum_{k=1}^{\sigma} \sum_{l=0}^{m_k - 1} \mathbf{Y}_{kl} \frac{1}{(s - s_k)^{l+1}}$$

where the matrices $\mathbf{Y}_{kl}$ have constant elements. The inverse transform of this sum gives

2 $$\exp(\mathbf{A}t) = \sum_{k=1}^{\sigma} \sum_{l=0}^{m_k - 1} \mathbf{Y}_{kl} \frac{t^l}{l!} e^{\lambda_k t}$$

This approach is satisfactory for very simple cases; however, when n is large, it is important to be able to organize the computation more systematically (especially for machine computation) and to have enough insight into the results to be able to check them. For this reason we present a more systematic approach to the question. The reader interested in only the practical computation of $\exp(\mathbf{A}t)$ is referred to *31* below.

Let us first deduce some results that follow from Cramer's rule.[1] Consider the matrix $\mathbf{\Phi}(t)$ (which is the state transition matrix with $t_0 = 0$) defined in *5.2.8*. Let $\hat{\mathbf{\Phi}}(s) = \mathcal{L}\{\mathbf{\Phi}(t)\}$, that is, $\hat{\mathbf{\Phi}}(s)$ is the matrix each element of which is the Laplace transform of the corresponding element of $\mathbf{\Phi}(t)$. Taking the $\mathcal{L}$ transform of both sides of *5.2.8*, we get

$$(s\mathbf{I} - \mathbf{A})\hat{\mathbf{\Phi}}(s) = \mathbf{I}$$

Since we know that each element of $\mathbf{\Phi}(t)$ can be bounded by $ke^{\sigma_0 t}$ for

[1] G. Birkhoff and S. MacLane, "A Survey of Modern Algebra," p. 306, The Macmillan Company, New York, 1953.

some k and σ_0, the above relation holds for all s such that Re $(s) > \sigma_0$ (*B.2.2* and *B.2.4*). Furthermore, this relation suggests to identify $\hat{\mathbf{\Phi}}(s)$ with the inverse of the matrix $(s\mathbf{I} - \mathbf{A})$. Since the determinant of this matrix is zero only for those s that are eigenvalues of $\mathbf{A}$, we are led to

$$\hat{\mathbf{\Phi}}(s) = (s\mathbf{I} - \mathbf{A})^{-1} \qquad \forall s \notin \mathcal{E}$$

where $\mathcal{E}$ is the set of all eigenvalues of $\mathbf{A}$. The first result we wish to consider is the

3 **Lemma** Let $\mathbf{A}$ be an $n \times n$ constant matrix; then

4
$$(s\mathbf{I} - \mathbf{A})^{-1} = \frac{\mathbf{B}(s)}{d(s)}$$

where[1]

5
$$d(s) = \det (s\mathbf{I} - \mathbf{A}) = s^n + d_1 s^{n-1} + \cdots + d_{n-1}s + d_n$$

and

6
$$\mathbf{B}(s) = s^{n-1}\mathbf{B}_0 + s^{n-2}\mathbf{B}_1 + \cdots + s\mathbf{B}_{n-2} + \mathbf{B}_{n-1}$$

with $\mathbf{B}_0, \mathbf{B}_1, \ldots, \mathbf{B}_{n-1}$ $n \times n$ matrices with constant elements.

Proof Cramer's rule establishes immediately that the (i,k) element of $(s\mathbf{I} - \mathbf{A})^{-1}$ is equal to $M_{ki}(s)/d(s)$, where $M_{ki}(s)$ is the cofactor of the (k,i) element of $(s\mathbf{I} - \mathbf{A})$. Now $M_{ki}(s)$ is $(-1)^{i+k}$ times the determinant of the matrix obtained from $(s\mathbf{I} - \mathbf{A})$ by deleting row k and column i; hence $M_{ki}(s)$ is a polynomial in s of degree $n - 1$ at most. Let $\mathbf{B}_l$ be the matrix whose (i,k) element is the coefficient of s^{n-1-l} in $M_{ki}(s)$ $(l = 0, 1, \ldots, n - 1)$; then *4* and *6* follow.

7 *Comment* This procedure for computing $\mathbf{B}(s)$ suggested by Cramer's rule is not a practical procedure when n is more than 2 or 3: it requires the evaluation of n^2 determinants of order $n - 1$, each of which requires the evaluation of $(n - 1)!$ products of $n - 1$ numbers, i.e., a total of $n^2(n - 1)(n - 1)!$ multiplications! In many instances, the problem of obtaining the polynomial $d(s)$ [let alone the matrix $\mathbf{B}(s)$] is not trivial. ◁

Let us establish a couple of facts which will turn out to be useful in the general development that follows.

8 **Corollary**

$$d_{n-1} = (-1)^{n-1} \sum_{i=1}^{n} A_{ii}$$

where A_{ii} is the cofactor of the element a_{ii} of $\mathbf{A}$.

[1] Usually one considers the characteristic polynomial $\Delta(s) = \det (\mathbf{A} - s\mathbf{I})$ instead of $d(s)$. Note that $\Delta(s) = (-1)^n d(s)$, since multiplying each row of $\mathbf{A} - s\mathbf{I}$ by -1 multiplies the determinant by $(-1)^n$.

Proof *5* implies that

$$d_{n-1} = \lim_{s \to 0} \frac{d}{ds} [d(s)]$$

Now $\frac{d}{ds} \det (s\mathbf{I} - \mathbf{A}) \Big|_{s=0}$ is a sum of n determinants whose ith term is the determinant obtained by differentiating the ith column of $(s\mathbf{I} - \mathbf{A})$ with respect to s and then setting s to zero; the resulting ith determinant is

$$i\text{th row} \rightarrow \begin{vmatrix} -a_{11} & -a_{12} & \cdots & 0 & -a_{1,i+1} & \cdots & -a_{1n} \\ -a_{21} & -a_{22} & \cdots & 0 & -a_{2,i+1} & \cdots & \cdot \\ \cdot & \cdot & \cdots & 0 & \cdot & \cdots & \cdot \\ \cdot & \cdot & \cdots & 1 & \cdot & \cdots & \cdot \\ & & & 0 & -a_{i+1,i+1} & & \\ & & & \cdot & & & \\ & & & \cdot & & & \\ & & & \cdot & & & \\ -a_{n1} & & \cdots & 0 & -a_{n,i+1} & \cdots & -a_{nn} \end{vmatrix}$$

$$\uparrow \; i\text{th column}$$

Clearly, this determinant is $(-1)^{n-1}A_{ii}$. Hence *8* follows. ◁

Let us consider *8* again. From *6*, $\mathbf{B}(0) = \mathbf{B}_{n-1}$, and since the (i,i) element of $\mathbf{B}_{n-1}$ is $(-1)^{n-1}A_{ii}$, we have immediately, by comparing *5* and *8*,

9
$$d_{n-1} = \operatorname{tr} \mathbf{B}(s) \Big|_{s=0}$$

In principle, Cramer's rule suffices to compute the state transition matrix $\mathbf{\Phi}(t)$. We proceed now to show (I) that there is a more convenient method for evaluating $\hat{\mathbf{\Phi}}(s) = (s\mathbf{I} - \mathbf{A})^{-1}$ than Cramer's rule and (II) that the usual method of expanding a rational function in partial fractions can readily be extended to matrices [such as $\hat{\mathbf{\Phi}}(s)$] whose elements are rational functions. This procedure will give us tools to establish some facts that can be proved only by a detailed study of linear transformations as in Appendices *C* and *D*.

The technique for computing the coefficient matrices $\mathbf{B}_i$ of $\hat{\mathbf{\Phi}}(s)$ is given in the form of a

10 **Theorem**[1] Given an $n \times n$ constant matrix $\mathbf{A}$, then the coefficients d_i of the polynomial $d(s)$ defined in *5* and the matrix coefficients $\mathbf{B}_j$ of the matrix polynomial $\mathbf{B}(s)$ defined in *6* are given by

[1] J. M. Souriau, *C. R. Acad. Sci. Paris,* vol. 227, pp. 1010, 1011, 1948. Also called Fadeeva's method, in V. N. Fadeeva, "Computational Methods in Linear Algebra," Dover Publications, Inc., New York, 1959.

11
$$\begin{aligned}
d_1 &= -\operatorname{tr}(\mathbf{A}) & \mathbf{B}_0 &= \mathbf{I}\\
d_2 &= -\tfrac{1}{2}\operatorname{tr}(\mathbf{B}_1\mathbf{A}) & \mathbf{B}_1 &= \mathbf{B}_0\mathbf{A} + d_1\mathbf{I}\\
d_3 &= -\tfrac{1}{3}\operatorname{tr}(\mathbf{B}_2\mathbf{A}) & \mathbf{B}_2 &= \mathbf{B}_1\mathbf{A} + d_2\mathbf{I}\\
&\cdots\cdots\cdots & &\cdots\cdots\cdots\\
d_k &= -\frac{1}{k}\operatorname{tr}(\mathbf{B}_{k-1}\mathbf{A}) & \mathbf{B}_k &= \mathbf{B}_{k-1}\mathbf{A} + d_k\mathbf{I}\\
&\cdots\cdots\cdots & &\cdots\cdots\cdots\\
d_{n-1} &= -\frac{1}{n-1}\operatorname{tr}(\mathbf{B}_{n-2}\mathbf{A}) & \mathbf{B}_{n-1} &= \mathbf{B}_{n-2}\mathbf{A} + d_{n-1}\mathbf{I}\\
d_n &= -\frac{1}{n}\operatorname{tr}(\mathbf{B}_{n-1}\mathbf{A}) & 0 &= \mathbf{B}_{n-1}\mathbf{A} + d_n\mathbf{I}
\end{aligned}$$

12 *Comment* This iterative technique for successively computing the d_i's and the $\mathbf{B}_i$'s determines the characteristic polynomial by *5* and $(s\mathbf{I} - \mathbf{A})^{-1}$ by *6* and *4*. Therefore, it is an alternative method to Cramer's rule. It is of interest to note that the last equation of *11* serves as a check: the extent to which $\mathbf{B}_{n-1}\mathbf{A}$ differs from the diagonal matrix $-d_n\mathbf{I}$ gives an estimate of the magnitude of the round-off errors.

Proof First clear the denominator in *4* and take into account *5* and *6*; then, for all s,

13
$$(s^{n-1}\mathbf{B}_0 + s^{n-2}\mathbf{B}_1 + \cdots + s\mathbf{B}_{n-2} + \mathbf{B}_{n-1})(s\mathbf{I} - \mathbf{A}) = (s^n + d_1 s^{n-1} + \cdots + d_{n-1}s + d_n)\mathbf{I}$$

This equality between matrix polynomials will hold if and only if coefficients of like powers of s are equal; i.e.,

14
$$\begin{aligned}
\mathbf{B}_0 &= \mathbf{I}\\
\mathbf{B}_1 &= \mathbf{B}_0\mathbf{A} + d_1\mathbf{I}\\
&\cdots\cdots\cdots\\
\mathbf{B}_k &= \mathbf{B}_{k-1}\mathbf{A} + d_k\mathbf{I}\\
&\cdots\cdots\cdots\\
\mathbf{B}_{n-1} &= \mathbf{B}_{n-2}\mathbf{A} + d_{n-1}\mathbf{I}\\
0 &= \mathbf{B}_{n-1}\mathbf{A} + d_n\mathbf{I}
\end{aligned}$$

If the coefficients $d_1, d_2, \ldots, d_n$ of $d(s)$ were known, *14* would constitute an algorithm for computing the $\mathbf{B}_k$'s. Suppose now that the d_i's are unknown. Let σ be an arbitrary complex number; then, remembering that $d(s + \sigma)$ is a polynomial in s of degree n, we have successively

$$\begin{aligned}
d(s+\sigma) &= \det((s+\sigma)\mathbf{I} - \mathbf{A})\\
&= d(\sigma) + s d'(\sigma) + \frac{s^2}{2!}d''(\sigma) + \cdots + \frac{s^n}{n!}d^{(n)}(\sigma)
\end{aligned}$$

From *9* the coefficient of s in the above relation is

$$\operatorname{tr} \mathbf{B}(s + \sigma) \Big|_{s=0} = \operatorname{tr} \mathbf{B}(\sigma)$$

In other words, changing σ into s,

$$d'(s) = \operatorname{tr} (\mathbf{B}(s))$$

Hence from *5* and *6*

$$\operatorname{tr} (s^{n-1}\mathbf{B}_0 + s^{n-2}\mathbf{B}_1 + \cdots + s\mathbf{B}_{n-2} + \mathbf{B}_{n-1}) = \sum_{k=0}^{n} k d_{n-k} s^{k-1} \qquad \forall s$$

where $d_0 = 1$. These polynomials in s will be equal if and only if the coefficients of like powers of s are equal; i.e.,

15
$$\begin{aligned} \operatorname{tr} (\mathbf{B}_0) &= n \\ \operatorname{tr} (\mathbf{B}_1) &= (n-1)d_1 \\ &\cdots\cdots \\ \operatorname{tr} (\mathbf{B}_k) &= (n-k)d_k \\ &\cdots\cdots \\ \operatorname{tr} (\mathbf{B}_{n-1}) &= d_{n-1} \end{aligned}$$

Let us take the trace of the matrix equations *14* and use $\mathbf{B}_0 = \mathbf{I}$:

16
$$\begin{aligned} \operatorname{tr} (\mathbf{B}_1) &= \operatorname{tr} (\mathbf{A}) + nd_1 \\ &\cdots\cdots \\ \operatorname{tr} (\mathbf{B}_k) &= \operatorname{tr} (\mathbf{B}_{k-1}\mathbf{A}) + nd_k \\ &\cdots\cdots \\ \operatorname{tr} (\mathbf{B}_{n-1}) &= \operatorname{tr} (\mathbf{B}_{n-2}\mathbf{A}) + nd_{n-1} \\ \mathbf{0} &= \operatorname{tr} (\mathbf{B}_{n-1}\mathbf{A}) + nd_n \end{aligned}$$

Subtracting each equation of *15* from the corresponding one of *16*, we get the first set of equations *11*. The second set is identical with *14*. Therefore, *11* is established. Q.E.D.

17 *Comment* In practical cases, it often happens that the numerator and denominator polynomials of a transfer function have common factors. In the present case where we deal with the matrix $(s\mathbf{I} - \mathbf{A})^{-1}$ whose elements are rational functions with the common denominator $d(s)$ (see *4*), a similar phenomenon may occur: it may happen that *all* elements of the matric polynomial $\mathbf{B}(s)$ have one or more factors in common with $d(s)$. When this happens, these common factors may be canceled from both the numerator and denominator of all the elements of $(s\mathbf{I} - \mathbf{A})^{-1}$ in *4*: more explicitly, since $\mathbf{B}(s)$ is an $n \times n$ matrix with polynomial elements, all the n^2 elements of $\mathbf{B}(s)/d(s)$ are rational functions in which the same common factors can be canceled.

This leads to a simplified expression, namely,

18 $$(s\mathbf{I} - \mathbf{A})^{-1} = \frac{\mathbf{P}(s)}{m(s)}$$

where the polynomial $m(s)$ and the matric polynomial $\mathbf{P}(s)$ are the result of these cancellations. Clearly, it is always true that $m(s)$ is a factor of $\Delta(s) = (-1)^n d(s)$ and degree $(m) \leq$ degree (Δ). When all possible cancellations common to all the n^2 rational functions that are elements of $(s\mathbf{I} - \mathbf{A})^{-1}$ have been performed, then the resulting denominator polynomial $m(s)$ is called the *minimal polynomial of the matrix* **A**. This minimal polynomial is identical with the minimal polynomial[1] defined in Appendix *D*, Sec. *2* (see in particular *D.2.1* to *D.2.6*). For our purposes here it is of interest to note that (I) $m(s)$ is the polynomial of least degree such that the matrix $m(\mathbf{A}) = \mathbf{0}$ and (II) every eigenvalue of **A** [that is, a zero of $\Delta(s)$] is a zero of $m(s)$. ◁

From the set of equations *11* the celebrated Cayley-Hamilton theorem is easily derived. Consider the system *11*: successively, from the bottom up, substitute for the successive $\mathbf{B}_k$'s the values assigned to them by the equation just above; then

$$\mathbf{0} = \mathbf{A}^n + d_1\mathbf{A}^{n-1} + \cdots + d_{n-1}\mathbf{A} + d_n\mathbf{I}$$

In other words, by referring to *5*,

$$d(\mathbf{A}) = \mathbf{0} \qquad \text{and} \qquad \Delta(\mathbf{A}) = \mathbf{0}$$

Therefore, we have the

19 **Cayley-Hamilton theorem** Let **A** be a matrix whose elements are real or complex numbers. Let the characteristic polynomial of **A** be denoted by Δ [that is, $\Delta(\lambda) = \det(\mathbf{A} - \lambda\mathbf{I})$]. Then the matrix $\Delta(\mathbf{A})$ is the zero matrix **0**. ◁

For the purpose of computing $(s\mathbf{I} - \mathbf{A})^{-1}$, Theorem *11* is a more efficient procedure than Cramer's rule because it requires only n matrix multiplications. The next step is to compute the inverse Laplace transform of $(s\mathbf{I} - \mathbf{A})^{-1}$. Since the n^2 elements of $(s\mathbf{I} - \mathbf{A})^{-1}$ are rational functions of s, the obvious procedure is to perform a partial-fraction expansion of each term of the matrix. Since from *18* every term of $(s\mathbf{I} - \mathbf{A})^{-1}$ has the same denominator, it seems reasonable to expect that this partial-fraction expansion can be carried out by matrix operations. That is, for the case of a simple pole at s_1, we can compute directly the matrix of residues at that pole, etc. The procedure is described by the

[1] In Appendix *D*, the minimal polynomial had to be labeled ψ in order to avoid conflicts of notation.

20 **Theorem** Let $\mathbf{A}$ be an arbitrary $n \times n$ matrix whose elements are real or complex numbers. Then

(I)

21 $$(s\mathbf{I} - \mathbf{A})^{-1} = \frac{\mathbf{P}(s)}{m(s)} = \sum_{k=1}^{\sigma} \sum_{l=0}^{m_k-1} \frac{\mathbf{Y}_{kl}}{(s - \lambda_k)^{l+1}}$$

where the minimal polynomial of $\mathbf{A}$ is written as

22 $$m(s) = \prod_{k=1}^{\sigma} (s - \lambda_k)^{m_k}$$

(II) In addition, for $k = 1, 2, \ldots,$

23 $$\sum_k \mathbf{Y}_{k0} = \mathbf{I}$$
$$\mathbf{Y}_{k,l+1} = \mathbf{Y}_{kl}(\mathbf{A} - \lambda_k\mathbf{I}) \qquad l = 0, 1, 2, \ldots, m_k - 2$$
$$\mathbf{Y}_{k,m_k-1}(\mathbf{A} - \lambda_k\mathbf{I}) = \mathbf{0}$$

(III) Finally,

24 $$\exp(\mathbf{A}t) = \sum_{k=1}^{\sigma} \sum_{l=0}^{m_k-1} \mathbf{Y}_{kl} \frac{t^l}{l!} e^{\lambda_k t} \qquad t \geq 0$$

25 *Comment* The matrices $\mathbf{Y}_{kl}$ defined in *21* are related to the $\mathbf{Z}_{kl}$ which appear in *1* as follows:

$$(l!)^{-1}\mathbf{Y}_{kl} = \mathbf{Z}_{kl}$$

Proof The proof of the theorem is very simple although somewhat detailed. The basic idea is to perform the partial-fraction expansion of $\mathbf{P}(s)/m(s)$ exactly as it is done in the scalar case. The next step is to study its properties.

(I) It is obvious from *21* and *22* that the elements of $(s\mathbf{I} - \mathbf{A})^{-1}$ have a pole of order $\leq m_k$ at λ_k. Let us therefore define the constant matrix $\mathbf{Y}_{k,m_k-1}$ by the relation

$$\mathbf{Y}_{k,m_k-1} \triangleq \lim_{s \to \lambda_k} \left[(s - \lambda_k)^{m_k} \frac{\mathbf{P}(s)}{m(s)} \right] \qquad k = 1, \ldots, \sigma$$

From *22* it follows that if we put $m(s) = (s - \lambda_k)^{m_k} p_k(s)$, we have $p_k(\lambda_k) \neq 0$ and

$$\mathbf{Y}_{k,m_k-1} = \frac{\mathbf{P}(\lambda_k)}{p_k(\lambda_k)}$$

This equality implies that

$$\mathbf{P}(s) - \mathbf{Y}_{k,m_k-1} p_k(s) = \mathbf{0} \qquad \text{for } s = \lambda_k$$

Therefore, the left-hand side is of the form

$$\mathbf{P}(s) - \mathbf{Y}_{k,m_k-1} p_k(s) = (s - \lambda_k)\mathbf{D}_1(s)$$

where each element of $\mathbf{D}_1(s)$ is obtained by factoring out $(s - \lambda_k)$ from the corresponding element of the left-hand side. Dividing this equation by $m(s)$ gives

$$(s\mathbf{I} - \mathbf{A})^{-1} = \frac{\mathbf{P}(s)}{m(s)} = \frac{\mathbf{Y}_{k,m_k-1}}{(s - \lambda_k)^{m_k}} + \frac{\mathbf{D}_1(s)}{(s - \lambda_k)^{m_k-1}p_k(s)} \qquad \forall s \notin \mathcal{E}$$

The last term of this expression is a matrix which has a pole of order $m_k - 1$ at λ_k. Therefore, define the constant matrix $\mathbf{Y}_{k,m_k-2}$ by the relation

$$\mathbf{Y}_{k,m_k-2} \triangleq \lim_{s \to \lambda_k} \left[(s - \lambda_k)^{m_k-1} \frac{\mathbf{D}_1(s)}{(s - \lambda_k)^{m_k-1}p_k(s)} \right]$$

Hence we can write

$$\frac{\mathbf{D}_1(s)}{(s - \lambda_k)^{m_k-1}p_k(s)} = \frac{\mathbf{Y}_{k,m_k-2}}{(s - \lambda_k)^{m_k-1}} + \frac{\mathbf{D}_2(s)}{(s - \lambda_k)^{m_k-2}p_k(s)} \qquad \forall s \notin \mathcal{E}$$

Iterating the procedure, we exhaust the pole at λ_k and get

$$(s\mathbf{I} - \mathbf{A})^{-1} = \sum_{l=0}^{m_k-1} \frac{\mathbf{Y}_{kl}}{(s - \lambda_k)^{l+1}} + \frac{\mathbf{C}_k(s)}{p_k(s)} \qquad \forall s \notin \mathcal{E}$$

where $\mathbf{C}_k(s)$ is the matrix with polynomials in s as elements resulting from these operations and, as above, $p_k(\lambda_k) \neq 0$. The summation in the above equation constitutes the contribution to $(s\mathbf{I} - \mathbf{A})^{-1}$ of the m_kth-order pole at λ_k.

Repeating this procedure at each eigenvalue of $\mathbf{A}$, one obtains *21*, thus establishing the statement (I) of the theorem.

(II) Define for $k = 1, 2, \ldots,$

26 $$\mathbf{Y}_k(s) = \sum_{l=0}^{m_k-1} \frac{\mathbf{Y}_{kl}}{(s - \lambda_k)^{l+1}}$$

Hence

27 $$(s\mathbf{I} - \mathbf{A})^{-1} = \sum_{k=1}^{\sigma} \mathbf{Y}_k(s)$$

and

28 $$\sum_{k=1}^{\sigma} \mathbf{Y}_k(s)(s\mathbf{I} - \mathbf{A}) = \mathbf{I}$$

Now

$$\begin{aligned}
\mathbf{Y}_k(s)(s\mathbf{I} - \mathbf{A}) &= \mathbf{Y}_k(s)[(s - \lambda_k)\mathbf{I} + (\lambda_k\mathbf{I} - \mathbf{A})] \\
&= \mathbf{Y}_{k0} + \frac{[\mathbf{Y}_{k1} + \mathbf{Y}_{k0}(\lambda_k\mathbf{I} - \mathbf{A})]}{(s - \lambda_k)} + \cdots \\
&\quad + \frac{[\mathbf{Y}_{kl+1} + \mathbf{Y}_{kl}(\lambda_k\mathbf{I} - \mathbf{A})]}{(s - \lambda_k)^{l+1}} + \cdots \\
&\quad + \frac{\mathbf{Y}_{k,m_k-1}(\lambda_k\mathbf{I} - \mathbf{A})}{(s - \lambda_k)^{m_k}}
\end{aligned}$$

If we substitute the above equation into *28*, we obtain a sum of the form

$$\sum_{k=1}^{\sigma} \sum_{l=0}^{m_k} \frac{\mathbf{N}_{kl}}{(s - \lambda_k)^l} = \mathbf{I} \qquad \forall s \notin \mathcal{E}$$

where the matrices $\mathbf{N}_{kl}$ have constant elements. Since this equality holds for all $s \notin \mathcal{E}$, each matrix $\mathbf{N}_{kl}$ must be zero for $l \geq 1$ and $\sum_{k=1} \mathbf{N}_{k0} = \mathbf{I}$. Hence,

$$\sum_{k=1}^{\sigma} \mathbf{Y}_{k0} = \mathbf{I}$$

$$\mathbf{Y}_{k,l+1} = \mathbf{Y}_{kl}(\mathbf{A} - \lambda_k\mathbf{I}) = \mathbf{Y}_{k0}(\mathbf{A} - \lambda_k\mathbf{I})^{l+1} \qquad l = 0, 1, 2, \ldots, m_k - 2 \quad k = 1, \ldots, \sigma$$

$$\mathbf{Y}_{k,m_k-1}(\mathbf{A} - \lambda_k\mathbf{I}) = \mathbf{0} \qquad k = 1, \ldots, \sigma$$

Thus *23* is established.

(III) This follows directly from taking the inverse transform of *21*.

29 *Comment* Observe that *23* implies that each term of *28* reduces as follows:

30
$$\mathbf{Y}_k(s)(s\mathbf{I} - \mathbf{A}) = \mathbf{Y}_{k0}$$

which is a constant matrix.

31 **Computational procedure** The computation of exp $(\mathbf{A}t)$ involves the following sequence of steps:

(I) Compute $\mathbf{B}(s)$ from $(s\mathbf{I} - \mathbf{A})$ by *11* or any other suitable method.

(II) Remove any known common factor between $d(s)$ and all the elements of $\mathbf{B}(s)$. The result is the numerator $\mathbf{P}(s)$ and the denominator $m(s)$. [If a common factor is overlooked, general theory shows that this will cause no erroneous result, only unnecessary computations (see *D.8.27*).] Let

$$m(s) = \prod_{k=1}^{\sigma} (s - \lambda_k)^{m_k}$$

(III) Compute the matrices $\mathbf{Y}_{k0}, \mathbf{Y}_{k1}, \ldots, \mathbf{Y}_{k,m_k-1}$, for

$$k = 1, 2, \ldots, \sigma$$

as follows: let $p_{ij}(s)$ be the (i,j) element of $\mathbf{P}(s)$; let $p_k(s)$ be defined by $m(s) = (s - \lambda_k)^{m_k}p_k(s)$ [note that $p_k(\lambda_k) \neq 0$]. Expand $p_{ij}(s)$ in Taylor series around λ_k, viz.:

$$\frac{p_{ij}(s)}{p_k(s)} = \alpha_0 + \alpha_1(s - \lambda_k) + \alpha_2(s - \lambda_k)^2 + \cdots + \alpha_{m_k-1}(s - \lambda_k)^{m_k-1} + \cdots$$

where

$$\alpha_l = \frac{1}{l!}\frac{d^l}{ds^l}\left[\frac{p_{ij}(s)}{p_k(s)}\right]\Bigg|_{s=\lambda_k}$$

Then $\mathbf{Y}_{k0}$ is the matrix whose (i,j) element is α_{m_k-1}
$\mathbf{Y}_{k1}$ is the matrix whose (i,j) element is α_{m_k-2}
$\mathbf{Y}_{k,m_k-1}$ is the matrix whose (i,j) element is α_0
(IV) Check that $\sum_{k=1} \mathbf{Y}_{k0} = \mathbf{I}$

$$\mathbf{Y}_{k,m_k-1}(\mathbf{A} - \lambda_k\mathbf{I}) = \mathbf{0} \qquad k = 1, 2, \ldots, \sigma$$

The knowledge of the $\mathbf{Y}_{kl}$'s leads directly to exp $(\mathbf{A}t)$ by *24*. ◁

This ends the description of the computational procedure for finding exp $(\mathbf{A}t)$ based on the Laplace transform method. However, let us use the results obtained above to establish a basic fact that can also be derived by techniques of linear algebra.

Consider the residue matrices $\mathbf{Y}_{k0}$. From *30* we have

$$\mathbf{Y}_k(s) = \mathbf{Y}_{k0}(s\mathbf{I} - \mathbf{A})^{-1} \qquad \forall s \notin \mathcal{E}$$

and, from *27*,

$$\mathbf{Y}_k(s) = \mathbf{Y}_{k0} \sum_{j=1}^{\sigma} \mathbf{Y}_j(s)$$

Hence, by rearranging terms,

$$(\mathbf{Y}_{k0} - \mathbf{I})\mathbf{Y}_k(s) + \mathbf{Y}_{k0} \sum_{j \neq k} \mathbf{Y}_j(s) = \mathbf{0} \qquad \forall s \notin \mathcal{E}$$

Since $\mathbf{Y}_j(s)$ is the contribution to $(s\mathbf{I} - \mathbf{A})^{-1}$ of the pole at λ_j, the left-hand side is a sum of terms of a partial-fraction expansion which is identically zero. Hence, with the notation of *26*, we get

$$\mathbf{Y}_{k0}^2 = \mathbf{Y}_{k0} \qquad k = 1, \ldots, \sigma$$
$$\mathbf{Y}_{k0}\mathbf{Y}_{j0} = 0 \qquad j, k = 1, \ldots, \sigma; j \neq k$$

The first of the relations above implies that $\mathbf{Y}_{k0}$ represents a projection (see *C.12.4*). We express these results in the form of a

32 **Theorem** The residue matrices $\mathbf{Y}_{10}, \mathbf{Y}_{20}, \ldots, \mathbf{Y}_{\sigma 0}$ at the poles of $(s\mathbf{I} - \mathbf{A})^{-1}$ represent *projections* with the following properties:

$$\sum_{k=0}^{\sigma} \mathbf{Y}_{k0} = \mathbf{I}$$
$$\mathbf{Y}_{k0}\mathbf{Y}_{j0} = \mathbf{0} \qquad k \neq j \quad ◁$$

In the language of linear vector spaces, this implies that the range spaces of the residue matrices constitute a partition of the state space into direct sum of subspaces (see *D.5.6* and *15*). In Appendix *D*, the projections $\mathbf{Y}_{k0}$ are denoted by $\mathbf{E}_k$ (*D.5.6*).

4 *Modes in linear time-invariant systems (distinct eigenvalues)*

Any book on classical mechanics includes a chapter dealing with the study of small vibrations of *conservative* systems about an equilibrium point. It is shown that once the equations are linearized, any (small) free motion of the system can be thought of as a superposition of modes of vibration. Each mode of vibration is sinusoidal and is characterized by its frequency. This interpretation of the free motions in terms of modes is extremely important for understanding many engineering vibration problems. It is, however, less well known that this concept of modes can be extended to any linear system without requiring that it be conservative. Therefore, the mode concept that is presented below will apply, for example, to any linear time-invariant differential system, be it electrical, mechanical, or electromechanical, reciprocal or nonreciprocal, lossy or lossless, passive or active.

As a first step we must introduce a few preliminary ideas. Although these ideas are presented much more thoroughly in Appendix *C*, the account that follows is sufficient for our present purposes.

Eigenvalues, eigenvectors, basis, reciprocal basis, spectral expansion

For simplicity, consider a *single-input* linear time-invariant system described by its state equation[1]

1 $$\dot{\mathbf{x}}(t) = \mathbf{A}\mathbf{x}(t) + \mathbf{a}u(t)$$

where $\mathbf{A}$ is an $n \times n$ constant matrix, $\mathbf{a}$ is a constant n-vector, $u(t)$ is the (scalar) input, and $\mathbf{x}(t)$ ranges over $\mathcal{C}^n$ (space of n-tuples of complex numbers; see *3.7.36*). Let us first define a number of terms.

2 **Definition** The number λ is said to be an *eigenvalue* of $\mathbf{A}$ if $\Delta(\lambda) = \det(\mathbf{A} - \lambda\mathbf{I}) = 0$. $\Delta(\lambda)$ is called the *characteristic polynomial*. It is a polynomial of degree n in λ; therefore, there are at most n distinct eigenvalues. When $\mathbf{A}$ is real and λ is a complex eigenvalue, then λ^* is also an eigenvalue because $\Delta(\lambda)$ is a polynomial with real coefficients.

3 **Assumption** Throughout Sec. *4* the eigenvalues of $\mathbf{A}$ are assumed to be *distinct;* they are denoted by $\lambda_1, \lambda_2, \ldots, \lambda_n$.

4 **Definition** Any nonzero vector $\mathbf{u}_i$ such that

5 $$\mathbf{A}\mathbf{u}_i = \lambda_i\mathbf{u}_i$$

is called an *eigenvector* of $\mathbf{A}$; more precisely, it is an eigenvector associated with the eigenvalue λ_i. ◁

[1] The vector $\mathbf{a}$ in *1* is used in a sense which is different from that of *5.2.20*.

Since the components of $\mathbf{u}_i$ are the solutions of the system of n linear homogeneous algebraic equations, they are determined only within a constant factor: if $\mathbf{u}_i$ is an eigenvector, so is $\alpha\mathbf{u}_i$, where α is an arbitrary constant. This constant factor α is usually so selected that the "length" of the eigenvector is unity: the eigenvector is then said to be *normalized.* More precisely, let $\mathbf{u}_i = (u_{1i},u_{2i}, \ldots ,u_{ni})$; then $\mathbf{u}_i$ is normalized if $\sum_{j=1}^{n} |u_{ji}|^2 = 1$.

6 **Lemma** If the eigenvalues of $\mathbf{A}$ are distinct (i.e., if Assumption *3* holds), the n eigenvectors of $\mathbf{A}$ are linearly independent and hence constitute a basis for the state space $\mathcal{C}^n$ of *1*.

Proof By contradiction. Suppose the eigenvectors are not linearly independent; then there are n constants, $\alpha_1, \alpha_2, \ldots, \alpha_n$, not all zero, such that $\sum_{k=1}^{n} \alpha_k \mathbf{u}_k = \mathbf{0}$. Let i be the index of one of the nonzero constants. Let us premultiply the above relation by the matrix

$$(\mathbf{A} - \lambda_1\mathbf{I})(\mathbf{A} - \lambda_2\mathbf{I}) \cdots (\mathbf{A} - \lambda_{i-1}\mathbf{I})(\mathbf{A} - \lambda_{i+1}\mathbf{I}) \cdots (\mathbf{A} - \lambda_n\mathbf{I})$$

Observe that the matrix $(\mathbf{A} - \lambda_i\mathbf{I})$ is not a factor of this product. Clearly, since the factors of this product commute and since

$$(\mathbf{A} - \lambda_k\mathbf{I})\mathbf{u}_k = 0$$

for $k = 1, \ldots, n$, the result is

$$\alpha_i(\lambda_i - \lambda_1)(\lambda_i - \lambda_2) \cdots (\lambda_i - \lambda_{i-1})(\lambda_i - \lambda_{i+1}) \cdots (\lambda_i - \lambda_n)\mathbf{u}_i = \mathbf{0}$$

Since $\mathbf{u}_i \neq \mathbf{0}$ by *4* and since, for any $k \neq i$, $\lambda_i - \lambda_k \neq 0$ by Assumption *3*, the above equality implies $\alpha_i = 0$. This contradicts the assumption; therefore, the $\mathbf{u}_i$'s are linearly independent. Since there are n distinct eigenvectors and since the state space of *1* is of dimension n, the set of vectors $\mathbf{u}_1, \mathbf{u}_2, \ldots, \mathbf{u}_n$ constitutes a basis for the state space (see *C.4.9* to *C.4.11*).

7 **Definition** Let $\mathbf{a} = (\alpha_1,\alpha_2, \ldots ,\alpha_n)$, $\mathbf{b} = (\beta_1,\beta_2, \ldots ,\beta_n)$. The scalar product of $\mathbf{a}$ and $\mathbf{b}$ is written $\langle \mathbf{a},\mathbf{b} \rangle$ and is by definition the complex number

8 $$\sum_{i=1}^{n} \alpha_i^* \beta_i = \langle \mathbf{a},\mathbf{b} \rangle$$

This form of scalar product is called the *complex scalar product.* In most cases the vectors we are considering have real components. However, there are still many instances in which they have complex components. For this reason it is more convenient to use the complex scalar product throughout.

9 **Definition** Given a basis $\{\mathbf{u}_i\}_{i=1}^n$, the set of vectors $\{\mathbf{v}_i\}_{i=1}^n$ is said to be the *reciprocal basis* of the basis $\{\mathbf{u}_i\}$ if

10 $$\langle \mathbf{v}_i, \mathbf{u}_j \rangle = \delta_{ij} \qquad i, j = 1, 2, \ldots, n$$

11 *Comment* (I) Given the n^2 components of $\mathbf{u}_1, \ldots, \mathbf{u}_n$, *10* is a set of n^2 linear algebraic equations in the n^2 components of $\mathbf{v}_1, \ldots, \mathbf{v}_n$. The system splits easily into n systems with n unknowns: consider $\mathbf{v}_1$; its n components are the solutions of the n linear algebraic equations $\langle \mathbf{v}_1, \mathbf{u}_i \rangle = \delta_{1i}$. The determinant of this system has as elements of its ith row the components of $\mathbf{u}_i$. Since the $\mathbf{u}_i$'s are linearly independent, the determinant is different from zero and hence $\mathbf{v}_1$ is uniquely determined: *any basis has one and only one reciprocal basis.*

(II) Consider the two-dimensional case illustrated in Fig. *5.4.1*. Let $\mathbf{u}_1$ and $\mathbf{u}_2$ be specified. The direction of $\mathbf{v}_1$ is determined by the fact that $\langle \mathbf{v}_1, \mathbf{u}_2 \rangle = 0$, that is, $\mathbf{v}_1$ is perpendicular to $\mathbf{u}_2$; its length is determined by $\langle \mathbf{v}_1, \mathbf{u}_1 \rangle = 1$.

(III) If all the vectors of the basis $\{\mathbf{u}_i\}_{i=1}^n$ are normalized to a unit length, then the $\mathbf{v}_i$'s will in general not be normalized.

(IV) If the basis $\{\mathbf{u}_i\}_{i=1}^n$ is orthonormal, that is, $\langle \mathbf{u}_i, \mathbf{u}_j \rangle = \delta_{ij}$, then the basis is itself its own reciprocal basis. ◁

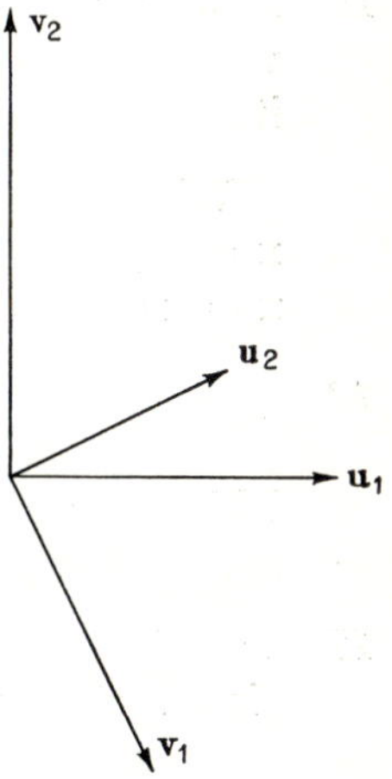

Fig. 5.4.1 The vectors $\mathbf{u}_1$, $\mathbf{u}_2$ constitute a basis for $\mathcal{C}^2$. The vectors $\mathbf{v}_1$, $\mathbf{v}_2$ are the vectors of the corresponding reciprocal basis.

The usefulness of the reciprocal basis follows from the

12 **Lemma** Let $\{\mathbf{u}_i\}_{i=1}^n$ be a basis. Let the (unique) representation of $\mathbf{x}$ be $\sum_{i=1}^n \xi_i \mathbf{u}_i$. Then

13 $$\xi_k = \langle \mathbf{v}_k, \mathbf{x} \rangle \qquad k = 1, 2, \ldots, n$$

and

14 $$\mathbf{x} = \sum_{i=1}^n \langle \mathbf{v}_i, \mathbf{x} \rangle \mathbf{u}_i$$

Proof *13* and *14* follow directly by taking the scalar product of $\mathbf{x} = \sum_{i=1}^n \xi_i \mathbf{u}_i$ with $\mathbf{v}_k$: that is, multiply the lth scalar equation by the complex conjugate of the lth component of $\mathbf{v}_k$ ($l = 1, 2, \ldots, n$) and sum; the result is *13* if we make use of *10*. *14* follows directly by substitution. ◁

At this point we need to introduce a convenient notation: consider the expression $\mathbf{a}\rangle\,\langle\mathbf{b}$ and call it a *dyad*. Given any vector $\mathbf{x}$, the product of the dyad $\mathbf{a}\rangle\langle\mathbf{b}$ and $\mathbf{x}$ is the vector $\mathbf{a}\langle\mathbf{b},\mathbf{x}\rangle$, that is, the vector $\mathbf{a}$ times the scalar $\langle\mathbf{b},\mathbf{x}\rangle$. The result may also be written $\langle\mathbf{b},\mathbf{x}\rangle\mathbf{a}$. Another way of looking at this definition is to observe that $\mathbf{a}\rangle\langle\mathbf{b}$, considered as a linear transformation operating on the vector $\mathbf{x}$, has a one-dimensional *range* (see *C.9.6*) which is spanned by $\mathbf{a}$, that is, all vectors $(\mathbf{a}\rangle\langle\mathbf{b})\mathbf{x}$ are proportional to $\mathbf{a}$. More precisely, we have the

15 **Definition** A *dyad* is a linear transformation denoted by $\mathbf{a}\rangle\langle\mathbf{b}$ which transforms the vector $\mathbf{x}$ into the vector $\langle\mathbf{b},\mathbf{x}\rangle\mathbf{a}$. In other words, it maps all vectors of $\mathcal{C}^n$ into vectors proportional to $\mathbf{a}$.

With the idea of the dyad in mind, if we look at *14* and note that $\mathbf{x} = \mathbf{I}\mathbf{x}$, then we may say that *14* has been obtained by multiplying $\mathbf{x} = \mathbf{x}$ on the left by

16
$$\mathbf{I} = \sum_{i=1}^{n} \mathbf{u}_i\rangle\langle\mathbf{v}_i$$

This point of view leads to a very important representation of any matrix $\mathbf{A}$ which satisfies Assumption *3*.

17 **Assertion** Let $\mathbf{A}$ have n distinct eigenvalues λ_i. Let $\{\mathbf{u}_i\}$ be its eigenvectors and $\{\mathbf{v}_i\}$ the reciprocal basis of the $\mathbf{u}_i$'s (see *9*). Then

18
$$\mathbf{A} = \sum_{i=1}^{n} \lambda_i\mathbf{u}_i\rangle\langle\mathbf{v}_i$$

Proof It follows by direct computation that for any $\mathbf{x}$, we have

$$\mathbf{A}\mathbf{x} = \mathbf{A}\left(\sum_{i=1}^{n} \langle\mathbf{v}_i,\mathbf{x}\rangle\mathbf{u}_i\right) = \sum_{i=1}^{n} \langle\mathbf{v}_i,\mathbf{x}\rangle\mathbf{A}\mathbf{u}_i$$
$$= \sum_{i=1}^{n} \langle\mathbf{v}_i,\mathbf{x}\rangle\lambda_i\mathbf{u}_i = \left(\sum_{i=1}^{n} \lambda_i\mathbf{u}_i\rangle\langle\mathbf{v}_i\right)\mathbf{x}$$

Note that in these steps we used *14*, the linearity of $\mathbf{A}$ and *5*. ◁

Expression *18* is called the *spectral representation* of $\mathbf{A}$: it gives a geometric interpretation of the effect of the linear transformation $\mathbf{A}$ on the vector $\mathbf{x}$. Consider all vectors with their components with respect to the basis $\{\mathbf{u}_i\}$; the ith component of $\mathbf{A}\mathbf{x}$ is equal to λ_i times the ith component of $\mathbf{x}$ ($i = 1, 2, \ldots, n$).

19 *Exercise* Define the product of two dyads by $(\mathbf{a}\rangle\langle\mathbf{b})(\mathbf{c}\rangle\langle d) = \langle\mathbf{b},\mathbf{c}\rangle\mathbf{a}\rangle\langle\mathbf{d}$. Let $\mathbf{E}_i = \mathbf{u}_i\rangle\langle\mathbf{v}_i$, in the notation of *10*. Show that $\mathbf{E}_i\mathbf{E}_j = \mathbf{0}$, $\mathbf{E}_i^2 = \mathbf{E}_i$, and $\sum_{i=1}^{n} \mathbf{E}_i = \mathbf{I}$.

20 *Exercise* Use the notation and assumptions of *17*. Let $\mathbf{P}$ be the matrix whose ith column is the components of $\mathbf{u}_i$. Let $\mathbf{Q}$ be the matrix whose ith row is the complex conjugates of the components of $\mathbf{v}_i$. (Remember we use the complex scalar product *7*.) (I) Show that $\mathbf{Q}\mathbf{P} = \mathbf{I}$.

(II) Show that both **P** and **Q** are nonsingular. (III) In the equation $\dot{\mathbf{x}} = \mathbf{A}\mathbf{x}$, change variable, put $\mathbf{x} = \mathbf{P}\mathbf{y}$. Put $\mathbf{y} = (\eta_1, \ldots, \eta_n)$. Show that the resulting equations are $\dot{\eta}_i = \lambda_i \eta_i$, $i = 1, 2, \ldots, n$. In other words if we pick as basis vectors the eigenvectors of **A**, the system $\dot{\mathbf{x}} = \mathbf{A}\mathbf{x}$ is diagonalized.

Mode interpretation of free motions

We now develop the mode interpretation of the free motions[1] of the system *1*; that is, throughout this subsection the input **u** is zero for all t. We use the same notation as above and we restrict ourselves to the case where Assumption *3* holds, that is, **A** has distinct eigenvalues.

To start the analysis, let us express the free motion $\mathbf{x}(t;\mathbf{x}_0,0)$ in terms of its components along the eigenvectors of **A**, viz.:

21 $$\mathbf{x}(t;\mathbf{x}_0,0) = \sum_{i=1}^{n} \alpha_i(t)\mathbf{u}_i \qquad -\infty < t < \infty$$

This is always possible, since by *6* the $\mathbf{u}_i$'s constitute a basis for the state space. Considering *21* at $t = 0$ and using *13*, we get

22 $$\alpha_i(0) = \langle \mathbf{v}_i, \mathbf{x}_0 \rangle \qquad i = 1, 2, \ldots, n$$

In order to calculate the functions $\alpha_i(t)$, substitute *21* into *1* and obtain successively

$$\sum_{i=1}^{n} \dot{\alpha}_i(t)\mathbf{u}_i = \mathbf{A}\left[\sum_{i=1}^{n} \alpha_i(t)\mathbf{u}_i\right] = \sum_{i=1}^{n} \alpha_i(t)\mathbf{A}\mathbf{u}_i$$
$$= \sum_{i=1}^{n} \alpha_i(t)\lambda_i\mathbf{u}_i$$

The resulting equality is one between two vectors, both expressed as a linear combination of the basis vectors. Since such representation is unique,

23 $$\dot{\alpha}_i(t) = \lambda_i \alpha_i(t) \qquad i = 1, 2, \ldots, n$$

The solution of *23* subject to the initial conditions *22* is

$$\alpha_i(t) = e^{\lambda_i t}\langle \mathbf{v}_i, \mathbf{x}_0 \rangle \qquad i = 1, 2, \ldots, n$$

Hence

24 $$\mathbf{x}(t;\mathbf{x}_0,0) = \sum_{i=1}^{n} e^{\lambda_i t}\langle \mathbf{v}_i, \mathbf{x}_0 \rangle \mathbf{u}_i$$

Comparing this with *5.2.11*, we are led to the expression

25 $$\exp(\mathbf{A}t) = \sum_{i=1}^{n} e^{\lambda_i t}\mathbf{u}_i\rangle\langle \mathbf{v}_i$$

[1] We could also call these free motions "zero-input state responses" since we are considering the case where the state itself is the output.

the *spectral representation* of exp ($\mathbf{A}t$). Note that in order to derive *25* we had to assume that $\mathbf{A}$ has distinct eigenvalues. In general, *25* holds if and only if $\mathbf{A}$ is a simple linear transformation (see *41* below).

26 *Exercise* Derive *24* by using the point of view of Comment *5.2.12*, especially Eq. *5.2.13*. ◁

We are now ready to introduce some interpretations. First let us define the concept of mode for systems satisfying Assumption *3*.

27 **Definition** The free system $\dot{\mathbf{x}} = \mathbf{A}\mathbf{x}$ will be said to have one and only one *mode* excited if its free motion (zero-input response) is of the form

$$\mathbf{x}(t) = \mathbf{c}e^{\lambda t}$$

where $\mathbf{c}$ is a constant vector and λ is a constant scalar. ◁

Clearly, since *24* represents all possible free motions of the system, the constant λ in the factor $e^{\lambda t}$ must be an eigenvalue of $\mathbf{A}$. To emphasize the ideas associated with the mode concept, we express them in terms of assertions.

28 **Assertion** Any free motion of the system *1* can be represented in one and only one way as the superposition of a certain number of modes of the system. ◁

The uniqueness follows from the fact that the eigenvectors of $\mathbf{A}$ constitute a basis for the state space (see *6*).

29 **Assertion** Each mode is excited independently of the other modes. ◁

Here by the "excitation" of the ith mode we mean $\langle \mathbf{v}_i, \mathbf{x}_0 \rangle$. The assertion follows immediately: the excitation of the ith mode is computed directly from $\mathbf{x}_0$ without regard to the values of $\langle \mathbf{v}_k, \mathbf{x}_0 \rangle$ for $k \neq i$. Another way of illustrating this fact is the following: suppose $\mathbf{x}_0 = \beta\mathbf{u}_i$, where β is a constant; geometrically the initial state is on the line supported by the ith eigenvector. By *10*, $\langle \mathbf{v}_k, \mathbf{x}_0 \rangle = 0$ for $k \neq i$; hence $\mathbf{x}(t;\beta\mathbf{u}_i,0) = \beta e^{\lambda_i t}\mathbf{u}_i$. Similarly, it is easy to find the initial state required for having the first mode excited by β_1, the second by β_2, . . . ; simply pick $\mathbf{x}_0 = \beta_1\mathbf{u}_1 + \beta_2\mathbf{u}_2 + \cdots + \beta_n\mathbf{u}_n$.

30 **Assertion** The excitation of each mode in the free motion depends only on the initial state. This is obvious from *24*. ◁

The interesting fact about these results is that the concept of modes applies to any linear time-invariant system with distinct eigenvalues. These concepts apply whether the system is active or passive, lossless or lossy, reciprocal or nonreciprocal. These concepts, of course, apply to a wide variety of physical systems: the only requirement is that the mathematical model satisfy *1* and *3*.

31 *Example*[1] We now illustrate these ideas by a simple example, namely, the *RC* circuit shown in Fig. *5.4.2*. Let us label the voltages across each

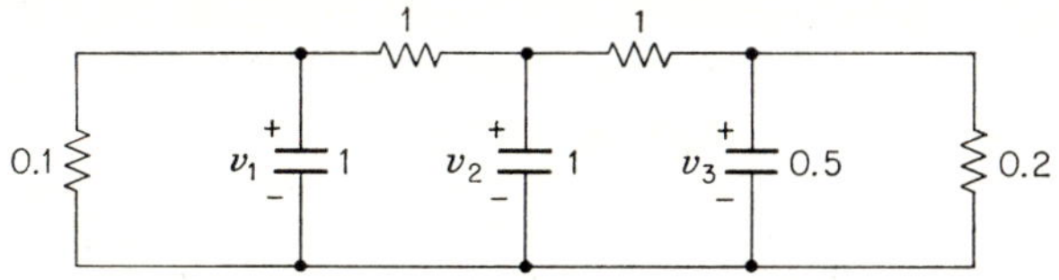

Fig. 5.4.2 Third-order system used in Example *5.4.31*. The element values are in mhos and darafs.

capacitance v_1, v_2, and v_3 and call $\mathbf{x}$ the vector whose components are v_1, v_2, and v_3. The three node equations read

$$\dot{v}_1 + 1.1v_1 - v_2 = 0$$
$$\dot{v}_2 - v_1 + 2v_2 - v_3 = 0$$
$$(\tfrac{1}{2})\dot{v}_3 - v_2 + 1.2v_3 = 0$$

In matrix form, after multiplying the last equation by 2, we have

$$\dot{\mathbf{x}} = \mathbf{A}\mathbf{x} \qquad \text{where } \mathbf{A} = \begin{bmatrix} -1.1 & 1 & 0 \\ 1 & -2 & 1 \\ 0 & 2 & -2.4 \end{bmatrix}$$

The third-degree characteristic equation $\det(\mathbf{A} - \lambda\mathbf{I}) = 0$ has three

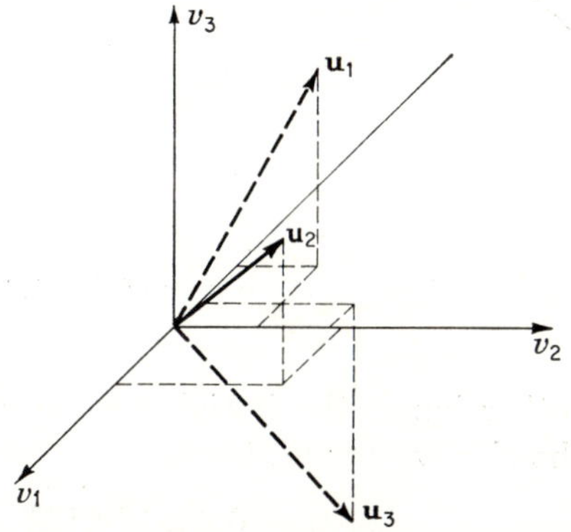

Fig. 5.4.3 Perspective view of the eigenvectors of the system shown in Fig. *5.4.2*.

roots: $\lambda_1 = -1.5878$, $\lambda_2 = -0.1126$, $\lambda_3 = -3.7996$. Solving Eq. *5* with these eigenvalues, we obtain the following normalized eigenvectors:

$$\mathbf{u}_1 = \begin{bmatrix} -0.6108 \\ 0.2980 \\ 0.7337 \end{bmatrix} \qquad \mathbf{u}_2 = \begin{bmatrix} 0.6064 \\ 0.5986 \\ 0.5235 \end{bmatrix} \qquad \mathbf{u}_3 = \begin{bmatrix} -0.2078 \\ 0.5609 \\ -0.8015 \end{bmatrix}$$

These eigenvectors are illustrated in Fig. *5.4.3*. If the initial conditions

[1] C. A. Desoer, Modes in Linear Circuits, *I.R.E. Trans. Circuit Theory*, vol. CT-7, no. 3, pp. 211–223, 1960.

are $v_1(0) = -0.2078$, $v_2(0) = 0.5609$, $v_3(0) = -0.8015$ (or any set of numbers proportional to the components of $\mathbf{u}_3$), the resulting behavior is

$$\begin{aligned} v_1(t) &= -0.2078e^{-3.7996t} \\ v_2(t) &= 0.5609e^{-3.7996t} \\ v_3(t) &= -0.8015e^{-3.7996t} \end{aligned}$$

These three waveforms are illustrated in Fig. *5.4.4*.

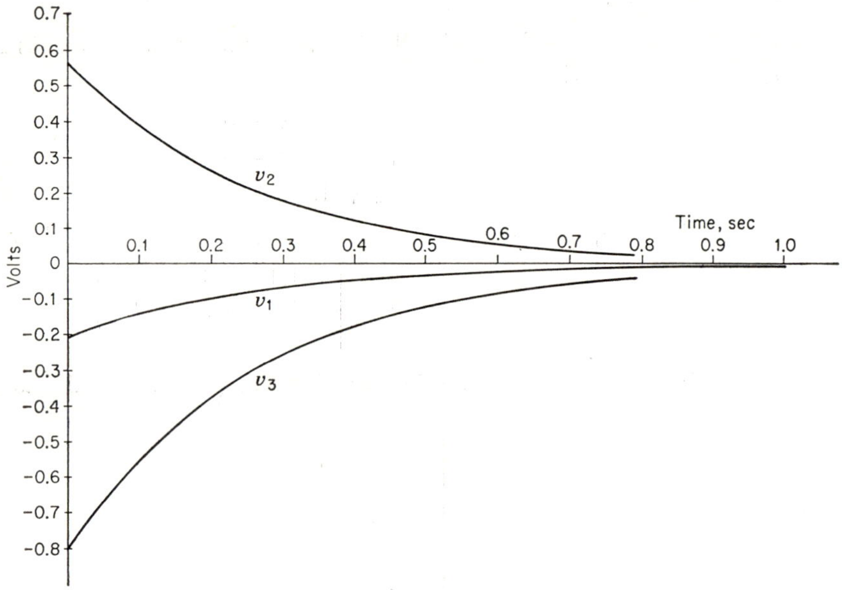

Fig. 5.4.4 Voltage waveforms in the system of Fig. *5.4.2* when only the third mode is excited.

If, on the other hand, the initial state vector $\mathbf{x}(0)$ had been proportional to the eigenvector $\mathbf{u}_2$, then

$$\begin{aligned} v_1(t) &= 0.6064e^{-0.1126t} \\ v_2(t) &= 0.5986e^{-0.1126t} \\ v_3(t) &= 0.5235e^{-0.1126t} \end{aligned}$$

Free motion (complex eigenvalues)

The analytical developments above apply in all cases where $\mathbf{A}$ is a constant matrix with distinct eigenvalues. In most applications the elements of $\mathbf{A}$, as well as those of $\mathbf{x}$, are real. However, the application of the formula *24* may lead to complex terms if some eigenvalues are complex. We propose here to consider this case in more detail.

We assume that $\mathbf{A}$ is real and $\mathbf{x}_0$ is real. Let $\lambda_1 = \alpha_1 + j\omega_1$ be a complex eigenvalue of $\mathbf{A}$. Since the characteristic polynomial det $(\mathbf{A} - \lambda\mathbf{I})$ has real coefficients, $\lambda_1^* = \alpha_1 - j\omega_1$ is also an eigenvalue. Let us call it λ_2. The eigenvector $\mathbf{u}_1$, corresponding to λ_1, has complex elements; we shall write it

$$\mathbf{u}_1 = \mathbf{u}_1' + j\mathbf{u}_1'' \qquad \mathbf{u}_1', \mathbf{u}_1'' \text{ real}$$

Similarly, $\mathbf{u}_2$, the eigenvector corresponding to λ_2, is of the form (see *5*)

$$\mathbf{u}_2 = \mathbf{u}_1' - j\mathbf{u}_1''$$

It is easily verified that $\mathbf{v}_1$ and $\mathbf{v}_2$, the elements of the reciprocal basis corresponding to $\mathbf{u}_1$ and $\mathbf{u}_2$, are complex conjugates of one another. Let us put

$$\mathbf{v}_1 = \tfrac{1}{2}\mathbf{v}_1' + j\tfrac{1}{2}\mathbf{v}_1'' \qquad \mathbf{v}_2 = \tfrac{1}{2}\mathbf{v}_1' - j\tfrac{1}{2}\mathbf{v}_1'' \qquad \mathbf{v}_1', \mathbf{v}_1'' \text{ real}$$

The relations

$$\langle \mathbf{v}_1,\mathbf{u}_1\rangle = 1 \qquad \langle \mathbf{v}_1,\mathbf{u}_2\rangle = 0$$

become four scalar relations involving real vectors $\mathbf{u}_1'$, $\mathbf{v}_1'$, $\mathbf{u}_1''$, and $\mathbf{v}_1''$:

$$\langle \mathbf{v}_1',\mathbf{u}_1'\rangle = 1 \qquad \langle \mathbf{v}_1'',\mathbf{u}_1'\rangle = 0$$
$$\langle \mathbf{v}_1',\mathbf{u}_1''\rangle = 0 \qquad \langle \mathbf{v}_1'',\mathbf{u}_1''\rangle = 1$$

With these notations, the first two terms of *24* become

32 $$e^{\alpha_1 t}[\langle \mathbf{v}_1',\mathbf{x}_0\rangle \cos \omega_1 t + \langle \mathbf{v}_1'',\mathbf{x}_0\rangle \sin \omega_1 t]\mathbf{u}_1' + e^{\alpha_1 t}[\langle \mathbf{v}_1'',\mathbf{x}_0\rangle \cos \omega_1 t - \langle \mathbf{v}_1',\mathbf{x}_0\rangle \sin \omega_1 t]\mathbf{u}_1''$$

or, in its equivalent polar form,

33 $$e^{\alpha_1 t}[\gamma \cos (\omega_1 t - \varphi)\, \mathbf{u}_1' - \gamma \sin (\omega_1 t - \varphi)\, \mathbf{u}_1'']$$

where γ and φ are given by

$$\gamma \cos \varphi = \langle \mathbf{v}_1',\mathbf{x}_0\rangle \qquad \gamma \sin \varphi = \langle \mathbf{v}_1'',\mathbf{x}_0\rangle$$

The bracketed quantity is the superposition of two sinusoidal oscillations along the directions $\mathbf{u}_1'$, $\mathbf{u}_1''$. The amplitude and phase of the oscillations depend only on the initial state $\mathbf{x}_0$. The whole motion occurs in the plane (more precisely, two-dimensional subspace), supported by $\mathbf{u}_1'$ and $\mathbf{u}_1''$. In order to better illustrate *32* and *33*, consider two special cases:

(I) $\mathbf{x}_0 = \mathbf{u}_1'$. Then $\gamma = 1$ and $\varphi = 0$.

$$\mathbf{x}(t) = e^{\alpha_1 t} \cos \omega_1 t\, \mathbf{u}_1' - e^{\alpha_1 t} \sin \omega_1 t\, \mathbf{u}_1''$$

(II) $\mathbf{x}_0 = \mathbf{u}_1''$. Then $\gamma = 1$ and $\varphi = \pi/2$.

$$\mathbf{x}(t) = e^{\alpha_1 t} \sin \omega_1 t\, \mathbf{u}_1' + e^{\alpha_1 t} \cos \omega_1 t\, \mathbf{u}_1''$$

These two cases are illustrated in Fig. *5.4.5*, where we have assumed $\alpha_1 < 0$ so that the state slides along the exponential spiral toward the origin.

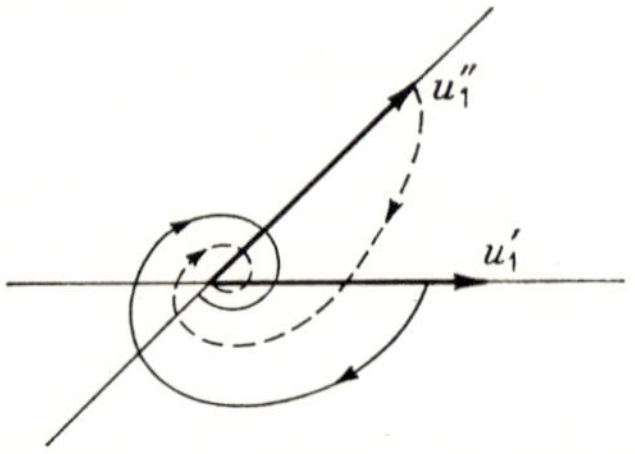

Fig. 5.4.5 Illustration of the spiral motions in the plane $\mathbf{u}_1'$, $\mathbf{u}_1''$ resulting from a mode with complex eigenvalue.

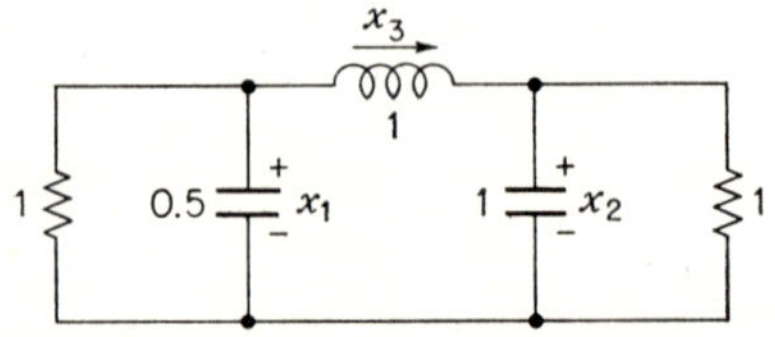

Fig. 5.4.6 Third-order system of Example *5.4.34*.

34 *Example*[1] Consider the *RLC* circuit shown in Fig. *5.4.6*. Writing two node equations and one loop equation, one gets

$$\begin{bmatrix} \dot{x}_1 \\ \dot{x}_2 \\ \dot{x}_3 \end{bmatrix} = \begin{bmatrix} -2 & 0 & -2 \\ 0 & -1 & +1 \\ 1 & -1 & 0 \end{bmatrix} \begin{bmatrix} x_1 \\ x_2 \\ x_3 \end{bmatrix}$$

where x_1, x_2, and x_3 are defined in the figure.

The matrix of this circuit has three eigenvalues, one real and a pair of complex conjugate ones. The corresponding eigenvectors are given in the table

$$\lambda_1 = -1.4534 \qquad \lambda_2 = -0.7733 + j1.468 \qquad \lambda_3 = \lambda_2^*$$

$$\mathbf{u}_1 = \begin{bmatrix} 0.834 \\ 0.503 \\ -2.228 \end{bmatrix} \qquad \mathbf{u}_2 = \begin{bmatrix} -0.100 - j0.646 \\ 0.268 + j0.322 \\ -0.4118 + j0.468 \end{bmatrix}$$

For example, if the initial conditions are along $\mathbf{u}_1$ [say, $x_1(0) = 0.834$, $x_2(0) = 0.503$, and $x_3(0) = -2.228$], only the exponentially damped mode is excited and the resulting voltages and currents are very similar to those of the *RC* circuit of the previous example. On the other hand, if the vector representing the initial conditions lies in the plane defined by $\mathbf{u}_2'$ and $\mathbf{u}_2''$ [where $\mathbf{u}_2' = (-0.100, 0.268, -0.4118)$ and $\mathbf{u}_2'' = (-0.646, 0.322, 0.468)$], then only the oscillatory mode is excited. Throughout the circuit the voltages and the currents are damped sinusoids of different amplitudes and phases, but they all have in common the decrement -0.7733 and the angular frequency 1.468 rad/sec. For such initial conditions, although this circuit is a three-reactance circuit, the voltages and currents have the simple waveforms of the standard *RLC* resonant circuit.

[1] *Ibid.*

Forced oscillations

Consider for simplicity a *single-input,* linear, time-invariant system represented by

$$\dot{\mathbf{x}} = \mathbf{A}\mathbf{x} + \mathbf{a}u(t)$$

where $\mathbf{a}$ is a constant vector and u is the input. From *5.2.21* the state of the system at time t, assuming that the state at $t = 0$ is $\mathbf{x}_0$, is given by

$$\mathbf{x}(t) = [\exp(\mathbf{A}t)]\mathbf{x}_0 + \int_0^t \{\exp[\mathbf{A}(t-\tau)]\}\mathbf{a}u(\tau)\,d\tau$$

Assuming that $\mathbf{A}$ has distinct eigenvalues, the spectral representation *25* is valid and

35 $$\mathbf{x}(t) = \sum_{i=1}^{n} e^{\lambda_i t}\mathbf{u}_i\left[\langle \mathbf{v}_i,\mathbf{x}_0\rangle + \int_0^t e^{-\lambda_i\tau}\langle \mathbf{v}_i,\mathbf{a}\rangle u(\tau)\,d\tau\right]$$

Thus, for any t, the state $\mathbf{x}(t)$ is a linear combination of the modes of the system; however, the time behavior of the coefficients of the linear combination is more complicated: compare the brackets of *35* and *24*. The excitation of the ith mode due to the input u is

$$e^{\lambda_i t}\langle \mathbf{v}_i,\mathbf{a}\rangle \int_0^t e^{-\lambda_i\tau}u(\tau)\,d\tau$$

For all t, this term is proportional to the constant $\langle \mathbf{v}_i,\mathbf{a}\rangle$ which is the component of $\mathbf{a}$ along $\mathbf{u}_i$ in the basis $\{\mathbf{u}_k\}$. Equation *35* and the above call for the following comments which we present in the form of an

36 **Assertion** Irrespective of the input u, each mode is excited independently of the others. ◁

The amount of excitation of any particular mode does not depend on the amount of excitation of the other modes.

37 **Assertion** The forcing vector $\mathbf{a}$ may be so chosen that, for all possible inputs u, a particular mode is never affected by the input. ◁

In such a case, its excitation, if any, is due exclusively to the initial state. For example, let $\langle \mathbf{v}_i,\mathbf{a}\rangle = 0$, that is, the vector $\mathbf{a}$ has no component along $\mathbf{u}_i$; then *35* shows that the input u will never affect the ith mode. We say that the *input u is uncoupled to the ith mode.*

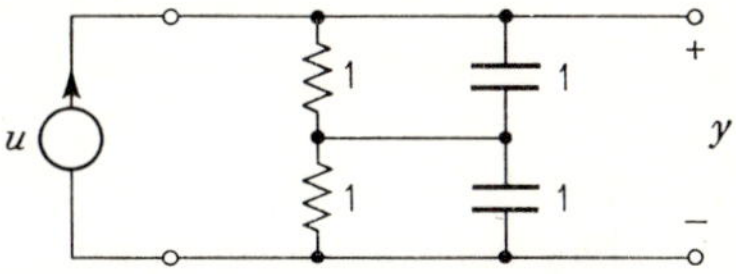

Fig. 5.4.7 Circuit considered in Exercise *5.4.38*. The element values are in ohms and farads.

38 *Exercise* Given the circuit shown in Fig. *5.4.7*, derive its equations in state form. Is the input coupled to all the modes of the circuit?

39 *Exercise* Suppose $\lambda_k = \alpha_k + j\omega_k$ and $\mathbf{u}_k = \mathbf{u}_k' + j\mathbf{u}_k''$, $\mathbf{u}_k'$ and $\mathbf{u}_k''$ real.

Show that if $\mathbf{a} = \mathbf{u}_k'$ and $\mathbf{x}_0 = \mathbf{0}$, then

$$\mathbf{x}(t) = \left[\int_0^t u(\tau)e^{\alpha_k(t-\tau)} \cos \omega_k(t-\tau)\, d\tau\right] \mathbf{u}_k' - \left[\int_0^t u(\tau)e^{\alpha_k(t-\tau)} \sin \omega_k(t-\tau)\, d\tau\right] \mathbf{u}_k''$$

and, if $\mathbf{a} = \mathbf{u}_k''$,

$$\mathbf{x}(t) = \left[\int_0^t u(\tau)e^{\alpha_k(t-\tau)} \sin \omega_k(t-\tau)\, d\tau\right] \mathbf{u}_k' + \left[\int_0^t u(\tau)e^{\alpha_k(t-\tau)} \cos \omega_k(t-\tau)\, d\tau\right] \mathbf{u}_k''$$

Resonance

Let the input be $u = e^{j\omega t}$. Assuming that the system is in the zero state at $t = 0$, we have

$$\mathbf{x}(t) = \sum_{i=1}^{n} \langle \mathbf{v}_i,\mathbf{a}\rangle \left[\int_0^t e^{j\omega(t-\tau)}e^{\lambda_i\tau}\, d\tau\right] \mathbf{u}_i = \sum_{i=1}^{n} \frac{\langle \mathbf{v}_i,\mathbf{a}\rangle}{\lambda_i - j\omega} (e^{\lambda_i t} - e^{j\omega t})\mathbf{u}_i$$

Suppose the system is asymptotically stable, that is, $\operatorname{Re} \lambda_i < 0$ for $i = 1, 2, \ldots, n$; then as $t \to \infty$, $e^{\lambda_i t} \to 0$.[1] Thus as $t \to \infty$, the motion of $\mathbf{x}(t)$ becomes asymptotic to the sinusoidal steady state

$$\mathbf{x}(t) \to \sum_{i=1}^{n} \frac{\langle \mathbf{v}_i,\mathbf{a}\rangle}{j\omega - \lambda_i} e^{j\omega t}\mathbf{u}_i \qquad \text{as } t \to \infty$$

The excitation of the ith mode has a peak amplitude:

$$\left| \frac{\langle \mathbf{v}_i,\mathbf{a}\rangle}{j\omega - \lambda_i} \right|$$

if λ_i is real. When λ_i is complex, we must consider both $\mathbf{u}_i$ and $\mathbf{u}_i^*$; the oscillation of the ith mode in the $\mathbf{u}_i'$, $\mathbf{u}_i''$ plane has an amplitude

$$\left| \frac{\langle \mathbf{v}_i,\mathbf{a}\rangle}{j\omega - \lambda_i} + \frac{\langle \mathbf{v}_i^*,\mathbf{a}\rangle}{j\omega - \lambda_i^*} \right|$$

Let $\lambda_i = \alpha_i + j\omega_i$. If $\omega \simeq \omega_i$ and if $\alpha_i \ll \omega_i$ (that is, if the ith mode has very little damping and if the input frequency is close to the frequency of the mode), then in the above expression the first term dominates and the amplitude of oscillation is approximately

$$\left| \frac{\langle \mathbf{v}_i,\mathbf{a}\rangle}{j\omega - \lambda_i} \right|$$

[1] Definition and characterization of asymptotic stability are found in Chap. 7, Secs. 2 and 3.

40 *Comment* We are thus led to the following conclusion. In the sinusoidal steady state the amplitude of oscillation of a particular mode depends on (I) the magnitude and direction of the forcing vector **a** and on (II) the distance between λ_i and $j\omega$ in the s plane. The first dependence is measured by $|\langle \mathbf{v}_i,\mathbf{a}\rangle|$; this scalar product is a measure of the coupling between the input and the ith mode. The second dependence is such that the amplitude of oscillation is inversely proportional to $|j\omega - \lambda_i|$, the distance in the s plane between the frequency of the excitation, $j\omega$, and that of the ith mode, $\alpha_i + j\omega_i$. This fact is well known from the Laplace transform approach: the residue at the pole $j\omega$ of the forcing function contains in its denominator the factor $|j\omega - \lambda_i|$.

Thus, in order to obtain the largest amplitude of oscillation in a particular mode, say, the ith, one should pick **a** so that $|\langle \mathbf{v}_i,\mathbf{a}\rangle|$ is as large as possible and use ω_i as the excitation frequency.

Remark on simple linear transformations[1]

In n-dimensional space, a matrix **A** represents, with respect to a given basis, a linear transformation which associates to each vector **x** a vector **y** given by $\mathbf{y} = \mathbf{A}\mathbf{x}$. In *C.18.6* we define a *simple* linear transformation as one whose eigenvectors constitute a basis for the space. From *6* it follows that a matrix whose eigenvalues are distinct represents a simple linear transformation. However, the class of simple linear transformations is much broader: it may include matrices with multiple eigenvalues; for example, when

$$\mathbf{A} = \begin{bmatrix} 1 & 0 \\ 0 & 1 \end{bmatrix}$$

$\det(\mathbf{A} - \lambda\mathbf{I}) = (\lambda - 1)^2$ and $\lambda = 1$ is a second-order zero of the characteristic equation. But the eigenvectors $\mathbf{u}_1 = (1,0)$, $\mathbf{u}_2 = (0,1)$ span the space.

The point we wish to make is the following: let $\lambda_1, \lambda_2, \ldots, \lambda_n$ be the (not necessarily distinct) eigenvalues of **A**. Let $\mathbf{u}_1, \mathbf{u}_2, \ldots, \mathbf{u}_n$ be the corresponding eigenvectors. If the eigenvectors constitute a basis for the space, then we still have the spectral representations

$$\mathbf{A} = \sum_{i=1}^{n} \lambda_i \mathbf{u}_i\rangle\ \langle\mathbf{v}_i$$

$$\exp(\mathbf{A}t) = \sum_{i=1}^{n} e^{\lambda_i t}\mathbf{u}_i\rangle\ \langle\mathbf{v}_i$$

where $\{\mathbf{v}_i, \mathbf{v}_2, \ldots, \mathbf{v}_n\}$ constitute the reciprocal basis of the basis $\{\mathbf{u}_1, \ldots, \mathbf{u}_n\}$. Therefore we have the

[1] For a more complete treatment see Appendix *C*, Sec. *18*.

41 **Assertion** The mode theory developed above is valid also when the matrix **A** represents a simple linear transformation. ◁

These statements are justified by the following observation. The only place where the assumption that the eigenvalues of **A** are distinct was used was in proving that the eigenvectors spanned the space. Under our present assumptions, this fact is true by assumption irrespective of whether the eigenvalues are distinct or not.

42 *Exercise* Let $\mathbf{A}^t$ be the transpose of **A**, that is, the matrix whose element of the ith row and jth column is a_{ji}. Show that, if **A** represents a simple linear transformation, $\mathbf{A}^t$ also represents a simple linear transformation; furthermore, that the spectral representations of $\mathbf{A}^t$ and $\exp(\mathbf{A}^t t)$ can be obtained from those of **A** and $\exp(\mathbf{A}t)$ by interchanging the $\mathbf{u}_i$'s and $\mathbf{v}_i$'s.

5 *Modes in linear time-invariant systems (general case)*

The mode picture in the case where **A** is not a simple linear transformation is very involved. In order to keep the discussion within reason, we assume that the reader is familiar with most of the ideas of Appendix *D*, Sec. *5*. This will allow us to describe the situation precisely and briefly. We shall consider only the free motion of the system described by

1 $$\dot{\mathbf{x}} = \mathbf{A}\mathbf{x} \qquad \mathbf{x}(0) = \mathbf{x}_0$$

If **A** is not a simple linear transformation, some zeros of the characteristic polynomial $\Delta(\lambda)$ are not simple. Let us label its *distinct* zeros (eigenvalues) as $\lambda_1, \lambda_2, \ldots, \lambda_\sigma$; clearly, $\sigma < n$. Let m_k be the multiplicity of λ_k as a zero of the *minimal* polynomial of **A** (see *D.2.1*) Let

2 $$\mathfrak{N}_k^l = \{\mathbf{x} \mid (\mathbf{A} - \lambda_k \mathbf{I})^l \mathbf{x} = 0\} \qquad l = 1, 2, \ldots; k = 1, \ldots, \sigma$$

$\mathfrak{N}_k^l$ is a linear vector space which is a subset of the state space $\mathcal{C}^n$. The following facts are either immediate or established in Appendix *D*.

(I) The subspaces $\mathfrak{N}_k^l$ are invariant subspaces of **A**; that is,

3 $$\mathbf{x} \in \mathfrak{N}_k^l \Rightarrow \mathbf{A}\mathbf{x} \in \mathfrak{N}_k^l$$

4 $$\text{(II)} \quad \mathfrak{N}_k^1 \subset \mathfrak{N}_k^2 \subset \cdots \subset \mathfrak{N}_k^{m_k} = \mathfrak{N}_k^{m_k+i} \qquad i = 1, 2, \ldots$$

For convenience, we write $\mathfrak{N}_k$ instead of $\mathfrak{N}_k^{m_k}$.

(III) The state space $\mathcal{C}^n$ is the direct sum of the $\mathfrak{N}_k$'s:

5 $$\mathcal{C}^n = \mathfrak{N}_1 \oplus \mathfrak{N}_2 \oplus \cdots \oplus \mathfrak{N}_\sigma$$

(IV) In the notation of *5.3.1*,

6 $$\mathbf{Z}_{kl}\mathbf{x} \in \mathfrak{N}_k \qquad \forall \mathbf{x} \in \mathcal{C}^n,\ l = 0, 1, \ldots, m_k - 1 \quad k = 1, 2, \ldots, \sigma$$

7 $$k \neq m \Rightarrow \mathbf{Z}_{kl}\mathbf{Z}_{mj} = 0 \qquad l = 0, 1, \ldots, m_k - 1 \quad j = 0, 1, \ldots, m_m - 1$$

From these properties of **A** we deduce the following conclusions.

8 **Assertion** If $\mathbf{x}_0 \in \mathfrak{N}_k$, then $\mathbf{x}(t) \in \mathfrak{N}_k$ for all $t > 0$; that is, if the initial state lies in subspace $\mathfrak{N}_k$, the state will remain in that subspace for all subsequent times.

This follows immediately from the fundamental formula for exp $(\mathbf{A}t)$ (*5.3.1*) and from *6* and *7*. Heuristically, it is obvious, because $\mathbf{x}_0 \in \mathfrak{N}_k \Rightarrow \mathbf{A}\mathbf{x}_0 \in \mathfrak{N}_k$; hence, by *1*, $\mathbf{x}_0 + \dot{\mathbf{x}}_0\, dt \in \mathfrak{N}_k$, since $\mathfrak{N}_k$ is itself a linear vector space, in fact, a subspace of the state space $\mathcal{C}^n$. Since this reasoning applies for all times, the conclusion follows.

9 **Assertion** Any initial state can be written in one and only one way as

$$\mathbf{x}_0 = \mathbf{y}_1 + \mathbf{y}_2 + \cdots + \mathbf{y}_\sigma$$

where $$\mathbf{y}_i \in \mathfrak{N}_i \qquad i = 1, 2, \ldots, \sigma$$

This follows immediately from *5*. Thus, the initial state can be considered as the superposition of σ initial states and, in view of *8*, the resulting free motion $\mathbf{x}(t)$ is a unique superposition of the free motions $\mathbf{y}_i(t)$, for all t.

Thus if we call $\mathbf{y}_i(t)$ the ith mode, we have the three properties: (I) any free motion can be represented in one and only one way as a superposition of a certain number of modes of the system, (II) each mode is excited independently of the other modes, and (III) the amount of excitation of each mode depends only on the initial conditions. These properties are identical with those valid for the case of simple operators; however, the time behavior of each mode is much more complicated. From *6* and the fundamental formula for exp $(\mathbf{A}t)$ we have

$$\mathbf{y}_i(t) = \sum_{l=0}^{m_i-1} t^l e^{\lambda_i t}\mathbf{Z}_{il}\mathbf{x}_0$$

The following special cases will illustrate the time behavior of the ith mode:

(I) If $\mathbf{y}_i(0) \in \mathfrak{N}_i^1$, then $\mathbf{y}_i(t) = e^{\lambda_i t}\mathbf{y}_i(0)$

(II) If $\mathbf{y}_i(0) \in \mathfrak{N}_i^2$, but $\mathbf{y}_i(0) \notin \mathfrak{N}_i^1$, then

$$\mathbf{y}_i(t) = e^{\lambda_i t}\mathbf{y}_i(0) + te^{\lambda_i t}(\mathbf{A} - \lambda_i\mathbf{I})\mathbf{y}_i(0)$$

(III) If $\mathbf{y}_i(0) \in \mathfrak{N}_i^{m_i}$, but $\mathbf{y}_i(0) \notin \mathfrak{N}_i^{m_i-1}$, then

$$\mathbf{y}_i(t) = e^{\lambda_i t}\mathbf{y}_i(0) + \sum_{k=1}^{m_i-1} t^k e^{\lambda_i t}(\mathbf{A} - \lambda_i \mathbf{I})^k \mathbf{y}_i(0)$$

Thus we arrive at the final

10 **Assertion** Any initial state that excites a free motion of the form $t^l e^{\lambda_i t}$ also excites motions of the form $t^{l-1}e^{\lambda_i t}, \ldots, te^{\lambda_i t}, e^{\lambda_i t}$. ◁

We conclude our discussion at this point. It is, however, obvious that the forced response can also be interpreted in terms of modes.

11 *Exercise* Consider the system $\dot{\mathbf{x}} = \mathbf{Ax} + \mathbf{a}u(t)$. Pick an input u and a constant forcing vector $\mathbf{a}$ such that the state of the system (assumed to be starting from rest at $t = 0$) is described by $\mathbf{x}(t) = t^N e^{\lambda_i t}\mathbf{d}$ for all $t > 0$, where $\mathbf{d}$ is some constant vector (not prescribed in advance) and N is some prescribed arbitrary positive integer. [*Hint:* Take $\mathbf{a}$ to be an eigenvector of $\mathbf{A}$ corresponding to λ_i and take u such that $\mathcal{L}\{u\} = N!/(s - \lambda_i)^N$.]

6 *Systems of differential equations*

In the preceding sections we focused our attention on the solution of first-order vector differential equations of the form $\dot{\mathbf{x}} = \mathbf{Ax} + \mathbf{B}u$. In this section, we shall consider vector differential equations (not necessarily state equations) of the more general form

1
$$\mathbf{L}(p)\mathbf{x}(t) = \mathbf{M}(p)u(t)$$

or more explicitly

2
$$\sum_{j=1}^{n} L_{ij}(p)x_j(t) = M_i(p)u(t) \qquad i = 1, 2, \ldots, n$$

where the $L_{ij}(p)$ and $M_i(p)$ are differential operators (*3.8.11*) with constant coefficients, u is a scalar input, and the dependent variable $\mathbf{x}(t)$ plays the role of a vector-valued output. In effect, *1* may be regarded as an input-output relation which differs from the relations considered in Chap. *4* in that $\mathbf{L}(p)$ and $\mathbf{M}(p)$ are matric rather than scalar differential operators. Equations of the form *1* are encountered quite frequently in practice as a result of direct application of the laws of physics. For example, in the analysis of linear time-invariant circuits Kirchhoff laws lead directly to a system of the form *1*; the components of $\mathbf{x}(t)$ are then either loop currents or node pair voltages.

In the analysis of the small vibrations of mechanical or electromechanical systems Lagrange's equations also lead to such equations.

Input-output-state relations

As usual, one of the first questions we raise in connection with *1* is: How can one associate a state vector with *1* and what is the corresponding input-output-state relation? To solve this problem, we can employ a slightly extended version of the Laplace transform technique described in Chap. *4*. A simple example will suffice to illustrate its use.

3 *Example* Consider the system

$$\begin{bmatrix} p^2+p+1 & 1 \\ p & p+2 \end{bmatrix} \begin{bmatrix} x_1(t) \\ x_2(t) \end{bmatrix} = \begin{bmatrix} p+1 \\ 1 \end{bmatrix} u(t)$$

On applying the Laplace transformation to both sides of this system, we obtain after a rearrangement of terms

4
$$\begin{bmatrix} s^2+s+1 & 1 \\ s & s+2 \end{bmatrix} \begin{bmatrix} X_1(s) \\ X_2(s) \end{bmatrix} = \begin{bmatrix} (s+1)U(s) + sx_1(0-) + \dot{x}_1(0-) + x_1(0-) - u(0-) \\ U(s) + x_1(0-) + x_2(0-) \end{bmatrix}$$

On inspection of these relations and recalling the procedure used in Chap. *4* (see Sec. *5*), it is evident that the 3-vector

$$\mathbf{z}(t) = \begin{bmatrix} x_1(t) \\ \dot{x}_1(t) + x_1(t) - u(t) \\ x_1(t) + x_2(t) \end{bmatrix}$$

qualifies as the state of the system at time t. Then, on solving *4* for $X_1(s)$ and $X_2(s)$, the desired input-output-state relation (in the Laplace transform form) is found to be

$$X_1(s) = \frac{N_1(s)}{D(s)} U(s) + \frac{s(s+2)z_1(0-) + (s+2)z_2(0-) - z_3(0-)}{D(s)}$$

$$X_2(s) = \frac{N_2(s)}{D(s)} U(s) + \frac{-s^2 z_1(0-) - sz_2(0-) + (s^2+s+1)z_3(0-)}{D(s)}$$

where $D(s) = (s^2+s+1)(s+2) - s$, $N_1(s) = (s^2+3s+1)$, and $N_2(s) = 1$.

The next step is to obtain the time responses x_1 and x_2 from the above Laplace transform equations. It is obvious that the value of the output at time t will consist of two terms, one depending on $u_{[0,t]}$ and the other on the initial state $\mathbf{z}(0-)$. The first term is the zero-state response of the system and the second is its zero-input response. (*Question:* What is the state equation satisfied by $\mathbf{z}$?)

The elimination method

A different approach, which has the advantage of being applicable not only to time-invariant but also to time-varying systems, is based on a reduction of the system of equations *1* to a triangular system which can be solved as a succession of scalar differential equations in a single unknown function. The reduction technique in question is analogous to the Gauss elimination procedure used in solving systems of linear algebraic equations. It is based on the following

5 **Theorem** Consider the system $\mathcal{S}$ of differential equations defined by *1* and the system $\mathcal{S}'$ obtained from $\mathcal{S}$ as follows: all the equations of $\mathcal{S}'$ are identical to their corresponding ones in $\mathcal{S}$ except for the kth equation, which is given by

$$\sum_{j=1}^{n} [L_{kj}(p) + M(p)L_{ij}(p)]x_j(t) = [F_k(p) + M(p)F_i(p)]u(t)$$

where $M(p)$ is an arbitrary differential operator, say $\sum_{k=0}^{N'} m_k p^k$, with the m_k's constant. Then $\hat{\mathbf{x}}(t) = (\hat{x}_1(t), \ldots, \hat{x}_n(t))$ is a solution of $\mathcal{S}$ if and only if $\hat{\mathbf{x}}(t)$ is a solution of $\mathcal{S}'$. $\triangleleft$

6 *Comment* The systems $\mathcal{S}$ and $\mathcal{S}'$ are equivalent in the sense that any solution of $\mathcal{S}$ is a solution of $\mathcal{S}'$ and conversely (see *2.5.4* and *3.9.1*). More specifically, the transformation of $\mathcal{S}$ into $\mathcal{S}'$ introduces no spurious solutions and hence $\mathcal{S} \equiv \mathcal{S}'$ rather than $\mathcal{S} \subset \mathcal{S}'$ (see *2.5.1*).

Proof (I) The direct statement $\Rightarrow$ is obvious from the relation between $\mathcal{S}$ and $\mathcal{S}'$. (II) The proof of $\Leftarrow$ is as follows: If $\hat{\mathbf{x}}(t)$ is a solution of $\mathcal{S}'$, then it is a solution of the system obtained from $\mathcal{S}'$ by replacing the kth equation of $\mathcal{S}'$ by a new one resulting from the addition to the kth equation of $\mathcal{S}'$ of the ith equation of $\mathcal{S}'$, multiplied by $-M(p)$. Note that this is the kth equation of $\mathcal{S}$. Thus $\hat{\mathbf{x}}(t)$ is a solution of $\mathcal{S}$.

7 **Corollary** The theorem is still true if the L_{ij}'s and M are linear time-varying differential operators. In other words, the theorem also applies to linear time-varying systems of differential equations.

8 **Corollary** The same statement holds if $\mathcal{S}'$ is obtained from $\mathcal{S}$ by multiplying the kth equation of $\mathcal{S}$ by a *nonzero constant*. Let us call *an elementary row operation* any operation on the system of equations of the kind described in Theorem *5* or Corollary *8*.

9 **Corollary** By a succession of elementary row operations, any two rows of the matrix $\mathbf{L}(p)$ may be interchanged. In other words, by a succession of elementary row operations, any two equations of the system $\mathcal{S}$ may be interchanged.

Proof Let us interchange the ith row with the first row. The procedure can be best explained by showing the successive states of the first column after each elementary row operation:

$$\begin{bmatrix} L_{11} \\ \cdot \\ \cdot \\ \cdot \\ L_{i1} \\ \cdot \end{bmatrix} \rightarrow \begin{bmatrix} L_{11} \\ \cdot \\ \cdot \\ \cdot \\ L_{11} + L_{i1} \\ \cdot \end{bmatrix} \rightarrow \begin{bmatrix} L_{11} \\ \cdot \\ \cdot \\ \cdot \\ -L_{11} - L_{i1} \\ \cdot \end{bmatrix} \rightarrow \begin{bmatrix} -L_{i1} \\ \cdot \\ \cdot \\ \cdot \\ -L_{11} - L_{i1} \\ \cdot \end{bmatrix}$$

$$\rightarrow \begin{bmatrix} L_{i1} \\ \cdot \\ \cdot \\ \cdot \\ -L_{11} - L_{i1} \\ \cdot \end{bmatrix} \rightarrow \begin{bmatrix} L_{i1} \\ \cdot \\ \cdot \\ \cdot \\ -L_{11} \\ \cdot \end{bmatrix} \rightarrow \begin{bmatrix} L_{i1} \\ \cdot \\ \cdot \\ \cdot \\ L_{11} \\ \cdot \end{bmatrix}$$

We shall now use a succession of elementary row operations to bring the system *1* to the required form. The procedure is described as a sequence of steps.

10 **Procedure**

Step 1. Among all elements of the first column (which consists of $L_{11}, L_{21}, \ldots, L_{n1}$), pick the polynomial which has the least degree and which is not identical to zero; by a succession of elementary row operations bring this polynomial to the (1,1) position.

Step 2. Let $\hat{L}_{k1}$ be the polynomial in the $(k,1)$ position resulting from step 1 ($k = 1, 2, \ldots, n$). Divide each polynomial $\hat{L}_{k1}$ by $\hat{L}_{11}$, viz.:

$$\hat{L}_{k1}(p) = q_{k1}\hat{L}_{11}(p) + r_{k1}(p)$$

where $d^0 r_{k1} < d^0 \hat{L}_{11}$.[1] For $k = 2, 3, \ldots, n$, subtract from the kth equation the first one multiplied by $q_{k1}(p)$. As a result of these elementary row operations, the first column is now $(\hat{L}_{11}, r_{21}, r_{31}, \ldots, r_{n1})$. The right-hand side of the kth equation is $[\hat{F}_k - q_{k1}(p)\hat{F}_1(p)]u(t)$, for $k = 1, 2, \ldots, n$.

Step 3. If all the remainders of the divisions of step 2 are identically zero, this cycle of elementary operations terminates and the matrix is in the form *11* below. If some remainders r_{k1} are not identically zero, bring to the (1,1) position the remainder which is of least degree not identically zero. Repeat step 2 until all the elements of the first column [except the one in the (1,1) position] are identically zero. These iterations require a finite number of operations, since each

[1] $d^0 r_{k1}$ = degree of the polynomial r_{k1}.

cycle of divisions reduces the degree of the polynomials by at least 1. The resulting matrix is of the form

11
$$\begin{bmatrix} L_{11} & L_{12} & L_{13} & \cdots & L_{1n} \\ 0 & L_{22} & L_{23} & \cdots & \cdot \\ \cdot & \cdot & \cdot & \cdot & \cdot \\ 0 & L_{2n} & L_{4n} & \cdots & L_{nn} \end{bmatrix}$$

where the L_{ij}'s are, in general, polynomials different from those of *1*.

Step 4. Perform the steps 1, 2, 3 on the $(n-1) \times (n-1)$ matrix formed by the rows and columns numbered 2 to n of *11*.

Step 5. Repeat step 4 on the successively obtained smaller matrices of order $n-2$, $n-3$, ... The final result is a system $\tilde{\mathfrak{S}}$ of equations of the form

12
$$\begin{bmatrix} \tilde{L}_{11} & \tilde{L}_{12} & \cdots & \cdot & \tilde{L}_{1n} \\ 0 & \tilde{L}_{22} & \cdots & \cdot & \tilde{L}_{2n} \\ \cdot & \cdot & \cdot & \cdot & \cdot \\ 0 & \cdot & & \tilde{L}_{n-1n-1} & \tilde{L}_{n-1n} \\ 0 & \cdot & & 0 & \tilde{L}_{nn} \end{bmatrix} \begin{bmatrix} \tilde{x}_1 \\ \tilde{x}_2 \\ \cdots \\ \tilde{x}_{n-1} \\ \tilde{x}_n \end{bmatrix} = \begin{bmatrix} \tilde{F}_1(p) \\ \tilde{F}_2(p) \\ \cdots \\ \cdot \\ \tilde{F}_n(p) \end{bmatrix} u(t)$$

where the $\tilde{x}_i$'s are identical to the x_i because these functions have not been affected by the sequence of elementary row operations.

13 **Conclusion** By a succession of elementary row operations, any system of differential equations such as *1* can be changed into an equivalent system which is in the triangular form *12*. ◁

The solution of *12* amounts to solving a succession of high-order differential equations in one unknown function: the nth equation determines $\tilde{x}_n$ and the $(n-1)$st then reads

$$\tilde{L}_{n-1,n-1}\tilde{x}_{n-1} + \tilde{L}_{n-1,n}\tilde{x}_n = F_{n-1}u$$

where both u and $\tilde{x}_n$ are known; therefore, it can be solved for $\tilde{x}_{n-1}$, etc.

14 *Comment* The procedure above which leads to *12* is also a convenient procedure for computing the characteristic polynomial of the system *1*. This characteristic polynomial is equal to the determinant, say Δ, of the matrix $\mathbf{L}(p)$ in *1*. It will be shown below that Δ is, within a constant factor, equal to the determinant of the matrix $(\tilde{L}_{ij}(p))$ in *12*. In other words,

$$\Delta = k \prod_{i=1}^{n} \tilde{L}_{ii}(p) \qquad k \text{ constant independent of } p$$

15 *Comment* The system *12* can be further simplified by the further use of the elementary row operation defined in *5*. Start by considering the second column: adding to the first row the second row multiplied

by a suitable factor, we can make the degree of $\tilde{L}_{12}$ become smaller than that of $\tilde{L}_{22}$. Next, operating on the third column, we make the degree of $\tilde{L}_{13}$ and $\tilde{L}_{23}$ smaller than that of $\tilde{L}_{33}$. Thus by a succession of elementary row operations the matrix $\tilde{L}$ may be brought into a form such that, for $k = 2, \ldots, n$, the degrees of $\tilde{L}_{1k}, \tilde{L}_{2k}, \ldots, \tilde{L}_{k-1,k}$ are smaller than the degree of $\tilde{L}_{kk}$. In particular, if $\tilde{L}_{kk}$ is a constant, all the elements of the kth column, except for the diagonal element, can be reduced to zero.

Matrix interpretation of the elimination method

The elementary row operation described in Theorem *5* is equivalent to multiplying *1* on the left by the matrix whose (α,β) element is $\delta_{\alpha,\beta} + \delta_{\alpha k}\delta_{\beta i}M(p)$, $\alpha,\beta = 1, 2, \ldots, n$, in other words, by a matrix all of whose elements are zero except the diagonal elements, which are equal to 1, and the element in position (k,i), which is equal to $M(p)$. Note that this matrix is obtained from the unit matrix by adding to its kth row the ith row multiplied by the differential operator $M(p)$, that is, by applying to the unit matrix the operation that we wish to apply to $\mathbf{L}$.

To multiply the kth row by a nonzero constant K, as in *8*, is equivalent to multiplying *1* on the left by the matrix whose (α,β) element is $\delta_{\alpha\beta} + \delta_{\alpha k}\delta_{\beta k}(K - 1)$, that is, by the diagonal matrix all of whose diagonal elements are unity except that in the kth row which is equal to K. Note again that this matrix is obtained from the unit matrix by multiplying its kth row by the constant K, that is, by applying to the unit matrix the operation that we wish to apply to $\mathbf{L}$.

In these two cases the determinants of the multiplying matrices are equal to 1 or K, respectively. Therefore $\det \mathbf{L}(p) = \gamma \det \tilde{\mathbf{L}}(p)$, where γ is a nonzero constant. Therefore, the determinant of the matrix $(\tilde{L}_{ij}(p))$ in *12* is equal to a nonzero constant times the determinant of the matrix $\mathbf{L}(p)$ of *1*.

16 *Example* An example is particularly useful to illustrate the simplicity of the procedure and to show how the required initial conditions are obtained. Consider the circuit shown in Fig. *5.6.1*. The voltage

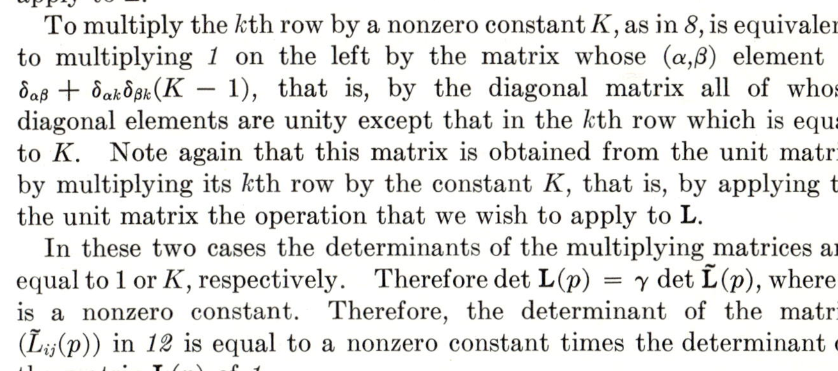

Fig. 5.6.1 Circuit studied in Example *5.6.16*. The element values are in ohms, farads, and henrys.

source e_1 is the input. Let us pick as variables the charges q_1, q_2 and the capacitor voltage v_1. If we write the current law for the parallel

RC circuit and the voltage laws for the loops defined in the figure, we obtain

17
$$\begin{bmatrix} p+1 & 0 & -2p \\ 0 & p^2+p+1 & -p \\ 0 & -p & p^2+2p+1 \end{bmatrix} \begin{bmatrix} v_1 \\ q_1 \\ q_2 \end{bmatrix} = \begin{bmatrix} 0 \\ e_1 \\ 0 \end{bmatrix}$$

This is a system of equations of the form *1.* Let us apply the elimination procedure *10.* The first column of *17* already fits the triangular pattern of *12*; we consider therefore the 2×2 submatrix formed by the second and third rows and columns. Step 1: interchange rows 2 and 3. Step 2: multiply row 2 by $(p+1)$ and add to row 3. Thus

$$\begin{bmatrix} p+1 & 0 & -2p \\ 0 & -p & (p+1)^2 \\ 0 & 1 & (p+1)^3 - p \end{bmatrix} \begin{bmatrix} v_1 \\ q_1 \\ q_2 \end{bmatrix} = \begin{bmatrix} 0 \\ 0 \\ e_1 \end{bmatrix}$$

Since the second column is not in the required form, repeat the procedure. Step 1: interchange rows 2 and 3. Step 2: multiply row 2 by p and add to row 3. Thus

18
$$\begin{bmatrix} p+1 & 0 & -2p \\ 0 & 1 & p^3+3p^2+2p+1 \\ 0 & 0 & p^4+3p^3+3p^2+3p+1 \end{bmatrix} \begin{bmatrix} v_1 \\ q_1 \\ q_2 \end{bmatrix} = \begin{bmatrix} 0 \\ e_1 \\ pe_1 \end{bmatrix}$$

This is the required form.

Now consider the initial conditions: The initial charges on the capacitors and the initial current in the inductances define the initial state of the system (see Chap. *4*, Sec. *6*). Thus the initial state prescribes $q_1(0)$, $\dot{q}_1(0)$, $q_2(0)$, $\dot{q}_2(0)$, $v_1(0)$ and the input specifies $e_{1[0,\infty)}$. Now the last equation in *18* requires $q_2(0)$, $\dot{q}_2(0)$, $\ddot{q}_2(0)$, and $\dddot{q}_2(0)$ in order to define $q_{2[0,\infty)}$ uniquely. In order to find $\ddot{q}_2(0)$ and $\dddot{q}_2(0)$, one has to go back to the original system, *17* in this case. From the second and third equations of *17* evaluated at $t = 0$, and from the initial state, one gets $\ddot{q}_1(0)$ and $\ddot{q}_2(0)$. Differentiating the third equation, one then gets $\dddot{q}_2(0)$. The solution of the last equation in *18* requires $e_1(0)$, which is known since $e_{1[0,\infty)}$ is given; thus the last equation defines q_2 uniquely. The solution q_2 and the input e_1 determine q_1 by the second equation of *18*. Finally, the first equation of *18* with the known initial condition $v_1(0)$ gives v_1.

7 *Solutions of the homogeneous system*

Consider a system described by the equations *5.6.1*, which represent a linear time-invariant system whose outputs are $x_1, x_2, \ldots, x_n$. Let us consider the zero-input response of this system, in other words, the differential equation

1 $$\mathbf{L}(p)\mathbf{x}(t) = \mathbf{0}$$

where $\mathbf{L}(p)$ is the matrix of differential operators defined in *5.6.1*. Suppose we write the state equations of this system and carry out a mode analysis of the free motions. Since each component of $\mathbf{x}(t)$ is a linear function of the state of the system, it will usually happen that some of the x_i's are uncoupled to some of the modes. Thus we may set ourselves the following problem: Find the differential operator of least order, Δ_i, such that all free motions of *1* are such that the ith component of the output satisfies the differential equation $\Delta_i(p)x_i = 0$. The purpose of this section is to answer the question without using the device of performing the mode analysis.

Let $\mathbf{\Gamma}(p)$ be the adjoint matrix of $\mathbf{L}(p)$; that is, by definition, $\Gamma_{ij}(p)$ is the cofactor of $L_{ji}(p)$.[1] Both $\mathbf{\Gamma}(p)$ and $\mathbf{L}(p)$ are matrices whose elements are polynomials in p. From elementary properties of determinants,

2 $$\mathbf{L}(p)\mathbf{\Gamma}(p) = \mathbf{\Gamma}(p)\mathbf{L}(p) = \Delta(p)\mathbf{I}$$

where $\Delta(p) = \det \mathbf{L}(p)$. As a first step we shall establish two facts which relate the columns of $\mathbf{\Gamma}$ with the solutions of *1*.

3 **Assertion** If s_1 is a simple zero of $\Delta(s)$, then any column of the matrix $\mathbf{\Gamma}(s_1)e^{s_1 t}$ is a solution of *1* and at least one column is a nontrivial solution of *1*.

Proof By direct calculation, using *2*, we get

$$\mathbf{L}(p)\mathbf{\Gamma}(s_1)e^{s_1 t} = e^{s_1 t}\mathbf{L}(s_1)\mathbf{\Gamma}(s_1) = e^{s_1 t}\,\Delta(s_1)\mathbf{I} = 0 \qquad \forall t$$

In other words, every column of $\mathbf{\Gamma}(s_1)e^{s_1 t}$ is a solution of *1*. It remains to show that some column of $\mathbf{\Gamma}(s_1)$ is not identically zero.

Now, from the rule of differentiation of a determinant, $\Delta^{(1)}(s) \triangleq \dfrac{d}{ds}\Delta(s)$ is a linear combination of all minors of $\Delta(s)$. Since s_1 is a simple zero of $\Delta(s)$, $\Delta^{(1)}(s_1) \neq 0$ and not all minors of $\Delta(s)$ are zero. In other words, not all columns of $\mathbf{\Gamma}(s_1)$ are identical to zero. Therefore, at least one column of $\mathbf{\Gamma}(s_1)e^{s_1 t}$ is a nontrivial solution of *1*.

4 **Assertion** Let $\mathbf{\Gamma}^{(k)}(s)$ be the kth derivative of the matrix $\mathbf{\Gamma}(s)$. If s_1 is a zero of order $m + 1$ of $\Delta(s)$, then any column of the matrix

5 $$e^{s_1 t}\left[\mathbf{\Gamma}^{(m)}(s_1) + \binom{m}{1}\mathbf{\Gamma}^{(m-1)}(s_1)t + \binom{m}{2}\mathbf{\Gamma}^{(m-2)}(s_1)t^2 + \cdots + \binom{m}{m-1}\mathbf{\Gamma}^{(1)}(s_1)t^{m-1} + \mathbf{\Gamma}(s_1)t^m\right]$$

[1] This adjoint matrix should not be confused with the matrix of the adjoint of the linear transformation represented by $\mathbf{L}$ (see *C.14.1*). Note that the cofactor of L_{ij} is $(-1)^{i+j}$ times the minor of L_{ij} and the minor of L_{ij} is the determinant of the matrix obtained by deleting the ith row and jth column of $\mathbf{L}$.

is a solution of *1* and at least one column of this matrix is not identically zero.

Proof Let us prove the assertion for $m = 1$, that is, s_1 is a double zero of $\Delta(s)$. Consider the matrix $\frac{\partial}{\partial s}[e^{st}\Gamma(s)]$; then successively we obtain

$$\begin{aligned}\mathbf{L}(p)\cdot\frac{\partial}{\partial s}[e^{st}\mathbf{\Gamma}(s)] &= \mathbf{L}\left(\frac{\partial}{\partial t}\right)\cdot\frac{\partial}{\partial s}[e^{st}\mathbf{\Gamma}(s)]\\ &= \frac{\partial}{\partial s}\cdot\mathbf{L}\left(\frac{\partial}{\partial t}\right)[e^{st}\mathbf{\Gamma}(s)]\\ &= \frac{\partial}{\partial s}[e^{st}\mathbf{L}(s)\mathbf{\Gamma}(s)]\\ &= \frac{\partial}{\partial s}[e^{st}\,\Delta(s)\mathbf{I}] = e^{st}\left[t+\frac{\partial}{\partial s}\right]\Delta(s)\mathbf{I}\\ &= e^{st}[t\,\Delta(s) + \Delta^{(1)}(s)]\mathbf{I}\end{aligned}$$

Since s_1 is a double zero of Δ, $\Delta(s_1) = \Delta^{(1)}(s_1) = 0$; and consequently the matrix $\frac{\partial}{\partial s}[e^{st}\mathbf{\Gamma}(s)]\Big|_{s=s_1}$ and every one of its column vectors satisfy *1*.

When $m > 1$, then every column of the $m + 1$ matrices

$$\frac{\partial^k}{\partial s^k}[e^{st}\mathbf{\Gamma}(s)]\Big|_{s=s_1} = e^{s_1 t}\left(t + \frac{\partial}{\partial s}\right)^k \mathbf{\Gamma}(s)\Big|_{s=s_1} \qquad k = 0, 1, \ldots, m$$

satisfies *1*. Next we show that not all columns of $\mathbf{\Gamma}(s_1), \mathbf{\Gamma}^{(1)}(s_1), \ldots, \mathbf{\Gamma}^{(m)}(s_1)$ are identically zero. If these $(m + 1)$ matrices were identically zero, then every minor of $\Delta(s)$, say, $q(s)$, would satisfy the conditions $q(s_1) = q^{(1)}(s_1) = \cdots = q^{(m)}(s_1) = 0$; hence, every minor would have $(s - s_1)^{m+1}$ as a factor and, by the argument used in the preceding proof, $\Delta(s)$ would have, at s_1, a zero of order greater than $m + 1$. This contradicts the definition of m, the order of the zero s_1 of Δ. Therefore, at least one column of the matrices $\mathbf{\Gamma}^{(k)}(s_1)$,

$$k = 0, 1, \ldots, m$$

is not identically zero. ◁

From *2* it follows immediately that if $\mathbf{x}(t)$ is a solution of *1*, any of its components, say, $x_i(t)$, satisfies the differential equation

6 $$\Delta(p)x_i(t) = 0$$

The converse, however, is not true, i.e., there are cases where some solutions of *6* are not the ith component of any solution, $\mathbf{x}(t)$, of *1*. To wit, let

7 $$\mathbf{L}(p) = \begin{bmatrix} p+1 & 0 & 0\\ 0 & p & 0\\ 0 & 0 & p\end{bmatrix}$$

Clearly, $x_2(t)$ and $x_3(t)$ are required by *1* to be constants. However, *6* reads in this case $p^2(p+1)x_i(t) = 0$, $i = 1, 2, 3$, and its linearly independent solutions are k_1, k_2t, k_3e^{-t}. In this case, k_2t and k_3e^{-t}, although solutions of *6*, cannot appear in the second and third components of any vector solution, $\mathbf{x}(t)$, of *1*. The following theorem gives the differential operator $\Delta_i(p)$ of least order such that the ith component of every solution $\mathbf{x}(t)$ of *1* satisfies $\Delta_i(p)x_i = 0$ and, conversely, every solution of that differential equation is the ith component of a vector solution of *1*.

8 **Theorem**[1] Let $\mathbf{x}(t) = (x_1(t), x_2(t), \ldots, x_n(t))$ be a solution of *1*. The differential equation of least order satisfied by its ith component $x_i(t)$, $i = 1, 2, \ldots, n$, is

9
$$\Delta_i(p)x_i(t) = 0$$

where $\Delta_i(s)$ is the quotient of $\Delta(s)$ by $g_i(s)$, the greatest common divisor of the ith row of $\mathbf{\Gamma}(s)$.

10 *Comment* $\Delta_i(s)$ is a polynomial in s. Consider the expansion of $\Delta(s)$ according to the elements of its ith column; the result is a linear combination of the elements of the ith row of $\mathbf{\Gamma}(s)$. Hence, by definition of $g_i(s)$, that polynomial can be factored.

Proof *Case 1* Let s_1 be a zero of order $m + 1$ of both $g_i(s)$ and $\Delta(s)$. Hence $\Delta_i(s_1) \neq 0$, $g_i(s_1) = g_1'(s_1) = \cdots = g_i^{(m)}(s_1) = 0$. All the elements of the ith row of the matrices in the sum *5* are identically zero because all the elements of the ith row of $\mathbf{\Gamma}(s)$ have the common factor $(s - s_1)^{m+1}$. The ith components of all vector solutions of *1* associated with s_1 are identically zero. Hence any zero of Δ which is not a zero of Δ_i plays no role in the ith component of any solution $\mathbf{x}(t)$ of *1*.

Case 2 Let s_2 be a zero of order $m + 1$ of $\Delta(s)$ but a zero of order $k + 1$ of $g_i(s)$ ($m \geq k$ by definition of g_i). It is therefore a zero of order $m - k$ of $\Delta_i(s)$. By the same argument as above, the ith rows of $\mathbf{\Gamma}(s_2)$, $\mathbf{\Gamma}^{(1)}(s_2), \ldots, \mathbf{\Gamma}^{(k)}(s_2)$ are identically zero. Furthermore, the ith row of $\mathbf{\Gamma}^{(k+1)}(s_2)$ cannot be identically zero, for otherwise $g_i(s)$ would contain the factor $(s - s_2)^{k+2}$. Let ${}^i\mathbf{\Gamma}(s)$ be the ith row of $\mathbf{\Gamma}(s)$. Hence any component of the row vector

$$e^{s_2t}\left[{}^i\mathbf{\Gamma}_{(s_2)}^{(m)} + \binom{m}{1}{}^i\mathbf{\Gamma}_{(s_2)}^{(m-1)}t + \cdots + \binom{m}{m-k-1}{}^i\mathbf{\Gamma}_{(s_2)}^{(k+1)}t^{m-k-1}\right]$$

is a solution of *1* and there is at least one vector solution of *1* whose ith component behaves like $(\alpha_0 + \alpha_1 t + \cdots + \alpha_{m-k-1}t^{m-k-1})e^{s_2t}$ with $\alpha_{m-k-1} \neq 0$. From these two cases, it follows that the order of the differential operator of *9* cannot be reduced further. ◁

[1] C. A. Desoer, and E. Polak, A Note on Lumped Time Invariant Systems, *I.R.E. Trans. Circuit Theory,* CT-9, no. 3, pp 282–283, September, 1962.

This concludes our study of linear time-invariant differential systems. The next chapter is devoted to linear time-varying systems.

REFERENCES AND SUGGESTED READING

1 Bellman, R.: "Stability Theory of Differential Equations," McGraw-Hill Book Company, Inc., New York, 1953.

2 Coddington, E. A., and N. Levinson: "Theory of Ordinary Differential Equations," McGraw-Hill Book Company, Inc., 1955.

3 Desoer, C. A.: Modes in Linear Circuits, *I.R.E. Trans. Circuit Theory,* vol. CT-7, no. 3, pp. 211–223, 1960.

4 Fadeeva, V. N.: "Computational Methods of Linear Algebra," Dover Publications, Inc., New York, 1959.

5 Gantmacher, F. R.: "The Theory of Matrices," vols. 1 and 2, Chelsea Publishing Company, New York, 1959.

6 Ince, E. L.: "Ordinary Differential Equations," Dover Publications, Inc., New York, 1956.

7 Kaplan, W.: "Operational Methods for Linear Systems," Addison-Wesley Publishing Company, Inc., Reading, Mass., 1962.

8 Kaplan, W.: "Ordinary Differential Equations," Addison-Wesley Publishing Company, Inc., Reading, Mass., 1958.

9 Lefschetz, S.: "Differential Equations: Geometric Theory," Interscience Publishers, Inc., New York, 1957.

10 Souriau, J. M.: Une méthode pour la décomposition spectrale et l'inversion des matrices, *C. R. Acad. Sci. Paris,* vol. 227, pp. 1010, 1011, 1948.

Linear time-varying differential systems 6

1 *Introduction*

In this chapter we investigate in detail the properties of the responses of linear time-varying systems. This chapter will, therefore, bear a strong resemblance to the preceding one. There will be, however, a striking difference: in the present chapter we have little to say about computational techniques because, except for a few special cases of some practical interest, the only effective means for computing the response of time-varying systems is the digital computer. In Sec. *2* we consider systems described by their equations in state form. This section bears great similarity to that of Chap. *5*, except for the fact that here the adjoint system (which is not even mentioned in Chap. *5*) plays a very important role. Sections *3* and *4* are devoted to the study of the relation between the impulse response of systems described by $L(p,t)x = u$ and $L(p,t)x = M(p,t)u$. In particular we investigate the relation between the impulse response and the basis functions of the operator $L(p,t)$ and its adjoint $L^*(p,t)$. Section *5* is devoted to the study of the tandem connection of two time-varying differential systems. In Sec. *6* the elimination procedure for solving high-order systems of differential equations (see Chap. *5*, Sec. *6*) is used to transform a time-varying system into one of simpler form. Finally, in Sec. *7* the very important practical case of periodically varying systems is considered, and the Poincaré-Lyapunov theory of such systems is developed.

2 *Linear time-varying systems described by their state equations*

As in the case of time-invariant systems, instead of considering a general differential system described by its state equations $\dot{\mathbf{x}} = \mathbf{A}(t)\mathbf{x} + \mathbf{B}(t)\mathbf{u}$

and $\mathbf{y} = \mathbf{C}(t)\mathbf{x} + \mathbf{D}(t)\mathbf{u}$, we shall restrict ourselves to the consideration of the first equation. To put it another way, we restrict ourselves to systems whose output is the state vector, for the reason that, once the properties of the trajectory in state space are known, the properties of an output $\mathbf{y}$ given by $\mathbf{C}(t)\mathbf{x} + \mathbf{D}(t)\mathbf{u}$ are easy to deduce. More specifically, in this section we shall investigate the zero-input response and the forced response of systems described by

$$\dot{\mathbf{x}}(t) = \mathbf{A}(t)\mathbf{x}(t) + \mathbf{B}(t)\mathbf{u}(t)$$

where $\mathbf{A}(t)$ is an $n \times n$ matrix of scalar functions assumed to be continuous for all t; similarly, $\mathbf{B}(t)$ is assumed to be an $n \times r$ continuous matrix, $\mathbf{x}(t)$ is the state vector, and $\mathbf{u}$ is the input.

Zero-input response (free motion)

First we consider the case where the input is identically zero; that is, we have, for all t,

1 $$\dot{\mathbf{x}}(t) = \mathbf{A}(t)\mathbf{x}(t)$$

We shall first derive a formula for $\mathbf{x}(t;\mathbf{x}_0,t_0)$, which is, by definition, the state of the system at time t given that the system was in state $\mathbf{x}_0$ at time t_0. Let us first quote a uniqueness result:

2 **Lemma**[1] For the linear system *1*, where the elements $a_{ij}(t)$ of $\mathbf{A}(t)$ are continuous on $[t_0,t_1]$, there is one and only one solution $\mathbf{x}(t;\mathbf{x}_0,t_0)$ of *1* on $[t_0,t_1]$ passing through the state $\mathbf{x}_0$ at time t_0. If $\mathbf{A}(t)$ is continuous for all $t \in (-\infty,\infty)$, then the system *1* has a unique solution $\mathbf{x}(t;\mathbf{x}_0,t_0)$ which is defined for all $t \in (-\infty,\infty)$ and which passes through $\mathbf{x}_0$ at t_0.

3 *Comment* In some applications, such as linear circuits with switches, linear pulse systems, etc., $\mathbf{A}(t)$ is not continuous. The following result is pertinent.[2] Let $\mathbf{A}(t)$ be an integrable function of t such that $\|\mathbf{A}(t)\| < \alpha(t)$ and $\int_{t_0}^{t_1} \alpha(t)\,dt < \infty$. Then there is a unique solution $\mathbf{x}(t;\mathbf{x}_0,t_0)$ over $[t_0,t_1]$ in the sense that $\mathbf{x}(t;\mathbf{x}_0,t_0)$ is continuous in t and satisfies the integral equation

$$\mathbf{x}(t;\mathbf{x}_0,t_0) = \mathbf{x}_0 + \int_{t_0}^{t} \mathbf{A}(\tau)\mathbf{x}(\tau;\mathbf{x}_0,t_0)\,d\tau$$

Sometimes[3] the theory of linear differential equations is developed for the case where $\mathbf{A}(t)$ is a *regulated* function of t; that is, each element $a_{ij}(t)$ of the matrix has at every point a right-hand limit and a left-

[1] E. A. Coddington and N. Levinson, "Theory of Ordinary Differential Equations," p. 20, McGraw-Hill Book Company, Inc., New York, 1955.

[2] *Ibid.*, p. 97, prob. 1.

[3] N. Bourbaki, "Fonctions d'une variable réelle," chap. IV, Hermann & Cie, Paris, 1961.

hand limit. This implies that, on every finite interval, the set of discontinuity points of $\mathbf{A}$ is countable and that $\mathbf{A}$ is integrable. Under these conditions, the differential equation *1* has a unique solution $\mathbf{x}(t;\mathbf{x}_0,t_0)$ which is a continuous function of t. It satisfies the above integral equation for every t but satisfies the differential equation *1* only for t outside an at most countable set of values of t. ◁

The expression for the zero-input response of the system *1* is given in the following

4 **Theorem** Let $\mathbf{A}(t)$ be an $n \times n$ matrix whose elements are continuous functions of time. Let $\mathbf{\Phi}(t,t_0)$ be the $n \times n$ matrix which is the solution of the equations

5 $$\frac{d}{dt}\mathbf{\Phi}(t,t_0) = \mathbf{A}(t)\mathbf{\Phi}(t,t_0) \qquad \mathbf{\Phi}(t_0,t_0) = \mathbf{I}$$

Then the solution to the equation

6 $$\dot{\mathbf{x}}(t) = \mathbf{A}(t)\mathbf{x}(t) \qquad \mathbf{x}(t_0) = \mathbf{x}_0$$

is denoted by $\mathbf{x}(t;\mathbf{x}_0,t)$ and is given by

7 $$\mathbf{x}(t;\mathbf{x}_0,t_0) = \mathbf{\Phi}(t,t_0)\mathbf{x}_0 \qquad \forall t,\ \forall \mathbf{x}_0$$

Proof By the definition of $\mathbf{\Phi}(t,t_0)$, observe that *7* reduces to $\mathbf{x}_0$ for $t = t_0$. Finally, on differentiating *7* we see that it satisfies the differential equation *6*.

8 **Definition** The matrix $\mathbf{\Phi}(t,t_0)$ is called the *state transition matrix*. ◁

Equation *7* can be interpreted as follows: $\mathbf{\Phi}(t,t_0)$ is a linear transformation which maps the state $\mathbf{x}_0$ at t_0 into the state $\mathbf{x}(t)$ at time t. This linear dependence of $\mathbf{x}(t)$ on $\mathbf{x}_0$ (exhibited by *7*) is a direct consequence of the zero-input linearity property of linear systems (see *3.4.17*). Another consequence is that all zero-input responses of *1* constitute an n-dimensional linear vector space (see *C.2.1* and *C.4.9*). Let $\mathcal{C}^n$ be the state space of *1*. These zero-input responses satisfy the requirements of additivity and homogeneity since, from *7*, $\mathbf{x}(t;\ \mathbf{x}_1 + \mathbf{x}_2,\ t_0) = \mathbf{x}(t;\mathbf{x}_1,t_0) + \mathbf{x}(t;\mathbf{x}_2,t_0)$, and $\mathbf{x}(t;\alpha\mathbf{x}_0,t_0) = \alpha\mathbf{x}(t;\mathbf{x}_0,t_0)$ for all $\mathbf{x}_1, \mathbf{x}_2, \mathbf{x}_0 \in \mathcal{C}^n$ and all complex numbers α. Let $\{\mathbf{a}_i\}_1^n$ be a basis for the state space $\mathcal{C}^n$ of *1* and let $\mathbf{x}_0 = \sum_1^n \xi_i\mathbf{a}_i$ be the (unique) representation of $\mathbf{x}_0$ in that basis; then the linearity of *7* with respect to $\mathbf{x}_0$ implies that

$$\mathbf{x}(t;\mathbf{x}_0,t_0) = \sum_{i=1}^n \xi_i\mathbf{x}(t;\mathbf{a}_i,t_0)$$

Thus any zero-input response of *1* can be expressed as a linear combination of n particular zero-input responses; hence, the dimension of the linear vector space is no greater than n. Let us pick as basis vectors the vectors $\mathbf{e}_i$ defined as follows: for $i = 1, 2, \ldots, n$, $\mathbf{e}_i$ has all its components equal to zero except the ith component, which is unity. From *7*, $\mathbf{x}(t;\mathbf{e}_i,t_0)$ is the ith column vector of $\mathbf{\Phi}(t,t_0)$. We shall show below (see *11*) that these n vectors are linearly independent for all t. Hence the dimension of the linear vector space of all zero-input responses of *1* is actually n.

Properties of $\mathbf{\Phi}(t,t_0)$

By analogy with the time-invariant case one might be tempted to write $\exp \int_{t_0}^{t} \mathbf{A}(\tau)\, d\tau$ for $\mathbf{\Phi}(t,t_0)$. In general this is erroneous because

$$\frac{d}{dt}[\mathbf{B}(t)]^2 = \dot{\mathbf{B}}(t)\mathbf{B}(t) + \mathbf{B}(t)\dot{\mathbf{B}}(t)$$

whereas the relation

$$\frac{d}{dt}[\mathbf{B}(t)]^2 = 2\mathbf{B}(t)\dot{\mathbf{B}}(t) = 2\dot{\mathbf{B}}(t)\mathbf{B}(t) \qquad \forall t$$

is true only if $\mathbf{B}(t)$ and $\dot{\mathbf{B}}(t)$ commute for all t. Thus referring to the series expansion definition of the exponential of a matrix (*5.2.3*), we have the

9 **Assertion** If, for all t, $\int_{t_0}^{t} \mathbf{A}(\tau)\, d\tau$ and $\mathbf{A}(t)$ commute, then

$$\mathbf{\Phi}(t,t_0) = \exp\left[\int_{t_0}^{t} \mathbf{A}(\tau)\, d\tau\right]$$

10 **Assertion** Let $\mathbf{\Phi}(t,t_0)$ be the state transition matrix of *1* defined by *5*. Then

$$\det \mathbf{\Phi}(t,t_0) = \exp\left[\int_{t_0}^{t} \alpha(\tau)\, d\tau\right]$$

where $\alpha(\tau) \triangleq \sum_{i=1}^{n} a_{ii}(\tau) \triangleq \operatorname{tr} \mathbf{A}(\tau)$.

The above assertion implies that

11 **Corollary** If $\mathbf{A}(t)$ is continuous over every finite time interval, then $\mathbf{\Phi}(t,t_0)$ is nonsingular for all finite t. (Therefore, for all t_0, the column vectors of $\mathbf{\Phi}(t,t_0)$ are linearly independent for all finite t.)

Proof of Assertion 10 Consider the time derivative of det $\mathbf{\Phi}$: it is a sum of n determinants $d_1, d_2, \ldots, d_n$, where d_i is the determinant all of whose elements are identical to those of $\mathbf{\Phi}$ except for the ith row, whose elements are the time derivatives of the corresponding ones in $\mathbf{\Phi}$. Let ϕ_{ij} be the general element of $\mathbf{\Phi}$. Then, by *5*

$$\dot{\phi}_{ij}(t) = \sum_{k=1}^{n} a_{ik}(t)\phi_{kj}(t)$$

Using some obvious row operations, one gets $d_i = a_{ii}(t) \det \mathbf{\Phi}$. Thus,

$$\frac{d}{dt}(\det \mathbf{\Phi}) = \left(\sum_{i=1}^{n} a_{ii}(t)\right) \det \mathbf{\Phi}$$

Since $\det \mathbf{\Phi}\Big|_{t_0} = \det \mathbf{\Phi}(t_0,t_0) = 1$, *10* follows by integrating this first-order differential equation in $\det \mathbf{\Phi}$.

12 **Assertion** Let $\mathbf{A}(t)$ be an $n \times n$ matrix whose elements are continuous functions of time. Then the solution of the matric differential equation

13 $$\dot{\mathbf{X}}(t) = \mathbf{A}(t)\mathbf{X}(t) \qquad \mathbf{X}(t_0) = \mathbf{C}$$

where $\mathbf{C}$ is an arbitrarily prescribed constant matrix, is given by

14 $$\mathbf{X}(t) = \mathbf{\Phi}(t,t_0)\mathbf{C}$$

Conversely, any solution of the above matric differential equation is of that form.

Proof Identical to that of Theorem *4* except for obvious modifications.

15 *Comment* Any nonsingular matrix which satisfies the matric differential equation *13* is called a *fundamental matrix* of the system *1*. Thus any fundamental matrix is of the form *14* for some nonsingular constant matrix $\mathbf{C}$. Also, the n columns of any fundamental matrix are n linearly independent solutions of *1*.

16 **Assertion** The state transition matrix has the group property

$$\mathbf{\Phi}(t_1,t_2)\mathbf{\Phi}(t_2,t_3) = \mathbf{\Phi}(t_1,t_3) \qquad \forall t_1, t_2, t_3$$

An immediate consequence of this property is

17 $$\mathbf{\Phi}(t_1,t_2)^{-1} = \mathbf{\Phi}(t_2,t_1)$$

Proof Exercise for the reader. (*Hint:* Use *12*.)

18 *Exercise* Show that the state transition matrix may be written as

$$\mathbf{\Phi}(t,t_0) = \mathbf{X}(t)\mathbf{X}^{-1}(t_0)$$

where $\mathbf{X}(t)$ is a nonsingular matrix which is differentiable for all t. [*Hint:* $\mathbf{X}(t) = \mathbf{\Phi}(t,0)$ and use *16*.]

Forced response

In many applications a linear system has an n-dimensional state space, has r inputs which are represented by the r-rowed vector $\mathbf{u}$, and has state equations of the form

$$\dot{\mathbf{x}}(t) = \mathbf{A}(t)\mathbf{x}(t) + \mathbf{B}(t)\mathbf{u}(t)$$

where $\mathbf{A}(t)$ and $\mathbf{B}(t)$ are known $n \times n$ and $n \times r$ matrices whose elements are continuous in t, $\mathbf{u}(t)$ is the input, and $\mathbf{x}(t)$ is the state.

For our purposes we adopt the simpler notation

19 $$\dot{\mathbf{x}}(t) = \mathbf{A}(t)\mathbf{x}(t) + \mathbf{a}(t)$$

where $\mathbf{a}(t) = \mathbf{B}(t)\mathbf{u}(t)$ is considered to be the input.

20 **Theorem** Consider the system *19* where $\mathbf{A}(t)$ is a continuous function of time. The solution that goes through the state $\mathbf{x}_0$ at t_0 is given by

21 $$\mathbf{x}(t;\mathbf{x}_0,t_0;\mathbf{a}_{[t_0,t]}) = \mathbf{\Phi}(t,t_0)\mathbf{x}_0 + \int_{t_0}^{t} \mathbf{\Phi}(t,\xi)\mathbf{a}(\xi)\,d\xi$$

Proof Immediate by direct verification of the initial condition and direct substitution into *19*. To effect this substitution note that

$$\begin{aligned}\frac{d}{dt}\int_{t_0}^{t} \mathbf{\Phi}(t,\xi)\mathbf{a}(\xi)\,d\xi &= \mathbf{\Phi}(t,t)\mathbf{a}(t) + \int_{t_0}^{t} \frac{d}{dt}\mathbf{\Phi}(t,\xi)\mathbf{a}(\xi)\,d\xi \\ &= \mathbf{a}(t) + \mathbf{A}(t)\int_{t_0}^{t} \mathbf{\Phi}(t,\xi)\mathbf{a}(\xi)\,d\xi\end{aligned}$$

22 *Comment* The notation $\mathbf{x}(t;\mathbf{x}_0,t_0;\mathbf{a}_{[t_0,t]})$ is used to exhibit the fact that the state at time t depends on the initial conditions $\mathbf{x}_0$, t_0 and on the values of the forcing vector $\mathbf{a}(\xi)$ over the interval $[t_0,t]$. Understanding of this triple dependence is the key to many control problems. Of course, later on we shall drop the unwieldy notation $\mathbf{x}(t;\mathbf{x}_0,t_0;\mathbf{a}_{[t_0,t]})$ and write $\mathbf{x}(t)$ for the state at time t.

23 *Comment* Let us give an interpretation of *21*. Suppose that the input $\mathbf{a}$ differs from zero only on the infinitesimal interval $[\tau, \tau + d\tau]$. Then, by *7*,

$$\mathbf{x}(t) = \mathbf{\Phi}(t,t_0)\mathbf{x}_0 \qquad \text{for } t_0 \leq t < \tau$$

but from *19*

$$\begin{aligned}\mathbf{x}(\tau + d\tau) &= \mathbf{x}(\tau) + \mathbf{A}(\tau)\mathbf{x}(\tau)\,d\tau + \mathbf{a}(\tau)\,d\tau \\ &= \mathbf{\Phi}(\tau + d\tau, t_0)\mathbf{x}_0 + \mathbf{a}(\tau)\,d\tau\end{aligned}$$

Therefore, considering the expression above for the initial state at time $\tau + d\tau$ and observing that the system is in free motion for $t > \tau + d\tau$, we have

$$\mathbf{x}(t) = \mathbf{\Phi}(t,t_0)\mathbf{x}_0 + \mathbf{\Phi}(t, \tau + d\tau)\mathbf{a}(\tau)\,d\tau \qquad t > \tau + d\tau$$

where we use the group property *16* to simplify the first term. Dropping second-order terms in $d\tau$, we get

$$\mathbf{x}(t) = \mathbf{\Phi}(t,t_0)\mathbf{x}_0 + \mathbf{\Phi}(t,\tau)\mathbf{a}(\tau)\,d\tau$$

Thus we see that the second term of *21* is simply the *superposition* of the effects *observed at time t* of the forcing function acting during $[\tau, \tau + d\tau]$. In the time-invariant case the superposition integral of *21* reduces to a convolution integral because $\mathbf{\Phi}(t,\tau) = \exp[\mathbf{A}(t - \tau)]$, that is, is a function of the difference $t - \tau$.

24 *Exercise* Consider a system described by *19* where

$$\mathbf{A}(t) = \begin{bmatrix} 2 & -e^{-t} \\ e^{-t} & 1 \end{bmatrix}$$

Show that $\mathbf{x}_1 = (e^{2t} \cos t, e^t \sin t)$, $\mathbf{x}_2 = (-e^{2t} \sin t, e^t \cos t)$ are two linearly independent solutions of *19* when $\mathbf{a} = \mathbf{0}$. Show that

$$\mathbf{\Phi}(t,t_0) = \begin{bmatrix} e^{2(t-t_0)} \cos (t - t_0) & -e^{(2t-t_0)} \sin (t - t_0) \\ e^{(t-2t_0)} \sin (t - t_0) & e^{(t-t_0)} \cos (t - t_0) \end{bmatrix}$$

Check that this matrix obeys Assertion *12* and the group property *16*. Write the general solution of *19* when $\mathbf{a} \neq \mathbf{0}$ for this $\mathbf{A}(t)$.

The adjoint system

In order to understand the structure of the properties of time-varying systems, it is very important to consider the adjoint system, which is a system closely related to the system *1*. In time-invariant systems, the adjoint system is never mentioned because to obtain the adjoint system one need only change t into $-t$: roughly speaking, in the time-invariant case, the adjoint of a given system is simply the given system running backward in time. For the time-invariant case, the two systems are different.

25 **Definition** The two systems of equations

26 $$\dot{\mathbf{x}} = \mathbf{A}(t)\mathbf{x}$$

and

27 $$\dot{\mathbf{y}} = -\mathbf{A}^*(t)\mathbf{y}$$

where $\mathbf{A}^*(t)$ is the conjugate transpose of $\mathbf{A}(t)$, are said to be *adjoint* to one another. ◁

In *27*, $\mathbf{A}^*(t)$ is the matrix representation of the adjoint linear transformation of the transformation represented by $\mathbf{A}(t)$ (see *C.14.1*).

The importance and usefulness of the adjoint system is based on the following lemma.

28 **Lemma** Let $\mathbf{x}(t)$ and $\mathbf{y}(t)$ be any solutions of *1* and *27*, respectively. Then

$$\langle \mathbf{x}(t),\mathbf{y}(t)\rangle \equiv \text{constant}$$

Furthermore, if we denote by $\mathbf{\Psi}(t,t_0)$ the state transition matrix of the adjoint system, that is,

29 $$\frac{d}{dt}\mathbf{\Psi}(t,t_0) = -\mathbf{A}^*(t)\mathbf{\Psi}(t,t_0) \qquad \mathbf{\Psi}(t_0,t_0) = \mathbf{I}$$

then

30 $$\mathbf{\Psi}^*(t,t_0)\mathbf{\Phi}(t,t_0) = \mathbf{I} \qquad \text{for all } t \text{ and } t_0$$

31 *Comment* Note that *30* and *17* imply that

32 $$\boldsymbol{\Psi}^*(t,t_0) = \boldsymbol{\Phi}(t,t_0)^{-1} = \boldsymbol{\Phi}(t_0,t) \qquad \forall t,t_0$$

Conversely, if *32* holds, then the corresponding systems are related to one another as *1* and *27*. For this reason *32* can be used as an alternative way of defining adjointness of systems of the form *1* (see also Theorem *6.3.16*).

Proof To prove *28*, note that

$$\begin{aligned}\frac{d}{dt}\langle \mathbf{x}(t),\mathbf{y}(t)\rangle &= \langle \dot{\mathbf{x}}(t),\mathbf{y}(t)\rangle + \langle \mathbf{x}(t),\dot{\mathbf{y}}(t)\rangle \\ &= \langle \mathbf{A}(t)\mathbf{x}(t),\mathbf{y}(t)\rangle + \langle \mathbf{x}(t),-\mathbf{A}^*(t)\mathbf{y}(t)\rangle \\ &= \langle \mathbf{A}(t)\mathbf{x}(t),\mathbf{y}(t)\rangle + \langle -\mathbf{A}(t)\mathbf{x}(t),\mathbf{y}(t)\rangle \\ &= \langle [\mathbf{A}(t) - \mathbf{A}(t)]\mathbf{x}(t),\mathbf{y}(t)\rangle = 0\end{aligned}$$

where we used the fact that $\langle \mathbf{u},\mathbf{A}\mathbf{v}\rangle = \langle \mathbf{A}^*\mathbf{u},\mathbf{v}\rangle$, $\forall \mathbf{u},\mathbf{v}$ (see *C.14.2*). *30* is proved in a similar manner: for $t = t_0$, the relation holds by the initial conditions imposed on $\boldsymbol{\Phi}$ and $\boldsymbol{\Psi}$. Differentiate the left-hand side of *30*, and use *29* and *5* to obtain

$$\begin{aligned}&\dot{\boldsymbol{\Psi}}^*(t,t_0)\boldsymbol{\Phi}(t,t_0) + \boldsymbol{\Psi}^*(t,t_0)\dot{\boldsymbol{\Phi}}(t,t_0) \\ &\qquad = \boldsymbol{\Psi}^*(t,t_0)\mathbf{A}(t)\boldsymbol{\Phi}(t,t_0) - \boldsymbol{\Psi}^*(t,t_0)\mathbf{A}(t)\boldsymbol{\Phi}(t,t_0) = \mathbf{0}\end{aligned}$$

Note that we also used the fact that $[\mathbf{A}^*(t)]^* = \mathbf{A}(t)$ (see *C.14.5*). Hence the left-hand side of *30* is always equal to $\mathbf{I}$.

33 *Comment* Let us interpret *30* geometrically. Let $\boldsymbol{\phi}_i(t,t_0)$ be the ith column of $\boldsymbol{\Phi}(t,t_0)$: $\boldsymbol{\phi}_i(t,t_0)$ is the solution of *1* such that all its components at t_0 are zero except the ith, which is unity. Let $\boldsymbol{\psi}_i(t,t_0)$ be the ith column of $\boldsymbol{\Psi}(t,t_0)$: $\boldsymbol{\psi}_i(t,t_0)$ is the solution of *27* such that all its components at t_0 are zero except the ith, which is unity. Remembering that we use the complex scalar product (see *5.4.7* and *C.5.1*), i.e., if $\mathbf{x} = (\xi_1, \ldots ,\xi_n)$ and $\mathbf{y} = (\eta_1, \ldots ,\eta_n)$, then $\langle \mathbf{x},\mathbf{y}\rangle = \sum_{i=1}^{n} \xi_i^*\eta_i$, let us observe that the (i,j) element of the product of *30* is the scalar product of $\boldsymbol{\psi}_i$ with $\boldsymbol{\phi}_j$; hence *30* implies

34 $$\langle \boldsymbol{\psi}_i(t,t_0),\boldsymbol{\phi}_j(t,t_0)\rangle = \delta_{ij} \qquad i, j = 1, 2, \ldots , n$$

Now, by *11*, the n column vectors of $\boldsymbol{\Phi}(t,t_0)$ are linearly independent for all t; that is, for all t, the set of vectors $\{\boldsymbol{\phi}_i(t,t_0)\}$ constitutes a basis for the state space $\mathcal{C}^n$. Thus *34* implies that (see *C.8.1* and *5.4.9*) for any t, the set of vectors $\{\boldsymbol{\psi}_i(t,t_0)\}$ constitutes the *reciprocal basis* of the basis $\{\boldsymbol{\phi}_i(t,t_0)\}$ of the state space $\mathcal{C}^n$ of *1*. Thus, for example, for any vector $\mathbf{c}$, for all t, and for all t_0

35 $$\mathbf{c} = \sum_{i=1}^{n} \langle \boldsymbol{\psi}_i(t,t_0),\mathbf{c}\rangle \boldsymbol{\phi}_i(t,t_0)$$

Therefore we also have

36 $$\mathbf{I} = \sum_{i=1}^{n} \boldsymbol{\phi}_i(t,t_0)\rangle \langle \boldsymbol{\psi}_i(t,t_0) \qquad \forall t,t_0$$

which is analogous to *5.4.16* and which occurred in the time-invariant case. In *36*, the vectors of the basis and those of the reciprocal basis vary with time.

37 **Remark: alternate forms for Eq.** ***21*** It is important to be able to write the expression for $\mathbf{x}(t;\mathbf{x}_0,t_0;\mathbf{a}_{[t_0,t]})$ in several ways. We have given the basic formula *21*. Let us now use successively *16, 17, 32,* and *36* to obtain four additional expressions for $\mathbf{x}$:

38 $$\mathbf{x}(t;\mathbf{x}_0,t_0;\mathbf{a}_{(t_0,t]}) = \begin{cases} \boldsymbol{\Phi}(t,t_0)\mathbf{x}_0 + \int_{t_0}^{t} \boldsymbol{\Phi}(t,\xi)\mathbf{a}(\xi)\,d\xi \\ \boldsymbol{\Phi}(t,t_0)\left[\mathbf{x}_0 + \int_{t_0}^{t} \boldsymbol{\Phi}(t_0,\xi)\mathbf{a}(\xi)\,d\xi\right] \\ \boldsymbol{\Phi}(t,t_0)\left\{\mathbf{x}_0 + \int_{t_0}^{t} [\boldsymbol{\Phi}(\xi,t_0)]^{-1}\mathbf{a}(\xi)\,d\xi\right\} \\ \boldsymbol{\Phi}(t,t_0)\left[\mathbf{x}_0 + \int_{t_0}^{t} \boldsymbol{\Psi}^*(\xi,t_0)\mathbf{a}(\xi)\,d\xi\right] \\ \sum_{i=1}^{n} \boldsymbol{\phi}_i(t,t_0)\left[\langle\boldsymbol{\psi}_i(t_0,t_0),\mathbf{x}_0\rangle + \int_{t_0}^{t} \langle\boldsymbol{\psi}_i(\tau,t_0),\mathbf{a}(\tau)\rangle\,d\tau\right] \end{cases}$$

39 *Comment* The last equation can be given an interpretation based on comment *33*: for any t, the response $\mathbf{x}(t)$ is a linear combination of the n linearly independent solutions $\boldsymbol{\phi}_i(t,t_0)$ of the homogeneous equation *1*. Each coefficient of this linear combination is a sum of two terms. The first term is a constant [$\langle\boldsymbol{\psi}_i(t_0,t_0),\mathbf{x}_0\rangle$, the ith component of $\mathbf{x}_0$] which depends only on the initial state $\mathbf{x}_0$; it measures the component of $\mathbf{x}_0$ along $\boldsymbol{\phi}_i(t_0,t_0)$. The second term of the sum is a time-varying term which takes into account the effect of the input $\mathbf{a}(t)$; it is the time integral over $[t_0,t]$ of the component of the input $\mathbf{a}(\tau)$ along the vector $\boldsymbol{\phi}_i(\tau,t_0)$ (compare the second term in the brackets of the last equation *38* and the expansion *35*). In other words, at every instant τ, one computes the components of $\mathbf{a}(\tau)$ along the instantaneous basis $\{\boldsymbol{\phi}_i(\tau,t_0)\}$, and the contribution of the input $\mathbf{a}(\tau)$ to the component of the state at time t along $\boldsymbol{\phi}_i(t,t_0)$, which is due to the input acting during the infinitesimal interval $(\tau, \tau + d\tau)$, is $\langle\boldsymbol{\psi}_i(\tau,t_0),a(\tau)\rangle\,d\tau$.

Another way of expressing this idea is the following. Let the response of *19* be written as $\mathbf{x}(t) = \sum_{i=1}^{n} \xi_i(t)\boldsymbol{\phi}_i(t,t_0)$, where the $\xi_i(t)$ are the components of $\mathbf{x}(t)$ with respect to the basis $\{\boldsymbol{\phi}_i(t,t_0)\}$ at time t. Then from the last equation *38* we see that

40 $$\dot{\xi}_i(t) = \langle\boldsymbol{\psi}_i(t,t_0),\mathbf{a}(t)\rangle \qquad \xi_i(t_0) = \langle\boldsymbol{\psi}_i(t_0,t_0),\mathbf{x}_0\rangle$$

Geometrically, the basis $\{\phi_i(t,t_0)\}$ is a "moving set of basis vectors" ("a moving reference frame," a physicist would say) because the vectors $\phi_i(t,t_0)$ change in both length and direction. If the input **a** is identically zero, then from the equation above the components ξ_i of $\mathbf{x}(t;\mathbf{x}_0,t_0)$ are constant! Also we could say that the set of basis vectors $\{\phi_i(t,t_0)\}$ "diagonalizes" the system in the sense that for $\mathbf{a} \equiv \mathbf{0}$ the equations with respect to that basis read $\dot{\xi}_i = 0$ or $\dot{\xi} = \mathbf{0}\xi$, where $\mathbf{0}$ is the zero matrix which is a special case of a diagonal matrix. In the time-invariant case, when **A** is a simple linear transformation (see *C.18.6*), we could pick a *fixed* basis with respect to which the equations of motion were diagonalized. In the time-varying case this can be done only by using a moving basis.

41 *Exercise* Write the solution of *19* as $\mathbf{x}(t;\mathbf{x}_0,t_0;\mathbf{a}_{[t_0,t]}) = \Sigma\xi_i(t)\phi_i(t;t_0)$. Find by direct substitution into *19* the differential equations obeyed by the $\xi_i(t)$'s and their initial conditions. (This is the method of variation of parameters.)

42 *Comment* The set of basis vectors $\{\phi_i(t,t_0)\}$ is not unique. For if **C** is an arbitrary nonsingular matrix and if $\hat{\mathbf{\Phi}}(t,t_0)$ and $\hat{\mathbf{\Psi}}(t,t_0)$ obey the same differential equations (*5* and *29*, respectively) but with the initial conditions $\hat{\mathbf{\Phi}}(t_0,t_0) = \mathbf{C}$, $\hat{\mathbf{\Psi}}(t_0,t_0) = \mathbf{C}^{-1}$, then *30* still holds and the $\{\psi_i(t,t_0)\}$ constitute for all t a reciprocal basis to the $\{\hat{\phi}_i(t,t_0)\}$. Thus equation *38* needs no modification if ^'s are put on all the ϕ_i's and ψ_i's.

43 *Exercise* For the time-varying case one may also introduce a *Green dyadic* and set

44
$$\begin{aligned} \mathbf{G}(t,\tau) &\triangleq \mathbf{\Phi}(t,\tau)1(t-\tau) \\ &= 1(t-\tau)\sum_{i=1}^{n} \phi_i(t,t_0)\rangle\langle\psi_i(t,t_0) \end{aligned}$$

Show that

$$\mathbf{x}(t;\mathbf{x}_0,t_0;\mathbf{a}_{[t_0,t]}) = \mathbf{\Phi}(t,t_0)\mathbf{x}_0 + \int_{t_0}^{t} \mathbf{G}(t,\tau)\mathbf{a}(\tau)\,d\tau$$

Comment on the relation between the nonanticipative character of the system *19* and the presence of $1(t-\tau)$ in *44*.

45 **Remark** *The use of the adjoint in engineering problems* In many communication and control problems, either linear or nonlinear, systems of time-varying differential equations appear naturally. Optimization problems, in particular, often require the consideration of the adjoint system. The basic reasons for this will become clear by considering in detail a specific though grossly simplified control problem. Suppose that the system we wish to control is described by a set of known nonlinear differential equations $\dot{\mathbf{x}} = \mathbf{f}(\mathbf{x},\mathbf{u},t)$. The n-vector **x** is the state, and the r-vector **u** is the control. We are given the initial state $\mathbf{x}_0$ and the initial time t_0. Let us denote by $\mathbf{x}(t_1)$ the state at a specified time t_1: $\mathbf{x}(t_1)$ is the state reached from $\mathbf{x}_0$ at time t_0 as a result of a control $\mathbf{u}_{[t_0,t_1]}$. The problem is to maximize the value that a given payoff function ϕ takes at $\mathbf{x}(t_1)$. Suppose that an approximate analysis or some intuitive argument suggests that a certain control $\hat{\mathbf{u}}_{[t_0,t_1]}$ would be a good first guess. Call $\hat{\mathbf{x}}(t_1)$ the resulting state at time t_1. Consider now the problem of improving the initial guess by a perturbation technique. If, starting from $\mathbf{x}_0$ at t_0, we apply a control $\hat{\mathbf{u}}_{[t_0,t_1]} + \delta\mathbf{u}_{[t_0,t_1]}$, then the corresponding solution of the nonlinear differential equations may be written as $\hat{\mathbf{x}}_{[t_0,t_1]} + \delta\mathbf{x}_{[t_0,t_1]}$. Furthermore, if $\mathbf{f}$ be sufficiently smooth and if $\delta\mathbf{u}_{[t_0,t_1]}$ is sufficiently small, then $\delta\mathbf{x}(t)$ will be small for all t in $[t_0,t_1]$. Under these assumptions, by performing a Taylor expansion of the original differential equation and by dropping the second-order terms, we arrive at the approximate differential equation for $\delta\mathbf{x}$:

$$\frac{d}{dt}\,\delta\mathbf{x} = \mathbf{A}(t)\delta\mathbf{x}(t) + \mathbf{B}(t)\delta\mathbf{u}(t) \qquad \delta\mathbf{x}(t_0) = \mathbf{0}$$

where $\mathbf{A}(t)$ is the $n \times n$ matrix whose i,j element is the partial derivative $\partial f_i/\partial x_j$ and $\mathbf{B}(t)$ is the $n \times r$ matrix whose i,k element is $\partial f_i/\partial u_k$. In both cases, the partial derivatives are evaluated along the original trajectory $\hat{\mathbf{x}}_{[t_0,t_1]}$ and the original control $\hat{\mathbf{u}}_{[t_0,t_1]}$; they are therefore known functions of time. The change in the state at time t_1 resulting from a small change of control $\delta\mathbf{u}_{[t_0,t_1]}$ is then given by

$$\delta\mathbf{x}(t_1) = \int_{t_0}^{t_1} \mathbf{\Phi}(t_1,\xi)\mathbf{B}(\xi)\delta\mathbf{u}(\xi)\,d\xi$$

This expression calls for a comment: the effect of $\delta\mathbf{u}_{[t_0,t_1]}$ on $\delta\mathbf{x}(t_1)$ depends on the behavior of $\mathbf{\Phi}(t_1,\xi)$ as a function of ξ, that is, as a function of its *second* argument. However, the defining equation of the state transition matrix, Eq. *5*, specifies the rate of change of $\mathbf{\Phi}$ with respect to its *first* argument. Equations *32* and *29* suggest that consideration of the state transition matrix of the adjoint system is called for because

$$\frac{d}{d\xi}\,\mathbf{\Phi}^*(t_1,\xi) = \frac{d}{d\xi}\,\mathbf{\Psi}(\xi,t_1) = -\mathbf{A}^*(\xi)\mathbf{\Psi}(\xi,t_1)$$

A little further analysis will, however, bring us directly to the adjoint system. Going back to the problem, recall that we should maximize ϕ at the final state. Again neglecting second-order terms, we see that the increase in ϕ is

$$\begin{aligned}\delta\phi &\triangleq \phi(\hat{\mathbf{x}}(t_1) + \delta\mathbf{x}(t_1)) - \phi(\hat{\mathbf{x}}(t_1))\\ &= \langle \mathbf{c},\delta\mathbf{x}(t_1)\rangle\end{aligned}$$

where the vector $\mathbf{c}$ is the gradient of ϕ evaluated at $\hat{\mathbf{x}}(t_1)$; in other words, **c** is a known vector. Now we obtain successively

$$\begin{aligned}\delta\phi &= \left\langle \mathbf{c}, \int_{t_0}^{t_1} \mathbf{\Phi}(t_1,\xi)\mathbf{B}(\xi)\delta\mathbf{u}(\xi)\, d\xi \right\rangle \\ &= \int_{t_0}^{t_1} \langle \mathbf{c}, \mathbf{\Phi}(t_1,\xi)\mathbf{B}(\xi)\delta\mathbf{u}(\xi)\rangle \, d\xi \\ &= \int_{t_0}^{t_1} \langle \mathbf{B}^*(\xi)\mathbf{\Phi}^*(t_1,\xi)\mathbf{c}, \delta\mathbf{u}(\xi)\rangle \, d\xi\end{aligned}$$

Since, by *32*, $\mathbf{\Phi}^*(t_1,\xi)\mathbf{c} = \mathbf{\Psi}(\xi,t_1)\mathbf{c}$, we see that this vector function of ξ is the solution of the adjoint differential equation $\dot{\mathbf{z}}(t) = -\mathbf{A}^*(t)\mathbf{z}(t)$ subject to the "initial" condition $\mathbf{z}(t_1) = \mathbf{c}$. (The adjoint system must be integrated backward in time starting at t_1 and ending at the prior time t_0!) Call $\boldsymbol{\lambda}(\xi)$ the solution just defined; then

$$\delta\phi = \int_{t_0}^{t_1} \langle \mathbf{B}^*(\xi)\boldsymbol{\lambda}(\xi), \delta\mathbf{u}(\xi)\rangle \, d\xi$$

Observe that the functions $\mathbf{B}^*_{[t_0,t_1]}$ and $\boldsymbol{\lambda}_{[t_0,t_1]}$ can be computed on the basis of the known data, so that the expression above allows us to compute $\delta\phi$ for any $\delta\mathbf{u}_{[t_0,t_1]}$ provided the latter is small. In order to keep $\delta\mathbf{u}_{[t_0,t_1]}$ small, it is convenient to require that

46 $$\int_{t_0}^{t_1} \langle \delta\mathbf{u}(\xi), \delta\mathbf{u}(\xi)\rangle \, d\xi = \eta$$

where η is a small number chosen in advance. With this constraint the problem of maximizing $\delta\phi$ is the same as that of maximizing

$$\frac{\delta\phi}{\eta} = \frac{\displaystyle\int_{t_0}^{t_1} \langle \mathbf{B}^*(\xi)\boldsymbol{\lambda}(\xi), \delta\mathbf{u}(\xi)\rangle \, d\xi}{\displaystyle\int_{t_0}^{t_1} \langle \delta\mathbf{u}(\xi), \delta\mathbf{u}(\xi)\rangle \, d\xi}$$

The denominator may be interpreted as the scalar product of the vector-valued function $\delta\mathbf{u}$, defined on $[t_0,t_1]$, with itself. Similarly, the numerator is the scalar product of $\mathbf{B}^*\boldsymbol{\lambda}$ with $\delta\mathbf{u}$ (see *C.5.1* and *C.5.3*). By the method used to prove *C.5.8*, it can easily be shown that Schwarz's inequality still holds in this case and that the above ratio will reach its maximum if and only if the change in control $\delta\mathbf{u}$ is given by

$$\delta\mathbf{u}(\xi) = k\mathbf{B}^*(\xi)\boldsymbol{\lambda}(\xi) \qquad t_0 \le \xi \le t_1$$

where the constant k is so chosen that the constraint *46* is satisfied.

Let us recapitulate the steps required to perform one iteration: (I) given $\hat{\mathbf{u}}_{[t_0,t_1]}$ and $\hat{\mathbf{x}}(t_1)$, calculate $\mathbf{A}(t)$, $\mathbf{B}(t)$, and $\mathbf{c}$; (II) integrate the adjoint differential equation to obtain $\boldsymbol{\lambda}(\xi)$; (III) determine $\delta\mathbf{u}$ by the expression above in which the unspecified constant k is determined through the constraint *46*.

The adjoint of a system represented by $L(p,t)y = u$

Consider the nth-order ordinary differential equation $L(p,t)y = u$ in which

47 $$L(p,t) = p^n + a_1(t)p^{n-1} + \cdots + a_{n-1}(t)p + a_n(t)$$

where p is the differential operator d/dt and $a_k(t) \in C^{n-k}$, $k = 1, 2, \ldots, n$. To start with, let us consider the homogeneous equation

48 $$p^n y + a_1(t)p^{n-1}y + \cdots + a_{n-1}(t)py + a_n(t)y = 0$$

If we use as state vector the vector $\mathbf{y} = (y,\dot{y},\ddot{y}, \ldots, y^{(n-1)})$, then we can write (see *4.3.33 et seq.*)

49 $$\dot{\mathbf{y}} = \mathbf{A}(t)\mathbf{y}$$

with

50 $$\mathbf{A}(t) = \begin{bmatrix} 0 & 1 & 0 & \cdots & \cdots & 0 \\ 0 & 0 & 1 & \cdots & \cdots & \cdots \\ \cdots & \cdots & 0 & 1 & \cdots & \cdots \\ \cdots & \cdots & \cdots & 0 & \cdots & \cdots \\ \cdots & \cdots & \cdots & \cdots & \cdots & 0 \\ 0 & 0 & \cdots & \cdots & 0 & 1 \\ -a_n & -a_{n-1} & \cdots & \cdots & -a_2 & -a_1 \end{bmatrix}$$

The adjoint state equation is, according to the Definition *27*,

$$\dot{\mathbf{z}} = -\mathbf{A}^*(t)\mathbf{z}$$

where the matrix $\mathbf{A}^*(t)$ is the conjugate transpose of the matrix $\mathbf{A}(t)$, namely,

51 $$-\mathbf{A}^*(t) = \begin{bmatrix} 0 & 0 & \cdots & \cdots & \cdots & 0 & a_n^* \\ -1 & 0 & \cdots & \cdots & \cdots & 0 & a_{n-1}^* \\ 0 & -1 & 0 & \cdots & \cdots & \cdots & \cdots \\ \cdots & 0 & -1 & 0 & \cdots & \cdots & \cdots \\ \cdots & \cdots & \cdots & \cdots & \cdots & \cdots & \cdots \\ \cdots & \cdots & \cdots & \cdots & 0 & 0 & \cdots \\ \cdots & \cdots & \cdots & \cdots & -1 & 0 & a_2^* \\ 0 & 0 & \cdots & \cdots & 0 & -1 & a_1^* \end{bmatrix}$$

Let the components of $\mathbf{z}$ be $z_1, z_2, \ldots, z_n$; then the adjoint system reads

52 $$\begin{aligned} p\,z_1 &= a_n^* z_n \\ p\,z_2 &= a_{n-1}^* z_n - z_1 \\ p\,z_3 &= a_{n-2}^* z_n - z_2 \\ &\cdots\cdots\cdots\cdots \\ p\,z_n &= a_1^* z_n - z_{n-1} \end{aligned}$$

If we multiply in the above system the kth equation by $(-1)^{k-1}p^{k-1}$ and add the resulting equations, we have

$$(-1)^n p^n z_n + (-1)^{n-1}p^{n-1}(a_1^* z_n) + \cdots + (-1)p(a_{n-1}^* z_n) + a_n^* z_n = 0$$

Therefore, it is reasonable to call the operator

53 $$L^*(p,t) \triangleq (-1)^n p^n \cdot + (-1)^{n-1}p^{n-1}(a_1^* \cdot) + \cdots + (-1)p(a_{n-1}^* \cdot) + a_n^* \cdot$$

the *adjoint of the operator*

54 $$L(p,t) \triangleq p^n \cdot + a_1(t)p^{n-1} \cdot + a_2(t)p^{n-1} \cdot + \cdots + a_{n-1}(t)p \cdot + a_n(t) \cdot$$

where the dots indicate the operand.

55 *Comment* It is very important to observe that the solution of $Ly = 0$ is the *first* component of the vector solution of *49*, whereas the solution $z(t)$ of $L^*z = 0$ is the nth component of the vector solution of

$$\dot{\mathbf{z}} = -\mathbf{A}^*(t)\mathbf{z}$$

This observation will be useful in some later developments.

56 *Exercise* Write $Ly = u$, with L given by *54*, in state equation form. Let $\mathbf{y} = (y, \dot{y}, \ldots, y^{(n-1)})$.
[*Ans.:* $\dot{\mathbf{y}} = \mathbf{A}(t)\mathbf{y} + \mathbf{b}u$, where $\mathbf{b} = (0,0,\ldots,0,1)$ and $\mathbf{A}(t)$ is given by *50*.]

57 *Exercise* Write $L^*z_n = r$, where L is given by *54*, in state equation form. Let $\mathbf{z} = (z_1, z_2, \ldots, z_n)$.
[*Ans.:* $\dot{\mathbf{z}} = -\mathbf{A}^*(t)\mathbf{z} - \mathbf{c}r$, where $\mathbf{c} = (1,0,\ldots,0)$ and $\mathbf{A}^*(t)$ is given by *51*.]

3 *System represented by $Ly = u$*

In this section we consider a single-input single-output linear time-varying differential system described by $Ly = u$.

The impulse response

We consider a nonanticipative system $\mathcal{S}_L$ represented by

1 $$L(p,t)y = \sum_{k=0}^{n} a_k(t)p^{n-k}y(t) = u$$

where y is the output and u the input. We assume that the $(n+1)$ functions $a_k(t)$ are continuous functions of time and that $a_0(t) \neq 0$ for all t. Since this representation directly relates the output y to the input u, it is natural to consider the impulse response of the system. Recall that we use the

2 **Definition** The unit-impulse response of the nonanticipative system $\mathcal{S}_L$ is denoted by $h(t,\xi)$; it is the zero-state response of the system at time t to a unit impulse applied at time ξ. ◁

Figure *6.3.1* illustrates the behavior of $h(t,\xi)$. Note that $h(t,\xi) = 0$ for all $t < \xi$, because the system is in the zero state at time $\xi-$ and

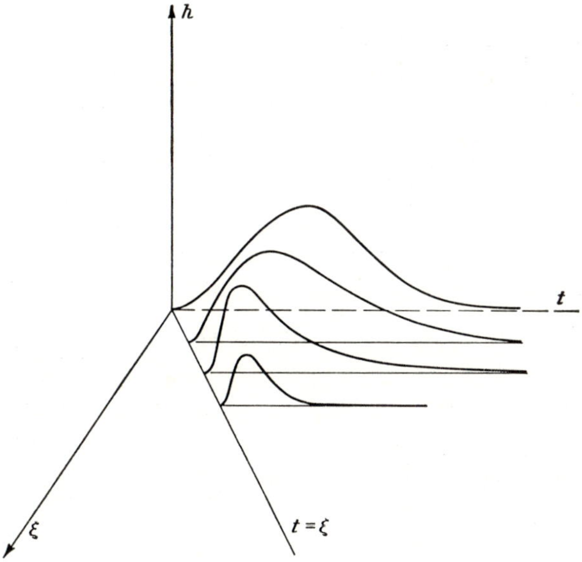

Fig. 6.3.1 The behavior of $h(t,\xi)$, the impulse response of the system represented by $L(p,t)y = u$.

because the system is nonanticipative [i.e., it does not respond prior to the application of the impulse (see *1.8.13*)]. The figure shows that the function $h(\cdot,0)$ delayed by one second is definitely different from the function $h(\cdot,1)$, thus exhibiting the fact that the properties of the system vary with time.

The relation between the impulse response and the differential equation *1* is given by the following

3 **Theorem** $h(t,\xi)$, the impulse response of the system $\mathcal{S}_L$ described by *1*, is such that[1]

4 (I) $$L_t h(t,\xi) = \delta(t - \xi) \qquad \text{for all } t \text{ and } \xi$$

(II) $$\left.\frac{\partial^k h(t,\xi)}{\partial t^k}\right|_{t=\xi+} = 0 \qquad k = 0, 1, \ldots, n-2$$

5 $$\left.\frac{\partial^{n-1} h(t,\xi)}{\partial t^{n-1}}\right|_{t=\xi+} = \frac{1}{a_0(\xi)}$$

[1] The subscript t attached to L in *4* is to specify that the differential operator p, which appears in the expression for L, operates on the variable t and not on the variable ξ.

(III)

6
$$h(t,\xi) = \frac{1(t-\xi)}{a_0(\xi)W(\xi)} \begin{vmatrix} \phi_1(\xi) & \phi_2(\xi) & \cdots & \phi_n(\xi) \\ \phi_1^{(1)}(\xi) & \phi_2^{(1)}(\xi) & \cdots & \phi_n^{(1)}(\xi) \\ \cdots & \cdots & \cdots & \cdots \\ \phi_1^{(n-2)}(\xi) & \phi_2^{(n-2)}(\xi) & \cdots & \phi_n^{(n-2)}(\xi) \\ \phi_1(t) & \phi_2(t) & \cdots & \phi_n(t) \end{vmatrix}$$

where $\phi_1, \phi_2, \ldots, \phi_n$ are n linearly independent solutions of $Ly = 0$ and $W(\xi)$ is called the Wronskian (*3.7.12*) of these solutions [i.e., the determinant whose ith column is $\phi_i(\xi), \phi_i^{(1)}(\xi), \ldots, \phi_i^{(n-2)}(\xi), \phi_i^{(n-1)}(\xi)$, for $i = 1, 2, \ldots, n$].

Proof From the definition of the impulse response, $h(t,\xi)$ satisfies $L_t h(t,\xi) = \delta(t-\xi)$ since the input u is $\delta(t-\xi)$ in this case. Furthermore, by definition of the impulse response, $h(t,\xi)$ is that solution of *4* such that the system is in its zero state at $t = \xi-$; therefore, $h(t,\xi)$ is the unique solution of $Ly = \delta(t-\xi)$ which satisfies the initial conditions

$$\left.\frac{d^k y}{dt^k}\right|_{t=\xi-} = 0 \qquad \text{for } k = 0, 1, 2, \ldots, n-1$$

Since the input is zero over the interval $(-\infty, \xi-)$, $y(t) = 0$ for all $t \in (-\infty, \xi)$. This justifies the presence of the factor $1(t-\xi)$ in *6*. In order to balance the unit impulse in the right-hand side of *4*, $a_0(t)\partial^n h(t,\xi)/\partial t^n$ must also behave like a unit impulse at $t = \xi$; therefore, $\partial^{n-1}h(t,\xi)/\partial t^{n-1}$ undergoes a jump of magnitude $1/a_0(\xi)$ at $t = \xi$.[1] Thus (I) and (II) are established. To prove (III), we show that *6* is an explicit expression of the solution of *4* with the initial conditions *5*. First compute $\partial^k h(t,\xi)/\partial t^k$ for some $t > \xi$; note that the kth derivative is obtained for $t > \xi$ by differentiating k times the last row of the determinant. Now for $t \to \xi+$ (in view of the continuity of $\phi_i^{(k)}$, $i = 1, 2, \ldots, n$, $k = 0, 1, \ldots, n-2$), the determinant vanishes since it has in the limit two identical rows. For $k = n-1$, in view of the continuity of the $\phi_i^{(n-1)}$, the determinant becomes identical to the Wronskian $W(\xi)$. Therefore, expression *6* satisfies the appropriate initial conditions at $\xi+$. Finally, it is a solution of $Ly = 0$ for $t > \xi$, since *6* is a linear combination of the $\phi_i(t)$. This completes the proof of (III).

7 *Comment* Note that *6* implies that $h(t,\xi)$ has the form

$$h(t,\xi) = 1(t-\xi) \sum_{i=1}^{n} \varphi_i(t)\alpha_i(\xi)$$

where $\alpha_i(\xi)$ is the cofactor of $\phi_i(\xi)$ in $W(\xi)$ divided by $a_0(\xi)W(\xi)$. It is easy to show (see Exercise *12*) that the $\alpha_i(t)$ are basis functions for the adjoint differential operator L^*.

[1] For a more detailed discussion of this point see *A.7.9* and *A.7.11*.

8 *Exercise* Let $a_0(t) = 1$ for all t. Prove Theorem *3* by considering the state equations of the system. [*Hint:* Use $(y,\dot{y}, \ldots, y^{(n-1)})$ as the state vector; show that the nth column of $\mathbf{\Phi}(t,\xi)$, for $t \geq \xi$, is $(h(t,\xi), h^{(1)}(t,\xi), \ldots, h^{(n-1)}(t,\xi))$, where the indicated differentiations are partial differentiations with respect to t.]

The basis functions

9 **Definition** Any set of n linearly independent solutions of $Ly = 0$ is called a set of *basis functions* of the operator L.

10 *Comment* The reason for calling such sets of solutions basis functions is that any such set constitutes a *basis* for the linear vector space whose elements are all the solutions of $Ly = 0$ (see *C.2.1* and *C.2.5*). Note also that this definition is equivalent to that given previously (see *3.6.39* and *3.6.41*).

Given the time-varying operator $L(p,t)$, there are no general methods for obtaining a set of basis functions. However, the converse problem —given n functions, find a linear differential operator L of order n such that these functions are basis functions of L—is easily solved.

11 **Theorem** Let $\phi_1(t), \phi_2(t), \ldots, \phi_n(t)$ be n functions having n continuous derivatives (i.e., belonging to the class C^n) over some interval $t_0 \leq t \leq t_1$. Let their Wronskian vanish nowhere in this interval. Then, these functions are n linearly independent solutions of the differential equation

$$W(\phi_1,\phi_2, \ldots, \phi_n, x) = 0$$

where W is the Wronskian of $\phi_1, \phi_2, \ldots, \phi_n, x$.
Proof Exercise for the reader.

12 *Exercise* Let the unit-impulse response be written as

$$h(t,\xi) = 1(t - \xi) \sum_{i=1}^{n} \phi_i(t)\alpha_i(\xi)$$

where the $\alpha_i(\xi)$ are defined by comparison with *6*. Let $a_0(t) = 1$, for all t, for convenience. From the above theorem show that $h(t,\xi)$, considered as a function of ξ, is such that

13 $$L_\xi^* h(t,\xi) = \delta(t - \xi)$$

14 $$\left.\frac{\partial^k h(t,\xi)}{\partial \xi^k}\right|_{\xi \to t-} = 0 \qquad k = 0, 1, \ldots, n-2$$
$$\left.\frac{\partial^{n-1} h(t,\xi)}{\partial \xi^{n-1}}\right|_{\xi \to t-} = (-1)^n$$

[*Hint:* Write the coefficient of $1(t - \xi)\phi_k(t)$ in *6* as $W_k(\xi)/W(\xi)$. Show that W_k/W are linearly independent solutions of $L^*y = 0$.]

The important facts concerning the forced response of the system *1* are summarized in the

15 **Theorem** The response of the system $\mathcal{S}_L$ described by *1* to the input $u_{[t_0,t]}$, given the initial state $y(t_0-)$, $\dot{y}(t_0-)$, . . . , $y^{(n-1)}(t_0-)$ of $\mathcal{S}_L$, is

$$y(t) = \sum_{i=1}^{n} \gamma_i\phi_i(t) + \int_{t_0}^{t} h(t,\xi)u(\xi)\, d\xi$$

where $\phi_1, \ldots, \phi_n$ are n basis functions of L and the constants $\gamma_1, \ldots, \gamma_n$ are determined by the prescribed initial state.
Proof Exercise for the reader.

The adjoint system

The following theorem exhibits a very important relation between the impulse response of the nonanticipative system $\mathcal{S}_L$ and that of the purely anticipative adjoint system $\mathcal{S}_L^*$. Observe the close analogy between the following and Remark *6.2.45*.

16 **Theorem** Consider the nonanticipative system $\mathcal{S}_L$ (see *1.8.13*) described by *1* [with $a_0(t) = 1$, $\forall t$; $a_k(t) \in C^{n-k}$] and whose impulse response is $h(t,\xi)$. Let $\mathcal{S}_L^*$ be a purely anticipative system (see *1.8.13*) described by

17 $$L^*z = u$$

Let $h^a(t,\xi)$ be the solution of $L_t^*h^a(t,\xi) = \delta(t-\xi)$ such that

$$\left.\frac{\partial^k h^a(t,\xi)}{\partial t^k}\right|_{t=\xi+} = 0 \qquad k = 0, 1, 2, \ldots, n-1$$

Then

18 $$h^a(t,\xi) = h^*(\xi,t)$$

Proof For $\mathcal{S}_L$ pick $(y,\dot{y}, \ldots, y^{n-1})$ as state vector. Then $\mathcal{S}_L$ is represented by the state equation

19 $$\dot{\mathbf{y}} = \mathbf{A}(t)\mathbf{y} + \mathbf{b}u$$

where $\mathbf{b} = (0,0, \ldots, 0,1)$. Let $\mathbf{\Phi}(t,\xi)$ be the state transition matrix of *19*. In order to represent the adjoint system $\mathcal{S}_L^*$ described by *17*, pick a state vector $\mathbf{z} = (z_1,z_2, \ldots, z_n)$ related to z as follows:

$$z_n = z \qquad z_{n-1} = a_1^*z_n - \dot{z}_n \qquad z_{n-2} = a_2^*z_n - \dot{z}_{n-1} \qquad \cdots$$
$$z_2 = a_{n-2}^*z_n - \dot{z}_3 \qquad z_1 = a_{n-1}^*z_n - \dot{z}_2$$

Then (see Exercise *6.2.57*)

20 $$\dot{\mathbf{z}} = -\mathbf{A}^*(t)\mathbf{z}(t) - \mathbf{c}u$$

where $\mathbf{c} = (1,0, \ldots ,0)$. Let $\mathbf{\Psi}(t,\xi)$ be the state transition matrix of *20*.

Let us make three observations: (I) In the case where $\mathbf{b} = \mathbf{c} = \mathbf{0}$, the homogeneous systems *19* and *20* are adjoints of one another (see *6.2.25*); hence we have, by *6.2.32*,

21 $$\mathbf{\Phi}(\xi,t) = \mathbf{\Phi}(t,\xi)^{-1} = \mathbf{\Psi}^*(t,\xi)$$

(II) From the definition of the impulse response and the choice of state vector in *19*, $h(t,\xi) = \phi_{1n}(t,\xi)$ for $t \geq \xi$. [Note that the function $\phi_{1n}(t,\xi)$ is not identically zero for $t < \xi$, but $h(t,\xi) = 0$ for such t's since $\mathcal{S}_L$ is nonanticipative.] (III) From the definition of $h^a(t,\xi)$ and from the choice of the state vector of *20*, we have $h^a(t,\xi) = \psi_{n1}(t,\xi)$ for all $t \leq \xi$.

Finally, by considering the $(1,n)$ element of the matrix equation *21*, we obtain $h^a(t,\xi)^* = h(\xi,t)$ for all $t \leq \xi$. Since both sides are identical to zero for $t > \xi$, the complex conjugate of the above equation gives *18*.

4 *System represented by $Lx = Mu$*

Consider a nonanticipative linear time-varying system $\mathcal{S}$ which has a single input u and a single output x. Let $\mathcal{S}$ be represented by a linear differential equation of the form

1 $$\mathcal{S}\colon\ u \to x \qquad\qquad L(p,t)x = M(p,t)u$$

where the symbol $\mathcal{S}\colon u \to x$ is used to indicate that u is the input and x is the output of $\mathcal{S}$. Let λ and μ be the orders of the differential operators L and M, respectively. Again we assume that the coefficients of the differential operator L satisfy the conditions $a_k \in C^{\lambda-k}$, $k = 0, 1, \ldots, \lambda$, and $a_0(t) \neq 0$ for all t.

By analogy with *6.3.9*, any set of λ linearly independent solutions of $L(p,t)y = 0$ will be referred to as a *set of basis functions of the system* $\mathcal{S}$. In addition, we define the set of zero-response functions of $\mathcal{S}$: any set of μ linearly independent solutions of $M(p,t)y = 0$ will be called a *set of zero-response functions of* $\mathcal{S}$. [Note that the basis functions as defined here coincide in meaning with those defined in Chaps. *3* and *4* (see *3.6.39*, *4.5.31*, and *4.5.50*) so long as $\mathcal{S}$ is a strictly proper system (*3.6.27*). If $\mathcal{S}$ is not strictly proper, then the basis functions of $\mathcal{S}$ (in the sense of *3.6.39*) will comprise the basis functions in the sense defined here together with a set of delta functions of orders up to and including $\mu - \lambda$]

2 *Exercise* Consider the time-invariant case. Suppose that the polynomials $L(s)$ and $M(s)$ have distinct roots. Call them s_i and z_j, respectively. Show that the sets $\{e^{s_i t}\}_1^\lambda$, $\{e^{z_j t}\}_1^\mu$ are respectively a set of

basis functions and a set of zero-response functions of $\mathcal{S}$. What happens if L or M has multiple zeros?

3 *Comment* If ψ is a nontrivial zero-response function of the system $\mathcal{S}$ defined by *1*, and if $\mu \geq 1$, then, in general, $1(t - t_0)\psi(t)$ is not a zero-response function of $\mathcal{S}$. Indeed by *6.3.15* the zero-state response of $\mathcal{S}$ to $1(t - t_0)\psi(t)$ is given by

4 $$x(t) = \int_{t_0}^{t} h(t,\xi)M(q,\xi)[1(\xi - t_0)\psi(\xi)]\, d\xi \qquad t \geq t_0$$

where $h(t,\xi)$ is the impulse response of the system $Ly = 0$ and $q = d/d\xi$. For simplicity let us consider an example: let $\mathcal{S}$ be specified by

$$(p^2 + a_1(t)p + 1)x(t) = (p + 1)u(t)$$

Clearly we can take ψ to be e^{-t}. For simplicity let $t_0 = 0$ and let $u(t) = 1(t)e^{-t}$. Then $M(p,t)u(t) = (p + 1)u(t) = \delta(t)e^{-t} = \delta(t)$ (see *A.5.12*). Therefore, from *4*, the zero-state response of that system to $1(t)e^{-t}$ is $h(t,0)$.

5 *Exercise* For the system and the input $u(t)$ just considered find the state $\mathbf{s}_0$ at time $0-$ such that $x(t;\mathbf{s}_0,0-;1(t)e^{-t}) = 0$ for all $t \geq 0$.

The relation between the impulse response of the system described by *1* and that described by *6.3.1* is given by the following

6 **Theorem** Consider the two systems

$$\mathcal{S}_M\colon\ u \to x \qquad\qquad L(p,t)x = M(p,t)u$$

7 $$\mathcal{S}_1\colon\ v \to x \qquad\qquad L(p,t)x = v$$

Let $h_M(t,\xi)$, $h_1(t,\xi)$ be the impulse response of $\mathcal{S}_M$ and $\mathcal{S}_1$, respectively. Then

8 $$h_M(t,\xi) = M^*(q,\xi)h_1(t,\xi)$$

where $M^*(q,\xi)$ is the adjoint operator of $M(q,\xi)$ and $q = d/d\xi$.

Proof From the definitions of $h_1(t,\xi)$ and $h_M(t,\xi)$, the response of $\mathcal{S}_M$, starting from the zero state at t_0-, may be written in two ways:

9 $$x(t) = \int_{t_0-}^{t} h_M(t,\xi)u(\xi)\, d\xi$$

10 $$x(t) = \int_{t_0-}^{t} h_1(t,\xi)M(q,\xi)u(\xi)\, d\xi$$

Since the state of $\mathcal{S}_M$ is specified at t_0- and since as a result of differentiations of the input u the integrand of *10* may include delta functions at t_0, we must carry out the integration from t_0- to t. By successive integrations by parts, we may bring *10* into the form

11 $$x(t) = \int_{t_0-}^{t} [M^*\!:(q,\xi)h_1(t,\xi)]u(\xi)\, d\xi$$

which by comparison with *9* leads to the conclusion that *8* holds because both *9* and *11* hold for all $u_{[t_0,t]}$. To illustrate the procedure, let $M(q,\xi) = \sum_{k=0}^{\mu} m_k(\xi)\, q^{\mu-k}$ and consider the contribution to *10* of the term $m_{\mu-1}(\xi)\, q$.

$$\int_{t_0-}^{t} h_1(t,\xi) m_{\mu-1}(\xi)\, q\, u(\xi)\, d\xi = [m_{\mu-1}(\xi) h_1(t,\xi) u(\xi)]\Big|_{t_0-}^{t} - \int_{t_0-}^{t+} q\, [m_{\mu-1}(\xi) h_1(t,\xi)] r(\xi)\, d\xi$$

Now $u(t_0-) = 0$ and $h_1(t,t) = 0$, by *6.3.5*; therefore,

$$\int_{t_0-}^{t} h_1(t,\xi) m_{\mu-1}(\xi)\, q\, r(\xi)\, d\xi = \int_{t_0-}^{t} -q\, [m_{\mu-1}(\xi) h_1(t,\xi)] r(\xi)\, d\xi$$

Observe that to the term $m_{\mu-1}(\xi)\, q$ of $M(q,\xi)$ there corresponds the term $-q\,[m_{\mu-1}(\xi)]$ of its adjoint $M^*(q,\xi)$. By performing the appropriate number of integrations by parts, a similar conclusion may be established for all terms of $M(q,\xi)$. In other words, this procedure transforms *10* into *11*. Therefore *8* is established. $\lhd$

Let us close this section by giving a realization of the system $\mathcal{S}_M$, which is similar to that given for the time-invariant case in *4.9.15*.

12 **Assertion** Consider the system $\mathcal{S}_M$ defined by

13
$$L(p,t)y = M(p,t)u$$

where

$$L(p,t) = \sum_{k=0}^{n} a_{n-k}(t) p^k \qquad M(p,t)u = \sum_{k=0}^{n} b_{n-k}(t) p^k$$

Let $a_0(t) = 1$ for all t (some of the $b_0, b_1, \ldots$ may be identically zero), so that the order of the system *13* is n. If, for $i = 0, 1, \ldots, n$, a_i and $b_i \in C^{n-i}$, then the system *13* has the following state equations:

14
$$\begin{bmatrix} \dot{x}_1 \\ \dot{x}_2 \\ \cdot \\ \cdot \\ \cdot \\ x_n \end{bmatrix} = \begin{bmatrix} 0 & 1 & 0 & \cdots & 0 \\ 0 & 0 & 1 & & \\ \cdots & \cdots & \cdots & \cdots & \cdots \\ \cdots & \cdots & \cdots & \cdots & 0 \\ \cdots & \cdots & \cdots & \cdots & 1 \\ -a_n & -a_{n-1} & \cdots & \cdots & -a_1 \end{bmatrix} \begin{bmatrix} x_1 \\ x_2 \\ \cdot \\ \cdot \\ \cdot \\ x_n \end{bmatrix} + \begin{bmatrix} \phi_1 \\ \phi_2 \\ \cdot \\ \cdot \\ \cdot \\ \phi_n \end{bmatrix} u$$

and

15
$$y = x_1 + \phi_0 u$$

where the ϕ_i may be computed successively in terms of the a_i's and b_i's by the relations

16
$$\phi_0(t) = b_0(t)$$
$$\phi_i(t) = b_i(t) - \sum_{r=0}^{i-1}\sum_{m=0}^{i-r}\binom{n+m-i}{n-i} a_{i-r-m}(t)p^m\phi_r(t)$$

Proof The proof is purely computational; so we shall only outline the steps: (I) use *14* and *15* to evaluate $y, \dot{y}, \ddot{y}, \ldots, y^{(n-1)}$: viz., $y = x_1 + \phi_0 u$, $\dot{y} = x_2 + \phi_1 u + p(\phi_0 u), \ldots$; (II) substitute in *13*; (III) reorder the summations so that both sides consist of weighted sums of successive derivatives of u; and (IV) equate coefficients of like derivatives of u. Then *16* follows.

17 *Comment* From *14* to *16* it is a simple matter to derive expressions for $x_1, x_2, \ldots, x_n$ in terms of u and y and their derivatives. The expres-

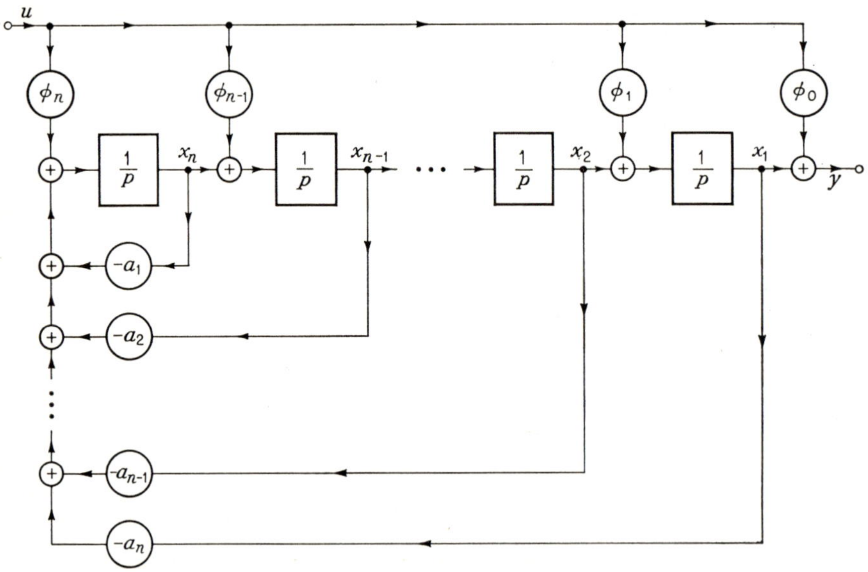

Fig. 6.4.1 Block-diagram representation of the system of Eqs. *6.4.14* which represent the system $L(p,t)y = M(p,t)u$. (Compare with Fig. *4.9.7*.)

sions are quite cumbersome even for small n (see *4.5.58*). An important point to note is that the state $\mathbf{x}$ is *not* given by $\mathbf{y} \triangleq (y,y^1, \ldots, y^{(n-1)})$ (see *3.7.33 et seq.*). Thus, while $\mathbf{y}$ has discontinuities at points where u has jumps, the state vector $\mathbf{x}$ is continuous. The state equations *14* lead to the realization shown in Fig. *6.4.1* in which there are n integrators and, in general, $2n$ time-varying scalors. A useful interpretation of the ϕ_i's is stated in the following

18 *Exercise* Suppose the system $\mathcal{S}_M$ is time-invariant; therefore, the a_i's and b_i's in *13* are constant. If any of the $b_i \neq 0$ $(i \geq 1)$, and if

$u(t) = 1(t)$, then the vector $\mathbf{y} = (y, y, \ldots, y^{n-1})$ is singular at $t = 0$. Using the techniques of Appendix A, show that

$$\mathbf{y}(0+) - \mathbf{y}(0-) = (\phi_0, \phi_1, \ldots, \phi_{n-1})$$

5 *Tandem connection*

First consider two time-invariant linear systems $\mathcal{A}_1$ and $\mathcal{A}_2$ described by

$$\mathcal{A}_1: \quad u \to y \qquad L_1(p)y = M_1(p)u$$
$$\mathcal{A}_2: \quad u \to y \qquad L_2(p)y = M_2(p)u$$

where L_1, L_2, M_1, and M_2 are time-invariant differential operators, i.e., polynomials in p with constant coefficients. We know that the tandem combination $\mathcal{A}_1\mathcal{A}_2$ ($\mathcal{A}_2$ preceding $\mathcal{A}_1$) is zero-state equivalent to $\mathcal{A}_2\mathcal{A}_1$ ($\mathcal{A}_1$ preceding $\mathcal{A}_2$) (see *3.6.26*) and that $\mathcal{A}_1\mathcal{A}_2$ is equivalent to $\mathcal{A}_2\mathcal{A}_1$ if and only if M_1 has no common factors with L_2 and M_2 has no common factors with L_1 (see *4.8.15*).

Let us now consider the time-varying case. Here the systems are characterized by

1 $$\mathcal{B}_1: \quad u \to y \qquad L_1(p,t)y = M_1(p,t)u$$
2 $$\mathcal{B}_2: \quad u \to y \qquad L_2(p,t)y = M_2(p,t)u$$

where L_1, L_2, M_1, and M_2 are time-varying differential operators, i.e., polynomials in p with coefficients that are known functions of time. Let the degrees of these polynomials be respectively λ_1, λ_2, μ_1, μ_2. We assume that the leading coefficients of L_1, L_2, M_1, and M_2 differ from zero for all t, so that the orders of these differential operators do not vary with time.

Consider now the tandem connection of $\mathcal{B}_1$ preceding $\mathcal{B}_2$, that is, the product $\mathcal{B}_2\mathcal{B}_1$. Let u be the input and y_i be the output of $\mathcal{B}_i$ in the tandem connection.

The equations of the tandem connection are then

3 $$L_1(p,t)y_1 = M_1(p,t)u$$
4 $$L_2(p,t)y_2 = M_2(p,t)y_1$$

We want to find the differential operators of least order, $L(p,t)$ and $M(p,t)$, such that the differential equation

5 $$L(p,t)y_2 = M(p,t)u$$

is satisfied by all the input-output pairs (u, y_2) of $\mathcal{B}_2\mathcal{B}_1$.

For the analysis to follow we introduce the following notations:

$\{\varphi_{i1}\}$, $i = 1, 2, \ldots, \lambda_1$, is a set of basis functions for $L_1(p,t)$.
$\{\varphi_{i2}\}$, $i = 1, 2, \ldots, \lambda_2$, is a set of basis functions for $L_2(p,t)$.
$\{\psi_{ij}\}$, $i = 1, 2, \ldots, \mu_j$, is a set of basis functions for $M_j(p,t)$, $j = 1, 2$.

Suppose we have found differential operators of minimal order $P_1(p,t)$ and $P_2(p,t)$ such that (note the order of the factors; see Comment *6.6.5*)

6 $$P_1(p,t)L_1(p,t) = P_2(p,t)M_2(p,t) \triangleq P(p,t)$$

where $P(p,t)$ is, by definition, the common value of the products. From this relation it is obvious that any basis function of $L_1(p,t)$ is a basis function of $P(p,t)$ and any basis function of $M_2(p,t)$ is a basis function of $P(p,t)$. Therefore, if we denote by π_{i1} and π_{i2} the basis functions of P_1 and P_2, respectively, we conclude that the basis functions of P_2 are

7 $$\pi_{i2} = M_2(p,t)\varphi_{i1} \qquad i = 1, 2, \ldots, \lambda_1$$

and those of P_1 are

8 $$\pi_{i1} = L_1(p,t)\varphi_{i2} \qquad i = 1, 2, \ldots, \lambda_2$$

9 **Assumption** It is assumed that the sets $\{\pi_{i1}\}$ and $\{\pi_{i2}\}$ are two sets of linearly independent functions.[1] ◁

This assumption justifies our calling these sets basis functions (see Definition *6.3.9*). Given these sets of basis functions of P_1 and P_2, it is easy to construct the differential operators P_1 and P_2 by use of *6.3.11*. Let us now operate on the left of *3* by P_1 and on the left of *4* by P_i. As a result of *6*, we obtain

$$P_2(p,t)L_2(p,t)y_2 = P_1(p,t)M_1(p,t)u$$

Therefore,

10 $$L(p,t) = P_2(p,t)L_2(p,t) \qquad M(p,t) = P_1(p,t)M_1(p,t)$$

We conclude this discussion by an

11 **Assertion** Given two time-varying differential systems $\mathcal{B}_1$ and $\mathcal{B}_2$ characterized by *1* and *2*, the product $\mathcal{B}_2\mathcal{B}_1$ (i.e., the tandem connection of $\mathcal{B}_1$ preceding $\mathcal{B}_2$) has all its input-output pairs (u,y_2) satisfying the

[1] Suppose Assumption *9* is not satisfied; suppose, in particular, that the π_{i2} are not linearly independent. We would like to point out the analogy between this situation and that of the time-invariant case where L_1 and M_2 have one or more factors in common. Indeed, if the π_{i2} are linearly dependent, there is a linear combination (with constant coefficients) of the π_{i2} that sums to zero for all t. Hence, by *7*, the same linear combination of the φ_{i1} is a solution of $M_2(p,t)z = 0$. Thus there is at least one basis function of $L_1(p,t)$ (the linear combination of the φ_{i1}) which is a zero-response function of $\mathcal{B}_2$, that is, is a solution of $M_2(p,t)z = 0$.

differential equation $L(p,t)y_2 = M(p,t)u$. If Assumption *9* is satisfied, L and M are given by *10*, where $P_1(p,t)$ and $P_2(p,t)$ are the differential operators which have $\{\pi_{i1}\}$ and $\{\pi_{i2}\}$ as sets of basis functions (see *7* and *8*). The arbitrary nonzero factors in P_1 and P_2 must be so selected that *6* holds.

6 *Systems of higher-order differential equations*

Let us consider a system described by a system of higher-order differential equations of the form

1 $$\mathbf{L}(p,t)\mathbf{x}(t) = \mathbf{M}(p,t)u(t)$$

or more explicitly

2 $$\sum_{j=1}^{n} L_{ij}(p,t)x_j(t) = M_i(p,t)u(t) \qquad i = 1, 2, \ldots, n$$

where the $L_{ij}(p,t)$ and $M_i(p,t)$ are differential operators with time-varying coefficients, u is a scalar input, and the dependent variable $\mathbf{x}$ is the vector-valued output. It is *assumed* that in each L_{ij} the non-identically zero coefficient of the highest power of p differs from zero for all t: thus, for each pair (i,j), the order of the differential operator does not change with t.

The first question one would like to answer in connection with *1* is: How can one associate a state vector with *1* and what is the corresponding input-output-state relation? In the case of time-invariant systems this question could easily be answered by the use of the Laplace transform technique. Since in the case of differential equations with time-varying coefficients the Laplace transformation does not lead to a set of linear *algebraic* equations in the transformed variable, it is no longer an effective tool for answering this question. Our first step will be to present the elimination method, which will bring the system *1* to a triangular form. Later we shall use this form to determine the state of the system.

The elimination method

The basic idea is identical with that of the method described in *5.6.10*: the idea is to reduce the system *1* to a form such that the matrix of the differential operators L_{ij} is triangular; consequently, the system can be solved as a succession of differential equations in one unknown function. The procedure is based on the

3 **Lemma** Let $L(p,t)$, $N(p,t)$ be two differential operators of order λ and ν, respectively, with $\lambda \geq \nu$; let $p = d/dt$. Let the coefficient of the highest power of p in L and N be distinct from zero for all t. Then there are two uniquely defined operators $M(p,t)$, $R(p,t)$, of orders μ and ρ, respectively, such that

4 $$L(p,t) = M(p,t)N(p,t) + R(p,t)$$

with $\lambda = \mu + \nu$ and $\rho < \nu$.

5 *Comment* The operation above is analogous to the "division" of $L(p,t)$ by $N(p,t)$; $M(p,t)$ is the "quotient" and $R(p,t)$ is the "remainder." It should be stressed, however, that, whereas time-invariant linear differential operators can be treated as if they were polynomials in the variable p, this is not the case for time-varying differential operators. The addition of time-varying differential operators is associative and commutative; the multiplication of time-varying differential operators is associative and distributive with respect to addition:

$$[L_1(p,t)L_2(p,t)]L_3(p,t) = L_1(p,t)[L_2(p,t)L_3(p,t)]$$

and

$$L_1(p,t)[L_2(p,t) + L_3(p,t)] = L_1(p,t)L_2(p,t) + L_1(p,t)L_3(p,t)$$

But the multiplication of such operators is *not commutative.* For example, let $L_1(p,t) = p$, $L_2(p,t) = e^{-t}p$; then $L_1L_2 = p(e^{-t}p) = e^{-t}p^2 - e^{-t}p$; $L_2L_1 = e^{-t}p^2$; hence $L_1L_2 \neq L_2L_1$. ◁

Proof To show that the coefficients $m_k(t)$ and $r_k(t)$ of the operators $M(p,t)$ and $R(p,t)$ defined in *4* are uniquely defined, let us consider the case where

$$L(p,t) = \sum_{k=0}^{4} a_{4-k}p^k \qquad N(p,t) = n_0p + n_1$$

Equation *4* gives

$$(m_0p^3 + m_1p^2 + m_2p + m_3)(n_0p + n_1) + r_0 = a_0p^4 + a_1p^3 + a_2p^2 + a_3p + a_4$$

A direct computation and identification of like powers of p gives

$$\begin{aligned}
n_0m_0 \qquad\qquad\qquad\qquad\qquad\qquad\qquad\qquad &= a_0\\
(\dot n_0 + n_1)m_0 + n_0m_1 \qquad\qquad\qquad\qquad &= a_1\\
(\ddot n_0 + \dot n_1)m_0 + (\dot n_0 + n_1)m_1 + n_0m_2 \qquad\qquad &= a_2\\
(\dddot n_0 + \ddot n_1)m_0 + (\ddot n_0 + \dot n_1)m_1 + (\dot n_0 + n_1)m_2 + n_0m_3 &= a_3\\
\dddot n_1m_0 + \ddot n_1m_1 + \dot n_1m_2 + n_1m_3 + r_0 &= a_4
\end{aligned}$$

By assumption $n_0(t) \neq 0$ for all t, the system may be solved from the top down by successive substitution for m_0, m_1, m_2, m_3, and r_0. Since the determinant of the system is $n_0^3(t)$, hence $\neq 0$ for all t, the solution is unique. This completes the proof of the assertion. ◁

As in *5.6.8* we define two types of *elementary row operations:*

(I) Replace in the system $\mathcal{S}$ of differential equations, given in *1*, the kth equation by the sum of the kth equation and the ith equation multiplied by $-M(p,t)$.

(II) Replace, in the system $\mathcal{S}$ of differential equations, the kth equations by K times the kth equation, where K is a nonzero constant.

As in *5.6.10*, we can show that by a succession of elementary row operations we can replace $\mathcal{S}$ by another system $\mathcal{S}'$ given by

6 $$\sum_{j=1}^{n} \tilde{L}_{ij}(p,t)x_j(t) = \tilde{F}_i(p,t)r(t) \qquad i = 1, 2, \ldots, n$$

where $\tilde{L}_{ij} \equiv 0$ for $i < j$. From the previously given arguments it follows that $\mathbf{x}(t) \equiv (x_1,x_2, \ldots ,x_n)$ is a solution of *1* if and only if it is a solution of *6*. In view of the triangular form of the matrix $(\tilde{L}_{ij})$, the system *6* can be solved by successive substitutions from the bottom equation up. At each step, one has to solve a linear differential equation with time-varying coefficients in one unknown.

Let us now consider the problem of assigning a state to the system *1*. As a first step let us make the following

7 *Comment* It is always possible to use elementary row operations to cause the system *6* to have the property that in the matric differential operator in the left-hand side, each diagonal element has an order larger than that of all other elements of the same column, i.e., for all j, the order of $\tilde{L}_{jj}(p,t)$ is larger than the order of $\tilde{L}_{ij}(p,t)$ for $i = 1, 2, \ldots, n$. The procedure is as follows: Consider the columns of the matric differential operator $\tilde{\mathbf{L}}$ of *6*, starting from the left. Suppose the jth column is the first column for which an off-diagonal element, say, $\tilde{L}_{i'j}$, has an order higher than that of the diagonal element L_{jj}. Since *6* is already in triangular form, $i' < j$. Let us now add to the i'th row the jth row premultiplied by a suitable $M(p,t)$ as in Lemma *3*: the order of the resulting (i',j) element is now smaller than that of $\tilde{L}_{jj}$. Since *6* is in the triangular form, this procedure does not affect the columns preceding the jth column; therefore, by successively operating on each column starting from the left, we shall end up with a matric differential operator $\hat{\mathbf{L}}$ in which each diagonal element has an order higher than that of all the elements of the same column.

Let us assume that the system is brought into this form and let us illustrate the selection of a state by considering a simple

8 *Example* Suppose that the methods described above lead to the following system of differential equations:

9 $$\begin{bmatrix} p^2 + \cos t\, p + 1 & p + 1 \\ 0 & p^2 + e^t p + 1 \end{bmatrix} \begin{bmatrix} x_1(t) \\ x_2(t) \end{bmatrix} = \begin{bmatrix} p^2 + p + 1 \\ 1 \end{bmatrix} u(t)$$

In the present case, the state at time t_0 is a collection of numbers such that the knowledge of the state at time t_0, the vector differential equation *9*, and the input $u_{(t_0,t]}$ is sufficient to determine the output over the interval $(t_0,t]$, that is, $x_1(\xi)$ and $x_2(\xi)$ for $t_0 < \xi \le t$. Recalling the definition of $p\,u(t)$ (see *3.8.4*), and assuming that $u_{(t_0,t]}$ is a time function, we obtain from *9* the system

10
$$\begin{bmatrix} p^2 + \cos t\, p + 1 & p + 1 \\ 0 & p^2 + e^t p + 1 \end{bmatrix} \begin{bmatrix} 1(t - t_0)x_1(t) \\ 1(t - t_0)x_2(t) \end{bmatrix} = \begin{bmatrix} p^2 + p + 1 \\ 1 \end{bmatrix} 1(t - t_0)u(t) + \begin{bmatrix} \varphi_1(t) \\ \varphi_2(t) \end{bmatrix}$$

where

$$\begin{aligned} \varphi_1(t) &= [x_1(t_0-) - u(t_0-)]\delta^{(1)}(t - t_0) + [x_1^{(1)}(t_0-) \\ &\qquad + \cos t_0\, x_1(t_0-) + x_2(t_0-) - u(t_0-) - u^{(1)}(t_0-)]\delta(t - t_0) \\ \varphi_2(t) &= x_2(t_0-)\delta^{(1)}(t - t_0) + [x_2^{(1)}(t_0-) + e^{t_0}x_2(t_0-)]\delta(t - t_0) \end{aligned}$$

It is easy to see that the vector

$$\boldsymbol{\alpha}(t_0) \triangleq \begin{bmatrix} x_1(t_0-) - u(t_0-) \\ x_1^{(1)}(t_0-) + \cos t_0\, x_1(t_0-) - u'(t_0-) - u(t_0-) \\ x_2(t_0-) \\ x_2^{(1)}(t_0-) \end{bmatrix}$$

qualifies for the state at time t_0 of the system described by *9*: in particular, given $\boldsymbol{\alpha}(t_0)$ and $u_{(t_0,t]}$, the uniqueness theorem of systems of differential equations guarantees that the output $\mathbf{x}_{(t_0,t]}$ is uniquely defined.

7 *Periodically varying systems*

Periodically varying linear systems are very important in engineering for two reasons: first, there are linear periodically varying systems in the real world, e.g., parametric amplifiers where the pump imposes a periodic variation on some circuit element; second, in the consideration of the stability of periodic solutions of nonlinear differential equations one has to consider the equations of the first variation which constitute a system of linear differential equations with periodically varying coefficients. In this case the period is the period of the oscillation whose stability is under study.

We consider in this section the free motion of a system described by

1
$$\dot{\mathbf{x}} = \mathbf{A}(t)\mathbf{x}$$

where $\mathbf{A}(t)$ depends continuously on t and is periodic with period T_0.

More precisely, $\mathbf{A}(t) = \mathbf{A}(t + T_0)$ for all t. Let us observe that if $\mathbf{\Phi}(t,t_0)$ is the state transition matrix of *1*, then $\mathbf{\Phi}(t + T_0, t_0)$ is also a fundamental matrix of *1*, that is, for all t its n columns are linearly independent solutions of *1* (see *6.2.15*). This is checked by direct verification:

$$\dot{\mathbf{\Phi}}(t + T_0, t_0) = \mathbf{A}(t + T_0)\mathbf{\Phi}(t + T_0, t_0) = \mathbf{A}(t)\mathbf{\Phi}(t + T_0, t_0)$$

Thus, $\mathbf{\Phi}(t + T_0, t_0)$ satisfies *1*. It is a fundamental matrix by *6.2.15*. The main result to be established is the

2 **Theorem** Let $\mathbf{A}(t)$ be continuous and periodic with period T_0. Then (I) the state transition matrix of *1* can always be written as

3 $$\mathbf{\Phi}(t,t_0) = \mathbf{P}(t,t_0) \exp [(t - t_0)\mathbf{B}] \qquad \mathbf{P}(t_0,t_0) = \mathbf{I}$$

where $\mathbf{P}(t,t_0)$ is a nonsingular periodic matrix with period T_0 and $\mathbf{B}$ is a constant matrix.

(II) By changing coordinates to a periodically varying system of coordinates such that the state is represented in the new system by $\mathbf{y}(t) = \mathbf{P}^{-1}(t,t_0)\mathbf{x}(t)$, the new representation of the state satisfies the equation

4 $$\dot{\mathbf{y}}(t) = \mathbf{B}\mathbf{y}(t)$$

which is an equation with constant coefficients.

Proof As we have seen, if $\mathbf{\Phi}(t,t_0)$ is the state transition matrix of the system *1*, then $\mathbf{\Phi}(t + T_0, t_0)$ is a fundamental matrix of *1*. Its ith column, $\boldsymbol{\phi}_i(t + T_C, t_0)$, is then of the form $\mathbf{\Phi}(t,t_0)\boldsymbol{\gamma}_i$, where $\boldsymbol{\gamma}_i$ is a suitably chosen constant vector since the vector $\boldsymbol{\phi}_i(t + T_0, t_0)$ can be represented as a linear combination of the basis vectors $\{\boldsymbol{\phi}_i(t,t_0)\}$. Let $\mathbf{C}$ be the matrix whose columns are the constant vectors $\boldsymbol{\gamma}_i$; $\mathbf{C}$ is nonsingular since it is equal to $\mathbf{\Phi}^{-1}(t,t_0)\mathbf{\Phi}(t + T_0, t_0)$, the product of two nonsingular matrices. Define $\mathbf{B}$ by $\exp (T_0\mathbf{B}) = \mathbf{C}$; $\mathbf{B}$ is a constant matrix. (However, if $\mathbf{C}$ is real, it does not follow that $\mathbf{B}$ is real.[1]) Now

$$\mathbf{\Phi}(t + T_0, t_0) = \mathbf{\Phi}(t,t_0)\mathbf{C} = \mathbf{\Phi}(t,t_0) \exp (T_0\mathbf{B})$$

Define

5 $$\mathbf{P}(t,t_0) \triangleq \mathbf{\Phi}(t,t_0) \exp [-(t - t_0)\mathbf{B}]$$

The matrix $\mathbf{P}(t,t_0)$ is nonsingular, since both $\mathbf{\Phi}(t,t_0)$ and $\exp [-(t - t_0)\mathbf{B}]$ are nonsingular. Furthermore $\mathbf{P}$ is periodic with period T_0 because,

[1] $\mathbf{B} = (1/T_0) \log \mathbf{C}$; now by *D.6.1*, if $\mathbf{C}$ has a negative real eigenvalue, say, λ_i, $\log \lambda_i$ is complex; hence $\mathbf{B}$ is complex.

for all t,

$$\begin{aligned}\mathbf{P}(t + T_0, t_0) &= \mathbf{\Phi}(t + T_0, t_0) \exp [-(t + T_0 - t_0)\mathbf{B}] \\ &= \mathbf{\Phi}(t + T_0, t_0)[\exp(-T_0\mathbf{B})] \{\exp [-(t - t_0)\mathbf{B}]\} \\ &= \mathbf{\Phi}(t,t_0) \{\exp [-(t - t_0)\mathbf{B}]\} \triangleq \mathbf{P}(t,t_0)\end{aligned}$$

From the last relation, *3* follows. Now let

6 $$\mathbf{Y}(t,t_0) \triangleq \mathbf{P}^{-1}(t,t_0)\mathbf{\Phi}(t,t_0)$$

By *5*,

$$\mathbf{Y}(t,t_0) = \exp [(t - t_0)\mathbf{B}]$$

Hence $\mathbf{Y}(t,t_0)$ is the solution of the equation $\dot{\mathbf{Y}} = \mathbf{BY}$ such that

$$\mathbf{Y}(t_0,t_0) = \mathbf{I}$$

Finally, let $\mathbf{z}(t)$ be any solution of *1*; then $\mathbf{z}(t) = \mathbf{\Phi}(t_0,t)\mathbf{z}(t_0)$, and using the definition of $\mathbf{y}(t)$, we observe that

$$\mathbf{y}(t) \triangleq \mathbf{P}^{-1}(t,t_0)\mathbf{z}(t) = \mathbf{P}^{-1}(t,t_0)\mathbf{\Phi}(t,t_0)\mathbf{z}(t_0) = \mathbf{Y}(t,t_0)\mathbf{z}(t_0)$$

is the solution of *4*, with the initial condition $\mathbf{y}(t_0) = \mathbf{z}(t_0)$.

7 *Comment* There is no general method for determining $\mathbf{P}(t,t_0)$ or $\mathbf{C}$, except numerical integration of *1* over an interval of length T_0.

8 *Comment* The eigenvalues of $\mathbf{B}$ are called *characteristic exponents* of $\mathbf{A}$, and the eigenvalues of $\mathbf{C}$ are called the *characteristic multipliers* of $\mathbf{A}$.

9 *Comment* The importance of the assertion (II) of *2* is that theoretical consideration of periodically varying systems often reduces to consideration of the time-invariant system *4*.

10 *Comment* Let $\mathbf{b}_1, \mathbf{b}_2, \ldots, \mathbf{b}_n$ be the columns of $\exp(t\mathbf{B})$ and let $\mathbf{p}_1, \ldots, \mathbf{p}_n$ be the complex conjugate of the rows of $\mathbf{P}(t)$ (for simplicity, we put $t_0 = 0$ and drop the second argument t_0). Then the ith column of $\mathbf{\Phi}(t)$ has the form

$$\begin{matrix}\langle \mathbf{p}_1(t),\mathbf{b}_i(t)\rangle \\ \langle \mathbf{p}_2(t),\mathbf{b}_i(t)\rangle \\ \cdot \\ \cdot \\ \cdot \\ \langle \mathbf{p}_n(t),\mathbf{b}_i(t)\rangle\end{matrix}$$

where the vectors $\mathbf{p}_i(t)$ are periodic in t and the $\mathbf{b}_i(t)$ have components that are linear combinations of $e^{\lambda_j t}, te^{\lambda_j t}, \ldots, t^{k_j-1}e^{\lambda_j t}, j = 1, 2, \ldots,$ where λ_j's are eigenvalues of $\mathbf{B}$ and k_j is the multiplicity of λ_j as a zero of the minimal polynomial of $\mathbf{B}$. Thus *3* imposes a strong constraint on the form of the solutions of *1*.

REFERENCES AND SUGGESTED READING

1 Bellman, R.: "Stability Theory of Differential Equations," McGraw-Hill Book Company, Inc., New York, 1953.

2 Bourbaki, N.: "Fonctions d'une variable réelle," chap. IV, Hermann & Cie, Paris, 1961.

3 Cesari, L.: "Asymptotic Behavior and Stability Problems in Ordinary Differential Equations," Springer Verlag, OHG, Berlin, 1959.

4 Coddington, E. A., and N. Levinson: "Theory of Ordinary Differential Equations," McGraw-Hill Book Company, Inc., New York, 1955.

5 Darlington, S.: Nonstationary Smoothing and Prediction Using Network Theory Concepts, *I.R.E. Trans. Circuit Theory,* vol. CT-6, pp. 1–13, 1959.

6 Ince, E. L.: "Ordinary Differential Equations," Dover Publications, Inc., New York, 1956.

7 Kaplan, W.: "Operational Methods for Linear Systems," Addison-Wesley Publishing Company, Inc., Reading, Mass., 1962.

8 Laning, J. H., Jr., and R. H. Battin: "Random Processes in Automatic Control," McGraw-Hill Book Company, Inc., New York, 1956.

9 Lefschetz, S.: "Differential Equations: Geometric Theory," Interscience Publishers, Inc., New York, 1957.

10 *I.R.E. Trans. Circuit Theory,* vol. CT-2, no. 1, 1955. (Special issue on linear time-varying systems.)

11. Zadeh, L. A.: Time-varying Networks, I, *Proc. I.R.E.*, vol. 49, pp. 1488-1503, October, 1961.

Stability of linear differential systems 7

1 Introduction

In Chaps. *5* and *6* we have investigated various basic properties of time-invariant and time-varying differential systems. In particular, we have seen that the state at time t is expressed in terms of the initial state $\mathbf{x}_0$ and the input $\mathbf{u}_{[t_0,t]}$ by an expression of the form

$$\mathbf{x}(t;\mathbf{x}_0,t_0;\mathbf{u}_{[t_0,t]}) = \mathbf{\Phi}(t,t_0)\mathbf{x}_0 + \int_{t_0}^{t} \mathbf{\Phi}(t,t')\mathbf{B}(t')\mathbf{u}(t')\,dt'$$

and we studied in detail the properties of the state transition matrix $\mathbf{\Phi}$, its relation to those of the adjoint system, and the relation between $\mathbf{\Phi}$ and the impulse response of the system. Now we turn to the study of the behavior of $\mathbf{x}$ as a function of t. In most practical systems, it is required that the system be "stable," i.e., when $\mathbf{u} = \mathbf{0}$, $\mathbf{x}(t)$ must remain bounded for all subsequent times; otherwise, some internal component of the system may burn out or break down. Section *2* will be devoted to the precise formulation of this concept of stability. We shall see that we shall have to distinguish between stability in the sense of Lyapunov and asymptotic stability. In Secs. *3* and *4* we shall present necessary and sufficient conditions for stability based on these concepts. In the case of time-varying systems, this characterization is based on the behavior of the state transition matrix which in most practical cases can be obtained only by numerical computation. For this reason we give in Sec. *5* some sufficiency tests for asymptotic stability which do not require the knowledge of $\mathbf{\Phi}(t,t_0)$. In Sec. *6* we develop the idea of reducible systems, systems that have the property that, when considered from a suitably chosen moving reference frame, they look like time-invariant systems. In the final section, we turn to another concept of stability. Roughly speaking, this concept is that a system is stable whenever any bounded input produces a bounded output; we must first formulate this idea precisely and then give its characterization. For the case of time-invariant systems whose output is actually the state of the system, the concept of asymptotic stability and the last concept are equivalent.

2 *Definition of stability based on the free motion of the state*

Let us consider a linear differential system $\mathcal{A}$ characterized by $\dot{\mathbf{x}} = \mathbf{A}\mathbf{x} + \mathbf{B}\mathbf{u}$. For simplicity we shall assume that the output of $\mathcal{A}$ is actually the state rather than a linear function of the state and the input. For the next three sections we shall restrict ourselves to consideration of the free motion of the state, i.e., we shall assume that $\mathbf{u}(t) = \mathbf{0}$ for all t, or equivalently we consider the system

1 $$\dot{\mathbf{x}}(t) = \mathbf{A}(t)\mathbf{x}(t)$$

where $\mathbf{A}$ is an $n \times n$ matrix whose elements are continuous functions of time on (t_0, ∞).

The first concept to be defined is that of *stability of the zero state* of $\mathcal{A}$ or, as it will occasionally be referred to, *zero-input stability* of $\mathcal{A}$, since it involves the free motion (zero-input response) of $\mathcal{A}$. Roughly speaking, the zero state $\mathbf{x} = \mathbf{0}$ is stable if, for every initial state $\mathbf{x}_0$ which is sufficiently close to $\mathbf{0}$, the corresponding free motion $\mathbf{x}(t;\mathbf{x}_0,t_0)$ remains close to zero for all $t \geq t_0$. A familiar example of a physical system which has a stable zero state [which is also its equilibrium state (*2.8.15*)] is a ball bearing in a teacup. There any sufficiently small displacement of the ball bearing from its equilibrium position results in a trajectory which remains close to the equilibrium position forever after.

2 **Definition** The zero state $\mathbf{x} = \mathbf{0}$ is said to be *stable in the sense of Lyapunov* or, equivalently, $\mathcal{A}$ is said to be *zero-input stable in the sense of Lyapunov*, if for any t_0 and any $\varepsilon > 0$, there is a $\delta > 0$ depending on ε and t_0 such that[1]

$$\|\mathbf{x}_0\| < \delta \Rightarrow \|\mathbf{x}(t;\mathbf{x}_0,t_0)\| < \varepsilon \qquad \forall t \geq t_0$$

For brevity, we shall write $\mathbf{x} = 0$ is stable i.s.L. (in the sense of Lyapunov).

If the zero state is stable i.s.L. and if, for every t_0 and every $\varepsilon > 0$ there is a $\delta > 0$ *independent of* t_0 such that the above inequality holds, then the zero state is said to be *uniformly stable i.s.L.* ◁

[1] In view of the fact that the norms in $\mathcal{C}^n$ are equivalent (*C.16.8 et seq.*), we have the freedom of picking the norm that is most convenient for the case at hand. Whenever in the course of a proof a specific norm is used, we shall so indicate.

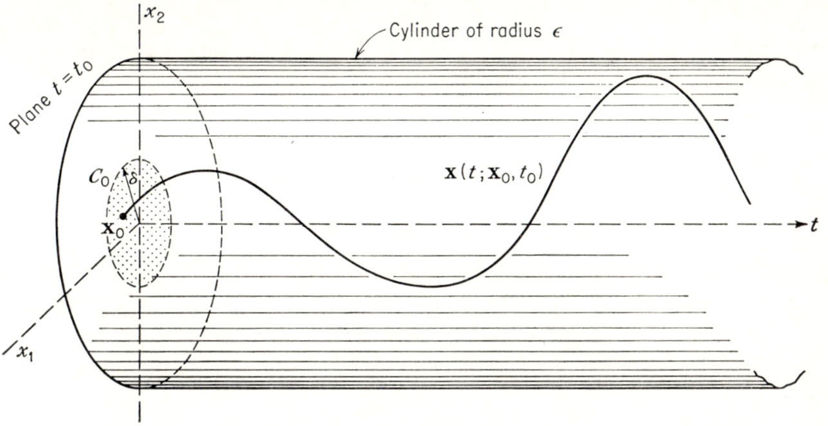

Fig. 7.2.1 The concept of stability in the sense of Lyapunov.

The concept of stability i.s.L. is illustrated in Fig. *7.2.1* for a second-order system: if $\mathbf{x}_0$ is in the interior of the circle C_0 of radius δ, then the trajectory $\mathbf{x}(t;\mathbf{x}_0,t_0)$ remains for all future times inside the cylinder of radius ε.

In most engineering applications we require that the response eventually go to zero. Since this requirement is not implied by the stability i.s.L., we must introduce a further

3 **Definition** The zero state $\mathbf{x} = 0$ is said to be *asymptotically stable* if (I) it is stable i.s.L. and (II) for any t_0 and for any $\mathbf{x}_0$ *sufficiently close to* $\mathbf{0}$, $\mathbf{x}(t;\mathbf{x}_0,t_0) \to \mathbf{0}$ as $t \to \infty$.

4 *Comment* The conditions (I) and (II) are equivalent to the following: For any $\varepsilon_1 > 0$ and $\varepsilon_2 > 0$ there are a $\delta > 0$ and a T sufficiently large that $\|\mathbf{x}_0\| < \delta$ implies that

$$\|\mathbf{x}(t;\mathbf{x}_0,t_0)\| < \varepsilon_1 \qquad \text{for all } t \geq t_0$$

and also that

$$\|\mathbf{x}(t;\mathbf{x}_0,t_0)\| < \varepsilon_2 \qquad \text{for all } t \geq t_0 + T$$

Both δ and T depend in general on t_0, ε_1, and ε_2.

5 *Comment* These concepts of the stability of the zero state consider exclusively the zero-input response (free motion) of the system. It should be stressed, in particular, that asymptotic stability of the zero state does not imply that any bounded input will produce a bounded output (see *7.7.2* and *7.7.13*).

6 **Remark: stability in the large** It is apparent that Definitions *2* and *3* may be applied to nonlinear systems, say, to a differential system defined by $\dot{\mathbf{x}} = \mathbf{f}(\mathbf{x})$, where $\mathbf{f}(\mathbf{0}) = \mathbf{0}$. In such cases the state $\mathbf{x} = \mathbf{0}$

is said to be an equilibrium state of the nonlinear system. There is, however, a considerable difference between the implications of the asymptotic stability of the equilibrium state of a nonlinear system and that of a linear system. In the case of linear systems the zero-input response is homogeneous (see *3.4.17*); this implies that if the zero state $\mathbf{x} = \mathbf{0}$ is asymptotically stable, then for any $\mathbf{x}_0$, the zero input response $\mathbf{x}(t;\mathbf{x}_0,t_0)$ is bounded and eventually becomes arbitrarily close to $\mathbf{0}$. More precisely, for any initial state $\mathbf{x}_0$, there is a positive number M (which depends on $\mathbf{x}_0$), such that

$$\|\mathbf{x}(t;\mathbf{x}_0,t_0)\| < M \qquad \forall t \geq t_0$$

and

$$\mathbf{x}(t;\mathbf{x}_0,t_0) \to \mathbf{0} \qquad \text{as } t \to \infty$$

This follows directly from the homogeneity of the zero-input response. We need only take $M = (\epsilon_1/\delta)\|\mathbf{x}_0\|$ in the expressions of Comment *4*. This fact is expressed as follows: If the linear system *1* has its zero state asymptotically stable, then it is *asymptotically stable in the large*. The expression "in the large" is used to indicate that the asymptotic stability holds for all initial states and is not restricted to states sufficiently close to the zero state. As a consequence we shall use the

7 **Definition** The linear system *1* is said to be *stable i.s.L.* (asymptotically stable) whenever its zero state is stable (asymptotically stable) i.s.L. ◁

For nonlinear systems, an equilibrium state may be asymptotically stable without being asymptotically stable in the large. This is illustrated by an example. Consider the second-order nonlinear system specified by the differential equations which define the motion of its state in terms of the polar coordinates of the state:

$$\dot{r} = r^2 - r \qquad r \geq 0$$
$$\dot{\theta} = r$$

The point $r = 0$, $\theta = 0$ is an equilibrium point since at $(r = 0, \theta = 0)$ the derivatives $\dot{r}$, $\dot{\theta}$ are both zero. Now, for any $r(t_0) < 1$ and any $\theta(t_0)$, the point $(r(t),\theta(t))$ spirals into the origin. Therefore, we say that the equilibrium point $(r = 0, \theta = 0)$ is asymptotically stable. However, for $r(t_0) > 1$ and arbitrary $\theta(t_0)$, the point $(r(t),\theta(t))$ spirals outward, and eventually the point gets arbitrarily far away from the origin. Therefore, the equilibrium point $(r = 0, \theta = 0)$ is not asymptotically stable in the large. This idea is sometimes expressed by saying that the asymptotic stability of $(r = 0, \theta = 0)$ is valid only *locally*.

3 *Characterization of stable systems*

In this section we state two theorems that characterize stable and asymptotically stable systems.

1 **Theorem** The system described by *7.2.1* is stable i.s.L. $\Leftrightarrow$ there exists a constant M, which may depend on t_0, such that

$$\|\mathbf{\Phi}(t,t_0)\| \leq M \qquad \forall t \geq t_0$$

Proof $\Leftarrow$ Observe that the zero-input response of *1* is always given by (see *6.2.7*)

$$\mathbf{x}(t;\mathbf{x}_0,t_0) = \mathbf{\Phi}(t,t_0)\mathbf{x}_0$$

Therefore (see *C.16.10*)

$$\begin{aligned}\|\mathbf{x}(t;\mathbf{x}_0,t_0)\| &\leq \|\mathbf{\Phi}(t,t_0)\|\,\|\mathbf{x}_0\| \\ &\leq M\|\mathbf{x}_0\|\end{aligned}$$

Thus

$$\|\mathbf{x}_0\| < \frac{\varepsilon}{M} \Rightarrow \|\mathbf{x}(t;\mathbf{x}_0,t_0)\| \leq \varepsilon \qquad \text{for all } t \geq t_0$$

and the system is stable i.s.L. in view of Definition *7.2.2*.

$\Rightarrow$ We prove it by contradiction. If $\mathbf{\Phi}$ is not uniformly bounded for $t \geq t_0$, there is at least one element, say, ϕ_{ik}, which takes arbitrarily large values as $t \to \infty$. Let us pick the initial state $\mathbf{x}_0$ to have all components zero except the kth component, which is equal to ξ. Then the ith component of the state vector $\mathbf{x}(t;\mathbf{x}_0,t_0)$ is

$$x_i(t) = \phi_{ik}(t)\xi$$

Now, however small ξ may be, since ϕ_{ik} takes arbitrarily large values at $t \to \infty$, $x_i(t)$ will also take arbitrarily large values as $t \to \infty$. Therefore, the system is unstable, which contradicts the hypothesis. Consequently, $\mathbf{\Phi}$ must be uniformly bounded for all $t \geq t_0$. ◁

To characterize the asymptotic stability, we have the

2 **Theorem** The linear system *7.2.1* is asymptotically stable if and only if (I) there is a constant M such that

$$\|\mathbf{\Phi}(t,t_0)\| \leq \text{M} \qquad \forall t \geq t_0$$

and (II)
$$\lim_{t\to\infty} \|\mathbf{\Phi}(t,t_0)\| = 0 \qquad \forall t_0$$

Proof The proof follows easily from that of *1* by a few easy modifications; it is therefore left as an exercise for the reader. ◁

We know that the state vector of a time-invariant system cannot grow faster than an exponential. For time-varying systems, this is still true provided $\mathbf{A}(t)$ remains bounded for all $t \geq t_0$. This fact is a consequence of the following

3 **Assertion** Any solution of *7.2.1* satisfies

$$\|\mathbf{x}(t;\mathbf{x}_0,t_0)\| \leq \|\mathbf{x}_0\| \exp\left[\int_{t_0}^{t} \|\mathbf{A}(\xi)\|\, d\xi\right] \qquad \forall t \geq t_0$$

Proof Let us integrate *6.2.1* between t_0 and t and obtain

$$\mathbf{x}(t;\mathbf{x}_0,t_0) = \mathbf{x}_0 + \int_{t_0}^{t} \mathbf{A}(\xi)\mathbf{x}(\xi;\mathbf{x}_0,t_0)\, d\xi$$

Let us take the norm of both sides and use the properties of the norm (*C.16.10* and *C.16.12*) to get

$$\|\mathbf{x}(t;\mathbf{x}_0,t_0)\| \leq \|\mathbf{x}_0\| + \int_{t_0}^{t} \|\mathbf{A}(\xi)\| \quad \|\mathbf{x}(\xi;\mathbf{x}_0,t_0)\|\, d\xi$$

Assertion *3* follows directly from this inequality by virtue of the inequality established in the next exercise.

4 *Exercise* *Bellman-Gronwall inequality* Let $k \geq 0$ and w be real-valued continuous functions on $[t_0, \infty)$ and let c be a constant. If

$$w(t) \leq c + \int_{t_0}^{t} k(\xi)w(\xi)\, d\xi \qquad \forall t \geq t_0$$

then

$$w(t) \leq c \exp \int_{t_0}^{t} k(\xi)\, d\xi \qquad \forall t \leq t_0$$

(*Hint:* Call the integral in the given inequality $y(t)$: then, $w(t) \leq y(t) + c$. Multiply this inequality by $k(t) \exp\left[-\int_{t_0}^{t} k(\xi)\, d\xi.\right]$ Observe that $\dot{y} = kw$; hence

$$\frac{d}{dt}\left\{y(t) \exp\left[-\int_{t_0}^{t} k(\xi)\, d\xi\right]\right\} \leq c \exp\left[-\int_{t_0}^{t} k(\xi)\, d\xi\right]$$

Integrate both sides of the inequality between t_0 and t, observe that $y(t_0) = 0$, and cancel the common exponential factor to finally obtain the claimed inequality.)

5 *Comment* If $\|\mathbf{A}(t)\| \leq M$ for all t, then

$$\|\mathbf{x}(t;\mathbf{x}_0,t_0)\| \leq \|\mathbf{x}_0\| \exp\left[M(t - t_0)\right] \qquad \forall \mathbf{x}_0, \forall t \geq t_0$$

4 *Special cases*

Linear time-invariant systems

The questions of stability of linear time-invariant systems are important not only because such systems occur so frequently in practice but also because their stability implies, in some cases, the local stability of the zero state of a nonlinear system. For example, the following theorem plays a very important role in the theory of nonlinear systems:[1]

1 **Theorem** Let $\dot{\mathbf{x}} = \mathbf{A}\mathbf{x} + \mathbf{f}(\mathbf{x},t)$, where $\mathbf{A}$ is a real constant matrix all of whose eigenvalues have negative real parts. Let $\mathbf{f}$ be real, continuous for small $\|\mathbf{x}\|$ and for all $t \geq 0$, and

$$\mathbf{f}(\mathbf{x},t) = o(\|\mathbf{x}\|) \qquad \|\mathbf{x}\| \to 0$$

uniformly in t, $t \geq 0$. Then the zero state, $\mathbf{x} = \mathbf{0}$, is asymptotically stable. $\triangleleft$

It is important to emphasize that this result concerning the nonlinear system is purely local: it asserts that states sufficiently close to $\mathbf{0}$ will remain close to $\mathbf{0}$ and approach $\mathbf{0}$ arbitrarily closely as $t \to \infty$.

The necessary and sufficient conditions for the stability and asymptotic stability of linear time-invariant systems follow easily from the general formula for $\mathbf{x}(t;\mathbf{x}_0,t_0)$ (see *5.2.9*) and from the fundamental formula for $\exp(\mathbf{A}t)$ (*5.3.1*). The characterization of stability i.s.L. is given by the

2 **Theorem** The system described by the state transition equation $\dot{\mathbf{x}} = \mathbf{A}\mathbf{x}$, where $\mathbf{A}$ is a constant matrix, is stable i.s.L. $\Leftrightarrow$ (I) all the eigenvalues of $\mathbf{A}$ have nonpositive real parts and (II) those eigenvalues of $\mathbf{A}$ that lie on the imaginary axis are simple zeros of the *minimal polynomial* of $\mathbf{A}$.

Proof From *5.2.11*, we have $\mathbf{x}(t;\mathbf{x}_0,t_0) = [\exp \mathbf{A}(t - t_0)]\mathbf{x}_0$. Therefore, we need only show that $\|\exp \mathbf{A}(t - t_0)\| < M$, $\forall t \geq t_0 \Leftrightarrow$ conditions (I) and (II) hold. For convenience we take $t_0 = 0$, which is justified since the system is time-invariant (*3.2.26*).

$\Leftarrow$ Consider the fundamental formula applied to $\exp(\mathbf{A}t)$ (*5.3.1*). We obtain successively (using *C.16.12*)

$$\begin{aligned} \|\exp(\mathbf{A}t)\| &= \left\| \sum_{k=1}^{\sigma} \sum_{l=0}^{m_k-1} t^l e^{\lambda_k t} \mathbf{Z}_{kl} \right\| \\ &\leq \sum_{k=1}^{\sigma} \sum_{l=0}^{m_k-1} t^l e^{\alpha_k t} \|\mathbf{Z}_{kl}\| \qquad t \geq 0 \end{aligned}$$

where $\alpha_k = \operatorname{Re}(\lambda_k)$, $k = 1, 2, \ldots, \sigma$.

Let

$$M = \sum_{k=1}^{\sigma} \sum_{l=0}^{m_k-1} \|\mathbf{Z}_{kl}\| \sup (t^l e^{\alpha_k t})$$

where the supremum is taken over $[0, \infty)$.

[1] E. A. Coddington and N. Levinson, "Theory of Ordinary Differential Equations," McGraw-Hill Book Company, Inc., New York, 1955.

If conditions (I) and (II) of the theorem are satisfied, then for all k, l, sup $(t^l e^{\alpha_k t})$ is finite. Since the matrices $\mathbf{Z}_{kl}$ are constant, the same is true of the numbers $\|\mathbf{Z}_{kl}\|$. Therefore M is a finite number and the stability i.s.L. follows.

$\Rightarrow$ Let us use a contradiction proof to establish that the stability i.s.L. implies (I) and (II). First suppose that some eigenvalue, say, λ_1, has a positive real part: that is, $\alpha_1 > 0$. Pick as an initial state $\mathbf{x}_0$ an eigenvector associated with the eigenvalue λ_1; then (*5.4.5*)

$$\mathbf{x}(t;\mathbf{u}_1,0) = e^{\lambda_1 t}\mathbf{u}_1$$

Since Re $(\lambda_1) = \alpha_1 > 0$, $\|\mathbf{x}(t;\mathbf{u}_1,0\| \to \infty$ exponentially, irrespective of how small $\|\mathbf{x}_0\|$ has been chosen. Therefore, stability i.s.L. implies that no eigenvalue of $\mathbf{A}$ can have a positive real part.

Suppose now that the minimal polynomial has a zero of order $m_1 > 1$ on the imaginary axis. Let λ_1 be this zero and $\lambda_1 = j\omega_1$. Since $m_1 > 1$ let us pick $\mathbf{x}_0 \in \mathfrak{N}_1^2$ (see *D.3.1*), but $\mathbf{x}_0 \notin \mathfrak{N}_1^1 \triangleq \mathfrak{N}(\mathbf{A} - \lambda_1\mathbf{I})$. Thus $(\mathbf{A} - j\omega_1\mathbf{I})\mathbf{x}_0 \neq \mathbf{0}$. The state at time t is then, by *D.6.2*,

$$\mathbf{x}(t;\mathbf{x}_0,0) = e^{j\omega_1 t}\mathbf{x}_0 + te^{j\omega_1 t}(\mathbf{A} - j\omega_1\mathbf{I})\mathbf{x}_0$$

Again $\|\mathbf{x}(t;\mathbf{x}_0,0)\| \to \infty$ (linearly with time in this case) irrespective of how small $\mathbf{x}_0$ has been chosen. Therefore stability i.s.L. implies that the minimal polynomial of $\mathbf{A}$ cannot have multiple zeros with zero real parts. ◁

The characterization of asymptotic stability is very similar and is stated as a

3 **Theorem** The system described by its state transition equation $\dot{\mathbf{x}} = \mathbf{A}\mathbf{x}$, where $\mathbf{A}$ is a constant matrix, is asymptotically stable if and only if all the eigenvalues of $\mathbf{A}$ have negative (<0) real parts.
Proof The proof is similar to that of *2* and is left as an exercise for the reader.

4 *Exercise* Prove Theorems *2* and *3* by using the Laplace transform to compute $\mathbf{x}(t;\mathbf{x}_0,t_0)$. (*Hint:* Use the basic property of the minimal polynomial given in *D.2.6*.)

Linear periodic systems

The study of the stability of linear periodic systems is important not only because such systems occur frequently in practice, say in parametric amplifiers, but also because the question of stability of periodic solutions of nonlinear systems reduces to that of linear periodic systems. More precisely, the relation is expressed by two fundamental theorems derived from the theory of nonlinear differential systems.[1] For the

[1] *Ibid.*

proofs of these theorems the reader is referred to the book indicated in the footnote.

5 **Theorem** Let the nonlinear system be $\dot{\mathbf{x}} = \mathbf{f}(\mathbf{x},t)$, where $\mathbf{f}$ is periodic in t. Suppose $\mathbf{s}(t)$ is a periodic solution of the nonlinear system which has the same period as $\mathbf{f}$. Then the solution $\mathbf{s}(t)$ is asymptotically stable as $t \to \infty$[1] if the characteristic exponents associated with the linear periodically varying equation

$$\dot{\mathbf{z}} = \mathbf{f}_x(\mathbf{s}(t),t)\mathbf{z}$$

have negative real parts. [$\mathbf{f}_x(\mathbf{s},t)$ is the Jacobian matrix of $\mathbf{f}$ evaluated at $(\mathbf{s},t)$, that is, the matrix whose (i,j) element is $\dfrac{\partial f_i}{\partial x_j}(\mathbf{s}(t),t)$.] ◁

The linear equation above is called the equation of the first variation with respect to the solution $\mathbf{s}(t)$.

When the right-hand side, $\mathbf{f}(\mathbf{x},t)$, does not depend explicitly on t, a difficulty arises from the fact that the system by itself does not have a preferred origin of time, so that if $\mathbf{s}(t)$ is a periodic solution of $\dot{\mathbf{x}} = \mathbf{f}(\mathbf{x})$ so is $\mathbf{s}(t + c)$, where c is an arbitrary constant. Under these circumstances Theorem *5* becomes[2]

6 **Theorem** Let $(n - 1)$ characteristic exponents of $\dot{\mathbf{z}} = \mathbf{f}_x(\mathbf{s}(t))\mathbf{z}$ have negative real parts. Let $\mathbf{y}(t)$ be a solution of $\dot{\mathbf{x}} = \mathbf{f}(\mathbf{x})$. Then there exists an $\varepsilon > 0$ such that whenever, for some t_1, t_2, $\|\mathbf{y}(t_1) - \mathbf{s}(t_2)\| < \varepsilon$, there also exists a constant ϕ such that, as $t \to \infty$, $\|\mathbf{y}(t) - \mathbf{s}(t + \phi)\| \to 0$.

Roughly speaking, whenever the system starts from a state sufficiently close to the periodic orbit, the resulting motion approaches that orbit as $t \to \infty$. In Theorems *5* and *6*, the stability of a periodic solution depends on the stability properties of a linear periodically varying system.

The specific results concerning linear periodic systems follow directly from the theory of linear periodic systems developed in Chap. *6*, Sec. *7*. The proofs of the results are left as an exercise for the reader.

7 **Assertion** The periodic system described by the state equation $\dot{\mathbf{x}} = \mathbf{A}(t)\mathbf{x}$, where $\mathbf{A}(t)$ is periodic with period T, is stable i.s.L. $\Leftrightarrow$ {(I) the real parts of the characteristic exponents of $\mathbf{A}(t)$ are nonpositive (≤ 0) and (II) those characteristic exponents that are purely imaginary are simple zeros of the minimal polynomial of $\mathbf{B}$}. [Note that $\mathbf{B}$ is defined in terms of $\mathbf{A}(t)$ by *6.7.3*.]

[1] That is, any solution which starts, at t_0, close to $\mathbf{s}(t_0)$ will eventually become arbitrarily close to the periodic motion $\mathbf{s}(t)$.

[2] Coddington and Levinson, *op. cit.*

8 **Assertion** The periodic system described by the state transition equation $\dot{\mathbf{x}} = \mathbf{A}(t)\mathbf{x}$ is *asymptotically stable* $\Leftrightarrow$ all the characteristic exponents have negative (<0) real parts.

9 **Assertion** For both time-invariant linear systems and periodically varying linear systems described by *7.2.1* asymptotic stability implies that $\|\mathbf{x}(t;\mathbf{x}_0,0)\| \to 0$ *exponentially;* more precisely, there is a constant C and a number $\eta > 0$ such that

$$\|\mathbf{x}(t;\mathbf{x}_0,0)\| \leq Ce^{-\eta t} \qquad \text{for all } t \geq 0$$

10 *Comment* This is not true for general time-varying systems; for example, the first-order system $\dot{x} = -x/2t$, $t \geq 1$, has a solution $s(t) = t^{-\frac{1}{2}}$. Therefore, the system is asymptotically stable but its solutions do not go to zero exponentially. In fact, the solution s goes to zero so slowly that $\int_1^\infty |s(t)|\, dt = \infty$.

Systems characterized by $Ly = u$

Let $\mathcal{A}$ be a system characterized by $L(p,t)y = u$, where $L(p,t)$ is an nth-order linear differential operator, and let

$$\mathbf{x}(t) = (y(t),y^{(1)}(t), \ldots ,y^{(n-1)}(t))$$

be its state vector at time t, with the corresponding state equations having the form *6.2.56*. From the general results which we established in the preceding discussion it is a simple matter to deduce a number of stability properties of the free motion of such systems. In what follows we shall merely state these properties in the form of assertions and leave the proofs as exercises for the reader.

11 **Assertion** The system characterized by $L(p,t) = u$ is stable i.s.L. (or equivalently, zero-input stable) $\Leftrightarrow$ for a set of basis functions $\phi_i(t)$ of the operator L there is a number M such that $|\phi_i^{(k)}(t)| \leq M$ for $i = 1, 2, \ldots, n$, for $k = 0, 1, \ldots, n-1$, and for all $t \geq t_0$.

12 **Assertion** The system described by $L(p,t)y = 0$ is *asymptotically stable* $\Leftrightarrow$ {for a set of basis functions $\phi_i(t)$ of the operator $L(p,t)$ (I) there is a number M such that $|\phi_i^{(k)}(t)| \leq M$ for $i = 1, 2, \ldots, n$, for $k = 0, 1, \ldots, n-1$, and for all $t \geq t_0$, and (II) $\lim_{t\to\infty} \phi_i^{(k)}(t) = 0$ for $i = 1, 2, \ldots, n$ and $k = 0, 1, \ldots, n-1$}.

13 **Assertion** The time-invariant system described by $L(p)y = 0$ is *stable* i.s.L. $\Leftrightarrow$ {(I) the zeros of the polynomial $L(s)$ have nonpositive real parts and (II) those zeros that lie on the imaginary axis are simple}.

14 **Assertion** The time-invariant system described by $Ly = 0$ is *asymptotically stable* $\Leftrightarrow$ all the zeros of the polynomial $L(s)$ have negative real parts. $\triangleleft$

In order to use Assertion *14*, it is not necessary to calculate the zeros of the polynomial $L(s)$; there are well-known necessary and sufficient conditions for the zeros of a polynomial to have negative real parts; these conditions are given in *9.4.22*.

15 **Assertion** The time-invariant system described by

$$\sum_{j=1}^{n} L_{ij}(p)x_j(t) = 0 \qquad i = 1, 2, \ldots, n$$

is *asymptotically stable* $\Leftrightarrow$ {all the zeros of $\Delta_0(s)$ have negative real parts, where Δ_0 is the quotient of Δ by g_0, which is the greatest common divisor of all the elements of the adjoint matrix $\mathbf{\Gamma}(s)$}.

Proof Follows directly from *5.7.8*.

5 *Some sufficient conditions for stability*

In this section we consider only systems of the form

1 $$\dot{\mathbf{x}} = \mathbf{A}(t)\mathbf{x}$$

where $\mathbf{A}(t)$ is continuous in t.

Let us first consider Wazewski's inequalities,[1] which bound the norm of free motions. More precisely we have the

2 **Theorem** Consider the system *1* and let $\mathbf{H}$ be the hermitian matrix defined by

3 $$\mathbf{H}(t) = \tfrac{1}{2}[\mathbf{A}(t) + \mathbf{A}^*(t)]$$

Let $\lambda(t)$ and $\Lambda(t)$ be, for each t, the smallest and largest eigenvalues of $\mathbf{H}(t)$. Then

4 $$\|\mathbf{x}_0\| \exp\left[\int_{t_0}^{t} \lambda(t')\,dt'\right] \le \|\mathbf{x}(t;\mathbf{x}_0,t_0)\| \le \|\mathbf{x}_0\| \exp\left[\int_{t_0}^{t} \Lambda(t')\,dt'\right]$$

where the norm is defined by $\|\mathbf{y}\|^2 = \langle \mathbf{y},\mathbf{y}\rangle$.

Proof Let us outline the proof. Consider $V(t) = \langle \mathbf{x}(t;\mathbf{x}_0,t_0), \mathbf{x}(t;\mathbf{x}_0,t_0)\rangle$. Then, by the definition of $\mathbf{H}$

$$\dot{V} = V\frac{2\langle \mathbf{x}(t),\mathbf{H}(t)\mathbf{x}(t)\rangle}{\langle \mathbf{x}(t),\mathbf{x}(t)\rangle}$$

[1] T. Wazewski, Sur la limitation des intégrales des systèmes d'equations différentielles lineáres ordinaires, *Studia Math.*, vol. 10, pp. 48–59, 1948.

where we have made an obvious simplification in the notation. From well-known properties of hermitian matrices (*C.19.15*),

$$\lambda(t)\langle \mathbf{y},\mathbf{y}\rangle \leq \langle \mathbf{y},\mathbf{H}(t)\mathbf{y}\rangle \leq \Lambda(t)\langle \mathbf{y},\mathbf{y}\rangle \qquad \text{for all } \mathbf{y}\text{'s}$$

Hence

$$2\lambda(t) \leq \frac{d}{dt}(\log V) \leq 2\Lambda(t)$$

Finally, *4* follows by integrating, taking the exponential of the integrals, and observing that $V = \|\mathbf{x}(t)\|^2$.

5 **Corollary** Using the notation of *2*, the system is stable i.s.L. if, for all t_0,

$$\limsup_{t\to\infty} \int_{t_0}^{t} \Lambda(t')\,dt' < \infty$$

6 **Corollary** All solutions of the system *1* are unstable if, for all t_0,

$$\limsup_{t\to\infty} \int_{t_0}^{t} \lambda(t')\,dt' = \infty \quad \triangleleft$$

There is a sharper result applicable only to systems for which $\mathbf{A}(t)$ is *normal.* $\mathbf{A}(t)$ is said to be normal if $\mathbf{A}^*(t)\mathbf{A}(t) = \mathbf{A}(t)\mathbf{A}^*(t)$ for all t. Equivalently, for each t, $\mathbf{A}(t)$ is unitarily similar to a diagonal matrix (see *C.19.1* and *C.19.11*). First let us introduce a

7 **Definition** For each $\mathbf{x}_i(t)$ which is a solution of *1*, we define a generalized characteristic exponent g_i of $\mathbf{A}(t)$:

$$g_i = \limsup_{t\to\infty} t^{-1} \log \|\mathbf{x}_i(t)\|$$

For the remainder of this section we shall use the norm defined by the scalar product $\|\mathbf{x}\|^2 = \langle \mathbf{x},\mathbf{x}\rangle$.

8 **Theorem**[1] Let $\mathbf{A}(t)$ be normal, continuous, and bounded for $t \geq t_0$. Let $R(t)$ and $r(t)$ be the maximum and the minimum of the real part of the eigenvalues of $\mathbf{A}(t)$. Then

9 $$\limsup_{t\to\infty} t^{-1} \int_{t_0}^{t} r(t')\,dt' \leq g_i \leq \limsup_{t\to\infty} t^{-1} \int_{t_0}^{t} R(t')\,dt'$$

for all generalized characteristic exponents of the system.

Proof Let $V(t) = \langle \mathbf{x}(t),\mathbf{x}(t)\rangle$, where $\mathbf{x}(t)$ is any solution. Then by *1* and using the definition of the adjoint (*C.14.1*)

$$\frac{\dot V(t)}{V(t)} = \frac{\langle \mathbf{Ax},\mathbf{x}\rangle + \langle \mathbf{x},\mathbf{Ax}\rangle}{\langle \mathbf{x},\mathbf{x}\rangle} = \frac{\langle \mathbf{Ax},\mathbf{x}\rangle + \langle \mathbf{A}^*\mathbf{x},\mathbf{x}\rangle}{\langle \mathbf{x},\mathbf{x}\rangle} \triangleq \tfrac{1}{2}\rho(t)$$

[1] L. Markus, Continuous Matrices and Stability of Differential Systems, *Math. Z.*, vol. 62, pp. 310–319, 1955.

Since the matrix $\mathbf{A}$ is normal, by *C.19.14*, $r(t) \leq \rho(t) \leq R(t)$; hence

$$\int_{t_0}^{t} r(t')\,dt' \leq \tfrac{1}{2}\log\frac{V(t)}{V(t_0)} \leq \int_{t_0}^{t} R(t')\,dt$$

Now, the generalized characteristic exponent of the solution $\mathbf{x}(t)$ is

$$g = \limsup_{t\to\infty} t^{-1}\log\|\mathbf{x}(t)\| = \tfrac{1}{2}\limsup_{t\to\infty} t^{-1}\log\frac{V(t)}{V(t_0)}$$

and *9* follows.

10 *Exercise* Use Theorem *8* to derive some sufficient conditions for stability (asymptotic stability) of the system *1* when $\mathbf{A}(t)$ is normal for all t. ◁

Finally, let us present a sufficient condition for stability which may be regarded either as an extension of Wazewski's theorem or an application of the Lyapunov second method[1] to linear systems.

11 **Theorem** Let $\mathbf{Q}$ be any constant self-adjoint positive definite matrix. Let μ be a positive constant (say, its smallest eigenvalue, *C.19.15*) such that

12 $$\langle \mathbf{z},\mathbf{Qz}\rangle \geq \mu\langle \mathbf{z},\mathbf{z}\rangle \qquad \forall \mathbf{z}$$

and let $m(t)$ be a continuous function of t such that

13 $$\langle \mathbf{z},(\mathbf{A}^*(t)\mathbf{Q} + \mathbf{QA}(t))\mathbf{z}\rangle \leq m(t)\langle \mathbf{z},\mathbf{z}\rangle \qquad \forall \mathbf{z},\ \forall t \geq t_0$$

If there is a constant M such that

14 $$\int_{t_0}^{t} m(t')\,dt' \leq M < \infty \qquad \forall t \geq t_0$$

and if

15 $$\int_{t_0}^{t} m(t')\,dt' \to -\infty \qquad \text{as } t \to \infty$$

then the system *1* is *asymptotically stable.*

Proof Let $\mathbf{x}(t)$ be any solution of *1*. Let $\phi(t) = \langle \mathbf{x}(t),\mathbf{Qx}(t)\rangle$; then by *1* and the definition of the adjoint

$$\dot{\phi}(t) = \langle \mathbf{x},(\mathbf{A}^*\mathbf{Q} + \mathbf{QA})\mathbf{x}\rangle$$

By *13* and *12*

$$\dot{\phi}(t) \leq m(t)\langle \mathbf{x},\mathbf{x}\rangle \leq \frac{m(t)}{\mu}\langle \mathbf{x},\mathbf{Qx}\rangle$$

In other words

$$\dot{\phi}(t) \leq \frac{m(t)}{\mu}\,\phi(t)$$

[1] J. La Salle and S. Lefschetz, "Stability by Lyapunov's Direct Method with Applications," Academic Press, Inc., New York, 1961. See also R. E. Kalman and J. Bertram, Control System Design via the Second Method of Lyapunov, *Trans. A.S.M.E., Ser. D*, vol. 82, no. 2, pp. 371–400, 1960.

and

$$\phi(t) \leq \phi(t_0) \exp\left[\int_{t_0}^{t} \frac{m(t')}{\mu}\, dt'\right] \qquad \forall t \geq t_0$$

This inequality together with condition *14* implies that $\phi(t)$ is bounded on $[t_0, \infty)$ by $\phi(t_0) \exp(M/\mu)$. Finally, with *15*, it implies that $\phi(t) \to 0$ as $t \to \infty$. These two facts and the positive definiteness of $\mathbf{Q}$ imply that $\|\mathbf{x}(t)\|$ is bounded on $[t_0, \infty)$ and $\|\mathbf{x}(t)\| \to 0$ as $t \to \infty$. Since $\mathbf{x}(t)$ was an arbitrary solution of *1*, these facts hold for all solutions of *1*; hence, *1* is asymptotically stable

6 *Reducible systems*

In Chap. *5*, Sec. *4*, we wrote the solution of the system $\dot{\mathbf{x}} = \mathbf{A}\mathbf{x}$ ($\mathbf{A}$ is a constant matrix with distinct eigenvalues) as a linear combination of the eigenvectors of $\mathbf{A}$ and obtained a considerable insight into the dynamics of the system. The new basis vectors $\mathbf{q}_i$ were not time-varying: their lengths and directions were fixed with respect to the original basis since they were eigenvectors of the constant matrix $\mathbf{A}$. Suppose we wish to extend this idea by letting the new basis vectors $\mathbf{q}_i$ be functions of time. In other words, we write $\mathbf{x}(t) = \sum_{i=1}^{n} \eta_i(t)\mathbf{q}_i(t)$, where the $\mathbf{q}_i(t)$ are known vector functions of time. Equivalently, if $\mathbf{y} = (\eta_1, \eta_2, \ldots, \eta_n)$, the above substitution may be written as

1 $$\mathbf{x}(t) = \mathbf{Q}(t)\mathbf{y}(t)$$

where $\mathbf{Q}(t)$ is the matrix whose *i*th column is $\mathbf{q}_i(t)$. In principle, once $\mathbf{Q}(t)$ is known, the stability of a solution $\mathbf{x}(t)$ can be inferred from the behavior of $\mathbf{y}(t)$. However, if no special conditions are imposed on $\mathbf{Q}(t)$, some difficulties may occur. For example, if, as $t \to \infty$, the length of one or more of the $\mathbf{q}_i(t)$ grows without bound, then even though $\mathbf{y}(t) \to \mathbf{0}$ it may happen that $\mathbf{x}(t)$ does not become arbitrarily small. Also if, say, as $t \to \infty$ the n vectors $\mathbf{q}_i(t)$ become coplanar (i.e., in the limit they all belong to the same hyperplane or belong to some subspace of $\mathcal{C}^n$ of dimension $< n - 1$), then it may happen that even though $\mathbf{x}(t) \to \mathbf{0}$, $\mathbf{y}(t)$ remains finite or even grows without bound. These considerations indicate that the transformation matrix $\mathbf{Q}(t)$ should be suitably restricted. For this purpose we introduce the

2 **Definition** The matrix $\mathbf{Q}(t)$ is said to represent a *Lyapunov transformation* (I) if $\mathbf{Q}(t)$ and $\dot{\mathbf{Q}}(t)$ are continuous and bounded on $[t_0, \infty)$ and

(II) if for some positive constant M

3 $$0 < M < |\det \mathbf{Q}(t)| \qquad \forall t \geq t_0$$

4 *Comment* It is easy to see that these conditions are equivalent to the following: (I) $\mathbf{Q}$ and $\dot{\mathbf{Q}}$ are continuous on $[t_0, \infty)$, (II) $\mathbf{Q}$ is nonsingular for all $t \geq t_0$, and (III) $\|\mathbf{Q}(t)\|$ and $\|\mathbf{Q}^{-1}(t)\|$ are bounded for all $t \geq t_0$.

5 **Assertion** The Lyapunov transformations constitute a group.
Proof First, if $\mathbf{Q}$ is a Lyapunov transformation, so is $\mathbf{Q}^{-1}(t)$. From *2* it is immediately apparent that the product of two such transformations is also a Lyapunov transformation.

6 *Exercise* Let $\mathbf{Q}(t)$ be an orthogonal transformation such that for all $t \geq t_0$ both $\mathbf{Q}$ and $\dot{\mathbf{Q}}$ are continuous. Show that it is a Lyapunov transformation.

7 *Exercise* Let $\mathbf{M}$ be a constant matrix. Let all the zeros of its minimal polynomial (*D.2.1*) be simple and purely imaginary. Show that $\exp(\mathbf{M}t)$ is a Lyapunov transformation.

Consider the system

8 $$\dot{\mathbf{x}}(t) = \mathbf{A}(t)\mathbf{x}(t)$$

where $\mathbf{A}(t)$ is continuous in t. Let $\mathbf{Q}(t)$ be a Lyapunov transformation and introduce the transformation *1*; then, by substituting in *8*,

$$\dot{\mathbf{Q}}\mathbf{y} + \mathbf{Q}\dot{\mathbf{y}} = \mathbf{A}\mathbf{Q}\mathbf{y}$$

and, since $\mathbf{Q}(t)$ is nonsingular,

$$\dot{\mathbf{y}} = \mathbf{Q}^{-1}(\mathbf{A}\mathbf{Q} - \dot{\mathbf{Q}})\mathbf{y}$$

If $\mathbf{Q}^{-1}(\mathbf{A}\mathbf{Q} - \dot{\mathbf{Q}}) = \mathbf{C}$ and if $\mathbf{C}$ is a constant matrix, then we say that the system *8* is *reducible*.

9 *Comment* Geometrically, with respect to the moving frame of reference defined by the column vectors of $\mathbf{Q}(t)$, any solution of *8* obeys the differential equation

10 $$\dot{\mathbf{y}} = \mathbf{C}\mathbf{y}$$

which has *constant* coefficients. Sometimes, it is said that *8* and *10* are *kinematically equivalent.*

11 *Comment* If $\mathbf{C} = \mathbf{0}$, the system is said to be *reducible to zero:* with respect to the moving frame of reference any of its solutions have constant components.

12 *Comment* Not all systems such as *8* are reducible. It is interesting, however, that from *6.7.2* it can be easily shown that

13 **Assertion** Every linear system of the form *8* which is periodically varying is reducible. ◁

A very important fact concerning reducible systems is that stated in the

14 **Theorem** If the system *8* is reducible to the system *10*, then the system *10* is stable (asymptotically stable, unstable) if and only if the system *8* is stable (asymptotically stable, unstable, respectively).
Proof Using Definition *2*, the proof is an easy exercise in norm comparison.

15 **Theorem** The system *8* is reducible to zero if and only if there is a constant M such that its state transition matrix satisfies

16 $$\|\mathbf{\Phi}(t,t_0)\| < M < \infty \qquad \|\mathbf{\Phi}(t,t_0)^{-1}\| < M < \infty \qquad \forall t_0, \forall t \geq t_0$$

Proof $\Rightarrow$ Let $\mathbf{Q}(t)$ be a Lyapunov transformation which reduces *8* to zero. Hence, $\dot{\mathbf{Q}}(t) = \mathbf{A}(t)\mathbf{Q}(t)$, and $\mathbf{Q}(t)$ is nonsingular. Hence by *6.2.15* there is a nonsingular constant matrix $\mathbf{K}$ such that $\mathbf{Q}(t) = \mathbf{\Phi}(t,t_0)\mathbf{K}$. Hence

$$\mathbf{\Phi}(t,t_0) = \mathbf{Q}(t)\mathbf{K}^{-1} \qquad \text{and} \qquad \mathbf{\Phi}(t,t_0)^{-1} = \mathbf{K}\mathbf{Q}^{-1}(t)$$

Therefore, by *C.16.13*, we have

$$\|\mathbf{\Phi}(t,t_0)\| \leq \|\mathbf{Q}(t)\|\|\mathbf{K}^{-1}\|$$

and

$$\|\mathbf{\Phi}(t,t_0)^{-1}\| \leq \|\mathbf{K}\|\|\mathbf{Q}(t)^{-1}\|$$

The inequalities *16* follow from the above inequalities and Comment *4*.
$\Leftarrow$ The state transition matrix $\mathbf{\Phi}(t,t_0)$ is continuous in t and

$$\dot{\mathbf{\Phi}}(t,t_0) = \mathbf{A}(t)\mathbf{\Phi}(t,t_0)$$

implies that $\dot{\mathbf{\Phi}}(t,t_0)$ is continuous. By *16*, the elements of both the matrix $\mathbf{\Phi}(t,t_0)$ and its inverse are bounded for $t \geq t_0$; it follows then by Cramer's rule that det $\mathbf{\Phi}(t,t_0)$ is bounded away from zero on $[t_0, \infty)$. By Definition *2*, $\mathbf{\Phi}(t,t_0)$ is a Lyapunov transformation. Finally, we obviously have

$$\mathbf{\Phi}(t,t_0)^{-1}(\dot{\mathbf{\Phi}}(t,t_0) - \mathbf{A}(t)\mathbf{\Phi}(t,t_0)) = \mathbf{0}$$

and hence *8* is reducible to zero.

17 *Exercise* Show that the following four conditions on the system *8* are equivalent:

(I) System *8* is reducible to zero.
(II) $\|\mathbf{\Phi}(t,t_0)\| < M < \infty$ for some M and $\forall t \geq 0$, $\forall t_0 \geq 0$.
(III) System *8* and its adjoint are stable.
(IV) $\|\mathbf{\Phi}(t,t_0)\| < M < \infty$, $\|\mathbf{\Phi}(t,t_0)^{-1}\| < M < \infty$ for some M and $\forall t \geq t_0$.

7 *Stability defined from the input-output point of view*

Definitions *7.2.2* and *7.2.3* are particularly well suited to the study of systems described by differential equations. However, they involve only the zero-input response (free motion) of these systems. In many applications, however, one is more interested in the input-output relationship than in the zero-input response of the system. Let us describe the input-output point of view. First consider the system described by

1 $$\dot{\mathbf{x}}(t) = \mathbf{A}(t)\mathbf{x}(t) + \mathbf{B}(t)\mathbf{u}(t)$$

where $\mathbf{A}(t)$ and $\mathbf{B}(t)$ are $n \times n$ and $n \times r$ matrices whose elements are known, bounded, continuous functions of time. The input is the r-rowed vector $\mathbf{u}$. For simplicity let us assume that the output is the state vector $\mathbf{x}$ rather than some linear function of $\mathbf{x}$ and $\mathbf{u}$.

The concept of stability we are about to define is usually referred to by the expression "any bounded input produces a bounded output." In order to distinguish this concept from the stability i.s.L. of linear differential systems defined in *7.2.2*, we shall refer to the new concept as stability b.i.b.o. It is defined as follows:

2 **Definition** The differential system *1* whose output is $\mathbf{x}$ and whose input is $\mathbf{u}$ is said to be *stable b.i.b.o.* if for all t_0, for all initial states $\mathbf{x}_0$, and for all bounded inputs $\mathbf{u}_{[t_0,\infty)}$, the output $\mathbf{x}(t;\mathbf{x}_0,t_0;\mathbf{u}_{[t_0,\infty)})$ is bounded on $[t_0, \infty)$. $\lhd$

More precisely, the concept of stability b.i.b.o. can be formulated as follows: Let $\mathcal{B}$ be the set of all bounded inputs $\mathbf{u}_{[t_0,\infty)}$, that is,

$$\mathcal{B} = \{\mathbf{u}_{[t_0,\infty)} | \text{ for some } M < \infty, \|\mathbf{u}(t)\| < M, \text{ for all } t \geq t_0\}$$

The system *1* is said to be stable b.i.b.o. if $\forall t_0$, $\forall \mathbf{x}_0 \in \mathcal{C}^n$, and $\forall \mathbf{u}_{[t_0,\infty)} \in \mathcal{B}$ there exists a finite number K independent of t_0, such that

$$\|\mathbf{x}(t;\mathbf{x}_0,t_0;\mathbf{u}_{[t_0,\infty)})\| < K \qquad \forall t \geq t_0$$

The bound K may depend on $\mathbf{x}_0$, on the bound M of the inputs, and on the characteristics of the system.

The stability b.i.b.o. is characterized by the following

3 **Theorem** Consider the differential system described by *1*: let $\mathbf{u}$ be its input and $\mathbf{x}$ be its output. Let $\mathbf{A}(t)$ and $\mathbf{B}(t)$ be continuous and bounded on any interval of the form $[t_0, \infty)$. This system is stable b.i.b.o. if and only if (I) there exists a number M_1 such that

4 $$\|\mathbf{\Phi}(t,t_0)\| < M_1 < \infty \qquad \forall t_0, \forall t \geq t_0$$

and (II) there exists a number M_2 such that

5 $$\int_{t_0}^{t} \|\mathbf{\Phi}(t,t')\mathbf{B}(t')\|_1 \, dt' < M_2 < \infty \qquad \forall t_0, \forall t \geq t_0$$

where, in the integrand of *5*, the subscript 1 indicates that the l_1 norm must be used. (The l_1 norm of a vector in $\mathcal{C}^n$ is $\|\mathbf{x}\|_1 = \sum_{i=1}^{n} |x_i|$; see *C.16.7, C.16.8 et seq.*)

6 *Comment* Instead of defining stability b.i.b.o. for all $\mathbf{x}_0$ (as in Definition *2*) we could have prescribed that $\mathbf{x}_0 = \mathbf{0}$. System *1* would then be called *zero-state stable b.i.b.o.* Theorem *3* would then have the following interpretation: System *1* is stable b.i.b.o. (in the sense of Definition *2* above) if and only if it is zero-state stable b.i.b.o. and uniformly zero-input stable i.s.L.

7 *Comment* Since in $\mathcal{C}^n$ all norms are equivalent, condition *4* holds whichever norm is used. This is not the case, however, for *5*, in which the l_1 norm must be used.

8 *Comment* Equation *4* by itself guarantees that, when $\mathbf{u} = \mathbf{0}$ for all t, the system is stable i.s.L. (see *7.3.2*). The additional condition *5* is required in order to force the matrix product $\mathbf{\Phi}(t,t')\mathbf{B}(t')$ to decrease sufficiently fast that the integrated effect of any bounded input remains bounded.

Proof $\Leftarrow$ The response of *1* is given by (*6.2.21*)

9 $$\mathbf{x}(t;\mathbf{x}_0,t_0;\mathbf{u}_{[t_0,t]}) = \mathbf{\Phi}(t,t_0)\mathbf{x}_0 + \int_{t_0}^{t} \mathbf{\Phi}(t,t')\mathbf{B}(t')\mathbf{u}(t') \, dt'$$

By simplifying the notation of the left-hand side in an obvious manner and using *C.16.12* and *C.16.13*, we obtain

10 $$\|\mathbf{x}(t)\| \leq \|\mathbf{\Phi}(t,t_0)\| \, \|\mathbf{x}_0\| + \int_{t_0}^{t} \|\mathbf{\Phi}(t,t')\mathbf{B}(t')\| \, \|\mathbf{u}(t')\| \, dt'$$

Since $\mathbf{u}_{[t_0,\infty)}$ is bounded, for some constant M_3, $\|\mathbf{u}(t)\| < M_3 < \infty$, $\forall t \geq t_0$. By *4*, *5*, and *10* we get successively

$$\|\mathbf{x}(t)\| < M_1\|\mathbf{x}_0\| + M_3 \int_{t_0}^{t} \|\mathbf{\Phi}(t,t')\mathbf{B}(t')\| \, dt'$$

$$\|\mathbf{x}(t)\| < M_1\|\mathbf{x}_0\| + M_2 M_3 \qquad \forall t \geq t_0$$

Hence *4* and *5* imply that the system is stable b.i.b.o.

$\Rightarrow$ First consider *9* with $\mathbf{u} = \mathbf{0}$. Clearly, for $\mathbf{x}$ to be bounded for all t and for all possible $\mathbf{x}_0$, *4* must hold; for otherwise from the definition of the norm of a linear transformation (*C.16.9*) there is an appropriate choice of $\mathbf{x}_0$ which would make $\mathbf{x}(t;\mathbf{x}_0,t_0;\mathbf{0})$ become arbitrarily large (see the proof of *7.3.1*). Thus stability b.i.b.o. implies *4*. In order to establish *5*, we consider the case $\mathbf{x}_0 = \mathbf{0}$; in other words, the first term on the right-hand side of *9* is zero in the proof that follows. Let us pick the l_1 norm: $\|\mathbf{x}\|_1 = \sum_{i=1}^{n} |x_i|$. The corresponding norm of a matrix $\mathbf{A}$ is then $\|\mathbf{A}\|_1 = \max_j \left(\sum_{i=1}^{n} |a_{ij}| \right)$, (see *C.16.17*). By assumption, $\mathbf{B}(t')$ is bounded for all t'. *4*, which we have just established, states that $\mathbf{\Phi}(t,t')$ is bounded for all $t \geq t'$. Therefore, the only way *5* may be violated is for the integral to become arbitrarily large as $t \to \infty$. For the purpose of a proof by contradiction, suppose that this is so: thus for any number N, however large, there is a T (which depends on N) such that

11
$$\int_{t_0}^{T} \|\mathbf{\Phi}(T,t')\mathbf{B}(t')\|_1 \, dt' > N$$

Let $\mathbf{\Phi}(T,t)\mathbf{B}(t) = \mathbf{\Psi}(T,t)$. *11* implies that for any N there is a T such that

$$\int_{t_0}^{T} \max_j \left(\sum_{i=1}^{n} |\psi_{ij}(T,t')| \right) dt' > N$$

This in turn implies that at least one component of $\mathbf{\Psi}$, say, the (i,j) component, has the property that $\int_{t_0}^{T} |\psi_{ij}(T,t')| \, dt'$ becomes arbitrarily large as $T \to \infty$. This is obvious because if it were not the case, we would have

$$\sup_{T\to\infty} \int_{t_0}^{T} |\psi_{kl}(T,t')| \, dt' < M < \infty \qquad \forall k, l, \text{ for some } M$$

and since $\mathbf{\Psi}$ has only nr components, this would contradict *11*. Therefore *11* implies that there is at least one component ψ_{ij} of $\mathbf{\Psi}$ such that given any number K, however large, there is a T' such that

$$\int_{t_0}^{T'} |\psi_{ij}(T',t)| \, dt > K$$

Pick the following bounded input: $u_l(t) = 0$, $\forall t$ and $\forall l \neq j$, and $u_j(t) = \operatorname{sgn} \psi_{ij}(T',t)$. Clearly from *9* we get

$$\|\mathbf{x}(T')\|_1 \geq |x_i(T')| = \int_{t_0}^{T'} |\psi_{ij}(T',t)| \, dt > K$$

This is a recipe for constructing a bounded input which will produce an output $\mathbf{x}(T')$ whose norm is larger than any preassigned number. In other words, the system is not stable b.i.b.o. This is a contradiction; therefore, stability b.i.b.o. implies *5*. ◁

12 *Comment* Throughout this section we have restricted ourselves to a differential system whose output is its state. In many practical instances, this is not so; usually, the output is a linear function of the state and the input. An analysis that would cover this general case

would be too involved. With the theory just developed, the reader will have no difficulty in using the specific features of the system under consideration to develop the necessary and sufficient conditions for the stability b.i.b.o. of that system.

13 *Exercise* Consider a linear time-invariant differential system defined by *1*. Suppose that $\mathbf{x}$ is the output of the system. Show that if such a system is asymptotically stable, then it is stable b.i.b.o. Show by an example that the converse is not true.

14 *Exercise* Give an example of a time-varying system which is asymptotically stable and which is not stable b.i.b.o. [*Hint:* Consider the scalar system $\dot{x} = -(1/2t)x + u(t)$, $t \geq 1$. $\Phi(t,t_0) = \sqrt{t_0/t}$.]

Theorem *3* has given a characterization of the stability b.i.b.o. If we put a few additional restrictions on $\mathbf{B}(t)$ of *1*, namely, that $\mathbf{B}(t)$ is a bounded $n \times n$ matrix whose inverse is bounded for all t, we can prove the rather surprising fact that the norm of the state transition matrix, $\|\mathbf{\Phi}(t,t_0)\|$, must decrease at least as fast as an exponential in $(t - t_0)$ in order that the system *1* be stable b.i.b.o., and conversely. We shall use a method of proof due to Kalman. More precisely, this fact is formulated as a

15 **Theorem** Consider the system *1*, where the input $\mathbf{u}$ and the output $\mathbf{x}$ are both n-vectors. Let the $n \times n$ matrices $\mathbf{A}(t)$ and $\mathbf{B}(t)$ be continuous and bounded for all t; that is,

16 $$\|\mathbf{B}(t)\|_1 \leq M_1 < \infty \qquad \forall t$$

Assume also that $\mathbf{B}(t)$ is nonsingular and that its inverse is bounded by M_2; that is,

17 $$\|\mathbf{B}^{-1}(t)\|_1 \leq M_2 < \infty \qquad \forall t$$

Then, system *1* is stable b.i.b.o. if and only if there are positive constants M_3 and M_4 such that

18 $$\|\mathbf{\Phi}(t,t_0)\|_1 \leq M_3 e^{-M_4(t-t_0)} \qquad \forall t_0,\ \forall t \geq t_0$$

Proof $\Leftarrow$ The response of the system to $\mathbf{u}_{[t_0,\infty)}$ starting from the state $\mathbf{x}_0$ at t_0 is given by *9*. As above, *9* implies *10*. If we call M the bound on the input $\mathbf{u}$, then by *16*, *18*, and the properties of the norm (see *C.16.10* and *C.16.13*), we get

$$\|\mathbf{x}(t;\mathbf{x}_0,t_0;\mathbf{u}_{[t,_0\infty)})\|_1 \leq M_3\|\mathbf{x}_0\|_1 + \int_{t_0}^{t} M_3 e^{-M_4(t'-t_0)} M_1 M \, dt'$$

In order to get an upper bound on the right-hand side, let us replace the upper limit of integration by ∞; then an elementary integration gives

$$\|\mathbf{x}(t;\mathbf{x}_0,t_0;\mathbf{u}_{[t_0,\infty)})\|_1 \leq M_3\|\mathbf{x}_0\| + \frac{M_1 M_3 M}{M_4} \qquad \forall t \geq t_0$$

The right-hand side is a constant independent of t_0 which plays the role of the constant K which was used in the definition of stability b.i.b.o. (see *2*). Therefore, *18* implies stability b.i.b.o.

$\Rightarrow$ The second part of the proof consists of two steps: First we use the assumed stability b.i.b.o. and the assumed properties of $\mathbf{B}(t)$ to prove inequality *22* below; next, using techniques due to Kalman, we show that *19*, which is a consequence of the stability b.i.b.o., and *22* together with the group property of the state transition matrix (see *6.2.16*) imply *18*.

By assumption the system is stable b.i.b.o. Therefore, by Theorem *3*, there is a constant M_5 such that

19 $$\|\mathbf{\Phi}(t,t_0)\|_1 \leq M_5 < \infty \qquad \forall t_0, \forall t \geq t_0$$

Now, for any pair t, t_0 with $t \geq t_0$, let us consider the zero-state response of *1*. By Theorem *3*, the stability b.i.b.o. of *1* implies that for some constant M_6,

20 $$\int_{t_0}^{t} \|\mathbf{\Phi}(t,t')\mathbf{B}(t')\mathbf{u}(t')\|_1 \, dt' \leq M_6 < \infty \qquad \forall t_0, \forall t \geq t_0$$

For all t', $\mathbf{B}(t')$ is an $n \times n$ nonsingular matrix which, by *17*, has a bounded inverse. Therefore, for any integer j, we can find a control $\mathbf{u}(t')$ such that $\|\mathbf{u}(t')\|_1 = M$ and such that the vector $\mathbf{B}(t')\mathbf{u}(t')$ has all its components equal to zero except the jth component. Call this jth component $b_j(t')$. Furthermore, by *17*, we have

$$|b_j(t')| \geq \|\mathbf{u}\|_1 \, \|\mathbf{B}^{-1}\|_1^{-1} \geq MM_2^{-1}$$

Now, for that $\mathbf{u}$,

21 $$\|\mathbf{\Phi}(t,t')\mathbf{B}(t')\mathbf{u}(t')\|_1 = \sum_k |\phi_{kj}(t,t')b_j(t')| \geq |\phi_{kj}(t,t')|MM_2^{-1}$$

Clearly, this inequality holds for all $k,j = 1, 2, \ldots, n$. Since $\|\mathbf{\Phi}(t,t_0)\|_1 = \max\limits_j \sum\limits_k |\phi_{kj}(t,t_0)|$ (see *C.16.17*), we obtain from *20* and *21*

22 $$\int_{t_0}^{t} \|\mathbf{\Phi}(t,t')\|_1 \, dt' \leq M_7 < \infty \qquad \forall t_0, \forall t \geq t_0$$

where M_7 can be taken to be nM_2M_6/M.

Now from *19* and *22* we have

$$\begin{aligned} \infty > M_5M_7 &\geq \int_{t_0}^{t} \|\mathbf{\Phi}(t,t')\|_1 \, \|\mathbf{\Phi}(t',t_0)\|_1 \, dt' \\ &\geq \int_{t_0}^{t} \|\mathbf{\Phi}(t,t')\mathbf{\Phi}(t',t_0)\|_1 \, dt' \geq \int_{t_0}^{t} \|\Phi(t,t_0)\|_1 \, dt' \end{aligned}$$

where we have used the properties of the norm and the group property of the state transition matrix. Thus, finally,

$$\|\mathbf{\Phi}(t,t_0)\|_1 \leq \frac{M_5 M_7}{t - t_0} \qquad \forall t_0,\ \forall t \geq t_0$$

Let us now define the time T by the condition

$$\frac{M_5 M_7}{T} = \frac{1}{2}$$

Consequently,

23 $$\|\mathbf{\Phi}(t_0 + T,\ t_0)\|_1 \leq \tfrac{1}{2} \qquad \forall t_0$$

By the group property of the state transition matrix, the properties of the norm, and *23*, we have

$$\begin{aligned}\|\mathbf{\Phi}(t_0 + kT,\ t_0)\|_1 &= \|\mathbf{\Phi}(t_0 + kT,\ t_0 + (k-1)T) \cdots \mathbf{\Phi}(t_0 + T,\ t_0)\|_1 \\ &\leq 2^{-k} \qquad \forall \text{ integers } k > 0,\ \forall t_0\end{aligned}$$

Consider an arbitrary $t \geq t_0$; let k be the largest integer in $(t - t_0)/T$; define τ by $t = t_0 + kT + \tau$. (By the definition of k, $0 \leq \tau < T$.) Using successively the group property of the state transition matrix, the properties of the norm *19*, and the above inequality we get

24 $$\begin{aligned}\|\mathbf{\Phi}(t,t_0)\|_1 &= \|\mathbf{\Phi}(t_0 + kT + \tau,\ \|t_0 + kT)\mathbf{\Phi}(t_0 + kT,\ \|t_0)\|_1 \\ &\leq \|\mathbf{\Phi}(t_0 + kT + \tau,\ \|t_0 + kT)\|_1 \ \ \|\mathbf{\Phi}(t_0 + kT,\ \|t_0)\|_1 \\ &\leq \frac{M_5}{2^k}\end{aligned}$$

where k is the largest integer in $(t - t_0)/T$.

This inequality has the following interpretation which is valid for all t_0: the scalar $\|\mathbf{\Phi}(t,t_0)\|_1$ (considered as a function of t) is bounded by M_5 over $[t_0,\ t_0 + T]$; by *24*, the same scalar is bounded by $M_5/2$ over $[t_0 + T,\ t_0 + 2T]$; more generally, for any positive integer k, that scalar is bounded by $M_5/2^k$ over $[t_0 + (k-1)T,\ t_0 + kT]$. Therefore if we define the positive number M_9 by $e^{-M_9 T} = \frac{1}{2}$ we have

$$\|\mathbf{\Phi}(t,t_0)\|_1 \leq M_5 e^{-M_9(t - t_0 - T)} \qquad \forall t_0,\ \forall t \geq t_0$$

This last inequality is of the same form as *18*. ◁

Let us now establish a connection between the concepts of stability based on the properties of the zero-input response (free motion) and those based on the input-output relationship. For this purpose we shall need a stability concept slightly more restrictive than those specified by *7.2.2* and *7.2.3*.

25 **Definition** The linear system $\dot{\mathbf{x}} = \mathbf{A}(t)\mathbf{x}$ is said to be *uniformly asymptotically stable* if (I) the system is uniformly stable i.s.L. and (II) for any fixed R, however large, $\|\mathbf{x}_0\| < R$ implies that $\mathbf{x}(t;\mathbf{x}_0,t_0) \to \mathbf{0}$ as $t \to \infty$, uniformly in t_0. ◁

More precisely, the conditions are: (I) for any $\varepsilon > 0$, there is a $\delta > 0$, independent of t_0, such that $\|\mathbf{x}_0\| \leq \delta \Rightarrow \|\mathbf{x}(t;\mathbf{x}_0,t_0)\| < \varepsilon$ $\forall t \geq t_0$, $\forall t_0$; and (II) for any $R > 0$ and any $\eta > 0$, there is a T, independent of t_0 (but which may depend on R and η), such that

$$\|\mathbf{x}_0\| \leq R \Rightarrow \|\mathbf{x}(t;\mathbf{x}_0,t_0)\| < \eta \qquad \forall t > t_0 + T, \forall t_0$$

26 *Comment* Since the system is linear, the uniform stability implies that any free motion is uniformly bounded. If we were to define the uniform asymptotic stability of the zero state for a nonlinear system, we would have to add the requirement that any free motion be uniformly bounded because, in the case of nonlinear systems, uniform stability of the zero state has only purely local implications (see Remark *7.2.6*).

27 *Comment* Let us exhibit an implication of the uniform asymptotic stability defined in *25*. Let us use l_1 norms and pick $R = 1$ and $\eta = \frac{1}{2}$. Since $\mathbf{x}(t;\mathbf{x}_0,t_0) = \mathbf{\Phi}(t,t_0)\mathbf{x}_0$, it follows from Definition *25* and from the definition of the norm of an L.T. that there exists a T', independent of t_0, such that

28 $$\|\mathbf{\Phi}(t_0 + T', t_0)\|_1 \leq \tfrac{1}{2} \qquad \forall t_0$$

As a first step, let us establish a preliminary

29 **Assertion** The linear system $\dot{\mathbf{x}} = \mathbf{A}(t)\mathbf{x}$ is uniformly asymptotically stable if and only if there exist two positive constants K_1 and K_2 such that

30 $$\|\mathbf{\Phi}(t,t_0)\|_1 \leq K_1 e^{-K_2(t-t_0)} \qquad \forall t \geq t_0, \forall t_0$$

Proof $\Rightarrow$ Uniform asymptotic stability implies uniform stability; hence, if we let $K_1 \triangleq \delta/\varepsilon$ and if, as in Comment *27*, we use l_1 norms, we get

$$\|\mathbf{\Phi}(t,t_0)\|_1 \leq K_1 \qquad \forall t \geq t_0, \forall t_0$$

Furthermore, uniform asymptotic stability implies *28*; hence by the last steps of the proof of Theorem *15*, *30* follows.

$\Leftarrow$ Follows immediately from *30* and $\mathbf{x}(t;\mathbf{x}_0,t_0) = \mathbf{\Phi}(t,t_0)\mathbf{x}_0$. $\triangleleft$

If now we combine the equivalences established by Theorem *15* and by Assertion *29*, we obtain the

31 **Theorem** Consider the linear system *1*, where the input $\mathbf{u}$ and the output $\mathbf{x}$ are both n-vectors. Let the matrices $\mathbf{A}(t)$ and $\mathbf{B}(t)$ be continuous and bounded for all t. Let us assume that *16* and *17* hold. Then, the linear system *1* is stable b.i.b.o. if and only if it is uniformly asymptotically stable.

REFERENCES AND SUGGESTED READING

1 Bellman, R.: "Stability Theory of Differential Equations," McGraw-Hill Book Company, Inc., New York, 1953.

2 Cesari, L.: "Asymptotic Behavior and Stability Problems in Ordinary Differential Equations," Springer Verlag OHG, Berlin, 1959.

3 Hahn, W.: "Theory and Application of the Direct Method of Lyapunov," Springer Verlag OHG, Berlin, 1959.

4 Kalman, R. E., and J. E. Bertram: Control System Analysis and Design via the Second Method of Lyapunov, *Trans. A.S.M.E., J. Basic Engrg.*, vol. 82, series D, 2, pp. 371–400, 1960.

5 LaSalle, J., and S. Lefschetz: "Stability by Lyapunov's Direct Method with Applications," Academic Press, Inc., New York, 1961.

6 Lefschetz, S.: "Differential Equations: Geometric Theory," Interscience Publishers, Inc., New York, 1957.

7 Sansone, G., and R. Conti: "Equazioni Differenziali Nonlineari," Edizioni Cremonese, Rome, 1956.

8 Bridgland, T. F., Jr.: Stability of Linear Transmission Systems, *Soc. Indust. Appl. Math. Rev.*, vol. 5, no. 1, pp. 7–32, 1963.

Impulse response of nondifferential linear systems

8

1 *Introduction*

In the preceding three chapters we have studied in detail the properties of both the zero-input and the zero-state response of linear, not necessarily time-invariant, differential systems. As was stated in Chap. 3 (*3.6.5*), the zero-state response of any nondifferential[1] nonanticipative linear system admits of the representation $\int_{t_0}^{t} h(t,\xi)u(\xi)\,d\xi$, where the impulse response $h(t,\xi)$ denotes the zero-state response of the system at time t to a unit impulse $\delta(t-\xi)$ applied at time ξ. In the case of an anticipative system (*1.8.13*), $h(t,\xi)$ does not vanish identically for $t < \xi$, as it does in the case of a nonanticipative system, and hence the expression for the zero-state response to $u \triangleq u_{(t_0,t]}$ must in general be written as $\int_{-\infty}^{+\infty} h(t,\xi)u(\xi)\,d\xi$.

Although the impulse response describes only the zero-state response of a system and hence can be used only to define the system to within zero-state equivalence, it is of considerable practical importance for a number of reasons. In the first place, in the case of nondifferential systems, the state space is, usually, infinite-dimensional and as a consequence it is usually much more difficult to characterize the system zero-input response than the zero-state response. [Among familiar examples of such systems are the ideal low-pass filter (see Chap. *9*, Sec. *5*), circuits containing transmission lines, ideal delay lines and wave guides, and control systems with transportation lags. In this connection it should be recalled that any nondifferential system whose impulse response is a reasonably well-behaved function of t and ξ can be approximated arbitrarily closely—in the sense of zero-state equivalence—by a differential system of sufficiently high order.

[1] The term "nondifferential" is used here in its wide sense to describe a system which is *not necessarily* differential, rather than one which is *not* differential. See *3.8.26* for the definition of a differential system.

See *9.16.2* for a more detailed discussion of this type of approximation in the case of time-invariant systems.] In the second place, there are many practical situations where the only data available about a system are measurements of its transfer function or its impulse response. (Knowing the impulse response, it is a routine matter to compute the transfer function, and vice versa, with the aid of a digital computer.) Frequently the impulse response is the only datum we are interested in because the rapid decay with time of the zero-input response makes its effect quite negligible in the analysis of many problems.

This is the motivation for focusing our attention in this chapter on those properties of nondifferential linear systems which involve only their impulse response. For simplicity, we shall restrict our attention to single-input single-output systems, since it is an easy matter to extend most of the results obtained under this assumption to multiple-input multiple-output systems.

In the following section, we consider the impulse response of the tandem connection of two time-varying systems $\mathcal{A}$ and $\mathcal{B}$ and show that, in general, $\mathcal{A}\mathcal{B}$ is not zero-state equivalent to $\mathcal{B}\mathcal{A}$. In Sec. *3*, we introduce a general definition of the adjoint of systems defined by their impulse response and show that it reduces to the conventional definition in the case of differential systems. In Sec. *4*, the problem of stability is met: Since the impulse response characterizes the system only as far as its zero-state response is concerned and since we do not have a state description of the system, we can define a stability only in terms of the zero-state response of the system to the class of all bounded inputs. Finally, an example shows that there is no simple relationship between this concept of stability and the concepts based on the state concept.

2 *Systems in tandem*

Consider two linear systems $\mathcal{A}$ and $\mathcal{B}$. Suppose that the only data available concerning the systems $\mathcal{A}$ and $\mathcal{B}$ are their respective impulse responses h_A and h_B. As in *1.10.23*, we denote by $\mathcal{B}\mathcal{A}$ the tandem connection of $\mathcal{A}$ and $\mathcal{B}$, with $\mathcal{A}$ preceding $\mathcal{B}$; thus $\mathcal{B}\mathcal{A}$ is the overall system obtained by connecting the output of $\mathcal{A}$ to the input of $\mathcal{B}$ (see Fig. *8.2.1*). Using the notation of the figure, the zero-state

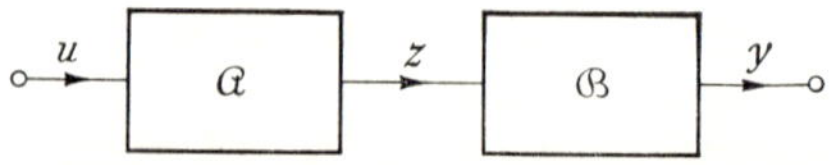

Fig. 8.2.1 The system $\mathcal{B}\mathcal{A}$, the tandem connection of $\mathcal{A}$ preceding $\mathcal{B}$.

response of these systems is given by

1 $$z(t) = \int_{-\infty}^{+\infty} h_A(t,\xi)u(\xi)\, d\xi$$

2 $$y(t) = \int_{-\infty}^{+\infty} h_B(t,\eta)z(\eta)\, d\eta$$

Let us eliminate z between *1* and *2* and interchange the order of integration;[1] we thereby obtain

3 $$y(t) = \int_{-\infty}^{+\infty} \left[\int_{-\infty}^{+\infty} h_B(t,\eta)h_A(\eta,\xi)\, d\eta \right] \mu(\xi)\, d\xi$$

The bracketed term, which is a function of t and ξ, is the weighting function which relates the zero-state response of $\mathcal{BA}$ to its input u; it is therefore the impulse response of $\mathcal{BA}$. We can therefore state the

4 **Assertion** The tandem connection $\mathcal{BA}$ (where $\mathcal{A}$ precedes $\mathcal{B}$) has an impulse response given by

5 $$h_{BA}(t,\xi) = \int_{-\infty}^{+\infty} h_B(t,\eta)h_A(\eta,\xi)\, d\eta$$

In the time-invariant case, *5* reduces to a convolution

$$h_{BA}(t-\xi) = \int_{-\infty}^{+\infty} h_B(t-\eta)h_A(\eta-\xi)\, d\eta$$

as can easily be seen by putting $t - \xi = \tau$ and $\eta - \xi \rightarrow \xi$; thus $h_{BA} = h_B * h_A$, where the $*$ denotes the operation of convolution (see *B.3.8*). From this relation (or also from its Laplace transform $H_{BA} = H_B H_A$) we immediately have $h_{BA} = h_{AB}$. Hence we state the

6 **Corollary** If $\mathcal{A}$ and $\mathcal{B}$ are time-invariant, the tandem connections $\mathcal{BA}$ and $\mathcal{AB}$ are zero-state equivalent:

$$\mathcal{AB} \doteq \mathcal{BA}$$

7 **Remark** For the time-varying case, $\mathcal{AB}$ is not zero-state equivalent to $\mathcal{BA}$. Consider the following example. Let

$$\mathcal{A}:\quad u \rightarrow z \qquad\qquad z(t) = p\,u(t)$$

$$\mathcal{B}:\quad v \rightarrow y \qquad\qquad y = tv$$

where the notation $u \rightarrow z$ indicates that u is the input and that z is the zero-state response due to u. Because we restrict ourselves to zero-state response, $\mathcal{A}$ defines a function which maps every u of the input function space into one and only one zero-state response z. Similarly, $\mathcal{B}$ maps the input v into the output y.

[1] The interchange of the order of integrations is not always permissible. Conditions under which the order of integration may be changed can be found in T. M. Apostol, "Mathematical Analysis," p. 448, Addison-Wesley Publishing Company, Inc., Reading, Mass., 1957.

The system $\mathcal{BA}$ (where $\mathcal{A}$ precedes $\mathcal{B}$) is characterized by

$$y_{BA} = t\, p\, u$$

whereas the system $\mathcal{AB}$ is characterized by

$$y_{AB} = p\,(tu) = u + t\, p\, u$$

8 *Exercise* Let $\mathcal{A}$ be time-varying and

$$h_A(t,\xi) = 1(t - \xi)(2 + \cos t) \exp\,[-(t - \xi)]$$

Let $\mathcal{B}$ be time-invariant and $h_B(t,\xi) = 1(t - \xi) - 1(t - \xi - 2)$. Show that $\mathcal{AB}$ and $\mathcal{BA}$ are not zero-state equivalent.

3 *Adjoint systems*

In this section we wish to develop the concept of the adjoint system of a linear system. This is an extension of the concept of the adjoint of a system of differential equations, which was introduced in *6.2.27* and *6.2.53*. The first step, however, is to introduce the idea of the zero-state inverse by the definition (compare with *2.10.6*):

1 **Definition** The system $\mathcal{A}^{-1}$ is said to be the *zero-state inverse* of the linear system $\mathcal{A}$ if, for all ξ, the zero-state response of $\mathcal{A}^{-1}\mathcal{A}$ and that of $\mathcal{A}\mathcal{A}^{-1}$ to a unit impulse applied at ξ are $\delta(t - \xi)$.

2 *Comment* Intuitively, we think of $\mathcal{A}^{-1}$ as canceling out the effect of $\mathcal{A}$: the system $\mathcal{A}$ spreads out the unit impulse along the time axis and $\mathcal{A}^{-1}$ reconcentrates the resulting output into a delta function. In practice, the concept of zero-state inverse is applied within an approximation for the following reason: The impulse response $h_A(t,\xi)$ of a nonanticipative system $\mathcal{A}$ usually differs from zero for $t > \xi$, that is, after the application of the impulse; thus the definition of the zero-state inverse requires $\mathcal{A}^{-1}$ to respond at time ξ to inputs which will be applied to $\mathcal{A}^{-1}$ at times later than ξ. In other words, if $\mathcal{A}$ is nonanticipative and if $h_A(t,\xi) \neq 0$ for some $t > \xi$, then $\mathcal{A}^{-1}$ will, in general, be anticipative and hence not physically realizable.[1] In practice, one builds systems that are approximately zero-state inverses of one another in the sense that the output of the tandem connection is a delayed delta function. For example, in amplitude-modulation communication systems any equalization network is, in this sense, the

[1] $\mathcal{A}^{-1}$ need not always be anticipative. For example, let $\mathcal{A}$ have $1/(s + 1)$ as a transfer function; then its zero-state inverse has $s + 1$ as a transfer function and $\delta^{(1)} + \delta$ as an impulse response. On the other hand, if $\mathcal{A}$ is an *RLC* circuit which contains delay lines, $\mathcal{A}^{-1}$ will then necessarily be anticipative, except, of course, for some trivial cases.

approximate zero-state inverse of that part of the system which requires equalization. Similarly, in a chirp radar system, the transmitter spreads out the radar pulse in time and the receiver is designed to reconcentrate it into a sharp pulse, ideally, a delta function.

3 **Remark** Definition *1* requires justification in that it is not a priori obvious (see *2.10.20*) that the same system $\mathcal{A}^{-1}$ may be connected either before $\mathcal{A}$ or after $\mathcal{A}$ and that the two resulting tandem connections are zero-state equivalent. Suppose that we have two linear systems $\mathcal{A}_R^{-1}$, $\mathcal{A}_L^{-1}$ such that $\mathcal{A}\mathcal{A}_R^{-1}$ ($\mathcal{A}_R^{-1}$ preceding $\mathcal{A}$) and $\mathcal{A}_L^{-1}\mathcal{A}$ ($\mathcal{A}$ preceding $\mathcal{A}_L^{-1}$) are zero-state equivalent to the identity operator. Consider then the tandem connections $\mathcal{A}_L^{-1}\mathcal{A}\mathcal{A}_R^{-1}$ (that is, $\mathcal{A}_R^{-1}$ precedes $\mathcal{A}$, which itself precedes $\mathcal{A}_L^{-1}$). Combine $\mathcal{A}$ and $\mathcal{A}_R^{-1}$; by definition this tandem connection is zero-state equivalent to the identity operator; hence $\mathcal{A}_L^{-1}\mathcal{A}\mathcal{A}_R^{-1}$ is zero-state equivalent to $\mathcal{A}_L^{-1}$. Similarly, by combining $\mathcal{A}_L^{-1}$ and $\mathcal{A}$ we can show that $\mathcal{A}_L^{-1}\mathcal{A}\mathcal{A}_R^{-1}$ is zero-state equivalent to $\mathcal{A}_R^{-1}$. Consequently, $\mathcal{A}_L^{-1}$ and $\mathcal{A}_R^{-1}$ are zero-state equivalent to one another. Another way of expressing this fact is to say that any left inverse of $\mathcal{A}$ (such as $\mathcal{A}_L^{-1}$) is necessarily zero-state equivalent to any right inverse of $\mathcal{A}$ (such as $\mathcal{A}_R^{-1}$).

4 *Exercise* Let $h(t,\xi)$ and $g(t,\xi)$ be respectively the impulse responses of a linear system $\mathcal{A}$ and its zero-state inverse $\mathcal{A}^{-1}$. Show that

$$\int_{-\infty}^{+\infty} g(t,\eta)h(\eta,\xi)\,d\eta = \delta(t-\xi) \qquad \forall t, \forall \xi$$

where the integral must be interpreted formally (see Appendix *A*). ◁

We now turn to the general definition of adjoint systems. In *6.2.27* we defined the adjoint of a set of differential equations. Using that definition, one could define the adjoint of a system. Such a definition would necessarily be restricted to differential systems. We shall introduce a definition of adjoint systems that is applicable to a much broader class of systems, namely, those characterized by their impulse response. We shall show that this more general definition reduces to the previous one for differential systems.

5 **Definition** Let the linear system $\mathcal{A}$ have a real-valued impulse response $h(t,\xi)$. The linear system $\mathcal{A}^{(a)}$ is said to be the *adjoint* of $\mathcal{A}$ if its impulse response $h^{(a)}(t,\xi)$ satisfies the relation

6 $$h^{(a)}(t,\xi) = h(\xi,t) \qquad \forall t, \forall \xi$$

where $h^{(a)}(t,\xi)$ is the response of $\mathcal{A}^a$ at time t to a unit impulse applied at time ξ. Thus *6* implies that $\mathcal{A}^{(a)}$ has a response at time t to a unit impulse applied at ξ equal to the response of $\mathcal{A}$ at time ξ to a unit impulse applied at time t. This definition implies that if $\mathcal{A}$ is nonanticipative [that is, $h(t,\xi) = 0$ for all $t < \xi$], then $\mathcal{A}^a$ is purely anticipative [that is, $h^{(a)}(t,\xi) = 0$ for all $t > \xi$.]

7 **Theorem** For any linear system, the zero-state inverse of its adjoint is zero-state equivalent to the adjoint of its zero-state inverse. Symbolically,

$$(\mathcal{A}^{(a)})^{-1} \doteq (\mathcal{A}^{-1})^{(a)}$$

Proof Let $h(t,\xi)$ and $g(t,\xi)$ be respectively the impulse response of $\mathcal{A}$ and that of $\mathcal{A}^{-1}$. Let us use the superscript (a) to denote the impulse responses of the adjoints. From the definition of inverse systems and *4* we have

$$\int_{-\infty}^{+\infty} h(\eta,\xi)g(t,\eta)\,d\eta = \delta(t-\xi) \qquad \forall t, \forall \xi$$

Using the defining relation *6* between $h^{(a)}$, $g^{(a)}$ and h, g, we get

$$\int_{-\infty}^{+\infty} h^{(a)}(\xi,\eta)g^{(a)}(\eta,t)\,d\eta = \delta(t-\xi) \qquad \forall t, \forall \xi$$

Interchanging t and ξ and observing that the delta function is unchanged, we get

$$\int_{-\infty}^{+\infty} h^{(a)}(t,\eta)g^{(a)}(\eta,\xi)\,d\eta = \delta(t-\xi) \qquad \forall t, \forall \xi$$

From *4*, this relation expresses the fact that $(\mathcal{A}^{-1})^{(a)}$ is the zero-state inverse of $\mathcal{A}^{(a)}$. ◁

As a first step in establishing the equivalence of the definitions of adjoint systems, consider a system $\mathcal{A}$ with input u and output y. Let its equation be

8 $$\mathcal{A}: \quad u \to y \qquad\qquad L(p,t)u = y$$

with

9 $$L(p,t) \triangleq \sum_{k=0}^{n} a_{n-k}(t)\,p^k$$

where $a_{n-k} \in C^k$, for $k = 0, 1, \ldots, n$. The impulse response of $\mathcal{A}$ is obviously

$$h(t,\xi) = \sum_{k=0}^{n} a_{n-k}(t)\delta^{(k)}(t-\xi)$$

where $\delta^{(k)}$ is the kth derivative of δ (see *A.2.5* and *A.5.4*). By Definition *5* the impulse response of the adjoint system $\mathcal{A}^{(a)}$ is given by

$$h^{(a)}(t,\xi) = \sum_{k=0}^{n} a_{n-k}(\xi)\delta^{(k)}(\xi-t)$$

Let z be the zero-state response of $\mathcal{A}^{(a)}$ to an input r; then

$$\begin{aligned} z(t) &= \int_{-\infty}^{+\infty} h^{(a)}(t,\xi)r(\xi)\,d\xi \\ &= \sum_{k=0}^{n} \int_{-\infty}^{+\infty} \delta^{(k)}(\xi - t)a_{n-k}(\xi)r(\xi)\,d\xi \\ &= \sum_{k=0}^{n} (-1)^k\, p^k\, (a_{n-k}(t)r(t)) \end{aligned}$$

where we used *A.2.11*. Thus, the output z of $\mathcal{A}^{(a)}$ may be obtained by operating on the input r by the differential operator L^*:

10 $\mathcal{A}^{(a)}\colon\ r \to z \qquad L^*(p,t)r = z$

where

11 $$L^*(p,t)r = \sum_{k=0}^{n} (-1)^k\, p^k\, (a_{n-k}(t)r)$$

Observe that the differential operator L^* is the adjoint of the differential operator L (see *6.2.53*).

Consider now the systems $\mathcal{A}^{-1}$ and $(\mathcal{A}^{(a)})^{-1}$, where $\mathcal{A}$ is defined by *8*. Let y be the input of $\mathcal{A}^{-1}$ and u its output. Then $\mathcal{A}^{-1}$ is characterized by (see Chap. *4*, Secs. *3* and *4*)

12 $\mathcal{A}^{-1}\colon\ y \to u \qquad L(p,t)u = y$

This result applied to the adjoint system $\mathcal{A}^{(a)}$ gives the following characterization for its inverse:

13 $(\mathcal{A}^{(a)})^{-1}\colon\ y \to v \qquad L^*(p,t)v = y$

Thus by Theorem *7*, the adjoint of the system *12* is the system *13*, where $L(p,t)$ and $L^*(p,t)$ are respectively given by *9* and *11*.

It is important to observe that in the system interpretation of the systems *12* and *13*, the system *12* in nonanticipative, whereas the system *13* is purely anticipative.

Finally, observing that this result coincides with that of *6.2.53*, we conclude with the

14 **Assertion** The general definition *5* of an adjoint system reduces to the usual one of *6.2.53* when applied to systems described by

$\mathcal{A}\colon\ u \to y \qquad L(p,t)y = u$

4 Zero-state stability

In Chap. 7, which was devoted to a detailed study of the stability of differential systems, three concepts of stability were introduced. Stability in the sense of Lyapunov and asymptotic stability are concepts which put restrictions on the motion of the *state* when the input is identically zero. The third concept, that of stability b.i.b.o., puts restrictions on the motion of the *state,* but the initial state is arbitrary and the input is only required to be bounded. The systems that we are considering in this chapter are specified only by their impulse response, so that we can only define a concept of stability which imposes restrictions on the *zero-state response* of the system. We shall require that the zero-state response remain bounded for all possible bounded inputs. For that reason we shall use the abbreviation b.i.b.o. We introduce the

1 **Definition** A linear system is said to be *zero-state stable (b.i.b.o.)* if its zero-state response to any bounded input u is bounded. ◁

To formulate this definition more precisely, let $A(t;\mathbf{0};u)$ be the zero-state response, at time t, of the system $\mathcal{A}$ to the input u. We do not write $u_{[t_0,\infty)}$, because we do not want to assume that the input was applied at some finite time t_0. With this notation the definition becomes

2 **Definition** $\mathcal{A}$ is said to be zero-state stable (b.i.b.o.) if and only if, for any input u such that $|u(t)| \leq M$, $\forall t \in (-\infty, \infty)$ and for some finite M, there is another constant M_0 (which depends on M) such that

3 $$|A(t;\mathbf{0};u)| \leq M_0 < \infty \qquad \forall t \in (-\infty, \infty)$$

4 *Comment* Whereas all the definitions of stability given in Chap. 7 apply only to differential systems which are nonanticipative, the present definition applies to anticipative systems just as well as to purely nonanticipative systems (see *1.8.13*).

5 *Exercise* Suppose the system is time-invariant and is defined by $h(t) = 1(t + 1) - 1(t - 1)$. Show that Definition *1* implies that it is zero-state stable b.i.b.o. In fact, show that M_0 can be taken to be $2M$.

6 *Exercise* Suppose again that the system is time-invariant and that $h(t) = 1(t) \cos \omega_0 t$. Show that it is not zero-state stable b.i.b.o. [*Hint:* Consider the input $1(t) \cos \omega_0 t$.]

Zero-state stable systems are characterized by the

7 **Theorem** Let $h(t,\xi)$ be the impulse response of the system $\mathcal{A}$. Then, $\mathcal{A}$ is zero-state stable b.i.b.o. if and only if there is a finite constant N such that

8 $$\int_{-\infty}^{+\infty} |h(t,\xi)|\, d\xi \leq N \qquad \forall t \in (-\infty, \infty)$$

Proof $\Leftarrow$ Consider an arbitrary bounded input u such that $|u(t)| < M$ for all $t \in (-\infty, \infty)$. The zero-state response of $\mathcal{A}$ to u is given by (see *3.6.3*)

9 $$A(t;\mathbf{0};u) = \int_{-\infty}^{+\infty} h(t,\xi)u(\xi)\, d\xi$$

We then obtain successively

$$|A(t;\mathbf{0};u)| \leq \int_{-\infty}^{+\infty} |h(t,\xi)u(\xi)|\, d\xi$$
$$< M \int_{-\infty}^{+\infty} |h(t,\xi)|\, d\xi$$

and, by *8*,

$$|A(t;\mathbf{0};u)| < MN \qquad \text{for all } t \in (-\infty, \infty)$$

Thus any bounded input u produces a zero-state response which is bounded for all t; hence $\mathcal{A}$ is zero-state stable b.i.b.o.

$\Rightarrow$ Let us start by an observation: the integral *8* defines a real-valued function f of t on $(-\infty, \infty)$. Now $\sup_{-\infty < t < \infty} |f(t)|$ is either finite or infinite.[1] If it is finite, we can take the supremum as the value for the constant N in *8*. The supremum is infinite if and only if the inequality *8* cannot hold for any finite N.

Thus for the purpose of a proof by contradiction, we assume that the system is zero-state stable b.i.b.o. and that *8* cannot hold for any finite N. In other words, we assume that

$$\sup_{-\infty < t < \infty} |f(t)| = \sup_{-\infty < t < \infty} \int_{-\infty}^{+\infty} |h(t,\tau)|\, d\tau = \infty$$

By definition of the supremum, it follows that given any positive M, however large, we can find a t_M in $(-\infty, \infty)$ such that

$$f(t_M) = \int_{-\infty}^{\infty} |h(t_M,\tau)|\, d\tau > M$$

To these numbers M and t_M, we associate a bounded input x_M defined by

$$x_M(\tau) = \operatorname{sgn} h(t_M,\tau) \qquad \forall \tau \in (-\infty, \infty)$$

[1] The supremum (abbreviated sup) of $|f(t)|$ on $(-\infty, \infty)$ is an upper bound b such that $b \geq |f(t)|$ for all t in $(-\infty, \infty)$ and such that no smaller number has the same property. Until recently, the supremum was called the least upper bound (abbreviated l.u.b.).

Then, by *8*, the output due to this input has the property that

$$y(t_M) > M$$

Therefore, we have shown that if the inequality *8* is not satisfied by any finite N, then, given any positive M, however large, we can construct a bounded input which will produce at a certain finite time t_M an output $y(t_M)$ which is larger than the given M. This clearly contradicts the assumption of stability b.i.b.o., which requires that any bounded input produce a bounded zero-state response.

10 *Example* Consider an infinite RC cable. Its input is a current source and its output is the voltage across the input terminals. It is well known that $h(t) = \sqrt{\frac{R}{\pi C t}}\, 1(t)$. Therefore, even though this system consists of a linear time-invariant dissipative network, it is not stable b.i.b.o., since the response to a unit step of current is a monotonically increasing voltage which tends to infinity as time increases indefinitely.

In some cases the concept of stability b.i.b.o. can easily be related to the concepts of stability introduced in Chap. *7*. For example, we have the

11 **Assertion** Let $\mathcal{Q}$ be a linear time-invariant system characterized by the differential equation $L(p)y = u$. Then the following statements are equivalent:

(I) $\mathcal{Q}$ is asymptotically stable (see *7.2.3*).
(II) The zeros of the polynomial $L(s)$ have negative (<0) real parts.
(III) The system $\mathcal{Q}$ is zero-state stable b.i.b.o. (see *2*).

Proof Exercise for the reader.

12 *Comment* In the time-invariant case, the system

$$\dot{\mathbf{x}} = \mathbf{Ax} + \mathbf{Bu} \qquad \mathbf{y} = \mathbf{Cx} + \mathbf{Du}$$

may be zero-state stable b.i.b.o. but not stable i.s.L. This fact becomes obvious if the system has an unstable mode and either the input is not coupled to that mode or that mode is not coupled to the output. A further discussion of these points will be given in Chap. *11*, because it requires a thorough definition of the notions of controllability (see *11.3.3*) and observability (see *11.4.1*). ◁

For the case of periodically varying systems we have the

13 **Assertion** Let $\mathcal{P}$ be a periodically varying system described by the differential equation $\dot{\mathbf{x}} = \mathbf{A}(t)\mathbf{x}(t) + \mathbf{u}(t)$, where $\mathbf{A}(t)$ is an $n \times n$ matrix continuous and periodic in t and where the input $\mathbf{u}(t)$ is an n-vector. Then the following statements are equivalent:

(I) The system $\mathcal{P}$ is asymptotically stable (see *7.2.3* and *7.4.8*).
(II) The characteristic exponents of the system have negative (<0) real parts (see *6.7.8*).
(III) The system $\mathcal{P}$ is zero-state stable b.i.b.o.

14 *Comment* For time-invariant systems, zero-state stability b.i.b.o. does not imply that the impulse response is bounded. Consider, for example, $h(t) = (1 - t)^{-1/2}$ for $0 \leq t < 1$ and $h(t) = 0$ for all other t. By *7*, such a system is zero-state stable b.i.b.o., but its impulse response is arbitrarily large as t approaches 1 from the left.

15 *Comment* In general for linear time-varying differential systems there is no simple relationship between zero-state stability b.i.b.o. and the concepts of stability, defined in Chap. *7*, which are based on the motion of the state.[1] The time-varying character of the system introduces an additional possibility which is over and above the problems of controllability and observability. Roughly the additional difficulty is the following. Suppose we have

$$\begin{aligned} \dot{\mathbf{x}} &= \mathbf{A}(t)\mathbf{x} + \mathbf{B}(t)\mathbf{u}(t) \\ \mathbf{y}(t) &= \mathbf{C}(t)\mathbf{x}(t) \end{aligned}$$

where $\mathbf{A}(t),\mathbf{B}(t)$ are bounded and continuous. Suppose the system is not stable i.s.L. and furthermore that, for some $\mathbf{x}_0$, $\|\mathbf{x}(t;\mathbf{x}_0,t_0;\mathbf{0})\| \to \infty$ as $t \to \infty$. Still, the system might be stable b.i.b.o. (see Chap. *7*, Sec. *7*) provided the matrix $\mathbf{C}(t)$ has a norm that decreases so rapidly that even though $\|\mathbf{x}(t;\mathbf{x}_0,t_0;\mathbf{0})\| \to \infty$, $\|\mathbf{C}(t)\mathbf{x}(t;\mathbf{x}_0,t_0,\mathbf{u})\|$ remains bounded for all $t \in [t_0, \infty)$ and for all bounded inputs $\mathbf{u}$. In such circumstances the system would obviously be also zero-state stable b.i.b.o. This idea is illustrated by the following

16 *Example* Consider the single-input single-output time-varying system $\mathcal{S}$ shown in *Fig. 8.4.1*. Calling x_0 the integrator output at time t_0, we

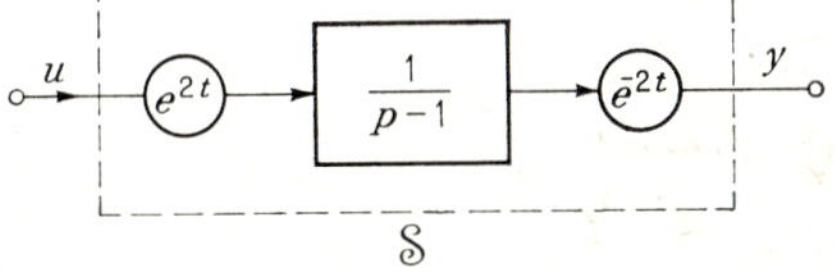

Fig. 8.4.1 System $\mathcal{S}$ considered in Example *8.4.16*.

obviously have

$$\begin{aligned} y(t;x_0,t_0;u_{(t_0,t]}) &= e^{-2t}e^{(t-t_0)}x_0 + e^{-2t}\int_{t_0}^{t} e^{\xi}e^{2(t-\xi)}u(t-\xi)\,d\xi \\ &= e^{-(t-t_0)}(e^{-2t_0}x_0) + \int_{t_0}^{t} e^{-\xi}u(t-\xi)\,d\xi \end{aligned}$$

[1] R. E. Kalman, On the Stability of Time-varying Systems, *I.R.E. Trans. Circuit Theory*, CT-9, no. 4, pp. 420–422, 1962.

The input-output relation of the system $\mathcal{S}$ is then seen to be identical to that of the time-invariant system $\mathcal{S}'$ shown in Fig. *8.4.2*, provided, of course, that the output at time t_0 of the integrator of the system $\mathcal{S}'$ is labeled $e^{-2t_0}x_0$. In other words, $\mathcal{S}$ and $\mathcal{S}'$ are equivalent systems. $\mathcal{S}'$ is stable i.s.L., asymptotically stable, stable b.i.b.o., and zero-state

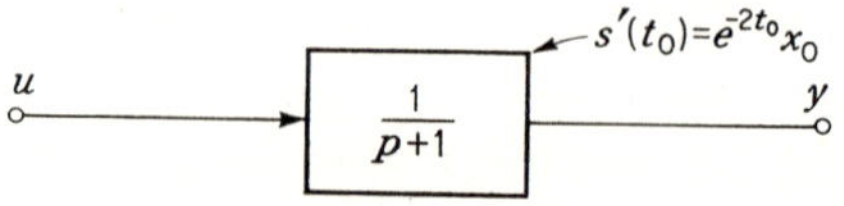

Fig. 8.4.2 The system $\mathcal{S}'$ shown here is equivalent to $\mathcal{S}$ shown in Fig. *8.4.1*.

stable b.i.b.o. On the other hand, $\mathcal{S}'$ is not even stable i.s.L. since

$$x(t;x_0,t_0;0) = e^{(t-t_0)}x_0$$

This concludes our consideration of systems specified by their impulse responses. In the next chapter we investigate in detail the properties of time-invariant systems from the transfer-function point of view.

REFERENCES AND SUGGESTED READING

1 Darlington, S.: Nonstationary Smoothing and Prediction Using Network Theory Concepts, *I.R.E. Trans. Circuit Theory,* vol. CT-6, pp. 1–13, 1959.

2 Laning, J. H., Jr., and R. H. Battin: "Random Processes in Automatic Control," McGraw-Hill Book Company, Inc., New York, 1956.

9 Transfer functions and their properties

1 *Introduction*

In the preceding chapters we have studied in detail the properties of the input-output relationships of linear systems. We now turn our attention to a very important tool in the analysis and design of linear systems: the transfer function. Since the transfer function is used most often in applications involving time-invariant linear systems, this chapter is devoted exclusively to time-invariant systems. From a purely mathematical point of view, the transfer function might be thought to be important because of its use in analyzing time-invariant differential systems and in solving integral equations of the form $y(t) = \int_0^\infty h(t - \xi)u(\xi)\, d\xi$. This point of view is valid as far as it goes, but it hides the main fact: the basic reason for the overwhelming importance of transfer functions is the fundamental theorem of the sinusoidal steady state (*9.4.17*), which states that any strictly stable transfer function can be determined by sinusoidal steady-state measurements. The practical importance of this fact arises from the availability of stable and accurate sinusoidal oscillators and of precise and convenient sinusoidal steady-state measurement equipment and techniques. In practice, it means that extremely complicated systems can be characterized by a number of simple measurements; indeed, in many practical instances, it would be extremely difficult and expensive to write the differential equations of the system under consideration. Consider, for example, a long-distance communications system with its dozens of amplifiers, filters, equalizers, modulators, etc. On the other hand, automatic measurement systems that determine in a few seconds the gain and phase [that is, $|H(j\omega)|$ and $\measuredangle H(j\omega)$] of such systems over a band of frequencies have been built. These facts of life explain why over the last decades engineers have developed so many analysis and synthesis techniques based on the concept of transfer function.

The purpose of this chapter is to present a number of basic properties of the transfer function that are useful for the intelligent application of the design techniques mentioned above. After a brief discussion of the realization of a system whose matrix transfer function is given, we consider the constraints imposed on transfer functions by certain physical characteristics of linear systems: stability, nonanticipative character, relation between the real and imaginary parts, uncertainty principle, dispersion of the impulse response. Next, we present a number of analysis techniques based on transfer functions: paired echoes, asymptotic relations between an impulse response and its transfer function, and the steady-state response to a periodic input. The third objective is to present methods for dealing with the stability problem for both single- and multiple-feedback-loop systems using exclusively data based on sinusoidal steady-state measurements. In order to deal with multiple-loop systems, we shall have to derive Mason's formula for the gain of a flow graph.

2 *Definition and basic relations*

In Chap. *3*, Sec. *6*, the notions of impulse response and transfer function have been defined to characterize the zero-state response of linear systems. In this chapter, we restrict ourselves to time-invariant systems; therefore (see *3.6.9*), the impulse response $h(\cdot)$ is the zero-state response of the system to $\delta(t)$, that is, to a unit impulse applied at $t = 0$. In this chapter we do not restrict ourselves to nonanticipative systems, and therefore it may happen that $h(t) \neq 0$ for some $t < 0$. For this reason, we introduce the following

1 **Definition** Let the system $\mathcal{A}$ be linear and time-invariant and let $h(\cdot)$ be its impulse response (*3.6.5*). Assume that h is Laplace transformable (unilateral transform if $\mathcal{A}$ is nonanticipative and bilateral if not). The relation

2 $$H(s) \triangleq \int_{-\infty}^{+\infty} e^{-st}h(t)\,dt$$

defines the *transfer function* H of $\mathcal{A}$.

3 *Comment* The transfer function H is defined only in the domain of convergence of the integral in *2*; in many practical instances it is possible to continue analytically the domain of definition of H to include the whole plane except the singular points of H. This extension is not always possible.

Alternatively, if (I) $\mathcal{A}$ has a ground state which coincides with its zero state [this implies that $\mathcal{A}$ is asymptotically stable (see *2.8.7* and *7.2.3*)] and (II) $\mathcal{A}$ is nonanticipative (*1.8.13*), then $H(s)$ can be expressed as follows:

$$H(s) \triangleq \frac{\text{steady-state response of } \mathcal{A} \text{ to } e^{st}}{e^{st}}$$

This relation is an immediate consequence of the definition of the steady-state response of $\mathcal{A}$. Thus by *2.8.20*

$$\begin{aligned}\text{Steady-state response of } \mathcal{A} \text{ to } e^{st} &= \lim_{t_0 \to -\infty} \int_{t_0}^{t} h(t-\xi)e^{-s\xi}\,d\xi \\ &= \lim_{\lambda \to \infty} e^{st} \int_0^{\lambda} h(\tau)e^{-s\tau}\,d\tau \\ &= e^{st}H(s)\end{aligned}$$

Note that the definition of $H(s)$ in terms of the steady-state response of $\mathcal{A}$ to e^{st} is not as general as *2*.

Finally, in the case of multiple-input multiple-output systems, the transfer function becomes a *matrix transfer function* $\mathbf{H}(s)$; its (i,j) element is the Laplace transform of $h_{ij}(t)$, the ith-output zero-state response due to a unit impulse applied at the jth input at time $t = 0$. ◁

If the transfer function $H(s)$ is given, then the impulse response can be calculated by the inversion integral, which reads

$$\frac{h(t+0) + h(t-0)}{2} = \frac{1}{2\pi j} \int_{c-j\infty}^{c+j\infty} H(s)e^{st}\,ds$$

where c is a real number which lies in the strip (or half-plane) of convergence of *2*. For more details the reader is referred to Appendix *B*. The property that makes the transfer function so useful in practice is contained in the

4 **Assertion** Let $\mathcal{A}$ be linear time-invariant and let it have $H(s)$ as a transfer function. If y is the zero-state response to the input u, then

5 $$Y(s) = H(s)U(s)$$

where Y and U are the Laplace transforms of y and u. [These transforms are bilateral if either $u(t) \not\equiv 0$ or $h(t) \not\equiv 0$ for some $t < 0$.]

Proof See *3.6.23*.

6 **Remark** It is of fundamental importance to constantly keep in mind that the transfer function of the system $\mathcal{A}$ *characterizes only its zero-state response.* In general, the transfer function gives only incomplete information regarding its zero-input response. For further elaboration of this fact, refer to Chap. *11*, Sec. *5*.

7 *Comment* The relationship between $H(s)$ and $h(t)$, which is expressed by *2* and *3*, deserves some comment in view of the importance of

approximations in engineering problems. Suppose that the defining integrals of transfer functions $H(s)$ and $H_1(s)$ converge for Re $s = 0$. Roughly speaking, the point we wish to make is that if two transfer functions $H_1(j\omega)$ and $H(j\omega)$ are approximately equal for all ω's, it does not follow that their impulse responses $h_1(t)$ and $h(t)$ are approximately equal for all t.

To exhibit this fact, let $H_1(j\omega) = H(j\omega) + \varepsilon(j\omega)$, where

$$\varepsilon(j\omega) = \begin{cases} \eta\left(1 - \dfrac{|\omega|}{\Omega}\right) e^{-j\omega t_0} & |\omega| \leq \Omega \\ 0 & |\omega| > \Omega \end{cases}$$

and the constant $\eta \ll 1$ and $t_0 > 0$. Clearly, $|\varepsilon(j\omega)| \ll 1$ for all $\omega \in (-\infty, \infty)$. A simple calculation shows that

$$h_1(t_0) - h(t_0) = \frac{\eta\Omega}{2\pi}$$

Therefore, by choosing Ω sufficiently large and keeping η very small and fixed, we can make this difference as large as we wish. Thus, although the "peak" difference between $H_1(j\omega)$ and $H(j\omega)$ remains constant and equal to the small number η, the "peak" difference between h_1 and h can be made arbitrarily large.

In general, however, we can write

$$|h_1(t) - h(t)| \leq \frac{1}{2\pi} \int_{-\infty}^{+\infty} |H_1(j\omega) - H(j\omega)|\ |e^{j\omega t}|\, d\omega$$

and

8 $$|h_1(t) - h(t)| \leq \frac{1}{2\pi} \int_{-\infty}^{+\infty} |H_1(j\omega) - H(j\omega)|\, d\omega \qquad \forall t \in (-\infty, \infty)$$

Thus if we measure the "distance" between H_1 and H by the integral $\int_{-\infty}^{+\infty} |H_1(j\omega) - H(j\omega)|\, d\omega$, then *8* gives an upper bound on $|h_1(t) - h(t)|$ over $(-\infty, \infty)$.

3 *Realization of a matrix transfer function*

In Chap. *4*, Sec. *9*, we considered the problem of representing a system specified by a rational transfer function as an interconnection of scalors, adders, and integrators. We now consider the corresponding problem: given a multiple-input multiple-output linear time-invariant system specified by its matrix transfer function $\mathbf{H}(s)$, find a realization of this system as an interconnection of scalors, adders, and differentiators.

We shall assume that all the elements of $\mathbf{H}(s)$ are rational functions of s. For simplicity, we shall further assume that all the elements of $\mathbf{H}(s)$ have simple poles: the extension of the method to the case where multiple poles are present does not require any additional devices. Its description in general terms would require cumbersome notation, however.

1 **Remark** For simplicity, we shall say the input $\mathbf{u}(s)$ when we mean the input whose Laplace transform is $\mathbf{u}(s)$, and we shall similarly express the output.

Special case Suppose $\mathbf{H}(s)$ is of the form $\mathbf{R}/(s - \lambda)$, where $\mathbf{R}$ is an $n \times n$ constant matrix which can be represented by a single dyad $\boldsymbol{\sigma}\rangle\langle\boldsymbol{\rho}$ (see *5.4.15*). By definition, the result of the dyad $\boldsymbol{\sigma}\rangle\langle\boldsymbol{\rho}$ operating on any vector $\mathbf{u}$ is the vector $\boldsymbol{\sigma}$ times the scalar $\langle\boldsymbol{\rho},\mathbf{u}\rangle$. The dyad $\boldsymbol{\sigma}\rangle\langle\boldsymbol{\rho}$ may also be thought of as the product of the column vector $\boldsymbol{\sigma}$ and the row vector $\boldsymbol{\rho}^*$, whose elements are the complex conjugates of those of $\boldsymbol{\rho}$. Thus, if we consider an input described by the vector $\mathbf{u}(s)$, the output

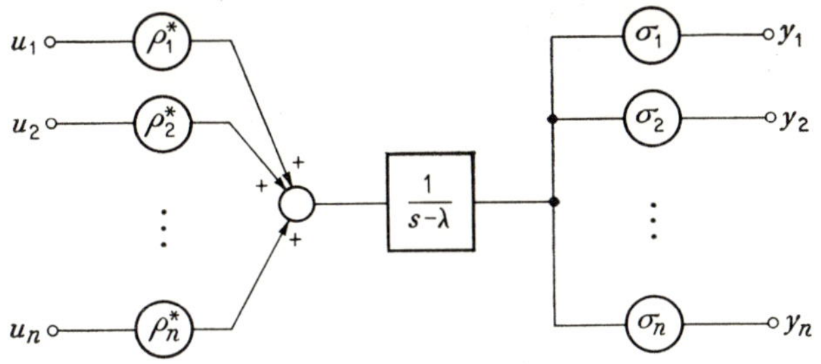

Fig. 9.3.1 Realization of the system represented by $\boldsymbol{\sigma}\rangle\langle\boldsymbol{\rho}/(s - \lambda)$.

will be the vector $\boldsymbol{\sigma}\langle\boldsymbol{\rho},\mathbf{u}(s)\rangle/(s - \lambda)$. This leads to the simple realization shown in Fig. *9.3.1* and the

2 **Assertion** If the transfer function $\mathbf{H}(s)$ is of the form $\mathbf{R}/(s - \lambda)$, where the constant matrix $\mathbf{R}$ is representable by a single dyad $\boldsymbol{\sigma}\rangle\langle\boldsymbol{\rho}$, then the system has a realization which requires only one integrator (see Fig. *9.3.1*).

3 *Exercise* Let $\mathbf{H}(s) = \mathbf{R}/d(s)$, where the constant matrix $\mathbf{R} = \boldsymbol{\sigma}\rangle\langle\boldsymbol{\rho}$ and $d(s)$ is a polynomial. Show that the realization of the system requires only that of the single-input single-output system whose transfer function is $1/d(s)$.

4 *Comment* The system shown in Fig. *9.3.1* exhibits very clearly the fact that for any real scalar k, $k\boldsymbol{\sigma}\rangle\langle k^{-1}\boldsymbol{\rho} = \boldsymbol{\sigma}\rangle\langle\boldsymbol{\rho}$. ◁

In order to proceed to the general case, we must first establish a lemma which gives the representation of any constant matrix $\mathbf{R}$ as a sum of dyads. Obviously if $\mathbf{R}$ is represented by a single dyad $\boldsymbol{\sigma}\rangle\langle\boldsymbol{\rho}$, the range of $\mathbf{R}$, $\mathcal{R}(\mathbf{R})$, is spanned by the vector $\boldsymbol{\delta}$ and hence is one-dimensional. If $\mathcal{R}(\mathbf{R})$ is of dimension c (that is, the matrix $\mathbf{R}$ has rank c) and is spanned by $\boldsymbol{\delta}_1, \boldsymbol{\delta}_2, \ldots, \boldsymbol{\delta}_c$, then $\mathbf{R}$ has a dyadic representation of the form $\sum_{i=1}^{c} \boldsymbol{\delta}_i\rangle\langle\boldsymbol{\varrho}_i$. Now the range of $\mathbf{R}$ is spanned by the column vectors of the matrix $\mathbf{R}$; therefore we may choose for the set $\{\boldsymbol{\delta}_i\}_1^c$ a maximal set of linearly independent column vectors of $\mathbf{R}$, that is, any set of linearly independent column vectors which spans $\mathcal{R}(\mathbf{R})$. This point of view naturally leads to the

5 **Lemma** Let $\mathbf{R}$ be an $n \times n$ matrix whose elements are complex numbers. Let $\tilde{\mathbf{R}}$ be any $n \times c$ matrix whose columns are a maximal set of linearly independent columns of $\mathbf{R}$. Let the $c \times n$ matrix $\mathbf{S}$ be defined by

6 $$\mathbf{R} = \tilde{\mathbf{R}}\mathbf{S}$$

Let, for $i = 1, 2, \ldots, n$, $\boldsymbol{\delta}_i$ be the ith column of $\tilde{\mathbf{R}}$ and $\boldsymbol{\varrho}_i$ be the complex conjugate ith row of $\mathbf{S}$; then

7 $$\mathbf{R} = \sum_{i=1}^{c} \boldsymbol{\delta}_i\rangle\langle\boldsymbol{\varrho}_i$$

Furthermore

8 $$\mathbf{S} = (\tilde{\mathbf{R}}^*\tilde{\mathbf{R}})^{-1}\tilde{\mathbf{R}}^*\mathbf{R}$$

where $\mathbf{R}^*$ is the conjugate transpose of $\mathbf{R}$.

9 *Comment* The elements of $\mathbf{S}$ defined in *6* may be interpreted as follows: the c elements of the kth column of $\mathbf{S}$ are the weighting coefficients defining the linear combination of the $\boldsymbol{\delta}_i$'s required to form the kth column of $\mathbf{R}$. This $\mathbf{S}$ is defined uniquely by *6*.

10 *Comment* It is not necessary to use columns of $\mathbf{R}$ for the basis vectors $\boldsymbol{\delta}_i$ since any basis of $\mathcal{R}(\mathbf{R})$ will do. Special choices of basis vectors may simplify greatly the inversion required by *8*.

Proof Let $\mathbf{x}$ be an arbitrary vector; then by *6*, $\mathbf{Rx} = \tilde{\mathbf{R}}\mathbf{Sx} = \tilde{\mathbf{R}}(\mathbf{Sx})$. Now $\mathbf{Sx}$ is a column vector whose components are $\langle\boldsymbol{\varrho}_i,\mathbf{x}\rangle$. Note that we use a complex scalar product (see *C.5.4*). Multiplying $\mathbf{Sx}$ by $\tilde{\mathbf{R}}$ on the left, we get $\Sigma\langle\boldsymbol{\varrho}_i,\mathbf{x}\rangle\boldsymbol{\delta}_i$. Since this holds for all $\mathbf{x}$, *7* follows.

From *6* we have

11 $$\tilde{\mathbf{R}}^*\mathbf{R} = (\tilde{\mathbf{R}}^*\tilde{\mathbf{R}})\mathbf{S}$$

Since the column vectors of $\tilde{\mathbf{R}}$ are linearly independent, with $\mathbf{z} \in \mathcal{C}^c$ we have $\mathbf{z} \neq 0 \Rightarrow \tilde{\mathbf{R}}\mathbf{z} \neq 0 \Rightarrow 0 < \langle\tilde{\mathbf{R}}\mathbf{z},\tilde{\mathbf{R}}\mathbf{z}\rangle = \langle\mathbf{z},\tilde{\mathbf{R}}^*\tilde{\mathbf{R}}\mathbf{z}\rangle$; that is, the hermitian matrix $\tilde{\mathbf{R}}^*\tilde{\mathbf{R}}$ is positive definite; hence it is nonsingular (see *C.19.19*). Multiplying *11* on the left by $(\tilde{\mathbf{R}}^*\tilde{\mathbf{R}})^{-1}$, we get *8*. ◁

We are now in a position to describe the main result of this section in a form of a

12 **Theorem** Let the system $\mathcal{A}$ be characterized by its transfer function matrix $\mathbf{H}(s)$. Suppose that all the elements of $\mathbf{H}(s)$ are rational functions and have simple poles. Let the partial-fraction expansion of $\mathbf{H}(s)$ be

13
$$\mathbf{H}(s) = \sum_{j=1}^{m} \frac{\mathbf{R}_j}{s - \lambda_j} + \mathbf{H}_0$$

where the residue matrices $\mathbf{R}_j$ and $\mathbf{H}_0$ are constant matrices. Let each residue matrix be expanded, as in Lemma *5*,

14
$$\mathbf{H}(s) = \sum_{j=1}^{m} \sum_{i=1}^{\alpha_j} \frac{\boldsymbol{\sigma}_i^{(j)}\rangle\langle\boldsymbol{\varrho}_i^{(j)}}{s - \lambda_j} + \mathbf{H}_0$$

Then the system $\mathcal{A}$, represented by $\mathbf{H}(s)$, is zero-state equivalent to the system $\mathcal{B}$ described by the following state equation:

15
$$\begin{bmatrix} \dot{\mathbf{x}}_1 \\ \dot{\mathbf{x}}_2 \\ \cdots \\ \dot{\mathbf{x}}_m \end{bmatrix} = \begin{bmatrix} \lambda_1\mathbf{I}_1 & \mathbf{0} & \cdots & \mathbf{0} \\ \mathbf{0} & \lambda_2\mathbf{I}_2 & \cdots & \cdots \\ \cdots & \cdots & \cdots & \cdots \\ \mathbf{0} & \cdots & \cdots & \lambda_m\mathbf{I}_m \end{bmatrix} \begin{bmatrix} \mathbf{x}_1 \\ \mathbf{x}_2 \\ \cdots \\ \mathbf{x}_m \end{bmatrix} + \begin{bmatrix} \mathbf{S}_1 \\ \mathbf{S}_2 \\ \cdots \\ \mathbf{S}_m \end{bmatrix} \mathbf{u}$$

16
$$\mathbf{y} = [\tilde{\mathbf{R}}_1 \quad \tilde{\mathbf{R}}_2 \quad \cdots \quad \tilde{\mathbf{R}}_m] \begin{bmatrix} \mathbf{x}_1 \\ \cdots \\ \mathbf{x}_m \end{bmatrix} + \mathbf{H}_0\mathbf{u}$$

where $\mathbf{I}_j$ is the unit matrix of order α_j and, as in Lemma *5*, $\mathbf{R}_j = \tilde{\mathbf{R}}_j\mathbf{S}_j$, $j = 1, 2, \ldots, m$. Furthermore, there is no system of order smaller than $n \triangleq \sum_{j=1}^{m} \alpha_j$ which is zero-state equivalent to $\mathcal{A}$.

Proof Let us establish the zero-state equivalence of $\mathcal{A}$ and $\mathcal{B}$. Since the $\mathbf{A}$ matrix of *15* is diagonal and since all matrices of *15* and *16* are partitioned in a conforming manner, the output $\mathbf{y}$ is easily written as the sum of the contribution of each block. Therefore, remembering that we consider only the zero-state response, we have successively

$$\mathbf{y}(t) = \sum_j \tilde{\mathbf{R}}_j \int_0^t e^{\lambda_j(t-\xi)}\mathbf{I}_j\mathbf{S}_j\mathbf{u}(\xi)\, d\xi + \mathbf{H}_0\mathbf{u}(t)$$

$$= \sum_j \tilde{\mathbf{R}}_j\mathbf{S}_j \int_0^t e^{\lambda_j(t-\xi)}\mathbf{u}(\xi)\, d\xi + \mathbf{H}_0\mathbf{u}(t)$$

$$\mathbf{y}(t) = \sum_j \mathbf{R}_j \int_0^t e^{\lambda_j(t-\xi)}\mathbf{u}(\xi)\, d\xi + \mathbf{H}_0\mathbf{u}(t)$$

Clearly this is exactly the expression that one would have obtained by starting from *13* and inverting the Laplace transforms by the convolution theorem. Therefore $\mathcal{A}$ and $\mathcal{B}$ are zero-state equivalent.

To show that $\mathcal{B}$ is of minimal order, let us show that the number of components of the state vector associated with λ_j cannot be less than α_j, the dimension of the range of $\mathbf{R}_j$. For all $\mathbf{u}(s)$ the contribution of the output associated with λ_j is $\mathbf{R}_j\mathbf{u}(s)/(s - \lambda_j)$. Now $\mathbf{R}_j$ has an α_j-dimensional range; therefore, the specification of the vector $\mathbf{R}_j\mathbf{u}(s)$ requires at least α_j scalars. [In the present case, where we use the vectors $\boldsymbol{\delta}_1^{(j)}, \ldots, \boldsymbol{\delta}_{\alpha_j}^{(j)}$ as a basis for $\mathcal{R}(\mathbf{R}_j)$, these scalars are $\langle \boldsymbol{\delta}_i^{(j)}, \mathbf{u}(s)\rangle$, $i = 1, 2, \ldots, \alpha_j$.] Therefore, the dimension of the state vector cannot be less than $\sum_{j=1}^{m} \alpha_j = n$ and $\mathcal{B}$ is of minimal order. Equivalently $\mathcal{B}$ is in reduced form (*2.2.11*).

17 *Example* Let the transfer function matrix be a 2×2 matrix. By expanding it in partial fractions, we get

$$\mathbf{H}(s) = \begin{bmatrix} \dfrac{2s+3}{s^2+3s+2} & \dfrac{6s^2+25s+23}{s^3+6s^2+11s+6} \\ \dfrac{1}{s+1} & \dfrac{3s+5}{s^2+3s+2} \end{bmatrix}$$

$$= \begin{bmatrix} \dfrac{1}{s+1} + \dfrac{1}{s+2} & \dfrac{2}{s+1} + \dfrac{3}{s+2} + \dfrac{1}{s+3} \\ \dfrac{1}{s+1} & \dfrac{2}{s+1} + \dfrac{1}{s+2} \end{bmatrix}$$

Thus

$$\mathbf{R}_1 = \begin{bmatrix} 1 & 2 \\ 1 & 2 \end{bmatrix} = \begin{bmatrix} 1 \\ 1 \end{bmatrix} [1 \quad 2]$$

$$\mathbf{R}_2 = \begin{bmatrix} 1 & 3 \\ 0 & 1 \end{bmatrix} = \begin{bmatrix} 1 \\ 0 \end{bmatrix} [1 \quad 0] + \begin{bmatrix} 3 \\ 1 \end{bmatrix} [0 \quad 1]$$

$$\mathbf{R}_3 = \begin{bmatrix} 0 & 1 \\ 0 & 0 \end{bmatrix} = \begin{bmatrix} 1 \\ 0 \end{bmatrix} [0 \quad 1]$$

Here the dyadic expansion of the $\mathbf{R}_i$'s is obtained by inspection using Comment *9*. The state equations of $\mathcal{B}$ are then

18
$$\dot{\mathbf{x}} = \begin{bmatrix} -1 & 0 & 0 & 0 \\ 0 & -2 & 0 & 0 \\ 0 & 0 & -2 & 0 \\ 0 & 0 & 0 & -3 \end{bmatrix} \mathbf{x} + \begin{bmatrix} 1 & 2 \\ 1 & 0 \\ 0 & 1 \\ 0 & 1 \end{bmatrix} \mathbf{u}$$

19
$$\mathbf{y} = \begin{bmatrix} 1 & 1 & 3 & 1 \\ 1 & 0 & 1 & 0 \end{bmatrix} \mathbf{x}$$

The realization of $\mathcal{B}$ is immediately obtained on the basis of *2* and Fig. *9.3.1*. It is shown in Fig. *9.3.2*. The scalors on the left of the

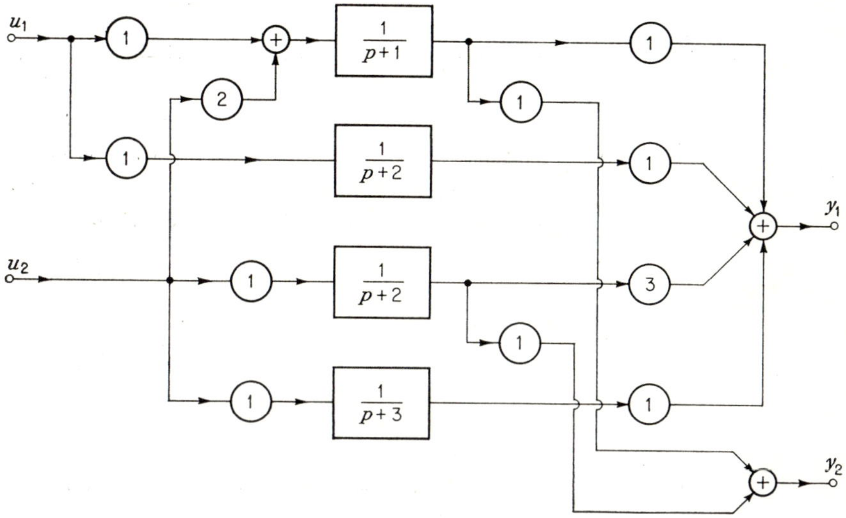

Fig. 9.3.2 Realization of the matrix transfer function of Example *9.3.17*.

integrators have multiplying factors which are the columns of the matrix multiplying $\mathbf{u}$ in *18*. Similarly, the scalors following the integrators have multiplying factors which are the rows of the matrix multiplying $\mathbf{x}$ in *19*. These two observations provide an easy check on the realization of $\mathcal{B}$.

4 *Stable transfer functions*

In Chap. *7*, we investigated the stability of differential systems, and in Chap. *8* we introduced the concept of zero-state stability for systems described by their impulse response. Here we consider exclusively time-invariant systems and we define stable and strictly stable transfer functions. The relationship between stable transfer functions and system stability is discussed below in Remark *15*.

Definition and characterization

1 **Definition** A transfer function $H(s)$ is said to be *stable* if (I) its impulse response $h(t)$ is bounded for all $t \in (-\infty, \infty)$ or if (II) h is the sum of such a bounded function and a countable linear combination of impulses such that the coefficients of the δ functions constitute an absolutely converging series.

2 *Example* For the circuit shown in Fig. *9.4.1*, $H(s) = 1/(s + 1)$ and $h(t) = 1(t)e^{-t}$; hence, $H(s)$ is stable.

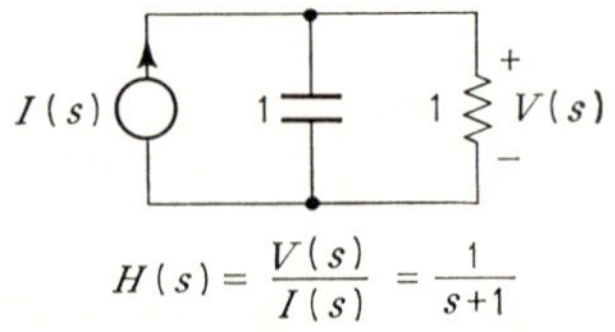

Fig. 9.4.1 Circuit whose transfer function is $H(s) = 1/(s + 1)$.

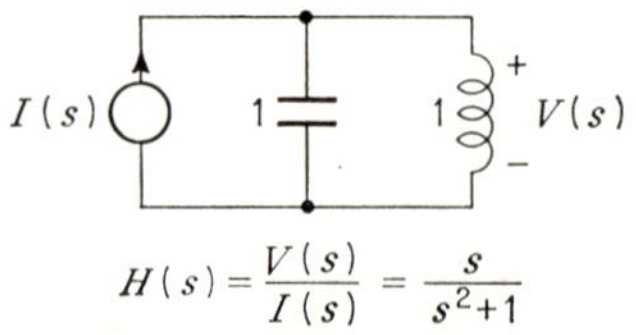

Fig. 9.4.2 Circuit whose transfer function is $H(s) = s/(s^2 + 1)$.

3 *Example* For the circuit shown in Fig. *9.4.2*, $H(s) = s/(s^2 + 1)$ and $h(t) = 1(t) \cos t$; again the transfer function is stable.

4 *Example* In the circuit shown in Fig. *9.4.3*, the transmission line is assumed to be lossless and to have a delay of 1 sec. The reflection

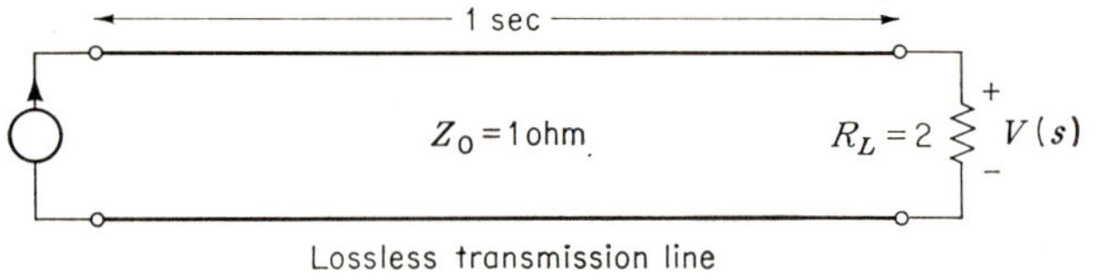

Fig. 9.4.3 Circuit whose impulse response is an infinite sum of impulses.

coefficient at the generator end is 1 and at the load end is $\frac{1}{3}$; therefore, the output voltage due to a unit impulse of current is

$$h(t) = \tfrac{4}{3}\left[\delta(t-1) + \sum_{n=1}^{\infty} \left(\tfrac{1}{3}\right)^n \delta(t - 1 - 2n)\right]$$

Again this particular transfer function is stable. Show that if the load resistance were -2 ohms, the transfer function would not be stable.

As an obvious consequence of the fact that, for any positive integer n,

5 $$\mathfrak{L}^{-1}\left(\frac{1}{(s-a)^n}\right) = 1(t)\,\frac{t^{n-1}}{(n-1)!}\,e^{at}$$

and of the partial-fraction expansion we have the

6 **Theorem** A rational transfer function is *stable* if and only if it has no poles inside the right half plane and its $j\omega$-axis poles (if any) are simple. [$H(s)$ is assumed to be nonanticipative.] ◁

From a practical point of view a more useful concept of stability is that of strict stability:

7 **Definition** A transfer function is said to be *strictly stable* if its zero-state response remains bounded for any bounded input. ◁

Therefore a linear time-invariant system that has a strictly stable transfer function is zero-state stable b.i.b.o. (*8.4.1*).

More precisely, a transfer function is said to be a strictly stable transfer function whenever any input x_i which satisfies $|x_i(t)| < M_i$, for all t and for some $M_i < \infty$, produces a zero-state response $x_0(t)$ which satisfies $|x_0(t)| < M_0$ for some finite M_0 and for all t. This bound M_0 depends on M_i and on the particular transfer function.

8 *Example* The transfer function $H(s) = 1/s$ is stable, since its impulse response $1(t)$ is uniformly bounded; but it is not strictly stable, since a bounded input $x_i(t) = 1(t)$ produces a zero-state response $x_0(t) = t1(t)$ which becomes arbitrarily large as $t \to \infty$.

9 *Example* There are system components that are not strictly stable in terms of certain variables; consider the conventional d-c motor which appears in every example of an electrical position servo. Its transfer function is $H(s) = \Theta(s)/I(s) = K/s(1 + Ts)$, where θ is the shaft position and i is the armature current. This transfer function is not strictly stable. Of course, the transfer function of the over-all position servo is obviously required to be strictly stable; hence, appropriate compensation and feedback networks are required.

10 *Example* The ideal low-pass filter, defined by

$$|H(j\omega)| = 1(\omega + 2\pi B) - 1(\omega - 2\pi B) \qquad \text{and} \qquad \measuredangle H(j\omega) = -j\omega t_0$$

has an impulse response $h(t) = 2B[\sin 2\pi B(t - t_0)]/2\pi B(t - t_0)$. Its transfer function is obviously stable, but a simple calculation will prove that it is not strictly stable. (See *12*.)

Again as a consequence of the partial-fraction expansion we have

11 **Theorem** A rational transfer function is strictly stable if and only if it has no poles inside the right half plane and no poles on the $j\omega$ axis.
Proof This follows immediately from the partial-fraction expansion of $H(s)$ and the convolution theorem. $\triangleleft$

A characterization of strictly stable transfer functions which applies to a much larger class of systems is the following

12 **Theorem** A transfer function $H(s)$ is strictly stable if and only if its impulse response $h(t)$ is such that

13
$$\int_{-\infty}^{+\infty} |h(t)|\, dt \triangleq A < \infty$$

that is, if and only if the area under the curve $|h(t)|$ is finite.
Proof $\Leftarrow$ Let us prove that if *13* holds, then $H(s)$ is strictly stable. Consider an arbitrary bounded input x_i; that is, there is a constant M_i such that

$$|x_i(t)| < M_i \qquad \forall t$$

The convolution theorem gives the zero-state response x_0 at time t due to the input x_i:

$$x_0(t) = \int_{-\infty}^{+\infty} x_i(t-\tau)h(\tau)\,d\tau$$

Hence

$$\begin{aligned} |x_0(t)| &= \left| \int_{-\infty}^{+\infty} x_i(t-\tau)h(\tau)\,d\tau \right| \\ &\leq \int_{-\infty}^{+\infty} |x_i(t-\tau)|\,|h(\tau)|\,d\tau \\ &< M_i \int_{-\infty}^{+\infty} |h(\tau)|\,d\tau \end{aligned}$$

Thus, in view of *13*, for all t,

$$|x_0(t)| < AM_i$$

That is, the output is uniformly bounded for all t.

$\Rightarrow$ It remains to prove that if $H(s)$ is strictly stable, then inequality *13* is satisfied. We shall establish this result by contradiction: we assume that *13* is not satisfied and then find a bounded input that will produce an output at time T which, for sufficiently large T, will be larger than any preassigned number.

Pick the input $\hat{x}_i$ defined as follows:

14 $$\hat{x}_i(t) = \begin{cases} \operatorname{sgn}[h(T-t)] & 0 \leq t \leq 2T \\ 0 & \text{elsewhere} \end{cases}$$

or equivalently, putting $T - t = \tau$,

$$\hat{x}_i(T-\tau) = \begin{cases} \operatorname{sgn}[h(\tau)] & -T \leq \tau \leq T \\ 0 & \text{elsewhere} \end{cases}$$

To this input $\hat{x}_i$ there corresponds an output $\hat{x}_0$ which at time T takes the value given by

$$\hat{x}_0(T) = \int_{-\infty}^{\infty} \hat{x}_i(T-\tau)h(\tau)\,d\tau = \int_{-T}^{T} h(\tau)\operatorname{sgn}[h(\tau)]\,d\tau$$

Thus,

$$\hat{x}_0(T) = \int_{-T}^{T} |h(\tau)|\,d\tau$$

The assumption that *13* does not hold means that given any positive number M, however large, we can find a value of T, say $T(M)$, such that

$$\int_{-T(M)}^{+T(M)} |h(\tau)|\,d\tau > M$$

The above two relations imply that if *13* is not satisfied, *14* gives a recipe for producing a bounded input such that the corresponding output becomes larger, in absolute value, then any preassigned M at time $T(M)$. In other words, if *13* does not hold, x_0 cannot be uniformly bounded for all bounded inputs. Hence if $H(s)$ is strictly stable, *13* must hold.

15 **Remark: transfer function stability versus other types of stability** Since the transfer function characterizes only the zero-state response of a system, one would expect that there are systems which have a strictly stable transfer function but which have a zero state that is unstable. Consider, for example, the system $\mathcal{S}$ shown in Fig. *9.4.4*. Its transfer function $H(s) = \frac{1}{2}$ and is strictly stable. The state is $\mathbf{x} = (x_1,x_2)$. It is obvious that any nonzero initial state such that $x_1(0) \neq x_2(0)$ will give rise to an exponentially increasing $\mathbf{x}(t;\mathbf{x}_0,0)$; in fact, it is easy to check that

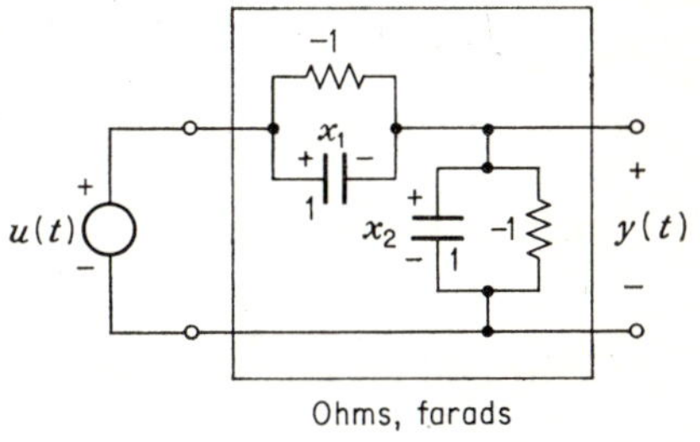

Fig. 9.4.4 Circuit whose transfer function is strictly stable but whose zero state is not asymptotically stable.

$$\mathbf{x}(t;\mathbf{x}_0,0) = \frac{e^t}{2}\begin{bmatrix} x_1(0) - x_2(0) \\ -x_1(0) + x_2(0) \end{bmatrix} \qquad \forall t > 0$$

Physically, an arbitrary small charge on either capacitor will produce an exponentially increasing output voltage. It should be stressed, however, that the voltage source, irrespective of its terminal voltage u, cannot excite the unstable mode of the system $\mathcal{S}$; thus in so far as $H(s)$ represents *only* the relation between the input u and the output y when the system is initially in the zero state, it is perfectly reasonable for the transfer function to be strictly stable. Needless to say, in engineering, our systems must have both an asymptotically stable zero state and a strictly stable transfer function.

The above example illustrates another point: the transfer function $H(s)$ becomes a constant independent of s if and only if the two RC parallel combinations have the same time constant (-1 sec in the present case); the slightest manufacturing deviation or change due to temperature or aging that makes the time constants different will cause $H(s)$ to be unstable. This shows the error that lies in the optimistic idea "let us cancel a right-half-plane pole (at $+1$ in the present case) by a zero at the same location." A further discussion of this subject using the concepts of controllability and observability is found in *11.5.10*.

A sufficient condition for strict stability is given by the following:

16 **Assertion** Let $H(s)$ be the transfer function. If $\int_{-\infty}^{+\infty} |H(j\omega)|^2\,d\omega < \infty$ and if $\int_{-\infty}^{+\infty} \left|\frac{d}{d\omega} H(j\omega)\right|^2 d\omega < \infty$, then $H(s)$ is strictly stable.

Proof By Parseval's theorem (*B.3.16*), the two assumptions imply that $\int_{-\infty}^{\infty} |h(t)|^2\,dt < \infty$, $\int_{-\infty}^{\infty} t^2|h(t)|^2\,dt < \infty$. Now

$$\int_{-\infty}^{+\infty} |h(t)|\,dt = \int_{-\infty}^{+\infty} (1+t^2)^{1/2}|h(t)|(1+t^2)^{-1/2}\,dt$$
$$\leq \left[\int_{-\infty}^{\infty} (1+t^2)|h(t)|^2\,dt\right]^{1/2}\left(\int_{-\infty}^{\infty} \frac{dt}{1+t^2}\right)^{1/2} < \infty$$

where the inequality follows from Schwarz's inequality (*C.5.8* and *9.8.12*). $H(s)$ is strictly stable by Theorem *12*.

Sinusoidal steady state

The main reason for the importance of transfer functions in engineering is that they can be measured easily and accurately. The basis of the measurements is the following

17 **Theorem** Let $\mathcal{S}$ be a single-input single-output linear nonanticipative time-invariant system whose transfer function is $H(s)$. Let $H(s)$ be strictly stable. Then the zero-state response to the input $1(t)\exp(j\omega t)$ becomes arbitrarily close to $H(j\omega)\exp(j\omega t)$ as $t \to \infty$. More precisely, given $\varepsilon > 0$, there is a time $T(\varepsilon)$ such that

18 $$|y(t; \mathbf{0}, 0; 1(t)\exp(j\omega t)) - H(j\omega)\exp(j\omega t)| < \varepsilon \qquad \forall t > T(\varepsilon)$$

In practical application, $h(t)$ is real; hence, $H(j\omega)^* = H(-j\omega)$ and consequently we have the

19 **Corollary** Let the assumptions of Theorem *17* hold and let $h(t)$ be real for all t; then

$$\left.\begin{array}{l} |y(t; \mathbf{0}, 0; 1(t)\cos\omega t) - |H(j\omega)|\cos[\omega t + \measuredangle H(j\omega)]| < \varepsilon \\ |y(t; \mathbf{0}, 0; 1(t)\sin\omega t) - |H(j\omega)|\sin[\omega t + \measuredangle H(j\omega)]| < \varepsilon \end{array}\right\} \forall t > T(\varepsilon)$$

The corollary follows immediately from $\exp(j\omega t) = \cos\omega t + j\sin\omega t$, the additivity property of the zero-state response (*3.4.1*) and *18*.

20 *Comment* $T(\varepsilon)$ sec after applying the excitation $1(t)\cos\omega t$, the amplitude of the response is within ε of that of the true steady state. Note that this $T(\varepsilon)$ is independent of the frequency of the excitation. Of course, if one knows more about the analytical properties of $H(s)$, one may obtain sharper estimates; see *9.13.11* for an example.

Proof of theorem Since $H(s)$ is strictly stable, $\int_0^{\infty} |h(t)|\,dt \triangleq A < \infty$ by *12*. Thus for any $\varepsilon > 0$, there is a $T(\varepsilon)$ such that $t > T(\varepsilon) \Rightarrow$

$\int_t^\infty |h(\xi)|\, d\xi < \varepsilon$. Consider now the zero-state response to $1(t) \exp (j\omega t)$; then

$$y(t; \mathbf{0}, 0; 1(t) \exp (j\omega t)) = \int_0^t h(\xi) \exp [j\omega(t - \xi)]\, d\xi$$
$$= [\exp (j\omega t)] \left[\int_0^\infty h(\xi)[\exp (-j\omega\xi)]\, d\xi - \int_t^\infty h(\xi)[\exp (-j\omega\xi)]\, d\xi \right]$$

Call $\eta_2(t)$ the second term in the outer brackets; then

21 $$y(t; \mathbf{0}, 0; 1(t) \exp (j\omega t)) - H(j\omega) \exp (j\omega t) = \eta_2(t) \exp (j\omega t)$$

Now

$$|\eta_2(t)| = \left| \int_t^\infty h(\xi)[\exp (-j\omega\xi)]\, d\xi \right| \leq \int_t^\infty |h(\xi)|\, d\xi$$

Thus

$$|\eta_2(t)| < \varepsilon \qquad \forall t < T(\varepsilon)$$

and, with this fact, *21* implies *18*.

Liénard and Chipart stability test

Theorem *11* states that a rational transfer function is strictly stable if and only if its denominator has all its roots inside the left half plane. It is important to be able to test a given transfer function for this condition. We shall now describe the Liénard and Chipart test, which is more efficient than the celebrated Hurwitz test.

Let the denominator of the given rational transfer function be

$$d(s) = a_0 s^n + a_1 s^{n-1} + \cdots + a_{n-1}s + a_n \qquad a_0 > 0$$

Let us define the Hurwitz determinants by the equality

$$\Delta_k = \begin{vmatrix} a_1 & a_3 & a_5 & \cdot & \cdot & \cdot & \cdot \\ a_0 & a_2 & a_4 & \cdot & \cdot & \cdot & \cdot \\ 0 & a_1 & a_3 & \cdot & \cdot & \cdot & \cdot \\ 0 & a_0 & a_2 & a_4 & \cdot & \cdot & \cdot \\ 0 & 0 & a_1 & a_3 & \cdot & \cdot & \cdot \\ 0 & 0 & a_0 & a_2 & \cdot & \cdot & \cdot \\ \cdot & \cdot & \cdot & \cdot & \cdot & \cdot & a_k \end{vmatrix} \qquad k = 1, 2, \ldots, n$$

where $a_k = 0$ for $k > n$. These determinants $\Delta_1, \Delta_2, \Delta_3, \ldots$ are called the *Hurwitz determinants*.

22 **Theorem**[1] The polynomial with real coefficients,

$$d(s) = a_0 s^n + a_1 s^{n-1} + \cdots + a_{n-1}s + a_n \qquad (a_0 > 0)$$

[1] The proof of this theorem is to be found in F. R. Gantmacher, "Theory of Matrices," vol. II, chap. XV, Chelsea Publishing Company, New York, 1959.

has all its roots with negative real parts if and only if any one of the following sets of conditions is satisfied:

(I) $a_n > 0, a_{n-2} > 0, \ldots ; \Delta_1 > 0, \Delta_3 > 0, \ldots$
(II) $a_n > 0, a_{n-2} > 0, \ldots ; \Delta_2 > 0, \Delta_4 > 0, \ldots$
(III) $a_n > 0, a_{n-1} > 0, a_{n-3} > 0, \ldots ; \Delta_1 > 0, \Delta_3 > 0, \ldots$
(IV) $a_n > 0, a_{n-1} > 0, a_{n-3} > 0, \ldots ; \Delta_2 > 0, \Delta_4 > 0 \ldots$

23 *Comment* Usually it is required that *all* the Hurwitz determinants be positive. However, this requirement is redundant because of the following fact: if $d(s)$ has all its coefficients positive and if the Hurwitz determinants of odd order are positive, then those of even order are also positive, and vice versa.

24 *Example* $a_0s^3 + a_1s^2 + a_2s + a_3 = 0 \qquad a_0 > 0$

If we take the second set of conditions, we get

$$a_1a_2 - a_0a_3 > 0 \qquad a_3 > 0, a_1 > 0$$

25 *Example* $a_0s^4 + a_1s^3 + a_2s^2 + a_3s + a_4 = 0 \qquad a_0 > 0$

If we take the first set of conditions, we get

$$a_4 > 0, a_2 > 0; a_1 > 0, a_1a_2a_3 - a_1^2a_4 - a_3^2a_0 > 0$$

Design considerations

In design problems it often occurs that we have to choose a parameter μ (say, the gain of some amplifier or some transducer) but the chosen μ must lead to a strictly stable transfer function. It is also usually the case that for some value of μ, say, μ_0, it is physically obvious or it is known that the transfer function is strictly stable. The problem is to devise an effective method for specifying the largest interval of the μ axis for which the transfer function is strictly stable. The answer is given by the

26 **Theorem** Given a system which has a rational transfer function whose coefficients are real and depend continuously on a parameter μ. Suppose that the denominator polynomial d has all its zeros with negative (<0) real parts when the parameter is equal to μ_0. Then, as μ moves away from μ_0, the first value of μ, say μ_1, at which a zero of d is on the $j\omega$ axis is such that $a_n(\mu_1) = 0$ or $\Delta_{n-1}(\mu_1) = 0$.

27 *Comment* In other words, we need only plot $a_n(\mu)$ and $\Delta_{n-1}(\mu)$ as a function of μ and detect the first time one of these curves touches the μ axis.

28 *Comment* The zeros of a polynomial are continuous functions of the coefficients; hence, in the present case, the zeros of d are continuous functions of μ. Thus the transfer function can become unstable only by some of its poles moving across the $j\omega$ axis. Therefore, Theorem *26*

allows us to assert that *for all* $\mu \in [\mu_0, \mu_1)$, *the transfer function of the system is strictly stable.*

Proof *Case 1* As μ varies away from μ_0, a real zero may move to the right and into the right half plane. The value of μ for which it is on the $j\omega$ axis corresponds to the fact that d has a zero at the origin of coordinates; hence, $a_n(\mu_1) = 0$.

Case 2 As μ varies away from μ_0, a pair of complex conjugate zeros may move into the right half plane. The value of μ at which they are on the $j\omega$ axis is such that $\Delta_{n-1}(\mu_1) = 0$. For that value of μ, d has a pair of pure imaginary zeros, say, $\pm j\omega_0$, and consequently $\text{Re}\,[d(j\omega_0)] = 0$ and $\text{Im}\,[d(j\omega_0)] = 0$. In other words, two polynomials easily derived from d have a common zero. Therefore, their discriminant is zero.[1] A few simple manipulations show that this discriminant is $a_n\Delta_{n-1}(\mu_1)$.

5 *The Paley-Wiener criterion*

The Paley-Wiener criterion imposes a fundamental restriction on all physical transfer functions. In order to describe it, let us give a definition: A transfer function is *causal* if its impulse response is identical to zero for $t < 0$. Roughly speaking, the Paley-Wiener criterion states that a transfer function $H(j\omega)$ is *causal* if and only if the inequality

$$\int_{-\infty}^{+\infty} \frac{|\log |H(j\omega)||}{1 + \omega^2}\, d\omega < \infty$$

is satisfied.

This requirement is not vacuous, as is shown by the following examples. (I) The ideal low-pass filter is defined by $H_L(j\omega) = e^{-j\omega\tau}$ for $|\omega| < 2\pi W$ and zero elsewhere; τ is the constant time delay. $H_L(j\omega)$ cannot be causal because $\log |H_L(j\omega)|$ is infinite in the intervals (W, ∞) and $(-\infty, -W)$. (II) No causal transfer function may be identical to zero in any interval (ω_0, ω_1) of positive length. It may, however, have a countable number of zeros on the $j\omega$ axis, as is shown by the example of Fig. *9.5.1*. At each resonance of the transmission line (and there is an infinite, but countable, number of resonances!) there is a zero of transmission. (III) The

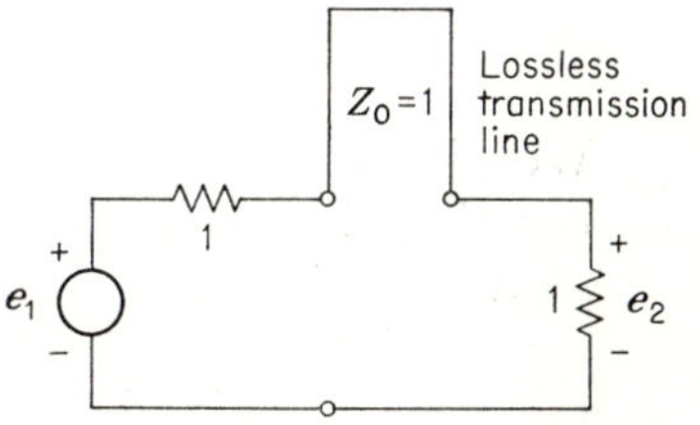

Fig. 9.5.1 Circuit whose transfer function has a countable number of zeros on the $j\omega$ axis.

[1] See, for example, M. Bocher, "Introduction to Higher Algebra," The MacMillan Company, New York, 1921.

Gaussian filter is defined by

$$H_G(j\omega) = (2\pi)^{-1/2} \exp\left(-j\omega\tau - \frac{\omega^2}{2}\right)$$

with τ as the constant time delay. It cannot be causal because $|\log |H_G(j)|| \sim \frac{\omega^2}{2}$ for large ω and the inequality above is not satisfied.

It is important to understand exactly what the Paley-Wiener criterion means in order not to draw false conclusions. An example in point is the Gaussian filter $H_G(j\omega)$. The criterion states that a system whose transfer function is *exactly* $H_G(j\omega)$, for all ω's, cannot be causal. It does not state that satisfactory approximations of $H_G(j\omega)$ will not be causal. For example, take $\tau = 0$ for simplicity and let

$$H_1(j\omega) = \begin{cases} \exp\left(-\frac{\omega^2}{2}\right) & \text{for } |\omega| \leq 4 \\ \left(\frac{4}{\omega}\right)^4 e^{-8} & \text{for } |\omega| > 4 \end{cases}$$

It satisfies the Paley-Wiener criterion. From a practical point of view, since e^{-8} is a small number, $|H_1(j\omega)|$ will usually be sufficiently Gaussian to satisfy ordinary engineering requirements.

Let us consider the ideal low-pass filter $H_L(j\omega)$ defined in the introduction. From the criterion we expect that its impulse response $h_L(t)$ will differ from zero for some $t < 0$; in fact,

$$h_L(t) = 2W \frac{\sin 2\pi W(t - t_0)}{2\pi W(t - t_0)}$$

for all t. Consider now a sequence of filters approximating the ideal low-pass filter such that (for simplicity we put $2\pi W = 1$ in the following)

$$|H_n(j\omega)|^2 = \frac{1}{1 + \omega^{2n}}$$

We pick the phase of $H_n(j\omega)$ so that it is a minimum phase transfer function, i.e., in this case the transfer function which satisfies the above condition and which has all its poles inside the left half plane has all its zeros at infinity. Clearly, for every integer n, however large, the criterion is satisfied and the corresponding impulse response $h_n(t) = 0$ for $t < 0$. Furthermore,

$$\lim_{n \to \infty} |H_n(j\omega)| = |H_L(j\omega)|$$

for all ω's. Thus comes the question: why do we have $h_n(t) = 0$ for all $t < 0$ and all n on the one hand and $h_L(t) \neq 0$ for negative values of t arbitrarily far away from $t = 0$ on the other? The answer lies in the fact that as $n \to \infty$, $|H_n(j\omega)| \to |H_L(j\omega)|$ but that $\measuredangle H_n(j\omega)$ does not tend to $\measuredangle H_L(j\omega)$. In fact, for each fixed ω, $\measuredangle H_n(j\omega) \to \infty$. This follows from the potential analogy[1] point of view: $\measuredangle H_n(j\omega)$ is zero at the origin and $-d/d\omega \measuredangle H_n(j\omega)$ is proportional to the E-field component normal to the $j\omega$ axis provided every pole of $H_n(s)$ is replaced by a line charge of 1 coulomb/m. Now the n poles of $H_n(s)$ are symmetrically distributed on the left half of the unit circle centered on the origin of the s plane. Clearly, the total charge on the left half of the unit circle is n coulombs/m and, as $n \to \infty$, it becomes infinite. Thus at every point of the $j\omega$ axis in $[-j,+j]$, $-d/d\omega \measuredangle H_n(j\omega)$ becomes infinite. In other words, as $n \to \infty$, the group delay (see Sec. *11*) through the filter grows without bound. All this amounts to the observation that given an amplitude $|H(j\omega)|$ that does not satisfy the Paley-Wiener criterion it is impossible to associate with it a finite phase $\phi(\omega)$ such that $|H(j\omega)|e^{-j\phi(\omega)}$ is a causal transfer function.

A precise statement of the Paley-Wiener criterion is as follows:

1 **Theorem** Let $h(t) = 0$ for $t < 0$. Define $H(j\omega)$ by

2 $$\int_0^\infty h(t)e^{-j\omega t}\,dt \triangleq H(j\omega)$$

If $|H(j\omega)| \triangleq A(\omega)$ satisfies

3 $$\int_{-\infty}^{+\infty} A(\omega)^2\,d\omega < \infty$$

and if $H(s)$ is analytic for $\sigma = 0$, then

4 $$\int_{-\infty}^{+\infty} \frac{|\log A(\omega)|}{1+\omega^2}\,d\omega < \infty$$

Conversely, if conditions *3* and *4* hold, then a phase $\phi(\omega)$ can be found such that $H(j\omega) = A(\omega)e^{j\phi(\omega)}$ is a causal transfer function [i.e., the impulse response of $H(j\omega)$ vanishes identically for $t < 0$].

5 *Comment* The assumption that $H(s)$ is analytic for $\sigma = 0$ is not necessary; it is introduced here to facilitate the proof by allowing us to drop some delicate passages to the limit of integrals ranging over an infinite domain. ◁

In order to facilitate the proof of the criterion, we introduce three preliminary results.

6 **Lemma** For ω and σ real and for $\sigma > 0$,

$$\frac{1}{\pi}\int_{-\infty}^{+\infty} \frac{\sigma}{\sigma^2 + (\omega - \omega')^2}\,d\omega' = 1$$

Proof This result is easily established by Cauchy's integral formula and is left as an exercise for the reader. ◁

[1] S. Seshu and N. Balabanian, "Linear Network Analysis," John Wiley & Sons, Inc., New York, 1959.

The next preliminary result required is *Jensen's formula*. Let us first recall a classical fact concerning harmonic functions. If $\psi(z)$ is analytic in the disk $|z| \leq R$ and if $\psi(z)$ has no zeros in that disk, $\log |\psi(z)|$ is harmonic in the disk. Integrating $[\log \psi(z)]/z$ over the circle of radius R we get, after taking real parts,

7 $$\log |\psi(0)| = \frac{1}{2\pi} \int_0^{2\pi} \log |\psi(Re^{i\theta})| \, d\theta$$

Let us observe that $\log |\psi(z)|$, being the real part of an analytic function, is a harmonic function. The above equality is often referred to as the *mean-value property of harmonic functions.* Assuming that these facts are known, we wish to prove the

8 **Lemma** If $f(z)$ is analytic in $|z| \leq R$, if $f(z)$ has zeros $a_1, a_2, \ldots, a_n$ in $|z| < R$ (multiple zeros being repeated), and if $f(0) \neq 0$, then

9 $$\log |f(0)| = \frac{1}{2\pi} \int_0^{2\pi} \log |f(Re^{i\theta})| \, d\theta - \sum_{i=1}^{n} \log \left(\frac{R}{|a_i|} \right)$$

Proof The function

$$F(z) = f(z) \prod_{i=1}^{n} \frac{R^2 - a_i^* z}{R(z - a_i)}$$

has the property that $|F(z)| = |f(z)|$ on $|z| = R$ and $F(z) \neq 0$ for $|z| \leq R$. Hence, from *7*,

$$\log |F(0)| = \frac{1}{2\pi} \int_0^{2\pi} \log |f(Re^{i\theta})| \, d\theta$$

and, using the first equation in the second, we get

$$\frac{1}{2\pi} \int_0^{2\pi} \log |f(Re^{i\theta})| \, d\theta = \log |f(0)| + \sum_{i=1}^{n} \log \left(\frac{R}{a_i} \right) \qquad \text{Q.E.D.}$$

Note If $f(0) = 0$, let $f(z) = b_n z^n + \cdots$ be its Taylor expansion about $z = 0$. Then apply the formula to $f(z)/b_n z^n$. The left-hand side of *9* then becomes

10 $$\log |b_n| + n \log R \quad \triangleleft$$

The next result we need makes use of the notion of convexity. By definition, e^x is a *convex* function because, for any two x, y

$$\exp [\alpha x + (1 - \alpha) y] \leq \alpha e^x + (1 - \alpha) e^y \qquad \text{for every } \alpha \in [0,1]$$

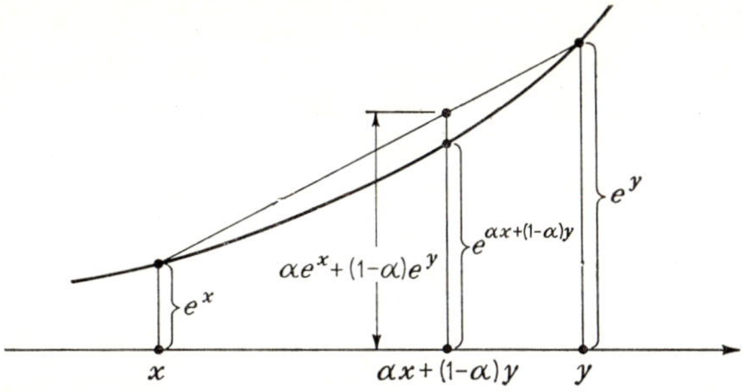

Fig. 9.5.2 Illustration exhibiting the convexity of the exponential function.

This inequality is illustrated in Fig. *9.5.2*. By induction it can be shown that if $w_1, w_2, \ldots, w_N$ are positive weights such that $\sum_1^N w_i = 1$, then

$$\exp\left(\sum_1^N w_i x_i\right) \leq \sum_{i=1}^N w_i e^{x_i}$$

and going over to the limit we get

11 **Lemma** If the weighting function $w(\omega')$ is such that $w(\omega') \geq 0$ for all ω' and $\int_{-\infty}^{+\infty} w(\omega')\, d\omega' = 1$, then

12
$$\exp\left[\int_{-\infty}^{+\infty} w(\omega')x(\omega')\, d\omega'\right] \leq \int_{-\infty}^{+\infty} w(\omega')\,\{\exp\,[x(\omega')]\}\, d\omega'$$

Proof of Theorem 1 $\Rightarrow$ By assumption, $H(s)$ is analytic for $\sigma = 0$. By *2*, $H(s)$ is analytic for all $\sigma \geq 0$. Map the right half s plane into the unit circle of the z plane by the transformation $z = (1 - s)/(1 + s)$; $H(s)$ becomes $\Phi(z) = \Phi(re^{j\theta})$. A simple computation gives [observe that $e^{j\theta} = (1 - j\omega)/(1 + j\omega)$; hence $\theta/2 = -\tan^{-1}\omega$]

13
$$\int_{-\pi}^{+\pi} |\Phi(e^{i\theta})|^2\, d\theta = 2\int_{-\infty}^{+\infty} \frac{|H(j\omega)|^2}{1+\omega^2}\, d\omega < \infty$$

where the inequality follows from assumption *3*.

It remains to establish *4*; observe that twice the L.H.S. of *4* is equal to $\int_{-\pi}^{+\pi} \log|\Phi(e^{j\theta})|\, d\theta$. Thus we must show that this integral is finite. For this purpose define over the interval $(-\pi,+\pi)$ the functions

$$\log^+ |\Phi(e^{j\theta})| = \max\,(0, \log|\Phi(e^{j\theta})|)$$

and
$$\log^- |\Phi(e^{j\theta})| = -\min\,(0, \log|\Phi(e^{j\theta})|)$$

Observe that, for all θ's, $\log^-|\Phi(e^{j\theta})| \geq 0$. From these definitions it follows that

$$\frac{1}{2\pi}\int_{-\pi}^{+\pi}\log|\Phi(e^{j\theta})|\,d\theta = \frac{1}{2\pi}\int_{-\pi}^{+\pi}\log^+|\Phi(e^{j\theta})|\,d\theta - \frac{1}{2\pi}\int_{-\pi}^{+\pi}\log^-|\Phi(e^{j\theta})|\,d\theta$$

and

$$\frac{1}{2\pi}\int_{-\pi}^{+\pi}\big|\log|\Phi(e^{j\theta})|\big|\,d\theta = \frac{1}{2\pi}\int_{-\pi}^{+\pi}\log^+|\Phi(e^{j\theta})|\,d\theta + \frac{1}{2\pi}\int_{-\pi}^{+\pi}\log^-|\Phi(e^{j\theta})|\,d\theta$$

From the fact that, for all real $x \geq 1$, $\log x < x^2$, it follows that

$$\frac{1}{2\pi}\int_{-\pi}^{+\pi}\log^+|\Phi(e^{j\theta})|\,d\theta \leq \frac{1}{2\pi}\int_{-\pi}^{+\pi}|\Phi(e^{j\theta})|^2\,d\theta < \infty$$

Now $\Phi(z)$ is analytic in the unit circle, and hence $\log \Phi(e^{i\theta})$ is harmonic. Therefore, we may apply Jensen's formula *9*. Let us use the two inequalities above in the following computation:[1]

$$\frac{1}{2\pi}\int_{-\pi}^{+\pi}\big|[\log|\Phi(e^{j\theta})|]\big|d\theta = \frac{1}{2\pi}\int_{-\pi}^{+\pi}\log^+|\Phi(e^{j\theta})|\,d\theta + \frac{1}{2\pi}\int_{-\pi}^{+\pi}\log^-|\Phi(e^{j\theta})|\,d\theta$$

14
$$= 2\frac{1}{2\pi}\int_{-\pi}^{+\pi}\log^+|\Phi(e^{j\theta})|\,d\theta - \frac{1}{2\pi}\int_{-\pi}^{+\pi}\log|\Phi(e^{j\theta})|\,d\theta$$

$$\leq \frac{1}{\pi}\int_{-\pi}^{+\pi}|\Phi(e^{j\theta})|^2\,d\theta - \log|\Phi(0)|$$

Therefore by *13*

$$\int_{-\pi}^{+\pi}\big|[\log|\Phi(e^{j\theta})|]\big|\,d\theta < \infty$$

and, finally,

$$\int_{-\infty}^{+\infty}\frac{\big|[\log|H(j\omega)|]\big|}{1+\omega^2}\,d\omega < \infty$$

This concludes the proof of the direct statement.

$\Leftarrow$ Let $s = \sigma + j\omega$. For $\sigma > 0$, define

15
$$\alpha(\sigma + j\omega) = \int_{-\infty}^{+\infty}\log A(\omega')\cdot\frac{1}{\pi}\frac{+\sigma}{\sigma^2 + (\omega - \omega')^2}\,d\omega'$$

This integral converges because inequality *4* is satisfied; hence, $\alpha(\sigma + j\omega)$ is a real finite harmonic function for all $\sigma > 0$. For $\sigma \to 0$, $\alpha(\sigma + j\omega) \to \log A(\omega)$ because the weighting function in *15* becomes $\delta(\omega - \omega')$. Thus $\alpha(j\omega)$ is the gain in nepers corresponding to $A(\omega)$. The next step is to construct the phase associated with $\alpha(j\omega)$. Let $\beta(\sigma + j\omega)$ be the harmonic function conjugate to $\alpha(\sigma + j\omega)$ in the

[1] The computation assumes that $\Phi(0) \neq 0$. If this is not the case, consider the Taylor expansion $\Phi(z) = b_n z^n + \cdots$. By *10*, the term $\log|\Phi(0)|$ in the right-hand side of *14* is replaced by $\log|b_n|$.

domain $\sigma > 0$. It is well known that $\beta(\sigma + j\omega)$ is determined only within an additive constant.[1] Since $\beta(j\omega)$ will be the phase associated with $\alpha(j\omega)$, take $\beta(0) = 0$. Therefore, for every point (σ,ω) such that $\sigma > 0$,

$$\beta(\sigma + j\omega) = \int_0^{\sigma+j\omega} \left(-\frac{\delta\alpha}{\delta\omega}\, d\sigma + \frac{\delta\alpha}{\delta\sigma}\, d\omega \right)$$

where the path of integration from 0 to $\sigma + j\omega$ must lie completely in the right half plane. The integral above is independent of the path because $\alpha(\sigma + j\omega)$ is harmonic.[1]

The functions $\theta(s) = \alpha(\sigma + j\omega) + j\beta(\sigma + j\omega)$ and $e^{\theta(s)}$ are analytic for all $\sigma > 0$. Observe that

$$\lim_{\sigma \to 0} e^{\alpha(\sigma+j\omega)} = A(\omega)$$

Since the exponential function e^z is a convex function of the real variable z, and since the weighting function

$$w(\omega') = \frac{1}{\pi} \frac{\sigma}{\sigma^2 + (\omega - \omega')^2}$$

is positive and $\int_{-\infty}^{+\infty} w(\omega')\, d\omega' = 1$, then by Lemma *11* it follows that

$$e^{\alpha(s)} \le \frac{1}{\pi} \int_{-\infty}^{+\infty} \frac{A(\omega')\sigma}{(\omega - \omega')^2 + \sigma^2}\, d\omega' \qquad \text{for all } \sigma > 0$$

and, with the Schwarz inequality,

$$\int_{-\infty}^{+\infty} e^{2\alpha(\sigma+i\omega)}\, d\omega$$
$$\le \frac{1}{\pi^2} \int_{-\infty}^{+\infty} d\omega \int_{-\infty}^{+\infty} \frac{A^2(\omega')\sigma}{(\omega - \omega')^2 + \sigma^2}\, d\omega' \int_{-\infty}^{+\infty} \frac{\sigma\, d\omega''}{(\omega - \omega'')^2 + \sigma^2}$$

The last integral is equal to π by Lemma *6*. Reordering the right-hand side, it becomes successively

$$\frac{1}{\pi} \int_{-\infty}^{+\infty} d\omega \int_{-\infty}^{+\infty} \frac{A^2(\omega')\sigma}{(\omega - \omega')^2 + \sigma^2}\, d\omega'$$
$$= \frac{1}{\pi} \int_{-\infty}^{+\infty} A^2(\omega')\, d\omega' \int_{-\infty}^{+\infty} \frac{\sigma\, d\omega}{(\omega - \omega')^2 + \sigma^2}$$
$$= \int_{-\infty}^{+\infty} A^2(\omega')\, d\omega'$$

Hence, for all $\sigma > 0$,

$$\int_{-\infty}^{+\infty} e^{2\alpha(\sigma+j\omega)}\, d\omega \le \int_{-\infty}^{+\infty} A^2(\omega')\, d\omega' < \infty$$

where the last inequality follows by assumption *3*.

[1] W. Kaplan, "Advanced Calculus," sec. 9.22, Addison-Wesley Publishing Company, Reading, Mass., 1952.

Since $|e^{\theta(s)}|^2 = e^{2\alpha(\sigma+j\omega)}$, the above inequality implies that $e^{\theta(s)}$ is integrable square along any parallel to the $j\omega$ axis for all $\sigma > 0$. Furthermore, it is analytic in the right half plane. Therefore,[1] there exists a function of time $h(t)$ (the corresponding impulse response) vanishing for $t < 0$ such that

$$e^{\theta(s)} = \lim_{\Omega \to \infty} \int_{-\Omega}^{+\Omega} h(t) e^{-(\sigma+j\omega)t}\, dt$$

We have, then, shown that if *3* and *4* hold, we can associate with $A(\omega)$ a phase $\beta(\omega)$ such that the transfer function $A(\omega)e^{j\beta(\omega)}$ has an impulse response which vanishes for all $t < 0$. Q.E.D.

6 *Relation between the real and imaginary parts of* $T(s)$

The purpose of this section is to establish a number of important relationships between the real part and the imaginary part of the values taken on the $j\omega$ axis by transfer functions and related functions. Bode's book (Ref. *1*) is the most complete reference on this subject. The proofs that follow are based on Cauchy's residue theorem and are due to Bode. In this section we make the following assumptions:

1 **Assumption** $T(s)$ is analytic for $\text{Re } s > 0$.

2 **Assumption** $[T(s)]^* = T(s^*)$ for $\text{Re } s \geq 0$. In particular, if we let $T(j\omega) = R(j\omega) + jI(j\omega)$, then $R(j\omega) = R(-j\omega)$ and $I(j\omega) = -I(-j\omega)$ for all real ω.

3 **Assumption** $T(s)$ may have a finite number of singularities on the $j\omega$ axis provided that at each such singularity, say $j\omega_0$, $(s - j\omega_0)T(s) \to 0$ as $s - j\omega_0 \to 0$.

4 **Assumption** As $|s| \to \infty$, $\dfrac{T(s)}{s} \to 0$ uniformly in $|\sphericalangle s| \leq \dfrac{\pi}{2}$.

5 *Comment* If $T(s)$ is a strictly stable rational transfer function and if $T(j\omega) \to 0$ as $|\omega| \to \infty$, Assumptions *1* to *4* are satisfied; the results that follow will therefore be applicable to the usual finite lumped linear systems. These assumptions, however, cover a broader class of functions. Consider

$$H(j\omega) = \frac{1}{\sqrt{1 - \omega^2} + j\omega}$$

Clearly, $|H(j\omega)| = 1$ for $-1 \leq \omega \leq 1$ and $|H(j\omega)|$ monotonically decreases to zero as $\omega \to \infty$; hence, H is the transfer function of a low-pass filter. Let $T(j\omega) = \log H(j\omega)$: $T(j\omega)$ satisfies the above assumptions and $R(j\omega)$ and $I(j\omega)$ are respectively the gain (in nepers) and the phase (in radians) of that low-pass filter.

[1] Ref. 6, p. 8; or Ref. 7, p. 214.

6 **Theorem** If $T(s)$ fulfills Assumptions *1* to *4* and if s_1 is inside the right half plane ($\operatorname{Re} s_1 > 0$), then

7 $$T(s_1) = \frac{2s_1}{\pi} \int_0^\infty \frac{R(j\omega)}{\omega^2 + s_1^2}\, d\omega$$

In other words, the knowledge of the real part of $T(s)$ along the $j\omega$ axis determines completely $T(s)$ inside the right half plane. This fact has been used for solving some time-domain approximation problems and for designing equalizers.

Proof The proof is a direct application of Cauchy's residue theorem. We shall take as contour the $j\omega$ axis from $+jR$ to $-jR$ and the right-half-plane semicircle of radius R, as shown in Fig. *9.6.1*. The integrand is chosen to be

$$\theta(s) = \frac{T(s)}{s - s_1} - \frac{T(s)}{s + s_1} = \frac{2s_1}{s^2 - s_1^2}\, T(s)$$

Assume first that $T(s)$ has no $j\omega$-axis singularities. Then $\theta(s)$ is analytic everywhere on the contour and inside it except for a pole at $s = s_1$, where its residue is $T(s_1)$; hence

$$T(s_1) = \frac{1}{2\pi j} \oint_C \frac{2s_1 T(s)}{s^2 - s_1^2}\, ds$$

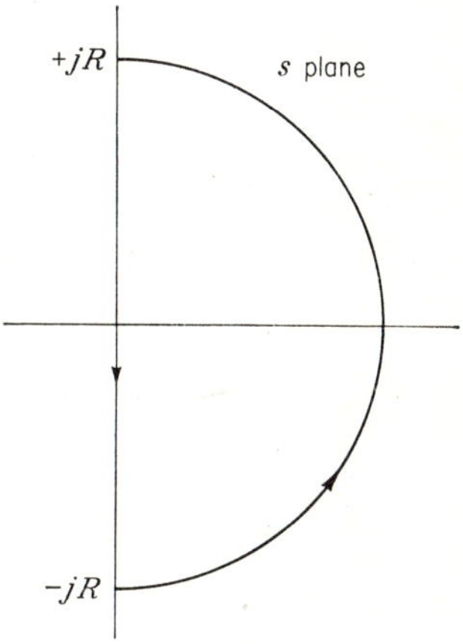

Fig. 9.6.1 Path of integration used in proving Theorem *9.6.6*.

The integral along the right-half-plane semicircle of radius R will go to zero as $R \to \infty$ because, as $|s| \to \infty$, $s\theta(s) \sim 2s_1 T(s)/s \to 0$ uniformly in $|\measuredangle s| \le \pi/2$. Thus

$$T(s_1) = \frac{-1}{2\pi j} \int_{-\infty}^{+\infty} \frac{2s_1[R(j\omega) + jI(j\omega)]}{-(\omega^2 + s_1^2)}\, j\, d\omega$$

or

$$T(s_1) = \frac{s_1}{\pi} \int_{-\infty}^{+\infty} \frac{R(j\omega) + jI(j\omega)}{\omega^2 + s_1^2}\, d\omega$$

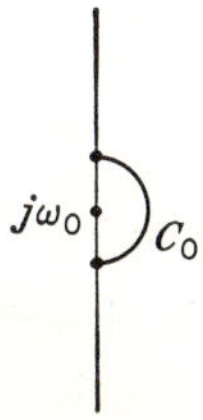

Fig. 9.6.2 Indentation used to bypass the singular point $j\omega_0$.

7 follows directly from the last equation by the symmetry properties mentioned in *2*.

If $T(s)$ has a singularity at $j\omega_0$, the contour C is modified by an indentation consisting of a semicircle C_0 of radius ρ centered on $j\omega_0$ (see Fig. *9.6.2*). The integral along C_0 vanishes, as $\rho \to 0$, because the length of the path of integration is $\pi\rho$ and, by *3*,

$$\lim_{\rho \to 0} \left[\rho \sup_{s \in C_0} \frac{T(s)}{s^2 - s_1^2} \right] = 0$$

Thus the $j\omega$ singularities will not contribute to the contour integral. This completes the proof of *7*.

8 **Theorem** If $T(s)$ satisfies Assumptions *1* to *4*, then, for any point $j\omega_1$ of the $j\omega$ axis,

9
$$I(j\omega_1) = \frac{2\omega_1}{\pi} \int_0^\infty \frac{R(j\omega) - R(j\omega_1)}{\omega^2 - \omega_1^2}\, d\omega$$

Proof Here we consider the following integrand:

$$\theta(s) = \frac{T(s) - R(j\omega_1)}{s - j\omega_1} - \frac{T(s) - R(j\omega_1)}{s + j\omega_1} = \frac{2j\omega_1[T(s) - R(j\omega_1)]}{s^2 + \omega_1^2}$$

To begin with, let $T(s)$ be analytic along the $j\omega$ axis; then $\theta(s)$ is analytic inside the right half plane and on the $j\omega$ axis except for two simple poles at $\pm j\omega_1$. Let us integrate $\theta(s)$ on the contour C shown in Fig. *9.6.3*: note the indentations C_1 and C_2 of the contour at the poles $j\omega_1$ and $-j\omega_1$. From Cauchy's integral theorem

$$\oint_C \theta(s)\, ds = 0$$

Since for $|s| \to \infty$, $s\theta(s) \sim T(s)/s \to 0$ uniformly in $|\measuredangle s| \leq \pi/2$, the contribution of the right semicircle of radius R vanishes as $R \to \infty$. Therefore, in the limit of $R \to \infty$,

$$-\!\!\!\!\!\!\int_{-\infty}^{+\infty} \frac{2\omega_1[T(j\omega) - R_1]}{\omega^2 - \omega_1^2}\, d\omega + \int_{C_1} \frac{T(s) - R_1}{s - j\omega_1}\, ds - \int_{C_2} \frac{T(s) - R_1}{s + j\omega_1}\, ds - \int_{C_1} \frac{T(s) - R_1}{s + j\omega_1}\, ds + \int_{C_2} \frac{T(s) - R_1}{s - j\omega_1}\, ds = 0$$

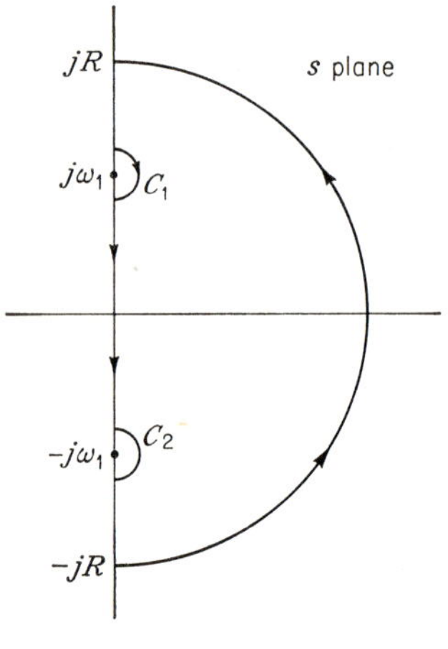

Fig. 9.6.3 Path of integration used in proving Theorem *9.6.8*.

where, for simplicity, R_1 is written for $R(j\omega_1)$. The horizontal bar across the first integral is used to indicate that the intervals of length 2ρ centered at $\pm j\omega_1$ are not included in the integration. Now as $\rho \to 0$, the last two integrals vanish, since their integrands are analytic in the neighborhoods of $-j\omega_1$ and $+j\omega_1$, respectively. Consider now the second integral: its integrand has a pole at the center of the semicircle C_1 and has the Laurent series expansion about $s = j\omega_1$,

$$\frac{T(s) - R_1}{s - j\omega_1} = \frac{jI(j\omega_1)}{s - j\omega_1} + \frac{d}{ds}[R(s) - R_1 + jI(s)]\Big|_{s=j\omega_1} + \frac{s - j\omega_1}{2!}\frac{d^2}{ds^2}[R(s) - R_1 + jI(s)]\Big|_{s=j\omega_1} + \cdots$$

Clearly, as $\rho \to 0$, only the first term contributes to the integral and

$$\int_{C_1} \frac{T(s) - R_1}{s - j\omega_1}\, ds \to jI(j\omega_1) \int_{C_1} \frac{ds}{s - j\omega_1} = -\pi I(j\omega_1)$$

The third integral gives the same answer. Therefore, as $\rho \to 0$, we get

$$2\pi I(j\omega_1) = \int_{-\infty}^{+\infty} \frac{2\omega_1[R(j\omega) - R(j\omega_1)]}{\omega^2 - \omega_1^2}\, d\omega + 2j\omega_1 \int_{-\infty}^{+\infty} \frac{I(j\omega)}{\omega^2 - \omega_1^2}\, d\omega$$

The last integral is zero, since $I(j\omega)$ is an odd function of ω by *2*. Since $R(j\omega)$ is an even function of ω, the first integrand remains finite at $\omega = \pm\omega_1$ and is an even function of ω; therefore, the integration need only be carried from 0 to ∞, and *9* follows.

If $T(s)$ has singularities on the $j\omega$ axis, the contour has to be appropriately indented and it can be shown, as in the preceding case, that the integrals along these indentations do not contribute to the integral. Q.E.D.

The next question that comes to mind is the following: Since $R(j\omega)$ determines the imaginary part, does the imaginary part determine the real part? Consider the example of $T(s)$ equal to an arbitrary real constant for all s. $I(j\omega)$ is identically zero, but $R(j\omega)$ is a nonzero constant. The answer to the question is, therefore, negative. The precise result is the following

10 **Theorem** If $T(s)$ satisfies Assumptions *1* to *3*, and if $T(s)$ is analytic at infinity with the Taylor expansion

$$T(s) = R_\infty + \frac{I_\infty}{s} + \frac{R_2}{s^2} + \cdots$$

then, for any point $j\omega_1$ of the $j\omega$ axis

11 $$R(j\omega_1) = R_\infty - \frac{2}{\pi} \int_0^\infty \frac{\omega I(j\omega) - \omega_1 I(j\omega_1)}{\omega^2 - \omega_1^2}\, d\omega$$

Proof Exercise for the reader. [*Hint:* Use the integrand

$$\theta(s) = \frac{T(s) - jI(j\omega_1)}{s - j\omega_1} + \frac{T(s) + jI(j\omega_1)}{s + j\omega_1}$$

It is the integration along the right half semicircle of radius R which contributes the term R_∞ in *11*.]

12 *Comment* In each of the above three formulas *7*, *9*, and *11*, the left-hand side is a *functional* of $R(\cdot)$, $R(\cdot) - R(j\omega_1)$, $I(\cdot) - j\omega_1 I(j\omega_1)$, respectively. [The dot $(\cdot)$ is used to indicate the variable $j\omega$ which runs from $-j\infty$ to $j\infty$.] For example, in *9*, $I(j\omega_1)$ depends on the values that $R(\cdot)$ takes over the whole $j\omega$ axis. The weighting function $2\omega_1/\omega^2 - \omega_1^2$ does give more emphasis to the values in the neighborhood of ω_1. However, it is instructive to consider a case where the value of $I(j\omega_1)$ is determined by the values of $R(j\omega)$ taken several decades away from ω_1. For example, consider a strictly stable amplifier with $H(s)$ as a transfer function. Take $T(s) = \log_e H(s)$; hence, $R(j\omega) = \log_e |H(j\omega)|$ and $I(j\omega) = \measuredangle H(j\omega)$. Let the gain $R(j\omega)$ be zero nepers for $\omega < \omega_0$ and be $-N$ nepers for $\omega > \omega_0$. Consider a frequency $\omega_1 \ll \omega_0$, say $10^{-3} \times \omega_0$. Then, since $\omega \gg \omega_1$ whenever $R(j\omega) \neq 0$, we obtain from *9*

$$I(j\omega_1) \simeq \frac{2\omega_1}{\pi} \int_0^\infty \frac{-N}{\omega^2}\, d\omega = \frac{-2N\omega_1}{\pi\omega_0} \qquad \text{rad}$$

Thus for any $\omega_1 \ll \omega_0$, the phase shift is linear with frequency but the slope of the phase curve depends on the amplifier gain many decades above the frequency ω_1.

13 *Comment* The three relations just established, *9* and *11* especially, are of great practical importance in the design of communication systems. For example, with $T(s)$ = log of a transfer function, it is often required to design a network that has a given gain $R(j\omega)$ and whose phase shift has a derivative almost constant through a certain frequency band: $dI(j\omega)/d\omega \simeq$ constant, i.e., the delay through the network is constant (*9.11.2*). The network is designed in two steps (see Fig. *9.6.4*): First a filter that provides the specified gain $R_F(j\omega) = R(j\omega)$

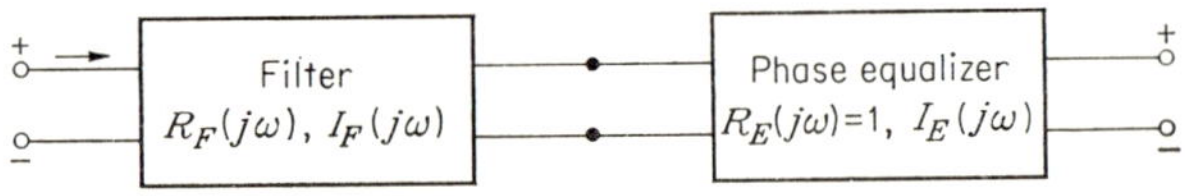

Fig. 9.6.4 Tandem connection of a filter and phase equalizer.

and the phase $I_F(j\omega)$ associated with $R(j\omega)$ as computed by *9* is designed. Next a phase equalizer whose gain $R_E(j\omega)$ is unity at all frequencies and whose phase $I_E(j\omega)$ when added to $I_F(j\omega)$ gives the required phase $I(j\omega)$ is designed. Formula *9* allows the designer to compute $I_F(j\omega)$ before the filter is built and therefore allows the design and testing of both filter and equalizer to proceed simultaneously. Another well-known example is the problem of shaping the loop gain of a single-loop-feedback amplifier. Bode's book is the best reference on this subject. Recently the extension to conditionally stable amplifiers has been carried out.[1] In physics, these relations are known as the

[1] J. Oizumi, M. Kimura, Design of Conditionally Stable Feedback System, *I.R.E. Trans. Circuit Theory*, vol. CT-4, no. 3, 1957.

Kramers-Kronig[1] relations or the dispersion relations; they relate, for example, the real and imaginary parts of the magnetic susceptibility χ and those of the dielectric constant ε. In nuclear theory these dispersion relations have recently received a great deal of attention.[2]

14 *Comment* In practice, it is convenient to plot curves on a log frequency scale. For this reason, a change of variable $u = \log(\omega/\omega_2)$, where ω_2 is a normalizing frequency, is introduced. The result is

$$I(j\omega_1) = \frac{1}{\pi}\int_0^\infty \frac{R(j\omega) - R(j\omega_1)}{\sinh u}\,du$$

or

$$I(j\omega_1) = \frac{1}{\pi}\int_{-\infty}^\infty \frac{dR}{du}\log\left|\coth\frac{u}{2}\right|du$$

This integral is of interest because it shows that $I(j\omega_1)$ is a weighted average of the slope of R when plotted on the log frequency scale $u = \log(\omega/\omega_1)$. A plot of the weighting function $\log\left|\coth\frac{u}{2}\right|$ is shown in Fig. *9.6.5*.

[1] See, for example, C. Kittel, "Elementary Statistical Physics," John Wiley & Sons, Inc., New York, 1958.

[2] A. Bohr, in W. E. Brittin (ed.), "Lectures in Theoretical Physics," vol. III, Interscience Publishers, Inc., New York, 1961.

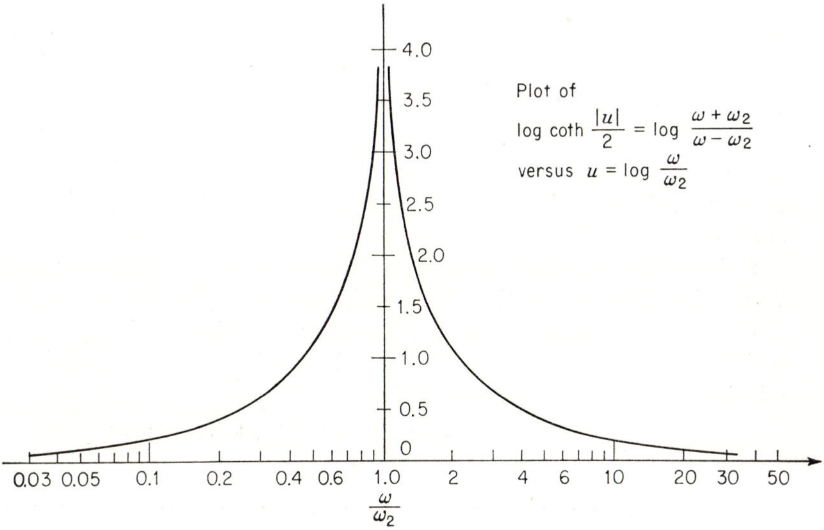

Fig. 9.6.5 Plot of $\log\coth\frac{|u|}{2} = \log\frac{\omega+\omega_2}{\omega-\omega_2}$ versus $u = \log\frac{\omega}{\omega_2}$.

7 *Minimum-phase transfer functions*

Consider the two rational transfer functions $H_1(s)$ and $H_2(s)$ that have the pole and zero configurations shown in Fig. *9.7.1*. They have iden-

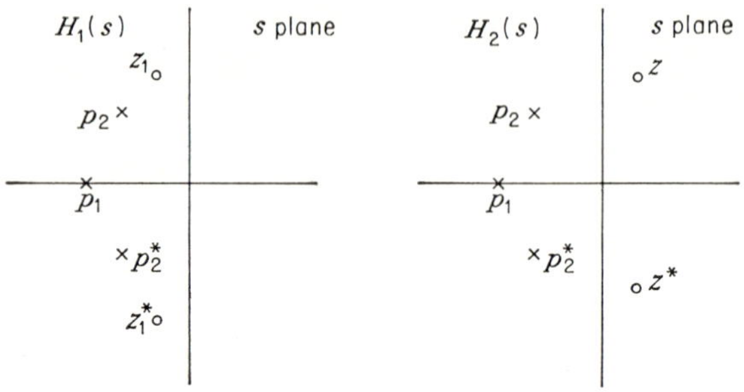

Fig. 9.7.1 Pole and zero locations of $H_1(s)$ and $H_2(s)$.

tical pole locations, but the zeros of $H_2(s)$ are the mirror images of those of $H_1(s)$ with respect to the $j\omega$ axis. Assume that $H_1(s)$ and $H_2(s)$ have the same constant factor; then from the potential analogy it is obvious that, for all ω's,

$$|H_1(j\omega)| = |H_2(j\omega)|$$

and for $\omega > 0$

$$\measuredangle H_1(j\omega) > \measuredangle H_2(j\omega)$$

They have the same gain, and the phase of H_2 lags that of H_1 at each positive frequency. It is clear also that, in general, if $H_1(s)$ and $H_2(s)$ are strictly stable rational transfer functions with the same gain at each frequency, then the one which has all its zeros in the left half plane will have at each frequency the least phase shift. Therefore, we say that the strictly stable rational transfer function $H(s)$ is *minimum phase* if and only if it has no finite zeros inside the right half plane.

It is clear also that for such a transfer function the new function $T(s) = \log H(s)$ may satisfy the requirements *9.6.1* to *9.6.4*. Therefore:

1 If $H(s)$ is a minimum-phase transfer function such that $\log H(s)$ satisfies *9.6.1* to *9.6.4*, then its phase is obtainable from its gain by the relation

$$\measuredangle H(j\omega_1) = \frac{2\omega_1}{\pi}\int_0^\infty \frac{\log|H(j\omega)| - \log|H(j\omega_1)|}{\omega^2 - \omega_1^2}\,d\omega$$

If $H(s)$ is a strictly stable rational transfer function and if it is not minimum phase, can its phase be computed by the contour integration technique used in the proof of *9.6.8*? The answer is yes, provided certain modifications are introduced: $T(s) = \log H(s)$ will have a branch point at each right-half-plane zero. For the case of Fig. *9.7.1*, the appropriate contour is shown in Fig. *9.7.2*: in order to remain on the same sheet of the Riemann surface of the integrand

$$\frac{2j\omega_1[\log H_2(s) - \log H_2(j\omega_1)]}{s^2 + \omega_1^2}$$

the branch cuts B_1 and B_2 had to be introduced. With respect to these branch-cut integrations it is straightforward to establish that (I) the numerator bracket of the integrand takes on the lower (outgoing) integration path of each cut a value larger by $2\pi j$ than the value it takes on the upper (incoming) integration path; (II) the integration around the almost complete circles centered on the zeros of $H_2(s)$ contribute nothing in the limit; and (III) a few standard manipulations lead to the answer

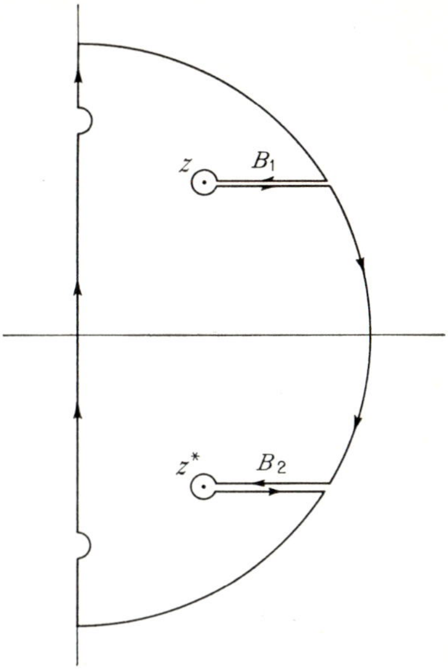

Fig. 9.7.2 Appropriate contour to compute the phase of H_2 from its gain along the $j\omega$ axis.

$$\measuredangle H_2(j\omega_1) = \frac{2\omega_1}{\pi}\int_0^\infty \frac{\log H(j\omega) - \log H(j\omega_1)}{\omega^2 - \omega_1^2}\,d\omega - 2\sum_i \tan^{-1}\frac{\sigma_{z_i}}{\omega_1 - \omega_{z_i}} - 2\sum_i \tan^{-1}\frac{\sigma_{z_i}}{\omega_1 + \omega_{z_i}}$$

where σ_{z_i} and ω_{z_i} are the real and imaginary parts of the first-quadrant zeros z_i of $H(s)$.

For the case where $H(s)$ is not rational, we adopt the following

2 **Definition** If $H(s)$ satisfies Assumptions *9.6.1* to *9.6.4* and if it satisfies the Paley-Wiener criterion, then it is said to be minimum phase if and only if its phase is given by

$$\measuredangle H(j\omega_1) = \frac{2\omega_1}{\pi}\int_0^\infty \frac{\log H(j\omega) - \log H(j\omega_1)}{\omega^2 - \omega_1^2}\,d\omega$$

3 *Comment* In evaluating experimental data it is strictly impossible to determine whether or not measured gain and phase curves are those of a minimum-phase transfer function. The reason is that no experimental data extend over all frequencies from 0 to ∞, and we have seen in *9.6.12* that the phase at some fixed frequency may depend significantly on the behavior of the gain at frequencies arbitrarily large compared with ω_1. Therefore, the only meaningful question is this: given gain and phase curves over a finite frequency band, and given an assumed gain over the remaining frequencies, do these gain and phase curves belong to a minimum-phase transfer function? This question can be answered simply by calculating the minimum phase associated with the given gain curve and comparing it with the given phase curve.

8 *Uncertainty principle*

It is a fact of experience that as the width of a pulse becomes smaller its bandwidth becomes larger in such a manner that the product of the two remains roughly constant. It is possible to give an exact and very general statement of this fact. It will turn out that the resulting inequality is exactly the same as that occurring in the statement of Heisenberg's uncertainty principle in quantum mechanics. The reason why this inequality occurs both in a time-frequency discussion and in a position-momentum discussion follows from the fact that in quantum mechanics the momentum is associated with the ϕ function that is the Fourier transform of the ψ function, which is related to the position.

We consider a real signal $h(t)$ and its Fourier transform

$$H(j\omega) = \int_{-\infty}^{+\infty} h(t)e^{-j\omega t}\, dt$$

We assume that the following inequalities are satisfied:

1 $$\|h\|^2 \triangleq \int_{-\infty}^{+\infty} h^2(t)\, dt < \infty$$

2 $$\bar{t} \triangleq \frac{\int_{-\infty}^{+\infty} th^2(t)\, dt}{\|h\|^2} < \infty$$

3 $$(\Delta t)^2 \triangleq \frac{1}{\|h\|^2} \int_{-\infty}^{+\infty} (t - \bar{t})^2 h^2(t)\, dt < \infty$$

If we think of $h(t)$ as a voltage across a 1-ohm load, $h^2(t)$ is the instantaneous power dissipated in that load. Therefore, $\|h\|^2$ represents the total energy dissipated in the load. Following Gabor, we shall call $\bar{t}$ the "epoch" of the signal: $\bar{t}$ is the center of gravity of the area under the curve $h^2(t)$. And $(\Delta t)^2$ is called the *dispersion* in time of the signal: it is a measure of how much the signal is spread about $\bar{t}$. More technically, Δt is the radius of gyration of the area under $h^2(t)$ about an axis going through $\bar{t}$. From another point of view, since the step response is the integral of the impulse response, Δt is a measure of the rise time of a linear system whose transfer function is $H(s)$. Similarly, we can define

4 $$\|H\|^2 = \int_{-\infty}^{+\infty} H(j\omega)H^*(j\omega)\,d\omega$$

5 $$\bar{\omega} \triangleq \frac{1}{\|H\|^2}\int_{-\infty}^{+\infty} \omega H(j\omega)H^*(j\omega)\,d\omega$$

6 $$(\Delta\omega)^2 = \frac{1}{\|H\|^2}\int_{-\infty}^{+\infty} \omega^2 H(j\omega)H^*(j\omega)\,d\omega$$

We assume that $\bar{\omega}$ and $(\Delta\omega)^2$ are finite. From Parseval's theorem,

$$\|h\|^2 = \frac{1}{2\pi}\|H\|^2$$

Since h is real, $H(-j\omega) = H(j\omega)^*$ for all real ω; hence $\bar{\omega} = 0$. Again $\Delta\omega$ is a measure of the bandwidth of the signal. The precise relationship between this bandwidth $\Delta\omega$ and the width Δt is the uncertainty principle, which may be expressed as the

7 **Theorem** If the signal $h(t)$ is such that the integrals *1* to *6* are finite, then

8 $$\Delta t \cdot \Delta\omega \geq \tfrac{1}{4}$$

Proof Define

$$H_1(j\omega) = H(j\omega)e^{+j\omega\bar{t}}$$

Then

$$|H_1(j\omega)|^2 = |H(j\omega)|^2 \qquad \text{and} \qquad \|h\|^2 = \|h_1\|^2$$

and since the Fourier transform is a special case of the bilateral Laplace transform (*B.2.12 et seq.*)

$$\mathcal{L}_{\mathrm{II}}^{-1}[H_1(s)] = h_1(t) = h(t + \bar{t})$$

From *3*

$$(\Delta t)^2 = \frac{1}{\|h\|^2}\int \tau^2 h_1^2(\tau)\,d\tau$$

Differentiating $H_1(j\omega) = \mathcal{L}_{\mathrm{II}}[h_1(t)]$, we get

$$j\frac{d}{d\omega}H_1(j\omega) = \int t h_1(t)e^{-j\omega t}\,dt$$

which means that the transform of $th_1(t)$ is $j\dfrac{d}{d\omega}H_1(j\omega)$. Parseval's theorem gives

$$\int t^2 h_1^2(t)\,dt = \frac{1}{2\pi}\int \frac{dH_1}{d\omega}\frac{dH_1^*}{d\omega}\,d\omega$$

and hence

9 $$(\Delta t)^2 = \frac{1}{\|H_1\|^2}\int \frac{dH_1}{d\omega}\frac{dH_1^*}{d\omega}\,d\omega$$

Consider the product of *6* and *9*. Since the product will not be affected if H_1 is multiplied by a constant, we may assume, without any loss of generality, that $\|H_1\|^2 = 1$. The result is

10 $$(\Delta\omega\,\Delta t)^2 = \int \omega^2 H_1 H_1^*\,d\omega \int \frac{dH_1}{d\omega}\frac{dH_1^*}{d\omega}\,d\omega$$

Now recall that the Schwarz inequality applied to complex functions states that

$$\int\psi_1\psi_1^*\,d\omega \int\psi_2\psi_2^*\,d\omega \geq \tfrac{1}{4}\left[\int(\psi_1\psi_2^* + \psi_2\psi_1^*)\,d\omega\right]^2$$

With $\psi_1 = \omega H_1$ and $\psi_2 = dH_1/d\omega$ the left-hand side of the above inequality is identical with the right-hand side of *10*; therefore

11 $$(\Delta t\,\Delta\omega)^2 \geq \tfrac{1}{4}N^2$$

where

$$N = \int \left(\omega H_1 \frac{dH_1^*}{d\omega} + \frac{dH_1}{d\omega}\,\omega H_1^*\right) d\omega$$

Upon integrating the first term by parts and observing that

$$\omega|H_1(j\omega)|^2 \to 0$$

as $\omega \to \pm\infty$, we get

$$N = -\int H_1(j\omega)H_1^*(j\omega)\,d\omega = -1$$

in view of the normalization assumption. Substituting this into *11* establishes the inequality *8*.

12 *Exercise* Prove Schwarz's inequality for complex functions. (*Hint:* Define the scalar product $\langle\psi_1,\psi_2\rangle \equiv \int\psi_1^*\psi_2\,d\omega$. Now, for all real λ, $\langle\psi_1 + \lambda\psi_2,\ \psi_1 + \lambda\psi_2\rangle = \langle\psi_1,\psi_1\rangle + \lambda^2\langle\psi_2,\psi_2\rangle + \lambda[\langle\psi_1,\psi_2\rangle + \langle\psi_2,\psi_1\rangle] \geq 0$, from which the desired inequality follows.)

9 *The dispersion of the unit-impulse response*

If the transfer function of a linear system has a constant gain $H(j\omega) = A$ and a phase linear with frequency $\phi(\omega) = -\omega\tau$ (that is, constant group delay τ), the impulse response is given by $A\delta(t - \tau)$. We know also that any system with only one of these types of ideal behavior no longer has an impulse response consisting of a single impulse; that is, if either or both of these properties are not fulfilled, the impulse response has a dispersion $(\Delta t)^2 > 0$. This intuitive idea can be given a precise expression by the following result.

1 **Theorem** If assumptions *9.8.1* to *9.8.3* hold and if we let

$$H(j\omega) = A(\omega)e^{j\phi(\omega)}$$

and $\phi(\omega) = -\omega\bar{t} + \Delta\phi(\omega)$, then

2 $$(\Delta t)^2 = \frac{1}{\|A\|^2}\int_{-\infty}^{+\infty} A^2(\omega)\left[\left(\frac{d}{d\omega}\log A\right)^2 + \left(\frac{d\Delta\phi}{d\omega}\right)^2\right] d\omega$$

where

$$\|A\|^2 = \int_{-\infty}^{+\infty} A^2(\omega)\, d\omega$$

3 *Comment* The energy density along the ω axis, $A^2(\omega)$, is a common weighting factor in the numerator. In addition, two terms contribute to $(\Delta t)^2$, one involving $(d \log A/d\omega)^2$, which is a measure of the *gain distortion* (the deviation of the gain from a constant), and the second involving $(d\Delta\phi/d\omega)^2$, which is a measure of the *group delay distortion.* Finally, this theorem implies the

4 **Assertion** Of all systems that have the same gain curve (i.e., the same $|H(j\omega)|$), the linear phase system has the least dispersion $(\Delta t)^2$.

Proof As a first step in the derivation, let us introduce $h_1(t)$ and $H_1(j\omega)$ as in *9.8.7*. With our present notations,

$$H_1(j\omega) = A(\omega)e^{j\Delta\phi(\omega)}$$

Furthermore, $h_1(t)$ has the same dispersion as $h(t)$, since it results from $h(t)$ by a translation in time of $\bar{t}$ sec. Observing that

$$\mathcal{L}[th_1(t)] = j\frac{d}{d\omega} H_1(j\omega)$$

we have, by Parseval's theorem,

5 $$(\Delta t)^2 = \frac{1}{\|H_1\|^2}\int_{-\infty}^{+\infty} \frac{d}{d\omega} H_1(j\omega) \cdot \frac{d}{d\omega} H_1^*(j\omega)\, d\omega$$

Since $\frac{d}{d\omega} H_1(j\omega) = A\left(\frac{1}{A}\frac{dA}{d\omega} e^{j\Delta\phi} + j\frac{d\Delta\phi}{d\omega} e^{j\Delta\phi}\right)$, we have

6 $$\frac{d}{d\omega} H_1(j\omega)\frac{d}{d\omega} H_1(j\omega)^* = A^2(\omega)\left[\left(\frac{d}{d\omega}\log A\right)^2 + \left(\frac{d\,\Delta\phi}{d\omega}\right)^2\right]$$

Finally,

7 $$\|H_1\|^2 = \int_{-\infty}^{+\infty} |H_1(j\omega)|^2\, d\omega = \int_{-\infty}^{+\infty} A^2(\omega)\, d\omega$$

and *2* follows from *5* to *7*.

10 *Moments*

Consider a linear time-invariant system $\mathcal{S}$ whose impulse response h is such that the integral $\int_{-\infty}^{+\infty} t^k h(t)\, dt$ exists for all integers $k \geq 0$. The value of the integral is called the *kth moment of h*, and we shall label it M_k; thus

1 $$M_k = \int_{-\infty}^{+\infty} t^k h(t)\, dt \qquad k = 0, 1, 2, \ldots$$

We shall now exhibit a relation between M_k and the system transfer function H. By definition

$$H(s) = \int_{-\infty}^{+\infty} h(t) e^{-st}\, dt$$

Let us proceed formally and differentiate k times under the integral sign. Thus, for s such that $\alpha < \operatorname{Re} s < \beta$, where α and β are the abscissae of absolute convergence of h (see *B.2.13*),

$$H^{(k)}(s) = \int_{-\infty}^{+\infty} (-t)^k h(t) e^{-st}\, dt$$

and, provided $\alpha < 0 < \beta$,

2 $$M_k = (-1)^k H^{(k)}(0) = \int_{-\infty}^{+\infty} t^k h(t)\, dt$$

From the Taylor series expansion of H about $s = 0$,

$$H(s) = \sum_{k=0}^{\infty} H^{(k)}(0) \frac{s^k}{k!}$$

and by *2*, we have

3 $$H(s) = \sum_{k=0}^{k} (-1)^k \frac{M_k}{k!} s^k$$

since $H(s)$ is analytic inside its strip of convergence. Therefore, the moments of h prescribe the behavior of H in the neighborhood of the origin. In those cases where, for example, H may be continued analytically in the whole plane, the knowledge of all the moments specifies H uniquely and, consequently, the function h itself. In general, however, the moments do not specify h uniquely.

4 *Exercise* Assume H to be analytic in a neighborhood of the origin. Let its Taylor expansion be written as *3* above. Is M_k the kth moment of h? [*Hint:* Consider the case $h(t) = 1(t)e^t$.]

The moments are of importance in engineering because the knowledge of, say, the first N moments of h describes h to a certain extent. For example, we may state the following

5 **Theorem** Let $\mathcal{S}_1$ ($\mathcal{S}_2$, respectively) be a linear time-invariant system with impulse response h_1 (h_2, respectively) and transfer function H_1 (H_2, respectively). Assume that the first N moments are equal. Then, for all t, the zero-state responses of $\mathcal{S}_1$ and $\mathcal{S}_2$ are equal provided their common input is a polynomial in t of degree smaller than or equal to N.

6 *Comment* We could say that the equalities

7 $$\int_{-\infty}^{+\infty} t^k h_1(t)\, dt = \int_{-\infty}^{+\infty} t^k h_2(t)\, dt = M_k \qquad k = 1, 2, \ldots, N$$

imply that the systems $\mathcal{S}_1$ and $\mathcal{S}_2$ are *zero-state equivalent* (*2.9.1*) over the class of polynomial inputs of degree $\leq N$.

Proof Let $\sum_{k=0}^{N} c_k t^k$ be the common input; then the outputs y_1 and y_2 are given by

$$y_i(t) = \int_{-\infty}^{+\infty} \sum_{0}^{N} c_k (t-\tau)^k h_i(\tau)\, d\tau \qquad i = 1, 2$$

$$= \sum_{k=0}^{N} c_k \sum_{\alpha=0}^{k} \binom{k}{\alpha} (-1)^\alpha t^{k-\alpha} \int_{-\infty}^{+\infty} \tau^\alpha h_i(\tau)\, d\tau$$

and using *7*,

$$y_i(t) = \sum_{k=0}^{T} c_k \sum_{k=0}^{k} \binom{\alpha}{k} (-1)^\alpha t^{k-\alpha} M_\alpha$$

The right-hand side is independent of i; thus the outputs y_1 and y_2 are identical for all t.

11 *Group delay*

The group delay—a concept closely related to the group velocity of the physicist—is a very basic concept which is very frequently used when requirements are set up for communication systems. In order to introduce the concept, let us consider the following situation. We are given a strictly stable linear system: it may be an amplifier, a filter, or a whole communication system. Let us excite it by a unit impulse $\delta(t)$; in other words, the input signal contains all frequencies with equal amplitude and with the same phase. The output is by definition the unit-impulse response $h(t)$. According to the Fourier integral theorem

1 $$h(t) = \frac{1}{2\pi} \int_{-\infty}^{+\infty} H(j\omega) e^{j\omega t} \, d\omega$$

That is, $h(t)$ is a superposition of sine waves of all frequencies. Let us ask the following question: If we consider only the contributions to $h(t)$ of the frequencies in the interval $(\omega_0 - \Delta, \omega_0 + \Delta)$, where Δ is very small and $\Delta \ll \omega_0$, is it possible to ascribe to this "group" of frequencies a time at which they appear at the output? The answer will turn out to be that it is possible to do so in a certain approximate sense, and the time required for the "group" of frequencies centered at ω_0 to emerge from the system is called the *group delay at* ω_0.

We shall now proceed with some analysis to answer the question fully. Let us exhibit in *1* the amplitude $A(\omega)$ and the phase $\phi(\omega)$ of the transfer function:

$$h(t) = \frac{1}{2\pi} \int_{-\infty}^{+\infty} A(\omega) e^{j[\omega t + \phi(\omega)]} \, d\omega$$

Let us now divide the frequency axis in equal segments of length Δ. Then

$$h(t) = \frac{1}{2\pi} \sum_{n=-\infty}^{+\infty} \int_{n\Delta}^{(n+1)\Delta} A(\omega) \{\exp j[\omega t + \phi(\omega)]\} \, d\omega$$

Suppose Δ is taken sufficiently small so that for all segments $(n\Delta, (n+1)\Delta)$ the amplitude $A(\omega)$ is almost constant. Also let $\omega_{0n} = (n + \frac{1}{2})\Delta$. Thus

$$h(t) \simeq \frac{1}{2\pi} \sum_{-\infty}^{+\infty} A(\omega_{0n}) \int_{n\Delta}^{(n+1)\Delta} \{\exp j[\omega t + \phi(\omega)]\} \, d\omega$$

Let us further require that Δ be sufficiently small that over each interval $(n\Delta, (n+1)\Delta)$ we may, with negligible error, replace $\phi(\omega)$ by

$$\phi(\omega_{0n}) + (\omega - \omega_{0n}) \left.\frac{d\phi}{d\omega}\right|_{\omega_{0n}}$$

Introducing this approximation, we get

$$h(t) \simeq \frac{1}{2\pi} \sum_{-\infty}^{+\infty} A(\omega_{0n}) e^{j[\omega_{0n}t - \phi(\omega_{0n})]} \int_{\omega_{0n}-\Delta/2}^{\omega_{0n}+\Delta/2} \{\exp j[(\omega - \omega_{0n})(t + \phi_n')]\} \, d\omega$$

where we wrote ϕ_n' for $\left.\frac{d\phi}{d\omega}\right|_{\omega_{0n}}$. Carrying out the integration, we get

$$h(t) \simeq \frac{\Delta}{2\pi} \sum_{-\infty}^{+\infty} A(\omega_{0n}) \frac{\sin (\Delta/2)(t + \phi_n')}{(\Delta/2)(t + \phi_n')} e^{j[\omega_{0n}t - \phi(\omega_{0n})]}$$

If we collect together the terms corresponding to ω_{0n} and $-\omega_{0n}$, then, taking into account the evenness of $A(\omega)$ and $d\phi/d\omega$, we get

$$h(t) \simeq \frac{\Delta}{\pi} \sum_{n=1}^{\infty} A(\omega_{0n}) \frac{\sin (\Delta/2)(t + \phi'_n)}{(\Delta/2)(t + \phi_n)} \cos [\omega_{0n}t - \phi(\omega_{0n})]$$

Each term of the sum can be interpreted as an amplitude-modulated cosine wave: $\cos [\omega_{0n}t - \phi(\omega_{0n})]$ is the carrier and the instantaneous amplitude is

$$\frac{\Delta}{\pi} A(\omega_{0n}) \frac{\sin (\Delta/2)(t + \phi'_n)}{(\Delta/2)(t + \phi'_n)}$$

The contribution of a single term is plotted in Fig. *9.11.1*. The shape of the curve suggests the following statement: the group of frequencies

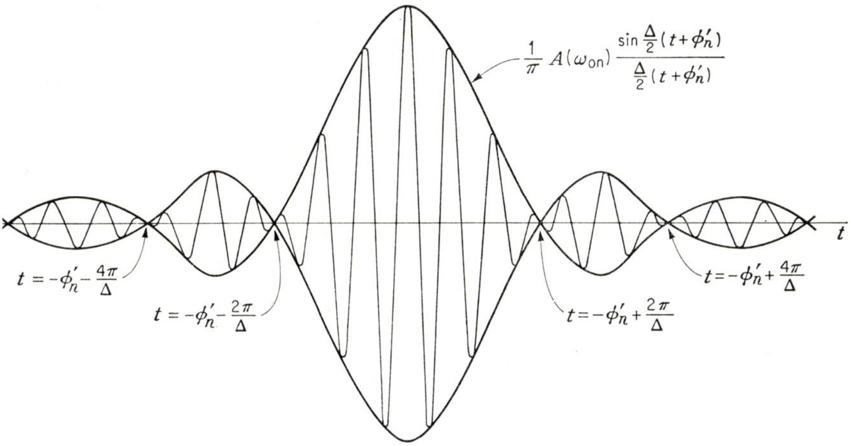

Fig. 9.11.1 Amplitude-modulated wave representing the contribution of the interval $(\omega_{0n} - \Delta/2, \omega_{0n} + \Delta/2)$.

centered around ω_{0n} are delayed by $-\left.\frac{d\phi}{d\omega}\right|_{\omega_{0n}}$ in the sense that their contribution to $h(t)$ is maximum at that instant.[1] For this reason, and in this sense, we say that

2 **Definition** $-\left.\frac{d\phi}{d\omega}\right|_{\omega_0}$ is the group delay of $H(j\omega) = A(\omega)e^{j\phi(\omega)}$ at the frequency ω_0.

Radio engineers talk about the *group velocity* for the excellent reason that they are interested in propagation through various media. If the

[1] To be accurate, we should take into account the phase of the carrier and say that the maximum occurs in the small interval $\left[-\frac{d\phi}{d\omega} - \frac{\pi}{2\omega_{0n}}, -\frac{d\phi}{d\omega} + \frac{\pi}{2\omega_{0n}}\right]$.

unit impulse of the preceding analysis is applied at $x = 0$ and $t = 0$ and the resulting phenomenon is observed at x_0, the appropriate transfer function is

$$A(\omega)e^{j[\omega t - \beta(\omega)x_0]}$$

Therefore, the phase shift $\phi(\omega)$ is now $-\beta(\omega)x_0$. The group delay is therefore $(d\beta/d\omega)x_0$, and the *group velocity* is $(d\beta/d\omega)^{-1} = d\omega/d\beta$.

The concept of group velocity is of great importance because in homogeneous media the group velocity is the velocity at which energy propagates. This concept appears in quantum mechanics and in the propagation of electromagnetic waves in wave guides, velocity waves in electron streams, and electromagnetic waves in ionized media (in fact, the group velocity explains the whistlers). In the case of gravity waves in deep water, the concept explains the great velocity of the tsunamis which periodically wash out a city or two on the shores of the Pacific.

The concept of group delay is also useful in the network approximation problem. For example, suppose we have to design a linear pulse amplifier. What criterion should one introduce: constant amplitude over the band of interest or something else? The group delay concept suggests that for a pulse amplifier all the frequencies of interest should have the same delay. This leads to an approximation problem as follows: $H(j\omega)$ should be such that $d\phi/d\omega$ is constant over the band of interest. This idea, coupled with the idea of Taylor series approximation, leads naturally to the maximally flat delay approximation which has recently been used with great success in many pulse systems. The concept of group delay is basic to the invention of the chirp radar.[1]

3 *Exercise* Can there exist a physical transfer function that has negative delay at some frequencies? (*Hint:* Use the potential analogy to construct a strictly stable rational transfer function that has negative group delay at some frequencies.) Compare the rate of change of $A(\omega)$ with that of $\phi(\omega)$ in the frequency interval where the delay is negative.

4 *Exercise* Physicists define the phase velocity v_p as $v_p = \lambda\nu = \omega/\beta$, where λ is the wavelength and ν the frequency. The group velocity v_g is $(d\beta/d\omega)^{-1}$. Show that

$$v_g = v_p - \lambda \frac{dv_p}{d\lambda} = v_p + \frac{1}{\lambda}\frac{dv_p}{d(1/\lambda)}$$

(*Hint:* $\beta = 2\pi/\lambda$. These relations are useful; for the phase velocity v_p is often plotted against the wavelength λ or the wave number $1/\lambda$.)

[1] J. R. Klauder, A. C. Price, S. Darlington, and W. J. Albersheim, The Theory and Design of Chirp Radars, *Bell System Tech. J.*, vol. 39, no. 4, pp. 745–808, 1960.

12 *Paired echoes*

Paired-echo theory

We wish to include a brief discussion of the theory of paired echoes for two reasons: first, it illuminates certain aspects of the relationship between a time function and its Fourier transform; second, it is the basis of the time-domain equalizer and the transversal filter which have found use in communication systems and in servomechanisms.

We shall be concerned with the following questions. Given a time function $h(t)$ with $H(j\omega)$ as its transform, what happens to the time function (I) if $|H(j\omega)|$ is changed but $\measuredangle H(j\omega)$ is left unchanged and (II) if $|H(j\omega)|$ is left unchanged but $\measuredangle H(j\omega)$ is changed? We can also think of $H(j\omega)$ as a transfer function and $h(t)$ as the corresponding impulse response. With this interpretation we ask what happens to the impulse response when we modify the gain $|H(j\omega)|$ of the system and leave the phase $\measuredangle H(j\omega)$ alone or vice versa. Let $H_1(j\omega)$ be the modified transforms.

Case 1 $H_1(j\omega) = H(j\omega)\,(1 + 2a \cos \omega\tau)$. Of course a must be real in order that $H_1(0)$ be real, and we assume $0 < |a| < \frac{1}{2}$. To be specific, we shall assume $a > 0$. Clearly

$$|H_1(j\omega)| = |H(j\omega)|\,(1 + 2a \cos \omega\tau) \qquad \text{and} \qquad \measuredangle H_1(j\omega) = \measuredangle H(j\omega)$$

Using the fact that

$$1 + 2a \cos \omega\tau = ae^{-j\omega\tau} + 1 + ae^{+j\omega\tau}$$

we get

1 $$\mathcal{L}_{\mathrm{II}}^{-1}\,[H_1(j\omega)] = ah(t - \tau) + h(t) + ah(t + \tau)$$

The new impulse response is equal to the original one plus two echoes: $ah(t - \tau)$ is the original impulse response multiplied by $a < \frac{1}{2}$ lagging $h(t)$ by τ sec. Similarly, $ah(t + \tau)$ leads it by τ sec. The pattern is illustrated by Fig. *9.12.1*.

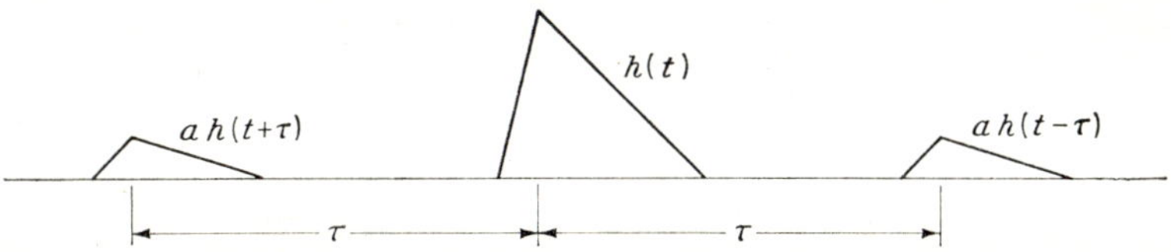

Fig. 9.12.1 Paired echoes created by a cosinusoidal perturbation of the gain.

Case 2 $H_1(j\omega) = H(j\omega) \exp(j\beta \sin \omega\tau)$, β real and constant. In this case,

$$|H_1(j\omega)| = |H(j\omega)|$$

and

$$\measuredangle H_1(j\omega) = \measuredangle H(j\omega) + \beta \sin \omega\tau$$

Therefore, β measures, in radians, the peak phase deviation from the original phase curve. Now consider the following Fourier series expansion:

$$\exp(j\beta \sin \theta) = \sum_{-\infty}^{+\infty} J_n(\beta) e^{jn\theta}$$

where $J_n(\cdot)$ is the nth Bessel function of the first kind. Hence

$$H_1(j\omega) = H(j\omega) \sum_{-\infty}^{+\infty} J_n(\beta) e^{jn\omega\tau}$$

Using an obvious notation, we get

2 $$h_1(t) = \sum_{n=-\infty}^{+\infty} J_n(\beta) h(t + n\tau)$$

Note that, for all integers n,

3 $$J_{-n}(\beta) = (-1)^n J_n(\beta)$$

In contrast with the preceding case, we have here an infinite number of echoes and the echoes have positive and negative weighting factors. Practically, however, for $\beta < \frac{1}{2}$ rad, a very good approximation is obtained by a few terms. This follows from the following expansion:

4 $$J_n(\beta) = \left(\frac{\beta}{2}\right)^n \sum_{l=0}^{\infty} \frac{(j\beta/2)^{2l}}{l!(n+l)!}$$
$$= \frac{1}{n!}\left(\frac{\beta}{2}\right)^n - \frac{1}{(n+1)!}\left(\frac{\beta}{2}\right)^{n+2} + \frac{1}{2!(n+2)!}\left(\frac{\beta}{2}\right)^{n+4} - \cdots$$

For example, if the peak phase deviation is 15°, $\beta \simeq 0.3$, $J_2(\beta) < 0.011$, and all terms beyond $|n| = 1$ are negligible. Hence

$$h_1(t) \simeq -J_1(\beta)h(t-\tau) + J_0(\beta)h(t) + J_1(\beta)h(t+\tau)$$

Again there are two "echoes," but their signs are reversed in the present case. In addition, the central term is multiplied by $J_0(\beta)$. This is illustrated in Fig. *9.12.2*.

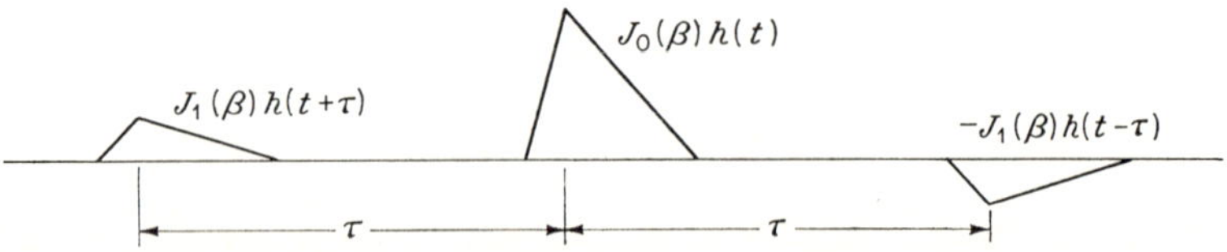

Fig. 9.12.2 Paired echoes created by a sinusoidal perturbation of the phase.

The transversal filter

In order to illustrate the concept, we shall briefly discuss the basic idea behind the transversal filter and the time-domain equalizer. The principal idea is to use a lossless delay line with $N + 1$ equally spaced taps. The delay between any two successive taps is τ sec. We assume that the delay line is matched at its end and that the energy derived from the main signal at each tap is a negligible portion of it. At each tap the signal is multiplied by a constant c_k, and if necessary its sign is changed by a phase inverter. The results of each multiplication are added together to form the output. Clearly, the transfer function is (except for a multiplicative constant independent of frequency which represents the "flat" attenuation through the system)

5 $$H(j\omega) = e^{-jN\omega\tau} \sum_{-N}^{+N} c_k e^{-jk\omega\tau} \qquad \text{with } c_0 = 1$$
$$= e^{-jN\omega\tau} \left(1 + \sum_1^N a_k \cos k\omega\tau + \sum_1^N b_k \sin k\omega\tau\right)$$

where we put

6 $$a_k = c_k + c_{-k} \qquad b_k = c_k - c_{-k} \qquad k = 1, 2, \ldots$$

The system is illustrated in Fig. *9.12.3*. If we set all the $b_k = 0$, we

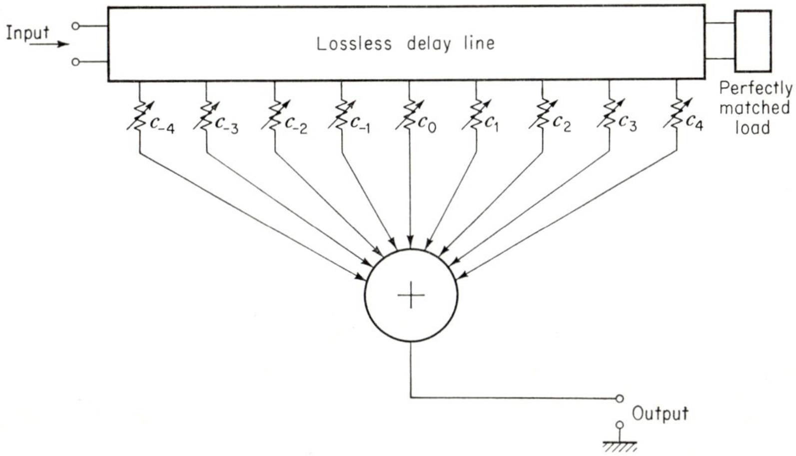

Fig. 9.12.3 A transversal filter.

get a transversal filter with a transfer function

$$H(j\omega) = e^{-jN\omega\tau} \left(1 + \sum_1^N a_k \cos k\omega\tau\right)$$

This is a *linear phase filter* whose gain is given by an N-term cosine Fourier series. The gain is adjustable.

In the time-domain equalizer case, let us assume $\varepsilon = \sum_1^N |c_k| \ll 1$.

Then

$$H(j\omega) \simeq e^{-jN\omega T}\left(1 + \sum_1^N a_k \cos k\omega T + j\sum_1^N b_k \sin k\omega T\right)$$

$$\simeq \left(1 + \sum_1^N a_k \cos k\omega T\right) \exp\left[-j\left(N\omega T - \sum_1^N b_k \sin k\omega T\right)\right] + 0(\varepsilon^2)$$

Therefore, provided the c_k's are kept sufficiently small, the equalizer provides an N-term cosine series for its gain and an N-term sine series for its phase. In addition, the adjustment of the gain can be made independently from that of the phase by *6*.

The reader should keep in mind that this elegant and simple behavior results from our assumptions: (I) a perfect delay line, (II) a perfect match at the end of the delay line, and (III) negligible loading by the taps. In practice, the delay line will be imperfectly uniform, lossy, and imperfectly matched.

13 *Asymptotic relations between $H(s)$ and $h(t)$*

Behavior of $h(t)$ for small t

It is easy to obtain a series expansion of $h(t)$ which has extensive use in evaluating $h(t)$ for small t. The advantage of this expansion is the ease with which a few terms of the series can be computed.

1 **Theorem** If $H(s)$ has a Taylor expansion about the point at infinity of the form

2 $$H(s) = \sum_{k=1}^{\infty} \frac{a_k}{s^k} \qquad \text{for } |s| > R > 0$$

then

3 $$h(t) = 1(t) \sum_{k=1}^{\infty} \frac{a_k}{(k-1)!}\, t^{k-1} \qquad \text{for all } t$$

where $\mathcal{L}\{h(t)\} = H(s)$.

4 *Comment* If $H(s)$ is rational, then $H(s)$ is analytic for $|s| > R_0$, where R_0 is the distance between the origin and the farthest pole of $H(s)$ and the Taylor series expansion of $H(s)$ about the point at infinity converges for all s such that $|s| > R_0$.

Proof First observe that, for the inversion integral, a suitable abscissa for the integration path is any $c > R$. With such a choice of c, we may replace $H(s)$ by its series expansion *2*, since the series expansion is valid along the whole path of integration. Furthermore, since inside its circle of convergence (which is $|s| > R$, in the present case) a power series converges uniformly, the summation and integration may be interchanged. Thus, successively

$$h(t) = \frac{1}{2\pi j} \int_{c-j\infty}^{c+j\infty} \left(\sum_{k=1}^{\infty} \frac{a_k}{s^k} \right) e^{st}\, ds = \sum_{k=1}^{\infty} \frac{a_k}{2\pi j} \int_{c-j\infty}^{c+j\infty} \frac{e^{st}}{s^k}\, ds$$

and $$h(t) = 1(t) \sum_{k=1}^{\infty} \frac{a_k}{(k-1)!}\, t^{k-1} \qquad \text{for all } t$$

5 *Comment* This result is very useful in practical applications for the following reasons: (I) In order to obtain the expansion in powers of $1/s$ of a rational transfer function, simply divide the denominator into the numerator. For example, let $H(s) = (s + 1)/(s^2 + s + 1)$:

$$\begin{array}{r|l}
 & \frac{1}{s} - \frac{1}{s^3} + \frac{1}{s^4} + \cdots \\ \hline
s^2 + s + 1 & s + 1 \\
 & s + 1 + \frac{1}{s} \\ \hline
 & -\frac{1}{s} \\
 & -\frac{1}{s} - \frac{1}{s^2} - \frac{1}{s^3} \\ \hline
 & \frac{1}{s^2} + \frac{1}{s^3} \\
 & \frac{1}{s^2} + \frac{1}{s^3} + \frac{1}{s^4} \\ \hline
 & -\frac{1}{s^4}
\end{array}$$

Hence $h(t) = 1(t)\left(1 - \frac{t^2}{2!} + \frac{t^3}{3!} + \cdots\right)$. (II) The power series in time is usually the fastest and most accurate way of obtaining the $h(t)$ for small t. (III) It is a useful check on numerical work when $h(t)$ is obtained by the partial-fraction-expansion method or by residues. ◁

The following result, which generalizes the preceding one, is often useful:

6 **Theorem** If, for $|s| > R$,

$$H(s) = \sum_{k=0}^{\infty} \frac{a_k}{s^{\alpha_k}} \qquad 0 < \alpha_0 < \alpha_1 < \cdots \to \infty$$

is an absolutely convergent series, then

$$h(t) = 1(t) \sum_{k=0}^{\infty} a_k \frac{t^{\alpha_k - 1}}{\Gamma(\alpha_k)} \qquad t \neq 0$$

where $\mathcal{L}\{h(t)\} = H(s)$ and the series converges absolutely.

Proof The derivation is identical to the preceding one.[1]

7 *Example* Let $H(s) = 1/s\sqrt{1+s}$. Using standard series expansions, we get successively, for $|s| > 1$,

$$H(s) = \frac{1}{s^{3/2}}\left(1 + \frac{1}{s}\right)^{-1/2} = \frac{1}{s^{3/2}} - \frac{1}{2s^{5/2}} + \frac{1 \times 3}{2 \times 4}\frac{1}{s^{7/2}} - \cdots$$

Hence, since $\Gamma(\frac{1}{2}) = \sqrt{\pi}$ and $\Gamma(z+1) = z\Gamma(z)$,

$$h(t) = 1(t)\left(\frac{2t^{1/2}}{\sqrt{\pi}} - \frac{2t^{3/2}}{3\sqrt{\pi}} + \frac{t^{5/2}}{5\sqrt{\pi}} - \cdots\right)$$

Asymptotic behavior for $t \to \infty$

To estimate the behavior of $h(t)$ for large t on the basis of $H(s)$ is very important in many applications, stability problems in particular. We shall give two results. The first one is well known and applies to rational $H(s)$; the second one is useful when $H(s)$ has more general singularities.

8 **Theorem** Assume that (I) $H(s)$ is a rational function of s that vanishes for $|s| \to \infty$ and is analytic for $\operatorname{Re} s > \alpha$, (II) $s_1, s_2, \ldots, s_n$ are the only poles of $H(s)$ in the strip $\alpha - \delta \leq \operatorname{Re} s \leq \alpha$, where $\delta > 0$, and (III) at each of these poles, the principal part of the Laurent series expansion of $H(s)$ is

$$\frac{b_1^{(k)}}{s - s_k} + \frac{b_2^{(k)}}{(s - s_k)^2} + \cdots + \frac{b_{l_k}^{(k)}}{(s - s_k)^{l_k}} \qquad k = 1, 2, \ldots, n$$

Then, as $t \to \infty$,

9 $$h(t) = 1(t)\left[\sum_{k=1}^{n}\left(b_1^{(k)} + b_2^{(k)}t + \cdots + \frac{b_{l_k}^{(k)}}{(k-1)!}t^{k-1}\right)e^{s_k t} + 0(e^{(\alpha-\delta)t})\right]$$

Proof The proof is very simple; so we shall only outline its steps: (I) consider the complete partial-fraction expansion of $H(s)$; (II) take its inverse transform; and (III) observe that, for any real ξ and η such that $\xi < \eta$, $e^{\xi t} + e^{\eta t} \sim e^{\eta t}$ as $t \to \infty$, from which *9* follows.

10 **Corollary** Let $H(s)$ be a rational transfer function. $H(s)$ is strictly stable if and only if all its poles are inside the left half plane, i.e., have negative (<0) real parts.

[1] G. Doetsch, "Handbuch der Laplace Transformation," vol. II, p. 175.

We shall now state, without proof, a result which is often useful when $H(s)$ is not rational. Doetsch (II, page 159) has the most complete discussion of this and many other asymptotic results.

11 **Theorem** Assume that (I) $H(s)$ is singular at s_0; (II) the path of the inversion integral can be deformed into the path C shown in Fig. *9.13.1*, where $\pi/2 < \phi < \pi$; (III) $H(s)$ is analytic on C; and (IV) the following asymptotic expansion is valid uniformly in the angle $|\measuredangle(s - s_0)| \le \phi$:

$$H(s) \sim \sum_{\nu=0}^{\infty} c_\nu (s - s_0)^{\gamma_\nu}$$

where $\operatorname{Re}\gamma_0 < \operatorname{Re}\gamma_1 < \operatorname{Re}\gamma_2 < \cdots$. Then, as $t \to \infty$,

$$h(t) \sim 1(t) e^{s_0 t} \sum_{\nu=0}^{\infty} \frac{c_\nu}{\Gamma(-\gamma_\nu)} t^{-1-\gamma_\nu}$$

where $\dfrac{1}{\Gamma(-\gamma_\nu)} = 0$ if γ_ν is a positive integer.

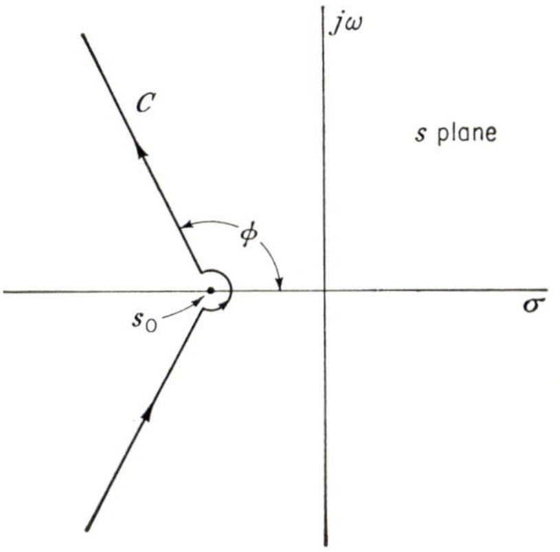

Fig. 9.13.1 Path C referred to in the statement of Theorem *9.13.11*.

12 *Example* $H(s) = \dfrac{1}{(s+2)\sqrt{s+1}}$. Since $H(s) \to 0$ as $|s| \to \infty$, uniformly in the angle $\measuredangle s$, by Jordan's lemma, the contour may be deformed as shown in Fig. *9.13.1*. Now

$$H(s) = \frac{1}{1 + s + 1} \frac{1}{\sqrt{s+1}} = \frac{1}{(s+1)^{1/2}} - (s+1)^{1/2} + (s+1)^{3/2} - \cdots$$

Hence

$$h(t) = 1(t) e^{-t} \left(- \frac{t^{-3/2}}{2\sqrt{\pi}} - \frac{t^{-5/2}}{\sqrt{\pi}} + \cdots \right)$$

13 *Comment* A word of warning should be inserted at this point: From the above it might appear that the only feature of $H(s)$ that counts as far as the behavior of $h(t)$ for $t \to \infty$ is concerned is the nature of the singularity that lies the farthest to the right in the s plane. This is not so; for such a statement to be true, $H(s)$ must behave appropriately at infinity. Observe that each of the last two theorems does impose conditions on the behavior of $H(s)$ as $|s| \to \infty$. A well-known example, due to Widder, is the following: $h(t) = 1(t) e^t \sin e^t$; that is, a signal whose amplitude and frequency increase exponentially with time. It can be shown, however, that $H(s) \triangleq \mathfrak{L}\{e^t \sin e^t\}$ has no singularities in the finite plane. It is, of course, singular at infinity, as required by Liouville's theorem; in fact, it is an essential singularity.

14 Steady-state response to a periodic input

Consider a strictly stable, causal, time-invariant system characterized by its transfer function $H(s)$. Let $H(s)$ be strictly stable and, in particular, be analytic for $\operatorname{Re} s \geq \sigma_2$, where $\sigma_2 < 0$. Let $H(s)$ be uniformly bounded as $|s| \to \infty$ and $\measuredangle s \in [\pi/2, 3\pi/2]$. At $t = 0$ the system is in the zero state and a periodic input $v(t)$ is applied: $v(t)$ satisfies the periodicity condition

1 $$v(t + T) = v(t) \qquad \text{for all } t \geq 0$$

Let us assume that $v(t)$ is Riemann integrable over the interval $[0,T]$. Since $H(s)$ is strictly stable, the response will eventually become periodic. We wish to evaluate this steady-state response. Let

2 $$v_0(t) = \begin{cases} v(t) & 0 \leq t < T \\ 0 & \text{elsewhere} \end{cases}$$

Consequently, $V_0(s)$ is an entire function of s: its singularities lie at infinity. It is well known that

$$V(s) \triangleq \int_0^\infty v(t)e^{-st}\,dt = \frac{V_0(s)}{1 - e^{-sT}}$$

Let us evaluate the response y of the system to the input $v(t)$ in the interval $(0,T)$. Since the system is causal, we have

3 $$y(t) = \mathcal{L}^{-1}\left\{H(s)\frac{V_0(s)}{1 - e^{-sT}}\right\} = \mathcal{L}^{-1}\{H(s)V_0(s)\} \qquad \text{for } 0 < t < T$$

Note that the second equality in *3* holds only because t is restricted to belonging to $(0,T)$. Furthermore, since $H(s)$ is analytic for $\operatorname{Re} s \geq \sigma_2$, and since $V_0(s)$ is an entire function,

4 $$y(t) = \frac{1}{2\pi j}\int_{\sigma_0 - j\infty}^{\sigma_0 + j\infty} H(s)V_0(s)e^{st}\,ds \qquad 0 < t < T,\ \sigma_0 \geq \sigma_2$$

Since $\sigma_2 < 0$ and since $1/(1 - e^{-sT})$ has simple poles located at $s = j2\pi n/T$, $n = \ldots, -1, 0, 1, \ldots$, the first inversion integral in *3* may be carried over any vertical path inside the right half plane:

$$y(t) = \frac{1}{2\pi j}\int_{\sigma_1 - j\infty}^{\sigma_1 + j\infty} \frac{H(s)V_0(s)}{1 - e^{-sT}}e^{st}\,ds \qquad \sigma_1 > 0$$

Let $\omega_n = (2n+1)\pi/T$, $n = \ldots, -1, 0, 1, \ldots$. The above equation may be rewritten as

$$y(t) = \frac{1}{2\pi j} \lim_{n\to\infty} \int_{\sigma_1 - j\omega_n}^{\sigma_1 + j\omega_n} \frac{H(s)V_0(s)e^{st}}{1-e^{-sT}}\, ds$$

$$= \frac{1}{2\pi j} \lim_{n\to\infty} \left(\int_{\sigma_1 - j\omega_n}^{\sigma_1 + j\omega_n} - \int_{\sigma_2 - j\omega_n}^{\sigma_2 + j\omega_n} + \int_{\sigma_2 - j\omega_n}^{\sigma_2 + j\omega_n} \right)$$

where σ_2 has been defined previously.

The integration paths are illustrated in Fig. *9.14.1*. The first two integrals may be combined to form a closed integration path by adding the paths joining $\sigma_2 - j\omega_n$ to $\sigma_1 - j\omega_n$ and $\sigma_1 + j\omega_n$ to $\sigma_2 + j\omega_n$. The integration along these paths tends to zero as $n \to \infty$ because (I) $e^{st}/(1 - e^{-sT})$ is uniformly bounded along the path, (II) $V_0(s) \to 0$ along the path by the Riemann-Lebesgue lemma, and (III) $H(s)$ is uniformly bounded on the path, by assumption. Let C_n denote the counterclockwise closed path described above; then

$$y(t) = \frac{1}{2\pi j} \int_{\sigma_2 - j\infty}^{\sigma_2 + j\infty} \frac{H(s)V_0(s)}{1-e^{-sT}} e^{st}\, ds + \frac{1}{2\pi j} \lim_{n\to\infty} \oint_{C_n} \frac{H(s)V_0(s)}{1-e^{-sT}} e^{sT}\, ds$$

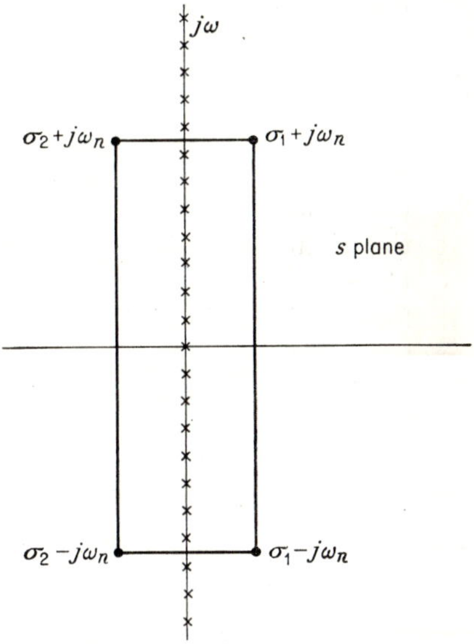

Fig. 9.14.1 Integration path used in the proof of Theorem *9.14.6*.

Since $\sigma_2 < 0$, the contour integration will contribute a Fourier series since the only singularities inside C_n are the simple poles of $1/(1 - e^{-sT})$. Therefore, this integral represents the periodic steady state, which we shall call $y_{ss}(t)$. From the equation above and *4* we have

5
$$y_{ss}(t) = \frac{1}{2\pi j} \int_{\sigma_0 - j\infty}^{\sigma_0 + j\infty} H(s)V_0(s)e^{st}\, ds - \frac{1}{2\pi j} \int_{\sigma_2 - j\infty}^{\sigma_2 + j\infty} \frac{H(s)V_0(s)e^{st}}{1-e^{-sT}}\, ds \qquad \text{for } 0 < t < T$$

where σ_0 is arbitrary except that $\sigma_0 \geq \sigma_2$. If σ_0 is taken equal to σ_2, this last equation can be rewritten by observing that

$$1 - \frac{1}{1-e^{-sT}} = \frac{e^{-sT}}{e^{-sT}-1} = \frac{1}{1-e^{sT}}$$

Hence,

$$y_{ss}(t) = \frac{1}{2\pi j}\int_{\sigma_2-j\infty}^{\sigma_2+j\infty} \frac{H(s)V_0(s)e^{st}}{1-e^{sT}}\,ds \qquad \text{for } 0 < t < T$$

We may state the result in the form of a

6 **Theorem** Given a linear, time-invariant, nonanticipative, system defined by its strictly stable transfer function $H(s)$, let the input $v(t)$ be periodic with period t. Then if $v(t)$ is Riemann integrable over $(0,T)$, if $H(s)$ is analytic for $\operatorname{Re} s \geq \sigma_2$, where $\sigma_2 < 0$, and if $H(s)$ is uniformly bounded as $|s| \to \infty$ for $\measuredangle s \in [\pi/2, 3\pi/2]$, then the periodic steady-state output $y_{ss}(t)$ is given by

7 $$y_{ss}(t) = \frac{1}{2\pi j}\int_{\sigma_2-j\infty}^{\sigma_2+j\infty} \frac{H(s)V_0(s)}{1-e^{sT}}\,e^{st}\,ds \qquad 0 < t < T$$

where $v_0(t)$ is defined in *2*.

8 *Comment* *7* describes $y_{ss}(t)$ *only* on the interval $(0,T)$. Furthermore, the right-hand side of *7* is *not* the inversion integral of $H(s)V_0(s)/(1-e^{sT})$ because that function has singularities on the $j\omega$ axis; therefore, its inversion integral should be carried over a vertical path inside the right half plane. Observe that $\sigma_2 < 0$; so the integration *7* is carried over a vertical path inside the left half plane.

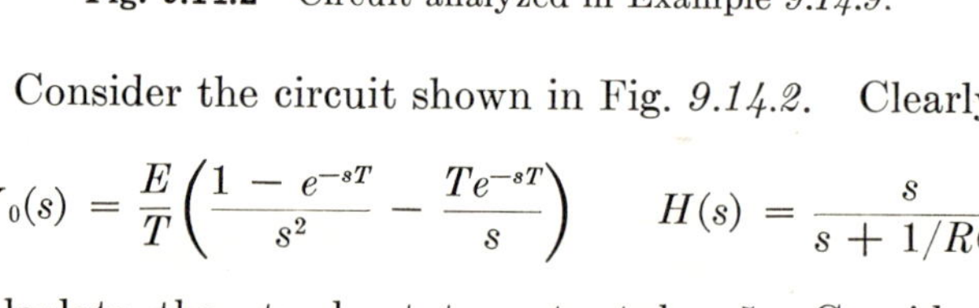

Fig. 9.14.2 Circuit analyzed in Example *9.14.9*.

9 *Example* Consider the circuit shown in Fig. *9.14.2*. Clearly,

$$V_0(s) = \frac{E}{T}\left(\frac{1-e^{-sT}}{s^2} - \frac{Te^{-sT}}{s}\right) \qquad H(s) = \frac{s}{s+1/RC}$$

Let us calculate the steady-state output by *5*. Consider the first integral:

$$H(s)V_0(s) = \underbrace{\frac{E}{T}\frac{1}{s(s+1/RC)}}_{\text{one}} - \underbrace{\frac{E}{T}\frac{e^{-sT}}{s(s+1/RC)}}_{\text{two}} - \underbrace{E\frac{e^{-sT}}{s+1/RC}}_{\text{three}}$$

Let us take $\sigma_0 = \sigma_2$; then $-1/RC < \sigma_0 < 0$. Consider term one: by Jordan's lemma (see *B.3.36*) we may close the integration path by a left-half-plane semicircle, and since the only enclosed pole is $s = -1/RC$, one gives $-(ERC/T)\exp[-(t/RC)]$. Terms two and three are such that Jordan's lemma requires that the integration path be closed *on the right*, because we consider values of t in $(0,T)$. The only enclosed pole is $s = 0$; hence, two and three give ERC/T and 0, respectively. Note the change of sign due to the fact that the integration is clockwise. Thus the first integral of *5* contributes

$$-\frac{ERC}{T}e^{-t/RC} + \frac{ERC}{T}$$

Now consider the second integral. Its integrand with sign reversed is

$$-\underbrace{\frac{E}{T}\frac{1}{s(s+1/RC)}}_{\text{four}} + \underbrace{\frac{E}{s+1/RC}\frac{e^{-sT}}{1-e^{-sT}}}_{\text{five}}$$

Jordan's lemma allows us to close the integration of four and five to the left. In both cases the only pole enclosed is at $-1/RC$. Hence the second integral contributes

$$\frac{ERC}{T}e^{-t/RC} - E\frac{e^{-t/RC}}{1-e^{-T/RC}}$$

The result is

$$y_{ss}(t) = \frac{ERC}{T} - E\frac{e^{-t/RC}}{1-e^{-T/RC}} \qquad 0 < t < T$$

15 *Signal-flow graphs*

In this section we define the signal-flow graph and prove the Mason formula for the gain of a signal-flow graph.

Definition of a linear signal-flow graph

The purpose of this section is to present precisely and completely some basic facts about linear signal-flow graphs. The reader is assumed to have a certain amount of familiarity with signal-flow graphs and their usual engineering applications. Instead of going through a long detailed example, let us assume that the reader has his pet example in mind and let us observe the following: (I) A signal-flow graph gives a graphical representation of the causal relationships between the variables chosen for the analysis of the system. It is the causal aspect of these relationships which makes the use of signal-flow graphs so intuitively appealing: the signal-flow graph gives a very satisfying graphical representation of these relationships. (II) For any given system, there are many ways of choosing variables and of writing equations. With each set of equations is associated a signal-flow graph, and thus a given system may be represented by many different signal-flow graphs. (III) It is of interest to note that the diagram of an analog-computer setup can with slight modification become a signal-flow graph. In fact the signal-flow graphs thus obtained are rather special, since their branches are restricted to have gains equal to $-1/s$ or -1 or a (where $0 \leq a \leq 1$), the corresponding analog-computer elements being, respectively, integrators, summing amplifiers, and potentiometers (scalors).

A (linear) signal-flow graph is a graphical means for representing linear equations. For convenience we write them in the form

1
$$\begin{array}{l} (1 - g_{11})x_1 - g_{12}x_2 - \cdots - g_{1n}x_n = f_1 \\ -g_{21}x_1 + (1 - g_{22})x_2 - \cdots - g_{2n}x_n = f_2 \\ \cdots\cdots\cdots\cdots\cdots\cdots\cdots\cdots \\ -g_{n1}x_1 - g_{n2}x_2 - \cdots + (1 - g_{nn})x_n = f_n \end{array}$$

The dependent variables x_i and the independent variables f_i (usually called inputs) are either functions of time or Laplace transforms of functions of time. The matrix elements g_{ij} are correspondingly either integrodifferential operators or transfer functions, i.e., functions of the complex frequency variable s. Thus, these equations may, for example, represent the familiar circuit equations resulting from Kirchhoff laws.

Let us rewrite the equations as follows:

2
$$x_i = \sum_{j=1}^{n} g_{ij}x_j + f_i \qquad i = 1, 2, \ldots, n$$

The relationship between this system of equations and the signal-flow graph is best explained by describing how one associates a signal-flow graph with the system of equations.

3 **Procedure: from the equations to the signal-flow graph** Carry out the following steps: (I) With each dependent variable x_i is associated one node. For convenience, the node associated with x_i is called the ith node. (II) With each input variable f_i is associated a node, a source node. (III) For all i,j if $g_{ij} \neq 0$, there is a directed branch that originates at node j and terminates at node i; this branch is assigned a gain g_{ij}. If $g_{ij} = 0$, no branch is introduced. (IV) A branch of unity gain joins the node associated with f_i to the ith node, $i = 1, 2, \ldots, n$.

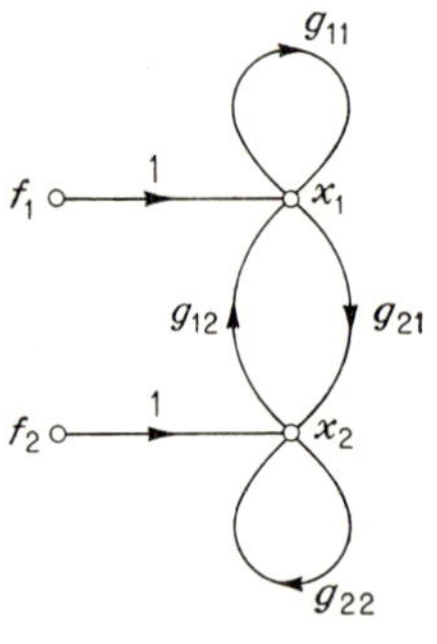

Fig. 9.15.1 Example of a signal-flow graph with two nodes.

The signal-flow graph associated with the system *1* for the case $n = 2$ is illustrated in Fig. *9.15.1*.

4 **Definition** A node on which no branch terminates is said to be a *source node.*

5 **Procedure: from the signal-flow graph to the equations** There is one equation associated with each node that is not a source node. The equation associated with the ith node is obtained by (I) considering all branches terminating on the ith node, (II) forming the product of the gain of each such branch with the node variable from which this branch originates, and (III) equating the sum of all these products to x_i. Thus we are led to the following

6 **Definition** A *signal-flow graph* is a weighted oriented graph which stands in a one-one correspondence (described by *3* and *5*) with a set of linear equations. ◁

Any system of linear algebraic equations such as *1* can be solved by systematic algebraic techniques, e.g., by the Gauss elimination method. However, in engineering it usually happens that the matrix $\mathbf{M}$ of the system *1* contains a large number of zero elements: it is this circumstance that makes the signal-flow-graph approach desirable and effective. Inspection of the signal-flow graph suggests the order in which the elimination should be carried out. The key operation is the node elimination, which will be discussed in the next subsection. It is also possible to solve directly for some x_k without successively eliminating all other nodes: this will be described in the subsection entitled The Gain of a Signal-flow Graph.

The node elimination

The node elimination technique consists in obtaining from the signal-flow graph G an equivalent signal-flow graph G' that has only $n - 1$ nodes. The equivalence means that the relationships between the $n - 1$ remaining variables expressed by G' are identical to those implied by G. By successively applying this operation to all the nodes except the kth node, we end up with the graph where each gain G_i is expressed in terms of the elements of the matrix $\mathbf{M}$, as in Fig. *9.15.2*.

The quantities G_i play a prominent role in signal-flow-graph theory. Refer to Fig. *9.15.2*, where G_i is the value of the output x_k when all inputs are set to zero except the ith input, which is set equal to unity. G_i is called the transmission from the ith input node to the kth node. Thus the

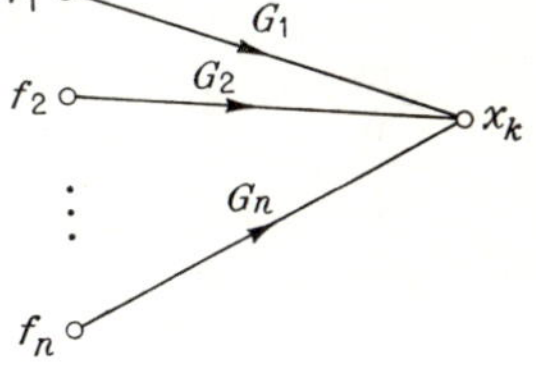

Fig. 9.15.2 Signal-flow graph resulting from the elimination of all nodes but node x_k.

7 **Definition** The graph transmission from node i to node j is the value of x_j when

a unit input is applied to node i and all other inputs are set equal to zero.

8 *Comment* Suppose there is a branch g_{ji} joining node i to node j; then we should stress the difference between the gain of the branch joining node i to node j and the graph transmission from node i to node j. It is obvious that the former is simply g_{ji}, whereas the latter usually depends on the gain of all the branches of the graph. The graph transmission is a property of the whole system. ◁

Let us describe the node elimination method by eliminating the ith node. For convenience define a *self-loop* as a single branch originating from and terminating at the same node.

Case 1 $g_{ii} = 0$, **that is, the** i**th node has no self-loop** The ith equation of *1* reads[1]

$$x_i = \sum_{j\neq i}' g_{ij}x_j + f_i$$

Substitution into any of the other equations gives

9 $$x_m = \sum_{\alpha\neq i}' g_{m\alpha}x_\alpha + g_{mi}x_i + f_m = \sum_{\alpha\neq i}' g_{m\alpha}x_\alpha + g_{mi}\sum_{\alpha\neq i} g_{i\alpha}x_\alpha + g_{mi}f_i + f_m$$

$$x_m = \sum_{\alpha\neq i} (g_{m\alpha} + g_{mi}g_{i\alpha})x_\alpha + g_{mi}f_i + f_m$$

$$m = 1, \ldots, i-1, i+1, \ldots, n$$

The operation is illustrated by Fig. *9.15.3*.

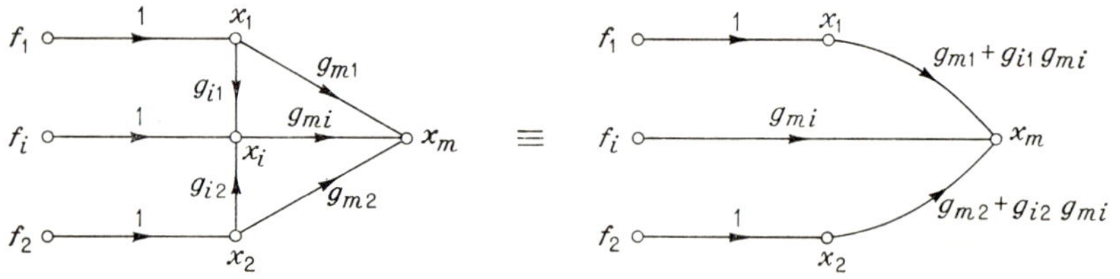

Fig. 9.15.3 Elimination of the ith node when $g_{ii} = 0$.

Case 2 $g_{ii} \neq 0$, **that is, the node to be eliminated has one self-loop** We may assume without restricting the generality of the discussion that $1 - g_{ii} \neq 0$. Indeed, if it were not possible to order the equations and the variables of the system *1* so that all diagonal elements $\neq 0$, then all products of the determinant expansion of **M** would be zero.[2] Since

[1] Unless otherwise indicated, the summation runs from 1 to n. The prime on the summation sign indicates that one value of the summation index is omitted.

[2] If one of them were not zero, each one of its factors would differ from zero. Consider one such nonzero product π. If we so reorder the columns that the element of π which is in the kth row of **M** ends up in the kth column, $k = 1, 2, \ldots, n$, we obtain a new determinant all of whose diagonal elements are different from zero. Thus such an ordering is impossible if and only if all products of the determinant expansion of **M** are zero. That is, det $\mathbf{M} = 0$; hence *1* does not specify the x_i uniquely.

$1 - g_{ii} \neq 0$, substitution in the mth equation gives

10 $$x_m = \sum_{\alpha \neq i}' \left(g_{m\alpha} + \frac{g_{mi} g_{i\alpha}}{1 - g_{ii}} \right) x_\alpha + \frac{g_{mi}}{1 - g_{ii}} f_i + f_m$$

$$m = 1, 2, \ldots, i - 1, i + 1, \ldots, n$$

The operation is illustrated by Fig. *9.15.4*. ◁

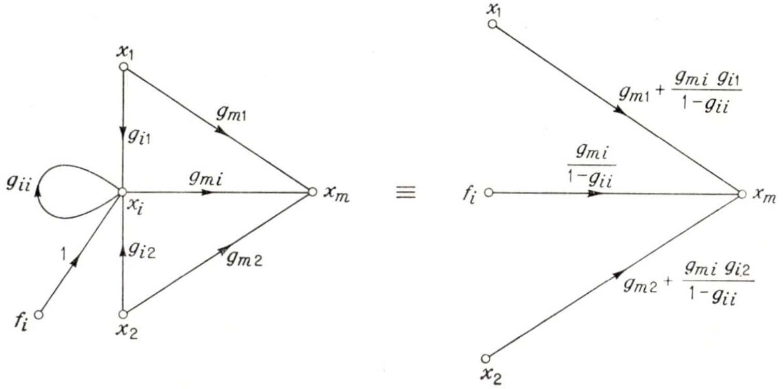

Fig. 9.15.4 Elimination of the ith node when $g_{ii} \neq 0$.

Since Case 1 is a special case of Case 2, we need formulate only one general statement.

11 **Node elimination rule** (refer to Fig. *9.15.5*) Suppose we wish to eliminate node i. With the exception of the possible self-loop of gain

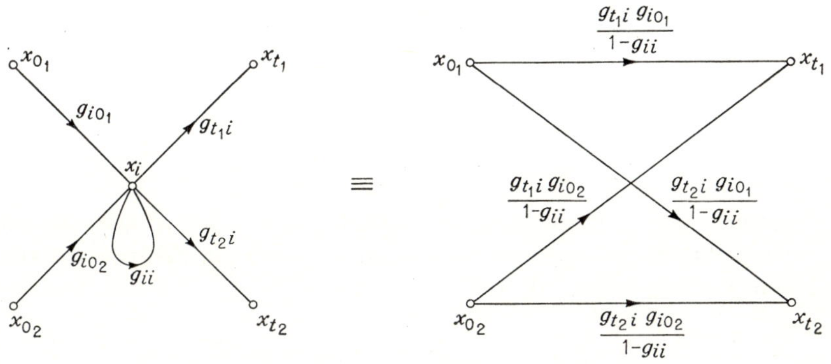

Fig. 9.15.5 Figure illustrating the node elimination rule.

g_{ii}, let $0_1, 0_2, \ldots, 0_k$ be the originating nodes of all the branches terminating at node i and let $t_1, t_2, \ldots, t_m$ be the terminating nodes of all the branches originating at node i. Then the new signal-flow graph is obtained from the given one as follows: (I) all branches originating from and/or terminating on node i are deleted, (II) each of the nodes $0_1, 0_2, \ldots, 0_k$ is connected to each one of the nodes

$t_1, t_2, \ldots, t_m$ by a new branch, and (III) the new branch connecting 0_α to t_β has a gain

$$\frac{g_{i0_\alpha} g_{t_\beta i}}{1 - g_{ii}} \qquad \alpha = 1, 2, \ldots, k; \beta = 1, 2, \ldots, m \quad \triangleleft$$

The diagrams of Fig. *9.15.6* show portions of a signal-flow graph the ith node of which is to be eliminated and the resulting graphs obtained by node elimination. The diagram shows *all* the branches connected to the ith node. If, in a particular instance, there are more branches terminating on or originating from the ith node than indicated on the graphs of Fig. *9.15.6*, the general rule just stated must be applied.

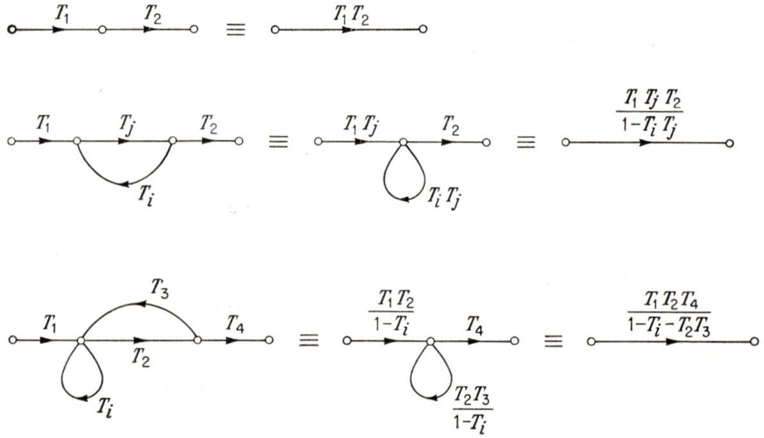

Fig. 9.15.6 Examples of node elimination in simple cases.

The gain of a signal-flow graph

Let us restrict ourselves to a single input f_1 and let us select an arbitrary variable, say, x_k, as output. We shall derive directly from the definition of a determinant the general expression for the gain of that signal-flow graph. For this purpose we first have to define a few topological concepts such as a connected subgraph, a loop, a forward path, and nontouching loops.

12 **Definition** A *subgraph* is said to be *connected* if, disregarding the branch directions, it is possible to go from any one node to any other node by only following branches of the subgraph.

13 **Definition** A *loop* (or, more precisely, a directed loop) is a connected subgraph whose branches $b_1, b_2, \ldots, b_l$ can be ordered in such a way that (I) the tip of b_k is the origin of b_{k+1}, $k = 1, 2, \ldots, l-1$; (II) the origin of b_1 is the tip of b_l; and (III) each node is encountered only once when the loop is traversed in the positive direction.

14 *Example* Figure *9.15.7a* is a loop; Fig. *9.15.7b* is not a loop because it violates (III); Fig. *9.15.7c* is not a loop because it violates (I).

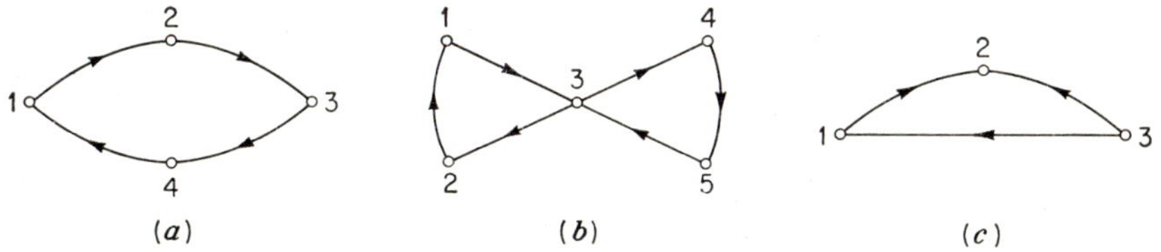

Fig. 9.15.7 Illustrations clarifying the concept of a loop.

15 **Definition** The *loop gain* of a loop is the product of the gains of the branches constituting that loop.

16 **Definition** A collection of loops is said to be *nontouching* if no two of the loops have either a branch or a node in common.

17 **Definition** A *directed path* from node α to node β is a connected subgraph such that (I) none of its branches terminates at node α and only one branch originates from it, (II) none of its branches originates from node β and only one branch terminates on it, and (III) all other branches connect at nodes in such a way that each node has only one incoming branch and one outgoing branch in the subgraph.

18 **Definition** A directed path from the input node f_1 to the output node k is called a *forward path.*

19 *Example* In Fig. *9.15.8a* is shown a directed path; however, *9.15.8b* violates (III), and *9.15.8c* violates (I) and (II).

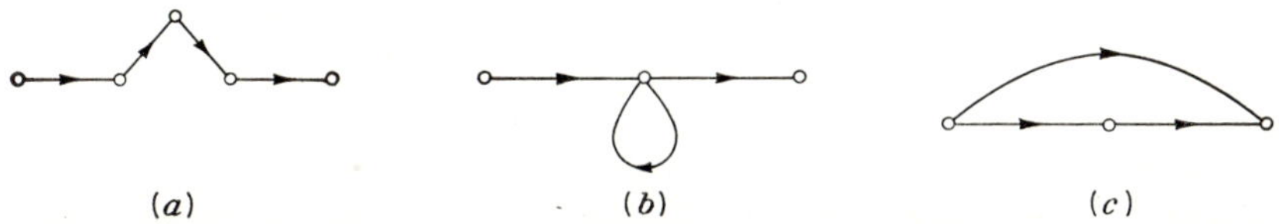

Fig. 9.15.8 Illustrations clarifying the concept of a directed path.

Now we are ready to state the main result in the form of a

20 **Theorem** The graph transmission G from input to output of a single-input signal-flow graph is given by Mason's formula

21
$$G = \frac{\Sigma G_k \Delta_k}{\Delta}$$

where the summation is taken over all forward paths from input to output, G_k is the gain of the kth forward path (i.e., the product of the gains of all branches constituting that forward path), and Δ is the determinant of the graph, given by

22 $$\Delta = 1 - \sum_m P_{m1} + \sum_k P_{k2} - \sum_l P_{l3} + \cdots$$

where P_{m1} is the loop gain of the mth loop and the summation is taken over all loops, P_{k2} is the product of the loop gains of the kth set of two nontouching loops and the summation is taken over all such pairs, P_{l3} is the product of the loop gains of the lth set of three nontouching loops and the summation is taken over all such triples, etc. Δ_k is the determinant of the graph obtained from the original graph by deleting all branches of the kth forward path and all branches having a node in common with it.

23 *Example* Consider the graph shown in Fig. *9.15.9.* It has four loops whose respective gains are g_{22}, g_{33}, $g_{41}g_{34}g_{13}$, $g_{21}g_{42}g_{34}g_{13}$. Consequently

$$\Delta = 1 - g_{22} - g_{33} - g_{41}g_{34}g_{13} - g_{21}g_{42}g_{34}g_{13} + g_{22}g_{33}$$

because g_{22} and g_{33} are the only nontouching loops.

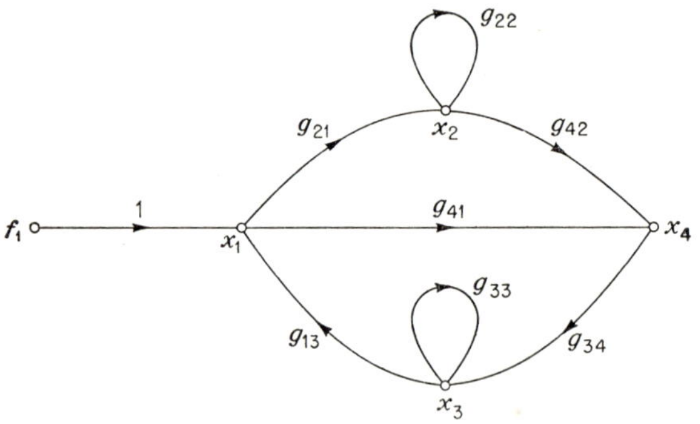

Fig. 9.15.9 Signal-flow graph analyzed in Example *9.15.23.*

There are two forward paths, and

$$G_1 = g_{41} \qquad \Delta_1 = 1 - g_{22} - g_{33} + g_{22}g_{33}$$
$$G_2 = g_{21}g_{42} \quad \Delta_2 = 1 - g_{33}$$

Hence

$$G = \frac{g_{41}(1 - g_{22} - g_{33} + g_{22}g_{33}) + g_{21}g_{42}(1 - g_{33})}{1 - g_{22} - g_{33} - g_{41}g_{34}g_{13} - g_{21}g_{42}g_{34}g_{13} + g_{22}g_{33}}$$

Proof The proof is based on elementary properties of determinants.

First, recall that the determinant of a matrix **M** is

$$\det \mathbf{M} = \sum_P \operatorname{sgn}(P) m_{i_1 1} m_{i_2 2} m_{i_3 3} \cdots m_{i_n n}$$

where $\sum_P$ denotes that the summation is taken over all the $n!$ permutations $i_1, i_2, \ldots, i_n$ of $1, 2, \ldots, n$ and sgn (P) is $+1$ or -1 depending on whether the permutation $i_1, i_2, \ldots, i_n$ is even or odd. Second, the permutation of two symbols of a permutation changes the sign of the permutation. Finally, if any two rows or columns of **M** are interchanged, the determinant of the resulting matrix is $-\det \mathbf{M}$.

Since f_1 is the only input, Eqs. *1* have zeros in their right-hand sides except the first one, which has f_1 as the right-hand side. The

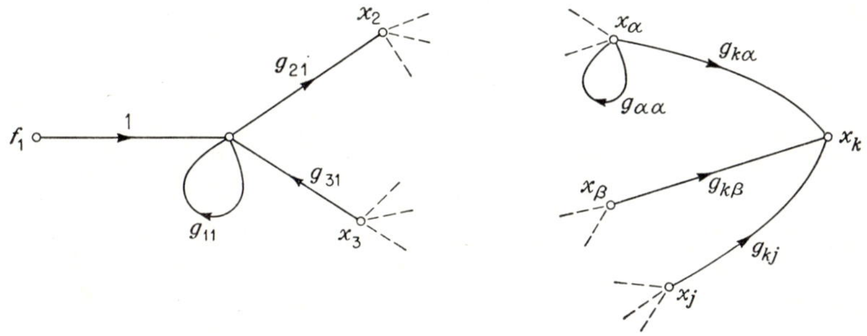

Fig. 9.15.10 Single-input signal-flow graph considered in the proof of Theorem *9.15.20*.

corresponding graph is shown in Fig. *9.15.10*. The output x_k of the graph is given by Cramer's rule:

24 $$x_k = \frac{N_{1k}}{\Delta}$$

where $\Delta = \det \mathbf{M}$ and N_{1k} is the determinant of the matrix obtained from **M** by replacing its kth column by a column vector whose first element is f_1 and all others are zero.

Evaluation of Δ

Let us expand Δ according to the definition of the determinant. Collecting terms, we get $\Delta = 1 + \sum_\alpha \Pi_\alpha$, where each Π_α is a product of branch gains. Consider a particular such product, say Π_β. Let us turn our attention to the topology of G_β, the subgraph formed by the branches whose gains appear in Π_β.

To start with, let us make three observations on the subgraph G_β and the gain product π_β.

(I) In G_β not more than one branch either originates from any given node or terminates at any given node. In other words, in G_β, the occurrences shown in Fig. *9.15.11* are excluded by this statement.

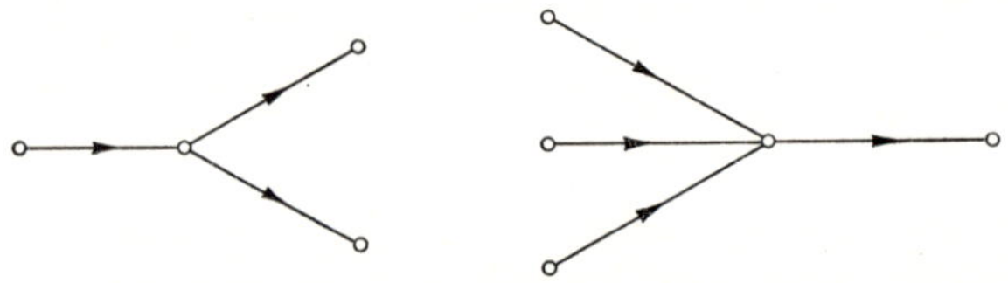

Fig. 9.15.11 Subgraphs that cannot occur in G_β.

This result follows directly from the definition of a determinant. Π_β is a product of branch gains with two properties: (*a*) it does not contain more than one element from each row of Δ and (*b*) it does not contain more than one element from each column of Δ. It follows from (*a*) that not more than one branch of G_β can terminate at any given node and from (*b*) that not more than one branch of G_β can originate from any given node.

(II) Π_β is a product of loop gains of nontouching loops. Note that, by definition, nontouching loops can have *neither* a branch *nor* a node in common. If G_β is not connected, we must show that it consists of nontouching loops. That the loops must be nontouching follows directly from (I). Therefore, if G_β is not connected, we must show that each connected subgraph of G_β is a loop; and if G_β is connected, we need only show that it is a loop. We establish this property by contradiction. Suppose that G_β consists of a directed path between some pair of nodes and some other nontouching configurations. Suppose, for simplicity, that the forward path is made of g_{21} and g_{32}, that is, originates from node 1 and terminates at node 3. Hence no branch of G_β terminates at 1 and no branch of G_β originates from 3. Hence the product cannot contain factors like g_{1k}, $k = 1, 2, \ldots, n$, and g_{k3}, $k = 1, 2, \ldots, n$; that is, no elements of row 1 and column 3 appear in G_β. The definition of a determinant implies, therefore, that Π_β includes the 1's which are part of the elements which are in the (1,1) and (3,3) location in the matrix $\mathbf{M}$. This implies a contradiction, because Π_β already includes g_{21} and g_{32} elements of column 1 and row 3 and hence cannot include, in addition, elements at locations (1,1) and (3,3). Therefore, if G_β is not connected, it consists of nontouching loops. If G_β is connected, it consists, obviously, of a single loop.

(III) The gain product in Π_β is assigned a negative sign if G_β consists of an odd number of loops and a positive sign if the number of loops is even. Recall that if, in a matrix, any two rows *and* any two columns are interchanged, the value of the determinant is unaffected. Observe the following: Suppose we are given two identical graphs G and G'. Suppose their nodes are identically numbered except that the nodes

i and j of G are labeled j and i, respectively, in G'. The matrix of the signal-flow graph G' can be obtained from that of G by interchanging rows i and j as well as columns i and j. Hence, the determinants are identical. Therefore, in evaluating the determinant of a signal-flow graph, we may arbitrarily number the nodes without affecting the determinant.

Suppose G_β consists of *two* nontouching loops. Let the nodes of each loop be numbered $1, 2, \ldots, l_1; l_1 + 1, \ldots, l_2$, respectively. The gain products associated with the loops are

$$g_{21}g_{32}g_{43} \cdots g_{l_1,l_1-1}g_{1l_1}g_{l_1+2,l_1+1} \cdots g_{l_2,l_2-1}g_{l_1+1,l_2}$$

This product will be assigned a sign according to the parity of the permutation P:

$$(2, 3, 4, \ldots, l_1, 1; l_1 + 2, l_1 + 3, \ldots, l_2, l_1 + 1)$$

Since the l_1th element, 1, can be brought to the position 1 by $l_1 - 1$ interchanges of adjacent symbols and since the $(l_1 + l_2)$nd element, $l_1 + 1$, can be brought to the $(l_1 + 1)$st position by $l_2 - 1$ interchanges of consecutive symbols, the sign of the permutation is

$$\operatorname{sgn} P = (-1)^{l_1-1}(-1)^{l_2-1}$$

Since, as elements of the system matrix, each of the g_{ik}'s has a minus sign, (see Eqs. *1*), the sign of the product Π_β is

$$(-1)^{l_1}(-1)^{l_2} \operatorname{sgn} P = (-1)^2$$

If G_β had consisted of λ loops, the reasoning above would have led to $(-1)^\lambda$. Therefore, any term of the expansion of Δ is either 1 or a product Π_α of loop gains of nontouching loops times $(-1)^{\lambda_\alpha}$, where λ_α is the number of loops corresponding to the product Π_α.

Conversely, any product of loop gains of nontouching loops of the graph G, with the appropriate sign, will appear as a term of the expansion of Δ. This follows directly from considerations almost identical with those outlined above.

Thus Eq. *22* is justified.

Evaluation of N_{1k}

Consider the matrix $\mathbf{M}'$ whose determinant appears as the numerator of *24*. Let $G_{M'}$ be the graph corresponding to the matrix $\mathbf{M}'$. $G_{M'}$ can be obtained from G by (I) deleting all branches that originate from the output node k and (II) inserting a branch of gain f_1 originating at k and terminating at 1. Since f_1 is the only nonzero element in the kth column of $\mathbf{M}'$, every nonzero product of the expansion of det $\mathbf{M}'$ includes the factor f_1. Since each such product corresponds to a set of nontouching loops, when f_1 is factored out, the remaining branch gains correspond to a set of directed loops and one *forward path* originating at the input node 1 and terminating at the output node k.

Let us consider a particular product Π'. Let the forward path of Π' contain $\lambda - 1$ branches and let its nodes be numbered $1, 2, \ldots, \lambda$, where 1 corresponds to the input node and λ to the output node. Such a renumbering does not affect the value of det $\mathbf{M}'$. The gain of the forward path is $g_{21}g_{32}g_{43} \cdots g_{\lambda,\lambda-1}$. Let us write the matrix $\mathbf{M}'$, after the reordering, putting the branch gains of the forward path in evidence.

$$\left[\begin{array}{cccccc:c}
\cdot & \cdot & \cdot & & & f_1 & \\
-g_{21} & & & & & \cdot & \\
& -g_{32} & & & & \cdot & \\
& & -g_{43} & & & \cdot & \\
& & & \cdot & & \cdot & \\
& & & & \cdot & \cdot & \\
& & & & -g_{\lambda,\lambda-1} & \cdot & \\
\hdashline
& & & & & & \\
& & & & & & \mathbf{M}'_\lambda \\
& & & & & &
\end{array}\right]$$

From the form of the matrix it is clear that the loop gain products appearing in Π' must come from the expansion of det $\mathbf{M}'_\lambda$. Consequently, if we consider in the expansion of det $\mathbf{M}'$ all the terms that have the product $f_1g_{21}g_{32} \cdots g_{\lambda,\lambda-1}$ as a common factor, we get

$$(-1)^{\lambda-1}f_1(-g_{21})(-g_{32})(-g_{43}) \cdots (-g_{\lambda,\lambda-1}) \det \mathbf{M}_\lambda$$

where $(-1)^{\lambda-1}$ comes from the fact that f_1 belongs to the row 1 and column λ of $\mathbf{M}'$. This product is thus

$$\underbrace{f_1g_{21}g_{32} \cdots g_{\lambda,\lambda-1}}_{\text{forward path gain}} \det \mathbf{M}'_\lambda$$

From a topological point of view, det $\mathbf{M}'_\lambda$ is the determinant associated with the signal-flow graph G_λ obtained from G by deleting all the nodes of the forward path and all the branches connected to this forward path. Therefore, any product Π' of the expansion of det $\mathbf{M}'$ is a term in an expression of the form $G_k\Delta_k$. Conversely, it is easy to establish that any expression of the form $G_k\Delta_k$ includes all terms of the expansion of det $\mathbf{M}'$ with the correct sign. Therefore, the formula det $\mathbf{M}' = \sum_k G_k\Delta_k$ is established.

25 *Exercise* Use Mason's formula (*21*) to compute the transfers function of the systems shown on Figs. *4.9.1* to *4.9.10* and Fig. *4.9.12*.

16 *Nyquist criterion*

In this section we shall consider exclusively linear time-invariant systems whose signal-flow graph is of the form shown in Fig. *9.16.1*. For obvious reasons such systems are called single-loop-feedback systems. Our purpose will be to characterize their stability in terms of the properties of the transfer functions F and H. Throughout this section we shall make the following

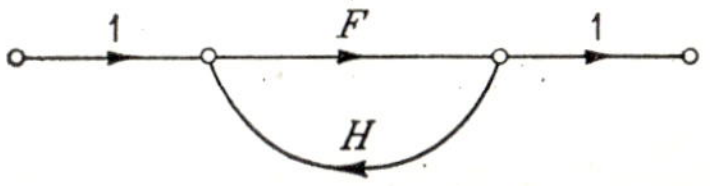

Fig. 9.16.1 Signal-flow-graph representation of a single-loop-feedback system; its transfer function is given by $G = F/(1 - FH)$.

1 **Assumption** The transfer functions F and H are rational functions of the complex frequency variable s.

2 *Comment* Of course, this assumption considerably simplifies the developments that follow. From an engineering point of view, it is certainly not a restriction to assume that the impulse response h of any transfer function that occurs in practice goes to zero at $t = \infty$ and belongs to $L^2(0, \infty)$; more precisely, that $\int_0^\infty |h(t)|^2\, dt < \infty$.[1] We assert that any such $h(t)$ can be approximated by an appropriately chosen linear combination of exponentials $\{e^{-(n+\frac{1}{2})t}\}_0^\infty$ and, therefore, the transfer function $H(s)$ can be approximated by an appropriately chosen linear combination of terms of the form $1/[s + (n + \frac{1}{2})]$. To be more explicit, let us show how this assertion follows from the Weierstrass approximation theorem. From the impulse response h, define a function f whose argument varies over (0,1), as follows:

$$h(t) = f(e^{-t})e^{-t/2}$$

[1] In many control applications, the transfer function has a pole at $s = 0$; in such a case, the impulse response can be written as a sum of two terms,

$$h(t) = a_0 1(t) + h_1(t)$$

where a_0 is the residue of $H(s)$ at $s = 0$. The function $h_1(t)$ satisfies the conditions above.

It is easy to check that the transformation $\xi = e^{-t}$ gives

$$\int_0^\infty |h(t)|^2\,dt = \int_0^1 |f(\xi)|^2\,d\xi$$

implying that $f \in L^2(0,1)$. Now, by the Weierstrass approximation theorem, f can be approximated (arbitrarily closely in the L^2 sense) by a polynomial $\sum_{n=0}^{N} \alpha_n x^n$, provided N is taken sufficiently large. Therefore, for N sufficiently large we can find a sequence of coefficients $\{\alpha_n\}$ such that $\int_0^\infty |h(t) - \sum_0^N \alpha_n[\exp -(n + \frac{1}{2})t]|^2\,dt$ is as small as we wish.

3 *Exercise* Show that if h is continuous on $[0, \infty)$ and goes to zero as $t \to \infty$, it can be approximated uniformly arbitrarily closely by a suitable linear combination of $\exp[-(n - \frac{1}{2})t]$, $n = 1, 2, \ldots$. (*Hint:* Use the following form of the Weierstrass approximation theorem: if f is continuous on $[0,1]$, then, for any $\varepsilon > 0$, there is a polynomial p such that

$$|f(x) - p(x)| < \varepsilon \qquad \forall x \in [0,1])$$

The first step in the analysis is to relate the strict stability of the gain of the single-loop-feedback system shown in Fig. *9.16.1* to the behavior of its loop gain FH. This is done by the following

4 **Theorem** Consider the single-loop-feedback system shown in Fig. *9.16.1*. If F and H are rational functions of s and if in the product FH the only cancellations that occur involve poles and zeros inside the left half plane, then the transfer function $G(s) = F/(1 - FH)$ is strictly stable if and only if $1 - FH$ has no zeros either on the $j\omega$ axis or in the right half plane.

5 *Comment* This theorem is intuitively reasonable: if F has a pole p_1 such that $\operatorname{Re} p_1 > 0$, a little analysis of the formula for G shows that G is well behaved in a small neighborhood of p_1 and $G(p_1) = -1/H(p_1)$. Similarly, if H has a pole at p_1, G is well behaved in a small neighborhood of p_1 and $G(p_1) = 0$.

Proof $\Rightarrow$ Use contradiction. Suppose that for some z_1 in the closed right half plane ($\operatorname{Re} z_1 \geq 0$), $1 - F(z_1)H(z_1) = 0$. By the second assumption, the last equality implies $F(z_1) \neq 0$; hence, G has a pole at z_1 and G is not strictly stable. This contradicts the hypothesis; therefore, $1 - FH$ cannot have a zero in the closed right half plane.

$\Leftarrow$ The only way G could have a pole p_0 in the closed right half plane (i.e., $\operatorname{Re} p_0 \geq 0$) is that either $1 - FH = 0$ at p_0 with $F(p_0) \neq 0$

or that F has a pole of order m at p_0 but $1 - FH$ has a pole of order $<m$. The first possibility is ruled out by the assumption of this sufficiency proof. The second possibility requires that there be a cancellation of factors of the form $(s - p_0)$ in the product FH. Since Re $p_0 \geq 0$, this is ruled out by the second assumption of the theorem. Hence G is strictly stable.

6 *Comment* The theorem makes no assumptions relative to the stability of either F or H. Furthermore, no practical system would ever violate the second assumption, namely, that there would be a right-half-plane pole canceled by a right-half-plane zero in the product FH. The reasons for this are well known and discussed in *11.5.10*. ◁

The most useful tool for the study of the loop gain FH is the

7 **Principle of the argument** Let C be a simple closed curve in the s plane. Let f be a rational function of s which has neither zeros nor poles on C. Let f map the curve C into the curve Γ of the f plane (see Fig. *9.16.2*). Then as the point s runs through the contour C in the

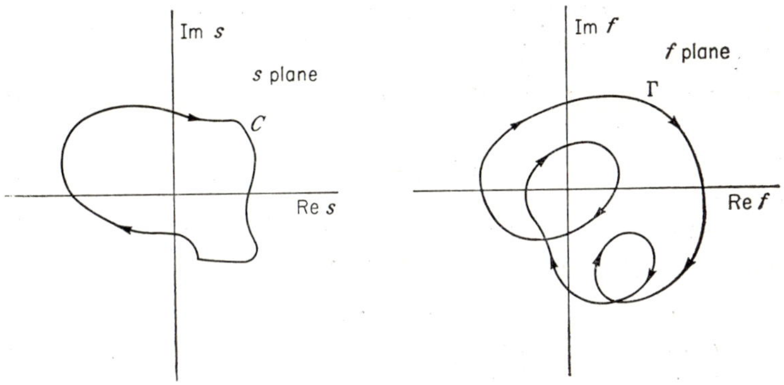

Fig. 9.16.2 Illustration of the principle of the argument: Γ is the f-plane curve which is the image under the mapping f of the s-plane curve C.

clockwise direction, the image of s, $f(s)$, follows Γ and encircles the origin of the f plane $Z - P$ times in the *clockwise* direction, where Z and P are the numbers of zeros and poles of f inside C, zeros and poles being counted with their proper multiplicities.

We shall not prove this result since it is classical.[1]

8 *Comment* For the purpose of investigating the stability of a transfer function, one uses the contour C_R, which consists of the j axis and the right-hand semicircle of infinite radius. Therefore, by *4*, any zero of $1 - FH$ inside C_R will make the transfer function G unstable.

[1] E. Hille, "Analytic Function Theory," Ginn and Co., Boston, 1959.

9 **Definition** The mapping of the contour C_R by a transfer function T is called the *Nyquist diagram* of T.

A Nyquist diagram of a simple low-pass amplifier is shown in Fig. *9.16.3*.

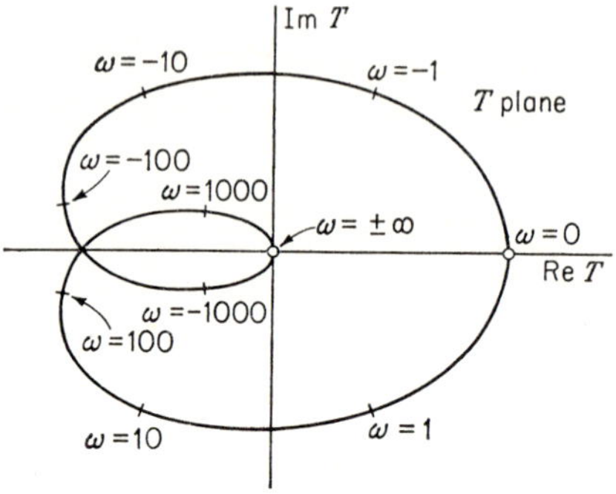

Fig. 9.16.3 Example of a Nyquist diagram (not to scale).

If we combine *4*, *7*, and *8* and observe that $1 - FH = 0$ whenever $FH = 1$, we obtain the

10 **Nyquist stability criterion** Let F and H be rational functions of s. Suppose that in the product FH the only cancellations that occur involve poles and zeros inside the left-hand plane. Suppose FH is stable: then the transfer function $G = F/(1 - FH)$ is strictly stable if and only if the Nyquist diagram of the loop gain FH does not encircle and does not go through the critical point (1,0) of the FH plane.

11 *Comment* In some cases a technical difficulty occurs in that FH has a pole (or zero) on C_R; in such a case the contour C_R is "indented" by a sufficiently small semicircle centered on the pole in such a way that the indented contour avoids the pole (see Fig. *9.6.2*).

12 *Comment* The importance of the Nyquist criterion lies in the following fact. All the assumptions of *10* can be tested experimentally: from *2* we know that we can always assume F and H to be rational. In *11.5.10* we show that if there are cancellations in FH involving right-half-plane zeros and poles, the system is unstable in practice; the stability of FH can be decided by observing the open-loop behavior of the system; and finally, the Nyquist diagram is obtained by plotting the result of sinusoidal steady-state measurements of the open-loop gain. In practice, there are simplifications: (I) $F(s)H(s) \to 0$ as $|s| \to \infty$; hence the right-half-plane semicircle of C_R maps into the origin of the FH plane, and (II) since $F(-j\omega)H(-j\omega) = [F(j\omega)H(j\omega)]^*$, the mapping of the negative part of the $j\omega$ axis is the mirror image of that of the positive part.

17 *Stability of multiple-loop systems*

Consider an example of a linear time-invariant multiple-feedback-loop system; in particular, consider the system shown in Fig. *9.17.1*. The

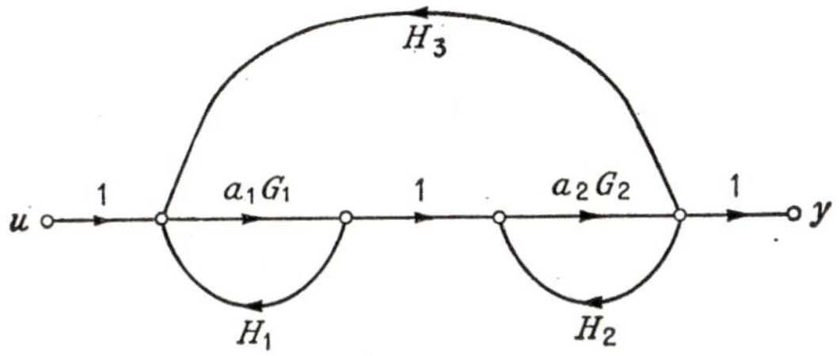

Fig. 9.17.1 Example of a multiple-feedback-loop system.

gain of the two forward branches is a_1G_1 and a_2G_2, where a_1 and a_2 are real numbers in [0,1] and G_1 and G_2 are functions of the complex variable s. The gain of the system from the input u to the output y is, according to Mason's formula *9.15.21*,

1 $$\frac{a_1a_2G_1G_2}{1 - a_1G_1H_1 - a_2G_2H_2 - a_1a_2G_1G_2H_3 + a_1a_2G_1G_2H_1H_2}$$

Suppose that while G_1, G_2, H_1, H_2, and H_3 are fixed functions of s, the parameters a_1 and a_2 are allowed to take any value in the square defined by $0 \leq a_1 \leq 1$, $0 \leq a_2 \leq 1$. The problem is to establish necessary and sufficient conditions for stability of the system irrespective of the values of a_1 and a_2 provided $0 \leq a_1 \leq 1$, $0 \leq a_2 \leq 1$.

The problem is important in many areas of engineering, in particular, in communications. It is often proposed to duplicate equipment to make the system immune to the failure or degradation of tubes, transistors, or other delicate devices. These questions have been considered in connection with active satellite repeaters, submarine cable repeaters, and, in general, any equipment whose maintenance is either impossible or prohibitively expensive. In repeaters it is proposed to duplicate the forward amplifiers in such a way that the failure or degradation of any one of them leaves the repeater performance within its specifications; this implies that the repeater is stable for all values of α_0 of the transistors lying between 0 and their nominal values.

Throughout the development of this section[1] we shall make the following

[1] The general method used in the following is originally due to L. Curtis of the Bell Telephone Laboratories. See also F. H. Blecher, Design Principles for Multiple Loop Transistor Feedback Amplifiers, *Proc. NEC,* vol. 13, pp. 19–34, 1957.

2 **Assumption** The system under consideration is a single-input single-output linear time-invariant system represented by a linear signal-flow graph.

3 **Assumption** All branches of the signal-flow graph are strictly stable.

4 **Assumption** There is a subset of l branches such that each branch of the subset has a transfer function proportional to one (real) parameter; the gain of the ith branch of the subset is proportional to a_i. These l parameters define a vector $\mathbf{a} = (a_1, a_2, \ldots, a_l)$. Since the components of $\mathbf{a}$ are subject to the inequalities $0 \leq a_i \leq 1$, the vector $\mathbf{a}$ is restricted to the cube

5 $$\mathcal{A} = \{\mathbf{a} | 0 \leq a_i \leq 1, i = 1, 2, \ldots, n\}$$

6 **Assumption** The gain of the system is strictly stable when $\mathbf{a} = \mathbf{0}$.

7 **Assumption** The zeros of the determinant $\Delta(s,\mathbf{a})$ of the signal-flow graph are continuous functions of $\mathbf{a}$.

8 *Comment* When all branch gains have their nominal values, $a_i = 1$ for $i = 1, 2, \ldots, l$. When aging, deterioration, or even failure occurs, some or all a_i decrease and may even become 0. Assumption *6* amounts to assuming that the gain of the system is strictly stable when all the elements of the subset have zero gain.

9 *Comment* Assumption *7* is always satisfied in practice: the implicit-function theorem[1] implies that *7* holds whenever each branch gain is the ratio of entire functions of s. In particular, if the system consists of lumped elements, each branch gain is a ratio of two polynomials. If in addition the system includes ideal transmission lines or transportation lags, the branch gains are ratios of polynomials in s, $e^{-T_1 s}$, $e^{-T_2 s}$, . . . , where T_1, T_2 are constants, and again *7* is satisfied.

The problem is to obtain necessary and sufficient conditions for the strict stability of the gain of the whole system. For this purpose we turn to the analysis of the problem. From Mason's formula (*9.15.21*) the gain of the system is of the form

$$G = \frac{\Sigma G_k \Delta_k}{\Delta(s,\mathbf{a})}$$

For each forward path, G_k and Δ_k are respectively a product and 1 minus an algebraic sum of products of branch gains; hence by *3*, the numera-

[1] T. M. Apostol, "Mathematical Analysis," p. 147, Addison-Wesley Publishing Company, Inc. Reading, Mass., 1957.

tor is analytic in the closed right half plane. Similarly, for all **a** in $\mathcal{A}$, $\Delta(s,\mathbf{a})$ is analytic in the closed right half plane. Thus,

10 **Assertion** The gain of the system is strictly stable for all $\mathbf{a} \in \mathcal{A}$ if and only if $\Delta(s,\mathbf{a})$ has no zeros in the closed right half plane for all $\mathbf{a} \in \mathcal{A}$. ◁

So the stability of the gain G depends on whether or not $\Delta(s,\mathbf{a})$ has right-half-plane zeros. Now, as in the Nyquist criterion case, we need only investigate the $j\omega$-axis behavior of $\Delta(s,\mathbf{a})$ in order to test for stability. More precisely, we have the

11 **Assertion** Let the system satisfy the assumptions *3* to *7*. If the gain of the system is unstable for some $\mathbf{a}_0 \in \mathcal{A}$, then there is a number $\lambda_0 \in (0,1]$ such that $\Delta(s,\lambda_0\mathbf{a}_0)$ has at least one $j\omega$-axis zero.
Proof Since the gain of the system is unstable at $\mathbf{a}_0$, *10* requires that at least one root of $\Delta(s,\mathbf{a}_0)$, say $s_1 = \sigma_1 + j\omega_1$, have a nonnegative real part σ_1. Note that σ_1 will vary with $\mathbf{a}$: $\sigma_1 = f(\mathbf{a})$, $f(\mathbf{a}_0) \geq 0$. Now introduce the function of the real variable λ defined by $\phi(\lambda) = f(\lambda\mathbf{a}_0)$. By *7*, $\phi(\lambda)$ is a continuous function of λ and $\phi(1) \geq 0$ but, by *6*, $\phi(0) < 0$. Therefore, there is a $\lambda_0 \in (0,1]$ such that $\phi(\lambda_0) = 0$; hence, at $\mathbf{a} = \lambda_0\mathbf{a}_0$, s_1 is on the $j\omega$ axis and $\Delta(s,\lambda_0\mathbf{a}_0)$ has a $j\omega$-axis zero.

12 **Corollary** The system is strictly stable if and only if $\Delta(s,\mathbf{a})$ has no $j\omega$-axis zeros for all $\mathbf{a} \in \mathcal{A}$. ◁

For convenience, let $\mathcal{R}^1$ be the set of all real numbers and define $T(s,\mathbf{a})$ by $\Delta(s,\mathbf{a}) = 1 - T(s,\mathbf{a})$. From Mason's formula (*9.15.21*), T is a linear combination of loop-gain products; hence, because of *4*, for any **a** and ω, the complex number $T(j\omega,\mathbf{a})$ can be determined experimentally by sinusoidal steady-state measurement of all the loop gains. In other words, the situation is analogous to the Nyquist case. Finally, if we observe that $\Delta(s,\mathbf{a}) = 0 \Leftrightarrow T(s,\mathbf{a}) = 1$, we may summarize the developments above by the

13 **Theorem** Let the system satisfy Assumptions *3* to *7*. The gain of the system is strictly stable for all $\mathbf{a} \in \mathcal{A}$ if and only if, for all $\omega \in \mathcal{R}^1$, the mapping of the cube $\mathcal{A}$ on the complex T plane by the function $T(j\omega,\cdot)$ does not include the point $1 + j0$. ◁

The problem now is to find an effective method of carrying out the mapping of $\mathcal{A}$ on the T plane for each $\omega \in \mathcal{R}^1$. Therefore, let us investigate some properties of the mapping.

Let the vector $\mathbf{a} \in \mathcal{A}$ and think of **a** as the point $(a_1,a_2, \ldots ,a_n)$ in $\mathcal{A}$. Since the cube $\mathcal{A}$ is convex, if a line is drawn through **a** parallel to the ith axis of coordinates, it will intersect the boundary of $\mathcal{A}$ at two points $\mathbf{a}'$ and $\mathbf{a}''$. Let $\mathbf{a}'$ (respectively, $\mathbf{a}''$) be the boundary point on the hyperplane $a_i = 0$ (respectively, $a_i = 1$). Observe that **a** belongs to the closed segment joining a′ to a″ (see Fig. *9.17.2*). Let $\mathbf{e}_i$ be the unit vector along the ith axis of coordinates; then

$$\mathbf{a} = \mathbf{a}' + a_i\mathbf{e}_i$$

$$\mathbf{a}'' = \mathbf{a}' + \mathbf{e}_i \quad \therefore a_i e_i = a_i(a'' - a')$$

or $a = a' + a_i(a'' - a') = a'(1 - a_i) + a_i a''$

Since each loop gain depends linearly on each of the branch gains of that loop, T is a multilinear function of the a_k's. More precisely, in

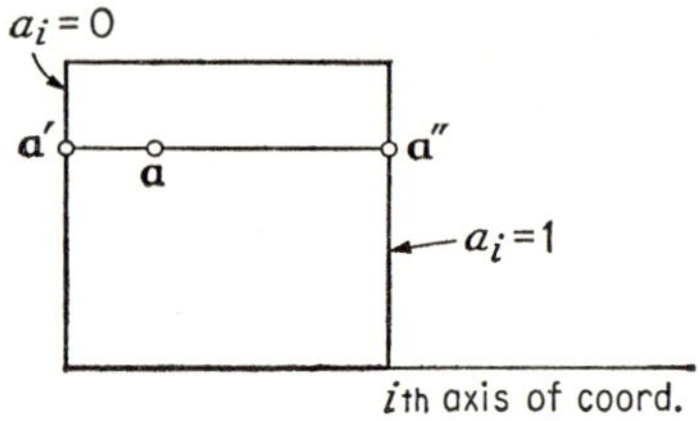

Fig. 9.17.2 Cross section of the n-dimensional cube $\mathcal{A}$ showing the segment a′a″.

the expression of T we can put any one of them in evidence, say, a_i, and write the complex number $T(j\omega,\mathbf{a})$ as $T(j\omega,\mathbf{a}) = A + a_iB$, where A and B depend on s, $a_1, a_2, \ldots, a_{i-1}, a_{i+1}, \ldots, a_n$. Observing that the ith components of a′ and a″ are 0 and 1, respectively, we have

$$T(j\omega,\mathbf{a}') = A \qquad T(j\omega,\mathbf{a}'') = A + B$$

Hence

$$T(j\omega,\mathbf{a}) = (1 - a_i)T(j\omega,\mathbf{a}') + a_iT(j\omega,\mathbf{a}'')$$

or

14 $$T(j\omega, \mathbf{a}' + a_i\mathbf{e}_i) = (1 - a_i)T(j\omega,\mathbf{a}') + a_iT(j\omega, \mathbf{a}' + \mathbf{e}_i)$$

Thus, in terms of mappings on the T plane, the point $T(j\omega, \mathbf{a}' + a_i\mathbf{e}_i)$, where $0 \leq a_i \leq 1$, lies on the segment joining $T(j\omega,\mathbf{a}')$ to $T(j\omega, \mathbf{a}' + \mathbf{e}_i)$. This is the fundamental property for the mapping of $\mathcal{A}$ into the complex T plane. Its geometric meaning can be stated as follows:

15 **Assertion** Any line segment of the cube $\mathcal{A}$ parallel to an axis of coordinates maps into a line segment on the T plane. The end points of the T-plane segment are the images of the end points of the segment in $\mathcal{A}$. ◁

Before deriving the general theorem, let us consider a simple example. Let $l = 2$; hence, the cube $\mathcal{A}$ reduces to a square with four vertices that are designated as $\mathbf{0}, \boldsymbol{\alpha}, \boldsymbol{\beta}, \boldsymbol{\gamma}$, as shown in Fig. *9.17.3*. We consider a fixed ω and wish to investigate the mapping of $\mathcal{A}$ on the complex T plane for that value of ω. Suppose for simplicity that $T(j\omega,\mathbf{0}) = 0$. In Fig. *9.17.4*, the images of the vertices are shown and are labeled with a primes in order to distinguish them from the vertices of $\mathcal{A}$. From *15* the mapping has the following properties:

16

$$\begin{aligned} &\text{Segment } [\mathbf{0},\boldsymbol{\alpha}] \to \text{segment } [\mathbf{0},\boldsymbol{\alpha}'] \\ &\text{Segment } [\boldsymbol{\alpha},\boldsymbol{\gamma}] \to \text{segment } [\boldsymbol{\alpha}',\boldsymbol{\gamma}'] \\ &\text{Segment } [\mathbf{0},\boldsymbol{\beta}] \to \text{segment } [\mathbf{0},\boldsymbol{\beta}'] \\ &\text{Segment } [\boldsymbol{\beta},\boldsymbol{\gamma}] \to \text{segment } [\boldsymbol{\beta}',\boldsymbol{\gamma}'] \end{aligned}$$

Consider now a segment parallel to the a_2 axis and of abscissa λ, $0 < \lambda < 1$. This segment is shown in Fig. *9.17.3*, and its vertices are labeled $\boldsymbol{\delta},\boldsymbol{\eta}$. Again, this segment maps onto the segment $[\boldsymbol{\delta}',\boldsymbol{\eta}']$ in Fig. *9.17.4*. It is clear that, as λ goes from 0 to 1, $[\boldsymbol{\delta},\boldsymbol{\eta}]$ sweeps through

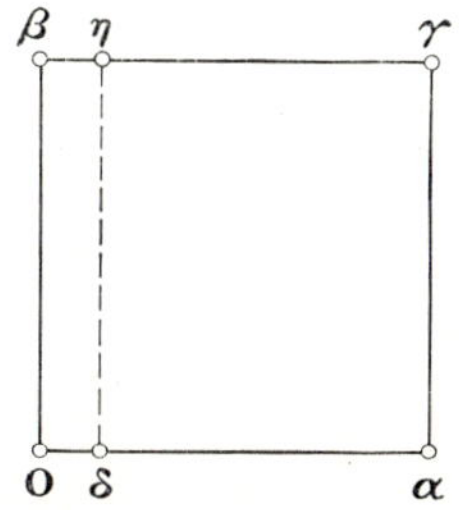

Fig. 9.17.3 Special case $l = 2$. The cube $\mathcal{A}$ reduces to a square.

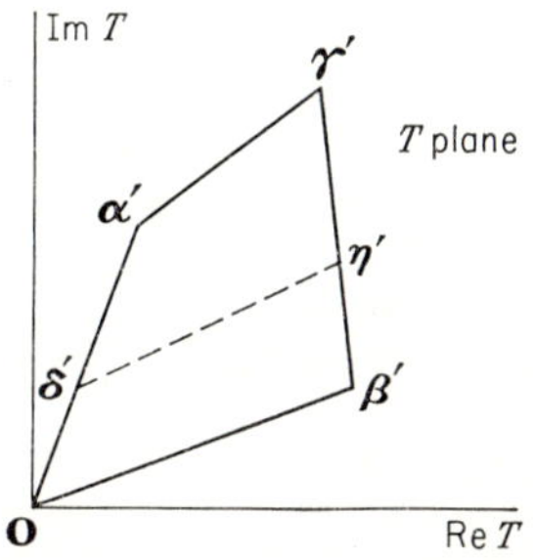

Fig. 9.17.4 T-plane image, for some fixed ω, of the square $\mathcal{A}$ of Fig. *9.17.3*.

the whole square $\mathcal{A}$ and $[\boldsymbol{\delta}',\boldsymbol{\eta}']$ sweeps through the whole quadrilateral $\mathbf{0}\boldsymbol{\alpha}'\boldsymbol{\gamma}'\boldsymbol{\beta}'$. Therefore, for that particular ω, the mapping of $\mathcal{A}$ requires only the plotting of the images of the vertices of $\mathcal{A}$. It is a fact that these happy circumstances will not always occur. Suppose the mapping of the vertices of $\mathcal{A}$ had been of the form shown in Fig. *9.17.5*. The statements *15* and *16* still hold: for small λ, the segment $[\boldsymbol{\delta}',\boldsymbol{\eta}']$ lies inside the quadrilateral $\mathbf{0}\boldsymbol{\alpha}'\boldsymbol{\gamma}'\boldsymbol{\beta}'$. However, for a larger λ, it maps onto

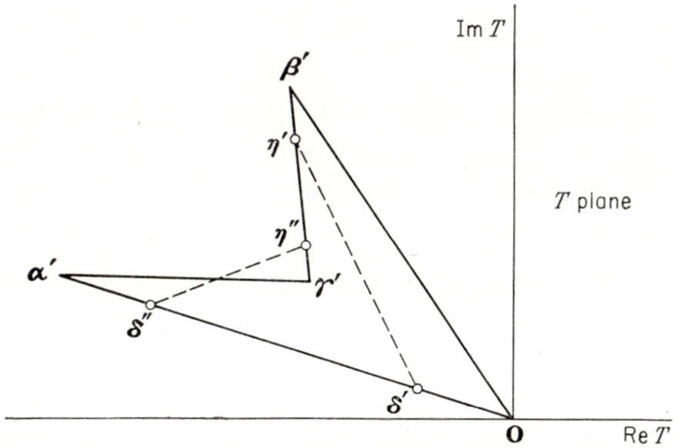

Fig. 9.17.5 T-plane image, for some other ω, of the square $\mathcal{A}$ of Fig. *9.17.3*.

$[\boldsymbol{\delta}'',\mathbf{n}'']$, some points of which are outside the quadrilateral! It is also easy to see that for all λ's it will be inside the triangle $\mathbf{0}\boldsymbol{\alpha}'\boldsymbol{\beta}'$. ← i.e. the convex hull

In order to proceed to the general case, we must introduce some new terms.

17 **Definition** For any $\omega \in \mathcal{R}^1$, let $V(\omega)$ be the images, on the T plane, of the 2^l vertices of $\mathcal{A}$. $V(\omega)$ is therefore a set of at most 2^l points, since the images of the vertices are not necessarily distinct.

18 **Definition** Let $X = \{\mathbf{x}_1,\mathbf{x}_2, \ldots ,\mathbf{x}_N\}$ be a finite set of points in the space $\mathcal{R}^N$. The *convex hull* of X, denoted by $[X]$, is the set

$$[X] = \left\{\mathbf{x} \middle| \mathbf{x} = \sum_{i=1}^{N} \lambda_i \mathbf{x}_i ; \lambda_i \geq 0 \text{ and } \sum_{i=1}^{N} \lambda_i = 1\right\}$$

For example, in Fig. *9.17.4*, the convex hull of $\{\mathbf{0},\boldsymbol{\alpha}',\boldsymbol{\beta}',\boldsymbol{\gamma}'\}$ is the quadrilateral defined by $\mathbf{0}$, $\boldsymbol{\alpha}'$, $\boldsymbol{\beta}'$, $\boldsymbol{\gamma}'$. On the other hand, in Fig. *9.17.5*, the convex hull of $\{\mathbf{0},\boldsymbol{\alpha}',\boldsymbol{\beta}',\boldsymbol{\gamma}'\}$ is the triangle defined by $\mathbf{0},\boldsymbol{\alpha}',\boldsymbol{\beta}'$. ◁

The following theorem states for the general case the situation suggested by these examples.

19 **Mapping theorem** For all $\omega \in \mathcal{R}^1$, the image of the cube $\mathcal{A}$ under the mapping $T(j\omega,\cdot)$ is contained in $[V(\omega)]$, the convex hull of $V(\omega)$ defined in *17*.

Proof The derivation is purely computational and is based on *14*. For simplicity we consider the case $l = 2$. From *14*, for all $\mathbf{a} = (a_1,a_2)$ in $\mathcal{A}$ we have

$$\begin{aligned} T(j\omega,\mathbf{a}) &= T(j\omega,\ a_1\mathbf{e}_1 + a_2\mathbf{e}_2) \\ &= (1 - a_2)T(j\omega,a_1\mathbf{e}_1) + a_2T(j\omega,\ a_1\mathbf{e}_1 + \mathbf{e}_2) \\ &= (1 - a_1)(1 - a_2)T(j\omega,\mathbf{0}) + a_1(1 - a_2)T(j\omega,\mathbf{e}_1) \\ &\qquad + (1 - a_1)a_2T(j\omega,\mathbf{e}_2) + a_1a_2T(j\omega,\ \mathbf{e}_1 + \mathbf{e}_2) \end{aligned}$$

This proves that, for all $\mathbf{a} \in \mathcal{A}$, the point $T(j\omega,\mathbf{a})$ belongs to the convex hull of the set $\{T(j\omega,\mathbf{0}),T(j\omega,\mathbf{e}_1),T(j\omega,\mathbf{e}_2),T(j\omega,\ \mathbf{e}_1 + \mathbf{e}_2)\}$, because each of the weighting factors is nonnegative and the sum of the weighting factors is equal to 1.

20 *Comment* From the example of Fig. *9.17.5* we know that there may be points in $[V(\omega)]$ that do not belong to the image of $\mathcal{A}$. We may finally state a sufficient condition for the stability of the system which follows directly from *13* and *19*.

21 **Theorem: sufficient condition for stability** Let the system satisfy assumptions *3* to *7*. If $[V(\omega)]$, the convex hull of $V(\omega)$, does not include the critical point $(1 + j0)$ for all $\omega \in \mathcal{R}^1$, then the gain of the system is strictly stable for all $\mathbf{a} \in \mathcal{A}$.

22 *Comment* Note that, first, this test is relatively easy to apply since it involves only 2^l Nyquist diagrams, one for each vertex of the cube $\mathcal{A}$, whereas the brute-force approach would require one Nyquist diagram for each point of $\mathcal{A}$. Second, *21* together with *13* is a quite powerful tool. Indeed, if for some ω, $[V(\omega)]$ includes the point $1 + j0$, *13* suggests that the shape of the mapping of $\mathcal{A}$ for that ω be sketched out in order to find out whether or not the system is actually unstable. Consider, for example, the case illustrated in Fig. *9.17.6*. The cube $\mathcal{A}$ is shown with its vertices **1,2,3,4,5,6,7,8**. On the T plane shown in Fig. *9.17.6* the points **1′,2′**, . . . ,**8′** are the mappings of the vertices

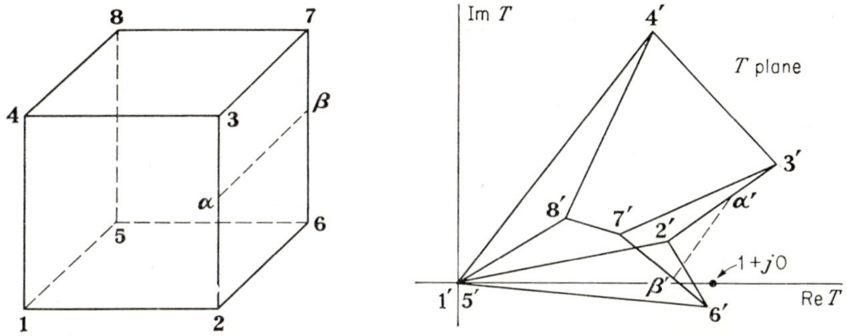

Fig. 9.17.6 Cube $\mathcal{A}$ and its T-plane image for some fixed ω.

of $\mathcal{A}$ for a particular ω, say, ω_0. It is clear that the critical point $(1 + j0)$ is inside $[V(\omega_0)]$. However, the critical point does not belong to the mapping of $\mathcal{A}$ at ω_0; all the faces of $\mathcal{A}$ map inside the polygon **1′,6′,2′,3′,4′** except the face **2,3,7,6**. The figure shows the mapping of the four edges of that face. Now consider parallels to the edge **3,7** such as the segment $\boldsymbol{\alpha\beta}$, which maps onto the segment $\boldsymbol{\alpha'\beta'}$. As $\boldsymbol{\alpha}$ moves along the edge **2,3**, the segment $\boldsymbol{\alpha'\beta'}$ slides in the T plane with $\boldsymbol{\alpha'}$ constrained to be on **2′3′** and $\boldsymbol{\beta'}$ constrained to remain on **6′7′**. The boundary of the mapping of the face **2,3,7,6** is the envelope generated by the motion of $\boldsymbol{\alpha'\beta'}$. It can be shown that it is an *arc of a parabola* and two straight-line segments. In practice, however, one need only trace a few positions of the segment $\boldsymbol{\alpha'\beta'}$ to decide whether or not the critical point is included in the mapping of $\mathcal{A}$.

23 *Exercise* Suppose that the vertices **2,3,6,7** of $\mathcal{A}$ map into the points $\mathbf{x}_2$, $\mathbf{x}_3$, $\mathbf{x}_6$, and $\mathbf{x}_7$ in the T plane shown in Fig. *9.17.7*. Segment $\boldsymbol{\alpha'\beta'}$ is

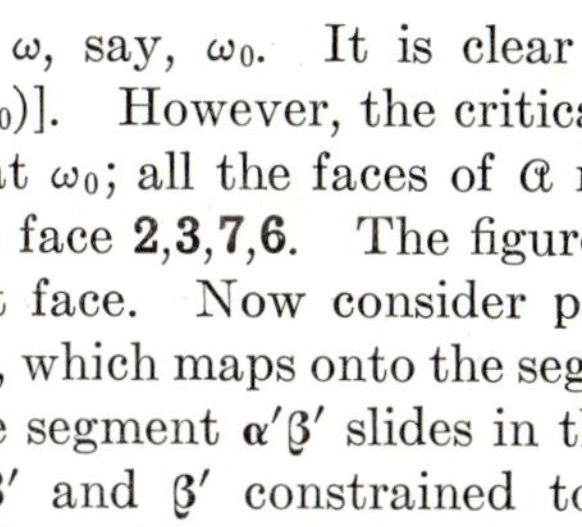

Fig. 9.17.7 Figure showing the segments $\mathbf{x}_2\mathbf{x}_3$ and $\mathbf{x}_6\mathbf{x}_7$ of Exercise *9.17.23*.

the segment joining $\boldsymbol{\alpha}' = \mathbf{x}_2 + \lambda\mathbf{x}_{23}$ and $\boldsymbol{\beta}' = \mathbf{x}_6 + \lambda\mathbf{x}_{67}$, where $\mathbf{x}_{23} = \mathbf{x}_3 - \mathbf{x}_2$, $\mathbf{x}_{67} = \mathbf{x}_7 - \mathbf{x}_6$, and $\lambda \in [0,1]$. Show that the envelope of that segment as λ traverses the interval $[0,1]$ is an arc of a parabola and two straight lines.

REFERENCES AND SUGGESTED READING

1 Bode, H. W.: "Network Analysis and Feedback Amplifier Design," D. Van Nostrand Company, Inc., Princeton, N.J., 1945.
2 Doetsch, G.: "Handbuch der Laplace Transformation," 3 volumes, Verlag Birkhäuser, Basel, 1950.
3 Doetsch, G.: "Guide to the Applications of Laplace Transforms," D. Van Nostrand Company, Inc., Princeton, N.J., 1961.
4 Guillemin, E. A.: "Mathematics of Circuit Analysis," John Wiley & Sons, Inc., New York, 1949.
5 Kaplan, W.: "Operational Methods for Linear Systems," Addison-Wesley Publishing Company, Inc., Reading, Mass., 1962.
6 Paley, R. E. A. C., and N. Wiener: Fourier Transforms in the Complex Domain, *Amer. Math. Soc. Colloq. Publ.*, vol. 19, New York, 1934.
7 Papoulis, A.: "The Fourier Integral and Its Applications," McGraw-Hill Book Company, Inc., New York, 1962.
8 Seshu, S., and N. Balabanian: "Linear Network Analysis," John Wiley & Sons, Inc., New York, 1959.
9 Van der Pol, B., and H. Bremmer: "Operational Calculus Based on the Two Sided Laplace Integral," Cambridge University Press, New York, 1950.
10 Widder, D. V.: "The Laplace Transform," Princeton University Press, Princeton, N.J., 1941.

Discrete-time systems

10

1 *Introduction*

So far we have considered exclusively systems in which the variables (input, output, state, etc.) can be observed at all instants of time. This is not always possible. Digital computers, for example, take in and print out data representing the values of variables at discrete instants of time. These discrete instants are taken sufficiently close that a simple interpolation would yield a satisfactory approximation for all intermediate values of time. In some other circumstances the system itself is not digital but observations can be made only periodically. A pulse radar, for example, obtains data from the target only at equally spaced instants of time. Another example comes from chemical engineering: it sometimes happens that an expensive analyzer is shared by several identical processes so that the output product of each process is analyzed only periodically.

Thus in general, when one or more variables can be observed only periodically and when the period of observation is sufficiently small that all variables can be reconstructed sufficiently accurately from their samples, it is an obvious simplification to write the equations of the system in terms of the sampled values of all the variables. In other words, we model the system by a *discrete-time system* (see *3.8.42*). This simplifies the analysis considerably, and, furthermore, it is precisely that which is done in numerical computation. We shall therefore restrict ourselves to *periodic sampling*, i.e., to cases where the samples are taken at equally spaced instants of time. We shall, furthermore, consider only *time-invariant* systems. This will be sufficient for our main purpose, which is to show the considerable extent to which the analysis of discrete systems follows that of continuous systems. The analysis and design of discrete control systems is a vast subject. Several books, listed in the references, have been written on the subject in the last few years, and the reader is referred to them if he wishes to pursue the subject beyond our brief treatment.

2 *Systems represented by their state equations*

Representation of discrete systems obtained by sampling a differential system

Suppose we are given a differential system $\mathcal{S}$ represented by its state equations

1 $$\dot{\mathbf{x}} = \mathbf{A}\mathbf{x} + \mathbf{B}\mathbf{u}$$

2 $$\mathbf{y} = \mathbf{C}\mathbf{x} + \mathbf{D}\mathbf{u}$$

where $\mathbf{A}$, $\mathbf{B}$, $\mathbf{C}$, and $\mathbf{D}$ are, respectively, $(n \times n)$, $(n \times r)$, $(p \times n)$, and $(p \times r)$ constant matrices. Suppose that the components of the input vector are sampled periodically and are constrained to be constant over each interval $(kT,(k+1)T)$, where $k = \ldots, -1, 0, 1, \ldots$. In Fig. *10.2.1*, this constraint is represented by the insertion of the sample-

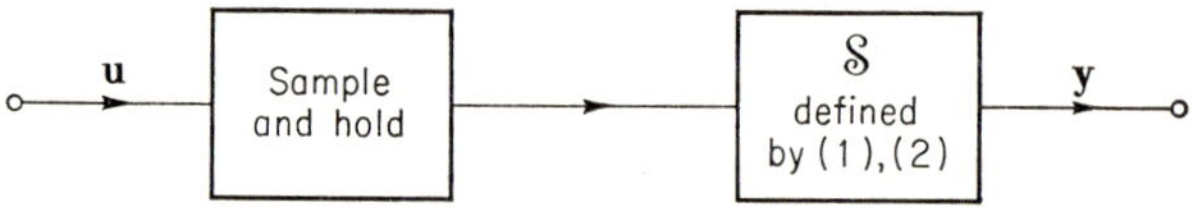

Fig. 10.2.1 The input u is sampled and held constant over the intervals $(kT, (k+1)T)$ by the sample-and-hold box before being applied to the system $\mathcal{S}$.

and-hold box between the input $\mathbf{u}$ and the system $\mathcal{S}$. By definition of the sample-and-hold box, if $\alpha(t)$ is its input, then its output is the piecewise constant function α_0 specified by

$$\alpha_0(t) = \alpha(kT) \qquad kT < t \leq (k+1)T, \text{ for all integers } k$$

We shall say that the input is sampled every T sec, and we shall refer to T as the *sampling period*. Thus the input is completely specified by the sequence of vectors $\{\mathbf{u}_k\}$, where $\mathbf{u}_k = \mathbf{u}(kT+)$. As stated earlier, in most applications the sampling period T is chosen sufficiently small that the interpolation of the sequences $\{\mathbf{x}_k\}$ and $\{\mathbf{y}_k\}$, where $\mathbf{x}_k = \mathbf{x}(kT+)$, $\mathbf{y}_k = \mathbf{y}(kT+)$, determines the functions $\mathbf{x}(t)$ and $\mathbf{y}(t)$ with sufficient accuracy for all t. For this reason, it is reasonable to seek the relations between the sequences $\{\mathbf{x}_k\}$ and $\{\mathbf{y}_k\}$ and the input sequence $\{\mathbf{u}_k\}$. It is most convenient to express these relations as recurrence relations giving $\mathbf{x}_{k+1}$ and $\mathbf{y}_{k+1}$ in terms of $\mathbf{x}_k$ and $\mathbf{u}_k$. Using *5.2.19* and defining

3 $$\mathbf{F} = \exp \mathbf{A}T$$

4 $$\mathbf{G} = \left(\int_0^T [\exp(\mathbf{A}\tau)]\, d\tau\right) \mathbf{B}$$

we immediately obtain

5 $$\mathbf{x}_{k+1} = \mathbf{F}\mathbf{x}_k + \mathbf{G}\mathbf{u}_k$$

6 $$\mathbf{y}_{k+1} = \mathbf{C}\mathbf{x}_{k+1} + \mathbf{D}\mathbf{u}_{k+1}$$

We can consider *5* and *6* as the state equations of a discrete-time system whose input, state, and output are respectively specified by the vector-valued sequences $\{\mathbf{u}_k\}$, $\{\mathbf{x}_k\}$, and $\{\mathbf{y}_k\}$. We shall refer to this system as the *discrete-time system 5* and *6*. Since **A**, **B**, **C**, and **D** are constant matrices, it is easy to verify that this system is linear and time-invariant (see *3.4.25* and *3.2.36*).

If we iterate *5*, then we can express $\mathbf{x}_k$ in terms of the initial state $\mathbf{x}_0$ and the sequence $\{\mathbf{u}_i\}_0^{k-1}$:

7 $$\mathbf{x}_k = \mathbf{F}^k\mathbf{x}_0 + \sum_{i=0}^{k-1} \mathbf{F}^i\mathbf{G}\mathbf{u}_{k-i-1} \qquad k = 0, 1, \ldots$$

The R.H.S. of *7* may also be written $\mathbf{x}(kT;\mathbf{x}_0,0;\{\mathbf{u}_i\}_0^{k-1})$ to emphasize that it is the state at time kT resulting from the initial state $\mathbf{x}_0$ at time 0 and the input defined by the sequence $\{\mathbf{u}_i\}_0^{k-1}$.

8 **Summary** Let the (continuous-time) system $\mathcal{S}$ be defined by its state equations *1* and *2*, where **A**, **B**, **C**, and **D** are constant matrices. Let the input be piecewise constant over the intervals $(kT,(k+1)T)$, $k = \ldots, -1, 0, 1, \ldots$; then the sampled values of the state vector, input, and output vector are related by *5* and *6*. $\mathbf{x}(kT;\mathbf{x}_0,0;\{\mathbf{u}_0\}_0^{k-1})$ is given by *7*.

9 *Exercise* Show that whatever **A** and $T > 0$ may be, **F** is always a nonsingular matrix.

10 *Example* For simplicity consider the single-input single-output differential system defined by

$$\begin{bmatrix} \dot{x}_1 \\ \dot{x}_2 \end{bmatrix} = \begin{bmatrix} -1 & 3 \\ 0 & -2 \end{bmatrix} \begin{bmatrix} x_1 \\ x_2 \end{bmatrix} + \begin{bmatrix} 0 \\ 1 \end{bmatrix} u$$

$$y = [1 \quad 1] \begin{bmatrix} x_1 \\ x_2 \end{bmatrix}$$

To obtain **F**, compute exp $(\mathbf{A}T)$ by the interpolation method of Appendix *D*, Sec. *8*; by integrating the result, **G** is easily obtained by *4*. The result is

$$\mathbf{x}_{k+1} = \begin{bmatrix} e^{-T} & 3(e^{-T} - e^{-2T}) \\ 0 & e^{-2T} \end{bmatrix} \mathbf{x}_k + \tfrac{1}{2} \begin{bmatrix} 3 - 6e^{-T} + 3e^{-2T} \\ 1 - e^{-2T} \end{bmatrix} u_k$$

$$y_{k+1} = [1 \quad 1]\mathbf{x}_{k+1}$$

Stability considerations

Consider now the discrete system *5* and *6* with its input, its state, and its output respectively specified by the sequences $\{\mathbf{u}_k\}$, $\{\mathbf{x}_k\}$, $\{\mathbf{y}_k\}$. In other words, we disregard the values of $\mathbf{x}(t)$ and $\mathbf{y}(t)$ for all t that are outside the set $\{kT\}$, where $k = \ldots, -1, 0, 1, \ldots$.

Let us first consider the idea of *asymptotic stability*. It will be recalled that asymptotic stability concerns itself with the state and not the output and that it imposes requirements only on the motion of the state in state space under zero-input conditions, in other words, under free motion (see *7.2.3*). By analogy with *7.2.4* we introduce the

11 **Definition** A linear time-invariant discrete system described by *5* is said to be *asymptotically stable* (in the large) if

(I) For any $M > 0$, there is a $\delta > 0$ such that

$$\|\mathbf{x}_0\| < \delta \Rightarrow \|\mathbf{x}(kT;\mathbf{x}_0,0;\mathbf{0})\| < M \qquad \text{for } k = 0, 1, 2, \ldots$$

(II) For all initial states $\mathbf{x}_0$, $\lim_{k\to\infty} \mathbf{x}(kT;\mathbf{x}_0,0;\mathbf{0}) = \mathbf{0}$. ◁

In other words, as in the continuous case, asymptotic stability implies that during any zero-input response (free motion) the state remains bounded and eventually tends to the zero state as $kT \to \infty$. As shown previously (*7.2.6*), the asymptotic stability of the zero state implies that the discrete system is asymptotically stable in the large.

12 **Theorem** The following three statements are equivalent:

(I) The linear time-invariant sampled system *5* is asymptotically stable (in the large).

(II) $\|\mathbf{F}^n\| \to 0$ as $n \to \infty$.

(III) All the eigenvalues of $\mathbf{F}$ are in the open disk $|\lambda| < 1$.

Proof Exercise for the reader. For the case where $\mathbf{F}$ has distinct eigenvalues use *C.18.18*; for the general case use *D.6.17*.

13 *Comment* If $\lambda_1, \lambda_2, \ldots, \lambda_\sigma$ are the eigenvalues of $\mathbf{A}$, then $e^{\lambda_1 T}, \ldots, e^{\lambda_\sigma T}$ are the eigenvalues of $\mathbf{F}$ (defined by *3*). This follows from the definition of a function of a matrix (see *D.4.1*). Therefore, the following three statements are equivalent:

(I) All eigenvalues of $\mathbf{F}$ are in the open disk $|\lambda| < 1$.

(II) All eigenvalues of $\mathbf{A}$ are inside the left half plane.

(III) The (continuous-time) system 1 is asymptotically stable (in the large).

In particular, the asymptotic stability of the sampled system *5* is equivalent to the asymptotic stability of the continuous system *1* from which it is obtained. Note that this equivalence holds for all possible sampling periods $T > 0$.

14 **Theorem** Let the discrete system be described by *5* and let the state $\mathbf{x}$ itself be the output. Then, if all the eigenvalues of $\mathbf{F}$ are in the open disk $|\lambda| < 1$ then, for all initial states, any bounded input $\{\mathbf{u}_i\}_0^\infty$ produces a bounded state vector.

Proof The input sequence is bounded; hence there is a finite number M such that $\|\mathbf{u}_i\| < M$ for $i = 0, 1, \ldots$. Then, by *7*, *C.16.4*, and *C.16.12* we obtain successively

15 $$\|\mathbf{x}_k\| \le \|\mathbf{F}^k\mathbf{x}_0\| + \sum_{i=0}^{k-1} \|\mathbf{F}^i\mathbf{G}\mathbf{u}_{k-i-1}\| \le \|\mathbf{F}^k\| \cdot \|\mathbf{x}_0\| + M\|\mathbf{G}\| \sum_{i=0}^{k-1} \|\mathbf{F}^i\|$$

Since the eigenvalues of $\mathbf{F}$ are in the open disk $|\lambda| < 1$, for some constant C we have

16 $$\|\mathbf{F}^i\| < C\rho^i \qquad i = 0, 1, 2, \ldots$$

where ρ is a real positive number smaller than 1 but larger than the absolute value of all the eigenvalues of $\mathbf{F}$. Therefore, *16* implies that

$$\|\mathbf{F}^k\| < C \qquad k = 0, 1, 2, \ldots$$

$$\sum_{i=0}^{\infty} \|\mathbf{F}^i\| < C(1 - \rho)^{-1}$$

Here, the first term of *15* is bounded, for all k, by $C\|\mathbf{x}_0\|$ and the second is bounded by $M\|\mathbf{G}\|C(1 - \rho)^{-1}$.

17 *Exercise* Show that if only the zero-state response of *5* were required to remain bounded for all bounded inputs, $\mathbf{F}$ might in some cases have eigenvalues larger than 1 in absolute value. [*Hint:* Let $\mathbf{F}$ have distinct eigenvalues $\lambda_1, \lambda_2, \ldots, \lambda_n$. Let λ_1 be the only eigenvalue outside the disk $|\lambda| < 1$ and let $|\lambda_1| > 1$. Pick $\mathbf{G}$ such that the range of $\mathbf{G}$, $\mathcal{R}(\mathbf{G})$, is spanned by the eigenvectors of $\lambda_2, \lambda_3, \ldots, \lambda_n$.]

3 *Transform theory of discrete systems*

In this section we shall consider discrete systems obtained by periodic-sampling continuous-time systems which are defined by their impulse response. Again we shall assume that the continuous-time systems are time-invariant. From Chap. *9* we would expect that the analysis can be considerably simplified by the introduction of a suitable transformation which, with respect to discrete systems, would play the role that the Laplace transform plays for continuous-time systems. The appropriate transform for the present case turns out to be the z transform. As a first step let us set up the model that we shall use throughout the section.

Impulse modulators and sampling

For simplicity, we consider a single-input single-output linear time-invariant system $\mathcal{S}$ represented by its impulse response h. We shall assume that the system is nonanticipative; therefore, $h(t) = 0$ for $t < 0$. The sampling will affect this continuous-time system $\mathcal{S}$ in the following manner (refer to Fig. *10.3.1*). Before being applied to the continuous-time system $\mathcal{S}$, the input u is periodically sampled with period T. Its value at the kth sampling instant, $u(kT)$, is used to amplitude-modulate a pulse generator whose output pulse is of height $u(kT)/\Delta$ and of width Δ. More precisely, the output pulse is given by

$$\Delta^{-1}[1(t - kT) - 1(t - kT - \Delta)]u(kT)$$

The zero-state response of the system $\mathcal{S}$ to this pulse can be shown to tend to $u(kT)h(t - kT)$ as $\Delta \to 0$; that is, in the limit of $\Delta \to 0$, the zero-state response of $\mathcal{S}$ is equal to its zero-state response to $u(kT)\delta(t - kT)$. This leads to the following idealization: Let us replace the pulse generator of Fig. *10.3.1* by an *impulse modulator*. By definition, an

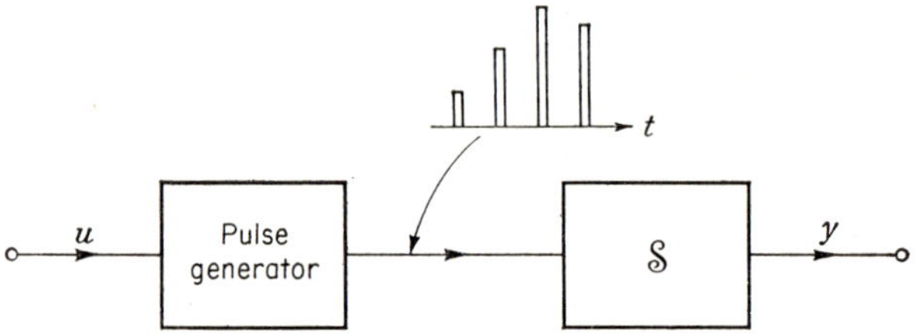

Fig. 10.3.1 The input u modulates the amplitude of the pulses produced by the pulse generator.

impulse modulator is characterized by its sampling period T; and if u is its input, its output is the sequence of impulses $\sum_{k=-\infty}^{\infty} u(kT)\delta(t - kT)$. This model is illustrated by Fig. *10.3.2*. The input u is fed into the

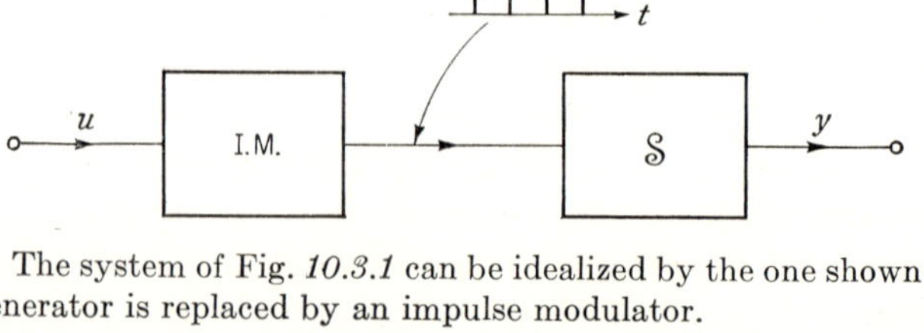

Fig. 10.3.2 The system of Fig. *10.3.1* can be idealized by the one shown here where the pulse generator is replaced by an impulse modulator.

impulse modulator (abbreviated as I.M.) whose output is the sequence of impulses just described. This sequence of impulses is the input of the continuous-time system $\mathcal{S}$; therefore, its zero-state output is given by

1
$$y(t) = \sum_{m=-\infty}^{+\infty} u(mT)h(t - mT) \qquad \forall t$$

Note that the output of $\mathcal{S}$ is defined for all t. If we restrict our attention to the sequence of values taken by y at the instants kT, then we obtain the relation

2
$$y_k = \sum_{m=-\infty}^{+\infty} u_m h_{k-m} \qquad k = \ldots, -1, 0, 1, \ldots$$

where $y_k = y(kT)$, $u_m = u(mT)$, $h_k = h(kT)$. Equation *2* is a linear relation between the *input sequence* $\{u_m\}$ and the (zero-state) *output sequence* $\{y_k\}$. In view of the close analogy between *2* and the convolution integral of continuous-time systems (*3.6.35*), the sequence $\{h_k\}$ is called the *weighting sequence*. In particular, if the input is applied at $t = 0$ and if the system is nonanticipative, the zero-state output sequence is given by

3
$$y_k = \sum_{m=0}^{\infty} u_m h_{k-m} \qquad k = 0, 1, \ldots$$

Because we restrict our attention to the sequence of values $y(kT)$ of the continuous-time output and because the impulse-modulator output depends only on the sequence of values, $u(kT)$, of the input, we think of the system shown in Fig. *10.3.2* as a *discrete system* whose input is completely characterized by the sequence $\{u_k\}$ and whose output is the sequence $\{y_k\}$. It is important to note that, in order to adopt this point of view, there is no need for an actual sampler to sample the output of $\mathcal{S}$. However, when systems are interconnected together, it will make a great difference whether there actually is or is not a sampler or impulse modulator at the output of $\mathcal{S}$ (see Remark *16*, below). For the remainder of this section, we shall consider the system of Fig. *10.3.2* as a linear discrete system transforming the input sequence $\{u_m\}$ into the output sequence $\{y_k\}$ by Eq. *3*.

4 *Comment* We have so far considered, and shall consider later in this section, only single-input single-output systems; the analysis that follows is also applicable to multiple-input and multiple-output systems. The extension is easy: it requires no new concepts but an elaborate notation.

z transform

In order to study the properties and relations between the input and output sequences of the system of Fig. *10.3.2*, we shall use the z transform. The definition and the principal theorems of the z transform are given in Appendix *B*, Sec. *4*. Let us, however, recall the

5 **Definition** The z transform of the function $u_{[0,\infty)}$ is the function $\tilde{U}$ of the complex variable z defined by

6
$$\tilde{U}(z) = \mathcal{Z}(u) = \sum_{n=0}^{\infty} u(nT)z^{-n}$$

where T is the sampling period of the impulse modulator.

7 **Remark** If u has a discontinuity at any of the sampling points, the defining relation *6* is ambiguous. We shall, therefore, restrict ourselves to the most important case that occurs in practice: we require all the functions of time which will be sampled to belong to the class A; that is, they are zero for $t < 0$, and if they are discontinuous at some sampling instant, then $u(nT-)$ and $u(nT+)$ must exist. In *6* we shall always take

8
$$u(nT) = u(nT+) \qquad n = 0, 1, \ldots$$

9 *Exercise* Establish the following z transforms. (*Hint:* Use *6* and sum the geometric series.)

Time function	*z transform*
$1(t)$	$1/(1 - z^{-1})$
e^{-at}	$1/(1 - z^{-1}e^{-aT})$
$e^{-\alpha t}\sin \omega t$	$ze^{-\alpha T}\sin \omega T(z^2 - 2e^{-\alpha T}z\cos \omega T + e^{-2\alpha T})^{-1}$

10 *Comment* $\tilde{U}(z)$ is defined by the power series in z^{-1} given in *6*. This power series converges for all z outside the circle $|z| = R_u$, where $R_u = \limsup_{n\to\infty} \sqrt[n]{|u(nT)|}$. (See Ref. *5*, page 141, and Ref. *6*, page 118.) Some functions of class A have a radius of convergence R_u that is infinite; for example, $u(t) = 1(t)\exp(t^4)$. We shall henceforth assume that each function of class A that we consider has a finite radius of convergence.

11 *Comment* If u is the input of the impulse modulator, its output is $\bar{u} = \sum_{k=0}^{\infty} u(kT)\delta(t - kT)$. This series of impulses has a Laplace transform $\bar{U}(s) = \sum_{k=0}^{\infty} u(kT)e^{-skT}$. Comparing *6* with this relation leads to the observation that $\tilde{U}(z)\Big|_{z=e^{sT}} = \bar{U}(s)$. In this sense, we might say

that the z plane, the complex plane over which $\tilde{U}$ is defined, is related to the s plane (of the Laplace transform of $\bar{u}$) by $z = e^{sT}$. The important fact to remember is that $\bar{U}(s)$ is *not* the Laplace transform of u; therefore, $\tilde{U}(z)$ is *not* obtained by taking the Laplace transform of u and then replacing s by $(1/T)\log z$. (See *19* and *22* below.)

12 **Theorem** Consider the system shown in Fig. *10.3.2*. Let $\tilde{H}(z)$ be the z transform of the impulse response h. Let y be the zero-state response due to the input u applied at $t = 0$. Then,

13 $$\tilde{Y}(z) = \tilde{H}(z)\tilde{U}(z) \qquad \text{for } |z| > \max(R_u, R_h)$$

14 *Comment* *13* is completely analogous to *3.6.23*, which relates the zero-state response, the impulse response, and the input of a continuous-time system. For this reason $\tilde{H}(z)$ is referred to as the *sampled transfer function* or the *z-transfer function*. It should be stressed that, in contradistinction to *3.6.23*, *13* does not define the zero-state response of the system shown in Fig. *10.3.2* for all t; it gives only the sequence $\{y_k\}$ of the output values at the sampling instants. If one wanted to obtain the output y for all t, then the usual Laplace transform technique (taking into account the presence of the impulse modulator) could be used. The same information can be obtained by the so-called modified z transform. A description of this technique is to be found in the books listed in the references.

Proof From Definition *5* we have

$$\tilde{H}(z)\tilde{U}(z) = \left(\sum_{\alpha=0}^{\infty} u_\alpha z^{-\alpha}\right)\left(\sum_{\beta=0}^{\infty} h_\beta z^{-\beta}\right) \qquad |z| > \max(R_h, R_u)$$

$$= \sum_{l=0}^{\infty}\left(\sum_{\alpha=0}^{l} u_\alpha h_{l-\alpha}\right) z^{-l}$$

and, with *3*,

$$\tilde{H}(z)\tilde{U}(z) = \sum_{l=0}^{\infty} y_l z^{-l} = \tilde{Y}(z) \qquad |z| > \max(R_h, R_u)$$

15 *Example* Consider the discrete system shown in Fig. *10.3.3*. Let us find the "sampled output" (i.e., the sequence $\{y(kT)\}$) caused by a unit step input. The system is known to be in the zero state at $t = 0$.

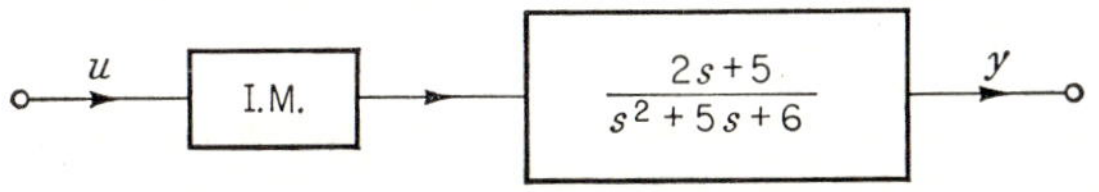

Fig. 10.3.3 System studied in Example *10.3.15*.

First, from $H(s)$ we obtain $h(t) = 1(t)(e^{-2t} + e^{-3t})$. Therefore,

$$\tilde{H}(z) = \sum_{k=0}^{\infty} (e^{-2kT} + e^{-3kT})z^{-k} = \frac{1}{1 - z^{-1}e^{-2T}} + \frac{1}{1 - z^{-1}e^{-3T}}$$

Since $\tilde{U}(z) = 1/(1 - z^{-1})$, the zero-state output sequence has a z transform given by

$$\begin{aligned}\tilde{Y}(z) &= \tilde{H}(z)\tilde{U}(z)\\ &= \frac{1}{1 - z^{-1}}\left(\frac{1}{1 - e^{-2T}} + \frac{1}{1 - e^{-3T}}\right) - \frac{1}{1 - z^{-1}e^{-2T}}\,\frac{1}{e^{2T} - 1}\\ &\qquad - \frac{1}{1 - z^{-1}e^{-3T}}\,\frac{1}{e^{3T} - 1}\end{aligned}$$

which has been obtained by partial-fraction expansion. The magnitude of the nth sample is (see Exercise *9*)

$$y_n = y(nT+) = \frac{1}{1 - e^{-2T}} + \frac{1}{1 - e^{-3T}} - \frac{e^{-n2T}}{e^{2T} - 1} - \frac{e^{-n3T}}{e^{3T} - 1}$$

The reader should sketch out the output y and observe that y is discontinuous at $t = nT$, $n = 0, 1, 2, \ldots$. This illustrates once more the importance of Remark *7* in the interpretation of the relationship between $\tilde{Y}(z)$ and $y_{[0,\infty)}$.

16 **Remark** When one interconnects several subsystems, it is very important to observe the position of the impulse modulators carefully before applying *13*. Consider the systems shown in Figs. *10.3.4* and

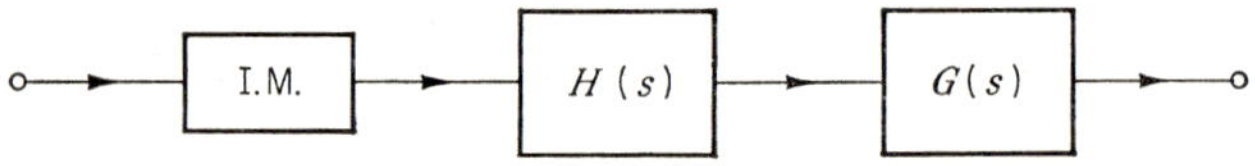

Fig. 10.3.4 The system G is directly connected to H.

10.3.5. (In both cases the impulse modulators are synchronized and have the same period.) Their respective z-transfer functions are $(\widetilde{GH})(z)$ and $\tilde{G}(z)\tilde{H}(z)$; these are different functions of z.

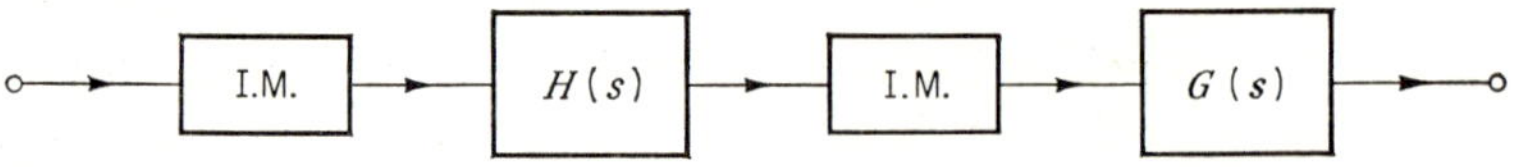

Fig. 10.3.5 The system G is connected to H through the impulse modulator. The two impulse modulators are synchronized and have the same period.

17 *Exercise* Let $G(s) = 1/(s + 1)$ and $H(s) = 1/(s + 2)$. Show that

$$(\widetilde{GH})(z) = \frac{e^{-T} - e^{-2T}}{(z - e^{-T})(z - e^{-2T})} \qquad \tilde{G}(z)\tilde{H}(z) = \frac{z^2}{(z - e^{-T})(z - e^{-2T})}$$

Sketch out the time functions at the input and output of the G and H boxes of Figs. *10.3.4* and *10.3.5*.

In Comment *11* we pointed out that the relation between $H(s)$ and $\tilde{H}(z)$ (respectively the Laplace transform of the function h and the z transform of the function h) is *not* merely a matter of substituting for s by some expression in z. In the case of rational functions, the relation between $H(s)$ and $\tilde{H}(z)$ is easily obtained through their partial-fraction expansions.

18 **Theorem** If $H(s)$ is a rational function of s, then $\tilde{H}(z)$ is also a rational function of z.

Proof If $H(s)$ is rational, $h(t)$ is a finite sum of terms of the form e^{-at}, $t^k e^{-at}$, where a may be complex and where k is an integer. Consider the contribution of the term e^{-at} in the defining relation *6*; then

$$\sum_{n=0}^{\infty} e^{-anT} z^{-n} = \sum_{n=0}^{\infty} (ze^{aT})^{-n} = \frac{1}{1-(ze^{aT})^{-1}} = \frac{z}{z-e^{-aT}} \qquad |z| > e^{-aT}$$

In other words, a simple pole of H at $s = -a$ contributes a simple pole of $\tilde{H}$ at $z = e^{-aT}$.

The term $t^k e^{-at}$ will occur in $h(t)$ when $H(s)$ has a pole of order $k+1$ at $s = -a$. Since

$$t^k e^{-at} = (-1)^k \frac{\partial^k}{\partial a^k} (e^{-at})$$

we have

$$\mathcal{Z}(t^k e^{-at}) \triangleq \sum_{n=0}^{\infty} (nT)^k e^{-anT} z^{-n} = (-1)^k \frac{\partial^k}{\partial a^k} \mathcal{Z}(e^{-at}) = (-1)^k \frac{\partial^k}{\partial a^k} \left(\frac{z}{z-e^{-aT}} \right)$$

It is clear that the right-hand side will be a rational function of z with a pole of order $k+1$ at $z = e^{-aT}$. Hence a pole of order $k+1$ of $H(s)$ at $s = a$ contributes a pole of order $k+1$ at $z = e^{-aT}$ in $\tilde{H}(z)$.

19 *Comment* The above proof shows that, given the partial-fraction expansion of $H(s)$, the expansion of $\tilde{H}(z)$ is immediately obtained by observing the following correspondences:

$$\frac{1}{s+a} \qquad \frac{z}{z-e^{-aT}}$$

$$\frac{1}{(s+a)^{k+1}} \qquad \frac{(-1)^k}{k!} \frac{\partial^k}{\partial a^k} \left(\frac{z}{z-e^{-aT}} \right)$$

20 **Corollary** If $H(s)$ has all its poles inside the left half plane, then $\tilde{H}(z)$ has all its poles in the open disk $|z| < 1$.

21 *Exercise* Give an example to show that the converse is false. [*Hint:* Consider $h(t) = e^t \sin 2\pi t$ and choose T appropriately.]

There is a useful relation between $H(s)$ and $\tilde{H}(z)$ which allows the direct determination of $\tilde{H}(z)$ when $H(s)$ is not rational. Let us express it in the form of a

22 **Theorem** Let f be a function belonging to class A (see Remark *7*) and be continuous at all points $t = kT$, $k = 1, 2, \ldots$. Let f have a Laplace transform F and let σ_f be its abscissa of convergence (see *B.2.6*). Let $\tilde{F}$ be the z transform of f. Then

23 $$\tilde{F}(e^{sT}) = \tfrac{1}{2}f(0+) + \frac{1}{2\pi j}\int_{c-j\omega}^{c+j\omega} \frac{F(p)}{1 - \exp[-T(s-p)]}\,dp \qquad \sigma_f < c < \operatorname{Re} s$$

24 *Comment* In the literature of z transforms many authors assume, often implicitly, that $F(s) \to 0$ at least as fast as $1/s^2$ as $s \to \infty$. Then f is continuous at 0 and the term $(\frac{1}{2})f(0+)$ does not appear in *23* since $f(0+) = 0$.

Proof By definition *6*, putting $z = e^{sT}$,

$$\tilde{F}(e^{sT}) = \sum_{n=0}^{\infty} f(nT)e^{-nsT}$$

The right-hand side of the above relation can be recognized to be the L transform of $f(t+)\left(\sum_{n=0}^{\infty} \delta(t - nT)\right)$, where the argument of f is $t+$ and not t (see Remark *7*). Since we shall evaluate $f(t)$ in terms of its $\mathcal{L}$ transform $F(s)$ and since the Laplace transformation inversion integral gives $\frac{1}{2}[f(t-) + f(t+)]$ (see *B.3.35*), let us rewrite the expression above in the following manner:

$$\begin{aligned}\tilde{F}(e^{sT}) &= \tfrac{1}{2}f(0+) + \sum_{n=0}^{\infty} \tfrac{1}{2}[f(nT-) + f(nT+)]e^{-nsT} \\ &= \tfrac{1}{2}f(0+) + \mathcal{L}\left\{\tfrac{1}{2}[f(t-) + f(t+)] \sum_{n=0}^{\infty} \delta(t - nT)\right\}\end{aligned}$$

Now

$$\mathcal{L}\left[\sum_{n=0}^{\infty} \delta(t - nT)\right] = \frac{1}{1 - e^{-sT}} \qquad \operatorname{Re} s > 0$$

and *23* follows by the complex convolution theorem (see *B.3.13*).

Relation to difference equations

In preceding chapters (*3.6.12* and *4.5.31*), we have seen that a system described by a differential equation with constant coefficients of the form $L(p)y = M(p)u$ is zero-state equivalent to the system described by the transfer function $H(s) = L(s)/M(s)$. We would expect that a similar situation exists for sampled systems; in fact, we have the

25 **Theorem** Let the linear discrete system $\mathcal{S}$ be characterized by its z-transfer function $\tilde{H}(z) = P(z)/Q(z)$, where P and Q are polynomials. Let

$$P(z) = \sum_{k=0}^{n} p_k z^{-k} \qquad \text{and} \qquad Q(z) = \sum_{k=0}^{m} q_k z^{-k}$$

Then the zero-state output sequence $\{y_0, y_1, \ldots\}$ of $\mathcal{S}$ is related to the input sequence $\{u_0, u_1, \ldots\}$ by the difference equation

26 $$\sum_{k=0}^{n} q_k y_{l-k} = \sum_{k=0}^{m} p_k u_{l-k} \qquad l = 0, 1, 2, \ldots$$

with the convention that all symbols with negative indices are zero.

Proof Since $\{y_n\}$ is a zero-state output sequence, we have by *12* $\tilde{Y}(z) = \tilde{H}(z)\tilde{U}(z)$; hence, $Q(z)\tilde{Y}(z) = P(z)\tilde{U}(z)$. Let us now express $\tilde{Y}$ and $\tilde{U}$ by their defining power series; see *6*. Since the last relation holds for all z such that $|z| > \max(R_u, R_y)$, the coefficients of like powers of z are equal. Considering the coefficients of z^{-l}, we get

$$q_0 y_l + q_1 y_{l-1} + \cdots + q_n y_{l-n} = p_0 y_l + p_1 u_{l-1} + \cdots + p_m u_{l-m}$$

Since this relation holds for all nonnegative integers l, *26* follows. ◁

The difference equation *26* has a representation in terms of adders, scalors, and delayors. The diagram is given by Fig. *10.3.6*. Note that

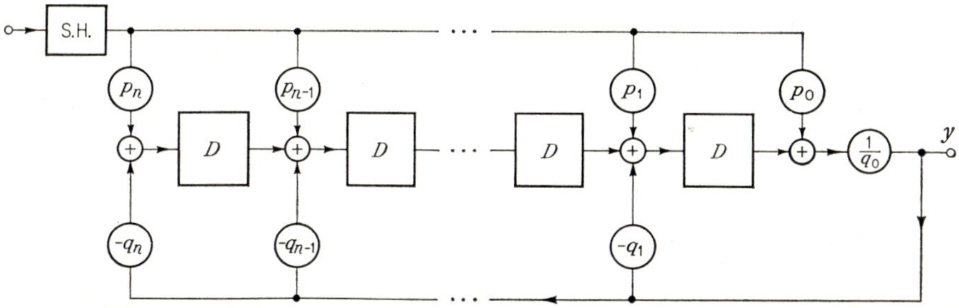

Fig. 10.3.6 Realization of the difference equation *10.3.26* by delayors, scalors, and a sample-and-hold box.

it is analogous to that given for differential systems. The only difference is that every integrator is now replaced by a delayor; furthermore in the discrete case there is a sample-and-hold circuit directly connected to the input. Compare Fig. *10.3.6* with Fig. *4.9.4*.

Stability considerations

The stability of a nonanticipative linear discrete system described by its z-transfer function $\tilde{H}(z)$ follows the same lines as that of the continuous system described by its transfer function $H(s)$ (Chap. *9*, Sec. *4*). Let us first introduce a

27 **Definition** Consider the system illustrated by Fig. *10.3.2*; its z-transfer function $\tilde{H}(z)$ is said to be *strictly stable* if any bounded input sequence produces a zero-state output sequence which is bounded. ◁

Strictly stable z-transfer functions and the usual transfer functions are characterized in a completely analogous manner. Indeed, we have the

28 **Theorem** The z-transfer function $\tilde{H}(z)$ of a nonanticipative discrete system is strictly stable if and only if its zero-state response sequence $\{h_n\}$ to the input 1, 0, . . . satisfies the condition

29
$$\sum_{n=0}^{\infty} |h_n| < \infty$$

30 *Comment* Condition *29* is the discrete form of condition *9.4.13* for the continuous case. Note, however, that, in general, *9.4.13* does not imply that $|h(t)| \to 0$ as $t \to \infty$, whereas *29* does imply that $|h_n| \to 0$ as $n \to \infty$.

31 *Comment* In *29*, the summation is carried from $n = 0$ to ∞ because the system is assumed nonanticipative. If the system were anticipative, we would have to sum from $-\infty$ to ∞ (compare with *9.4.13*).

Proof Exercise for the reader. (*Hint:* Use the proof of *9.4.12* as a guide.)

32 *Exercise* (For the reader interested in teratology!) Find a nonanticipative linear time-invariant system whose impulse response satisfies $\int_0^\infty |h(t)|\,dt < \infty$ and is such that, if sampled periodically with the appropriate period, it will have a zero-state output sequence $\{y_n\}$ with $y_n \to \infty$ in response to the input $\{1,0, \ldots ,0, \ldots\}$. *Hint:* Let $h(t)$ be zero everywhere except when $t \in (n - n^{-3}, n + n^{-3})$, in which case

$$h(t) = n \exp \{n^{-6}[t - (n - n^{-3})]^{-1}[t - (n + n^{-3})]^{-1}\} \qquad n = 2, 3, \ldots$$

For the case of rational transfer functions, we have the

33 **Theorem** Let the z-transfer function $\tilde{H}(z)$ be a rational function of z. Then $\tilde{H}(z)$ is strictly stable if and only if all the poles of $\tilde{H}$ are in the open disk $|z| < 1$.

Proof Exercise for the reader. (*Hint:* $\Rightarrow$ Use contradiction and invoke the inversion theorem *B.4.8*. $\Leftarrow$ Use the inversion theorem.)

34 *Caveat* If, as in Fig. *10.3.2*, the discrete system $\mathcal{S}$, with z-transfer function $\tilde{H}(z)$, is obtained by sampling the continuous system Σ whose transfer function is $H(s)$, then the strict stability of $\tilde{H}(z)$ does not imply that of $H(s)$. This follows from the observation that the strict stability of $\tilde{H}(z)$ imposes a boundedness requirement on the zero-state output sequence $\{y_k\}$; it does not require the output $y(t)$ to be bounded for all $t \in (0, \infty)$. Consider the following example: $h(t) = 1(t)[e^{-t} + e^t \sin (\pi t/T)]$. Clearly the transfer function $H(s)$ is unstable, since $h(t)$ increases exponentially with t (see *9.4.1*). However, if the sampling period is T,

the sequence $\{h_n\}$ is $\{e^{-nT}, n = 0, 1, 2, \ldots\}$ and $\tilde{H}(z)$ is strictly stable. In Sec. *2*, where the systems were described by their state equations, asymptotic stability of the sampled system was equivalent to that of the continuous system from which the sampled system was obtained (see *10.2.12*). This again stresses the importance of the state concept and that the attempt to think of linear systems exclusively in terms of transfer functions leads to difficulties.

By Theorem *33* the problem of checking the strict stability of a rational z-transfer function $\tilde{H}(z)$ reduces to finding conditions under which its denominator polynomial has all its zeros in the open disk $|z| < 1$. The classical test for this condition is the celebrated Schur-Cohn criterion. Recently, however, Jury has simplified the test considerably. Let us now describe the *Jury stability test*.

35 **Theorem**[1] Let f be a polynomial with real coefficients.

36 $$f(z) = a_n z^n + a_{n-1}z^{n-1} + \cdots + a_1 z + a_0 \quad a_n > 0$$

Define the triangular matrices

37 $$\mathbf{X}_k = \begin{bmatrix} a_0 & a_1 & \cdots & & a_{k-1} \\ 0 & a_0 & a_1 & & \cdots \\ \cdots & \cdots & \cdots & \cdots & \cdots \\ 0 & \cdots & \cdots & & a_1 \\ 0 & 0 & \cdots & 0 & a_0 \end{bmatrix} \qquad \mathbf{Y}_k = \begin{bmatrix} a_{n-k+1} & \cdots & a_{n-1} & a_n \\ \cdots & \cdots & a_n & 0 \\ \cdots & \cdots & \cdots & \cdots \\ a_{n-1} & a_n & 0 & 0 \\ a_n & 0 & 0 & 0 \end{bmatrix}$$
$$k = 1, 2, \ldots, n$$

Define the numbers A_k and B_k by the relations

38 $$\begin{aligned} \det(\mathbf{X}_k + \mathbf{Y}_k) &= A_k + B_k \\ \det(\mathbf{X}_k - \mathbf{Y}_k) &= A_k - B_k \end{aligned} \qquad k = 1, 2, \ldots, n$$

Then the polynomial f has all its zeros in the open disk $|z| < 1$ if and only if

39

n even		*n odd*	
$f(1) > 0,$	$f(-1) > 0$	$f(1) > 0,$	$f(-1) < 0$
$\lvert A_1\rvert < B_1$			
$\lvert A_3\rvert < \lvert B_3\rvert,$	$B_3 < 0$	$\lvert A_2\rvert > \lvert B_2\rvert,$	$A_2 < 0$
$\lvert A_5\rvert < \lvert B_5\rvert,$	$B_5 > 0$	$\lvert A_4\rvert > \lvert B_4\rvert,$	$A_4 > 0$
$\lvert A_7\rvert < \lvert B_7\rvert,$	$B_7 < 0$	$\lvert A_6\rvert > \lvert B_6\rvert,$	$A_6 < 0$
$\vdots$	$\vdots$	$\vdots$	$\vdots$
$\lvert A_{n-1}\rvert < \lvert B_{n-1}\rvert,$	$B_{n-1} \begin{cases} <0 \text{ if } n = 4k \\ >0 \text{ otherwise} \end{cases}$ (k is an integer)	$\lvert A_{n-1}\rvert > \lvert B_{n-1}\rvert,$	$A_{n-1} \begin{cases} >0 \text{ if } n = 4k+1 \\ <0 \text{ otherwise} \end{cases}$

[1] E. I. Jury, A Simplified Stability Criterion for Linear Discrete Systems, *Proc. I.R.E.*, vol. 50, no. 6, pp. 1493–1500, and no. 9, p. 1973, 1962. See also *I.R.E. Trans. Automatic Control*, vol. AC-7, no. 4, pp. 51–55, 1962.

40 *Comment* A systematic way of obtaining A_k and B_k from their defining equations *34* is as follows: (I) put a ^ on all the elements of $\mathbf{Y}_k$; then expand det $(\mathbf{X}_k + \mathbf{Y}_k)$. The result is a polynomial in the a_i and $\hat{a}_i$. (II) A_k is the sum of all the monomial terms which include an *even* number of $\hat{a}_i$, and B_k is the sum of all those which include an *odd* number of $\hat{a}_i$.

41 *Example* For $n = 2$ the conditions are (by assumption $a_2 > 0$)

$$|a_0| < a_2 \qquad a_0 + a_1 + a_2 > 0 \qquad a_0 - a_1 + a_2 > 0$$

Observe that the last two inequalities allow us to replace the first one by $a_0 < a_2$.

For $n = 3$ the conditions are (by assumption $a_3 > 0$)

$$|a_0| < a_3 \qquad a_0^2 - a_3^2 < a_0a_2 - a_1a_3 \qquad \sum_0^3 a_i > 0 \qquad \sum_0^3 (-1)^i a_i < 0$$

42 *Comment* In the design of sampled systems one encounters the same problem as that solved by *9.4.22* and *9.4.24*. For the sampled systems with rational z-transfer functions whose denominator $f(z,\mu)$ depends on a parameter μ, the end point μ_1 of the stability interval is given[1] by either $f(1,\mu_1) = 0$, $f(-1,\mu_1) = 0$, or $A_{n-1}(\mu_1) - B_{n-1}(\mu_1) = 0$.

REFERENCES AND SUGGESTED READING

1 Bridgland, T. F., Jr.: A Linear Algebraic Formulation of the Theory of Sampled Data Control, *J. Soc. Indust. Appl. Math.*, vol. 7, no. 4, pp. 431–446, 1959.

2 Jury, E. I.: "Sampled Data Control Systems," John Wiley & Sons, Inc., New York, 1958.

3 Ragazzini, J. R., and G. F. Franklin: "Sampled-data Control Systems," McGraw-Hill Book Company, Inc., New York, 1958.

4 Tou, J. T.: "Digital and Sampled-data Control Systems," McGraw-Hill Book Company, Inc., New York, 1959.

5 Ahlfors, L. V.: "Complex Analysis," McGraw-Hill Book Company, Inc., New York, 1953.

6 Hille, E.: "Analytic Function Theory," vol. 1, Ginn and Company, Boston, 1959.

[1] *Ibid.*

Controllability and observability 11

1 Introduction

In this chapter we shall apply the technical background provided by the preceding chapters to a study of an aspect of the behavior of linear systems which is of particular importance in control theory. For simplicity, we shall focus our attention on time-invariant systems.

It will be recalled (*3.6.1*) that the response of any linear system comprises two parts: the zero-input response and the zero-state response, with the latter characterized by the transfer function of the system. As we have seen in Chap. *5*, the zero-input response can be described in terms of modes of various types and may exhibit rather complex behavior. For example let the matrix **A** of the linear time-invariant system described by

1
$$\dot{\mathbf{x}} = \mathbf{A}\mathbf{x} + \mathbf{B}\mathbf{u}$$
$$\mathbf{y} = \mathbf{C}\mathbf{x} + \mathbf{D}\mathbf{u}$$

have distinct eigenvalues. In this case, the simple concept of modes of oscillations is applicable (see *5.4.27*). We know from experience that some systems have modes that are uncoupled to the input or the output or both. Thus for this special case we see that we have a four-way classification of the modes: (I) coupled to both the input and the output, (II) coupled to the input but not the output, (III) coupled to the output but not to the input, and (IV) coupled to neither. Is this classification valid in general? We shall see that it is approximately valid, but in order to be able to develop the theory, we must first consider in detail the dual concepts of controllability and observability.[1]

[1] The theory of controllability and observability of linear systems is due largely to R. E. Kalman. (See Refs. *1*, *3*, and *4*). As pointed out in *11.3.11* and *11.4.3*, the concepts of controllability and observability are closely related to those of "strong connectedness" and "reduced system," which were introduced by E. F. Moore in the context of finite-state systems.

Once they are available, we shall develop the canonical decomposition of the state space. In terms of this canonical decomposition, the question raised above is easily answered.

2 *Exercise* This simple exercise is intended for the student who has had no experience in the ideas mentioned above. Consider the circuit shown in Fig. *11.1.1*. Write the state equations of the circuit in the

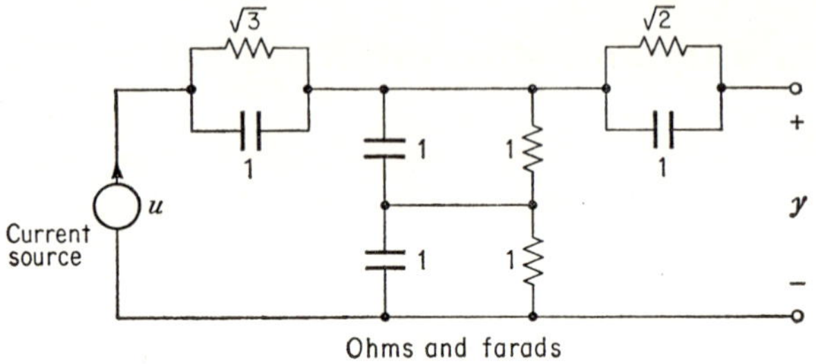

Fig. 11.1.1 Circuit to be analyzed in Exercise *11.1.2*.

form *1*. Analyze the circuit and exhibit the four-way classification of the modes.

2 *Impulse and doublet responses of a single-input system*

Consider a single-input linear time-invariant system described by its state equation

$$\dot{\mathbf{x}} = \mathbf{A}\mathbf{x} + \mathbf{b}u$$

where $\mathbf{A}$ and $\mathbf{b}$ are $(n \times n)$ and $(n \times 1)$ constant matrices. The output is the state vector $\mathbf{x}$ itself. The *unit-impulse response* of the system is, by definition, the zero-state response to a unit impulse applied at $t = 0$. By *6.2.21* we have

$$\mathbf{x}(t;\mathbf{0},0-;\delta(t)) = \int_{0-}^{t} \{\exp[\mathbf{A}(t-\tau)]\}\mathbf{b}\delta(\tau)\,d\tau \qquad t > 0$$

By the sifting property of δ (see *A.2.11*), this expression becomes

$$\exp[\mathbf{A}(t-\tau)]\Big|_{\tau=0}\mathbf{b} = [\exp(\mathbf{A}t)]\mathbf{b} \qquad t > 0$$

Since the impulse response is identically zero for $t < 0$, we finally write

1 $$\mathbf{x}(t;\mathbf{0},0-;\delta(t)) = [\exp(\mathbf{A}t)]\mathbf{b}1(t) \quad \forall t$$

Thus in state space the trajectory of the impulse response is not continuous at $t = 0$; it undergoes a "jump" $\mathbf{b}$ at $t = 0$.

Similarly, the *doublet response* is the zero-state response to a doublet applied at $t = 0$. It is given by

2 $$\mathbf{x}(t;\mathbf{0},0-;\delta^{(1)}(t)) = \mathbf{A}[\exp(\mathbf{A}t)]\mathbf{b}1(t) + \mathbf{b}\delta(t) \qquad \forall t$$

Let us justify *2* by three observations: (I) *2* indicates that, for $t < 0$, the response is identically zero as it should be. (II) For $t > 0$, *2* reduces to[1]

$$\mathbf{A}[\exp(\mathbf{A}t)]\mathbf{b} = [\exp(\mathbf{A}t)]\mathbf{A}\mathbf{b}$$

which is a solution of $\dot{\mathbf{x}} = \mathbf{A}\mathbf{x}$ as it should be for $t > 0$. (III) Finally, by inserting *2* in the system equation and performing the differentiation in the distribution sense, it is found that $\dot{\mathbf{x}} - \mathbf{A}\mathbf{x} = \mathbf{b}\delta^{(1)}(t)$.

The zero-state response to the kth derivative of δ is by definition $\mathbf{x}(t;\mathbf{0},0-;\delta^{(k)}(t))$ and, in fact,

3 $$\mathbf{x}(t;\mathbf{0},0-;\delta^{(k)}(t)) = \mathbf{A}^k[\exp(\mathbf{A}t)]\mathbf{b}1(t) + \sum_{i=1}^{k} \mathbf{A}^{(k-i)}\mathbf{b}\delta^{(i-1)}(t) \qquad \forall t$$

This can be verified in three steps, as above. Note that for $t > 0$

$$\mathbf{x}(t;\mathbf{0},0-;\delta^{(k)}(t)) = \mathbf{A}^k[\exp(\mathbf{A}t)]\mathbf{b} \qquad t > 0$$

If we compare *1* and *3*, we see that (see *A.5.20*)

$$\mathbf{x}(t;\mathbf{0},0-;\delta^{(k)}(t)) = \frac{d^k}{dt^k}\mathbf{x}(t;\mathbf{0},0-;\delta(t)) \qquad \forall t$$

provided the derivatives are taken in the distribution sense. Let us next prove a lemma which will be necessary for the following developments.

4 **Lemma** Let $\{\varphi_i\}$ be a set of N complex-valued continuous functions of t defined over $[0,T]$. Let $\mathbf{G}$ be the $N \times N$ matrix whose (i,j) element is $g_{ij} = (\varphi_i,\varphi_j) \triangleq \int_0^T \varphi_i^*(t)\varphi_j(t)\,dt$. Then the functions φ_i are linearly independent if and only if $\det \mathbf{G} \neq 0$.

5 *Comment* The continuous functions φ_i may be considered as vectors of an obviously defined linear vector space: $\det \mathbf{G}$ is called the *Gram determinant* of the vectors $\varphi_1, \varphi_2, \ldots, \varphi_N$.

6 *Comment* Since $g_{ij} = g_{ji}^*$ $(\forall i,j)$, $\mathbf{G}$ is a *hermitian matrix* (*C.14.8*). Furthermore, $\mathbf{G}$ is *nonnegative definite;* indeed, let $\boldsymbol{\alpha} = (\alpha_1, \ldots, \alpha_N)$ be an N-tuple of complex numbers: then with $\psi = \sum_{i=1}^{N} \alpha_i\varphi_i$

[1]From *5.2.3*, the matrices $\mathbf{A}$ and $\exp(\mathbf{A}t)$ commute.

7
$$\|\psi\|^2 = (\psi,\psi) = (\Sigma\alpha_i\varphi_i,\Sigma\alpha_j\varphi_j) = \Sigma\alpha_i^*\alpha_j(\varphi_i,\varphi_j)$$
$$= \langle\boldsymbol{\alpha},\mathbf{G}\boldsymbol{\alpha}\rangle \geq 0$$

where $\langle\cdot,\cdot\rangle$ is the usual complex scalar product over $\mathcal{C}^N$ (see *C.5.4*).
Proof $\Rightarrow$ The φ_i are linearly independent. Therefore, for all $\boldsymbol{\alpha} \in \mathcal{C}^N$, ψ is not identically zero; hence $\|\psi\|^2 = \int_0^T |\psi(t)|^2\, dt > 0$ and

$$\langle\boldsymbol{\alpha},\mathbf{G}\boldsymbol{\alpha}\rangle > 0 \qquad \forall\boldsymbol{\alpha} \in \mathcal{C}^N$$

Consequently $\det \mathbf{G} \neq 0$ (see *C.19.19*).

$\Leftarrow$ By hypothesis $\det \mathbf{G} \neq 0$. For the purpose of a proof by contradiction, suppose the φ_i's were linearly dependent. Then for some $\boldsymbol{\alpha} \neq \mathbf{0}$, say, $\boldsymbol{\alpha}_0$, the corresponding ψ would be identical to zero over $[0,T]$. Hence, by *7*, $\langle\boldsymbol{\alpha}_0,\mathbf{G}\boldsymbol{\alpha}_0\rangle = 0$. Therefore, the smallest eigenvalue of $\mathbf{G}$ is zero (*C.19.15*) and consequently $\det \mathbf{G} = 0$. This is a contradiction.

3 *Controllability*

Consider a multiple-input multiple-output linear time-invariant system $\mathcal{S}$ described by

1
$$\dot{\mathbf{x}} = \mathbf{A}\mathbf{x} + \mathbf{B}\mathbf{u}$$
2
$$\mathbf{y} = \mathbf{C}\mathbf{x} + \mathbf{D}\mathbf{u}$$

where **A**, **B**, **C**, and **D** are, respectively, $(n \times n)$, $(n \times r)$, $(p \times n)$, $(p \times r)$ constant matrices. The n-vector $\mathbf{x}$ is the state of the system (*4.5.19*), the r-vector $\mathbf{u}$ is the input, and the p-vector $\mathbf{y}$ is the output of $\mathcal{S}$.

Speaking intuitively, we shall say that the system $\mathcal{S}$ is *controllable* if, knowing the matrices **A** and **B** and the state of $\mathcal{S}$ at t_0, $\mathbf{x}_0$, we can construct an input $\mathbf{u}_{[t_0,t_0+T]}$ which will bring the state of $\mathcal{S}$ to the zero state $\mathbf{0}$ at time $t_0 + T$. Since the system $\mathcal{S}$ is time-invariant, we may take $t_0 = 0$ without loss of generality (*3.2.36*). More formally, we have the

3 **Definition** The system $\mathcal{S}$ defined by *1* is said to be *controllable* if and only if the following is true for all $\mathbf{x}_0 \in \mathcal{C}^n$: given that the system is in state $\mathbf{x}_0$ at time 0, then for some *finite* $T > 0$, there is an input $\mathbf{u}_{[0,T]}$ such that

4
$$\mathbf{x}(T;\mathbf{x}_0,0;\mathbf{u}_{[0,T]}) = \mathbf{0} \quad \triangleleft$$

The controllability of the system $\mathcal{S}$ is characterized by the following

5 **Theorem** The system $\mathcal{S}$ described by *1* is *controllable* if and only if the column vectors of the matrix

6
$$\mathbf{Q} \triangleq [\mathbf{B},\mathbf{AB}, \ldots ,\mathbf{A}^{n-1}\mathbf{B}]$$

where $\mathbf{B}$, $\mathbf{AB}$, . . . , $\mathbf{A}^{n-1}\mathbf{B}$ denote the columns of $\mathbf{Q}$, span the state space $\mathcal{C}^n$ of $\mathcal{S}$.

7 *Comment* This condition can also be expressed as follows: The matrix $\mathbf{Q}$ has rank n; among the nr columns of $\mathbf{Q}$ there is a set of n linearly independent vectors which, therefore, constitute a basis for $\mathcal{C}^n$.

Proof $\Rightarrow$ The system is controllable by hypothesis; therefore, any state $\mathbf{x}_0$ can be transferred to the zero state in a finite time; thus,

8
$$\mathbf{0} = [\exp (\mathbf{A}T)\mathbf{x}_0] + \int_0^T \{\exp [\mathbf{A}(T - \tau)]\}\mathbf{Bu}(\tau)\, d\tau$$

Multiplying *8* by $\exp(-\mathbf{A}T)$ on the left and remembering that $\exp(-\mathbf{A}\tau) = \sum_{k=0}^{n-1} \mathbf{A}^k\alpha_k(-\tau)$ (*D.8.12*), we have

$$-\mathbf{x}_0 = \sum_{k=0}^{n-1} \mathbf{A}^k\mathbf{B} \int_0^T \alpha_k(-\tau)\mathbf{u}(\tau)\, d\tau$$

Thus any $\mathbf{x}_0$ that can be transferred to $\mathbf{0}$ in a finite time is a linear combination of the column vectors of $\mathbf{Q}$. Hence, controllability of $\mathcal{S}$ implies that the columns of $\mathbf{Q}$ span the state space $\mathcal{C}^n$.

$\Leftarrow$ In order to give a constructive proof and not bog down in notational difficulties, consider the case where $n = 3$ and $r = 2$. By assumption, $\mathcal{R}(\mathbf{Q}) = \mathcal{C}^n$. Suppose, for example, that $\mathcal{C}^n$ is spanned by $\mathbf{a}_1$,$\mathbf{a}_2$, and $\mathbf{a}_3$, where, say, $\mathbf{a}_1$ is the second column of $\mathbf{B}$, $\mathbf{a}_2$ is the first column of $\mathbf{AB}$, and $\mathbf{a}_3$ is the second column of $\mathbf{A}^2\mathbf{B}$. Consider a control $\mathbf{u}_{[0,T]} = \sum_{i=1}^{3} \gamma_i\mathbf{u}_i(t)$, where the constants γ_i will be chosen later, $\mathbf{u}_1(t) = (0,\delta(t))$, $\mathbf{u}_2(t) = (\delta^{(1)}(t),0)$, $\mathbf{u}_3(t) = (0,\delta^{(2)}(t))$, and T is an arbitrary finite positive number. If $\mathbf{u}_{[0,T]}$ drives $\mathbf{x}_0$ to $\mathbf{0}$, then

$$\mathbf{x}(T;0,0;\mathbf{u}_{[0,T]}) = \mathbf{0} = [\exp (\mathbf{A}T)]\mathbf{x}_0 + [\exp (\mathbf{A}T)] \sum_{i=1}^{3} \gamma_i\mathbf{a}_i$$

where we have used the definitions of the $\mathbf{a}_i$'s and *11.2.3*. Since the $\mathbf{a}_i$'s constitute a basis for $\mathcal{C}^n$, if we pick the γ_i's to be the components of $-\mathbf{x}_0$ with respect to that basis, then $\mathbf{u}_{[0,T]} = \Sigma\gamma_i\mathbf{u}_i(t)$ will transfer $\mathbf{x}_0$ at time 0 to $\mathbf{0}$ at time T.

9 *Exercise* Show that $\mathcal{S}$ is controllable if and only if any state $\mathbf{x}$ is reachable from the origin in a finite time.

10 *Exercise* Show that $\mathcal{S}$ is controllable if and only if any state $\mathbf{x}_1$ is reachable from any initial state $\mathbf{x}_0$.

11 *Comment* Note that this implies that, in the case of linear systems, the concept of a controllable system coincides with the concept of a "strongly connected" system, which was introduced by E. Moore (see *2.4.17*).

12 *Exercise* Show that the system $\mathcal{S}$ described by

$$(a_0p^n + \cdots + a_n)y = (b_0p^m + \cdots + b_m)u \qquad m \leq n$$

or, more compactly, $L(p)y = M(p)u$, is controllable if and only if $L(p)$ and $M(p)$ have no factors in common. [*Hint:* Note that if $L(p)$ and $M(p)$ have a factor in common, then $\mathcal{S}$ is zero-state equivalent to a system $\mathcal{S}'$ of lower order than n. If all the states in the state spaces of $\mathcal{S}$ and $\mathcal{S}'$ are reachable from the origin, then by *2.9.5* $\mathcal{S}$ and $\mathcal{S}'$ would be equivalent, which contradicts *4.2.12*.]

13 **Assertion** The subspace spanned by the column vectors of the matrix $\mathbf{Q}$ defined in *6* is an invariant subspace of $\mathbf{A}$ (see *C.13.1*).
Proof Exercise for the reader. (*Hint:* Use the Cayley-Hamilton theorem, *5.13.19*.)

It is useful to apply the concept of controllability to a particular state and introduce the

14 **Definition** A state $\mathbf{x}_1$ of the system $\mathcal{S}$ described by *1* is said to be *controllable* if and only if, for some finite T, there exists a control $\mathbf{u}_{[0,T]}$ such that

$$\mathbf{x}(T;\mathbf{x}_1,0;\mathbf{u}_{[0,T]}) = \mathbf{0}$$

15 **Corollary** A state of the system $\mathcal{S}$ is controllable if and only if it belongs to $\mathcal{R}(\mathbf{Q})$, the subspace spanned by the column vectors of $\mathbf{Q}$.
Proof Exercise for the reader.

The corollary states that $\mathcal{R}(\mathbf{Q})$ is the subspace of all controllable states. Let us exhibit the meaning of this fact in terms of special representations of $\mathcal{S}$. Let $\mathcal{R}(\mathbf{Q})$ be of dimension $\gamma < n$. Of course, $\mathcal{C}^n = \mathcal{R}(\mathbf{Q}) \oplus \mathcal{R}(\mathbf{Q})^\perp$. Let us pick a new basis for $\mathcal{C}^n$ as follows: the first γ basis vectors constitute a basis for $\mathcal{R}(\mathbf{Q})$ and the $(n - \gamma)$ remaining ones constitute a basis for $\mathcal{R}(\mathbf{Q})^\perp$. Let us use a prime to indicate the matrices representing $\mathbf{x}$, $\mathbf{A}$, $\mathbf{B}$, and $\mathbf{u}$ in the new basis. We shall show that we have

16
$$\mathbf{A}' = \begin{bmatrix} \mathbf{A}'_{11} & \mathbf{A}'_{12} \\ \mathbf{0} & \mathbf{A}'_{22} \end{bmatrix} \qquad \mathbf{B}' = \begin{bmatrix} \mathbf{B}'_1 \\ \mathbf{0} \end{bmatrix}$$

where $\mathbf{A}'_{11}$ is $\gamma \times \gamma$, $\mathbf{A}'_{12}$ is $\gamma \times (n - \gamma)$, and $\mathbf{A}'_{22}$ is $(n - \gamma) \times (n - \gamma)$.

$\mathbf{B}_1'$ is a $\gamma \times r$ matrix. Equation *1* can be partitioned into two vector equations if we write $\mathbf{x}' = \mathbf{x}_1' + \mathbf{x}_2'$, where $\mathbf{x}_1' \in \mathcal{R}(\mathbf{Q})$ and $\mathbf{x}_2' = \mathcal{R}(\mathbf{Q})^\perp$:

17
$$\begin{aligned}\dot{\mathbf{x}}_1' &= \mathbf{A}_{11}'\mathbf{x}_1' + \mathbf{A}_{12}'\mathbf{x}_2' + \mathbf{B}_1'\mathbf{u}' \\ \dot{\mathbf{x}}_2' &= \qquad\quad \mathbf{A}_{22}'\mathbf{x}_2'\end{aligned}$$

The lower submatrix of $\mathbf{B}'$ in *16* is zero because the columns of $\mathbf{B}$ are necessarily in $\mathcal{R}(\mathbf{Q})$ by the definition *6* of $\mathbf{Q}$. The lower left-hand submatrix of $\mathbf{A}'$ in *16* is zero because, by *13*, $\mathcal{R}(\mathbf{Q})$ is an invariant subspace under $\mathbf{A}$, that is, $\mathbf{x}' \in \mathcal{R}(\mathbf{Q}) \Rightarrow \mathbf{A}'\mathbf{x}' \in \mathcal{R}(\mathbf{Q})$. Thus if $\mathbf{x}' = (\mathbf{x}_1',\mathbf{0})$, then for all possible $\mathbf{x}_1'$, $\mathbf{A}'\mathbf{x}' = (\mathbf{A}_{11}'\mathbf{x}_1',\mathbf{A}_{21}'\mathbf{x}_1') = (\mathbf{A}_{11}'\mathbf{x}_1',\mathbf{0})$, and therefore $\mathbf{A}_{21}' = \mathbf{0}$.

18 *Comment* Equation *17* shows that whenever $\gamma < n$, any state vector $\mathbf{x}'$ can be decomposed in a sum $\mathbf{x}_1' + \mathbf{x}_2'$, where $\mathbf{x}_2'$ is unaffected by the input $\mathbf{u}$, whatever it may be. Furthermore, $\mathbf{x}'$ is controllable if and only if it is of the form $\mathbf{x}_1' + \mathbf{0}$. This leads to the representation of Fig. *11.3.1*. It is of interest to note that both $\mathbf{x}'(t)$ and $\mathbf{x}_1'(t)$ depend

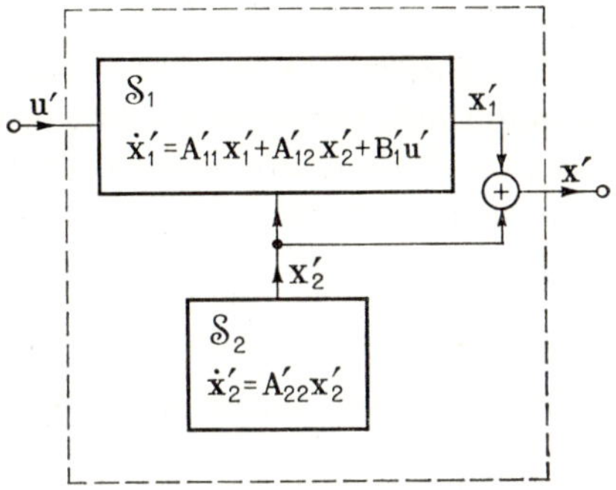

Fig. 11.3.1 Direct sum decomposition of the state space $\mathcal{C}^n$ in controllable states $\mathbf{x}_1'$ and uncontrollable states $\mathbf{x}_2'$.

(for $t > 0$) on $\mathbf{x}_2'(0)$. The transfer function of the system shown in Fig. *11.3.1* depends only on $\mathbf{A}_{11}'$ and $\mathbf{B}_1'$ because the transfer function depends only on the zero-state response and $\mathbf{x}_2'(0) = \mathbf{0} \Rightarrow \mathbf{x}_2'(t) = \mathbf{0}$ for all t.

4 *Observability*

The idea of observability is closely related to that of controllability. Controllability means that knowing the initial state and the matrices characterizing the system, we can find an input which brings this state to the zero state in a finite time. Observability means that the knowledge of the matrices characterizing the system and the zero-input response $\mathbf{y}_{[0,T]}$ over a finite interval is sufficient to determine uniquely the initial state of the system. More precisely, we have the

1 **Definition** The system $\mathcal{S}$ defined by *11.3.1* and *11.3.2* is said to be *observable* if and only if, for some $T > 0$ and for all initial states $\mathbf{x}(0)$, the knowledge of **A**, **C**, and the zero-input response $\mathbf{y}_{[0,T]}$ suffices to determine the initial state $\mathbf{x}(0)$.

2 *Comment* Since throughout this section we assume that the matrices **A**, **B**, **C**, and **D** are known, we could just as well have said that the system is observable if the input $\mathbf{u}_{[0,T]}$ and the output $\mathbf{y}'_{[0,T]}$ resulting from $\mathbf{u}_{[0,T]}$ suffice to determine $\mathbf{x}(0)$. Indeed, the contribution to the output due to the observed input, namely,

$$\mathbf{C}\int_0^t \{\exp[\mathbf{A}(t-\tau)]\}\,\mathbf{B}\mathbf{u}(\tau)\,d\tau + \mathbf{D}\mathbf{u}(t)$$

may always be subtracted from the observed output $\mathbf{y}'_{[0,T]}$ to obtain the output under free motion.

3 *Comment* From the foregoing comment it follows that the concept of observability is essentially the same as that of initial state determinability (*1.8.12 et seq.*). It should be noted that, by Corollary *3.7.55*, any system that is in reduced form is observable. Thus, in the case of linear systems, the concepts of "observable system" and "system in reduced form" are coextensive. (This is not true, however, in the case of nonlinear systems.) Note that if, in the definition of observability, we merely required that the initial state be determinable to within equivalent states, then every linear system would be observable.

The property of observability can be characterized in a manner similar to that of controllability (see *11.3.5*). We have the

4 **Theorem** The system $\mathcal{S}$ described by *11.3.1* and *11.3.2* is *observable* if and only if the np columns of the matrix

$$\mathbf{P} \triangleq [\mathbf{C}^*,\mathbf{A}^*\mathbf{C}^*, \ldots ,\mathbf{A}^{*(n-1)}\mathbf{C}^*]$$

span the state space $\mathcal{C}^n$. (The superscript * indicates the conjugate transpose of the unstarred matrix.)

Proof The zero-input response is $\mathbf{y}(t) = \mathbf{C}[\exp(\mathbf{A}t)]\mathbf{x}_0$, and by *D.8.12*,

5 $$\mathbf{y}(t;\mathbf{x}_0,0;\mathbf{0}) = \sum_{k=0}^{m-1} \alpha_k(t)\mathbf{C}\mathbf{A}^k\mathbf{x}_0$$

where m is the degree of the minimal polynomial of **A**.

$\Rightarrow$ Use contradiction. Suppose the columns of $\mathbf{P}$ do not span $\mathcal{C}^n$. Then there exists a nonzero vector $\boldsymbol{\eta} \in \mathcal{C}^n$ which is orthogonal to all the columns of $\mathbf{P}$. This orthogonality may be expressed by

$$\mathbf{0} = (\mathbf{A}^{*k}\mathbf{C}^*)^*\boldsymbol{\eta} = \mathbf{C}\mathbf{A}^k\boldsymbol{\eta} \qquad k = 0, 1, \ldots, n-1$$

where we have used $(\mathbf{AB})^* = \mathbf{B}^*\mathbf{A}^*$ (see *C.14.4*). From *5*,

$$\mathbf{y}(t;\boldsymbol{\eta},0;\mathbf{0}) = \mathbf{0}$$

for all t; thus, any initial state proportional to η produces a zero-input response which is identical to zero. In other words, there are nonzero initial states which cannot be determined by observing only the zero-input response. This contradicts the assumption that $\mathcal{S}$ is observable. Hence, the columns of $\mathbf{P}$ span $\mathcal{C}^n$.

$\Leftarrow$ Let $\mathbf{c}_j^*$ be the jth column of $\mathbf{C}^*$ (consequently $\mathbf{c}_j$ is the jth row of $\mathbf{C}$). Remembering that we use the complex scalar product, the jth component of $\mathbf{y}$ is

$$y_j(t) = \langle \mathbf{c}_j^*, [\exp(\mathbf{A}t)]\mathbf{x}_0\rangle = \sum_{k=1}^{m-1} \alpha_k(t)\langle \mathbf{c}_j^*, \mathbf{A}^k\mathbf{x}_0\rangle$$

or

6 $$y_j(t) = \sum_{k=0}^{m-1} \alpha_k(t)\langle \mathbf{A}^{*k}\mathbf{c}_j^*, \mathbf{x}_0\rangle$$

where we have used the definition of the adjoint (*C.14.2*). Now, from *D.8.13*, the functions α_k are linearly independent over $[0,T]$. Taking the scalar product of *6* with α_i (as in Lemma *11.2.4*) we obtain the system of equations

$$(\alpha_i, y_j) = \sum_{k=0}^{m-1} (\alpha_i, \alpha_k)\langle \mathbf{A}^{*k}\mathbf{c}_j^*, \mathbf{x}_0\rangle \qquad \begin{array}{l} i = 0, 1, \ldots, m-1 \\ j = 1, 2, \ldots, p \end{array}$$

For each j, this is a set of m equations in the m scalars $\langle \mathbf{A}^{*k}\mathbf{c}_j^*, \mathbf{x}_0\rangle$,

$$k = 0, 1, \ldots, m-1$$

and the determinant of the system is different from zero by *11.2.4*. Solving these p systems yields the mp scalars $\langle \mathbf{A}^{*k}\mathbf{c}_j^*, \mathbf{x}_0\rangle$. Now, by assumption, among the mp vectors $\mathbf{A}^{*k}\mathbf{c}_j^*$ there is a basis for $\mathcal{C}^n$; therefore, $\mathbf{x}_0$ is uniquely specified, since we know the scalar product of $\mathbf{x}_0$ with n basis vectors of $\mathcal{C}^n$ (*C.8.5*).

7 *Comment* The proof of Theorem *4* shows that if the system is observable in an interval $[0,T]$, then it is observable in any interval $[0,T_1]$, where $0 < T_1$.

8 *Comment* If $\mathcal{R}(\mathbf{P})$, the subspace spanned by the column vectors of $\mathbf{P}$, is a proper subset of $\mathcal{C}^n$, the proof of *4* shows that $\mathbf{y}_{[0,T]}$ is not sufficient to specify $\mathbf{x}_0$ uniquely. Now, *5* shows that the knowledge of $\mathbf{y}_{[0,T]}$ is equivalent to that of the *mp* scalar products $\langle \mathbf{A}^{*k}\mathbf{c}_j^*, \mathbf{x}_0 \rangle$. The *mp* vectors $\mathbf{A}^{*k}\mathbf{c}_j^*$ span $\mathcal{R}(\mathbf{P})$. Consider the orthogonal decomposition of the state space $\mathcal{C}^n = \mathcal{R}(\mathbf{P}) \oplus \mathcal{R}(\mathbf{P})^\perp$ and the corresponding decomposition of $\mathbf{x}_0$, $\mathbf{x}_0 = \mathbf{x}_1' + \mathbf{x}_2'$, where $\mathbf{x}_1' \in \mathcal{R}(\mathbf{P})$, $\mathbf{x}_2' \in \mathcal{R}(\mathbf{P})^\perp$. All the scalar products $\langle \mathbf{A}^{*k}\mathbf{c}_j^*, \mathbf{x}_0 \rangle$ determine $\mathbf{x}_1'$ uniquely but give no information whatsoever about $\mathbf{x}_2'$. In other words, we have the

9 **Assertion** When $\mathcal{R}(\mathbf{P})$ is a proper subset of $\mathcal{C}^n$, the output $\mathbf{y}_{[0,T]}$ determines only the orthogonal projection of $\mathbf{x}_0$ on $\mathcal{R}(\mathbf{P})$. ◁

If the initial state $\mathbf{x} \in \mathcal{R}(\mathbf{P})^\perp$, then the corresponding zero-input response $\mathbf{y}_{[0,T]}$ is identically zero; in that sense, $\mathbf{x}$ is said to be *unobservable*. Furthermore,

10 **Assertion** The set of all *unobservable* states of the system $\mathcal{S}$ is the subspace $\mathcal{R}(\mathbf{P})^\perp$.

It is of interest to note that $\mathcal{R}(\mathbf{P})$ is not necessarily invariant under $\mathbf{A}$ but that we have the

11 **Assertion** The subspace $\mathcal{R}(\mathbf{P})^\perp$ is invariant under $\mathbf{A}$.
Proof Exercise for the reader. [*Hint:* Use the definition of $\mathcal{R}(\mathbf{P})^\perp$ and the Cayley-Hamilton theorem.] ◁

The close relationship between the concept of controllability and that of observability is clearly exhibited by the

12 **Duality theorem of Kalman** Consider the system $\mathcal{S}$ defined by *11.3.1* and *11.3.2*: $\mathbf{x}$ is an n-vector, $\mathbf{u}$ an r-vector, and $\mathbf{y}$ a p-vector. Consider also the system Σ defined by

13 $$\dot{\boldsymbol{\xi}} = \mathbf{A}^*\boldsymbol{\xi} + \mathbf{C}^*\mathbf{v}$$

14 $$\boldsymbol{\eta} = \mathbf{B}^*\boldsymbol{\xi} + \mathbf{D}^*\mathbf{v}$$

where the state $\boldsymbol{\xi}$ is an n-vector, the input $\mathbf{v}$ is a p-vector, and the output $\boldsymbol{\eta}$ is an r-vector. The system $\mathcal{S}$ is controllable (observable) if and only if the system Σ is observable (controllable, respectively).
Proof Immediate consequence of Theorem *11.3.5* and *4*.

15 *Exercise* The purpose of this exercise is to relate the concept of observability and that of state equivalence (see *2.2.1*). Consider the differential system described by *11.3.1* and *11.3.2*. Show that if the system is observable, any two distinct points of its state space are not equivalent states. Prove the converse.

5 *Canonical decomposition of the state space of* $\mathcal{S}$

We have shown that intrinsically associated with the system $\mathcal{S}$, defined by (*11.3.1* and *11.3.2*), there are two invariant subspaces: $\mathcal{R}(\mathbf{Q})$, the subspace of all controllable states, and $\mathcal{R}(\mathbf{P})^{\perp}$, the subspace of all unobservable states. Our purpose is to use these subspaces to create a direct sum decomposition of the whole state space $\mathcal{C}^n$. Let $\mathfrak{M}_1$ be the subspace defined by

1 $$\mathfrak{M}_1 = \mathcal{R}(\mathbf{Q}) \cap \mathcal{R}(\mathbf{P})^{\perp}$$

$\mathfrak{M}_1$ is also an invariant subspace under $\mathbf{A}$. It is uniquely defined by *1*. Define $\mathfrak{M}_2$, $\mathfrak{M}_3$, and $\mathfrak{M}_4$ by the following relations:

2 $$\mathcal{R}(\mathbf{Q}) = \mathfrak{M}_1 \oplus \mathfrak{M}_2$$

3 $$\mathcal{R}(\mathbf{P})^{\perp} = \mathfrak{M}_1 \oplus \mathfrak{M}_3$$

4 $$\mathcal{C}^n = \mathfrak{M}_1 \oplus \mathfrak{M}_2 \oplus \mathfrak{M}_3 \oplus \mathfrak{M}_4$$

Clearly, $\mathfrak{M}_1$ is the subspace of the states that are both controllable and unobservable. Any state in $\mathfrak{M}_2$ is controllable, but only its projection on $\mathcal{R}(\mathbf{P})$ is observable. Any state in $\mathfrak{M}_3$ is unobservable, but it is also not controllable. Any state in $\mathfrak{M}_4$ is uncontrollable, and its projection on $\mathcal{R}(\mathbf{P})^{\perp}$ is unobservable.

It is important to note that the subspaces $\mathfrak{M}_2$, $\mathfrak{M}_3$, and $\mathfrak{M}_4$ are not uniquely defined. However, as long as they satisfy the conditions *2* to *4* the above statements hold.

Let us now use the direct sum decomposition *4* to write the equations of the system. More precisely, we use a basis of $\mathcal{C}^n$ which is the union of bases for $\mathfrak{M}_1$, $\mathfrak{M}_2$, $\mathfrak{M}_3$, $\mathfrak{M}_4$. In this new basis, the state is $\boldsymbol{\xi}_1 + \boldsymbol{\xi}_2 + \boldsymbol{\xi}_3 + \boldsymbol{\xi}_4$, where $\boldsymbol{\xi}_i \in \mathfrak{M}_i$. Equations *11.3.1* and *11.3.2* become in these new coordinates

5 $$\begin{bmatrix} \dot{\boldsymbol{\xi}}_1 \\ \dot{\boldsymbol{\xi}}_2 \\ \dot{\boldsymbol{\xi}}_3 \\ \dot{\boldsymbol{\xi}}_4 \end{bmatrix} = \begin{bmatrix} \bar{\mathbf{A}}_{11} & \bar{\mathbf{A}}_{12} & \bar{\mathbf{A}}_{13} & \bar{\mathbf{A}}_{14} \\ \mathbf{0} & \bar{\mathbf{A}}_{22} & \mathbf{0} & \bar{\mathbf{A}}_{24} \\ \mathbf{0} & \mathbf{0} & \bar{\mathbf{A}}_{33} & \bar{\mathbf{A}}_{34} \\ \mathbf{0} & \mathbf{0} & \mathbf{0} & \bar{\mathbf{A}}_{44} \end{bmatrix} \begin{bmatrix} \boldsymbol{\xi}_1 \\ \boldsymbol{\xi}_2 \\ \boldsymbol{\xi}_3 \\ \boldsymbol{\xi}_4 \end{bmatrix} + \begin{bmatrix} \bar{\mathbf{B}}_1 \\ \bar{\mathbf{B}}_2 \\ \mathbf{0} \\ \mathbf{0} \end{bmatrix} \mathbf{v}$$

6 $$\mathbf{n} = [\mathbf{0} \quad \bar{\mathbf{C}}_2 \quad \mathbf{0} \quad \bar{\mathbf{C}}_4] \begin{bmatrix} \boldsymbol{\xi}_1 \\ \boldsymbol{\xi}_2 \\ \boldsymbol{\xi}_3 \\ \boldsymbol{\xi}_4 \end{bmatrix} + \mathbf{D}\mathbf{v}$$

The pattern of the zero submatrices in *5* and *6* is justified as follows. $\mathcal{R}(\mathbf{Q}) = \mathfrak{M}_1 \oplus \mathfrak{M}_2$ is invariant under $\mathbf{A}$ (see *11.3.13*); therefore, the submatrices in the positions (3,1), (3,2), (4,1), and (4,2) are zero matrices in this basis (see *C.13.3*). $\mathcal{R}(\mathbf{P})^\perp = \mathfrak{M}_1 \oplus \mathfrak{M}_3$ is invariant under $\mathbf{A}$ (see *11.4.11*); hence the submatrices in the positions (2,1), (4,1), (2,3), and (4,3) are zero matrices. By *2* and *4*, all states in $\mathfrak{M}_3 \oplus \mathfrak{M}_4$ are outside $\mathcal{R}(\mathbf{Q})$ and hence are uncontrollable; consequently, the bottom two submatrices of $\mathbf{B}$ in *5* are zero matrices. From *11.4.10* and *11.4.11*, any initial state in the invariant subspace $\mathcal{R}(\mathbf{P})^\perp = \mathfrak{M}_1 \oplus \mathfrak{M}_3$ will produce a zero-input response which is identical to zero; hence, the first and third submatrices of $\mathbf{C}$ in *6* are zero.

The decomposition *5*, *6* leads to the block diagram shown in Fig. *11.5.1*. This decomposition, though not unique, can be said to be *canonical* because it is based on the intrinsic properties of the linear transformations $\mathbf{A}$, $\mathbf{B}$, $\mathbf{C}$.

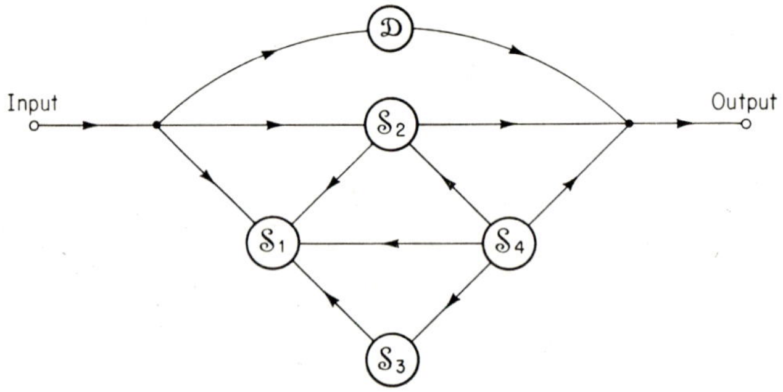

Fig. 11.5.1 Canonical decomposition of the state space $\mathbb{C}^n$ into the direct sum of four subspaces. With each subspace is associated a subsystem $\mathcal{S}_i$, $i = 1, 2, 3, 4$.

7 *Comment* Suppose we consider Eqs. *5* and *6* and try to construct an example of the system. Can the matrices $\bar{\mathbf{A}}_{ij}$, $\bar{\mathbf{B}}_i$, $\bar{\mathbf{C}}_j$ be chosen arbitrarily? The answer is no. The matrices $\bar{\mathbf{A}}_{ij}$, $\bar{\mathbf{B}}_i$ must be such that

$$\mathcal{R}(\mathbf{Q}) = \mathfrak{M}_1 \oplus \mathfrak{M}_2$$

and the matrices $\bar{\mathbf{A}}_{ij}$, $\bar{\mathbf{C}}_j$ must be such that $\mathcal{R}(\mathbf{P})^\perp = \mathfrak{M}_1 \oplus \mathfrak{M}_3$.

8 *Comment* The (matrix) transfer function of the system of Fig. *11.5.1* will characterize only the properties of $\mathcal{S}_2$ (that part of the system that is both observable and controllable) and D. $\mathcal{S}_1$ does not affect the transfer function because all states of $\mathcal{S}_1$ are unobservable. $\mathcal{S}_3$ and $\mathcal{S}_4$ do not affect the transfer function because all states of $\mathcal{S}_3$ and $\mathcal{S}_4$ are uncontrollable and because the transfer function is defined on the basis of the *zero-state response* of the system; indeed, when the system starts from the zero state, $\boldsymbol{\xi}_3 = \boldsymbol{\xi}_4 = \mathbf{0}$, for all t, irrespective of the input $\mathbf{u}$.

9 *Example* The purpose of this example is to illustrate the concepts of controllability and observability in a simple situation and also to show

how it can throw some light on certain puzzling situations. Consider the linear feedback system shown in Fig. *11.5.2*. It is a single-input

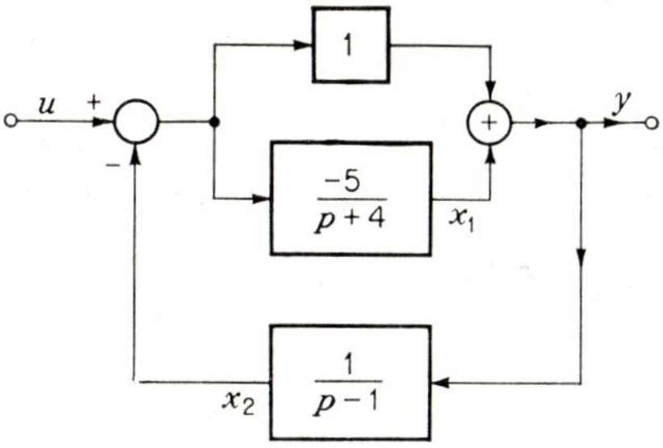

Fig. 11.5.2 Linear feedback system considered in Example *11.5.9*.

(u), single-output (y) system. Let us pick x_1 and x_2 as the components of the state vector. It is immediately found that

10 $$\begin{bmatrix}\dot{x}_1\\ \dot{x}_2\end{bmatrix} = \begin{bmatrix}-4 & 5\\ 1 & 0\end{bmatrix}\begin{bmatrix}x_1\\ x_2\end{bmatrix} + u\begin{bmatrix}-5\\ 1\end{bmatrix}$$

11 $$y = u + \langle \mathbf{c},\mathbf{x}\rangle$$

where $\mathbf{c} = (1,-1)$.

Referring to the figure, the transfer function of the two forward paths combined is easily seen to be $G(s) = (s-1)/(s+4)$. That of the feedback branch is $H(s) = 1/(s-1)$. Therefore, the transfer function of the system is

$$\frac{G}{1+GH} = \frac{(s-1)/(s+4)}{1+(s-1)^{-1}(s-1)/(s+4)} = \frac{s-1}{s+5}$$

12 Hence the transfer function of the system is strictly stable (see *9.4.7*).

From *10* we find that the eigenvalues and eigenvectors are[1] $\lambda_1 = 1$, $\mathbf{u}_1 = (1,1)$; $\lambda_2 = -5$, $\mathbf{u}_2 = (5,-1)$. The system has an eigenvalue with a positive real part ($\lambda_1 = 1$), and

13 Therefore the system *10* is unstable i.s.L (see *7.2.2* and *7.4.2*).

The apparent contradiction between *12* and *13* is easily resolved by noting that the output y depends on $\mathbf{x}$ through the scalar product $\langle \mathbf{c},\mathbf{x}\rangle$ and since $\langle \mathbf{c},\mathbf{u}_1\rangle = 0$,

14 The output y is uncoupled to the unstable mode of *10* and *11*.

Equivalently, the subspace of unobservable states, $\mathcal{R}(\mathbf{P})^{\perp}$, is spanned by $\mathbf{u}_1$.

It is of interest to follow around the loop the physical events: suppose there is some initially stored energy in the black box in the feedback

[1] For the eigenvectors, we picked a normalization which suited our numerical convenience.

branch; then x_2 will increase exponentially at the rate e^t, and x_1 will also include a term which increases as e^t. However, it is easy to check that these two terms will exactly cancel at the summing point in the forward path, and consequently no such exponentially increasing term will appear in the output y.

Another interesting fact is the following: Suppose that at $t = 0$ the system is in the zero state. For any input $u_{[0,T]}$, the resulting state is

$$\mathbf{x}(T;\mathbf{0},0;u_{[0,T]}) = \int_0^T \{\exp[\mathbf{A}(T-\tau)]\}\mathbf{a}u(\tau)\,d\tau$$
$$= \mathbf{a}\int_0^T \alpha_1(T-\tau)u(\tau)\,d\tau + \mathbf{A}\mathbf{a}\int_0^T \alpha_2(T-\tau)u(\tau)\,d\tau$$

where we used the fact that $\exp(\mathbf{A}t) = \alpha_1(t)\mathbf{I} + \alpha_2(t)\mathbf{A}$ and we called $\mathbf{a}$ the forcing vector of *10*. In the present case, both $\mathbf{a}$ and $\mathbf{Aa}$ are easily verified to be proportional to $\mathbf{u}_2$; therefore, the only states reachable from the zero state are the states in the one-dimensional subspace spanned by $\mathbf{u}_2$. In other words,

15 The input is uncoupled to the unstable mode and therefore the unstable mode can be excited only if initially the system is not in the zero state.

The canonical structure of the system shown in Fig. *11.5.2* is given, for the particular choice of state vector, by Fig. *11.5.3*.

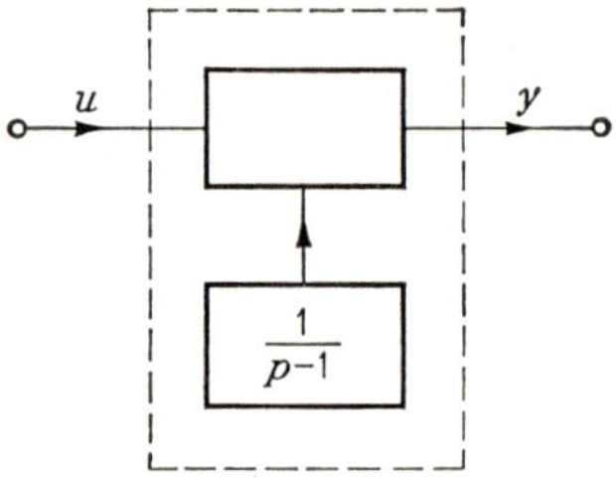

Fig. 11.5.3 Canonical decomposition of the system of Fig. *11.5.2*.

16 **Remark** Observe that if the two forward paths of the system of Fig. *11.5.3* are joined together, the resulting gain is $(s-1)/(s+4)$. Thus the right-half-plane zero of the forward path cancels the right-half-plane pole of the feedback path. The usual admonition given to beginners to the effect that they should never design a system such that a right-half-plane pole is canceled by a right-half-plane zero can now be justified in two ways: (I) The boundedness of y is based on the *exact* cancellation of rising exponentials at the summing node in the forward path; the cancellation will never be exact in practice because physical components cannot be specified with arbitrary precision and because of aging, temperature variations, etc., of the same components. (II) Even if the cancellation did occur exactly as on paper, any noise or other spurious input to the black box in the feedback loop would create two exponentially increasing outputs x_1 and x_2. This instability, excitable by any suitably placed noise source, demonstrates also the worthlessness of such cancellation schemes. ◁

Finally, it should be stressed that the particular canonical structure one obtains for the system of Fig. *11.5.2* depends on the choice of the state vector (see Exercise *18*).

17 *Exercise* Apply the theory of Secs. *2* and *3* to verify that the system of Fig. *11.5.2* is neither controllable nor observable.

18 *Exercise* Consider the system of Fig. *11.5.2*. Choose y and x_2 as state variables. Find the state equations and discuss the stability, observability, and controllability of the system. Give its canonical structure.

19 *Exercise* Call $\mathcal{A}$ the system shown in Fig. *11.5.1* and described by Eqs. *5* and *6*. Call $\mathcal{A}_{24}$ the system obtained from $\mathcal{A}$ by deleting the subsystems $\mathcal{S}_1$ and $\mathcal{S}_3$. Call $\mathcal{A}_2$ the system obtained by deleting from $\mathcal{A}$ the subsystems $\mathcal{S}_1$, $\mathcal{S}_3$, and $\mathcal{S}_4$. Show that (I) $\mathcal{A}_{24} \doteq \mathcal{A}_2 \doteq \mathcal{A}$, (II) $\mathcal{A}_2$ is not zero-input equivalent to $\mathcal{A}_{24}$, (III) $\mathcal{A}_2$ is not zero-input equivalent to $\mathcal{A}$, and (IV) $\mathcal{A}_{24}$ is zero-state and zero-input equivalent to $\mathcal{A}$. (*Hint:* See Definitions *2.9.1* and *2.9.2*.)

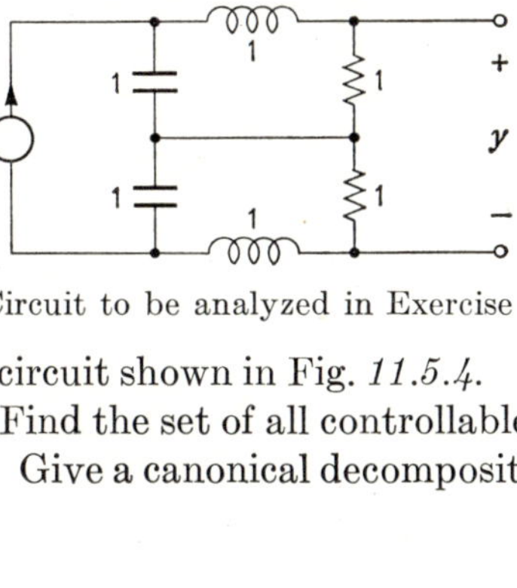

Fig. 11.5.4 Circuit to be analyzed in Exercise *11.5.20*.

20 *Exercise* Consider the circuit shown in Fig. *11.5.4*. It is a single-input single-output system. Find the set of all controllable states and the set of unobservable states. Give a canonical decomposition for the system.

6 *Alternate characterization of controllability*

Let us consider the interpretation of Theorem *11.3.5* in terms of the Jordan canonical form of the system. Using the Jordan form will allow us to actually compute the control required to bring any state to **0**. For simplicity we shall consider a single-input system described by

1 $$\dot{\mathbf{x}} = \mathbf{A}\mathbf{x} + \mathbf{b}u$$

where **A** is an $n \times n$ constant matrix and **b** is an n-vector. u is the scalar input.

2 *Exercise* Show that if the minimum polynomial of **A** is of degree $k \leq n - 1$, then the system *1* is not controllable. (*Hint:* Use *11.3.5* and note that since k is the degree of the minimal polynomial, $\mathbf{A}^k$ is a linear combination of $\mathbf{I}, \mathbf{A}, \ldots, \mathbf{A}^{k-1}$.)

Let us change variables: let $\mathbf{x} = \mathbf{Ty}$, where the transformation matrix $\mathbf{T}$ is such that $\mathbf{T}^{-1}\mathbf{AT} = \mathbf{J}$ and $\mathbf{J}$ is the Jordan canonical form of $\mathbf{A}$ (see *D.5.21*). With $\mathbf{e} = \mathbf{T}^{-1}\mathbf{b}$, *1* becomes

3 $$\dot{\mathbf{y}} = \mathbf{Jy} + \mathbf{e}u$$

The first characterization of controllability for the case where the eigenvalues of $\mathbf{A}$ are distinct is given by the

4 **Theorem** Let $\mathbf{A}$ have distinct eigenvalues; hence $\mathbf{J} = \text{diag}\,(\lambda_1, \ldots, \lambda_n)$. Then the system *1* is controllable if and only if all the components of $\mathbf{e} = \mathbf{T}^{-1}\mathbf{d}$ are different from zero.

5 *Comment* In terms of the mode interpretation of the dynamics of the system (Chap. *5*, Sec. *4*), the conditions are equivalent to the following: The (scalar) input must be coupled to all the modes of the system.

Proof $\Rightarrow$ The ith component of the vector equation *3* is $\dot{y}_i = \lambda_i y_i + e_i u$. Suppose that one component of $\mathbf{e}$, say, e_k, is zero. Then the kth component of $\mathbf{y}$ is $y_k(t) = y_k(0)e^{\lambda_k t}$, which differs from zero for all *finite* t once $y_k(0) \neq 0$. Therefore, if $e_k = 0$, any state $\mathbf{y}$ such that $y_k(0) \neq 0$ cannot be brought to $\mathbf{0}$ in a finite time.

$\Leftarrow$ Pick an arbitrary time $T > 0$. Any input $u_{[0,T]}$ which brings $\mathbf{y}$ to $\mathbf{0}$ in T sec satisfies the n conditions

6 $$\int_0^T e^{-\lambda_i \xi} u(\xi)\, d\xi = \frac{-y_i(0)}{e_i} \qquad i = 1, 2, \ldots, n$$

It is well known that the functions $\{e^{-\lambda_i \xi}\}$ are linearly independent over any finite interval if and only if the λ_i's are pairwise distinct. Therefore, in view of our hypothesis, the functions $\{e^{-\lambda_i \xi}\}$ are linearly independent over $[0,T]$. If we put $u(\xi) = \Sigma \eta_j \exp\,(-\lambda_j^* \xi)$, the n conditions *6* will give a set of n linear algebraic equations in n unknowns,

7 $$\sum_{j=1}^{n} (\varphi_i, \varphi_j)\eta_j = \frac{y_i(0)}{e_i} \qquad i = 1, 2, \ldots, n$$

where $\varphi_i(\xi) = \exp\,(-\lambda_i^* \xi)$. By Lemma *11.2.4*, the determinant of the system *7* is nonzero and the solution is unique. The solution of *7* specifies a control $u_{[0,T]}$ which will bring the state $\mathbf{y}(0)$ to $\mathbf{0}$ in T sec.

8 *Exercise* If to the control u just computed one would add any function v such that $(\varphi_i, v) = 0$ for all i, the control $u + v$ would still bring the state $\mathbf{y}(0)$ to zero in T sec. ◁

Suppose now that the eigenvalues of $\mathbf{J}$ are not distinct: let the matrix $\mathbf{J}$ consist of α Jordan blocks of respective size $r_1, r_2, \ldots, r_\alpha$; let the corresponding eigenvalues be $\lambda_1, \lambda_2, \ldots, \lambda_\alpha$. Note that, in general, these λ's are not necessarily distinct, since the same eigenvalue may be associated with two or more Jordan blocks (*D.5.21*). For example, suppose $r_1 = 1$, $r_2 = 2$, $r_3 = 3$. The matrix is then

9
$$\begin{bmatrix} \lambda_1 & 0 & 0 & 0 & \cdot & 0 \\ 0 & \lambda_2 & 1 & \cdot & \cdot & \cdot \\ 0 & 0 & \lambda_2 & \cdot & \cdot & \cdot \\ \cdot & \cdot & \cdot & \lambda_3 & 1 & 0 \\ \cdot & \cdot & \cdot & 0 & \lambda_3 & 1 \\ 0 & \cdot & \cdot & 0 & 0 & \lambda_3 \end{bmatrix}$$

The block-diagram representation corresponding to this matrix with the notation of *2* is shown in Fig. *11.6.1*. The general characterization of controllability is contained in the

10 **Theorem** With the notation just described, the system *1* is controllable if and only if (I) the eigenvalues $\lambda_1, \ldots, \lambda_\alpha$ are pairwise distinct and

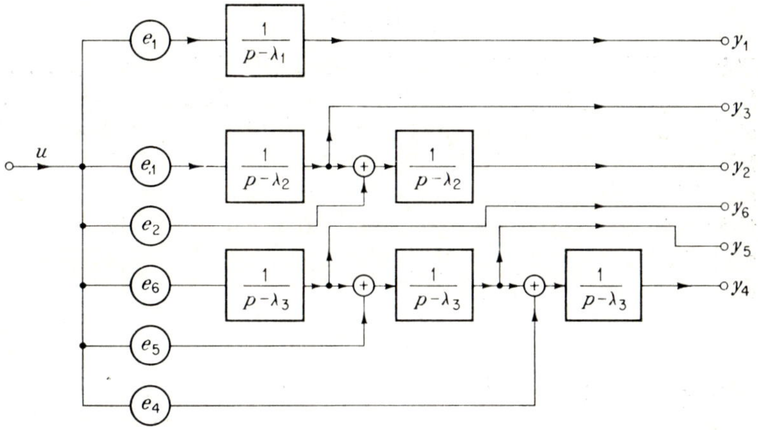

Fig. 11.6.1 Realization of the system whose matrix **A** is given by *11.6.9*.

(II) all the components of **e** that are in the last row of each Jordan block are different from zero.

Proof $\Rightarrow$ The system is assumed to be controllable. Condition (II) is easily shown to hold by the use of the same argument as that used in the proof of *4*. Suppose condition (I) were violated; say, $\lambda_i = \lambda_j$. Considering the components of the vector equation *3* corresponding to the last row of the ith and jth Jordan blocks, we get

$$\dot{y}_i = \lambda_i y_i + e_i u \qquad \dot{y}_j = \lambda_i y_j + e_j u$$

If we put $z = e_j y_i - e_i y_j$, we see that

$$\dot{z} = \lambda_i z$$

Therefore z is independent of the input and $z(t) = z(0) \exp(\lambda_i t) \neq 0$ for all finite t provided $z(0) \neq 0$. Therefore, it is obvious that there are noncontrollable states. This is a contradiction; hence condition (II) must hold.

$\Leftarrow$ Let us establish controllability by actually computing a control $u_{[0,T]}$ which brings any state $\mathbf{y}(0)$ to $\mathbf{0}$ in T sec. By *5.2.17*, any such control is characterized by

$$-\mathbf{y}(0) = \int_0^T [\exp(-\mathbf{J}\xi)]\, \mathbf{e}u(\xi)\, d\xi$$

Suppose, for simplicity, that $\mathbf{J}$ is given by *9*; then

$$\boldsymbol{\varphi} \triangleq [\exp(-\mathbf{J}\xi)]\mathbf{e} = \begin{bmatrix} e_1 \exp(-\lambda_1\xi) \\ e_2 \exp(-\lambda_2\xi) - e_3\xi \exp(-\lambda_2\xi) \\ e_3 \exp(-\lambda_2\xi) \\ e_4 \exp(-\lambda_3\xi) - e_5\xi \exp(-\lambda_3\xi) \\ \qquad + e_6(\xi^2/2) \exp(-\lambda_3\xi) \\ e_5 \exp(-\lambda_3\xi) - e_6\xi \exp(-\lambda_3\xi) \\ e_6 \exp(-\lambda_3\xi) \end{bmatrix}$$

It is a well-known fact that for all nonnegative integers $r_1, r_2, \ldots, r_\alpha$, the functions $\{\xi^{k_i} \exp(-\lambda_i\xi)\}$, where $k_i = 1, 2, \ldots, r_i$, $i = 1, 2, \ldots, \alpha$, are linearly independent over any finite interval $[0,T]$ if and only if the λ_i's are pairwise distinct. By condition (I) this is the case, and by condition (II) $e_1 \neq 0$, $e_3 \neq 0$, $e_6 \neq 0$; therefore, the six components of the vector $\boldsymbol{\varphi}$ displayed above are linearly independent. Consequently, a control of the form $\Sigma\eta_j\varphi_j^*$ (where φ_j is the jth componet of $\boldsymbol{\varphi}$) can be computed as in the proof of *4*.

7 *Controllability of linear time-varying systems*

In order to check the controllability of a time-invariant system, one need only find the rank of a matrix with constant elements (*11.3.5*). In the time-varying case, Kalman et al. have shown that it is still possible to check the controllability by evaluating the rank of a matrix with constant elements.

We consider the system $\mathcal{S}$ defined by

1 $$\dot{\mathbf{x}} = \mathbf{A}(t)\mathbf{x} + \mathbf{B}(t)\mathbf{u}(t)$$

where $\mathbf{A}(t)$ and $\mathbf{B}(t)$ are $n \times n$ and $n \times r$ matrices of continuous elements. By analogy with the time-invariant case (*11.3.3*), we introduce the

2 **Definition** The system $\mathcal{S}$ defined by *1* is said to be *controllable at time* t_0 if and only if the following is true for all states $\mathbf{x}_0$ at time t_0: Given that the system is in state $\mathbf{x}_0$ at t_0, then, for some finite time t_1, there exists an input $\mathbf{u}_{[t_0,t_1]}$ such that

$$\mathbf{x}(t_1;\mathbf{x}_0,t_0;\mathbf{u}_{[t_0,t_1]}) = \mathbf{0}$$

3 *Comment* More informally, the system is controllable at t_0 if any possible state at t_0 can be brought to the zero state in a finite time.

4 *Comment* In the time-invariant case we could say "the system $\mathcal{S}$ is controllable." Here we have to say "controllable at time t_0" because the properties of the system vary with time so that it may happen that the system is controllable up to some time T and not controllable forever after. It is obvious, however, that the definition implies that if the system is *controllable at time* T, it is controllable at time t_0 for all $t_0 < T$. ◁

Let us state a result analogous to *11.3.10*.

5 **Assertion** The system $\mathcal{S}$ is controllable at time t_0 if and only if any state $\mathbf{x}_0$ at time t_0 can be transferred to any state $\mathbf{x}_1$ at time t_1 (where $t_1 < \infty$ and depends on $\mathbf{x}_0$, $\mathbf{x}_1$, and t_0).
Proof Exercise for the reader. [*Hint:* The state transition matrix $\mathbf{\Phi}(t_1,t_2)$ is nonsingular for all t_1,t_2.] ◁

Let us now turn to the test for controllability: define

6 $$\mathbf{W}(t_0,t) = \int_{t_0}^{t} \mathbf{\Phi}(t_0,t')\mathbf{B}(t')\mathbf{B}^*(t')\mathbf{\Phi}^*(t_0,t')\,dt'$$

where the superscript * denotes the conjugate transpose. $\mathbf{W}(t_0,t)$ is a hermitian matrix; in particular, if $\mathbf{A}(t)$ and $\mathbf{B}(t)$ are real, $\mathbf{W}$ is real and symmetrical.

7 **Theorem** The system $\mathcal{S}$ defined by *1* is *controllable at time* t_0 if and only if, for some finite t_1, the matrix $\mathbf{W}(t_0,t_1)$ defined by *6* is positive definite.

8 *Comment* For any fixed t_1, $\mathbf{W}(t_0,t_1)$ is a matrix whose elements are numbers; furthermore, since $\mathbf{W}$ is hermitian, it is easy to check its positive definiteness by diagonalizing it and requiring that all the resulting diagonal elements be positive (see *C.19.18*).
Proof For purposes of the proof let us pick t_1 to be the smallest t for which $\mathbf{W}$ has maximal rank.

$\Leftarrow$ The assumption is that $\mathbf{W}(t_0,t_1)$ is positive definite: consequently it is nonsingular. Consider the control $\mathbf{u}'_{[t_0,t_1]}$ defined by

$$\mathbf{u}'(t) = -\mathbf{B}^*(t)\mathbf{\Phi}^*(t_0,t)\mathbf{W}^{-1}(t_0,t_1)\mathbf{x}_0$$

Now since

$$\mathbf{x}(t_1;\mathbf{x}_0,t_0;\mathbf{u}'_{[t_0,t_1]}) = \mathbf{\Phi}(t_1,t_0)\mathbf{x}_0 + \int_{t_0}^{t_1} \mathbf{\Phi}(t_1,\tau)\mathbf{B}(\tau)\mathbf{u}'(\tau)\,d\tau$$

it is easy to check, using *6.2.16*, that $\mathbf{u}'_{[t_0,t_1]}$ transfers $(\mathbf{x}_0,t_0)$ to $(\mathbf{0},t_1)$.

$\Rightarrow$ The assumption is that $\mathcal{S}$ is controllable at t_0. First observe that *6* requires that $\mathbf{W}$ be nonnegative definite. Indeed, let $\boldsymbol{\xi}$ be any state. We have to show that the quadratic form $\langle\boldsymbol{\xi},\mathbf{W}\boldsymbol{\xi}\rangle \geq 0$ for all $\boldsymbol{\xi}$. From *6*, we obtain successively

$$\begin{aligned}\langle \xi, \mathbf{W}\xi \rangle &= \int_{t_0}^{t_1} \langle \xi, \mathbf{\Phi}\mathbf{B}\mathbf{B}^*\mathbf{\Phi}^*\xi \rangle \, dt \\ &= \int_{t_0}^{t_1} \langle \mathbf{B}^*\mathbf{\Phi}^*\xi, \mathbf{B}^*\mathbf{\Phi}^*\xi \rangle \, dt \\ &= \int_{t_0}^{t_1} \|\mathbf{B}^*\mathbf{\Phi}^*\xi\|^2 \, dt \geq 0 \qquad \forall \xi\end{aligned}$$

It remains, therefore, to show that $\mathbf{W}$ is nonsingular. We shall prove this by contradiction in two steps. (I) Suppose $\mathbf{W}$ is singular; then there is a state $\mathbf{n} \neq \mathbf{0}$ such that $\langle \mathbf{n}, \mathbf{W}\mathbf{n} \rangle = 0$. Define, then, $\mathbf{u}''_{[t_0,t_1]}$ by $\mathbf{u}''(t) = -\mathbf{B}^*(t)\mathbf{\Phi}^*(t_0,t)\mathbf{n}$; $\mathbf{u}''(t)$ is continuous in t. Then

$$\begin{aligned}\int_{t_0}^{t_1} \|\mathbf{u}''(t)\|^2 \, dt &= \int_{t_0}^{t_1} \langle \mathbf{B}^*\mathbf{\Phi}^*\mathbf{n}, \mathbf{B}^*\mathbf{\Phi}^*\mathbf{n} \rangle \, dt \\ &= \int_{t_0}^{t_1} \langle \mathbf{n}, \mathbf{\Phi}\mathbf{B}\mathbf{B}^*\mathbf{\Phi}^*\mathbf{n} \rangle \, dt \\ &= \left\langle \mathbf{n}, \int_{t_0}^{t_1} \mathbf{\Phi}\mathbf{B}\mathbf{B}^*\mathbf{\Phi}^* \, dt \, \mathbf{n} \right\rangle = \langle \mathbf{n}, \mathbf{W}\mathbf{n} \rangle = 0\end{aligned}$$

Thus $\mathbf{u}''(t) = \mathbf{0}$ for all t in $[t_0,t_1]$.

(II) The system is completely controllable; hence by *5* there is a control $\mathbf{u}^{(1)}_{[t_0,t_1]}$ which transfers the state $\mathbf{n} \neq \mathbf{0}$ at t_0 to $\mathbf{0}$ at t_1. That is, the control $\mathbf{u}^{(1)}_{[t_0,t_1]}$ satisfies the condition (see *6.2.21*)

9 $$\mathbf{n} = -\int_{t_0}^{t_1} \mathbf{\Phi}(t_0,t)\mathbf{B}(t)\mathbf{u}^{(1)}(t) \, dt$$

Now, the contradiction is obtained by the following calculations:

$$\begin{aligned}0 &= \int_{t_0}^{t_1} \langle \mathbf{u}''(t), \mathbf{u}^{(1)}(t) \rangle \, dt = -\int_{t_0}^{t_1} \langle \mathbf{B}^*(t)\mathbf{\Phi}^*(t_0,t)\mathbf{n}, \mathbf{u}^{(1)} \rangle \, dt \\ &= -\int_{t_0}^{t_1} \langle \mathbf{n}, \mathbf{\Phi}(t_0,t)\mathbf{B}(t)\mathbf{u}^{(1)}(t) \rangle \, dt \\ &= -\left\langle \mathbf{n}, \int_{t_0}^{t_1} \mathbf{\Phi}(t_0,t)\mathbf{B}(t)\mathbf{u}^{(1)}(t) \, dt \right\rangle\end{aligned}$$

Hence, by *9*, $0 = \langle \mathbf{n}, \mathbf{n} \rangle = \|\mathbf{n}\|^2$. This is a contradiction, since in (I) we had $\mathbf{n} \neq \mathbf{0}$ by assumption.

REFERENCES AND SUGGESTED READING

1 Kalman, R. E.: On the General Theory of Control Systems, *Proc. First Intl. Cong. Automatic Cont., Moscow, 1960*, vol. 1, pp. 481–493, 1961, Butterworth & Co. (Publishers), Ltd., London.

2 Bertram, J. E., and P. E. Sarachik: On Optimal Computer Control, *Proc. First Intl. Cong. Automatic Cont., Moscow, 1960*, vol. 1, pp. 419–423, 1961, Butterworth & Co. (Publishers), Ltd., London.

3 Kalman, R. E., Y. C. Ho, and K. S. Narendra: Controllability of Linear Dynamical Systems, in "Contributions to Differential Equations," vol. 1, Interscience Publishers, Inc., New York, 1962.

4 Kalman, R. E.: Canonical Structure of Linear Dynamical Systems, *Proc. Natl. Acad. Sci. U.S.A.*, vol. 48, no. 4, pp. 596–600, 1962.

Delta functions and distributions

A

1 *Introduction*

Dirac delta functions, or simply *delta functions,* denote singular "functions" having the form of impulses. Strictly speaking, they are not functions in the accepted mathematical sense, and they cannot be treated with complete rigor within the framework of classical analysis.

Distributions, on the other hand, are "generalized functions" in a sense which will be defined in Sec. *4*. Thus, distributions subsume both ordinary functions[1] and delta functions as special cases. The theory of distributions, which is due to L. Schwartz, provides a framework within which the concept of a delta function and operations on delta functions can be given a precise meaning. Thus, familiarity with the theory of distributions is indispensable for one who is not satisfied with the conventional formalistic way of manipulating delta functions and their derivatives.

Since we make considerable use of delta functions and their properties in the main body of this text, a summary of the principal definitions and main properties of delta functions is given in Sec. *2*.[2] Later sections constitute a self-contained and compact exposition of the theory of distributions in which the properties deduced in a purely formal way in Sec. *2* are established in a rigorous fashion.

2 *Delta functions*

Intuitively, a delta function is an idealization of a very narrow pulse or impulse occurring at, say, $t = 0$, with a finite total area which, for

[1] Not every ordinary function is a distribution. Specifically, as will be indicated in *A.6.1*, a distribution is essentially a derivative (of some finite order) of a continuous function. (See also Example *A.4.3*.)

[2] A detailed exposition of the properties of delta functions and a historical account of the development of the concept of a delta function and related notions can be found in Ref. *1*.

convenience, is normalized to 1. Thus, formally, $\delta(t) = 0$ for all $t \neq 0$, $\delta(0) = \infty$, and $\int_{-\infty}^{\infty} \delta(t)\,dt = 1$. Furthermore, if $\delta(t)$ is a narrow pulse of area 1, then $g(t)\delta(t)$,[1] where g is any function which is "well behaved" in the neighborhood of $t = 0$, will be a narrow pulse of area $g(0)$. Thus, intuitively, $\delta(t)$ should have the property expressed in a formal way by the relation

1 $$\int_{-\infty}^{\infty} \delta(t)g(t)\,dt = g(0)$$

This suggests that $\delta(t)$ be defined as a "limit" of an approximating sequence of functions $f(t,\lambda)$ such that, for any "well-behaved" g,

$$\lim_{\lambda \to 0} \int_{-\infty}^{\infty} f(t,\lambda)g(t)\,dt = g(0)$$

There are many approximating sequences which have this property (see *A.4.6* and Ref. *1*). A simple one is the sequence of pulses of width λ and height $1/\lambda$ centered on $t = 0$, in which case $f(t,\lambda)$ is expressed by

$$f(t,\lambda) = \frac{1}{\lambda}\left[1\left(t + \frac{\lambda}{2}\right) - 1\left(t - \frac{\lambda}{2}\right)\right]$$

Now if $\delta(t) = \lim_{\lambda \to 0} f(t,\lambda)$, then formally

2 $$\delta(t) = \lim_{\lambda \to 0} \frac{1}{\lambda}\left[1\left(t + \frac{\lambda}{2}\right) - 1\left(t - \frac{\lambda}{2}\right)\right] = \frac{d}{dt}1(t)$$

To say that $\delta(t)$ is the derivative of a unit step function does not constitute a satisfactory definition of a delta function in the context of differential calculus, since $\frac{d}{dt}1(t)$ has no existence in the conventional sense of the derivative. Nonetheless, if a relation such as *2* is interpreted in the sense of Definition *6* given below, then *2* offers a convenient point of departure if not a rigorous definition of $\delta(t)$.

3 *Note* When the derivative of a time function, say, v, is expressed in terms of delta functions and is interpreted in the sense of Definition *6*, then dv/dt or, in symbolic form, $p\,v$ is said to be the derivative of v in the *distribution sense.* (A rigorous definition of this term is given in *A.5.4*.)

With these remarks as a background, the definition of $\delta(t)$ may be worded as follows:

[1] In this section we are not using our notation for time functions (*1.3.3*) in a consistent manner, for reasons explained in *1.3.3*. Thus, $g(t)\delta(t)$ should be interpreted as the time function $g(\cdot)\delta(\cdot)$ rather than as the value of this time function at time t.

4 **Definition** The zero-order delta function, or simply delta function, $\delta(t)$, is the derivative of $1(t)$ in the distribution sense; i.e.,

$$\delta(t) = \frac{d}{dt} 1(t)$$

More generally, $\delta^{(n)}(t)$, the delta function of order n, is the $(n + 1)$st derivative of $1(t)$ in the distribution sense or, recursively, the first derivative of $\delta^{(n-1)}(t)$. ◁

Thus

5 $$\delta^{(n)}(t) = \frac{d^{n+1}}{dt^{n+1}} 1(t) = \frac{d}{dt} \delta^{(n-1)}(t) \qquad n = 1, 2, \ldots$$

Since pointwise equality of functions is not applicable to delta functions, we must define what is meant by the equality of expressions involving delta functions. This is done in the

6 **Definition** Let $\Delta_1(t)$ and $\Delta_2(t)$ be two expressions involving delta functions and ordinary time functions. Then $\Delta_1(t)$ and $\Delta_2(t)$ are identical (or equal), written as $\Delta_1(t) \equiv \Delta_2(t)$ or, more simply, $\Delta_1(t) = \Delta_2(t)$, if and only if

7 $$\int_{-\infty}^{\infty} \Delta_1(t)g(t)\, dt = \int_{-\infty}^{\infty} \Delta_2(t)g(t)\, dt$$

for all time functions g for which the integrals in question exist. (Compare with *A.5.1*.) In this sense, the equality $\delta(t) = \frac{d}{dt} 1(t)$ has the meaning

$$\int_{-\infty}^{\infty} \delta(t)g(t)\, dt = \int_{-\infty}^{\infty} g(t) d1(t) = g(0)$$

which agrees with *1*. ◁

By treating $\delta(t)$ and its derivatives as if they were ordinary time functions, the following basic properties of delta functions can readily be deduced from Definition *4*.

8 **Property**

$$\delta(t - t_0) = \frac{d}{dt} 1(t - t_0) \qquad \forall t_0$$

$\delta(t - t_0)$ is usually referred to as a delta function which *occurs* at $t = t_0$.

9 **Property**

$$\delta^{(n)}(-t) = (-1)^n \delta^{(n)}(t)$$

[*Hint:* Express $\delta(t)$ as a sum of an even function and an odd function and then use *5* and the fact that the derivative of an odd function is even and, conversely, the derivative of an even function is odd.]

10 **Property** The key property of delta functions is the so-called *sifting property* expressed by

11 $$\int_{-\infty}^{\infty} \delta^{(n)}(t-\xi)g(\xi)\,d\xi = g^{(n)}(t) \qquad \text{for all finite } t$$

where g is any function which is n-fold differentiable in the usual sense. If g is not such a function, but its nth derivative is defined in the distribution sense, then *11* should be interpreted as an identity in the sense of Definition *6*. ◁

Property *10* can readily be established by repeated integration by parts. For example, on replacing the infinite limits in *11* by finite limits $-a$, a [which does not change the value of the integral so long as $|t| < a$, since $\delta^{(n)}(t-\xi)$ vanishes for $\xi \neq t$], we have for $\delta(t)$

$$\int_{-a}^{a} \delta(t-\xi)f(\xi)\,d\xi = -f(\xi)1(t-\xi)\Big]_{-a}^{a} + \int_{-a}^{a} 1(t-\xi)\dot{f}(\xi)\,d\xi \qquad -a<t<a$$

$$= f(t)$$

and similarly for higher-order delta functions.

A special instance of *11* is the relation

$$\int_{-\infty}^{\infty} \delta^{(n)}(t-\xi)\delta^{m}(\xi)\,d\xi = \delta^{(m+n)}(t)$$

12 **Remark** In the main body of the text, we frequently use the symbol p to denote the operator which, acting on g, yields the derivative of g in the distribution sense. Thus on the basis of *11* we can write

13 $$p^n u = \int_{-\infty}^{\infty} \delta^{(n)}(t-\xi)u(\xi)\,d\xi$$

14 **Property** (see Exercise *A.5.19*)

15 $$f(t)\delta^{(n)}(t) = \sum_{0}^{n} (-1)^k \binom{n}{k} f^{(k)}(0)\delta^{(n-k)}(t)$$

where the equality should be understood in the sense of Definition *6*. To demonstrate this property, it is sufficient to multiply both sides of *15* by $g(t)$ and verify that *6* holds.

16 **Property** (See also *A.7.14*.) Let $r(t)$ be a function with simple zeros at $t_1, t_2, \ldots$. Then

17 $$\delta(r(t)) = \sum_i \frac{1}{|\dot{r}(t_i)|}\delta(t-t_i)$$

18 $$\delta^{(1)}(r(t)) = \sum_i \frac{1}{\dot{r}(t)|\dot{r}(t_i)|}\delta^{(1)}(t-t_i)$$

Hint: Write

19 $$1(r(t)) = \sum_i \varepsilon_i 1(t - t_i)$$

where $\varepsilon_i = 1$ if $\dot{r}(t_i) > 0$ and $\varepsilon_i = -1$ if $\dot{r}(t_i) < 0$. Then, by *4*,

$$\delta(r(t)) = \frac{d}{d(r(t))} 1(r(t))$$
$$= \frac{1}{\dot{r}(t)} \frac{d}{dt} 1(r(t))$$

and the rest follows from *19* and *18*.

A special instance of *16* is the simple relation

$$\delta(at) = \frac{1}{|a|} \delta(t) \qquad \forall a$$

Thus concludes our brief discussion of those properties of delta functions which are relevant to the material in the main body of the text.

3 *Testing functions*

A distribution is a generalization of a function.[1] Distributions are very useful in engineering because many operations which have no meaning when applied to certain functions are quite meaningful when applied to distributions. For example, within the framework of distributions, any function encountered in applications (steps, rectangular pulses, etc.) may be differentiated as many times as we wish; any convergent series of functions may be differentiated term by term and the sum of the derivatives is the derivative of the sum, etc. In other words, distribution theory provides a framework in which the sometimes illegitimate (in the usual sense) operations that engineers like to perform become legitimate and meaningful; furthermore, the theory provides a sure way to carry them out and to interpret them.

Since a distribution is a generalization of the concept of function, let us review what is meant by a function. With any *function* is associated a set A of elements called the *domain* of the function and another set B called the *range;* the function associates with each element of the domain A one and only one element of the range. Thus a function is defined by all the ordered pairs $(x,f(x))$, where $x \in A$ and $f(x)$ is the unique element

[1] The reader should be warned that distributions (in the sense used here) have nothing to do with probability distributions and distribution functions encountered in probability. Distributions are sometimes called generalized functions or symbolic functions. None of these terminologies is above criticism; so we have used Schwartz's terminology.

of B associated with x by the function f. $f(x)$ is called the value of f at x. For example, let the values of f be given by $f(x) = 1 + x + x^2$, where A is the set of all real numbers and so is B. This formula specifies a rule for computing $f(x)$ for each $x \in A$; in other words, it specifies all ordered pairs $(x,f(x))$ for $x \in A$.

There are other ways of defining functions. Consider, for example, a periodic function ψ of period 2π which can be expanded in Fourier series. In principle, at least, it is equivalent to knowing all the ordered pairs $(t,\psi(t))$ for $t \in [-\pi,\pi]$ or all of its Fourier coefficients $\ldots, c_{-1}, c_0, c_1, \ldots$; these are given by

1 $$c_k = \frac{1}{2\pi}\int_{-\pi}^{\pi} \psi(t)\,[\exp(-jkt)]\,dt \qquad k = \ldots, -1, 0, 1, \ldots$$

The point is that *1* sets up a one-one correspondence between the functions ψ and the sequences $\{\ldots, c_{-1}, c_0, c_1, \ldots\}$. We may interpret *1* as defining a functional[1] associated with ψ, which assigns to each function $\rho_k = \exp(-jkt)$ the number c_k. Observe also that this functional is linear. The function ψ is completely defined by the values taken by the functional (associated with ψ) at the functions ρ_k, $k = \ldots, -1, 0, 1, \ldots$. This leads to the point of view that ψ could just as well be defined by all the ordered pairs (c_k,ρ_k). The usefulness of this point of view is that even though with each periodic function ψ (which has a Fourier series) is associated a unique set of ordered pairs (c_k,ρ_k), there are sets of ordered pairs to which there do not correspond periodic function. If $c_k = 1$ for all k, one has the set of ordered pairs associated with $2\pi \sum_{-\infty}^{+\infty} \delta(t - 2n\pi)$, which is not a function.

It is by generalizing this approach to the definition of functions that L. Schwartz has constructed the theory of distributions. The first step will be to specify which functions will play the role of the exponentials $\exp(-jkt)$. These will be the so-called testing functions. We shall define these first before giving the formal definition of a distribution. We shall next define operations on distributions, and, finally, we shall give applications.

For our purposes it is sufficient to consider distributions in one variable. The extension to several variables does not involve any new ideas. On this topic the reader should consult Ref. *2*.

2 **Definition** A *testing function* is a real-valued function of the real variable which may be differentiated an arbitrary number of times and which is identical to zero outside a finite interval.

We shall denote testing functions by φ and the variable by t. The closure of the set of all points at which $\varphi \neq 0$ is called the *support of* φ;

[1] A functional is a function whose range is the set of all real (or complex) numbers.

or equivalently, the support of φ consists of (I) the set of all points at which $\varphi(t) \neq 0$ and (II) the limit points of that set. In other words, if φ is a testing function, $\varphi^{(n)}(t)$ exists for all t and all n, and φ is identical to zero outside its support.

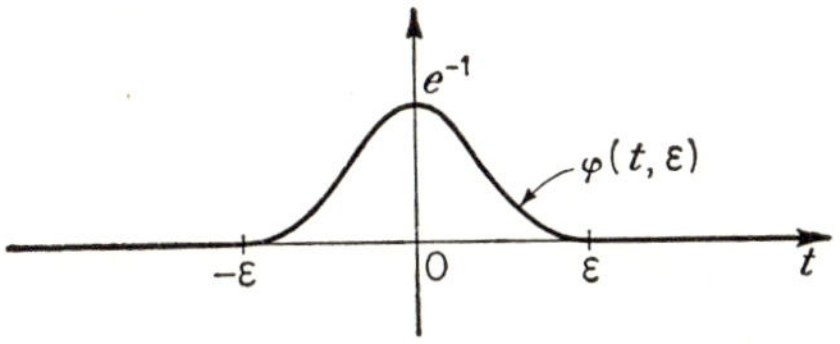

Fig. A.3.1 The testing function $\varphi(t,\varepsilon)$.

3 *Example* A well-known example of a testing function is illustrated in Fig. *A.3.1*. By definition

4
$$\varphi(t,\varepsilon) = \begin{cases} \exp\left(-\dfrac{\varepsilon^2}{\varepsilon^2 - t^2}\right) & |t| < \varepsilon \\ 0 & |t| \geq \varepsilon \end{cases}$$

The support of $\varphi(t,\varepsilon)$ is $[-\varepsilon,\varepsilon]$.

5 *Example* Given any continuous function, say, f, identically zero outside a finite interval and any testing function φ, the function ψ defined by

6
$$\psi(t) = \int_{-\infty}^{+\infty} f(\tau)\varphi(t-\tau)\,d\tau \qquad -\infty \leq t \leq \infty$$

is a testing function. The conditions on f imply that *6* is always the integral of a continuous function which differs from zero only on a finite interval; therefore, it is well defined for each t. Since φ can be differentiated arbitrarily often, it is easy to see that ψ has the same property. Finally, since both f and φ are zero outside some finite interval, so is ψ. Hence ψ is a testing function.

7 *Example* If f is any function that can be differentiated arbitrarily often (for example, a polynomial in t) and if φ is a testing function, then so is $f\varphi$.

8 *Example* The testing function $\rho(t,\varepsilon)$, shown in Fig. *A.3.2*, is defined by

9
$$\rho(t,\varepsilon) = k\int_{-1}^{+1} \varphi(t-\tau,\varepsilon)\,d\tau \qquad \varepsilon < 1$$

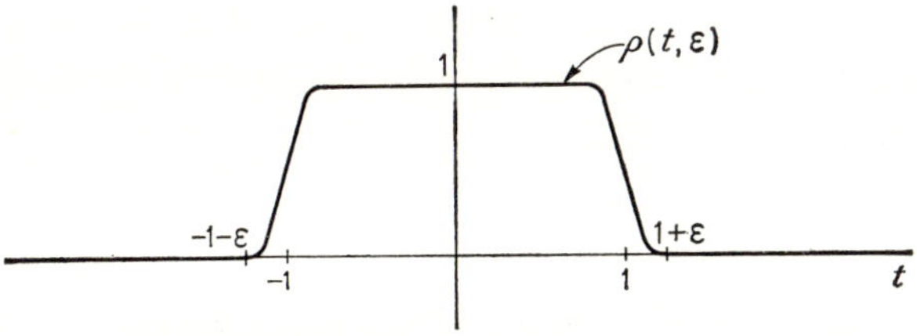

Fig. A.3.2 The testing function $\rho(t,\varepsilon)$.

where the constant k is so selected that $\rho(0,\varepsilon) = 1$. Note that $\rho(t,\varepsilon) = 1$ for all $t \in [-1 + \varepsilon, 1 - \varepsilon]$.

10 *Example*

11
$$\varphi_n(t) = \frac{t^n}{n!}\rho(t,\varepsilon) \qquad \varepsilon < 1$$

Observe that φ_n satisfies the conditions

12
$$\frac{d^k}{dt^k}\varphi_n(t)\bigg|_{t=0} = \delta_{kn} \qquad \begin{array}{l} k = 0, 1, 2, \ldots \\ n = 0, 1, 2, \ldots \end{array}$$

where δ_{kn} is the Kronecker symbol. Furthermore,

13
$$\varphi_n(t) = \frac{t^n}{n!} \qquad \text{for } -1 + \varepsilon \leq t \leq 1 - \varepsilon$$

4 *Definition of distributions*

Let $\mathfrak{D}$ be the set of all testing functions. $\mathfrak{D}$ is a linear vector space because whenever $\varphi_1 \in \mathfrak{D}$ and $\varphi_2 \in \mathfrak{D}$, then $\varphi_1 + \varphi_2 \in \mathfrak{D}$ and $a\varphi_1 \in \mathfrak{D}$ for any number a (see *C.2.1*).

In a very rough way and following the analogy with the definition of the periodic function by its Fourier coefficients, a distribution will be defined as an object specified by all the values taken by a linear functional at all the testing functions $\varphi \in \mathfrak{D}$.

As a first step toward a more rigorous definition, let us define the notion of convergence in the space of testing function $\mathfrak{D}$. We say that a sequence of testing functions $\{\varphi_n\}_1^\infty$ converges to zero if (I) all φ_n's are identically zero outside some fixed interval independent of n and (II) each φ_n, as well as all its derivatives, tends uniformly to zero. Observe that the convergence to zero is required for derivatives of all orders and that the uniformity (in t) of the convergence is required only separately for each order.[1]

The reason for defining the convergence of testing functions is that it enables us to define a continuous functional on the space $\mathfrak{D}$ of testing functions: such a functional f is said to be *continuous* if the convergence

[1] Another way of expressing this idea is the following: let each testing function of the sequence $\{\varphi_j\}_0^\infty$ have its support in some finite interval K. We shall say that the φ_j tend to zero if, for each m, the numbers $N_m(\varphi_j)$ tend to zero as $j \to \infty$, where

$$N_m(\varphi) \triangleq \sup_{k \leq m} [\sup_{t \in K} |\varphi^{(k)}(t)|]$$

of any sequence of testing functions φ_n to φ implies the convergence of the numbers $f(\varphi_n) \to f(\varphi)$, where $f(\varphi_n)$ is the value taken by the functional f at φ_n.

Let $\mathfrak{D}_K$ be the subspace of $\mathfrak{D}$ containing all testing functions whose support is contained in the closed and bounded interval K.

1 **Definition** A *distribution* is a linear functional defined on the space $\mathfrak{D}$ of testing functions and such that, for all closed and bounded intervals K, the restriction of the functional to $\mathfrak{D}_K$ is continuous. We shall denote distributions by T, F, S, ... Let us denote by $\langle T,\varphi\rangle$ the number which is the value that the distribution T takes at φ. Thus T is a linear mapping of $\mathfrak{D}$ into the set of all real (or complex) numbers. Linearity implies that

2
$$\begin{cases} \langle T, \varphi_1 + \varphi_2\rangle = \langle T,\varphi_1\rangle + \langle T,\varphi_2\rangle & \forall \varphi_1,\varphi_2 \in \mathfrak{D} \\ \langle T,a\varphi_1\rangle = a\langle T,\varphi_1\rangle & \forall \text{ numbers } a \end{cases}$$

The continuity of T and the restriction of the testing functions to $\mathfrak{D}_K$ implies that if the $\{\varphi_n\}_1^\infty$ have their supports in a fixed closed and bounded interval K and if the φ_j's as well as all their derivatives converge uniformly to zero, then the numbers $\langle T,\varphi_n\rangle \to 0$. We shall not pursue the requirements of continuity further.

3 *Example* Any real-valued function f integrable (in the Lebesgue sense) over any finite interval can be identified with a distribution; indeed, from the definition of the testing functions, the integral

$$\int_{-\infty}^{+\infty} f(t)\varphi(t)\,dt$$

is a continuous linear functional on the space $\mathfrak{D}$. Thus with any such function f we can associate a distribution F such that

$$\langle F,\varphi\rangle = \int_{-\infty}^{+\infty} f(t)\varphi(t)\,dt \qquad \forall \varphi \in \mathfrak{D}$$

4 *Example* Distributions include more general objects than functions. For example, the relations

5
$$\langle T,\varphi\rangle = \varphi(0) \qquad \forall \varphi \in \mathfrak{D}$$

define the distribution δ of Dirac (the unit impulse of electrical engineers) by its sifting property over the space of testing functions. The relations

$$\langle S,\varphi\rangle = \varphi''(0) \qquad \forall \varphi \in \mathfrak{D}$$

define a linear functional on $\mathfrak{D}$; hence, it defines a distribution S which we shall later identify with δ'', the second derivative of the distribution δ.

We shall see later that the class of distributions includes all the functions which are integrable over every finite interval as well as all their derivatives (to be taken in the distribution sense when they do not exist in the sense of the calculus).

6 *Comment* It is very important to appreciate the difference between a function of the real variable t and a distribution. Whereas a function f associates with every real number t one and only one value $f(t)$, it is usually not possible to associate a "value" to a distribution for each t. The following is an example where it is possible. Let T be a distribution such that $\langle T,\varphi\rangle = 0$ for all those φ which are zero outside $[-1,1]$. Then it is natural to say that $T = 0$ on $(-1,1)$. Similarly, the distribution R such that $\langle R,\varphi\rangle = 0$ for all φ's is the zero distribution.

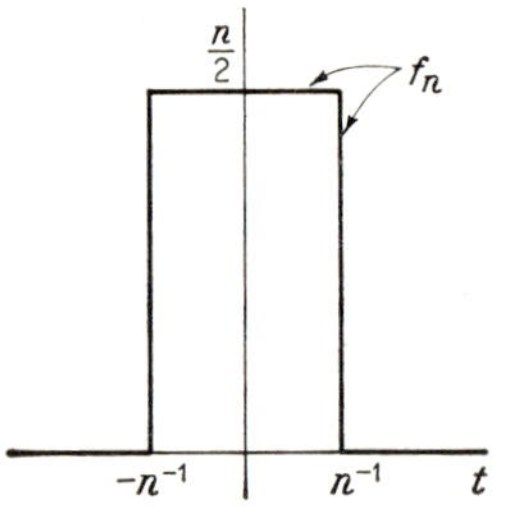

Fig. A.4.1 Example of a function that tends to δ and for which $\lim_{n\to\infty} f_n(0) = \infty$.

Consider the distribution δ. It is defined by its sifting property *5*. Now, many sequences of functions have this property; for example, consider the functions illustrated by Figs. *A.4.1* to *A.4.3* and defined by

$$f_n(t) = \begin{cases} \dfrac{n}{2} & |t| < n^{-1} \\ 0 & \text{elsewhere} \end{cases}$$

$$g_n(t) = \begin{cases} 2n^2 t & 0 \le t < \dfrac{1}{2n} \\ 2n^2(n^{-1} - t) & \dfrac{1}{2n} < t \le \dfrac{1}{n} \\ 0 & t > \dfrac{1}{n} \\ g_n(t) = g_n(-t) & \forall t \end{cases}$$

$$h_n(t) = \begin{cases} -\dfrac{n}{10} & |t| < \dfrac{1}{n} \\ \dfrac{3n}{5} & \dfrac{1}{n} < |t| < \dfrac{2}{n} \\ 0 & \text{elsewhere} \end{cases}$$

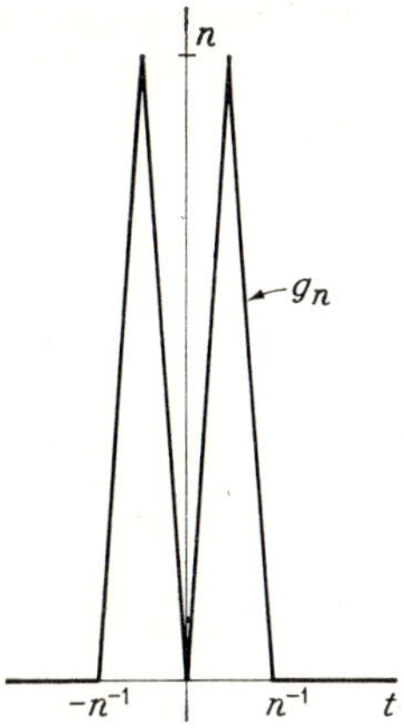

Fig. A.4.2 Example of a function that tends to δ and for which $\lim_{n\to\infty} g_n(0) = 0$.

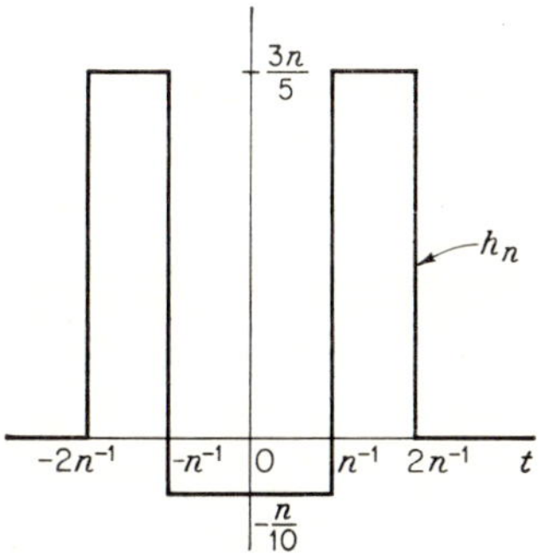

Fig. A.4.3 Example of a function that tends to δ and for which $\lim_{n\to\infty} h_n(0) = -\infty$

These three sequences of functions have the property that

$$\lim_{n\to\infty} \langle f_n,\varphi\rangle = \lim_{n\to\infty} \int_{-\infty}^{+\infty} f_n(t)\varphi(t)\,dt = \varphi(0) \qquad \forall\varphi \in \mathfrak{D}$$

$$\lim_{n\to\infty} f_n(t) = 0 \qquad \forall t \neq 0$$

and similar relations for g_n and h_n. Thus in the limit, these sequences behave like δ. It is of interest to note that for $t = 0$,

$$\lim f_n(0) = \infty \qquad \lim g_n(0) = 0 \qquad \lim h_n(0) = -\infty$$

These facts emphasize the point that although it is meaningful to state that the distribution δ is zero in any closed interval excluding the origin, it is impossible to assign a "value" to δ at $t = 0$.

5 *Operations on distributions*

Let S, T denote distributions: they are linear continuous functionals over $\mathfrak{D}$ and they are defined by all the numbers $\langle S,\varphi\rangle$, $\langle T,\varphi\rangle$ for all $\varphi \in \mathfrak{D}$.

1 **Definition: equality** Two distributions S, T are equal if

$$\langle T,\varphi\rangle = \langle S,\varphi\rangle \qquad \forall\varphi \in \mathfrak{D}$$

2 **Definition: addition** The distribution $S + T$ is that linear functional which with each $\varphi \in \mathfrak{D}$ associates the number

$$\langle S + T,\ \varphi\rangle \triangleq \langle S,\varphi\rangle + \langle T,\varphi\rangle$$

3 **Definition: multiplication by a scalar** c The distribution cT is defined by

$$\langle cT,\varphi\rangle \triangleq c\langle T,\varphi\rangle \qquad \forall \varphi \in \mathfrak{D}$$

The product of two distributions cannot be defined satisfactorily in general; however, we shall give later a restricted definition of interest in practice (see *10* below).

4 **Definition: differentiation** The derivative of a distribution S is the distribution S' defined by

$$\langle S',\varphi\rangle \triangleq -\langle S,\varphi'\rangle \qquad \forall \varphi \in \mathfrak{D}$$

When the distribution is the linear functional associated with a differentiable function $s(t)$, *4* is simply the rule of integration by parts,

$$\int_{-\infty}^{+\infty} s'(t)\varphi(t)\,dt = s(t)\varphi(t)\Big|_{-\infty}^{+\infty} - \int_{-\infty}^{+\infty} s(t)\varphi'(t)\,dt$$

where the integrated term is identically zero in view of the properties of the testing functions.

For derivatives of higher order we have

5 $$\langle S^{(k)},\varphi\rangle = (-1)^k\langle S,\varphi^{(k)}\rangle \qquad \forall \varphi \in \mathfrak{D}$$

Note that if $\varphi \in \mathfrak{D}$, then for any positive integer k, $\varphi^{(k)} \in \mathfrak{D}$; hence *4* and *5* have the very useful and important consequence:

6 **Theorem** Any distribution T may be differentiated as many times as we wish; that is, the derivative of any distribution always exists (in the distribution sense) and is a distribution.

7 *Example* Associated with the unit step $1(t)$ is the distribution T defined by

$$\langle T,\varphi\rangle = \int_0^{\infty} 1(t)\varphi(t)\,dt = \int_0^{\infty} \varphi(t)\,dt$$

Although $\frac{d}{dt}1(t)$ does not exist at $t = 0$ in the calculus sense, we may define T' by *4*:

$$\langle T',\varphi\rangle = -\int_0^{\infty} \varphi'(t)\,dt = -\varphi(t)\Big|_0^{\infty} = \varphi(0)$$

Therefore

8 $$\left[\frac{d}{dt}1(t)\right] = \delta(t)$$

That is, the derivative of $1(t)$ *in the distribution sense* is the Dirac delta function (see *A.4.5*).

The product of two distributions cannot be defined satisfactorily in general. Some restrictions are required; to present them we give the following

9 **Definition** An (ordinary) function f is said to *belong to* C^m if $f, f', \ldots, f^{(m)}$ are continuous functions of t. (*Note:* $f', \ldots, f^{(m)}$ are the first m derivatives of f in the calculus sense.)

10 **Definition** A *distribution* is said to be of *order m* if m is the least integer for which it is the $(m + 1)$st derivative (in the distribution sense) of a function integrable over any finite interval.

11 *Example* $\delta(t)$ is the derivative (in the distribution sense) of $1(t)$, and hence it is of order 0. $\delta'(t)$ is $\left[\dfrac{d^2}{dt^2} 1(t)\right]$; hence it is of order 1.

In some cases, distributions of finite order have an interesting integral representation. First a definition: a distribution T is said to be of *finite support* and its support is said to be in $(-a,a)$ if the interval $(-a,a)$ is such that, for all testing functions φ whose support is outside $(-a,a)$, $\langle T,\varphi\rangle = 0$. The representation theorem is as follows (Ref. 5, vol. 2, p. 31): Let T be of finite support and let its support be in $(-a,a)$. Let T be of finite order m. Then $\langle T,\varphi\rangle$ is given in terms of the Stieltjes integral

$$\langle T,\varphi\rangle = \int_{-a}^{+a} \varphi^{(m)}(x)\, d\mu(x) \qquad \forall \varphi \in \mathfrak{D}$$

where μ is a function of bounded variation.

12 **Definition: product of a distribution by a function** If T is a distribution of order $\leq m$ and if $f \in C^m$, then the product of T by f is the distribution defined by

13 $$\langle fT,\varphi\rangle = \langle T,f\varphi\rangle \qquad \forall \varphi \in \mathfrak{D}$$

Since T is of order $\leq m$, no trouble will arise from the fact that the functional has to be evaluated at $f\varphi$ rather than φ even though $f\varphi$ is not a testing function. The reason is that T is of order $\leq m$ and $f\varphi \in C^m$.

14 *Example* Let T be the distribution δ (of order 0); hence, in order for the product fT to be meaningful, f need only be continuous:[1]

$$\langle f\delta,\varphi\rangle \triangleq \langle \delta,f\varphi\rangle = f(0)\varphi(0)$$

Hence, if f is continuous at 0, we have the distribution equality: $f\delta = f(0)\delta$.

[1] Compare this result with the usual definition of δ in physics texts: δ is a "function" such that $\delta(t) = 0$ for all $t \neq 0$ and $\int_{-\infty}^{+\infty} \delta(t)\psi(t)\, dt = \psi(0)$ for all functions ψ continuous at 0.

15 *Exercise* Show that $t\delta(t)$ is the zero distribution.

16 *Example* Let T be δ' (of order 1); hence, f must have a continuous first derivative. The product of δ' and f is defined by

$$\langle f\delta',\varphi\rangle \triangleq \langle \delta',f\varphi\rangle$$

and by the definition of the derivative (*4*)

17 $$\langle \delta',f\varphi\rangle = -\left\langle \delta, \frac{d}{dt}(f\varphi)\right\rangle = -f'(0)\varphi(0) - f(0)\varphi'(0)$$

where the last equality follows from the definition of the distribution δ. Referring to the definition of the product of a distribution by a scalar and to the definition of δ and δ', we recognize that the last right-hand side can be written as $-f'(0)\langle\delta,\varphi\rangle + f(0)\langle\delta',\varphi\rangle$. Hence if $f \in C^1$,

18 $$f(t)\delta'(t) = -f'(0)\delta(t) + f(0)\delta'(t)$$

which is an equality between distributions (to be interpreted by *1*).

19 *Exercise* Show that if $f \in C^{n+1}$ in some neighborhood of zero,

$$f(t)\delta^n(t) = \sum_{k=0}^{n} (-1)^k \binom{n}{k} f^{(k)}(0)\delta^{n-k}(t)$$

and that the nth derivative of $f\delta$ is $f(0)\delta^{(n)}$.

20 *Exercise* Let $f \in C^n$ over $(-\infty,\infty)$. Let $g(t) = f(t)1(t)$. Show that, in the sense of distributions,

$$\begin{aligned} g^{(1)}(t) &= f^{(1)}(t)1(t) + f(0)\delta(t) \\ g^{(2)}(t) &= f^{(2)}(t)1(t) + f'(0)\delta(t) + f(0)\delta'(t) \\ &\vdots \\ g^{(k)}(t) &= f^{(k)}(t)1(t) + \sum_{i=1}^{k} f^{(k-i)}(0)\delta^{(i-1)}(t) \qquad k \le n \end{aligned}$$

6 *Further properties*

We state here without proof some properties of distributions which illustrate their nature and their convenience in applications.

1 **Theorem** Over any finite interval, a distribution T is the derivative of some finite order of some continuous function.[1] This property does

[1] L. Schwartz, "Théorie des distributions," vol. 1, p. 82, Hermann & Cie., Paris, 1957.

not extend to infinite intervals. Consider, for example,

$$\langle T,\varphi\rangle = \sum_{k=1}^{\infty} \varphi^{(k)}(k)$$

The notion of convergence of distributions is defined as follows:[1]

2 **Definition** The sequence of distributions $\{T_n\}_1^\infty$ is said to converge to the linear functional T if

$$\lim_{n\to\infty} \langle T_n,\varphi\rangle = \langle T,\varphi\rangle \qquad \forall \varphi \in \mathfrak{D}$$

It can be shown that the functional T is a distribution; this fact is not obvious because, in general, a limit of a sequence of continuous functionals is not a continuous functional.

3 **Theorem** Every distribution is the limit, in the sense of distributions, of a sequence of infinitely differentiable functions.

For example, $\delta(t) = \lim_{\sigma\to 0} (1/\sqrt{2\pi}\,\sigma) \exp(-t^2/2\sigma^2)$.

4 **Theorem** If $T_n \to T$, $R_n \to R$ (in the sense of distributions), and the numbers $a_n \to a$, then

$$T'_n \to T' \qquad R_n + T_n \to R + T \qquad a_n T_n \to aT$$

This implies, in particular, that every series of functions may be differentiated term by term any number of times; however, the sum must then be interpreted in the sense of distributions. For example, any Fourier series, such as $\sum_{-\infty}^{+\infty} c_n e^{jnt}$, where the c_n's behave for n large like a fixed finite power of n, converges in the sense of distributions because it is the finite-order derivative of a Fourier series that converges in the usual sense.

7 *Applications*

1 **Problem: equating distributions** Determine the constants α, β, γ, η such that

2 $$\alpha\delta(t) + \beta\delta'(t) + \gamma\delta(t-2) + \eta\delta''(t) = 3\delta(t) + 4\delta'(t)$$

Solution This expression is an equality of distributions. It does not mean that for each t the left-hand side takes on the same value as the

[1] *Ibid.*, chap. 3. sec 3.

right-hand side, because distributions are not functions of t. This equality must be interpreted by *A.5.1*. In other words, *2* means that

3 $$\langle\alpha\delta(t),\varphi\rangle + \langle\beta\delta'(t),\varphi\rangle + \langle\gamma\delta(t-2),\varphi\rangle + \langle\eta\delta''(t),\varphi\rangle = \langle 3\delta(t),\varphi\rangle + \langle 4\delta'(t),\varphi\rangle$$

which is an equality between numbers and which must hold for all $\varphi \in \mathfrak{D}$. Replace $\varphi(t)$ in *3* by $\varphi_0(t)$ defined in *A.3.11*. Then in view of *A.4.5*, *A.5.3*, *A.5.4*, and *A.3.10*, we obtain successively

$$\begin{aligned}\langle\alpha\delta(t),\varphi_0\rangle + \langle\beta\delta'(t),\varphi_0\rangle + \langle\gamma\delta(t-2),\varphi_0\rangle + \langle\eta\delta''(t),\varphi_0\rangle &\\ = \alpha\langle\delta(t),\varphi_0\rangle + \beta\langle\delta'(t),\varphi_0\rangle + \gamma\langle\delta(t-2),\varphi_0\rangle + \eta\langle\delta''(t),\varphi_0\rangle &\\ = \alpha\varphi_0(0) + \beta(-\varphi_0'(0)) + \gamma\varphi_0(2) + \eta\varphi_0''(0) &\\ = \alpha &\end{aligned}$$

Similarly, the right-hand side of *3* with φ replaced by φ_0 is equal to 3. Thus $\alpha = 3$. Replacing $\varphi(t)$ in *3* by $\varphi_1(t)$ defined in *A.3.11* gives $\beta = 4$. Replacing $\varphi(t)$ by $\varphi_0(t-2)$ leads to $\gamma = 0$, and replacing $\varphi(t)$ by $\varphi_2(t)$ leads to $\eta = 0$. Thus we have

$$3\delta(t) + 4\delta'(t) = 3\delta(t) + 4\delta'(t)$$

which is obviously an equality between distributions according to *A.5.1*. Thus we have the

4 **Assertion** Two linear combinations of $\delta, \delta', \ldots, \delta^{(k)}$ with *constant* coefficients are equal if and only if the coefficients of derivatives of the same order are equal.

5 *Caveat* The reader should notice the importance of the assumption that α, β, γ are constants. Indeed, *A.5.18* implies that

$$(1 + \sin t)\delta'(t) \neq \delta'(t)$$

because, in fact,

$$(1 + \sin t)\delta'(t) = \delta'(t) - \delta(t)$$

6 **Problem Differentiating a discontinuous function** Find the first and second derivatives of the function f, illustrated by Fig. *A.7.1* and defined by

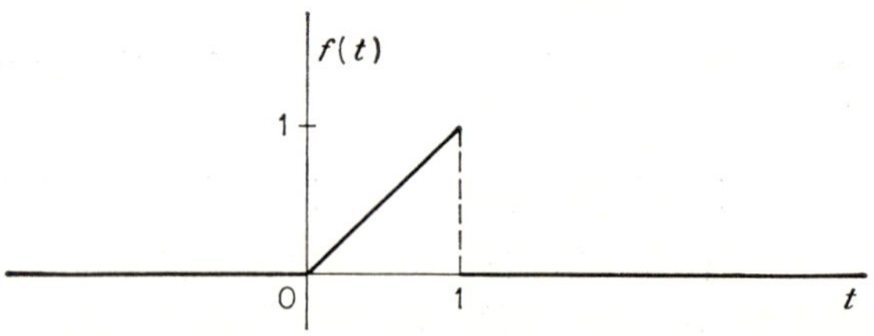

Fig. A.7.1 The function f which is the subject of Problem *A.7.6*.

7 $$f(t) = t1(t) - (t-1)1(t-1) - 1(t-1)$$

Solution f is obviously not differentiable at $t = 0$ and at $t = 1$. Consider f as a distribution; then its derivative, in the distribution sense, is obtained by differentiating term by term; thus

$$f' = t\delta(t) + 1(t) - (t-1)\delta(t-1) - 1(t-1) - \delta(t-1)$$

which, by *A.5.15*, reduces to

8 $$f' = 1(t) - 1(t-1) - \delta(t-1)$$
$$f'' = \delta(t) - \delta(t-1) - \delta'(t-1)$$

Let us verify that *8* actually is the derivative of f. For all $\varphi \in \mathfrak{D}$, we have

$$\begin{aligned} \langle f',\varphi\rangle &= -\langle f,\varphi'\rangle && \text{by } A.5.4 \\ &= \int_{-\infty}^{+\infty} f(t)\varphi'(t)\,dt && \text{because } f \text{ is an integrable function } A.4.3 \\ &= -\int_0^1 t\varphi'(t)\,dt && \text{by } 7 \\ &= \int_0^1 \varphi(t)\,dt - \varphi(1) && \text{by integration by parts} \\ &= \langle (1(t) - 1(t-1) - \delta(t-1)), \varphi(t)\rangle \end{aligned}$$

Since this equality holds for all $\varphi \in \mathfrak{D}$, *8* is a valid equality between distributions (*A.5.1*).

9 **Problem Determining the impulse response of a time-invariant differential system** Consider the differential system defined by

10 $$\dot{y} + y = \ddot{u} + \dot{u} + u$$

By *4.5.50*, its impulse response is the solution of the differential equation such that $y(0-) = u(0-) = \dot{u}(0-) = 0$ and $u = \delta(t)$.
Solution $\delta(t) = 0$ for all $t \neq 0$ in the sense that $\langle\delta(t),\varphi\rangle = 0$ for any testing function φ whose support does not include 0. Hence for $t \neq 0$, $\dot{y} + y = 0$. That is, y satisfies the homogeneous equation $\dot{y} + y = 0$; therefore

$$y(t) = y(0+)e^{-t} \qquad \forall t > 0$$
$$y(t) = 0 \qquad \forall t < 0$$

since $y(0-) = 0$.

The problem, then, is to evaluate $y(0+)$ and determine the behavior (in the distribution sense) of y at 0.

In view of *4*, since the right-hand side of *10* is $\ddot{\delta} + \dot{\delta} + \delta$, y must include both δ and $\dot{\delta}$. By *4*, it does not include any higher derivative of δ. Hence, put

$$y = \alpha\delta'(t) + \beta\delta(t) + \gamma 1(t)e^{-t} \qquad \forall t$$

where α, β, γ are unknown *constants.* Taking derivatives in the distribution sense, we have (using *A.5.10*),

$$y + \dot{y} = \alpha\delta'' + (\beta + \alpha)\delta'(t) + \beta\delta(t) + \gamma e^{-t}\Big|_{t=0}\delta(t) + \gamma 1(t)(e^{-t} - e^{-t})$$
$$= \delta'' + \delta' + \delta$$

Noting that the coefficients of the δ, δ', δ'' are constants, we invoke *4* and get

$$\alpha = 1 \qquad \beta + \alpha = 1 \qquad \beta + \gamma = 1$$

that is,
$$y(t) = \delta'(t) + 1(t)e^{-t} \qquad \forall t$$

11 **Problem Determining the impulse response of a time-varying differential system** Consider the time-varying differential system defined by

12
$$(1 + e^{-t})\dot{y} + \cos ty = 2\dot{u} + u$$

By *4.5.58*, its impulse response is the solution of *12* such that

$$y(0-) = u(0-) = 0 \qquad u = \delta(t)$$

Solution Let $\psi(t)$ be the solution of the homogeneous equation such that $\psi(0) = 1$. Since the right-hand side of *12* has only δ and δ', let us try

$$y = \alpha\delta(t) + \gamma 1(t)\psi(t)$$

where α and γ are unknown constants. By direct substitution we have

$$\alpha(1 + e^{-t})\dot{\delta}(t) + [\gamma(1 + e^{-t}) + \alpha\cos t]\delta(t)$$
$$+ \gamma 1(t)[(1 + e^{-t})\dot{\psi} + \cos t\psi] = 2\dot{\delta} + \delta$$

The last brackets are identical to zero from the definition of ψ. Using *A.5.10* and *A.5.13*, we have

$$2\alpha\dot{\delta}(t) + (2\gamma + 2\alpha)\delta(t) = 2\dot{\delta} + \delta$$

Hence, since α, γ are constants, by *4* we get

$$\alpha = 1 \qquad \gamma = -\tfrac{1}{2} \qquad \text{and} \qquad y(t) = \delta(t) - \tfrac{1}{2}1(t)\psi(t)$$

13 *Comment* Consider the system described by $L(p,t)y = M(p,t)u$, where L and M are differential operators of order n and m, respectively. Then, it follows immediately from the method of solution of Problem *11* that $n < m$ implies that the impulse response will contain derivatives of δ; in other words, for $n < m$, the system is improper (see *3.6.27*).

14 **Problem** Let $g(t)$ be a real-valued function of t having a continuous derivative for all t. Let $g(t) \neq 0$ for all $t \neq t_0$ and $g(t_0) = 0$ with $g'(t_0) \neq 0$. Find a convenient expression for $\delta[g(t)]$.

Solution Let us give an interpretation to the distribution $\delta[g(t)]$ by considering the integral

$$\int_{-\infty}^{+\infty} \delta[g(t)]\varphi(t)\,dt \triangleq \langle \delta[g(t)],\varphi(t) \rangle$$

From the properties of the distribution δ and of g, the integral depends only on the behavior of φ in an arbitrarily small neighborhood of t_0. Pick $\varepsilon > 0$ small enough that g' does not change sign in the interval $[t_0 - \varepsilon, t_0 + \varepsilon]$. Hence g defines a one-one mapping of $[t_0 - \varepsilon, t_0 + \varepsilon]$ onto $[\eta_1,\eta_2]$, where $\eta_1 = g(t_0 - \varepsilon)$, $\eta_2 = g(t_0 + \varepsilon)$. Let $u = g^{-1}$ be the inverse mapping, that is, u maps $[\eta_1,\eta_2]$ onto $[t_0 - \varepsilon, t_0 + \varepsilon]$ and

$$u[g(t)] = t \qquad \forall\, t \in [t_0 - \varepsilon, t_0 + \varepsilon]$$

and

$$u(0) = t_0$$

For convenience set $y = g(t)$. We obtain successively

$$\begin{aligned}\langle \delta(g(t)),\varphi(t) \rangle &= \int_{t_0-\epsilon}^{t_0+\epsilon} \delta(g(t))\varphi(t)\,dt \\ &= \int_{\eta_1}^{\eta_2} \delta(y)\varphi(u(y))u'(y)\,dy\end{aligned}$$

Note that $\varphi(u(y))u'(y)$ is a continuous function of y at $y = 0$ because the continuity of g' implies that of u' by the implicit-function theorem. If $g' > 0$ in $[t_0 - \varepsilon, t_0 + \varepsilon]$, then $\eta_2 > \eta_1$ and

$$\langle \delta(g(t)),\varphi(t) \rangle = \varphi(u(0))u'(0) = \frac{\varphi(t_0)}{g'(t_0)}$$

If $g' < 0$ in $[t_0 - \varepsilon, t_0 + \varepsilon]$, then $\eta_2 < \eta_1$ and

$$\langle \delta(g(t)),\varphi(t) \rangle = -\varphi(u(0))u'(0) = \frac{-\varphi(t_0)}{g'(t_0)}$$

Hence, in both cases,

$$\langle \delta(g(t)),\varphi(t) \rangle = \frac{\varphi(t_0)}{|g'(t_0)|}$$

or, equivalently,

$$\delta(g(t)) = \frac{\delta(t - t_0)}{|g'(t_0)|}$$

REFERENCES AND SUGGESTED READING

1 Van der Pol, B., and H. Bremmer: "Operational Calculus," Cambridge University Press, New York, 1955.

2 Schwartz, L.: "Théorie des distributions," vols. 1 and 2, Hermann & Cie., Paris, 1951, 1957.
3 Schwartz, L.: "Méthodes mathématiques pour les sciences physiques," Hermann & Cie., Paris, 1961.
4 Beckenbach, E. F. (ed.): "Modern Mathematics for the Engineer," Second Series, chap. 1, From Delta Functions to Distributions, by A. Erdelyi, McGraw-Hill Book Company, Inc., New York, 1961.
5 Gelfand, I. M., and G. E. Schilow: "Verallgemeinerte Funktionen und das Rechnen mit Ihnen," vols. 1 and 2, Deutscher Verlag der Wissenschaften, Berlin, 1960, 1962.

Laplace transformation and z transformation

B

1 Introduction

This appendix starts with a brief exposition of the Laplace transformation and is concerned primarily with those aspects of this vast subject which are relevant to the material in the main body of the text. Since there are many readily available books on the Laplace transformation, ranging all the way from exhaustive and rigorous treatments to cookbook-type presentations, we shall merely state the principal definitions and some of the main results and add, when necessary, a few clarifying comments. For proofs the reader is referred to the texts listed at the end of this appendix. The last section gives the definition and the main properties of the z transformation.

2 Basic concepts and definitions of the Laplace transformation

1 **Remark** In the following discussion, s stands for a complex variable ranging over the complex s plane. The real and imaginary parts of s are denoted by Re $s \triangleq \sigma$ and Im $s \triangleq \omega$, respectively. A time function is denoted by $f(t)$, or f for short, and its Laplace transform is written as $F(s)$ or $\mathcal{L}\{f(t)\}$. Note that we do not adhere to the convention *1.3.3* for reasons explained in *1.3.3*. Integrals are understood in the Lebesgue sense unless otherwise indicated.

Stated informally, the basic idea behind the Laplace transformation is that of resolution of a time function $f(t)$ into a continuum of "elementary" time functions $\{e^{st}\}$, where s is a complex variable taking values on a line parallel to the axis of imaginaries in the s plane. Thus, informally, the Laplace transform of $f(t)$ is a function $F(s)$ such that $F(s)\,ds$ represents the content in $f(t)$ of elementary time functions $e^{\lambda t}$ whose λ lies in the interval $(s, s + ds)$.

This idea underlies the definitions of the unilateral (one-sided) and the bilateral (two-sided) Laplace transforms given below.

2 **Definition** The *unilateral Laplace transform* of a time function $f(t)$ defined on the interval $[0, \infty)$ is denoted by $F(s)$ or $\mathcal{L}\{f(t)\}$ and is defined by the integral

3 $$F(s) = \int_0^\infty f(t)e^{-st}\,dt$$

for all values of s for which *3* exists (converges).

A basic result concerning the existence of $F(s)$ is contained in the following

4 **Theorem** If $F(s)$ exists for some s, say, $s_0 \triangleq \sigma_0 + j\omega_0$, then it exists for all s such that $\text{Re}\, s > \text{Re}\, s_0$. Furthermore, $F(s)$ is analytic in the domain defined by $\text{Re}\, s > \text{Re}\, s_0$. (See Ref. *1*, vol. 1, page 35 *et seq.*)

5 *Comment* The theorem makes no assertion concerning the convergence of *3* for s such that $\text{Re}\, s = \text{Re}\, s_0$. The reason for it is that, in general, *3* may converge for some points on the vertical line passing through the point s_0 and may diverge for other points on this line.

The greatest lower bound on $\text{Re}\, s_0$ for which *3* converges is denoted by α and is called the *abscissa of convergence* for f. More compactly, we have the

6 **Definition** Abscissa of convergence for $f \triangleq \alpha \triangleq$ infimum of σ over all s for which *3* converges. The domain defined by $\text{Re}\, s > \alpha$ constitutes the *domain of existence* (or *convergence*) of $F(s)$. The infimum of σ over all s for which *3* converges *absolutely* [i.e., with the integrand taken as $|f(t)e^{-st}|$] is the *abscissa* of *absolute convergence* for f and is denoted by α_a.

7 *Example* The abscissa of convergence for the exponential function e^{at}, where a is a real constant, is a. The abscissa of convergence for e^{-t^2} is $-\infty$. The function e^{t^2} is an example of a function which is not Laplace transformable since it has no finite abscissa of convergence.

8 **Remark** In many of the applications that are of interest to us the function $f(t)$ may contain delta functions at $t = 0$. In such cases, there are two alternative ways of resolving the ambiguity at $t = 0$. One is to extend the lower limit of integration to $0-$ in order to include the delta functions in question within the limits of integration, and the other is to leave out the delta functions by starting the integration at $0+$. Thus, under these two alternatives we have, respectively,

9 $$F(s) = \int_{0-}^\infty f(t)e^{-st}\,dt$$

and

10 $$F_+(s) = \int_{0+}^\infty f(t)e^{-st}\,dt$$

We use the symbol $F_+(s)$ to identify the latter because we shall be using for the most part *9* rather than *10*. Thus, when we speak of the unilateral Laplace transform of $f(t)$, *9* will be understood unless explicitly stated to the contrary.

We turn next to bilateral Laplace transforms.

11 **Definition** The *bilateral Laplace transform* of a time function $f(t)$ defined on $(-\infty, \infty)$ is denoted by $\bar{F}(s) = \mathfrak{L}_{\text{II}}\{f(t)\}$ [or simply $F(s)$ where no confusion with the unilateral Laplace transform is likely to arise] and is defined by the integral

12
$$\bar{F}(s) = \mathfrak{L}_{\text{II}}\{f(t)\} = \int_{-\infty}^{\infty} f(t)e^{-st}\,dt$$

for all values of s for which *12* exists (converges). ◁

By considering the two halves of $f(t)$ separately, it is easy to verify the truth of the following

13 **Assertion** If the abscissae of convergence for $1(t)f(t)$ and $1(t)f(-t)$ are α and β, respectively, and if $\beta \geq \alpha$, then $\bar{F}(s)$ exists and is analytic throughout the region (called the *strip of convergence*) defined by $\alpha < \text{Re } s < \beta$.

14 *Example* For the function $f(t)$ defined by

$$f(t) = 1(-t)e^t + 1(t)e^{-2t}$$

the strip of convergence is given by $-2 < \text{Re } s < 1$.

It is easy to show that a sufficient condition for the existence of a strip of convergence is that $f(t)$ be of *exponential class* in the sense that $|f(t)| \leq Ce^{\alpha t}$ for $t \geq 0$ and $|f(t)| \leq Ce^{\beta t}$ for $t < 0$, with $\alpha < \beta$ and C a finite constant (see Ref. *1*, vol. 1, page 407).

The *Fourier transform* of $f(t)$ may be regarded as a special case of the bilateral transform of $f(t)$. Specifically, if the axis of imaginaries falls within the strip of convergence, then we can set $s = j\omega$ and define

15
$$\mathfrak{F}\{f(t)\} \triangleq \text{Fourier transform of } f(t) \triangleq \int_{-\infty}^{\infty} f(t)e^{-j\omega t}\,dt$$

For simplicity, $\bar{F}(j\omega)$ will usually be written as $F(j\omega)$.

3 *Basic properties of Laplace transforms*

In what follows, we list without proof some of the main properties of unilateral and bilateral Laplace transforms. Whenever we do not differentiate between them, the implication is that the property in question is possessed by both. It is understood that the equalities involving time functions hold for all t (more strictly, *almost* all t) over which the functions are defined and that the equalities involving Laplace transforms hold for s which fall within their common domain of existence.

1 **Property: linearity** The Laplace transformation is a linear operation in the sense that, for all constants a, b and all time functions f_1, f_2

2 $$\mathcal{L}\{af_1(t) + bf_2(t)\} = a\mathcal{L}\{f_1(t)\} + b\mathcal{L}\{f_2(t)\}$$

over the range of definition of f_1 and f_2, provided the Laplace transforms on the left-hand side and the right-hand side of *2* exist. [This condition implies that the strip of convergence for $af_1 + bf_2$ is $\max(\alpha_1,\alpha_2) < \operatorname{Re} s < \min(\beta_1,\beta_2)$, where α_1, β_1 and α_2, β_2 define the strips of convergence for f_1 and f_2, respectively.]

3 **Property: translation theorem for time functions** For all real non-negative constants a

4 $$\mathcal{L}\{f(t-a)\} = e^{-as}\mathcal{L}\{f(t)\}$$

Note that for bilateral transforms, the restriction $a > 0$ is unnecessary. If $a < 0$, then for unilateral transforms *4* should be replaced by

5 $$\mathcal{L}\{f(t-a)\} = e^{-as}\left[\mathcal{L}\{f(t)\} - \int_0^{-a} f(t)e^{-st}\,dt\right]$$

6 **Property: translation theorem for unilateral Laplace transforms** Let $F(s)$ be the unilateral Laplace transform of $f(t)$. Then for any constant a (real or complex)

7 $$\mathcal{L}\{e^{-at}f(t)\} = F(s+a)$$

The same result holds for bilateral transforms, with the strip of convergence for $e^{-at}f(t)$ being that for $f(t)$ shifted to the left by amount $\operatorname{Re} a$. (See Ref. *2*, page 39.)

A key property of the Laplace transform involves the convolution of two time functions, defined as follows.

8 **Definition** Let f_1 and f_2 be defined on $(-\infty,\infty)$. Then the *convolution* of f_1 and f_2 is a time function denoted by $f_1 * f_2$, with the value of $f_1 * f_2$ at time t given by the integral

9 $$f_1 * f_2(t) = \int_{-\infty}^{\infty} f_1(t)f_2(t-\xi)\,d\xi$$

if it exists for all t. The convolution of f_1 and f_2 is a commutative operation in the sense that $f_1 * f_2 = f_2 * f_1$ for all t. ◁

If f_1 and f_2 are one-sided, i.e., vanish for $t < 0$, then their convolution is given by

10 $$f_1 * f_2(t) = \int_0^t f_1(\xi) f_2(t - \xi)\, d\xi$$

For the convolution of one-sided f_1 and f_2 we have the following

11 **Property: convolution theorem for time functions** The unilateral Laplace transform of the convolution of $f_1(t)$ and $f_2(t)$ is given by

12 $$\mathcal{L}\{f_1 * f_2\} = F_1(s) F_2(s)$$

More specifically, let $\mathcal{L}\{f_1(t)\}$ and $\mathcal{L}\{f_2(t)\}$ converge absolutely for some s, say s_0. Then (I) $f_1 * f_2$ exists for almost all $t \geq 0$, (II) $\mathcal{L}\{f_1 * f_2\}$ converges for s_0, and (III) *12* holds for s_0 and all s such that $\text{Re}\, s > \text{Re}\, s_0$. (See Ref. *1*, vol. 1, p. 124.)

The same result holds for bilateral transforms, provided $f_1 * f_2$ exists and there is a common strip of convergence defined by $\max(\alpha_1, \alpha_2) < \text{Re}\, s < \min(\beta_1, \beta_2)$ (see Ref. *2*, page 39).

13 **Property: convolution theorem for Laplace and Fourier transforms** The unilateral Laplace transform of the product of $f_1(t)$ and $f_2(t)$ is given by the following integral in the s plane:

14 $$\mathcal{L}\{f_1(t) f_2(t)\} = \frac{1}{2\pi j} \int_{\sigma_1 - j\infty}^{\sigma_1 + j\infty} F_1(\lambda) F_2(s - \lambda)\, d\lambda$$

where the integration is carried out along the vertical line $\text{Re}\, s = \sigma_1$, with σ_1 and s in *14* constrained by two conditions: (I) $\sigma_1 > \alpha_1$ (abscissa of convergence for f_1) and (II) $\text{Re}\, s - \sigma_1 > \alpha_2$ (abscissa of convergence for f_2). (See Ref. *1*, vol. 1, page 414.)

The same result holds for bilateral transforms, with the strip of convergence for $f_1(t)f_2(t)$ defined by $\alpha_2 + \sigma_1 < \text{Re}\, s < \beta_2 + \sigma_1$, $\alpha_1 < \sigma_1 < \beta_1$, where α_1, β_1 and α_2, β_2 define the strips of convergence for f_1 and f_2, respectively. (See Ref. *2*, page 47.) An immediate consequence of this property of bilateral transforms is the following: If $f_1(t)$ and $f_2(t)$ have Fourier transforms $F_1(j\omega)$ and $F_2(j\omega)$, respectively, then the Fourier transform of the product $f_1(t)f_2(t)$ is given by the convolution of $F_1(j\omega)$ and $F_2(j\omega)$; that is,

15 $$\mathcal{F}\{f_1(t) f_2(t)\} = \frac{1}{2\pi} \int_{-\infty}^{\infty} F_1(j\omega') F_2(j\omega - j\omega')\, d\omega'$$

From this relation it readily follows that f_1, f_2 and their respective Fourier transforms have the

16 **Property: Parseval's formula** If f_1 and f_2 are square integrable, that is,

$$\int_{-\infty}^{\infty} |f_i(t)|^2\,dt < \infty \qquad i = 1, 2$$

then
$$\int_{-\infty}^{\infty} f_1(t)f_2(t)\,dt = \frac{1}{2\pi}\int_{-\infty}^{\infty} F_1^*(j\omega)F_2(j\omega)\,d\omega$$

where $F_1^*(j\omega)$ is the complex conjugate of $F_1(j\omega)$. (See Ref. *3*, page 48, and Ref. *4*, page 27.)

Laplace transforms of derivatives and integrals

17 **Remark** In what follows, $F(s)$ stands for the unilateral Laplace transform defined by *9* and p denotes the derivative operator in the distribution sense (see *A.2.3* and *A.5.4*). $f^{(n)}(0-)$ stands for the nth derivative of $f(t)$ evaluated at $0-$ and analogously for $f^{(n)}(0+)$.

The following formulae summarize those properties of the Laplace transforms which are of frequent use in the solution of differential and integrodifferential equations.

18 **Property: formula for the Laplace transform of nth derivative** We have

19 $$\mathcal{L}\{p^n f(t)\} = s^nF(s) - s^{n-1}f(0-) - \cdots - f^{(n-1)}(0-) \qquad n = 0, 1, 2, \ldots$$

which reduces to

20 $$\mathcal{L}\{p^n f(t)\} = s^nF(s)$$

if $f(t)$ and its derivatives vanish for $t < 0$. It should be noted that in terms of $F_+(s)$ (see *B.2.10*), we have

21 $$\mathcal{L}\{p^n f(t)\} = s^nF_+(s) - s^{n-1}f(0+) - \cdots - f^{(n-1)}(0+) \qquad n = 0, 1, 2, \ldots$$

Note that it is this expression rather than *19* that is found in most classical texts on the Laplace transformation. We use for the most part *19* because it is better adapted to the needs of system analysis.

22 **Remark** Equation *19* can be expressed in a more compact form by using the initial-value derivative operator p_0 defined by

$$p_0^n f(t) = f^{(n)}(0-) \qquad n = 0, 1, 2, \ldots$$

Then *19* becomes simply

$$\mathcal{L}\{p^n f(t)\} = s^nF(s) - \left(\frac{s^n - p_0^n}{s - p_0}\right)f(t)$$

This relation easily generalizes to differential operators of the form

$$L(p) = a_n p^n + \cdots + a_0$$

Thus

23 $$\mathfrak{L}\{L(p)f(t)\} = L(s)F(s) - \left(\frac{L(s) - L(p_0)}{s - p_0}\right)f(t)$$

24 **Remark** A special case of *18* which is of frequent use in dealing with linear time-invariant systems is the following: Suppose that $f(t)$ is given by the convolution of one-sided functions h and u,

$$f(t) = \int_0^t h(t - \xi)u(\xi)\, d\xi$$

and $L(p)$ is a differential operator. Then

25 $$\mathfrak{L}\{L(p)f(t)\} = L(s)U(s)H(s)$$

where $U(s) \triangleq \mathfrak{L}\{u(t)\}$ and $H(s) \triangleq \mathfrak{L}\{h(t)\}$. This follows at once from *20* and *11*.

26 **Property: Laplace transform of nth-order integral** Let $\frac{1}{p^n} f(t)$, $n \geq 0$, denote the nth-order indefinite integral of $f(t)$, with $f^{(-n)}(0-)$ denoting the initial value of $\frac{1}{p^n} f(t)$ at $0-$. Then, if the Laplace transform of $\frac{1}{p^n} f(t)$ exists, we have

27 $$\mathfrak{L}\left\{\frac{1}{p^n} f(t)\right\} = s^{-n}F(s) + s^{-n}f^{(-1)}(0-) + \cdots + s^{-1}f^{(-n)}(0-)$$

28 **Property** For all points within the domain of existence of $F(s)$ (unilateral or bilateral) we have

29 $$\mathfrak{L}\{t^n f(t)\} = (-1)^n \frac{d^n}{ds^n} F(s) \qquad n = 0, 1, 2, \ldots$$

Initial- and final-value theorems for unilateral transforms

30 **Theorem** If $\lim_{t \to 0} f(t)$ exists as t approaches zero from the right, then

31 $$f(0+) = \lim_{s \to \infty} sF(s)$$

Comment It should be noted that the right-hand member of *31* may exist without the existence of $f(0+)$. For this reason it is important to know that $f(0+)$ exists before using *31*.

32 **Theorem** If $\lim_{t\to\infty} f(t)$ exists, then

33
$$\lim_{t\to\infty} f(t) = \lim_{s\to 0} sF(s)$$

Comment The comment for *30* applies to this theorem also.

Inversion formulae for Laplace transforms

Of basic importance in the theory of the Laplace transformation is the so-called inversion formula which yields the expression for $f(t)$ in terms of $F(s)$. For the one-sided Laplace transform we have the

34 **Theorem** Let $F(s) = \mathcal{L}\{f(t)\}$, with $f(t) = 0$ for $t < 0$, and let α_a be the abscissa of absolute convergence of f. If f is of bounded variation in the neighborhood of t, then,

35
$$\frac{f(t+0)+f(t-0)}{2} = \frac{1}{2\pi j}\int_{c-j\infty}^{c+j\infty} F(s)e^{st}\,ds \qquad \forall c > \alpha_a$$

where the integral is to be interpreted as $\lim_{\Omega\to\infty} \int_{c-j\Omega}^{c+j\Omega}$. ◁

In the case of the two-sided Laplace transformation, formula *35* holds for all $\alpha < c < \beta$, where α and β are the abscissae of absolute convergence of $1(t)f(t)$ and $1(t)f(-t)$, respectively.

In many instances, the inversion formula *35* can be calculated by contour integration provided the open path of integration can be closed. The *Jordan lemma* gives a set of sufficient conditions for closing the integration path (see Ref. *1*, vol. 1, page 224).

36 **Theorem** Let C_l be the left-half-plane semicircle of radius R centered on the origin: $C_l = \left\{s \,\middle|\, |s| = R \text{ and } \frac{\pi}{2} \le \measuredangle s \le \frac{3\pi}{2}\right\}$.

If, as $R \to \infty$, $F(s) \to 0$ uniformly on C_l, then

37
$$\lim_{R\to\infty} \int_{C_l} F(s)e^{st}\,ds = 0 \qquad \forall t > 0$$

A similar statement holds for $C_r = \left\{s \,\middle|\, |s| = R, -\frac{\pi}{2} \le \measuredangle s \le \frac{\pi}{2}\right\}$, but then *37* holds for all $t < 0$.

38 *Exercise* Let $P(s)$ and $Q(s)$ be polynomials such that the degree of Q is larger than the degree of P; let $t_0 > 0$. Show that

$$\mathcal{L}^{-1}\left(\frac{P(s)}{Q(s)}e^{-st_0}\right) = 0 \qquad \text{for all } t < t_0$$

This concludes our résumé of those properties of the Laplace transforms which are of relevance to the material in the main body of the text.

4 *z transforms*

In the case of time-invariant discrete-time systems, the z transform plays a role similar to that of the Laplace transform. In this section we define the z transform and cite its principal properties.

1 **Definition** Given a sequence of scalars $\{f_n\}_0^\infty$, the z transform of this sequence is the function of the complex variable z defined by

2 $$\tilde{F}(z) = \sum_{n=0}^{\infty} f_n z^{-n}$$

3 *Comment* The z transform can be considered as an operator mapping sequences of scalars into functions of the complex variable.

4 *Comment* In most engineering applications, one considers a function f which satisfies $f(t) = 0$ for $t < 0$ and which is either continuous at $t = nT$, $n = 0, 1, \ldots,$ or is such that $f(nT-)$ and $f(nT+)$ both exist for all nonnegative integers n. To the function f one associates "by sampling with period T" the sequence $\{f_n\}$, where $f_n = f(nT+)$. One refers to the z transform of the sequence $\{f_n\}$ as the "z transform of the function f." In symbols, $\tilde{F}(z) \triangleq Z\{f(t)\}$, with

5 $$\tilde{F}(z) \triangleq \sum_{k=0}^{\infty} f(kT+)z^{-k} \quad \triangleleft$$

The following properties of the z transform follow directly from the fact that $\tilde{F}(z)$ is a power series in z^{-1}.

6 **Property**[1] (I) The radius of convergence R_f of the power series *2* is given by

$$R_f = \limsup_{n\to\infty} \sqrt[n]{|f_n|}$$

(II) The power series *2* converges absolutely for all $z \ni |z| > R_f$ and uniformly for all $z \ni |z| > R$, where R is any radius larger than R_f.

(III) $\tilde{F}(z)$ is an analytic function of z in the domain $|z| > R_f$.

(IV) Derivatives of $\tilde{F}$ of any order can be obtained by differentiating the series term by term provided $|z| > R_f$.

[1] E. Hille, "Analytic Function Theory," vol. 1, secs. 5.4 and 5.5, Ginn and Company, Boston, 1959.

Properties of the z transform

We shall state the properties of the z transform by using the engineering terminology indicated in *4*. We *assume* throughout that the functions that we consider satisfy the conditions stated in *4* and have a z transform with a finite radius of convergence.

7 **Linearity theorem**

$$\mathcal{Z}(\alpha f) = \alpha \mathcal{Z}(f) \qquad \forall \text{ complex numbers } \alpha, \forall |z| > R_f$$
$$\mathcal{Z}(f + g) = \mathcal{Z}(f) + \mathcal{Z}(g) \qquad \forall |z| > \max (R_f, R_g)$$

Proof Refer to Definition *1* and *5*.

8 **Inversion theorem**

9
$$f(nT) = \frac{1}{2\pi j} \int_\Gamma \tilde{F}(z) z^{n-1}\, dz \qquad n = 0, 1, \ldots$$

where Γ is any simple closed rectifiable curve enclosing the origin and lying outside the circle $|z| = R > R_f$. The integration is performed in the counterclockwise direction.

10 *Comment* This theorem implies that, provided $R_f < \infty$, the z transform is a one-one mapping of the space of sequences into the set of functions which are analytic in some neighborhood of infinity.

Proof Start from the R.H.S. of *9*. Replace $\tilde{F}$ by its defining series *2*. Integrate term by term, which is legitimate by *6*(II). Finally, use the residue theorem to obtain *9*.

From definition *5* we immediately obtain

11 **Initial-value theorem**

$$f(0+) = \lim_{z \to \infty} \tilde{F}(z)$$

12 **Final-value theorem** If the sequence $\{f(nT+)\}_0^\infty$ tends to a limit, say, f_∞, then

(I) $R_f = 1$ and $\tilde{F}(z)$ is analytic outside the circle $|z| = 1$.

(II) $\lim_{z \to 1} (z - 1)\tilde{F}(z) = f_\infty$, z real and $z > 1$.

Proof To prove (I), use *6*(I) and observe that $f_n \to f_\infty$ implies that $\limsup_{n \to \infty} \sqrt[n]{|f_n|} = 1$.

To prove (II), observe that $f_n \to f_\infty$ means that for any $\varepsilon > 0$ there is a number N (which depends on ε) such that $n \geq N$ implies $|f_n - f_\infty| < \varepsilon$. Therefore,

$$\sum_{k=0}^{\infty} f_{N+k} z^{-(N+k)} = z^{-N} \sum_{k=0}^{\infty} f_{N+k} z^{-k}$$
$$= z^{-N} \left\{ \sum_{k=0}^{\infty} (f_{N+k} - f_\infty) z^{-k} + \sum_{k=0}^{\infty} f_\infty z^{-k} \right\} \qquad |z| > 1$$

Since $|z| > 1$, the second sum is $f_\infty(1 - z^{-1})^{-1}$. Let the first sum be called $R(\varepsilon,z)$, i.e.,

13 $$R(\varepsilon,z) \triangleq \sum_{k=0}^{\infty} (f_{N+k} - f_\infty)z^{-k}$$

Thus

$$\sum_{k=0}^{\infty} f_{N+k}z^{-(N+k)} = \frac{z^{-N}}{1 - z^{-1}} f_\infty + z^{-N}R(\varepsilon,z)$$

Now

$$\begin{aligned}\tilde{F}(z) &= \sum_{n=0}^{N-1} f_n z^{-n} + \sum_{k=0}^{\infty} f_{N+k}z^{-(N+k)} \\ &= \sum_{n=0}^{N-1} f_n z^{-n} + \frac{z^{-N}}{1 - z^{-1}} f_\infty + z^{-N}R(\varepsilon,z) \qquad |z| > 1\end{aligned}$$

Therefore,

14 $$(z - 1)\tilde{F} = (z - 1) \sum_{n=0}^{N-1} f_n z^{-n} + z^{-N+1} f_\infty + (z - 1)z^{-N}R(\varepsilon,z) \qquad |z| > 1$$

Now, restricting z to be real, we obtain from *13*

$$\begin{aligned}|(z - 1)z^{-N}R(\varepsilon,z)| &= \left| (z - 1)z^{-N} \sum_{k=0}^{\infty} (f_{N+k} - f_\infty)z^{-k} \right| \qquad |z| > 1 \\ &< (z - 1)z^{-N} \sum_{k=0}^{\infty} \varepsilon z^{-k} \qquad z \text{ real}, z > 1\end{aligned}$$

and

15 $$|(z - 1)z^{-N}R(\varepsilon,z)| < \varepsilon z^{(1-N)} \qquad z \text{ real}, z > 1$$

Now let $z \to 1$, with z real and $z > 1$. The first term of *14* tends to zero; the second tends to f_∞; and, by *15*, the last one is smaller in absolute value than any ε. Hence, (II) follows.

16 **Shifting theorem** If $\tilde{F}(z)$ is the z transform of the sequence $\{f_0,f_1, f_2, \ldots\}$, then $z^{-1}\tilde{F}(z)$ is the z transform of sequence $\{0,f_0,f_1,f_2, \ldots\}$.
Proof Immediate from Definition *1*.

REFERENCES AND SUGGESTED READING

1 Doetsch, G.: "Handbuch der Laplace Transformation," vols. 1 to 3, Verlag Birkhauser, Basel, 1950.
2 Van der Pol, B., and H. Bremmer: "Operational Calculus," Cambridge University Press, New York, 1955.
3 Goldberg, R. R.: "Fourier Transforms," Cambridge University Press, New York, 1961.

4 Papoulis, A.: "The Fourier Integral and Its Applications," McGraw-Hill Book Company, Inc., New York, 1962.

5 Jury, E. I.: "Sampled Data Control Systems," John Wiley & Sons, Inc., New York, 1958.

6 Ragazzini, J. R., and G. F. Franklin: "Sampled-data Control Systems," McGraw-Hill Book Company, Inc., New York, 1958.

7 Tou, J. T.: "Digital and Sampled-data Control Systems," McGraw-Hill Book Company, Inc., New York, 1959.

8 Bridgland, T. F., Jr.: A Linear Algebraic Formulation of the Theory of Sampled Data Control, *J. Soc. Indust. Appl. Math.*, vol. 7, no. 4, pp. 431–446, 1959.

9 Doetsch, G.: "Guide to the Applications of Laplace Transforms," D. Van Nostrand Company, Inc., Princeton, N.J., 1961.

C

Vectors and linear transformations

1 *Introduction*

The purpose of this appendix is twofold: first, to collect a number of results necessary for developments in the text and, second, to give a rapid, yet logical, derivation of results of interest in the theory and the applications of linear systems. In order to keep the size of this appendix within reason, we have assumed that the reader has some knowledge of vector and matrix algebra; more specifically, we have assumed that he knows the basic properties of determinants (including Cramer's rule) and the basic facts of matrix algebra (addition, multiplication, and inversion). This allows us to shorten our presentation, to leave a number of derivations as exercises, and, for example, to mention in a comment the inverse of a matrix before we encounter the definition of the inverse in formal presentation.

The next thirteen sections of the appendix are devoted to the definition of the basic concepts and the derivation of the associated fundamental facts. The final six sections make full use of the preceding ones and are devoted to the derivation of those results that are used most frequently in the text. Since the text itself provides ample motivation for this appendix (and the large number of cross references from the text to this appendix proves this point) we have eliminated almost all motivational introductions. We have, however, added a last section which reviews the reasons for the appearance of the adjoint in so many applications.

Most of the material presented here is well known and is available in numerous books, a number of which are listed as references. On the other hand, the section on pseudo inverses is new.

2 *Linear vector space*

1 **Definition** A (*linear*) *vector space* $\mathfrak{X}$ is a set of elements called *vectors* satisfying the following conditions:

(I) For every pair of vectors $\mathbf{x}$ and $\mathbf{y}$ in $\mathfrak{X}$ there corresponds a unique vector $\mathbf{x} + \mathbf{y}$ in $\mathfrak{X}$, called the sum of $\mathbf{x}$ and $\mathbf{y}$; this operation of addition has the following properties:

(*a*) $\mathbf{x} + \mathbf{y} = \mathbf{y} + \mathbf{x}$
(*b*) $\mathbf{x} + (\mathbf{y} + \mathbf{z}) = (\mathbf{x} + \mathbf{y}) + \mathbf{z}$
(*c*) There is a unique vector $\mathbf{0}$ such that $\mathbf{x} + \mathbf{0} = \mathbf{x}$ for all $\mathbf{x} \in \mathfrak{X}$.
(*d*) For every vector $\mathbf{x} \in \mathfrak{X}$ there is a vector $-\mathbf{x} \in \mathfrak{X}$ such that

$$\mathbf{x} + (-\mathbf{x}) = \mathbf{0}$$

(II) To every scalar α and vector $\mathbf{x} \in \mathfrak{X}$ there is a vector $\alpha\mathbf{x} \in \mathfrak{X}$, called the *product* of α and $\mathbf{x}$. It has the following properties:

(*a*) $\alpha(\beta\mathbf{x}) = (\alpha\beta)\mathbf{x}$ (*b*) $1\mathbf{x} = \mathbf{x}$ for every $\mathbf{x}$
(*c*) $\alpha(\mathbf{x} + \mathbf{y}) = \alpha\mathbf{x} + \alpha\mathbf{y}$ (*d*) $(\alpha + \beta)\mathbf{x} = \alpha\mathbf{x} + \beta\mathbf{x}$

In applications, α will usually be a real number; however, there are always instances where α is complex. For this reason we shall henceforth develop the theory under the assumption that the scalars are complex numbers. Technically, such linear vector spaces are called linear vector spaces over the field of complex numbers.

2 *Example* In physics all the forces applied at a point constitute a vector space: the vectors are ordered triples of real numbers, which are their components with respect to a system of coordinates; the scalars are real numbers. They may be complex when the phasor notation is used. The ith component is then $f_i(t) = \text{Re}\,[F_i \exp(j\omega t)]$, where $|F_i|$ is the amplitude and $\measuredangle F_i$ is the phase of the sinusoidal component $f_i(t)$.

3 *Example* In most of our applications, we shall consider vectors in n-dimensional space. These vectors may be considered as defined by an ordered n-tuple of complex numbers, e.g., $\mathbf{x} = (\xi_1, \xi_2, \ldots, \xi_n)$, where the ξ_i's are called the components of $\mathbf{x}$ with respect to a set of basis vectors. We shall henceforth call this space $\mathfrak{C}^n$. Let

$$\mathbf{y} = (\eta_1, \eta_2, \ldots, \eta_n)$$

The sum will be defined as $\mathbf{x} + \mathbf{y} = (\xi_1 + \eta_1, \ldots, \xi_n + \eta_n)$ and the product as $\alpha\mathbf{x} = (\alpha\xi_1, \alpha\xi_2, \ldots, \alpha\xi_n)$. The zero vector is

$$\mathbf{0} = (0,0, \ldots, 0)$$

4 *Example* In signal analysis, the vectors are "signals," i.e., functions of time f subject to $\int_{-\infty}^{+\infty} |f(t)|^2\,dt < \infty$. The class of all such functions of time constitutes a vector space. The scalars are either real or complex numbers.

5 *Example* All the solutions of the linear homogeneous differential equation

$$\frac{d^2y}{dt^2} + 3\frac{dy}{dt} + 2y = 0$$

are of the form $\alpha e^{-t} + \beta e^{-2t}$, where α and β are scalars (possibly complex). They constitute a linear vector space.

6 *Example* If $\mathbf{a}$ and $\mathbf{b}$ are two vectors in the plane, the set of all vectors of the form $\alpha\mathbf{a} + \beta\mathbf{b}$, with $|\alpha| \leq 1$ and $|\beta| \leq 1$, is not a vector space. (State why.)

7 *Example* The set of all complex-valued continuous functions over the interval [0,1] is a linear vector space.

8 *Example* The set of all solutions $\mathbf{E}(x,y,z,t)$, $\mathbf{H}(x,y,z,t)$ of Maxwell's equations for a cavity resonator with perfectly conducting walls constitutes a linear vector space. (*Hint:* $\nabla \times \mathbf{E} = -\mu\, \partial\mathbf{H}/\partial t, \nabla \times \mathbf{H} = \varepsilon\, \partial\mathbf{E}/\partial t$, and, on the boundary, $\mathbf{n} \times \mathbf{E} = \mathbf{0}$ and $\mathbf{n} \cdot \mathbf{H} = \mathbf{0}$.)

3 *Linear dependence*

1 **Definition** The vectors $\mathbf{a}_1, \mathbf{a}_2, \ldots, \mathbf{a}_k$ are said to be *linearly independent* if any relation of the form

2 $$\sum_{i=1}^{k} \alpha_i \mathbf{a}_i = \mathbf{0}$$

implies that $\alpha_i = 0$ for $i = 1, 2, \ldots, k$.

The vectors $\mathbf{a}_1, \mathbf{a}_2, \ldots, \mathbf{a}_k$ are *linearly dependent* if there is a set of scalars $\{\alpha_i\}$ not all zero, such that

3 $$\sum_{i=1}^{k} \alpha_i \mathbf{a}_i = \mathbf{0} \qquad \text{with} \sum_{i=1}^{n} |\alpha_i| > 0$$

4 *Example* In $\mathcal{C}^3$ $\mathbf{a}_1 \triangleq (1,0,0)$, $\mathbf{a}_2 \triangleq (0,1,0)$, $\mathbf{a}_3 \triangleq (0,0,1)$ are linearly independent. If $\mathbf{a}_4 \triangleq (1,3,0)$, then $\mathbf{a}_1$, $\mathbf{a}_2$, and $\mathbf{a}_4$ are linearly dependent.

5 *Exercise* Let $\mathfrak{X}$ be the linear vector space of all complex-valued functions of t, $-\infty < t < \infty$, with the usual definitions of addition and multiplication. Show that the vectors ϕ_i, $i = 1, 2, \ldots, N$, defined by $\phi_i(t) = e^{\lambda_i t}$ are linearly independent if and only if the exponents λ_i are pairwise distinct, that is, $\lambda_i \neq \lambda_j$ for all i, j.

6 **Theorem** The vectors $\mathbf{a}_1, \mathbf{a}_2, \ldots, \mathbf{a}_n$ are linearly dependent if and only if one of them, say, $\mathbf{a}_k$, $1 \leq k \leq n$, can be expressed as a linear combination of the others.

Proof Follows directly from the definitions.

7 **Theorem** The vectors $\mathbf{a}_1, \mathbf{a}_2, \ldots, \mathbf{a}_n$ of $\mathcal{C}^n$ are linearly independent if and only if

$$\det \mathbf{A} = 0$$

where $\mathbf{A}$ is the matrix whose ith column is made of the components of $\mathbf{a}_i$.

If the vectors $\mathbf{a}_1, \mathbf{a}_2, \ldots, \mathbf{a}_r$ of $\mathcal{C}^n$, where $r < n$, are linearly independent, then

$$\text{rank}\ (\mathbf{A}) = r$$

i.e., at least one $(r \times r)$ minor of $\mathbf{A}$ is not zero and all minors of higher order are zero.

Proof These statements follow directly from a well-known property of determinants: a determinant is equal to zero if and only if there is a linear combination of its column vectors that is equal to the zero vector.

4 *Bases*

1 **Definition** A set $\mathcal{B}$ of vectors of a linear vector space $\mathcal{X}$ is said to be a *basis* of $\mathcal{X}$ if (I) the vectors of $\mathcal{B}$ are linearly independent and (II) every vector of $\mathcal{X}$ can be expressed as a linear combination of the vectors of $\mathcal{B}$.

2 **Definition** A set of vectors is said to *span* a vector space $\mathcal{X}$ if any vector in $\mathcal{X}$ can be expressed as a linear combination of vectors of that set.

Given any set of vectors $\mathbf{x}_1, \mathbf{x}_2, \ldots, \mathbf{x}_m$, the set of all vectors of the form

$$\sum_{i=1}^{m} \alpha_i \mathbf{x}_i$$

is itself a vector space: it is the vector space spanned by $\mathbf{x}_1, \mathbf{x}_2, \ldots, \mathbf{x}_m$. We denote it by $\mathcal{S}[\mathbf{x}_1,\mathbf{x}_2, \ldots ,\mathbf{x}_m]$.

3 *Example* The vectors $\mathbf{x}_1 = (1,0,0)$, $\mathbf{x}_2 = (0,1,0)$, $\mathbf{x}_3 = (0,0,1)$, and $\mathbf{x}_4 = (1,1,0)$ span the three-dimensional euclidean space, but they do not constitute a basis. The first three vectors constitute a basis. In fact, $\mathbf{x}_3$ with any two of the remaining vectors constitutes a basis.

4 **Theorem** If $\mathbf{a}_1, \mathbf{a}_2, \ldots, \mathbf{a}_n$ form a basis of the vector space $\mathcal{X}$, then every vector $\mathbf{x} \in \mathcal{X}$ has a *unique* representation of the form

5 $$\mathbf{x} = \sum_{i=1}^{n} \xi_i \mathbf{a}_i$$

The numbers $\xi_1, \xi_2, \ldots, \xi_n$ are called the *components* of $\mathbf{x}$ with respect to the basis $\mathbf{a}_1, \mathbf{a}_2, \ldots, \mathbf{a}_n$.

Proof By definition of the basis, for any $\mathbf{x}$ there exists a representation such as *5*. It is unique because if there were two distinct representations of that form, then by subtraction we would get

$$\sum_{i=1}^{n} (\xi_i - \xi_i')\mathbf{a}_i = \mathbf{0}$$

where the ξ_i''s are the coefficients of the assumed second representation. Now, the linear independence of the set $\{\mathbf{a}_1,\mathbf{a}_2, \ldots ,\mathbf{a}_n\}$ implies that $\xi_i - \xi_i' = 0$ for $i = 1, 2, \ldots , n$. In other words, the representations are identical.

6 **Corollary** If $\mathcal{A} = \{\mathbf{a}_1,\mathbf{a}_2, \ldots ,\mathbf{a}_n\}$ constitutes a basis of $\mathcal{X}$, if $\mathbf{x} \in \mathcal{X}$, and if $\mathbf{x} \neq \mathbf{0}$, then $\{\mathbf{x},\mathbf{a}_1, \ldots ,\mathbf{a}_n\}$ are linearly dependent and for some k, $1 \leq k \leq n$, the set $\mathcal{B} = \{\mathbf{x},\mathbf{a}_1, \ldots ,\mathbf{a}_{k-1},\mathbf{a}_{k+1}, \ldots ,\mathbf{a}_n\}$ is a basis of $\mathcal{X}$.

Proof Since $\mathcal{A}$ is a basis, we have, for any $\mathbf{x} \in \mathcal{X}$,

7 $$\mathbf{x} = \sum_{i=1}^{n} \xi_i\mathbf{a}_i \qquad \text{or} \qquad \mathbf{x} - \sum_{1}^{n} \xi_i\mathbf{a}_i = \mathbf{0}$$

where $\sum_{1}^{n} |\xi_i| > 0$. From the last equality, it follows that $\mathbf{x}, \mathbf{a}_1, \ldots , \mathbf{a}_n$ are linearly dependent.

Now to show that $\mathcal{B}$ is a basis of $\mathcal{X}$, let k, $1 \leq k \leq n$, be an index such that $\xi_k \neq 0$; then

8 $$\mathbf{a}_k = \frac{1}{\xi_k}\mathbf{x} - \sum_{i=1}^{n} \left(\frac{\xi_i}{\xi_k} - \delta_{ik}\right) \mathbf{a}_i$$

Now since $\mathcal{A}$ is a basis, for any $\mathbf{y} \in \mathcal{X}$,

$$\mathbf{y} = \sum_{i=1}^{n} \eta_i\mathbf{a}_i$$

and, upon substituting *8* for $\mathbf{a}_k$, we get a relation of the form

$$\mathbf{y} = \xi\mathbf{x} + \sum_{\substack{j=1 \\ j \neq k}}^{n} \xi_j'\mathbf{a}_j$$

In other words, the set $\mathcal{B} = \{\mathbf{x},\mathbf{a}_1, \ldots ,\mathbf{a}_{k-1},\mathbf{a}_{k+1}, \ldots ,\mathbf{a}_n\}$ spans $\mathcal{X}$. To show that this set of vectors is linearly independent, we use contradiction. We suppose they are not independent; then we use *7* to show that this would imply that the set $\mathcal{A}$ would be linearly dependent. Thus, the set $\mathcal{B}$ spans $\mathcal{X}$ and is linearly independent. Therefore, it is a basis.

9 **Definition** A vector space $\mathfrak{X}$ is said to be *a finite-dimensional vector space* if it has a finite basis. If the number of elements in its basis is n, it is designated by $\mathcal{C}^n$.

10 **Theorem** Any two bases of a finite-dimensional vector space have the same number of elements. This theorem justifies the definition of the *dimension.*

11 **Definition** The number of elements in a basis of the vector space $\mathfrak{X}$ is called the *dimension* of $\mathfrak{X}$.

Proof The theorem is proved by contradiction. Let

$$\mathcal{A} = \{\mathbf{a}_1, \ldots, \mathbf{a}_n\} \qquad \text{and} \qquad \mathcal{B} = \{\mathbf{b}_1, \mathbf{b}_2, \ldots, \mathbf{b}_m\}$$

be bases of $\mathfrak{X}$. Suppose that $n < m$. By the corollary, for some k, $1 \leq k \leq n$, $\mathcal{A}_1 = \{\mathbf{b}_1, \mathbf{a}_1, \ldots, \mathbf{a}_{k-1}, \mathbf{a}_{k+1}, \ldots, \mathbf{a}_n\}$ is a basis of $\mathfrak{X}$. Applying it once more for some j, $1 \leq j \leq n$ and $j \neq k$,

$$\mathcal{A}_2 = \{\mathbf{b}_1, \mathbf{b}_2, \mathbf{a}_1, \ldots, \mathbf{a}_{j-1}, \mathbf{a}_{j+1}, \ldots, \mathbf{a}_{k-1}, \mathbf{a}_{k+1}, \ldots, \mathbf{a}_n\}$$

is a basis of $\mathfrak{X}$. Note that since $\mathbf{b}_2$ and $\mathbf{b}_1$ are elements of the same basis $\mathcal{B}$, when $\mathbf{b}_2$ is expressed as a linear combination of the elements of $\mathcal{A}_1$, at least one of the $\mathbf{a}_i$'s appears in the linear combination with a coefficient $\neq 0$; for otherwise the elements of $\mathcal{B}$ would be linearly dependent, which would contradict the assumption that $\mathcal{B}$ is a basis. After n applications of Corollary *6*, we arrive at the conclusion that $\mathcal{A}_n = \{\mathbf{b}_1, \mathbf{b}_2, \ldots, \mathbf{b}_n\}$ is a basis of $\mathfrak{X}$. Therefore, $\mathbf{b}_m$ can be expressed as a linear combination of $\mathbf{b}_1, \mathbf{b}_2, \ldots, \mathbf{b}_n$, that is, the vectors $\mathbf{b}_1, \mathbf{b}_2, \ldots, \mathbf{b}_n, \mathbf{b}_m$ are linearly dependent. This contradicts the fact that $\mathcal{B}$ is a basis. Therefore, $n = m$.

12 *Example* In Example *C.2.3* the vectors $\mathbf{u}_i = (\delta_{i1}, \delta_{i2}, \ldots, \delta_{in})$ constitute a basis for $\mathcal{C}^n$. (The Kronecker symbol $\delta_{ij} \triangleq 0$ whenever $i \neq j$ and $\delta_{ij} \triangleq 1$ when $i = j$.)

13 *Example* In Example *C.2.5*, the solutions e^{-t} and e^{-3t} constitute a basis for the space of solutions.

14 *Exercise* Consider the space of all band-limited functions f: that is, the functions whose Fourier transform vanishes outside the interval $(-W, W)$, where W is the bandwidth, in cycles per second. Interpret the sampling theorem to conclude that (I) these band-limited functions form a linear vector space $\mathfrak{X}$ and (II) a basis of $\mathfrak{X}$ is the denumerable set of functions

$$\phi_n(t) = \frac{\sin(2\pi Wt - n\pi)}{2\pi Wt - n\pi} \qquad n = \ldots, -1, 0, 1 \ldots$$

15 *Exercise* In $\mathcal{C}^n$ any set of n linearly independent nonzero vectors is a basis.

5 *Scalar product*

1 **Definition** The *scalar product* of two vectors $\mathbf{x}$ and $\mathbf{y} \in \mathfrak{X}$ is a complex number, denoted by $\langle \mathbf{x},\mathbf{y} \rangle$, having the properties that for all $\mathbf{x}$, $\mathbf{y}$ in $\mathfrak{X}$ and all complex numbers α_1, α_2

(I) $\langle \mathbf{x},\mathbf{y} \rangle^* = \langle \mathbf{y},\mathbf{x} \rangle$

(II) $\langle \alpha_1\mathbf{x}_1 + \alpha_2\mathbf{x}_2, \mathbf{y} \rangle = \alpha_1^*\langle \mathbf{x}_1,\mathbf{y} \rangle + \alpha_2^*\langle \mathbf{x}_2,\mathbf{y} \rangle$

(III) $\langle \mathbf{x},\mathbf{x} \rangle > 0$ for all $\mathbf{x} \neq \mathbf{0}$

2 *Comment* First, (I) implies that $\langle \mathbf{x},\mathbf{x} \rangle$ is real. Second, (I) and (II) imply that $\langle \mathbf{x},\mathbf{0} \rangle = \langle \mathbf{0},\mathbf{x} \rangle^* = 0\langle \mathbf{x},\mathbf{x} \rangle^* = 0$. Third, (I) and (II) imply that $\langle \mathbf{x},\alpha\mathbf{y} \rangle = \alpha\langle \mathbf{x},\mathbf{y} \rangle$.

3 *Example* In Example *C.2.7* show that $\int_0^1 x(t)^*y(t)\,dt$ satisfies the three requirements for being called a scalar product.

4 *Example* In $\mathcal{C}^n$ let $\mathbf{x} = (\xi_1,\xi_2, \ldots ,\xi_n)$ and $\mathbf{y} = (\eta_1,\eta_2, \ldots ,\eta_n)$; then $\langle \mathbf{x},\mathbf{y} \rangle = \sum_{i=1}^{n} \xi_i^*\eta_i$. This choice of scalar product is equivalent to the assumption that the ξ_i's and the η_i's are the coefficients of the representation of $\mathbf{x}$ and $\mathbf{y}$, respectively, in terms of an *orthonormal basis* $(\mathbf{u}_1,\mathbf{u}_2, \ldots ,\mathbf{u}_n)$, that is, a basis such that

$$\langle \mathbf{u}_i,\mathbf{u}_j \rangle = \delta_{ij} \qquad i,j = 1, 2, \ldots , n$$

This scalar product is called the *complex scalar product* in $\mathcal{C}^n$.

The scalar product gives a natural norm on the vector space: the *norm* of $\mathbf{x}$ is defined by

5 $$\|\mathbf{x}\| = \sqrt{\langle \mathbf{x},\mathbf{x} \rangle}$$

A more complete discussion of norms is to be found in Sec. *16*.

6 *Comment* If the vector $\mathbf{x}$ in $\mathcal{C}^n$ has real components ξ_i, with respect to an orthonormal basis, its norm is given by

7 $$\|\mathbf{x}\| = \sqrt{\sum_{i=1}^{n} \xi_i^2}$$

which is the *euclidean length* of the vector. The term "length" is justified because for $n = 3$, $\|\mathbf{x}\|$ is actually the length of $\mathbf{x}$.

8 **Schwarz inequality** For all $\mathbf{x},\mathbf{y} \in \mathfrak{X}$,

9 $$|\langle \mathbf{x},\mathbf{y} \rangle| \leq \|\mathbf{x}\| \; \|\mathbf{y}\|$$

where the equality holds if and only if $\mathbf{x} = \mu\mathbf{y}$ for some scalar μ.

Proof For $\mathbf{x},\mathbf{y} \in \mathfrak{X}$ and any scalar λ, (II) and (III) of Definition *1* imply that

$$\langle \mathbf{x} + \lambda\mathbf{y}, \mathbf{x} + \lambda\mathbf{y} \rangle = \langle \mathbf{x},\mathbf{x} \rangle + \lambda^*\langle \mathbf{y},\mathbf{x} \rangle + \lambda\langle \mathbf{x},\mathbf{y} \rangle + \lambda\lambda^*\langle \mathbf{y},\mathbf{y} \rangle \geq 0$$

where the equals sign holds if and only if $\mathbf{x} + \lambda\mathbf{y} = \mathbf{0}$. Since for $\mathbf{y} = \mathbf{0}$, *8* is obviously true, let us consider the case where $\mathbf{y} \neq \mathbf{0}$. Picking $\lambda = -\langle\mathbf{y},\mathbf{x}\rangle/\langle\mathbf{y},\mathbf{y}\rangle$, we get, using (I),

$$\langle\mathbf{x},\mathbf{x}\rangle\langle\mathbf{y},\mathbf{y}\rangle \geq \langle\mathbf{x},\mathbf{y}\rangle\langle\mathbf{y},\mathbf{x}\rangle = |\langle\mathbf{x},\mathbf{y}\rangle|^2$$

Finally, *9* follows from the definition of the norm *5*.

10 **Definition** $\mathbf{x}$ is said to be *orthogonal to* $\mathbf{y}$ if $\langle\mathbf{x},\mathbf{y}\rangle = 0$.

11 *Exercise* If $\mathbf{x} \in \mathcal{C}^n$ is orthogonal to all vectors of some basis of $\mathcal{C}^n$, then $\mathbf{x} = \mathbf{0}$. (*Hint:* Let $\{\mathbf{u}_i\}$ be a basis of $\mathcal{C}^n$ and let $\mathbf{x} = \sum_{i=1}^{n} \xi_i\mathbf{u}_i$. Take the scalar product of both sides by $\mathbf{u}_k$, $k = 1, \ldots, n$.)

12 **Definition** A *subspace* $\mathcal{B}$ of a finite dimensional space $\mathcal{X}$ is a set of vectors of $\mathcal{C}^n$ such that if $\mathbf{x}$ and $\mathbf{y}$ are in $\mathcal{B}$, then for all α and β, $\alpha\mathbf{x} + \beta\mathbf{y} \in \mathcal{B}$.

13 *Example* In the ordinary three-dimensional space with $\mathbf{u}_1$, $\mathbf{u}_2$, $\mathbf{u}_3$ as a basis, the plane which contains $\mathbf{u}_1$ and $\mathbf{u}_2$ is a subspace of the whole space. Also, $\mathbf{u}_1$ and $\mathbf{u}_2$ span that subspace.

14 *Example* If e^{-t}, e^{-2t}, e^{-3t} are three solutions of a linear differential equation $Ly = 0$, then all solutions of the form $\alpha e^{-t} + \beta e^{-et}$ constitute a subspace of the vector space of solutions.

6 *The Schmidt orthonormalization procedure*

The Schmidt orthonormalization procedure solves the following problem: given a basis $\{\mathbf{a}_1,\mathbf{a}_2, \ldots ,\mathbf{a}_n\}$ of $\mathcal{C}^n$, find an orthonormal basis of $\mathcal{C}^n$, that is, a basis $\{\mathbf{v}_1,\mathbf{v}_2, \ldots ,\mathbf{v}_n\}$ such that $\langle\mathbf{v}_i,\mathbf{v}_j\rangle = \delta_{ij}$, $i,j = 1, 2, \ldots , n$.

Let us first describe the mechanics of the solution: at each step we construct a basis vector $\mathbf{u}_k$ of an intermediate basis and by normalization of $\mathbf{u}_k$ we get $\mathbf{v}_k$, the vector of the desired orthonormal basis.

1

$$\begin{aligned}
\mathbf{u}_1 &= \mathbf{a}_1 & \mathbf{v}_1 &= \mathbf{u}_1/\|\mathbf{u}_1\| \\
\mathbf{u}_2 &= \mathbf{a}_2 - \langle\mathbf{v}_1,\mathbf{a}_2\rangle\mathbf{v}_1 & \mathbf{v}_2 &= \mathbf{u}_2/\|\mathbf{u}_2\| \\
\mathbf{u}_3 &= \mathbf{a}_3 - \langle\mathbf{v}_1,\mathbf{a}_3\rangle\mathbf{v}_1 - \langle\mathbf{v}_2,\mathbf{a}_3\rangle\mathbf{v}_2 & \mathbf{v}_3 &= \mathbf{u}_3/\|\mathbf{u}_3\| \\
&\cdots\cdots\cdots & &\cdots\cdots \\
\mathbf{u}_r &= \mathbf{a}_r - \sum_{k=1}^{r-1} \langle\mathbf{v}_k,\mathbf{a}_r\rangle\mathbf{v}_k & \mathbf{v}_r &= \mathbf{u}_r/\|\mathbf{u}_r\| \\
&\cdots\cdots\cdots & &\cdots\cdots \\
\mathbf{u}_n &= \mathbf{a}_n - \sum_{k=1}^{n-1} \langle\mathbf{v}_k,\mathbf{a}_n\rangle\mathbf{v}_k & \mathbf{v}_n &= \mathbf{u}_n/\|\mathbf{u}_n\|
\end{aligned}$$

It is obvious that the construction guarantees that $\|\mathbf{v}_k\| = 1$ for all k's. Direct verification shows that $\langle \mathbf{v}_k, \mathbf{u}_r \rangle = 0$ for all $k < r$. Since this is true for all r, we have $\langle \mathbf{v}_i, \mathbf{v}_j \rangle = 0$ for all $i \neq j$. Therefore, the basis $\{\mathbf{v}_1, \mathbf{v}_2, \ldots, \mathbf{v}_n\}$ is *orthonormal* since

2 $$\langle \mathbf{v}_i, \mathbf{v}_j \rangle = \delta_{ij} \qquad i,j = 1, 2, \ldots, n$$

Let us consider the geometric aspect of the process: $\mathbf{v}_1$ is the unit vector along $\mathbf{a}_1$; $\mathbf{u}_2$ is the vector perpendicular to $\mathbf{a}_1$ whose length is the distance from the tip of $\mathbf{a}_2$ to the line supporting $\mathbf{v}_1$ (Fig. *C.6.1*); $\mathbf{u}_3$ is

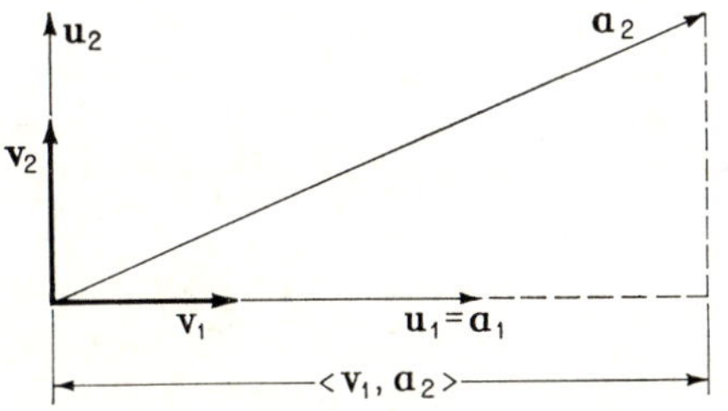

Fig. C.6.1 Schmidt orthonormalization. Relation between the vectors $\mathbf{a}_1$ and $\mathbf{a}_2$ of the original basis and the vectors $\mathbf{v}_1$ and $\mathbf{v}_2$ of the new orthonormal basis.

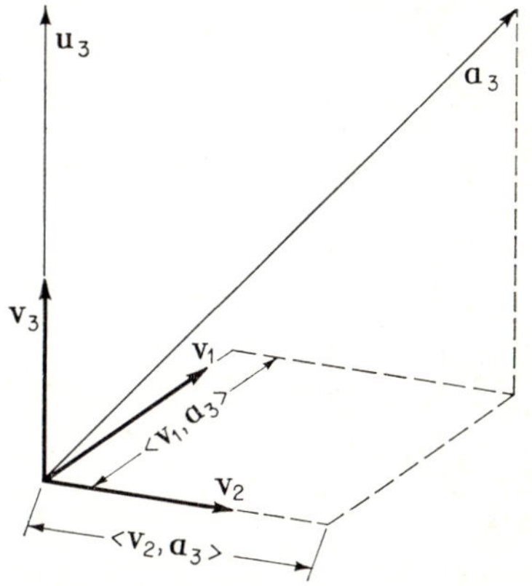

Fig. C.6.2 Schmidt orthonormalization; the figure illustrates the projection of $\mathbf{a}_3$ on the plane supported by $\mathbf{v}_1$ and $\mathbf{v}_2$ and the resulting definition of $\mathbf{u}_3$. The vectors $\mathbf{v}_1$, $\mathbf{v}_2$, and $\mathbf{v}_3$ belong to the new orthonormal basis.

parallel to the perpendicular dropped from the tip of $\mathbf{a}_3$ on the plane spanned by $\mathbf{a}_1$, $\mathbf{a}_2$ (Fig. *C.6.2*, where the process is shown in perspective); $\mathbf{u}_r$ is parallel to the perpendicular dropped from the tip of $\mathbf{a}_r$ to the subspace spanned by $\mathbf{a}_1, \mathbf{a}_2, \ldots, \mathbf{a}_{r-1}$. Observe that this subspace is also spanned by $\mathbf{v}_1, \mathbf{v}_2, \ldots, \mathbf{v}_{r-1}$; hence, any vector in that subspace is of the form $\sum_{i=1}^{r-1} \alpha_i \mathbf{v}_i$ and any such vector is also orthogonal to $\mathbf{u}_r$ since $\langle \mathbf{u}_r, \mathbf{v}_i \rangle = 0$ for $i = 1, 2, \ldots, r-1$.

7 *Orthogonal projections*

1 **Theorem** Let $\mathfrak{M}$ be a subspace of $\mathcal{C}^n$. Then any vector $\mathbf{x}$ not in $\mathfrak{M}$ can be represented in one and only one way as

2 $$\mathbf{x} = \mathbf{y} + \mathbf{y}_p$$

where $\mathbf{y} \in \mathfrak{M}$ and $\mathbf{y}_p = \mathbf{x} - \mathbf{y}$ is orthogonal to all vectors of $\mathfrak{M}$. (See Fig. C.7.1.)

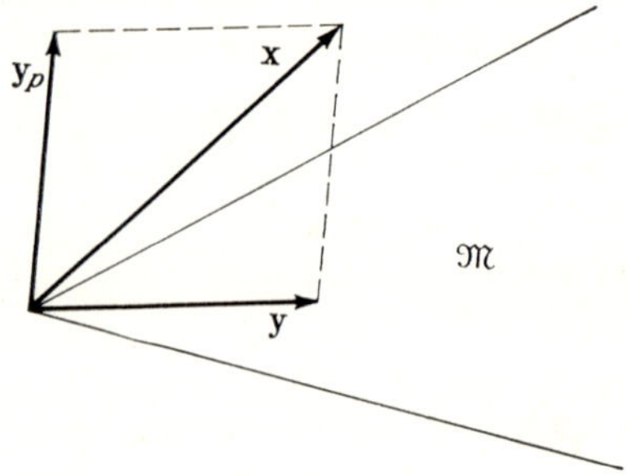

Fig. C.7.1 Orthogonal projection of **x** on the subspace $\mathfrak{M}$ (shown here as a plane), resulting in two vectors $\mathbf{y} \in \mathfrak{M}$ and $\mathbf{y}_p \perp \mathfrak{M}$.

3 **Definition** **y** is said to be the *orthogonal projection of* **x** *on* $\mathfrak{M}$.

Proof Pick a basis for $\mathfrak{M}$ and adjoin to it linearly independent vectors of $\mathcal{C}^n$ so that the resulting set is a basis for $\mathcal{C}^n$. Using these vectors, by the Schmidt process we obtain an orthonormal basis for $\mathcal{C}^n$:

4 $$\mathbf{u}_1, \mathbf{u}_2, \ldots, \mathbf{u}_s, \mathbf{u}_{s+1}, \ldots, \mathbf{u}_n$$

where $\mathbf{u}_1, \mathbf{u}_2, \ldots, \mathbf{u}_s$ span $\mathfrak{M}$. Thus, for any $\mathbf{x} \in \mathcal{C}^n$

5 $$\mathbf{x} = \sum_{i=1}^{s} \xi_i \mathbf{u}_i + \sum_{k=s+1}^{n} \xi_k \mathbf{u}_k$$

Let the first sum be **y** and the second be $\mathbf{y}_p$. Clearly, $\mathbf{y} \in \mathfrak{M}$. Since $\mathbf{y}_p = \mathbf{x} - \mathbf{y}$ is a linear combination of $\mathbf{u}_{s+1}, \ldots, \mathbf{u}_n$, and since the basis $\{\mathbf{u}_i\}$ is an orthonormal basis, $\mathbf{y}_p$ is orthogonal to $\mathbf{u}_1, \mathbf{u}_2, \ldots, \mathbf{u}_s$ and hence to every vector of $\mathfrak{M}$, because $\mathfrak{M} = \mathcal{S}[\mathbf{u}_1, \ldots, \mathbf{u}_s]$. It remains to show that the representation is unique. Suppose it were not, i.e., suppose there is another representation of **x** such that $\mathbf{x} = \mathbf{y}' + \mathbf{y}'_p$, where $\mathbf{y}' \in \mathfrak{M}$ and $\mathbf{y}'_p$ is orthogonal to all the vectors of $\mathfrak{M}$. Then $\mathbf{x} = \mathbf{y}' + \mathbf{y}'_p$ with $\mathbf{y} \neq \mathbf{y}'$. By subtracting this second representation from *2*, we have

6 $$(\mathbf{y} - \mathbf{y}') + (\mathbf{y}_p - \mathbf{y}'_p) = \mathbf{0}$$

and, by scalar multiplication,[1]

$$\langle \mathbf{y} - \mathbf{y}', \mathbf{y} - \mathbf{y}' \rangle + \langle \mathbf{y} - \mathbf{y}', \mathbf{y}_p - \mathbf{y}'_p \rangle = 0$$

[1] Since we are using the complex scalar product (see *C.5.4*), the order of the factors in a scalar product is important. In the case at hand we have to have $\mathbf{y} - \mathbf{y}'$ as the left-hand member in both scalar products. To see the point, imagine Eq. *6* written out as n scalar equations relative to some orthonormal basis. Now multiply each scalar equation by the complex conjugate of the corresponding component of $\mathbf{y} - \mathbf{y}'$ (relative to the same basis!). Add the results and observe that they amount to the sum of the two scalar products with their factors in the order shown above.

Now $\mathbf{y}_p$ and $\mathbf{y}'_p$ are orthogonal to all vectors of $\mathfrak{M}$ and hence to $\mathbf{y} - \mathbf{y}'$; therefore, the second term is zero and the above equality then implies $\mathbf{y} = \mathbf{y}'$. Hence, the representation is unique.

Let $\mathfrak{M}_0 = \mathcal{S}[\mathbf{u}_{s+1},\mathbf{u}_{s+2}, \ldots ,\mathbf{u}_n]$. Then, every vector of $\mathfrak{M}_0$ is orthogonal to every vector of $\mathfrak{M}$. Furthermore, any vector in $\mathcal{C}^n$ orthogonal to every vector in $\mathfrak{M}$ is a vector of $\mathfrak{M}_0$. Hence $\mathfrak{M}_0$ is the set of all vectors of $\mathcal{C}^n$ that are orthogonal to all vectors of $\mathfrak{M}$. $\mathfrak{M}_0$ is said to be the *orthogonal complement* of $\mathfrak{M}$. It is denoted by $\mathfrak{M}^\perp$.

In order to further exhibit the relation between the subspaces $\mathfrak{M}$ and $\mathfrak{M}_0$, we need to introduce a concept that will be of great usefulness later on.

7 **Definition** Let $\mathfrak{M}$ and $\mathfrak{N}$ be two subspaces of a vector space $\mathfrak{X}$. $\mathfrak{X}$ is said to be the *direct sum of* $\mathfrak{M}$ *and* $\mathfrak{N}$, written $\mathfrak{M} \oplus \mathfrak{N} = \mathfrak{X}$, if any $\mathbf{x} \in \mathfrak{X}$ may be written in one and only one way as $\mathbf{x} = \mathbf{y} + \mathbf{z}$, where $\mathbf{y} \in \mathfrak{M}$ and $\mathbf{z} \in \mathfrak{N}$.

8 *Example* Let $\mathbf{e}_1$, $\mathbf{e}_2$, and $\mathbf{e}_3$ be a basis for $\mathcal{C}^3$. Let $\mathfrak{M} = \mathcal{S}[\mathbf{e}_1,\mathbf{e}_2]$ and $\mathfrak{N} = \mathcal{S}[\mathbf{e}_3]$, that is, $\mathfrak{M}$ is the plane spanned by $\mathbf{e}_1$ and $\mathbf{e}_2$ and $\mathfrak{N}$ is the straight line spanned by $\mathbf{e}_3$. Clearly, $\mathcal{C}^3 = \mathfrak{M} \oplus \mathfrak{N}$. ◁

Going back to the subspaces $\mathfrak{M}$ and $\mathfrak{M}_0$ of $\mathcal{C}^n$ defined in Theorem *1*, we see that their relation can be expressed by

$$\mathcal{C}^n = \mathfrak{M} \oplus \mathfrak{M}_0 \qquad \mathfrak{M} \perp \mathfrak{M}_0$$

By the second expression, $\mathbf{y} \in \mathfrak{M} \Rightarrow \langle \mathbf{y},\mathbf{y}_0\rangle = 0, \ \forall \mathbf{y}_0 \in \mathfrak{M}_0$.

9 **Corollary** With the notation of Theorem *1*, we have

10 $$\dim \mathcal{C}^n = \dim \mathfrak{M} + \dim \mathfrak{M}_0$$

Proof Exercise for the reader. (*Hint:* Look at the proof of *1*.)

11 **Theorem** Let $\mathfrak{M}$ be a subspace of $\mathcal{C}^n$; then $\mathbf{y}$ is the orthogonal projection of $\mathbf{x}$ on $\mathfrak{M}$ if and only if

12 $$\|\mathbf{x} - \mathbf{y}\| = \min_{\mathbf{z} \in \mathfrak{M}} \|\mathbf{x} - \mathbf{z}\|$$

Proof Use the orthonormal basis *4* and the representation *5*. For any $\mathbf{z}$ in $\mathfrak{M}$, let $\mathbf{z} = \sum_{i=1}^{s} \zeta_i \mathbf{u}_i$. Hence

$$\mathbf{x} - \mathbf{z} = \sum_{i=1}^{s} (\xi_i - \zeta_i)\mathbf{u}_i + \sum_{k=s+1}^{s} \xi_k \mathbf{u}_k$$

and

$$\|\mathbf{x} - \mathbf{z}\|^2 = \sum_{i=1}^{s} |\xi_i - \zeta_i|^2 + \sum_{k=s+1}^{n} |\xi_k|^2$$

Clearly, $\|\mathbf{x} - \mathbf{z}\|$ is minimum if and only if $\xi_i = \zeta_i$, $i = 1, 2, \ldots, s$; that is, $\mathbf{z} = \mathbf{y}$.

8 *Reciprocal basis*

1 **Definition** Let $\mathcal{B} = \{\mathbf{u}_1,\mathbf{u}_2, \ldots ,\mathbf{u}_n\}$ be a basis of $\mathcal{C}^n$. The reciprocal basis to $\mathcal{B}$ is the set of vectors $\{\mathbf{v}_1,\mathbf{v}_2 \ldots ,\mathbf{v}_n\}$ such that

2 $$\langle \mathbf{v}_i,\mathbf{u}_j\rangle = \delta_{ij} \qquad i,j = 1, 2, \ldots , n \quad \triangleleft$$

In other words, $\langle \mathbf{v}_1,\mathbf{u}_1\rangle = 1$ and $\mathbf{v}_1$ is orthogonal to $\mathbf{u}_2, \mathbf{u}_3, \ldots , \mathbf{u}_n$. The vectors $\mathbf{v}_1, \ldots , \mathbf{v}_n$ constitute a basis because (I) they are linearly independent and (II) they span the space $\mathcal{C}^n$. To establish (I) suppose they were not linearly independent; then $\Sigma_k\alpha_k\mathbf{v}_k = \mathbf{0}$ for some α_k's not all zero. From *2* we get, by taking the scalar product of the last vector equation with $\mathbf{u}_i$, $i = 1, 2, \ldots , n$,

$$\sum_{k=1}^{n} \alpha_k\langle \mathbf{u}_i,\mathbf{v}_k\rangle = \alpha_i\langle \mathbf{u}_i,\mathbf{v}_i\rangle = \alpha_i = 0 \qquad i = 1, 2, \ldots , n$$

Hence, $\alpha_i = 0$ for all i's, which contradicts the assumption that the $\mathbf{v}_i$'s were linearly dependent. (II) Any vector $\mathbf{x} \in \mathcal{C}^n$ can be expressed as a linear combination of the $\mathbf{v}_i$'s, viz.,

$$\mathbf{x} = \sum_{i=1}^{n} \langle \mathbf{u}_i,\mathbf{x}\rangle \mathbf{v}_i$$

To establish this relationship, let $\mathbf{y} \triangleq \mathbf{x} - \sum_{i=1}^{n} \langle \mathbf{u}_i,\mathbf{x}\rangle \mathbf{v}_i$. Taking the scalar product with $\mathbf{u}_k$ and using *2*, we get, as above, $\langle \mathbf{u}_k,\mathbf{y}\rangle = 0$, $k = 1, 2, \ldots , n$; hence, $\mathbf{y} = \mathbf{0}$ by Exercise *C.5.11*. The reciprocal basis to $\mathbf{u}_1$, $\mathbf{u}_2$ is illustrated in Fig. *C.8.1*.

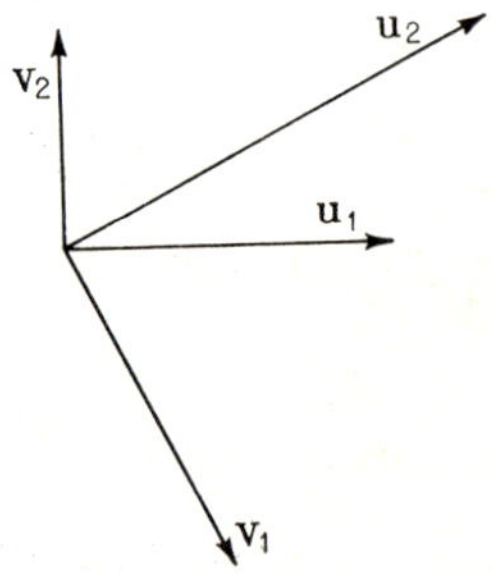

Fig. C.8.1 Relation between a basis $\{\mathbf{u}_1,\mathbf{u}_2\}$ for the plane and its reciprocal basis.

The main use of the reciprocal basis is the convenient expression for the components of any $\mathbf{x}$ in $\mathcal{C}^n$ relative to the basis $\{\mathbf{u}_1,\mathbf{u}_2, \ldots ,\mathbf{u}_n\}$:

3 $$\mathbf{x} = \sum_{i=1}^{n} \langle \mathbf{v}_i,\mathbf{x}\rangle \mathbf{u}_i \qquad \forall \mathbf{x} \in \mathcal{C}^n$$

4 *Exercise* Let $\{\mathbf{u}_1, \mathbf{u}_2, \ldots, \mathbf{u}_n\}$ be an arbitrary basis of $\mathcal{C}^n$. Let u_{ij}, $j = 1, 2, \ldots, n$, be the n components of $\mathbf{u}_i$. The matrix $\mathbf{U} = (u_{ij})$ has as its ith column the components of $\mathbf{u}_i$. Let $\mathbf{V}$ be the matrix whose jth row consists of the complex conjugate of the components of $\mathbf{v}_j$, where $\{\mathbf{v}_1, \ldots, \mathbf{v}_n\}$ is the reciprocal basis of $\{\mathbf{u}_1, \ldots, \mathbf{u}_n\}$. Show that

$$\mathbf{VU} = \mathbf{I} \qquad \text{or equivalently} \qquad \mathbf{V} = \mathbf{U}^{-1}$$

In other words, the computation of a reciprocal basis amounts to a matrix inversion.

5 *Exercise* Let $\{\mathbf{v}_1, \mathbf{v}_2, \ldots, \mathbf{v}\}$ be a basis for $\mathcal{C}^n$. Show that the n scalars $\langle \mathbf{v}_i, \mathbf{x} \rangle$ uniquely define $\mathbf{x}$. (*Hint:* Use *3*.)

9 *Linear transformation*

The principal reason for learning the basic facts and the terminology of linear vector spaces is to provide a framework to discuss the linear transformations, or linear operators as they are often called. We shall give a very general definition of linear transformations in order to emphasize the fact that they appear in almost all branches of engineering. Very soon, however, we shall restrict ourselves to linear transformations whose domain and whose range are an n-dimensional linear vector space.

1 **Definition** A *linear transformation* (or a linear operator) $\mathcal{L}$ is a function whose domain is a linear vector space $\mathfrak{X}$ and whose range is in a linear vector space $\mathcal{Y}$ such that for any $\mathbf{x}_1$ and $\mathbf{x}_2$ in $\mathfrak{X}$ and any scalar α

(I) $\qquad \mathbf{y}_1 = \mathcal{L}\mathbf{x}_1 \Rightarrow \alpha \mathbf{y}_1 = \mathcal{L}(\alpha \mathbf{x}_1)$

(II) $\qquad \mathbf{y}_i = \mathcal{L}\mathbf{x}_i, \; i = 1, 2 \Rightarrow \mathcal{L}(\mathbf{x}_1 + \mathbf{x}_2) = \mathbf{y}_1 + \mathbf{y}_2$

2 *Example* Let $\mathbf{x}$ be an n-rowed column vector whose elements are complex numbers and let $\mathcal{L}$ be represented by $\mathbf{A}$, an $m \times n$ matrix of complex numbers; $\mathbf{y}$ is the m-rowed column vector

$$\mathbf{Ax} = \mathbf{y}$$

Here $\mathfrak{X} = \mathcal{C}^n$ is the usual n-dimensional vector space and $\mathcal{Y}$ an m-dimensional vector space. If $m = n$, $\mathcal{Y}$ is usually identified with $\mathfrak{X}$. In many applications where $m < n$, $\mathcal{Y}$ is considered to be a subspace of $\mathfrak{X}$.

3 *Example* Let $\mathfrak{X}$ be the set of all bounded integrable functions over $(-\infty, +\infty)$; let $\mathcal{Y}$ be identical to $\mathfrak{X}$. The transformation is defined in terms of a function $h(t)$ such that $\int_{-\infty}^{+\infty} |h(\tau)| \, d\tau < \infty$:

$$y(t) = \int_{-\infty}^{+\infty} h(t - \tau) x(\tau) \, d\tau \qquad -\infty < t < \infty$$

x is the input and y is the zero-state response of a linear time-invariant system whose impulse response is $h(t)$.

4 *Example* $\mathcal{X}$ is the set of all functions having continuous derivatives in $(-\infty,+\infty)$. $\mathcal{Y}$ is the set of continuous functions in $(-\infty,\infty)$. The linear transformation is defined by

$$y = \frac{d}{dt} x$$

5 *Example* $\mathcal{X}$ is the set of all periodic functions with the same period $T = 2\pi/\omega_0$, with continuous derivative. Thus for any $x \in \mathcal{X}$,

$$x(t) = \sum_{-\infty}^{+\infty} \xi_n \exp(jn\omega_0 t) \qquad -\frac{\pi}{\omega_0} \leq t \leq \frac{\pi}{\omega_0}$$

where the ξ_n's are the components of x with respect to the orthogonal basis $\{\exp(jn\omega_0 t)\}_{-\infty}^{+\infty}$. Let y be the steady-state output of a given linear time-invariant system when x is the input; then, we know that (see *9.4.17*)

$$y(t) = \sum_{-\infty}^{+\infty} H(jn\omega_0)\xi_n \exp(jn\omega_0 t)$$

where $H(j\omega)$ is the transfer function of the system and H is assumed to be strictly stable. As x ranges over $\mathcal{X}$, y will range over a linear vector space $\mathcal{Y}$. Thus, the linear system defines a linear transformation $\mathfrak{N}$ on the inputs x to obtain the outputs y.

6 **Definition** The *range of a linear transformation* $\mathcal{A}$ is the set $\mathcal{R}(\mathcal{A})$ defined by

$$\mathcal{R}(\mathcal{A}) = \{\mathbf{y} \in \mathcal{Y} | \mathbf{y} = \mathcal{A}\mathbf{x} \text{ for some } \mathbf{x} \in \mathcal{X}\}$$

In other words, $\mathcal{R}(\mathcal{A})$ is the image of the vector space $\mathcal{X}$ under the linear transformation $\mathcal{A}$.

7 **Definition** The *null space of an L.T.*[1] $\mathcal{A}$ is the set $\mathfrak{N}(\mathcal{A})$ defined by

$$\mathfrak{N}(\mathcal{A}) = \{\mathbf{x} \in \mathcal{X} | \mathcal{A}\mathbf{x} = \mathbf{0}\}$$

That is, $\mathfrak{N}(\mathcal{A})$ is the set of all vectors of $\mathcal{X}$ that $\mathcal{A}$ maps into the zero vector of $\mathcal{Y}$.

8 **Remark** It is very important to realize that *in general* $\mathcal{R}(\mathcal{A})$ and $\mathfrak{N}(\mathcal{A})$ *may have nonzero vectors in common.* Take $\mathcal{C}^2$ and $\mathbf{A} = \begin{bmatrix} 0 & 1 \\ 0 & 0 \end{bmatrix}$: since $\mathbf{A}^2 = \begin{bmatrix} 0 & 0 \\ 0 & 0 \end{bmatrix}$, any vector $\mathbf{x} = (\alpha,0)$ is in both $\mathcal{R}(\mathbf{A})$ and $\mathfrak{N}(\mathbf{A})$. Indeed, $\mathbf{x}$ is the image of $\mathbf{z} = (0,\alpha)$ under $\mathbf{A}$ (that is, $\mathbf{x} = \mathbf{A}\mathbf{z}$) and $\mathbf{A}\mathbf{x} = \mathbf{0}$.

[1] L.T. is an abbreviation for linear transformation.

9 *Exercise* Show that $\mathcal{R}(\mathcal{A})$ and $\mathcal{N}(\mathcal{A})$ are linear vector spaces.

10 **Definition** The dimension of the subspace $\mathcal{R}(\mathcal{A})$ is called the *rank* of the linear transformation $\mathcal{A}$. The dimension of $\mathcal{N}(\mathcal{A})$ is called the *nullity* of the linear transformation $\mathcal{A}$.

10 *Representation of an linear transformation in* $\mathcal{C}^n$

Let $\mathcal{L}$ be an L.T. mapping of the n-dimensional space $\mathcal{C}^n$ into the m-dimensional space $\mathcal{C}^m$. We shall show that with respect to a specified basis in $\mathcal{C}^n$ and a specified basis in $\mathcal{C}^m$ the linear transformation $\mathcal{L}$ is represented by a matrix **L**. The matrix **L** representing the L.T. depends on the particular bases chosen. This dependence of the representation of the L.T. on the basis is analogous to the dependence of the components of a vector on the basis. For example, a vector **a** in a plane is represented by the ordered pair (0,1) with respect to a first basis and is represented by $(1/\sqrt{2}, 1/\sqrt{2})$ with respect to another basis.

Let $\{\mathbf{u}_1,\mathbf{u}_2, \ldots ,\mathbf{u}_n\}$ be a basis for $\mathcal{C}^n$ and $\{\mathbf{v}_1, \ldots ,\mathbf{v}_m\}$ be a basis for $\mathcal{C}^m$. Consider the arbitrary vector $\mathbf{x} \in \mathcal{C}^n$; then $\mathbf{x}$ has a unique representation in terms of the basis $\{\mathbf{u}_1, \mathbf{u}_2, \ldots ,\mathbf{u}_n\}$,

1 $$\mathbf{x} = \sum_{i=1}^{n} \xi_i \mathbf{u}_i$$

Consider the vector $\mathcal{L}\mathbf{u}_i$. It belongs to $\mathcal{C}^m$; hence, it has a unique representation in terms of the basis $\{\mathbf{v}_1,\mathbf{v}_2, \ldots ,\mathbf{v}_m\}$. Define the scalars λ_{ji} by

2 $$\mathcal{L}\mathbf{u}_i = \sum_{j=1}^{m} \lambda_{ji}\mathbf{v}_j \qquad i = 1, 2, \ldots , n$$

By direct application of the definition of an L.T.,

$$\mathbf{y} = \mathcal{L}\mathbf{x} = \mathcal{L}\Big(\sum_{i=1}^{n} \xi_i\mathbf{u}_i\Big) = \sum_{i=1}^{n} \xi_i\mathcal{L}\mathbf{u}_i = \sum_{i=1}^{n} \xi_i \Big(\sum_{j=1}^{m} \lambda_{ji}\mathbf{v}_j\Big)$$

where we have used *1*, the properties (I) and (II) of *C.9.1*, and *2*. Finally,

$$\mathbf{y} = \sum_{j=1}^{m} \Big(\sum_{i=1}^{n} \lambda_{ji}\xi_i\Big) \mathbf{v}_j$$

Thus the components η_j of **y** relative to the basis $\mathbf{v}_1, \mathbf{v}_2, \ldots , \mathbf{v}_m$ of $\mathcal{C}^m$ are obtained in terms of the components of **x** relative to the basis

$\mathbf{u}_1, \ldots, \mathbf{u}_n$ of $\mathcal{C}^n$ by the operation

3 $$\eta_j = \sum_{i=1}^{n} \lambda_{ji}\xi_i$$

Thus we are led to state the

4 **Theorem** Given a basis $\{\mathbf{u}_1,\mathbf{u}_2, \ldots, \mathbf{u}_n\}$ of $\mathcal{C}^n$ and a basis $\{\mathbf{v}_1, \ldots, \mathbf{v}_m\}$ of $\mathcal{C}^m$ the L.T. $\mathcal{L}$ mapping $\mathcal{C}^n$ into $\mathcal{C}^m$ is represented by a matrix $\mathbf{L} = (\lambda_{ji})$ whose ith column consists of the components of $\mathcal{L}\mathbf{u}_i$ relative to the basis $\{\mathbf{v}_1, \ldots, \mathbf{v}_m\}$. ◁

Frequently we shall drop the distinction between the L.T. $\mathcal{L}$ and its matrix representation $\mathbf{L}$ when we are studying a specific problem in which the bases are fixed once and for all throughout the problem.

5 *Exercise* The purpose of this exercise is to show that the definitions of matrix addition and matrix multiplication follow naturally from the consideration of addition and composition of L.T. Let $\mathcal{C}^n$ be a vector space with basis $\{\mathbf{u}_1,\mathbf{u}_2, \ldots, \mathbf{u}_n\}$. Let $\mathcal{A}$ and $\mathcal{B}$ be two L.T.'s on $\mathcal{C}^n$ into itself. *The sum* $\mathcal{A} + \mathcal{B}$ is defined as the L.T. which maps any vector $\mathbf{x} \in \mathcal{C}^n$ into the vector $\mathcal{A}\mathbf{x} + \mathcal{B}\mathbf{x}$. *The product* $\mathcal{A}\mathcal{B}$ is the L.T. which maps the vector $\mathbf{x} \in \mathcal{C}^n$ into the vector $\mathbf{z} \in \mathcal{C}^n$, where $\mathbf{z}$ is defined as follows: let $\mathbf{y} = \mathcal{B}\mathbf{x}$; then $\mathbf{z} = \mathcal{A}\mathbf{y}$. This may be written as $\mathbf{z} = \mathcal{A}(\mathcal{B}\mathbf{x})$. Show that if, with respect to the given basis of $\mathcal{C}^n$, the L.T. $\mathcal{A}$ and $\mathcal{B}$ are respectively represented by the matrices $\mathbf{A} = (\alpha_{ij})$ and $\mathbf{B} = (\beta_{ij})$, then

(I) $\mathcal{A} + \mathcal{B}$ is represented by the matrix $\mathbf{A} + \mathbf{B} = (\alpha_{ij} + \beta_{ij})$.

(II) $\mathcal{A}\mathcal{B}$ is represented by the matrix $\mathbf{AB} = \left(\sum_{k=1}^{n} \alpha_{ik}\beta_{kj}\right)$.

6 *Exercise* Using the notation of Exercise *5*, construct an example in $\mathcal{C}^2$ showing that

(I) $\mathbf{AB} \neq \mathbf{BA}$.

(II) $\mathbf{CD} = \mathbf{0}$, where $\mathbf{C} \neq \mathbf{0}$, $\mathbf{D} \neq \mathbf{0}$, and $\mathbf{0}$ is the zero matrix, namely, the matrix all of whose elements are zero.

7 **Remark** (I) of Exercise *6* shows that matrix multiplication is not commutative; (II) shows that in matrix algebra there are divisors of zero and consequently the cancellation law of ordinary algebra does not apply to matrices. In other words, $\mathbf{AB} = \mathbf{AC}$ and $\mathbf{A} \neq \mathbf{0}$ do not imply $\mathbf{B} = \mathbf{C}$.

8 *Exercise* Let $\mathbf{A} = \begin{bmatrix} 0 & 1 & 0 \\ 0 & 0 & 1 \\ 0 & 0 & 0 \end{bmatrix}$. Show that $\mathbf{A}^3$ is the zero matrix. (Again, $\mathbf{A}$ is a nonzero matrix and $\mathbf{A} \cdot \mathbf{A} \cdot \mathbf{A}$ is the zero matrix.)

9 **Theorem** Let $\mathcal{A}$ be an L.T. of $\mathcal{C}^n$ into $\mathcal{C}^n$. Let the $n \times n$ matrix $\mathbf{A}$ be the representation of $\mathcal{A}$ with respect to some basis of $\mathcal{C}^n$. Then

$$\text{rank of } \mathcal{A} \triangleq \dim \mathcal{R}(\mathcal{A}) = \text{rank } (\mathbf{A})$$

That is, the rank of the linear transformation $\mathcal{A}$ is the rank of the matrix **A** which represents the linear transformation with respect to some basis.
Proof This assertion follows directly from Theorem *4*, the definition of the rank of $\mathcal{A}$, and standard properties of determinants. It is therefore left as an exercise for the reader.

11 *Matrix representation of an L.T. and changes of bases*

Let, for simplicity, $\mathcal{L}$ be an L.T. of $\mathcal{C}^n$ into $\mathcal{C}^n$. Let $\{\mathbf{u}_1,\mathbf{u}_2, \ldots ,\mathbf{u}_n\}$ be a basis of $\mathcal{C}^n$ and $\mathbf{L} = (\lambda_{ij})$ be the corresponding representation of $\mathcal{L}$; that is, from *C.10.2*,

1
$$\mathcal{L}\mathbf{u}_i = \sum_{j=1}^{n} \lambda_{ji}\mathbf{u}_j$$

Let $\{\hat{\mathbf{u}}_1,\hat{\mathbf{u}}_2, \ldots ,\hat{\mathbf{u}}_n\}$ be a new basis of $\mathcal{C}^n$. The problem is to find $\hat{\mathbf{L}}$, the matrix representation of $\mathcal{L}$ with respect to the new basis.

Expand the new basis vectors in terms of the old:

2
$$\hat{\mathbf{u}}_i = \sum_{\alpha=1}^{n} c_{\alpha i}\mathbf{u}_\alpha$$

That is, the ith column of the matrix $(c_{\alpha i})$ consists of the components of $\hat{\mathbf{u}}_i$ relative to the old basis.

Now consider an arbitrary vector $\mathbf{x} \in \mathcal{C}^n$. The components of $\mathbf{x}$ are defined as

$$\mathbf{x} = \sum_{\alpha} \xi_\alpha \mathbf{u}_\alpha = \sum_{\beta} \hat{\xi}_\beta \hat{\mathbf{u}}_\beta$$

Hence, from *2*,

$$\mathbf{x} = \sum_{\beta} \hat{\xi}_\beta \sum_{\alpha} c_{\alpha\beta}\mathbf{u}_\alpha = \sum_{\alpha} \Big(\sum_{\beta} c_{\alpha\beta}\hat{\xi}_\beta\Big) \mathbf{u}_\alpha$$

and by the uniqueness of the representation (see *C.4.4*)

$$\xi_\alpha = \sum_{\beta=1}^{n} c_{\alpha\beta}\hat{\xi}_\beta \qquad \alpha = 1, 2, \ldots , n$$

Hence, we conclude:

3 The column vector of the old components (ξ_α) of $\mathbf{x}$ is the product of the matrix $\mathbf{C} = (c_{\alpha\beta})$ and the column vector of the new components $(\hat{\xi}_\beta)$ of $\mathbf{x}$, where $\mathbf{C}$ is the matrix defined in *2*.

Let us now compute the relation between $\mathbf{L}$ and $\hat{\mathbf{L}}$. From the definition of $\hat{\mathbf{L}}$ (by analogy with *1*) we get, using *2*,

4
$$\mathcal{L}\hat{\mathbf{u}}_\alpha = \sum_{\beta} \hat{\lambda}_{\beta\alpha}\hat{\mathbf{u}}_\beta = \sum_{\beta} \hat{\lambda}_{\beta\alpha} \Big(\sum_{\gamma} c_{\gamma\beta}\mathbf{u}_\gamma\Big)$$
$$= \sum_{\gamma} \Big(\sum_{\beta} \hat{\lambda}_{\beta\alpha}c_{\gamma\beta}\Big) \mathbf{u}_\gamma$$

Using *2*, rewrite $\mathcal{L}\hat{\mathbf{u}}_\alpha$ in a different way and use *1* to get

5
$$\mathcal{L}\hat{\mathbf{u}}_\alpha = \mathcal{L}\left(\sum_\eta c_{\eta\alpha}\mathbf{u}_\eta\right) = \sum_\eta c_{\eta\alpha}\left(\sum_\gamma \lambda_{\gamma\eta}\mathbf{u}_\gamma\right) = \sum_\gamma \left(\sum_\eta c_{\eta\alpha}\lambda_{\gamma\eta}\right)\mathbf{u}_\gamma$$

Comparing *4* and *5*,

$$\sum_\eta c_{\eta\alpha}\lambda_{\gamma\eta} = \sum_\beta \hat{\lambda}_{\beta\alpha}c_{\gamma\beta} \qquad \alpha,\gamma = 1, 2, \ldots, n$$

Let $\mathbf{C}$ be the matrix $(c_{\alpha\beta})$; then

$$\mathbf{LC} = \mathbf{C\hat{L}}$$

These results can be condensed in the following

6 **Theorem** Let $\mathcal{L}$ be a linear transformation of $\mathcal{C}^n$ into $\mathcal{C}^n$. Let $\{\mathbf{u}_1, \ldots, \mathbf{u}_n\}$ and $\{\hat{\mathbf{u}}_1, \ldots, \hat{\mathbf{u}}_n\}$ be respectively the old and the new bases of $\mathcal{C}^n$. Let $\mathbf{C}$ be the matrix $(c_{\alpha\beta})$ expressing the new basis vectors in terms of the old ones:

$$\hat{\mathbf{u}}_i = \sum_{k=1}^{n} c_{ki}\mathbf{u}_k$$

If $\mathbf{L}$ and $\hat{\mathbf{L}}$ are the matrix representations of $\mathcal{L}$ with respect to the corresponding bases, then

7
$$\hat{\mathbf{L}} = \mathbf{C}^{-1}\mathbf{LC}$$

If $\mathbf{x} \in \mathcal{C}^n$ and if

8
$$\mathbf{x} = \sum_{k=1}^{n} \xi_k\mathbf{u}_k = \sum_{i=1}^{n} \hat{\xi}_i\hat{\mathbf{u}}_i$$

then

9
$$\xi_k = \sum_{j=1}^{n} c_{kj}\hat{\xi}_j \qquad k = 1, 2, \ldots, n$$

10 *Exercise* Show that $\mathbf{C}$ is nonsingular.

12 *Direct sums and projections*

Let us recall (see *C.5.12*) that a *subspace* $\mathcal{B}$ of a finite dimensional space $\mathcal{C}^n$ is a set of vectors in $\mathcal{C}^n$ such that if $\mathbf{x} \in \mathcal{B}$ and $\mathbf{y} \in \mathcal{B}$, then for all α's and β's, $\alpha\mathbf{x} + \beta\mathbf{y} \in \mathcal{B}$. Also recall that (see *C.7.7*) the space $\mathcal{C}^n$ is said to be the *direct sum* of two subspaces $\mathfrak{M}_1$ and $\mathfrak{M}_2$ if $\mathfrak{M}_1$ and $\mathfrak{M}_2$ are subspaces of $\mathcal{C}^n$ and if every $\mathbf{x} \in \mathcal{C}^n$ has one and only one representation as the sum of a vector in $\mathfrak{M}_1$ and a vector in $\mathfrak{M}_2$. This is written as

1
$$\mathcal{C}^n = \mathfrak{M}_1 \oplus \mathfrak{M}_2$$

From the definition we have for every **x** a unique representation of the form

2 $$\mathbf{x} = \mathbf{x}_1 + \mathbf{x}_2 \qquad \text{with } \mathbf{x}_i \in \mathfrak{M}_i \quad i = 1, 2$$

3 **Theorem** $\mathcal{C}^n = \mathfrak{M}_1 \oplus \mathfrak{M}_2 \Leftrightarrow \mathfrak{M}_1$ and $\mathfrak{M}_2$ have only the zero vector in common and any vector **x** in $\mathcal{C}^n$ can be represented as the sum of a vector in $\mathfrak{M}_1$ and a vector in $\mathfrak{M}_2$.

Proof Exercise for the reader.

4 **Definition** Let $\mathcal{C}^n = \mathfrak{M}_1 \oplus \mathfrak{M}_2$ and, $\forall \mathbf{x} \in \mathcal{C}^n$, let $\mathbf{x} = \mathbf{x}_1 + \mathbf{x}_2$, where $\mathbf{x}_i \in \mathfrak{M}_i$, $i = 1, 2$; then the transformation **E** such that

5 $$\mathbf{E}\mathbf{x} = \mathbf{x}_1$$

is called the *projection on* $\mathfrak{M}_1$ *along* $\mathfrak{M}_2$.

6 *Exercise* Verify that **E** is a *linear* transformation.

7 *Comment* It is important to be aware of two subtle points: (I) the projection just defined is not necessarily an orthogonal projection as was the case in Sec. *7*; (II) in order to completely define the projection, we must say the projection on $\mathfrak{M}_1$ *along* $\mathfrak{M}_2$. This point is illustrated in Fig. *C.12.1a*, where $\mathcal{C}^n$ is the plane and $\mathfrak{M}_1$ is the ξ axis. Figure *C.12.1a* shows a possible choice for $\mathfrak{M}_2$ and Fig. *C.12.1b* shows another. Note the difference between the projections of **x** on $\mathfrak{M}_1$ in the two cases.

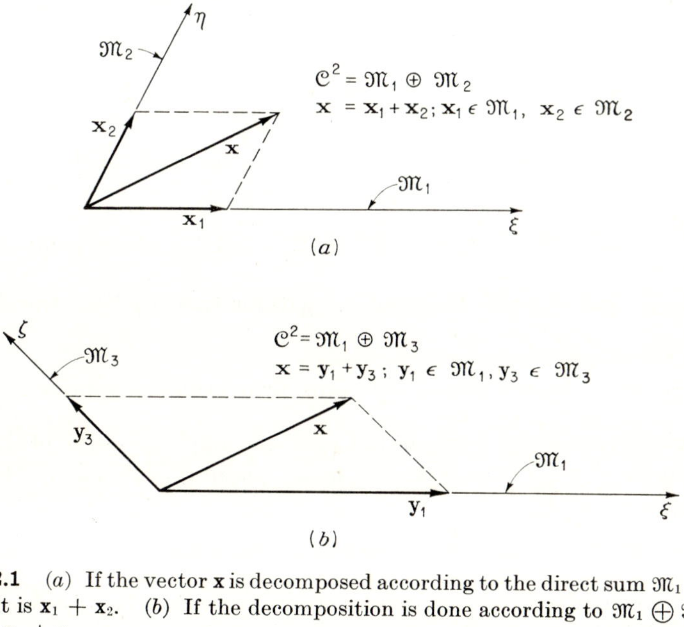

Fig. C.12.1 (*a*) If the vector **x** is decomposed according to the direct sum $\mathfrak{M}_1 \oplus \mathfrak{M}_2$, the result is $\mathbf{x}_1 + \mathbf{x}_2$. (*b*) If the decomposition is done according to $\mathfrak{M}_1 \oplus \mathfrak{M}_3$, the result is $\mathbf{y}_1 + \mathbf{y}_3$.

8 *Exercise* Let $\mathbf{E}$ be the projection on $\mathfrak{M}_1$ along $\mathfrak{M}_2$. Show that

$$\mathfrak{M}_1 = \mathcal{R}(\mathbf{E}) \triangleq \{\mathbf{x}|\mathbf{Ex} = \mathbf{x}\}$$
$$\mathfrak{M}_2 = \mathfrak{N}(\mathbf{E}) \triangleq \{\mathbf{y}|\mathbf{Ey} = \mathbf{0}\}$$

9 **Theorem** The L.T. $\mathbf{E}$ is the projection on $\mathfrak{M}_1$ along $\mathfrak{M}_2$ if and only if

10 $$\mathbf{E} = \mathbf{E}^2$$

where $\mathfrak{M}_1 = \mathcal{R}(\mathbf{E})$ and $\mathfrak{M}_2 = \mathfrak{N}(\mathbf{E})$.

Proof $\Rightarrow$ By Definition *4*, $\mathcal{C}^n = \mathfrak{M}_1 \oplus \mathfrak{M}_2$. Therefore, for any $\mathbf{x}$ in $\mathcal{C}^n$ there is a unique decomposition $\mathbf{x} = \mathbf{x}_1 + \mathbf{x}_2$, with $\mathbf{x}_i \in \mathfrak{M}_i$. The decomposition of $\mathbf{x}_1$ is $\mathbf{x}_1 + \mathbf{0}$. Therefore, by the definition of $\mathbf{E}$,

$$\mathbf{E}^2\mathbf{x} = \mathbf{E}(\mathbf{Ex}) = \mathbf{Ex}_1 = \mathbf{x}_1 = \mathbf{Ex} \qquad \forall \mathbf{x} \in \mathcal{C}^n$$

and *10* follows.

$\Leftarrow$ We have the algebraic fact that

$$\mathbf{x} = \mathbf{Ex} + (\mathbf{I} - \mathbf{E})\mathbf{x}$$

Clearly $\mathbf{Ex} \in \mathfrak{M}_1$. We assert that $(\mathbf{I} - \mathbf{E})\mathbf{x} \in \mathfrak{M}_2$; indeed, the assumption $\mathbf{E} = \mathbf{E}^2$ implies that $\mathbf{E}(\mathbf{I} - \mathbf{E})\mathbf{x} = (\mathbf{E} - \mathbf{E}^2)\mathbf{x} = \mathbf{0}$, that is, $(\mathbf{I} - \mathbf{E})\mathbf{x} \in \mathfrak{N}(\mathbf{E}) \triangleq \mathfrak{M}_2$. Thus any $\mathbf{x}$ in $\mathcal{C}^n$ can be written as a sum of a vector in $\mathfrak{M}_1$ and a vector in $\mathfrak{M}_2$. By *3* it remains to show that $\mathfrak{M}_1$ and $\mathfrak{M}_2$ have only the vector $\mathbf{0}$ in common. Suppose $\mathbf{y} \in \mathfrak{M}_1$ and $\mathbf{y} \in \mathfrak{M}_2$. Then, by *8*, $\mathbf{Ey} = \mathbf{y}$ and $\mathbf{Ey} = \mathbf{0}$; hence, $\mathbf{y} = \mathbf{0}$. Therefore,

$$\mathfrak{M}_1 \oplus \mathfrak{M}_2 = \mathcal{C}^n$$

and $\mathbf{E}$ is the projection on $\mathfrak{M}_1$ along $\mathfrak{M}_2$.

11 **Theorem** Let $\mathbf{E}$ be the projection on $\mathfrak{M}_1$ along $\mathfrak{M}_2$ where $\mathfrak{M}_1 \oplus \mathfrak{M}_2 = \mathcal{C}^n$. *E* is an *orthogonal* projection if and only if

$$\mathbf{E}^2 = \mathbf{E} \qquad \text{and} \qquad \mathbf{E}^* = \mathbf{E} \qquad (\mathbf{E}^* \text{ is the adjoint of } \mathbf{E})$$

Proof Exercise for the reader. (Refer to *C.14.1* for the definition of $\mathbf{E}^*$.)

A useful representation of some projections is obtained by the *dyad.*

12 **Definition** A *dyad* is an L.T., or an algebraic sum of L.T.[1] of the form $\mathbf{A} = \mathbf{a}\rangle\langle\mathbf{b}$ (where $\mathbf{a}$ and $\mathbf{b}$ are vectors), defined by the following relation:

13 $$\mathbf{Ax} = \langle\mathbf{b},\mathbf{x}\rangle\mathbf{a} \qquad \text{for all } \mathbf{x} \in \mathcal{C}^n$$

[1] In cases where the dyad consists of only one such term, it is sometimes referred to as a "one-term dyad."

In other words, **A** maps any vector **x** into a vector proportional to **a**; clearly, then, $\mathcal{R}(\mathbf{a}\rangle\langle\mathbf{b}) = \mathcal{S}[\mathbf{a}]$. The rank of $\mathbf{a}\rangle\langle\mathbf{b}$ is unity and its nullity is $n - 1$ because $\mathfrak{N}(\mathbf{a}\rangle\langle\mathbf{b})$ is the hyperplane orthogonal to **b**, that is, $\{\mathbf{x}|\langle\mathbf{b},\mathbf{x}\rangle = 0\}$.

14 *Exercise* Establish that the matrix representation of a one-term dyad $\mathbf{a}\rangle\langle\mathbf{b}$ may be described as follows. Consider an arbitrary orthonormal basis $\mathcal{B}$: let (α_i) and (β_j) be the components of **a** and **b** with respect to the basis $\mathcal{B}$; then the (i,k) element of the matrix representing $\mathbf{a}\rangle\langle\mathbf{b}$ is $\alpha_i\beta_k^*$.

15 *Exercise* Consider two dyads $\mathbf{A} = \mathbf{a}\rangle\langle\mathbf{b}$, $\mathbf{B} = \mathbf{c}\rangle\langle\mathbf{d}$. Show that $\mathbf{AB} = \langle\mathbf{b},\mathbf{c}\rangle\,\mathbf{C}$, where $\mathbf{C} = \mathbf{a}\rangle\langle\mathbf{d}$.

16 **Theorem** Let $\mathcal{B} = \{\mathbf{u}_1,\mathbf{u}_2, \ldots ,\mathbf{u}_n\}$ be a basis and $\{\mathbf{v}_1, \ldots ,\mathbf{v}_n\}$ be the reciprocal basis of $\mathcal{B}$. For $i = 1, 2, \ldots , n$ let

$$\mathfrak{M}_i = \mathcal{S}[\mathbf{u}_i] \qquad \text{and} \qquad \mathfrak{M}_i' = \mathcal{S}[\mathbf{u}_1,\mathbf{u}_2, \ldots ,\mathbf{u}_{i-1},\mathbf{u}_{i+1}, \ldots ,\mathbf{u}_n]$$

Then

$$\mathbf{E}_i = \mathbf{u}_i\rangle\langle\mathbf{v}_i \qquad i = 1, 2, \ldots , n$$

is the projection on $\mathfrak{M}_i$ along $\mathfrak{M}_i'$.

Proof By direct computation, using Exercise *15* and $\langle\mathbf{u}_i,\mathbf{v}_i\rangle = 1$, we get $\mathbf{E}_i^2 = \mathbf{E}_i$; hence, $\mathbf{E}_i$ is a projection. We also have $\mathcal{R}(\mathbf{E}_i) = \mathcal{S}[\mathbf{u}_i] = \mathfrak{M}_i$ and $\mathfrak{N}(\mathbf{E}_i) = \{\mathbf{y}|\langle\mathbf{v}_i,\mathbf{y}\rangle = 0\}$. Now, from the definition of the reciprocal basis, that set is spanned by $\mathbf{u}_1, \ldots , \mathbf{u}_{i-1}, \mathbf{u}_{i+1}, \ldots , \mathbf{u}_n$; hence $\mathfrak{N}(\mathbf{E}_i) = \mathfrak{M}_i'$. Therefore, $\mathbf{E}_i$ is the projection on $\mathfrak{M}_i$ along $\mathfrak{M}_i'$.

17 **Corollary** With the same notation as above, with **I** being the identity transformation

18
$$\mathbf{I} = \sum_{i=1}^{n} \mathbf{u}_i\rangle\langle\mathbf{v}_i$$

and, for all $\mathbf{x} \in \mathcal{C}^n$,

19
$$\mathbf{x} = \Big(\sum_{i=1}^{n} \mathbf{u}_i\rangle\langle\mathbf{v}_i\Big)\,\mathbf{x} = \sum_{i=1}^{n} \langle\mathbf{v}_i,\mathbf{x}\rangle\mathbf{u}_i$$

Proof Exercise for the reader.

20 **Corollary** The projections $\mathbf{E}_i$ defined in Theorem *16* are such that

$$\mathbf{E}_i\mathbf{E}_k = \delta_{ik}\mathbf{E}_i \qquad i,k = 1, 2, \ldots , n$$

$$\sum_{i=1}^{n} \mathbf{E}_i = \mathbf{I}$$

21 **Corollary** Let $\mathbf{A} = (a_{ij})$ be the matrix representation of an L.T. $\mathcal{A}$ relative to the basis $\mathcal{B} \triangleq \{\mathbf{u}_1,\mathbf{u}_2, \ldots ,\mathbf{u}_n\}$. Let $\mathbf{v}_1, \ldots , \mathbf{v}_n$ be the reciprocal basis to $\mathcal{B}$. Then a dyad representation of $\mathcal{A}$ is

22
$$\sum_{i=1}^{n} \Big(\mathbf{u}_i\rangle\langle\sum_{j=1}^{n} a_{ij}^*\mathbf{v}_j\Big)$$

Proof Clearly *22* is a dyad representation, Call it $\mathbf{B}$. It remains to show that $\mathbf{B}$ represents the L.T. $\mathcal{A}$. In particular, we need only show that

$$\mathbf{B}\mathbf{u}_k = \mathbf{A}\mathbf{u}_k \qquad k = 1, 2, \ldots, n$$

By *22*, and since $\langle \mathbf{v}_j, \mathbf{u}_k \rangle = \delta_{jk}$ by definition of a reciprocal basis, we have

$$\mathbf{B}\mathbf{u}_k = \sum_{i=1}^{n} a_{ik}\mathbf{u}_i \qquad k = 1, 2, \ldots, n$$

That is, $\mathbf{B}\mathbf{u}_k$ has $(a_{1k}, a_{2k}, \ldots, a_{nk})$ as components with respect to the basis $\mathcal{B}$.

By *C.10.4* we also have

$$\mathbf{A}\mathbf{u}_k = \sum_{i=1}^{n} a_{ik}\mathbf{u}_i \qquad k = 1, 2, \ldots, n$$

Hence the corollary is established.

13 *Invariant subspaces*

1 **Definition** $\mathfrak{M}$ is said to be an *invariant subspace under the L.T.* $\mathbf{A}$ if $\mathbf{x} \in \mathfrak{M} \Rightarrow \mathbf{A}\mathbf{x} \in \mathfrak{M}$.

2 *Exercise* Show that the sets $\mathcal{C}^n$, $\mathcal{R}(\mathbf{A})$, $\mathfrak{N}(\mathbf{A})$, $\left\{\mathbf{x} \middle| \sum_{k=0}^{N} \alpha_k \mathbf{A}^k \mathbf{x} = \mathbf{0}\text{, where the } \alpha_k\text{'s are numbers and } N \text{ is an arbitrary integer}\right\}$ are invariant subspaces under $\mathbf{A}$.

3 **Theorem** Let $\mathcal{C}^n = \mathfrak{M} \oplus \mathfrak{N}$. Let $\mathfrak{M}$ be invariant under $\mathbf{A}$ and have dimension k; then the L.T. $\mathbf{A}$ has a matrix representation of the form

$$4 \qquad \begin{matrix} & \overbrace{}^{k} & \\ k\{ & \left[\begin{array}{c|c} \mathbf{P} & \mathbf{R} \\ \hline \mathbf{0} & \mathbf{Q} \end{array}\right] & \}\, n-k \\ & & \underbrace{}_{n-k} \end{matrix}$$

provided a suitable basis of $\mathcal{C}^n$ is used.

Proof Let $\{\mathbf{u}_1, \mathbf{u}_2, \ldots, \mathbf{u}_k\}$ be a basis for $\mathfrak{M}$ and $\{\mathbf{u}_{k+1}, \mathbf{u}_{k+2}, \ldots, \mathbf{u}_n\}$ be a basis for $\mathfrak{N}$. Then $\{\mathbf{u}_1, \mathbf{u}_2, \ldots, \mathbf{u}_n\}$ is a basis for $\mathcal{C}^n$. The zero submatrix in *4* follows immediately from the fact that for any

$$\mathbf{x} \in \mathfrak{M} \rightarrow \mathbf{x} = \sum_{1}^{k} \xi_i \mathbf{u}_i$$

for some ξ_i's and $\mathbf{Ax} = \sum_1^k \eta_i \mathbf{u}_i$, since $\mathfrak{M}$ is invariant under **A**. (The η_i's are appropriate linear combinations of the ξ_k's.)

5 **Corollary** If $\mathcal{C}^n = \mathfrak{M} \oplus \mathfrak{N}$ and both $\mathfrak{M}$ and $\mathfrak{N}$ are invariant under **A**, then the L.T. **A** has a representation of the form

$$\left[\begin{array}{c|c} \mathbf{P} & \mathbf{0} \\ \hline \mathbf{0} & \mathbf{Q} \end{array}\right]$$

6 *Comment* These two results form the basis for the diagonalization of matrices. (See also *D.5.21*.)

14 *Adjoint transformation*

The concept of the adjoint of an L.T. plays a very important role in many problems. It is used throughout the text and is basic to the developments of the rest of this appendix. Section *20* is, in fact, devoted to the review of those problems in which the adjoint is important.

1 **Definition** Let **A** be an L.T. on $\mathcal{C}^n$.[1] The adjoint transformation $\mathbf{A}^*$ of **A** is a transformation such that

2 $$\langle \mathbf{Ax},\mathbf{y} \rangle = \langle \mathbf{x},\mathbf{A}^*\mathbf{y} \rangle \qquad \forall \mathbf{x},\mathbf{y} \in \mathcal{C}^n$$

3 *Exercise* Verify that the transformation $\mathbf{A}^*$ defined by *2* is linear. [*Hint:* Consider $\langle \mathbf{x},\mathbf{A}(\mathbf{y}_1 + \mathbf{y}_2) \rangle$ $\forall \mathbf{x},\mathbf{y}_1,\mathbf{y}_2$; use *1* and the linearity of the scalar product (see *C.5.1*).]

4 *Exercise* Show that $(\mathbf{AB})^* = \mathbf{B}^*\mathbf{A}^*$; $(\mathbf{A} + \mathbf{B})^* = \mathbf{A}^* + \mathbf{B}^*$; $(\alpha\mathbf{A})^* = \alpha^*\mathbf{A}^*$, where α is any scalar.

5 **Theorem** $(\mathbf{A}^*)^* = \mathbf{A}$.

Proof Interchange the order of the factors in the scalar products of *2*:

$$\langle \mathbf{y},\mathbf{Ax} \rangle = \langle \mathbf{A}^*\mathbf{y},\mathbf{x} \rangle \qquad \forall \mathbf{x},\mathbf{y} \in \mathcal{C}^n$$

Apply the property defined by *2* to the right-hand side of the above relation and get

$$\langle \mathbf{y},\mathbf{Ax} \rangle = \langle \mathbf{y},(\mathbf{A}^*)^*\mathbf{x} \rangle \qquad \forall \mathbf{x},\mathbf{y} \in \mathcal{C}^n$$

Hence

$$\langle \mathbf{y},[\mathbf{A} - (\mathbf{A}^*)^*]\mathbf{x} \rangle = 0 \qquad \forall \mathbf{x},\mathbf{y} \in \mathcal{C}^n$$

[1] Strictly speaking, we should use the notation $\mathcal{A}$ for the L.T. and **A** for its matrix representation with respect to some basis. However, to adhere strictly to this notation becomes very cumbersome. In the following we shall often use the same notation for an L.T. and its matrix representation.

Since this equality holds for all **y**'s and **x**'s, it holds in particular if **y** is successively taken to be the vectors of a basis of $\mathcal{C}^n$. Consequently, by *C.5.11*,

$$[\mathbf{A} - (\mathbf{A}^*)^*]\mathbf{x} = \mathbf{0} \qquad \forall \mathbf{x} \in \mathcal{C}^n$$

Therefore, the L.T. $\mathbf{A} - (\mathbf{A}^*)^*$ is the L.T. that maps any **x** into **0**; in other words, $\mathbf{A} = (\mathbf{A}^*)^*$.

6 **Theorem** Let $\mathfrak{A}$ be an L.T. in $\mathcal{C}^n$ and **A** its matrix representation relative to an orthonormal basis. The matrix representation of the adjoint transformation $\mathfrak{A}^*$ (relative to the same basis) is the conjugate transpose of the matrix **A**.

Proof By definition, $\langle \mathfrak{A}\mathbf{x},\mathbf{y}\rangle = \langle \mathbf{x},\mathfrak{A}^*\mathbf{y}\rangle$ for all $\mathbf{x},\mathbf{y} \in \mathcal{C}^n$. Let $\mathbf{u}_1, \ldots, \mathbf{u}_n$ be the orthonormal basis. Let $\mathbf{x} = \sum_j \xi_j \mathbf{u}_j$ and $\mathbf{y} = \sum_i \eta_i \mathbf{u}_i$. Then, by the linearity of the scalar product, we have

$$\left\langle \sum_j \mathfrak{A}\xi_j\mathbf{u}_j, \sum_i \eta_i\mathbf{u}_i \right\rangle = \sum_{ij} \xi_j^*\eta_i \langle \mathfrak{A}\mathbf{u}_j,\mathbf{u}_i\rangle$$

$$\left\langle \sum_j \xi_j\mathbf{u}_j, \sum_i \mathfrak{A}^*\eta_i\mathbf{u}_i \right\rangle = \sum_{i,j} \xi_j^*\eta_i \langle \mathbf{u}_j,\mathfrak{A}^*\mathbf{u}_i\rangle$$

From the definition of $\mathfrak{A}^*$, the right-hand sides of these two equations are equal for all possible choices of ξ_j and η_i. Hence

$$\langle \mathbf{u}_j,\mathfrak{A}^*\mathbf{u}_i\rangle = \langle \mathbf{u}_i,\mathfrak{A}\mathbf{u}_j\rangle^* \qquad i,j = 1, 2, \ldots, n$$

Since the $\mathbf{u}_i$'s are orthonormal, the basis coincides with its reciprocal basis and the left-hand side is the jth component of $\mathfrak{A}^*\mathbf{u}_i$, that is, it is the element of the jth row and ith column of the matrix $\mathbf{A}^*$. The right-hand side is the complex conjugate of the (i,j) element of the matrix **A**. Therefore, if $\mathbf{A} = (a_{ij})$ and $\mathbf{A}^* = (b_{ij})$,

$$b_{ij} = a_{ji}^* \qquad i,j = 1, 2, \ldots, n$$

7 **Definition** Let $\mathfrak{A}$ be an L.T. on $\mathcal{C}^n$. If $\mathfrak{A} = \mathfrak{A}^*$, then $\mathfrak{A}$ is said to be *self-adjoint.*

8 *Comment* Let **A** be the matrix representing the self-adjoint L.T. $\mathfrak{A}$ with respect to an orthonormal basis. Then, by *6*, $a_{ij} = a_{ji}^*$ for

$$i,j = 1, 2, \ldots, n$$

The matrix **A** is said to be *hermitian.*

9 *Exercise* Show that if **A** is self-adjoint, $\langle \mathbf{x},\mathbf{A}\mathbf{x}\rangle$ is real, $\forall \mathbf{x} \in \mathcal{C}^n$.

10 *Exercise* Let **A** be an arbitrary $n \times n$ matrix. Show that $\mathbf{A}^*\mathbf{A}$ is hermitian. Show that it is a nonnegative definite hermitian matrix. (A hermitian matrix **H** is said to be nonnegative definite if $\langle \mathbf{x},\mathbf{H}\mathbf{x}\rangle \geq 0$, $\forall \mathbf{x} \in \mathcal{C}^n$.)

15 *Systems of linear equations*

Consider the system of linear equations

1 $$\sum_{j=1}^{n} a_{ij}x_j = b_i \qquad i = 1, 2, \ldots, n$$

where the matrix $\mathbf{A} = (a_{ij})$ and the column vector $\mathbf{b}$ are given.[1] The problem is to settle the following questions: When does the system have a solution, and if it has a solution, is it unique? If it is not, how does one obtain all of its solutions? To settle these questions, it is expedient to first consider a theorem listing relevant properties of $\mathbf{A}$ and then deduce their consequences relative to the system *1*.

The reader might be astonished that we shall not go into the usual long-detailed study of determinants, minors, etc. The reason is that in the solution of systems of equations involving more than, say, five equations, determinants are very seldom used. In fact, their principal use is as a tool to prove theorems. The main drawback of determinants is that the determinant of an $n \times n$ matrix consists of an algebraic sum of $n!$ terms and that each term is a product of n elements of the determinant. If $n = 20$, this is 2.4329×10^{18} terms, and if the computation of one term requires one microsecond, the $n!$ terms require about 10^5 years! For this reason modern computer programs solve systems of linear equations either by the Gauss elimination procedure or by iteration.

To proceed further, we must recall the

2 **Definition** The *rank of an L.T.* is the dimension of its range space $\mathcal{R}(\mathbf{A})$; that is, it is the maximum number of linearly independent vectors $\mathbf{y}_1, \mathbf{y}_2, \ldots, \mathbf{y}_r$ that can be written as $\mathbf{y}_i = \mathbf{A}\mathbf{x}_i$ for some $\mathbf{x}_i \in \mathcal{C}^n$. The *nullity of an L.T.* is the dimension of its null space; i.e., it is the maximum number of linearly independent vectors $\mathbf{x}_1, \mathbf{x}_2, \ldots, \mathbf{x}_\nu$ such that $\mathbf{A}\mathbf{x}_i = \mathbf{0}$, $i = 1, 2, \ldots, \nu$.

3 *Exercise* Let $\mathbf{A}$ be a matrix representing an L.T. $\mathcal{A}$ in $\mathcal{C}^n$. Show that the rank of the L.T. $\mathcal{A}$ just defined is equal to the rank of the matrix $\mathbf{A}$ representing $\mathcal{A}$. (An $n \times n$ matrix $\mathbf{A}$ is of rank r if and only if it has at least one minor of order r which is not zero and all its minors of order greater than r are zero.)

Theorems *4* and *6* below have two interpretations: (I) in abstract language, $\mathbf{A}$ is an L.T. of $\mathcal{C}^n$ into itself, $\mathbf{x}$ and $\mathbf{b}$ are elements of $\mathcal{C}^n$. (II) In the language of linear algebraic equations, $\mathbf{A}$ is an $n \times n$ matrix of complex numbers, and $\mathbf{x}$ and $\mathbf{b}$ are n-tuples of complex numbers. The matrix $\mathbf{A}$ is the representation of some L.T. relative to some (usually unspecified) orthonormal basis.

[1] *1* might suggest that we assumed as many equations as unknowns. This is not so, because we may always write a matrix with some of its rows or columns filled with zeros. Therefore, the following discussion is completely general.

4 **Theorem** Let $\mathbf{A}$ be a linear transformation in $\mathcal{C}^n$; then

(I) $\mathcal{C}^n = \mathcal{R}(\mathbf{A}) \oplus \mathcal{N}(\mathbf{A}^*)$.

(II) $\mathcal{N}(\mathbf{A}^*)$ is the orthogonal complement of $\mathcal{R}(\mathbf{A})$.

(III) $\dim \mathcal{R}(\mathbf{A}) + \dim \mathcal{N}(\mathbf{A}^*) = n$.

(IV) $\dim \mathcal{R}(\mathbf{A}) = \dim \mathcal{R}(\mathbf{A}^*)$ and $\dim \mathcal{N}(\mathbf{A}) = \dim \mathcal{N}(\mathbf{A}^*)$: that is, $\mathbf{A}$ and $\mathbf{A}^*$ have the same rank and nullity.

(V) $\mathbf{A}$ is a one-one mapping of $\mathcal{R}(\mathbf{A}^*)$ onto $\mathcal{R}(\mathbf{A})$.

Proof To prove (I) and (II) let $\mathbf{y} \in \mathcal{R}(\mathbf{A})^\perp$ [that is, the orthogonal complement of $\mathcal{R}(\mathbf{A})$; *C.7.6 et seq.*]. Since $\mathbf{A}(\mathbf{A}^*\mathbf{y}) \in \mathcal{R}(\mathbf{A})$ and since $\mathbf{y} \in \mathcal{R}(\mathbf{A})^\perp$, we have $0 = \langle \mathbf{y},\mathbf{A}\mathbf{A}^*\mathbf{y}\rangle = \langle \mathbf{A}^*\mathbf{y},\mathbf{A}^*\mathbf{y}\rangle = \|\mathbf{A}^*\mathbf{y}\|^2 = 0$; thus $\mathbf{A}^*\mathbf{y} = \mathbf{0}$ and $\mathbf{y} \in \mathcal{N}(\mathbf{A}^*)$. Thus we have proved that $\mathbf{y} \in \mathcal{R}(\mathbf{A})^\perp \Rightarrow \mathbf{y} \in \mathcal{N}(\mathbf{A}^*)$. Suppose that $\mathbf{z} \in \mathcal{N}(\mathbf{A}^*)$; then, for all $\mathbf{x}$, $0 = \langle \mathbf{A}^*\mathbf{z},\mathbf{x}\rangle = \langle \mathbf{z},\mathbf{A}\mathbf{x}\rangle$, that is, $\mathbf{z}$ is orthogonal to all vectors in $\mathcal{R}(\mathbf{A})$. Hence we have proved that $\mathbf{z} \in \mathcal{N}(\mathbf{A}^*) \Rightarrow \mathbf{z} \in \mathcal{R}(\mathbf{A})^\perp$. Therefore, $\mathcal{N}(\mathbf{A}^*) = \mathcal{R}(\mathbf{A})^\perp$; i.e., (II) is proved, and (I) follows from the fact that $\mathcal{C}^n = \mathcal{R}(\mathbf{A}) \oplus \mathcal{R}(\mathbf{A})^\perp$.

(III) From (I) it follows that any basis of $\mathcal{R}(\mathbf{A})$ together with any basis of $\mathcal{N}(\mathbf{A}^*)$ constitutes a basis for $\mathcal{C}^n$. Since the dimension of a subspace is the number of vectors in any of its bases, (III) follows.

(IV) Let $\{\mathbf{u}_1,\mathbf{u}_2, \ldots ,\mathbf{u}_n\}$ be a basis of $\mathcal{C}^n$ such that $\{\mathbf{u}_{r+1}, \ldots ,\mathbf{u}_n\}$ is a basis for $\mathcal{N}(\mathbf{A})$; then for any $\mathbf{x} = \sum_{i=1}^{n} \xi_i\mathbf{u}_i$,

$$\mathbf{A}\mathbf{x} = \sum_{1}^{n} \xi_i\mathbf{A}\mathbf{u}_i = \sum_{1}^{r} \xi_k\mathbf{A}\mathbf{u}_k$$

Therefore, $\mathcal{R}(\mathbf{A})$ is spanned by $\{\mathbf{A}\mathbf{u}_1,\mathbf{A}\mathbf{u}_2, \ldots ,\mathbf{A}\mathbf{u}_r\}$; hence, $\dim \mathcal{R}(\mathbf{A}) \leq r$ and $\dim \mathcal{R}(\mathbf{A}) \leq n - \dim \mathcal{N}(\mathbf{A})$. By applying this inequality to $\mathbf{A}^*$, we get $\dim \mathcal{R}(\mathbf{A}^*) \leq n - \dim \mathcal{N}(\mathbf{A}^*)$, which together with (III) gives

$$\dim \mathcal{R}(\mathbf{A}^*) \leq \dim \mathcal{R}(\mathbf{A})$$

We again apply this inequality to $\mathbf{A}^*$ and use the fact that $(\mathbf{A}^*)^* = \mathbf{A}$ to get

$$\dim \mathcal{R}(\mathbf{A}) \leq \dim \mathcal{R}(\mathbf{A}^*)$$

Therefore, $\dim \mathcal{R}(\mathbf{A}) = \dim \mathcal{R}(\mathbf{A}^*)$; hence, by *2*, $\mathbf{A}$ and $\mathbf{A}^*$ have the same rank. This fact together with (III) implies that they also have the same nullity. (V) Apply (I) to $\mathbf{A}^*$; thus $\mathcal{C}^n = \mathcal{R}(\mathbf{A}^*) \oplus \mathcal{N}(\mathbf{A})$. Therefore, if $\{\mathbf{u}_1,\mathbf{u}_2, \ldots ,\mathbf{u}_r\}$ is a basis for $\mathcal{R}(\mathbf{A}^*)$ and $\{\mathbf{u}_{r+1}, \ldots ,\mathbf{u}_n\}$ is a basis for $\mathcal{N}(\mathbf{A})$, $\{\mathbf{u}_1, \ldots ,\mathbf{u}_n\}$ is a basis for $\mathcal{C}^n$. For any $\mathbf{x} \in \mathcal{C}^n$ we have a unique decomposition

5 $$\mathbf{x} = \sum_{i=1}^{n} \xi_i\mathbf{u}_i = \mathbf{y} + \mathbf{z}$$

where $\mathbf{y} = \sum_{1}^{r} \xi_i\mathbf{u}_i$ and $\mathbf{z} = \sum_{r+1}^{n} \xi_i\mathbf{u}_i$, $\mathbf{y} \in \mathcal{R}(\mathbf{A}^*)$ and $\mathbf{z} \in \mathcal{N}(\mathbf{A})$, and also $\mathbf{A}\mathbf{x} = \mathbf{A}\mathbf{y}$. Thus, any vector $\mathbf{A}\mathbf{x}$ in $\mathcal{R}(\mathbf{A})$ is the mapping of a vector

$\mathbf{y} \in \mathcal{R}(\mathbf{A}^*)$, that is, $\mathbf{A}$ maps $\mathcal{R}(\mathbf{A}^*)$ *onto* $\mathcal{R}(\mathbf{A})$. Now, r is by assumption the dimension of $\mathcal{R}(\mathbf{A}^*)$, but by (IV) it is also that of $\mathcal{R}(\mathbf{A})$; hence, the r vectors $\{\mathbf{Au}_1, \ldots ,\mathbf{Au}_r\}$ spanning $\mathcal{R}(\mathbf{A})$ constitute a basis for $\mathcal{R}(\mathbf{A})$. For any $\mathbf{x} \in \mathcal{C}^n$ the relation between $\mathbf{Ax}$ and the vector $\mathbf{y}$ defined in *5* is one-one, since they both have $(\xi_1,\xi_2, \ldots ,\xi_r)$ as components relative to the basis $\{\mathbf{Au}_1, \ldots ,\mathbf{Au}_r\}$ and $\{\mathbf{u}_1, \ldots ,\mathbf{u}_r\}$, respectively. Hence $\mathbf{A}$ is a one-one mapping of $\mathcal{R}(\mathbf{A}^*)$ onto $\mathcal{R}(\mathbf{A})$. Q.E.D.

With these properties of $\mathbf{A}$ at our fingertips, we return to our equations and state the pertinent results in the form of a

6 **Theorem** Let $\mathbf{A}$ be a linear transformation on $\mathcal{C}^n$ and $\mathbf{b} \in \mathcal{C}^n$. Then we assert that

(I) The system of equations

7
$$\mathbf{Ax} = \mathbf{b}$$

has a solution if and only if $\mathbf{b}$ is orthogonal to all solutions of $\mathbf{A}^*\mathbf{y} = \mathbf{0}$, that is, if $\mathbf{b} \in \mathcal{N}(\mathbf{A}^*)^\perp = \mathcal{R}(\mathbf{A})$.

(II) The system *7* has a *unique solution* if and only if

8
$$\mathbf{Ay} = \mathbf{0}$$

has no nontrivial solutions.

(III) For all $\mathbf{b} \in \mathcal{C}^n$ the system *7* has a *unique solution* if and only if the rank of $\mathbf{A}$ is n.

(IV) If the rank of $\mathbf{A}$ is r and $r < n$ and if $\mathbf{b} \in \mathcal{R}(\mathbf{A})$, then the *most general solution* of *7* is

9
$$\mathbf{x} = \mathbf{x}_0 + \mathbf{y}$$

where $\mathbf{x}_0 \in \mathcal{R}(\mathbf{A}^*)$ and $\mathbf{y}$ is any vector of $\mathcal{N}(\mathbf{A})$. Furthermore,

10
$$\|\mathbf{x}\|^2 = \|\mathbf{x}_0\|^2 + \|\mathbf{y}\|^2$$

where the norm is defined by the scalar product (see *C.5.5*).

11 *Comment* In case (IV), among all the solutions of *7*, $\mathbf{x}_0$ is that which has the minimum length.

Proof (I) By the definition of $\mathcal{R}(\mathbf{A})$ (see *C.9.6*), *7* has a solution if and only if $\mathbf{b} \in \mathcal{R}(\mathbf{A})$. By conclusion (I) of Theorem *4*, this is equivalent to $\mathbf{b}$ belonging to the orthogonal complement of $\mathcal{N}(\mathbf{A}^*)$, that is, $\mathbf{b}$ is orthogonal to every solution of $\mathbf{A}^*\mathbf{y} = \mathbf{0}$. (II) Use contradiction. To prove $\Rightarrow$, suppose *8* has a nonzero solution $\mathbf{y}_0$. If *7* has a solution, say, $\mathbf{x}_0$, then $\mathbf{x}_0 + \mathbf{y}_0$ is also a solution of *7*. That is, the solution of *7* is not unique. Therefore, if *7* has a unique solution, *8* cannot have a nontrivial solution. To prove $\Leftarrow$, suppose the solution of *7* were not unique; then the difference $\mathbf{x}_0 - \mathbf{x}_0'$ between two solutions would be a nontrivial solution of *8*.

(III) By conclusion (III) of Theorem *4*, "rank of $\mathbf{A}$ is equal to n" is equivalent to "dimension of $\mathcal{N}(\mathbf{A}^*)$ is zero"; hence, by Theorem

4, conclusion (IV), dim $\mathfrak{N}(\mathbf{A}) = 0$. Therefore, "rank of $\mathbf{A}$ is n" is equivalent to "$\mathbf{y} = \mathbf{0}$ is the only solution of *8*." (IV) By Theorem *4*, conclusion (V), if $\mathbf{b} \in \mathcal{R}(\mathbf{A})$, there is a unique $\mathbf{x}_0 \in \mathcal{R}(\mathbf{A}^*)$ such that

$$\mathbf{A}\mathbf{x}_0 = \mathbf{b}$$

For any $\mathbf{y} \in \mathfrak{N}(\mathbf{A})$, we have $\mathbf{A}(\mathbf{x}_0 + \mathbf{y}) = \mathbf{b}$. Finally, by Theorem *4*, conclusion (II), $\mathcal{R}(\mathbf{A}^*)$ is the orthogonal complement of $\mathfrak{N}(\mathbf{A})$; hence, $\langle \mathbf{x}_0,\mathbf{y} \rangle = 0$. Let $\mathbf{x} = \mathbf{x}_0 + \mathbf{y}$; then upon expanding the equality $\langle \mathbf{x}_0 + \mathbf{y}, \mathbf{x}_0 + \mathbf{y} \rangle = \langle \mathbf{x},\mathbf{x} \rangle = \|\mathbf{x}\|^2$ we get *10*. Q.E.D.

12 *Exercise* Let $\mathbf{A}$ be an L.T. on $\mathcal{C}^n$ defined by its matrix representation (a_{ij}). For all $i,j = 1, 2, \ldots, n$, let $\mathbf{a}_j$ be the vector whose elements are $(a_{1j},a_{2j}, \ldots ,a_{nj})$ and let $\boldsymbol{\alpha}_i$ be the vector whose elements are $(a_{i1},a_{i2}, \ldots ,a_{in})$: $\mathbf{a}_j$ is the jth column vector of $\mathbf{A}$ and $\boldsymbol{\alpha}_i$ is the ith row vector of $\mathbf{A}$. Show that (I) $\mathcal{R}(\mathbf{A}) = \mathcal{S}[\mathbf{a}_1,\mathbf{a}_2, \ldots ,\mathbf{a}_n]$, that is, the range of $\mathbf{A}$ is spanned by its columns; (II) $\mathcal{R}(\mathbf{A}^*) = \mathcal{S}[\boldsymbol{\alpha}_1^*,\boldsymbol{\alpha}_2^*, \ldots ,\boldsymbol{\alpha}_n^*]$, where $\boldsymbol{\alpha}_i^*$ is the vector whose components are $(a_{i1}^*,a_{i2}^*, \ldots ,a_{in}^*)$; (III) the system of n linear homogeneous equations $\mathbf{A}\mathbf{x} = \mathbf{0}$ is equivalent to $\langle \boldsymbol{\alpha}_i^*,\mathbf{x} \rangle = 0$, $i = 1, 2, \ldots, n$. (IV) Use (II) and (III) to show that $\mathfrak{N}(\mathbf{A}) = \{\mathbf{x} | \mathbf{A}\mathbf{x} = \mathbf{0}\}$ is the orthogonal complement of $\mathcal{R}(\mathbf{A}^*)$.

Let the rank of $\mathbf{A}$ be n. Then from (III), Theorem *6*, the L.T. $\mathbf{A}$ defines for each $\mathbf{b}$ in $\mathcal{C}^n$ one and only one $\mathbf{x}$ such that $\mathbf{A}\mathbf{x} = \mathbf{b}$. In other words, $\mathbf{A}$ is a one-one mapping of $\mathcal{C}^n$ onto itself. Hence an inverse mapping exists. This inverse mapping is itself a linear transformation called the *inverse transformation* and is denoted $\mathbf{A}^{-1}$. We write $\mathbf{x} = \mathbf{A}^{-1}\mathbf{b}$.

13 **Definition** Any L.T. that has an inverse is called *nonsingular*. A *singular* L.T. is one that has no inverse. By *6*, any singular L.T. has a rank $< n$.

14 *Exercise* Apply Theorem *4* to a hermitian matrix $\mathbf{H}$ (see *C.14.8*). Show, in particular, that $\mathcal{C}^n = \mathcal{R}(\mathbf{H}) \oplus \mathfrak{N}(\mathbf{H})$ and $\mathcal{R}(\mathbf{H}) \perp \mathfrak{N}(\mathbf{H})$. [This implies that $\mathbf{H}$ is a one-one mapping of $\mathcal{R}(\mathbf{H})$ onto itself.]

16 *Norms*

Norm of a vector

Given a linear vector space, the idea of the norm of a vector $\mathbf{x}$ is a generalization of the idea of length. There are, however, several ways in which a norm may be defined, but in any case the norm of $\mathbf{x}$ is a real number, denoted by $\|\mathbf{x}\|$, which must satisfy the following postulates.

1 $\|\mathbf{x}\| \geq 0$ for all $\mathbf{x} \in \mathcal{X}$

2 $\|\mathbf{x}\| = 0 \Leftrightarrow \mathbf{x} = \mathbf{0}$

3 $\|a\mathbf{x}\| = |a|\ \|\mathbf{x}\|$ for all numbers a and vectors $\mathbf{x} \in \mathcal{X}$

4 $\|\mathbf{x} + \mathbf{y}\| \leq \|\mathbf{x}\| + \|\mathbf{y}\|$ for all pairs $\mathbf{x},\mathbf{y} \in \mathcal{X}$

This last requirement is called the *triangle inequality.*

Examples of norms in $\mathcal{C}^n$[1]

Let $\mathbf{x}$ have components $\xi_1, \xi_2, \ldots, \xi_n$ with respect to some orthonormal basis $\mathbf{u}_1, \mathbf{u}_2, \ldots, \mathbf{u}_n$.

5 $$\|\mathbf{x}\|_2 = \sqrt{\langle \mathbf{x},\mathbf{x} \rangle} = \sum_{i=1}^{n} |\xi_i|^2$$

Let $\mathbf{x}$ have components $\zeta_1, \zeta_2, \ldots, \zeta_n$ with respect to some arbitrary basis $\mathbf{v}_1, \mathbf{v}_2, \ldots, \mathbf{v}_n$.

6 $$\|\mathbf{x}\|_\infty = \max_i |\zeta_i|$$

7 $$\|\mathbf{x}\|_1 = \sum_{i=1}^{n} |\zeta_i|$$

8 *Exercise* Show that the norms *5* to *7* satisfy the four postulates.

The principal use of norms is to define the concept of convergence. Given a norm $\|\cdot\|$, the sequence of vectors $\{\mathbf{x}_n\}$ is said to converge to some vector $\mathbf{x}$ if the sequence of numbers $\|\mathbf{x}_n - \mathbf{x}\| \to 0$. Thus the norm ties back the concept of convergence of vectors to that of numbers. This type of convergence is referred to as "convergence in the norm topology" or "convergence in the strong topology."

It can be shown that, in the n-dimensional space $\mathcal{C}^n$, the three norms defined by *5* to *7* are *equivalent;* i.e., any sequence which converges in the sense of any one of these norms converges in the sense of the others.

Norm of a linear transformation

The concept of norm can be extended to linear transformations. Consider an L.T. $\mathbf{A}$ of $\mathcal{C}^n$ into $\mathcal{C}^n$ and a norm $\|\cdot\|$ on $\mathcal{C}^n$. For any $\mathbf{x}$, we may consider the numbers $\|\mathbf{Ax}\|$ and $\|\mathbf{x}\|$. It is obvious that $\|\mathbf{Ax}\|$ will usually depend on the direction of $\mathbf{x}$ and will always depend on its "length" $\|\mathbf{x}\|$; thus, a natural way to define the norm of $\mathbf{A}$, denoted by $\|\mathbf{A}\|$, is

9 $$\|\mathbf{A}\| = \inf \{k \mid \|\mathbf{Ax}\| \leq k\|\mathbf{x}\| \quad \forall \mathbf{x} \in \mathcal{C}^n\}$$

[1] The subscripts 2, ∞, and 1 are used to distinguish the several norms.

It is immediately apparent that

10 $$\|\mathbf{Ax}\| \leq \|\mathbf{A}\| \; \|\mathbf{x}\| \qquad \forall \mathbf{x} \in \mathcal{C}^n$$

and

11 $$\|\mathbf{A}\| = \sup_{\|\mathbf{x}\|=1} \|\mathbf{Ax}\|$$

In other words, $\|\mathbf{A}\|$ is the maximum "length" of the vectors $\mathbf{Ax}$ when $\mathbf{x}$ ranges over the unit sphere S in $\mathcal{C}^n$. The unit sphere is defined by $S = \{\mathbf{x}:\|\mathbf{x}\| = 1\}$.

The norms of L.T. of $\mathcal{C}^n$ into $\mathcal{C}^n$ have the following properties:

12 $$\|\mathbf{A} + \mathbf{B}\| \leq \|\mathbf{A}\| + \|\mathbf{B}\|$$

13 $$\|\mathbf{AB}\| \leq \|\mathbf{A}\| \; \|\mathbf{B}\|$$

The proof of *12* and *13* follow from the following observations: $\forall \mathbf{x}$ in $\mathcal{C}^n$,

$$\|(\mathbf{A} + \mathbf{B})\mathbf{x}\| = \|\mathbf{Ax} + \mathbf{Bx}\| \leq \|\mathbf{Ax}\| + \|\mathbf{Bx}\| \leq (\|\mathbf{A}\| + \|\mathbf{B}\|) \; \|\mathbf{x}\|$$
$$\|\mathbf{ABx}\| \leq \|\mathbf{A}\| \; \|\mathbf{Bx}\| \leq \|\mathbf{A}\| \; \|\mathbf{B}\| \; \|\mathbf{x}\|$$

The convergence of a sequence of L.T. is defined as follows: the sequence of L.T. $\{\mathbf{A}_n\}$ converges to an L.T. $\mathbf{A}$ if the sequence of numbers $\|\mathbf{A}_n - \mathbf{A}\| \to 0$.

14 *Exercise* Prove that if $\|\mathbf{A}\| < 1$, then $\sum_{k=0}^{\infty} \mathbf{A}^k$ converges to $(\mathbf{I} - \mathbf{A})^{-1}$. (*Hint:* Consider the sequence $\mathbf{B}_n$, where $\mathbf{B}_n = \sum_{k=0}^{n} \mathbf{A}^k$. Use *12* and *13* to establish convergence. Then use $\mathbf{B}_n - \mathbf{AB}_n = \mathbf{I} - \mathbf{A}^{n+1}$ and $\mathbf{A}^k \to \mathbf{0}$ as $k \to \infty$.)

15 *Exercise* Prove that for any $\mathbf{A}$, $\sum_{k=0}^{\infty} \frac{1}{k!} \mathbf{A}^k$ converges.

16 *Exercise* Let $\{\mathbf{u}_i\}^n$ be an orthonormal basis. Let (a_{ij}) represent $\mathbf{A}$ in that basis. With norm *5* show that

$$\|\mathbf{A}\|_2 \leq \Big(\sum_{i,j=1}^{n} |a_{ij}|^2\Big)^{\frac{1}{2}}$$

17 *Exercise* Let $\{\mathbf{u}_i\}$ be a basis and (a_{ij}) represent $\mathbf{A}$ in that basis. With norm *6*, show that $\|\mathbf{A}\|_\infty = \max_i \Big(\sum_{j=1}^{n} |a_{ij}|\Big)$. With norm *7*, show that $\|\mathbf{A}\|_1 = \max_j \Big(\sum_{i=1}^{n} |a_{ij}|\Big)$.

17 *Pseudo inverse of a matrix*

The pseudo inverse of a matrix is a generalization of the notion of inverse that has already found numerous applications. Let us recall that if the $n \times n$ matrix $\mathbf{A}$ is nonsingular, then for any $\mathbf{y} \in \mathcal{C}^n$ there is one and only one $\mathbf{x} \in \mathcal{C}^n$ such that $\mathbf{Ax} = \mathbf{y}$. Furthermore, the relationship between $\mathbf{y}$ and $\mathbf{x}$ is linear and defines the inverse of $\mathbf{A}$: $\mathbf{x} = \mathbf{A}^{-1}\mathbf{y}$.

Let $\mathbf{A}$ be singular (that is, its null space contains nonzero vectors and the rank of $\mathbf{A}$ is smaller than n). Given any $\mathbf{y} \in \mathcal{C}^n$, the equation $\mathbf{Ax} = \mathbf{y}$ either has no solutions [when $\mathbf{y} \notin \mathcal{R}(\mathbf{A})$; see *C.15.6*] or has an infinite number of solutions [because to any solution we may add a nonzero vector of $\mathcal{N}(\mathbf{A})$ and obtain another solution]. These two features of the problem suggest that we should reformulate the problem in order that it will always have one and only one solution. First, we should not ask for $\mathbf{Ax}$ to be equal to $\mathbf{y}$ and, second, among all the solutions, we should pick the one having the least norm. A possible reformulation is the following: given $\mathbf{A}$ (possibly singular), for any $\mathbf{y} \in \mathcal{C}^n$, find the vector $\mathbf{x}_0$ of least norm[1] which minimizes $\|\mathbf{Ax} - \mathbf{y}\|$. We can describe $\mathbf{x}_0$ as follows. Project $\mathbf{y}$ orthogonally on $\mathcal{R}(\mathbf{A})$; let $\mathbf{y}_p$ be its projection; then $\mathbf{y}_p \in \mathcal{R}(\mathbf{A})$ (see *C.7.1*). There is one and only one vector in $\mathcal{R}(\mathbf{A}^*)$, say, $\mathbf{x}_0$, such that $\mathbf{Ax}_0 = \mathbf{y}_p$ (see *C.15.4*). Finally, from *C.7.11* and *C.15.10* it follows that $\|\mathbf{Ax}_0 - \mathbf{y}\| = \min_{\mathbf{x} \in \mathcal{C}^n} \|\mathbf{Ax} - \mathbf{y}\|$. It can further be shown that the relation between the solution $\mathbf{x}_0$ of the modified problem and $\mathbf{y}$ is linear: $\mathbf{x}_0 = \mathbf{A}^\dagger\mathbf{y}$, where $\mathbf{A}^\dagger$ is said to be the pseudo inverse of $\mathbf{A}$. Obviously, if $\mathbf{A}$ is nonsingular, $\mathbf{A}^\dagger = \mathbf{A}^{-1}$. Note that $\mathbf{A}^\dagger$ maps $\mathcal{C}^n$ onto $\mathcal{R}(\mathbf{A}^*)$ and that the null space of $\mathbf{A}^\dagger$ is $\mathcal{R}(\mathbf{A})^\perp = \mathcal{N}(\mathbf{A}^*)$.

Definition of the pseudo inverse

We could develop the theory from this point of view. It is, however, more efficient to use the insight of the above discussion to set up a formal definition of the pseudo inverse and easily derive its properties.

1 **Definition** Let $\mathbf{A}$ be an L.T. mapping $\mathcal{C}^n$ into $\mathcal{C}^n$. The pseudo inverse of $\mathbf{A}$ is denoted by $\mathbf{A}^\dagger$ and satisfies the conditions

$$\begin{aligned} &\text{(I)}\quad \mathbf{A}^\dagger\mathbf{Ax} = \mathbf{x} && \forall \mathbf{x} \in \mathcal{N}(\mathbf{A})^\perp = \mathcal{R}(\mathbf{A}^*) \\ &\text{(II)}\quad \mathbf{A}^\dagger\mathbf{z} = 0 && \forall \mathbf{z} \in \mathcal{R}(\mathbf{A})^\perp = \mathcal{N}(\mathbf{A}^*) \\ &\text{(III)}\quad \mathbf{A}^\dagger(\mathbf{y} + \mathbf{z}) = \mathbf{A}^\dagger\mathbf{y} + \mathbf{A}^\dagger\mathbf{z} && \forall \mathbf{y} \in \mathcal{R}(\mathbf{A}),\ \forall \mathbf{z} \in \mathcal{R}(\mathbf{A})^\perp \end{aligned}$$

2 *Comment* (I) implies that the L.T. $\mathbf{A}^\dagger\mathbf{A}$ restricted to $\mathcal{R}(\mathbf{A}^*) = \mathcal{N}(\mathbf{A})^\perp$ is the identity mapping. Conditions (I), (II), and (III) define $\mathbf{A}^\dagger$

[1] We use here the norm defined by the scalar product $\|\mathbf{x}\|^2 = \langle \mathbf{x},\mathbf{x} \rangle$.

completely: (I) specifies the values of $\mathbf{A}^\dagger$ on $\mathcal{R}(\mathbf{A})$, (II) specifies them on $\mathcal{R}(\mathbf{A})^\perp$, and (III) specifies them on $\mathcal{R}(\mathbf{A}) \oplus \mathcal{R}(\mathbf{A})^\perp = \mathcal{C}^n$.

3 **Corollary** $\mathbf{A}^\dagger$ defined by *1* is a linear transformation.

Proof The corollary follows immediately from the computation $\mathbf{A}^\dagger\mathbf{u}$, where $\mathbf{u}$ is an arbitrary vector of $\mathcal{C}^n$. Since $\mathcal{C}^n = \mathcal{R}(\mathbf{A}) \oplus \mathcal{R}(\mathbf{A})^\perp$, let $\mathbf{u} = \mathbf{y} + \mathbf{z}$, where $\mathbf{y} \in \mathcal{R}(\mathbf{A})$, $\mathbf{z} \in \mathcal{R}(\mathbf{A})^\perp$. By (III) and (II) we get

$$\mathbf{A}^\dagger\mathbf{u} = \mathbf{A}^\dagger\mathbf{y} + \mathbf{A}^\dagger\mathbf{z} = \mathbf{A}^\dagger\mathbf{y}$$

Now from *C.15.4*, there is one and only one $\mathbf{x}_1 \in \mathcal{R}(\mathbf{A}^*)$ such that $\mathbf{A}\mathbf{x}_1 = \mathbf{y}$. Finally, from (I) $\mathbf{A}^\dagger\mathbf{A}\mathbf{x}_1 = \mathbf{x}_1$; that is, $\mathbf{A}^\dagger\mathbf{u} = \mathbf{x}_1$. It is obvious from the above constructions that this relation between $\mathbf{u}$ and $\mathbf{x}_1$ is homogeneous and additive. Hence $\mathbf{A}^\dagger$ is an L.T. ◁

From these considerations the following result is obvious.

4 **Corollary**

$$\mathcal{R}(\mathbf{A}^\dagger) = \mathcal{R}(\mathbf{A}^*) = \mathcal{N}(\mathbf{A})^\perp$$
$$\mathcal{N}(\mathbf{A}^\dagger) = \mathcal{N}(\mathbf{A}^*) = \mathcal{R}(\mathbf{A})^\perp$$

Definition *1* and Corollaries *3* and *4* are illustrated in Fig. *C.17.1*. The two sets on the left of this figure represent the orthogonal decomposition of $\mathcal{C}^n$ in $\mathcal{N}(\mathbf{A}) \oplus \mathcal{N}(\mathbf{A})^\perp$, and the two sets on the right represent

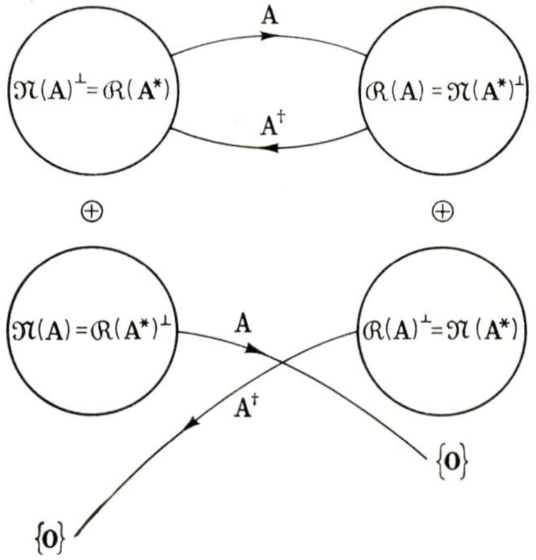

Fig. C.17.1 Illustration of

$$\mathcal{C}^n = \mathcal{N}(\mathbf{A}) \oplus \mathcal{N}(\mathbf{A})^\perp = \mathcal{R}(\mathbf{A}) \oplus \mathcal{R}(\mathbf{A})^\perp$$

and the corresponding mapping properties of $\mathbf{A}$ and its pseudo inverse $\mathbf{A}^\dagger$.

the decomposition of $\mathcal{C}^n$ in $\mathcal{R}(\mathbf{A}) \oplus \mathcal{R}(\mathbf{A})^\perp$. By definition, $\mathbf{A}$ maps $\mathcal{N}(\mathbf{A})$ into $\{\mathbf{0}\}$ and $\mathbf{A}^\dagger$ maps $\mathcal{N}(\mathbf{A}^*) = \mathcal{R}(\mathbf{A})^\perp$ into $\{\mathbf{0}\}$. Also, $\mathbf{A}^\dagger$ restricted to $\mathcal{R}(\mathbf{A})$ [which is a mapping denoted by $\mathbf{A}^\dagger/\mathcal{R}(\mathbf{A})$] is the inverse of $\mathbf{B}$, where by definition, $\mathbf{B}$ is the one-one linear mapping which is $\mathbf{A}$ restricted to $\mathcal{R}(\mathbf{A}^*) = \mathcal{N}(\mathbf{A})^\perp$.

Properties of the pseudo inverse

5 **Theorem**

(I) $\mathbf{A}^\dagger\mathbf{A}$ is the orthogonal projection of $\mathcal{C}^n$ onto $\mathcal{R}(\mathbf{A}^*) = \mathcal{N}(\mathbf{A})^\perp$

(II) $(\mathbf{A}^\dagger)^\dagger = \mathbf{A}$

(III) $\mathbf{A}^\dagger\mathbf{A}\mathbf{A}^\dagger = \mathbf{A}^\dagger$

(IV) $\mathbf{A}\mathbf{A}^\dagger\mathbf{A} = \mathbf{A}$

(V) $\mathbf{A}\mathbf{A}^\dagger$ is the orthogonal projection of $\mathcal{C}^n$ onto $\mathcal{R}(\mathbf{A}) = \mathcal{N}(\mathbf{A}^*)^\perp$

6 *Comment* (I) and (V) imply that $(\mathbf{A}^\dagger\mathbf{A})^* = \mathbf{A}^\dagger\mathbf{A}$ and $(\mathbf{A}\mathbf{A}^\dagger)^* = \mathbf{A}\mathbf{A}^\dagger$. Penrose[1] has shown that these two relations, together with (III) and (IV), considered as matric equations in $\mathbf{A}^\dagger$, define a unique matrix which he calls the pseudo inverse of $\mathbf{A}$.

Proof (I) For any $\mathbf{x}$ in $\mathcal{C}^n$ consider the orthogonal decomposition $\mathbf{x} = \mathbf{x}_1 + \mathbf{x}_2$ with $\mathbf{x}_1 \in \mathcal{N}(\mathbf{A})^\perp$ and $\mathbf{x}_2 \in \mathcal{N}(\mathbf{A})$. Then, by the definition of $\mathcal{N}(\mathbf{A})$ and *1*, $\mathbf{A}^\dagger\mathbf{A}\mathbf{x} = \mathbf{A}^\dagger\mathbf{A}\mathbf{x}_1 = \mathbf{x}_1$ and the assertion follows.

(II) This fact is obvious from the symmetry of the definition and of Fig. *C.17.1*. It is easily proved with the help of *1* and *4*.

(III) Let $\mathbf{y} \in \mathcal{C}^n$ and $\mathbf{y} = \mathbf{y}_1 + \mathbf{y}_2$, with $\mathbf{y}_1 \in \mathcal{R}(\mathbf{A})$ and $\mathbf{y}_2 \in \mathcal{R}(\mathbf{A})^\perp$. From *1*, $\mathbf{A}^\dagger\mathbf{y} = \mathbf{A}^\dagger\mathbf{y}_1 \in \mathcal{N}(\mathbf{A})^\perp$. From (I),

$$\mathbf{A}^\dagger\mathbf{A}(\mathbf{A}^\dagger\mathbf{y}) = \mathbf{A}^\dagger\mathbf{A}(\mathbf{A}^\dagger\mathbf{y}_1) = \mathbf{A}^\dagger\mathbf{y}_1 = \mathbf{A}^\dagger\mathbf{y}$$

Thus $\mathbf{A}^\dagger\mathbf{A}\mathbf{A}^\dagger\mathbf{y} = \mathbf{A}^\dagger\mathbf{y}$ for all $\mathbf{y} \in \mathcal{C}^n$.

(IV) Put $\mathbf{A}^\dagger = \mathbf{B}$. Use (II) and (III) to get (IV).

(V) Let $\mathbf{y} \in \mathcal{C}^n$ and let $\mathbf{y} = \mathbf{y}_1 + \mathbf{y}_2$, with $\mathbf{y}_1 \in \mathcal{R}(\mathbf{A})$ and $\mathbf{y}_2 \in \mathcal{R}(\mathbf{A})^\perp$; then $\mathbf{A}\mathbf{A}^\dagger\mathbf{y} = \mathbf{A}\mathbf{A}^\dagger\mathbf{y}_1 = \mathbf{y}_1$, where the last equality follows from (IV). Thus, for any $\mathbf{y} \in \mathcal{C}^n$, $\mathbf{A}\mathbf{A}^\dagger\mathbf{y} = \mathbf{y}_1$, where $\mathbf{y}_1$ is the orthogonal projection of $\mathbf{y}$ on $\mathcal{R}(\mathbf{A})$.

7 **Theorem** $(\mathbf{A}^*)^\dagger = (\mathbf{A}^\dagger)^*$.

Proof By *5*, $\mathbf{A}\mathbf{A}^\dagger = (\mathbf{A}\mathbf{A}^\dagger)^* = \mathbf{A}^{\dagger*}\mathbf{A}^*$; hence $\mathbf{A}^{\dagger*}\mathbf{A}^*$ is by *5*(V) the orthogonal projection of $\mathcal{C}^n$ onto $\mathcal{R}(\mathbf{A})$. On the other hand, *5*(I) applied to $\mathbf{A}^*$ asserts that $\mathbf{A}^{*\dagger}\mathbf{A}^*$ is the orthogonal projection of $\mathcal{C}^n$ onto $\mathcal{R}(\mathbf{A}^{**}) = \mathcal{R}(\mathbf{A})$. Hence, $\mathbf{A}^{*\dagger}\mathbf{A}^* = \mathbf{A}^{\dagger*}\mathbf{A}^*$, or, equivalently,

$$\mathbf{A}^{*\dagger}\mathbf{x} = \mathbf{A}^{\dagger*}\mathbf{x}$$

for all $\mathbf{x} \in \mathcal{R}(\mathbf{A}^*)$. It remains to show that the same relation holds for all $\mathbf{x} \in \mathcal{N}(\mathbf{A})$. Applying *1* to $\mathbf{A}^\dagger$, we have $\mathcal{N}(\mathbf{A}^{\dagger*}) = \mathcal{N}(\mathbf{A}^{\dagger\dagger}) = \mathcal{N}(\mathbf{A})$, where the last equality follows from *5*(II). But

$$\mathcal{N}(\mathbf{A}^{*\dagger}) = \mathcal{N}(\mathbf{A}^{**}) = \mathcal{N}(\mathbf{A})$$

Hence $\mathbf{A}^{*\dagger}\mathbf{x} = \mathbf{A}^{\dagger*}\mathbf{x} = \mathbf{0}$ for all $\mathbf{x} \in \mathcal{N}(\mathbf{A})$. ◁

The approximating property of $\mathbf{A}^\dagger$ mentioned in the introduction can be derived from the definition.

[1] R. Penrose, A Generalized Inverse for Matrices, *Proc. Cambridge Phil. Soc.*, vol. 51, part 3, pp. 406–413, 1955.

8 **Theorem** Let $\mathbf{y} \in \mathcal{C}^n$ and $\mathbf{x}_0 = \mathbf{A}^\dagger\mathbf{y}$. Then

9 $$\|\mathbf{A}\mathbf{x}_0 - \mathbf{y}\| \leq \|\mathbf{A}\mathbf{x} - \mathbf{y}\| \qquad \forall \mathbf{x} \in \mathcal{C}^n$$

and $\mathbf{x}_0$ is the vector of least norm which satisfies *9*. (The norm is defined by $\|\mathbf{y}\|^2 = \langle \mathbf{y},\mathbf{y}\rangle$.)

Proof Let $\mathbf{y} = \mathbf{y}_1 + \mathbf{y}_2$, with $\mathbf{y}_1 \in \mathcal{R}(\mathbf{A})$ and $\mathbf{y}_2 \in \mathcal{R}(\mathbf{A})^\perp$. Then

$$\|\mathbf{A}\mathbf{x}_0 - \mathbf{y}\| = \|\mathbf{A}\mathbf{A}^\dagger\mathbf{y} - \mathbf{y}\| = \|\mathbf{y}_1 - (\mathbf{y}_1 + \mathbf{y}_2)\| = \|\mathbf{y}_2\|$$

For any $\mathbf{x} \in \mathcal{C}^n$; $\mathbf{A}\mathbf{x} \in \mathcal{R}(\mathbf{A})$, hence

$$\|\mathbf{A}\mathbf{x} - \mathbf{y}\|^2 = \|\mathbf{A}\mathbf{x} - \mathbf{y}_1 - \mathbf{y}_2\|^2 = \|\mathbf{A}\mathbf{x} - \mathbf{y}_1\|^2 + \|\mathbf{y}_2\|^2$$

Hence *9* follows from these two equalities. Finally, to show that $\mathbf{x}_0$ is the vector of least norm satisfying *9*, observe the following. If $\mathbf{x}_1 \in \mathfrak{N}(\mathbf{A})$, $\mathbf{A}\mathbf{x}_0 = \mathbf{A}(\mathbf{x}_0 + \mathbf{x}_1)$. And, since $\mathbf{x}_0 \in \mathcal{R}(\mathbf{A}^\dagger) = \mathcal{R}(\mathbf{A}^*) = \mathfrak{N}(\mathbf{A})^\perp$, $\|\mathbf{x}_0 + \mathbf{x}_1\|^2 = \|\mathbf{x}_0\|^2 + \|\mathbf{x}_1\|^2$; hence, $\|\mathbf{x}_0\| \leq \|\mathbf{x}_0 + \mathbf{x}_1\|$.

10 *Comment* In the language of linear algebraic equations, Theorem *8* may be rephrased as follows.

(I) If $\mathbf{y} \in \mathcal{R}(\mathbf{A})$, $\mathbf{A}\mathbf{x} = \mathbf{y}$ has a solution and any of its solutions is of the form

$$\mathbf{x} = \mathbf{A}^\dagger\mathbf{y} + (\mathbf{I} - \mathbf{A}^\dagger\mathbf{A})\mathbf{z}$$

where $\mathbf{z}$ is an arbitrary vector of $\mathcal{C}^n$. [Note that *5*(I) implies that $\mathbf{I} - \mathbf{A}^\dagger\mathbf{A}$ is the orthogonal projection of $\mathcal{C}^n$ on $\mathfrak{N}(\mathbf{A})$; hence, $(\mathbf{I} - \mathbf{A}^\dagger\mathbf{A})\mathbf{z}$ is an arbitrary vector of $\mathfrak{N}(\mathbf{A})$.]

(II) If $\mathbf{y} \notin \mathcal{R}(\mathbf{A})$, the equation $\mathbf{A}\mathbf{x} = \mathbf{y}$ has no solutions. The vector $\mathbf{x}_0 = \mathbf{A}^\dagger\mathbf{y}$ is the vector of least norm which when reinserted into the equation gives the least error $\mathbf{e} \triangleq \mathbf{A}\mathbf{x} - \mathbf{y}$ (in the sense of *9*). ◁

Now we prove a polar decomposition theorem which is useful in getting a geometric insight into the effect of a general L.T. mapping $\mathcal{C}^n$ into $\mathcal{C}^n$.

11 **Theorem** If $\mathbf{A}$ is a nonsingular L.T. of $\mathcal{C}^n$ into $\mathcal{C}^n$, there is a unique decomposition of $\mathbf{A}$,

12 $$\mathbf{A} = \mathbf{U}\mathbf{H}$$

where $\mathbf{H}$ is positive definite hermitian and $\mathbf{U}$ is unitary, that is, $\mathbf{U}^* = \mathbf{U}^{-1}$. If $\mathbf{A}$ is singular, then we have a decomposition

13 $$\mathbf{A} = \mathbf{U}\mathbf{H}$$

where both $\mathbf{H}$ and $\mathbf{U}$ are singular, $\mathbf{H}$ is hermitian nonnegative definite, and

14 $$\mathbf{H}^2 = \mathbf{A}^*\mathbf{A} \qquad \mathbf{U} = \mathbf{A}\mathbf{H}^\dagger \qquad \mathbf{U}^* = \mathbf{U}^\dagger$$

Proof Let us prove the second statement. We anticipate a little and use *C.19.1*, *C.19.2*, and *C.19.7*. Since $\mathbf{A}^*\mathbf{A}$ is hermitian (self-adjoint) and nonnegative definite (see *C.14.10*), $\mathbf{A}^*\mathbf{A} = \sum_{i=1}^{n} \lambda_i \mathbf{E}_i$, with $\lambda_i \geq 0$. Define the matrix $\mathbf{H}$ by $\mathbf{H} \triangleq \Sigma \sqrt{\lambda_i}\, \mathbf{E}_i$. $\mathbf{H}$ is then hermitian and is such that $\mathbf{A}^*\mathbf{A} = \mathbf{H}^2$; thus $\mathcal{R}(\mathbf{H}) = \mathcal{R}(\mathbf{H}^*)$ and $\mathbf{H}\mathbf{H}^\dagger = \mathbf{H}^\dagger\mathbf{H}$. Since $\mathbf{H}^\dagger$ maps $\mathcal{C}^n$ onto $\mathcal{R}(\mathbf{H})$, we have $\mathbf{H}^\dagger = \sum_{\lambda_i > 0} \lambda_i^{-1/2}\mathbf{E}_i$. Now define $\mathbf{U}$ by $\mathbf{U} = \mathbf{A}\mathbf{H}^\dagger$. Since $\mathbf{H}^\dagger$ commutes with $\mathbf{H}$ and since $\mathbf{H}$ is hermitian, we have

$$\mathbf{U}^*\mathbf{U} = (\mathbf{A}\mathbf{H}^\dagger)^*(\mathbf{A}\mathbf{H}^\dagger) = \mathbf{H}^{\dagger *}\mathbf{A}^*\mathbf{A}\mathbf{H}^\dagger = \mathbf{H}^{\dagger *}\mathbf{H}\mathbf{H}\mathbf{H}^\dagger = \mathbf{H}^\dagger\mathbf{H}\mathbf{H}^\dagger\mathbf{H} = \mathbf{H}^\dagger\mathbf{H}$$

By *5*(I), $\mathbf{U}^*\mathbf{U} = \mathbf{H}^\dagger\mathbf{H}$ is the orthogonal projection of $\mathcal{C}^n$ onto $\mathcal{R}(\mathbf{H})$: $\mathbf{U}^*\mathbf{U}\mathbf{x} = \mathbf{x}$ for all $\mathbf{x} \in \mathcal{R}(\mathbf{H})$. However, by considering Fig. *C.17.1*, it is easy to show that $\mathcal{R}(\mathbf{H}) = \mathcal{R}(\mathbf{A}^*) = \mathcal{R}(\mathbf{U}^*)$; thus $\mathbf{U}^*\mathbf{U}\mathbf{x} = \mathbf{x}$ for all $\mathbf{x}$ in $\mathcal{R}(\mathbf{U}^*)$. Finally, for all $\mathbf{y} \in \mathcal{R}(\mathbf{U})^\perp = \mathfrak{N}(\mathbf{U}^*)$, $\mathbf{U}^*\mathbf{y} = \mathbf{0}$; and by *1*, $\mathbf{U}^* = \mathbf{U}^\dagger$. The decomposition *12* follows directly from the definition of $\mathbf{U}$: $\mathbf{U}\mathbf{H} = \mathbf{A}\mathbf{H}^\dagger\mathbf{H} = \mathbf{A}$ because $\mathbf{H}^\dagger\mathbf{H}$ is the orthogonal projection of $\mathcal{C}^n$ on $\mathcal{R}(\mathbf{A}^*)$.

The proof for the case where $\mathbf{A}$ is nonsingular is a special case of the preceding one.

The calculation of $\mathbf{A}^\dagger$

15 **Theorem** Let $\mathbf{S}$ be the hermitian nonnegative definite matrix defined by

16
$$\mathbf{S} = \mathbf{A}^*\mathbf{A}$$

Then

17
$$\mathbf{A}^\dagger = \mathbf{S}^\dagger\mathbf{A}^*$$

Comment This theorem implies that the computation of the pseudo inverse of any $n \times n$ matrix $\mathbf{A}$ reduces to the computation of the pseudo inverse of a hermitian nonnegative definite $n \times n$ matrix $\mathbf{S}$.

Proof If $\mathbf{x} \in \mathfrak{N}(\mathbf{A}^*)$, then, by *1*, $\mathbf{A}^\dagger\mathbf{x} = \mathbf{0}$ and also $\mathbf{S}^\dagger\mathbf{A}^*\mathbf{x} = \mathbf{0}$. Now let $\mathbf{y} \in \mathcal{R}(\mathbf{A})$. By *C.15.4* there is a unique $\mathbf{x} \in \mathcal{R}(\mathbf{A}^*)$ such that $\mathbf{A}\mathbf{x} = \mathbf{y}$. On the one hand, $\mathbf{A}^\dagger\mathbf{y} = \mathbf{A}^\dagger\mathbf{A}\mathbf{x} = \mathbf{x}$ by (I) of Definition *1*. On the other hand, from *C.15.4*, $\mathbf{A}^*\mathbf{A}$ is a one-one map of $\mathcal{R}(\mathbf{A}^*)$ onto itself and, by *1*, for any $\mathbf{x} \in \mathcal{R}(\mathbf{A}^*)$, $(\mathbf{A}^*\mathbf{A})^\dagger\mathbf{A}^*\mathbf{A}\mathbf{x} = \mathbf{x}$, or $(\mathbf{A}^*\mathbf{A})^\dagger\mathbf{A}^*\mathbf{y} = \mathbf{x} = \mathbf{A}^\dagger\mathbf{y}$. Thus for all $\mathbf{y} \in \mathcal{R}(\mathbf{A})$, $(\mathbf{A}^*\mathbf{A})^\dagger\mathbf{A}^*\mathbf{y} = \mathbf{A}^\dagger\mathbf{y}$. Thus the linear operators on both sides of *17* give the same result when they operate on vectors of $\mathcal{R}(\mathbf{A})$ and $\mathfrak{N}(\mathbf{A}^*)$. Since $\mathcal{C}^n = \mathcal{R}(\mathbf{A}) \oplus \mathcal{R}(\mathbf{A})^\perp$ and $\mathfrak{N}(\mathbf{A}^*) = \mathcal{R}(\mathbf{A})^\perp$, *17* is established.

18 *Exercise* Let $\mathbf{A} = \text{diag}(a_{11}, a_{22}, \ldots, a_{nn})$. Show that

$$\mathbf{A}^\dagger = \text{diag}(\alpha_{11}, \alpha_{22}, \ldots, \alpha_{nn})$$

where $\alpha_{ii} = a_{ii}^{-1}$ if $a_{ii} \neq 0$ and $\alpha_{ii} = 0$ if $a_{ii} = 0$. Give the geometric interpretation.

19 *Exercise* Let $\mathbf{B}$ be hermitian. Let $\mathbf{U}$ be the matrix whose ith column is the normalized ith eigenvector of $\mathbf{B}$. Then $\mathbf{U}$ is unitary; that is, $\mathbf{U}^*\mathbf{U} = \mathbf{I}$. Let $\mathbf{B} = \mathbf{U}\mathbf{\Lambda}\mathbf{U}^*$, where $\mathbf{\Lambda} = \text{diag}(\lambda_1, \lambda_2, \ldots, \lambda_n)$, and then show that $\mathbf{B}^\dagger = \mathbf{U}\mathbf{\Lambda}^\dagger\mathbf{U}^*$.

20 *Exercise*[1] Suppose $\mathbf{A}$ is of rank $r < n$. Suppose we know the $n \times r$ matrix $\mathbf{B}$ and $r \times n$ matrix $\mathbf{C}$, both of rank r, which have the property that

$$\mathbf{A} = \mathbf{BC}$$

Show that

$$\mathbf{A}^\dagger = \mathbf{C}^*(\mathbf{CC}^*)^{-1}(\mathbf{B}^*\mathbf{B})^{-1}\mathbf{B}^*$$

21 *Exercise*[2] Suppose $\mathbf{A}$ has its last $n - r$ columns identical to zero and that its first r columns are linearly independent. Let $\mathbf{N}$ be the $n \times r$ matrix consisting of these r columns. Then

$$\mathbf{A}^\dagger = \mathbf{K}(\mathbf{N}^*\mathbf{N})^{-1}\mathbf{N}^*$$

where $\mathbf{K}$ is an $n \times r$ matrix whose first r rows constitute a unit matrix and all other rows are identical to zero.

18 *Simple L.T.*

1 **Definition** Let $\mathbf{A}$ be an $n \times n$ matrix whose elements are real or complex numbers. An *eigenvalue of* $\mathbf{A}$ is a number λ, possibly complex, such that the polynomial $\Delta(\lambda) \equiv \det(\mathbf{A} - \lambda\mathbf{I}) = 0$. The polynomial $\Delta(\lambda)$ is called the *characteristic polynomial.* We may also define the eigenvalues in terms of L.T. without recourse to any matrix representation: an eigenvalue of an L.T. $\mathbf{A}$ is a number λ such that the L.T. $\mathbf{A} - \lambda\mathbf{I}$ is singular, i.e., there is at least one vector $\mathbf{x} \neq \mathbf{0}$ such that $(\mathbf{A} - \lambda\mathbf{I})\mathbf{x} = \mathbf{0}$.

2 *Exercise* Show that, in $\mathcal{C}^n$, the two definitions are equivalent.

3 *Exercise* Show directly that the zeros of $\Delta(\lambda)$ are independent of the basis used to obtain the matrix representation of the L.T. (*Hint:* Use

[1] T. N. E. Greville, Some Applications of the Pseudo Inverse of a Matrix, *Soc. Indust. Appl. Math. Rev.*, vol. 2, no. 1, pp. 15–22, 1960.

[2] J. B. Rosen, The Gradient Projection Method for Nonlinear Programming, *J. Soc. Indust. Appl. Math.*, vol. 8, no. 1, pp. 181–217, 1960.

C.11.7) and the fact that the determinant of a product of $n \times n$ matrices is the product of their respective determinants.)

4 **Definition** An *eigenvector* of $\mathbf{A}$ is a nonzero vector $\mathbf{x}$ such that $\mathbf{Ax} = \lambda\mathbf{x}$. Clearly, since $(\mathbf{A} - \lambda\mathbf{I})\mathbf{x} = \mathbf{0}$ is implied, with $\mathbf{x} \neq \mathbf{0}$, the number λ is necessarily an eigenvalue. $\mathbf{x}$ is said to be an eigenvector associated with the eigenvalue λ. This is equivalent to saying $\mathbf{x}$ is in the null space of the L.T. $\mathbf{A} - \lambda\mathbf{I}$.

Note that if $\mathbf{x}$ is an eigenvector, then for any scalar $k \neq 0$, $k\mathbf{x}$ is also an eigenvector. Eigenvectors will usually be so scaled that $\|\mathbf{x}\| = 1$; in such case the eigenvector is said to be *normalized.*

5 *Exercise* Show that $\mathfrak{N}(\mathbf{A} - \lambda_i\mathbf{I}) \triangleq \{\mathbf{x} | (\mathbf{A} - \lambda_i\mathbf{I})\mathbf{x} = \mathbf{0}\}$ is an invariant subspace under $\mathbf{A}$.

6 **Definition** A *simple L.T.* is one whose eigenvectors span the space.

7 *Comment* Not all L.T. are simple; for example, let

8 $$\mathbf{A} = \begin{bmatrix} 1 & 1 \\ 0 & 1 \end{bmatrix}$$

Then $\det(\mathbf{A} - \lambda\mathbf{I}) = (\lambda - 1)^2$ and $\lambda = 1$ is the only eigenvalue. There is only one (normalized) eigenvector, namely, $\mathbf{u}_1 = \begin{bmatrix} 1 \\ 0 \end{bmatrix}$. Obviously, it cannot span $\mathcal{C}^2$.

It should be stressed that the eigenvectors spanning the space need not correspond to distinct eigenvalues. For example, let $\mathbf{A} = \begin{bmatrix} 1 & 0 \\ 0 & 1 \end{bmatrix}$. Its eigenvalues are both equal to 1 and the eigenvectors are $\mathbf{u}_1 = \begin{bmatrix} 1 \\ 0 \end{bmatrix}$ and $\mathbf{u}_2 = \begin{bmatrix} 0 \\ 1 \end{bmatrix}$. $\mathbf{u}_1$ and $\mathbf{u}_2$ span $\mathcal{C}^2$; hence, $\mathbf{A}$ is simple. In *5.4.6* we have shown that if $\mathbf{A}$ has distinct eigenvalues, then the eigenvectors of $\mathbf{A}$ span the space $\mathcal{C}^n$. Consequently, the class of all simple L.T. is larger than the class of all L.T. that have distinct eigenvalues.

9 **Corollary** $\mathbf{A}$ is a simple L.T. if and only if there is a basis with respect to which matrix representation of $\mathbf{A}$ is diagonal.

Proof $\Rightarrow$ $\mathbf{A}$ is simple; then there are n eigenvectors $\mathbf{u}_1, \mathbf{u}_2, \ldots, \mathbf{u}_n$ (with associated eigenvalues $\lambda_1, \lambda_2, \ldots, \lambda_n$) which constitute a basis $\mathcal{B}$ for $\mathcal{C}^n$. Hence $\mathbf{Au}_i = \lambda_i\mathbf{u}_i$, $i = 1, 2, \ldots, n$. Therefore, by *C.10.4*, the representation of $\mathbf{A}$ in that basis is a matrix whose ith column has all its components equal to zero except the ith, which is λ_i. Thus, $\mathbf{A} = \text{diag}(\lambda_1,\lambda_2, \ldots ,\lambda_n)$.

$\Leftarrow$ By assumption, $\mathbf{A}$ is diagonal; let $\mathbf{A} = \text{diag}\,(\lambda_1,\lambda_2, \ldots ,\lambda_n)$. If $\mathbf{u}_i$ is the ith basis vector for this matrix representation, we have

$$\mathbf{A}\mathbf{u}_i = \lambda_i \mathbf{u}_i \qquad i = 1, 2, \ldots , n$$

Hence, $\mathbf{u}_1, \mathbf{u}_2, \ldots , \mathbf{u}_n$ are eigenvectors and they span the space. Therefore, by definition $\mathbf{A}$ is a simple L.T.

The effect of a simple L.T. $\mathbf{A}$ on a vector $\mathbf{x}$ is illustrated in Fig. *C.18.1*.

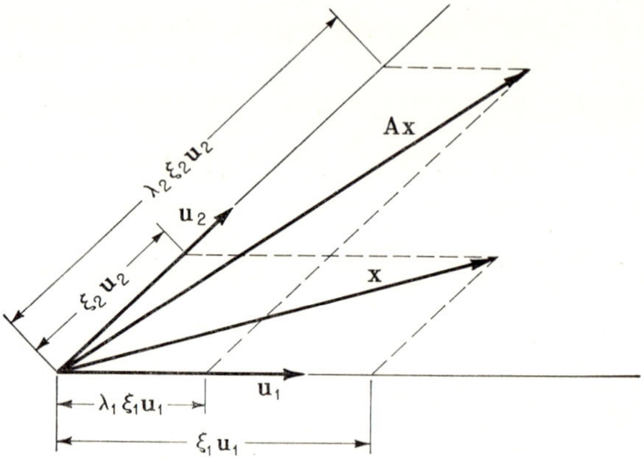

Fig. C.18.1 Relation between $\mathbf{x}$ and $\mathbf{A}\mathbf{x}$ through the spectral representation of $\mathbf{A}$.

10 *Exercise* Let $\mathbf{A}$ be an L.T. on $\mathcal{C}^n$. Then λ_i is an eigenvalue of $\mathbf{A}$ if and only if λ_i^* is an eigenvalue of $\mathbf{A}^*$. (*Hint:* Use an orthonormal basis and take the conjugate transpose of the matrix $\mathbf{A} - \lambda\mathbf{I}$.)

11 **Theorem** Let $\mathbf{A}$ be a simple L.T. on $\mathcal{C}^n$. Let $\mathbf{u}_1, \mathbf{u}_2, \ldots , \mathbf{u}_n$ be a set of eigenvectors of $\mathbf{A}$ which constitutes a basis $\mathcal{B}$ of $\mathcal{C}^n$. Then, the reciprocal basis of $\mathcal{B}$ $\{\mathbf{v}_1, \ldots ,\mathbf{v}_n\}$ is constituted by eigenvectors of $\mathbf{A}^*$; and if the eigenvalue associated with $\mathbf{u}_i$ is λ_i, that associated with $\mathbf{v}_i$ is λ_i^*.

Proof By definition of a reciprocal basis: $\langle \mathbf{u}_i,\mathbf{v}_j\rangle = \delta_{ij}$, $i,j = 1, \ldots , n$. Now on the one hand, for all i, j's

$$\langle \mathbf{v}_j,\mathbf{A}\mathbf{u}_i\rangle = \langle \mathbf{v}_j,\lambda_i\mathbf{u}_i\rangle = \lambda_i\langle \mathbf{v}_j,\mathbf{u}_i\rangle = \lambda_i\delta_{ij} = \lambda_j\delta_{ij}$$

and on the other hand, using the definition of the adjoint (see *C.14.1*)

$$\lambda_j\delta_{ij} = \langle \mathbf{v}_j,\mathbf{A}\mathbf{u}_i\rangle = \langle \mathbf{A}^*\mathbf{v}_j,\mathbf{u}_i\rangle = \langle \mathbf{u}_i,\mathbf{A}^*\mathbf{v}_j\rangle^*$$

Hence, for all i, j's

$$\langle \mathbf{u}_i,\mathbf{A}^*\mathbf{v}_j\rangle = \lambda_j^*\delta_{ij}$$

Since the $\mathbf{u}_i$'s constitute a reciprocal basis of the $\mathbf{v}_j$'s, we see that

$$\mathbf{A}^*\mathbf{v}_j = \lambda_j^*\mathbf{v}_j \qquad j = 1, 2, \ldots , n$$

12 **Corollary** If $\mathbf{A}$ is a simple L.T. on $\mathcal{C}^n$, $\mathbf{A}^*$ is also a simple L.T. on $\mathcal{C}^n$.

13 **Theorem** Let $\mathbf{A}$ be a simple L.T. on $\mathcal{C}^n$. Let $\{\mathbf{u}_1, \ldots, \mathbf{u}_n\}$ be a set of eigenvectors of $\mathbf{A}$ constituting a basis $\mathcal{B}$ of $\mathcal{C}^n$. Let the $\mathbf{v}_j$'s be the reciprocal basis to $\mathcal{B}$. If $\mathbf{E}_i = \mathbf{u}_i\rangle\langle\mathbf{v}_i$, $i = 1, 2, \ldots, n$, then

14 $$\mathbf{A} = \sum_{i=1}^{n} \lambda_i \mathbf{E}_i = \sum_{i=1}^{n} \lambda_i \mathbf{u}_i\rangle\langle\mathbf{v}_i$$

and

15 $$\mathbf{A}^* = \sum_{i=1}^{n} \lambda_i^* \mathbf{E}_i^* = \sum_{i=1}^{n} \lambda_i^* \mathbf{v}_i\rangle\langle\mathbf{u}_i$$

Proof $\forall \mathbf{x} \in \mathcal{C}^n$ we have a unique representation by *C.4.4* and *C.8.3*:

$$\mathbf{x} = \sum_{i=1}^{n} \zeta_i \mathbf{u}_i = \sum_{i=1}^{n} \langle\mathbf{v}_i,\mathbf{x}\rangle\mathbf{u}_i$$

Hence

$$\mathbf{A}\mathbf{x} = \sum_{i=1}^{n} \langle\mathbf{v}_i,\mathbf{x}\rangle\mathbf{A}\mathbf{u}_i = \sum_{i=1}^{n} \lambda_i\langle\mathbf{v}_i,\mathbf{x}\rangle\mathbf{u}_i$$

from which *14* follows. The proof of *15* follows a similar computation.

16 *Comment* The formulae *14* and *15* are the main reason for introducing simple L.T. These formulae give an easily visualized interpretation of the relation between $\mathbf{x}$ and $\mathbf{A}\mathbf{x}$. Indeed, *14* shows that the simple L.T. $\mathbf{A}$ is completely described by a basis made up of its eigenvectors $\{\mathbf{u}_1, \ldots, \mathbf{u}_n\}$ and the corresponding eigenvalues λ_i. For any $\mathbf{x}$ consider the expansion of $\mathbf{x}$ along the eigenvectors $\mathbf{u}_1, \ldots, \mathbf{u}_n$ together with the corresponding components (see Fig. *C.18.1*). $\mathbf{A}\mathbf{x}$ is the vector having each one of its components multiplied by the corresponding eigenvalue. This very simple visualization of the effect of operating on $\mathbf{x}$ by $\mathbf{A}$ is the geometrical equivalent to choosing $\{\mathbf{u}_1,\mathbf{u}_2, \ldots, \mathbf{u}_n\}$ as basis vectors and observing that with respect to that basis $\mathbf{A}$ is represented by the diagonal matrix, diag $(\lambda_1,\lambda_2, \ldots, \lambda_n)$. An analogous interpretation holds for *15*.

Anticipating a bit what is to follow (*D.4.1* in particular), let us observe that if the function f is analytic in an open set containing the eigenvalues λ_i of $\mathbf{A}$, then $f(\mathbf{A})$ is *defined* by

17 $$f(\mathbf{A}) = \sum_{i=1}^{n} f(\lambda_i)\mathbf{E}_i = \sum_{i=1}^{n} f(\lambda_i)\mathbf{u}_i\rangle\langle\mathbf{v}_i$$

In particular,

18 $$\mathbf{A}^k = \sum_{i=1}^{n} \lambda_i^k \mathbf{u}_i\rangle\langle\mathbf{v}_i$$

19 $$e^{\mathbf{A}t} = \sum_{i=1}^{n} e^{\lambda_i t}\mathbf{u}_i\rangle\langle\mathbf{v}_i$$

20 **Corollary** If f is an analytic function in an open set containing the eigenvalues λ_i of a simple L.T. $\mathbf{A}$ on $\mathcal{C}^n$ then the L.T. $f(\mathbf{A})$ has the same eigenvectors as $\mathbf{A}$ and its corresponding eigenvalues are $f(\lambda_i)$.

21 *Exercise* Let $\mathbf{A}$ be simple. Show that $\mathcal{C}^n = \mathcal{R}(\mathbf{A}) \oplus \mathcal{N}(\mathbf{A})$. Compare with *C.15.4*.

19 *Normal L.T.*

In this section we shall consider a subclass of simple L.T., the normal transformation. It will show that for any normal matrix there is an *orthonormal* set of eigenvectors which span the space.

1 **Definition** An L.T. $\mathcal{A}$ is called *normal* if $\mathcal{A}^*\mathcal{A} = \mathcal{A}\mathcal{A}^*$. If $\mathbf{A}$ is the matrix representation of a normal L.T. $\mathcal{A}$ relative to any basis, then $\mathbf{A}^*\mathbf{A} = \mathbf{A}\mathbf{A}^*$. If the basis is orthonormal, then the matrix $\mathbf{A}$ commutes with its *conjugate transpose* $\mathbf{A}^*$.

In order to visualize the scope of normal L.T., consider the following

2 *Exercise* A matrix is normal if it belongs to one of the following classes: (I) hermitian ($\mathbf{A}^* = \mathbf{A}$), (II) skew hermitian ($\mathbf{A}^* = -\mathbf{A}$), (III) real symmetric ($a_{ij} = a_{ji} = a_{ij}^*$), (IV) real skew symmetric ($a_{ij} = -a_{ji} = a_{ij}^*$),

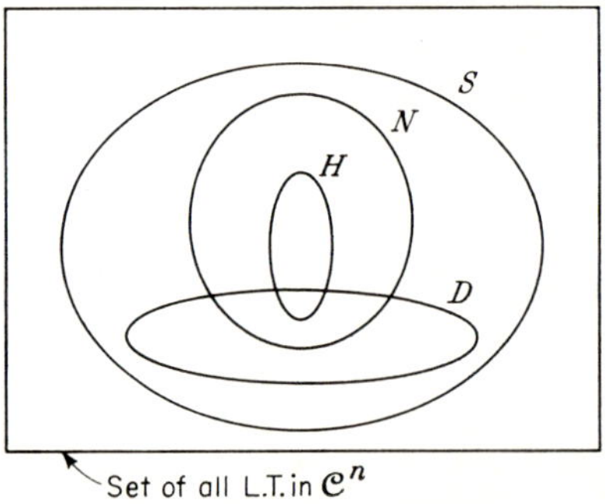

Fig. C.19.1 Inclusion relations between the set of all L.T. in $\mathcal{C}^n$; the set of all simple (S), normal (N), and hermitian (H), L.T.; and the L.T. that have distinct eigenvalues (D).

(V) unitary ($\mathbf{AA}^* = \mathbf{I}$), (VI) orthogonal (all a_{ij}'s real, and $\mathbf{AA}^* = \mathbf{I}$), and (VII) diagonal ($a_{ij} = 0$, if $i \neq j$). We therefore arrive at the following classification of L.T. in $\mathcal{C}^n$. Consider Fig. *C.19.1*. The rectangle represents the set of all L.T. in $\mathcal{C}^n$. The set S is the set of all simple L.T., and the sets D, N, H respectively represent the set of all L.T. with distinct eigenvalues, the set of all normal L.T., and the set of all hermitian L.T. We have $H \subset N \subset S$ and $D \subsetneq S$. However, there are L.T. in D that are not in N, for example, the matrix

$$\begin{bmatrix} 1 & 2 \\ 0 & 3 \end{bmatrix}$$

3 **Theorem** Let $\mathbf{A}$ be normal, $\mathbf{Au} = \lambda\mathbf{u} \Leftrightarrow \mathbf{A}^*\mathbf{u} = \lambda^*\mathbf{u}$.
Proof By assumption, $(\mathbf{A} - \lambda\mathbf{I})\mathbf{u} = \mathbf{0}$. Since $\mathbf{A}$ is normal and since λ is a number, $\mathbf{A} - \lambda\mathbf{I} \triangleq \mathbf{B}$ is also normal (note that $\mathbf{B}^* = \mathbf{A}^* - \lambda^*\mathbf{I}$). Hence,

$$\begin{aligned} \|\mathbf{Bu}\|^2 &= \langle \mathbf{Bu},\mathbf{Bu}\rangle = \langle \mathbf{u},\mathbf{B}^*\mathbf{Bu}\rangle = \langle \mathbf{u},\mathbf{BB}^*\mathbf{u}\rangle \\ &= \langle \mathbf{B}^*\mathbf{u},\mathbf{B}^*\mathbf{u}\rangle = \|\mathbf{B}^*\mathbf{u}\|^2 \end{aligned}$$

Hence $\mathbf{Bu} = \mathbf{0}$ if and only if $\mathbf{B}^*\mathbf{u} = \mathbf{0}$.

4 **Corollary** The eigenvalues of a hermitian matrix are real.

5 *Exercise* Prove Corollary *4* directly from $\mathbf{A} = \mathbf{A}^*$. (*Hint*: Consider $\langle \mathbf{u},\mathbf{Au}\rangle = \lambda\langle \mathbf{u},\mathbf{u}\rangle$ and its conjugate. Use the properties of the scalar product and the definitions of $\mathbf{A}^*$ and of a hermitian matrix.)

6 **Theorem** Let $\mathbf{A}$ be normal; then the eigenvectors corresponding to distinct eigenvalues are orthogonal. That is, if $\mathbf{Au}_1 = \lambda_1\mathbf{u}_1$ and $\mathbf{Au}_2 = \lambda_2\mathbf{u}_2$ with $\lambda_1 \neq \lambda_2$, then $\langle \mathbf{u}_1,\mathbf{u}_2\rangle = 0$.
Proof From the assumption and Theorem *3*, $\mathbf{u}_1$ is also an eigenvector of $\mathbf{A}^*$ and $\mathbf{A}^*\mathbf{u}_1 = \lambda_1^*\mathbf{u}_1$. Therefore, $\langle \mathbf{u}_2,\mathbf{A}^*\mathbf{u}_1\rangle = \lambda_1^*\langle \mathbf{u}_2,\mathbf{u}_1\rangle$. Hence, taking the complex conjugate of this relation and using the definition of $\mathbf{A}^*$, $\langle \mathbf{u}_1,\mathbf{Au}_2\rangle = \lambda_1\langle \mathbf{u}_1,\mathbf{u}_2\rangle$. From the assumption, $\langle \mathbf{u}_1,\mathbf{Au}_2\rangle = \lambda_2\langle \mathbf{u}_1,\mathbf{u}_2\rangle$. Therefore, by subtraction $0 = (\lambda_1 - \lambda_2)\langle \mathbf{u}_1,\mathbf{u}_2\rangle$, from which the conclusion follows, since $\lambda_1 \neq \lambda_2$. ◁

The most important result concerning normal L.T. is their *spectral representation*, which may be stated in the form of

7 **Theorem** Let $\mathbf{A}$ be *normal*. Let $\lambda_1, \lambda_2, \ldots, \lambda_\sigma$ be its distinct eigenvalues. Then

(I) $\mathcal{C}^n = \mathfrak{N}(\mathbf{A} - \lambda_1\mathbf{I}) \oplus \mathfrak{N}(\mathbf{A} - \lambda_2\mathbf{I}) \oplus \cdots \oplus \mathfrak{N}(\mathbf{A} - \lambda_\sigma\mathbf{I})$ with

$$\mathfrak{N}(\mathbf{A} - \lambda_i\mathbf{I}) \perp \mathfrak{N}(\mathbf{A} - \lambda_j\mathbf{I})$$

whenever $i \neq j$.
(II) Let $\mathbf{E}_i$ be the orthogonal projection of $\mathcal{C}^n$ on $\mathfrak{N}(\mathbf{A} - \lambda_i\mathbf{I})$,

$$i = 1, 2, \ldots, \sigma$$

Then

8
$$\begin{aligned} \mathbf{E}_i^2 = \mathbf{E}_i &= \mathbf{E}_i^* & i &= 1, 2, \ldots \\ \mathbf{E}_i\mathbf{E}_j &= \mathbf{0} & i &\neq j \end{aligned}$$

and

9
$$\mathbf{A} = \sum_{i=1}^{\sigma} \lambda_i\mathbf{E}_i$$

(III) There exists a set of *orthonormal* eigenvectors which span the space; hence

10 $$\mathbf{A} = \sum_{i=1}^{n} \lambda_{\nu_i}\mathbf{u}_i\rangle\,\langle\mathbf{u}_i$$

where λ_{ν_i} is the eigenvalue associated with the eigenvector $\mathbf{u}_i$ and the integers $\nu_i, \ldots, \nu_n$ belong to the set $\{1,2, \ldots ,\sigma\}$.

11 *Comment* (I) implies that the eigenvectors (associated with possibly multiple eigenvalues of **A**) span the space $\mathcal{C}^n$; hence, any normal transformation is a simple transformation. Furthermore, if in each

$$\mathfrak{N}(\mathbf{A} - \lambda_i\mathbf{I})$$

one picks an orthonormal basis (whose elements are of necessity eigenvectors associated with the same eigenvalue λ_i), then the union of all these bases constitutes an orthonormal basis for $\mathcal{C}^n$ made up of eigenvectors of **A**. With respect to such an orthonormal basis the matrix representation of the L.T. $\mathfrak{A}$ is diagonal.

12 *Comment* Equations *9* and *10* should be compared with *C.18.14* and *C.18.15*. For the special class of normal transformations, the projections $\mathbf{E}_i$ are now *orthogonal projections;* and since the basis $\{\mathbf{u}_1,\mathbf{u}_2, \ldots ,\mathbf{u}_n\}$ is orthonormal, one need not in *10* use the reciprocal basis as one had to in *C.18.14*.

13 *Comment* Equations *9* and *10* are applicable in the case of self-adjoint L.T. in more general spaces with the only modification that the summation is over an infinite, not necessarily denumerable, set. (See, e.g., Refs. *2*, *10*, *12*, and *14*.) ◁

Before proving Theorem *7*, let us deduce two important consequences.

14 **Theorem** If **A** is normal and if R, r respectively denote the largest and smallest real parts of the eigenvalues of **A**, then

$$2r \le \frac{\langle\mathbf{x},\mathbf{A}\mathbf{x}\rangle + \langle\mathbf{x},\mathbf{A}^*\mathbf{x}\rangle}{\langle\mathbf{x},\mathbf{x}\rangle} \le 2R \qquad \forall\mathbf{x} \in \mathfrak{X}$$

Proof Exercise for the reader. (*Hint:* Use *8* and *9*.)

15 **Corollary** If **A** is hermitian, then

$$\lambda \le \frac{\langle\mathbf{x},\mathbf{A}\mathbf{x}\rangle}{\langle\mathbf{x},\mathbf{x}\rangle} \le \Lambda$$

where Λ, λ are the largest and the smallest eigenvalues of **A**, respectively.
Proof **A** is hermitian and hence is normal. However, by *4*, all its eigenvalues are real. Hence *14* ⇒ *15*.

16 *Exercise* Let $\mathbf{A}$ be any $n \times n$ matrix. Using the norm defined by $\|\mathbf{x}\|^2 = \langle \mathbf{x},\mathbf{x} \rangle$, show that $\|\mathbf{A}\| \leq \operatorname{tr}(\mathbf{A}^*\mathbf{A})$. [*Hint:* Observe that $\mathbf{A}^*\mathbf{A}$ is hermitian and nonnegative; also $\operatorname{tr}(\mathbf{A}^*\mathbf{A})$ is the sum of the eigenvalues of $\mathbf{A}^*\mathbf{A}$.]

17 *Exercise* Let the n-tuples $\mathbf{a}_i = (a_{1i}, a_{2i}, \ldots, a_{ni})$ be the components of n linearly independent vectors with respect to some orthonormal basis. Show that $\mathbf{A} = (\alpha_{ij})$, where $\alpha_{ij} = \langle \mathbf{a}_i, \mathbf{a}_j \rangle$, is hermitian and positive definite (that is, $\langle \mathbf{x}, \mathbf{A}\mathbf{x} \rangle > 0$ for all $\mathbf{x} \neq \mathbf{0}$). *Hint:* For all $\boldsymbol{\xi} \neq \mathbf{0}$,

$$\Big\| \sum_i \xi_i \mathbf{a}_i \Big\|^2 = \sum_{i,j} \xi_i^* \xi_j \alpha_{ij} > 0$$

18 *Exercise* A hermitian matrix is positive definite if and only if all its eigenvalues are positive (>0). (*Hint:* Use *15* and the definition of an eigenvector.)

19 *Exercise* If $\mathbf{A}$ is a positive definite hermitian matrix, then $\mathbf{A}$ is nonsingular (that is, $\det \mathbf{A} \neq 0$). Furthermore, using the representation *10*, show that the inverse of $\mathbf{A}$ is given by

$$\mathbf{A}^{-1} = \sum_{i=1}^{n} \lambda_{\nu_i}^{-1} \mathbf{u}_i \rangle \langle \mathbf{u}_i$$

Let us leave the proof of Theorem *7* as a graded set of exercises.

20 *Exercise* Prove that $\mathfrak{N}(\mathbf{A} - \lambda_i \mathbf{I}) \perp \mathfrak{N}(\mathbf{A} - \lambda_j \mathbf{I})$ for $i \neq j$ follows directly from Theorem *6*. Show that this also implies $\mathbf{E}_i \mathbf{E}_j = \mathbf{0}$ for $i \neq j$, since $\mathfrak{R}(\mathbf{E}_k) = \mathfrak{N}(\mathbf{A} - \lambda_k \mathbf{I})$ for all k's.

21 *Exercise* Let $\mathbf{A}$ be normal; let λ_i be an eigenvalue of $\mathbf{A}$; then $\mathfrak{N}(\mathbf{A} - \lambda_i \mathbf{I})$ and its orthogonal complement are invariant subspaces under $\mathbf{A}$. *Hint:* Observe that $\mathbf{A}$ commutes with $\mathbf{A} - \lambda_i \mathbf{I}$ and that

$$\mathbf{A}^*(\mathbf{A} - \lambda_i \mathbf{I}) = \mathbf{A}\mathbf{A}^* - \lambda_i \mathbf{A}^* = (\mathbf{A} - \lambda_i \mathbf{I})\mathbf{A}^*$$

22 *Exercise* Show that the dimension of $\mathfrak{N}(\mathbf{A} - \lambda_i \mathbf{I})$ is equal to the multiplicity of λ_i as a zero of $\Delta(\lambda)$. (*Hint:* For conciseness let

$$\mathfrak{N} \triangleq \mathfrak{N}(\mathbf{A} - \lambda_i \mathbf{I})$$

and $\mathfrak{N}^\perp$ be its orthogonal complement. Both $\mathfrak{N}$ and $\mathfrak{N}^\perp$ are invariant under $\mathbf{A}$. Let $\mathbf{A}_1$ (and $\mathbf{A}_2$) be the L.T. a restricted to the domain $\mathfrak{N}$ (and $\mathfrak{N}^\perp$), respectively. Let d be the dimension of $\mathfrak{N}$. Then $\mathbf{A}_1$ ($\mathbf{A}_2$, respectively) is representable by a $d \times d$ matrix [respectively, $(n - d) \times (n - d)$ matrix]. Since $\mathfrak{N}$ and $\mathfrak{N}^\perp$ are invariant,

$$\Delta(\lambda) \triangleq \det(\mathbf{A} - \lambda \mathbf{I}) = \det(\mathbf{A}_1 - \lambda \mathbf{I}) \det(\mathbf{A}_2 - \lambda \mathbf{I})$$

Since $\mathbf{A}_1$ is restricted to $\mathfrak{N}$, each vector of any basis of $\mathfrak{N}$ is an eigenvector of $\mathbf{A}$ associated with the same eigenvalue λ_i; hence,

$$\det(\mathbf{A}_1 - \lambda_i \mathbf{I}) = (\lambda_i - \lambda)^d$$

From the definition of $\mathfrak{N}^\perp$, $\det(\mathbf{A}_2 - \lambda_i\mathbf{I}) \neq 0$; hence λ_i is a root of multiplicity d of $\Delta(\lambda)$.)

23 *Exercise* Use Theorem *6* and Exercise *20* to show that

$$\mathfrak{N}(\mathbf{A} - \lambda_1\mathbf{I}) \oplus \mathfrak{N}(\mathbf{A} - \lambda_2\mathbf{I}) \oplus \cdots \oplus \mathfrak{N}(\mathbf{A} - \lambda_\sigma\mathbf{I}) = \mathcal{C}^n$$

20 *Comment on the adjoint*

The adjoint of an L.T. $\mathbf{A}$ is a concept that has appeared repeatedly throughout the text and throughout this appendix. We propose here to exhibit together some of the facts concerning the adjoint that cause it to appear in so many apparently unrelated instances.

Looking at $\mathbf{A}$ as a mapping of $\mathcal{C}^n$ into $\mathcal{C}^n$, we have the following facts[1] (see *C.15.4*):

1 $$\mathcal{C}^n = \mathcal{R}(\mathbf{A}^*) \overset{\perp}{\oplus} \mathfrak{N}(\mathbf{A}) = \mathcal{R}(\mathbf{A}) \overset{\perp}{\oplus} \mathfrak{N}(\mathbf{A}^*)$$

2 $\mathbf{A}$ is a one-one mapping of $\mathcal{R}(\mathbf{A}^*)$ onto $\mathcal{R}(\mathbf{A})$

These facts are illustrated by Fig. *C.17.1* and are basic to the understanding of the problems involved in the solution of a set of linear algebraic equations and in the definition of the pseudo inverse of $\mathbf{A}$.

If, in addition, $\mathbf{A}$ is a simple L.T., then we have the spectral representation (see *C.18.13*)

3 $$\mathbf{A} = \sum_{i=1}^{n} \lambda_i\mathbf{u}_i\rangle\langle\mathbf{v}_i$$

where $\mathbf{u}_i$ is an eigenvector of $\mathbf{A}$ associated with the eigenvalue λ_i; $\mathbf{v}_i$ is the ith vector of the basis reciprocal to the basis $\{\mathbf{u}_1,\mathbf{u}_2, \ldots ,\mathbf{u}_n\}$; and $\mathbf{v}_i$ is an eigenvector of $\mathbf{A}^*$ associated with the eigenvalue λ_i^* (see *C.18.11*).

Considering now differential equations of the form $\dot{\mathbf{x}} = \mathbf{A}\mathbf{x} + \mathbf{a}$, we assume that $\mathbf{A}$ is constant. If $\mathbf{A}$ is a simple L.T., the Green dyadic is given by (see *5.2.24*)

4 $$\begin{aligned}\mathbf{G}(t,\xi) &= 1(t-\xi) \sum_{i=1}^{n} e^{\lambda_i(t-\xi)}\mathbf{u}_i\rangle\langle\mathbf{v}_i \\ &= 1(t-\xi) \sum_{i=1}^{n} e^{\lambda_i t}\mathbf{u}_i\rangle\langle e^{-\lambda_i^*\xi}\mathbf{v}_i\end{aligned}$$

Let us recall that t is the instant of observation and ξ is the instant at which the input is applied. (t and ξ play along the time axis the

[1] The notation $\overset{\perp}{\oplus}$ is used to indicate at the same time that the direct sum of $\mathcal{R}(\mathbf{A}^*)$ and $\mathfrak{N}(\mathbf{A})$ is $\mathcal{C}^n$ and that $\mathfrak{N}(\mathbf{A})$ is the orthogonal complement of $\mathcal{R}(\mathbf{A}^*)$.

same role as the observation point and the source point in the Green function of problems involving, for example, the Laplace or Helmholtz partial differential equation.) The behavior of **G** with respect to t is determined by the dynamics of the free system $\dot{\mathbf{x}} = \mathbf{A}\mathbf{x}$. Indeed, $e^{\lambda_i t}\mathbf{u}_i$ is one of its solutions; it is, in fact, a rather special one, since $\mathbf{u}_i$ is an eigenvector of **A** associated with the eigenvalue λ_i. The behavior of **G** with respect to ξ is determined by the dynamics of the adjoint system $d\mathbf{z}/d\xi = -\mathbf{A}^*\mathbf{z}$; indeed, $e^{-\lambda_i^* t}\mathbf{v}_i$ is one of its solutions and $\mathbf{v}_i$ is an eigenvector of $\mathbf{A}^*$ associated with the eigenvalue λ_i^* (of $\mathbf{A}^*$).

In the time-varying case, the Green dyadic becomes (see *6.2.44*)

$$\mathbf{G}(t,\xi) = -1(t-\xi)\sum_{i=1}^{n} \boldsymbol{\phi}_i(t,t_0)\rangle\,\langle\boldsymbol{\psi}_i(\xi,t_0)$$

Let us recall that $\boldsymbol{\phi}_i(t,t_0)$ is the ith column of the state transition matrix $\boldsymbol{\Phi}(t,t_0)$ and, hence, is a solution of $\dot{\mathbf{x}} = \mathbf{A}(t)\mathbf{x}$. $\boldsymbol{\psi}_i(\xi,t_0)$ is the ith column vector of the state transition matrix $\boldsymbol{\Psi}(\xi,t_0)$ of the adjoint problem $d\mathbf{z}/d\xi = -\mathbf{A}^*(\xi)\mathbf{z}$. So again the behavior of the Green dyadic **G** with respect to its first argument, the instant of observation t, is governed by the dynamics of the free system. The behavior of **G** with respect to its second argument ξ, the instant at which the input is applied, is governed by the dynamics of the adjoint system. Since in many engineering problems it is the second behavior that is of interest, it is natural that the adjoint system plays a key role in the solution of the problem. (See, for example, Remark *6.2.45*.)

REFERENCES AND SUGGESTED READING

1 Bellman, R.: "Introduction to Matrix Analysis," McGraw-Hill Book Company, Inc., New York, 1960.
2 Dunford, N., and J. T. Schwartz: "Linear Operators. Part I: General Theory," Interscience Publishers, Inc., New York, 1958.
3 Finkbeiner, D. T.: "Introduction to Matrices and Linear Transformations," W. H. Freeman and Company, San Francisco, 1960.
4 Frazer, R. A., W. J. Duncan, and A. R. Collar: "Elementary Matrices," Cambridge University Press, New York, 1938.
5 Gantmacher, F. R.: "The Theory of Matrices," vols. 1 and 2, Chelsea Publishing Company, New York, 1959.
6 Halmos, P. R.: "Finite Dimensional Vector Spaces," 2d ed., D. Van Nostrand Company, Inc., Princeton, N.J., 1950.
7 Hamburger, H. L., and M. E. Grimshaw: "Linear Transformations in n Dimensional Space," Cambridge University Press, New York, 1951.
8 Hoffman, K., and R. Kunze: "Linear Algebra," Prentice-Hall, Inc., Englewood Cliffs, N.J., 1961.
9 MacDuffee, C. C.: "The Theory of Matrices," Chelsea Publishing Company, New York, 1956.

10 Morse, P. M., and H. Feshbach: "Methods of Theoretical Physics," vols. 1 and 2, McGraw-Hill Book Company, Inc., New York, 1953.

11 Perlis, S.: "Theory of Matrices," Addison-Wesley Publishing Company, Inc., Reading, Mass., 1952.

12 Riesz, F., and B. Sz.-Nagy: "Functional Analysis," Frederick Ungar Publishing Co., New York, 1955.

13 Stewart, F. M.: "Introduction to Linear Algebra," D. Van Nostrand Company, Inc., Princeton, N.J., 1963.

14 Taylor, A. E.: "Introduction to Functional Analysis," John Wiley & Sons, Inc., New York, 1958.

D

Function of a matrix

1 *Introduction*

The purpose of this appendix is to present the definition of a function of a matrix and to derive computational techniques. In Sec. *2* we introduce the concept of the multiplicity of an eigenvalue (i.e., the order of the eigenvalue as a zero of the minimal polynomial). In Sec. *3* we define the index of an eigenvalue and show that the multiplicity of an eigenvalue is always equal to its index. With these results at hand, the definition of a function of a matrix follows quite naturally. In the next two sections we develop and use the geometric insight built up previously to obtain the fundamental formula for a function of a matrix. Section *8* is devoted to methods of computation.

2 *Minimal polynomial and multiplicity of an eigenvalue*

We know that the polynomial $\Delta(\lambda) = \det(\mathbf{A} - \lambda\mathbf{I})$ is called the *characteristic polynomial of* **A** and that its zeros are the *eigenvalues*. We shall denote them by $\lambda_1, \lambda_2, \ldots$. Given the polynomial

$$p(\lambda) = \sum_{k=0}^{N} a_k \lambda^k$$

we *define* $p(\mathbf{A})$ as the matrix equal to the polynomial $\sum_{k=0}^{N} a_k \mathbf{A}^k$, where $\mathbf{A}^0 = \mathbf{I}$. This definition implies that if two polynomials p and q are equal, that is, $p(\lambda) = q(\lambda)$ for all λ, then $p(\mathbf{A}) = q(\mathbf{A})$.

1 **Definition** The minimal polynomial of the matrix **A** is the polynomial $\psi(\lambda)$ of least degree such that $\psi(\mathbf{A}) = \mathbf{0}$ and the coefficient of the highest power of λ is unity.

2 **Assertion** If λ_i is an eigenvalue of $\mathbf{A}$, then $\psi(\lambda_i) = 0$.
Proof Exercise for the reader. [*Hint:* Let $\mathbf{u}_i \neq \mathbf{0}$ be an eigenvector of $\mathbf{A}$ associated with the eigenvalue λ_i; compute $\psi(\mathbf{A})\mathbf{u}_i$.] ◁

Consequently, we shall use throughout the notation

$$\psi(\lambda) = \prod_{i=1}^{\sigma} (\lambda - \lambda_i)^{m_i}$$

where the product is carried over the σ distinct eigenvalues of $\mathbf{A}$. The exponent m_i is called the *multiplicity of the eigenvalue* λ_i.[1]

3 *Example*

$$\text{For } \mathbf{A} = \begin{bmatrix} 2 & 0 \\ 0 & 2 \end{bmatrix}, \Delta(\lambda) = (\lambda - 2)^2, \psi(\lambda) = \lambda - 2$$

$$\text{For } \mathbf{A} = \begin{bmatrix} 1 & 1 & 0 \\ 0 & 1 & 0 \\ 0 & 0 & 1 \end{bmatrix}, \Delta(\lambda) = -(\lambda - 1)^3, \psi(\lambda) = (\lambda - 1)^2$$

4 *Example* The matrices

$$\begin{bmatrix} 2 & 0 & 0 & 0 \\ 0 & 2 & 0 & 0 \\ 0 & 0 & 2 & 0 \\ 0 & 0 & 0 & 2 \end{bmatrix}, \begin{bmatrix} 2 & 1 & 0 & 0 \\ 0 & 2 & 0 & 0 \\ 0 & 0 & 2 & 1 \\ 0 & 0 & 0 & 2 \end{bmatrix}, \begin{bmatrix} 2 & 0 & 0 & 0 \\ 0 & 2 & 1 & 0 \\ 0 & 0 & 2 & 1 \\ 0 & 0 & 0 & 2 \end{bmatrix}, \begin{bmatrix} 2 & 1 & 0 & 0 \\ 0 & 2 & 1 & 0 \\ 0 & 0 & 2 & 1 \\ 0 & 0 & 0 & 2 \end{bmatrix}$$

all have a common characteristic polynomial $(\lambda - 2)^4$ but have, respectively, $\lambda - 2$, $(\lambda - 2)^2$, $(\lambda - 2)^3$, $(\lambda - 2)^4$ as minimal polynomials.

5 *Exercise* If $\mathbf{T}$ is a nonsingular matrix and if ψ is the minimal polynomial of $\mathbf{A}$, then ψ is also the minimal polynomial of $\mathbf{T}^{-1}\mathbf{A}\mathbf{T}$. [*Hint:* For any integer k, $\mathbf{T}^{-1}\mathbf{A}^k\mathbf{T} = (\mathbf{T}^{-1}\mathbf{A}\mathbf{T})^k$.]

6 *Exercise* Show that $\Delta(\lambda) = \psi(\lambda) \cdot$ [greatest common divisor of all minors of order $(n - 1)$ of $(\mathbf{A} - \lambda\mathbf{I})$]. [*Hint:* Consider the matrix $\mathbf{B}(\lambda)$ such that $b_{ik}(\lambda) =$ cofactor of the (k,i) element of $\mathbf{A} - \lambda\mathbf{I}$. Observe that $\mathbf{B}(\lambda)(\mathbf{A} - \lambda\mathbf{I}) = \Delta(\lambda)\mathbf{I}$. Cancel common factors, etc.]

7 *Exercise* Let $1(t)$ be the unit step function; let $\mathbf{f}$ be an arbitrary constant vector. Assume $\mathbf{x}(0) = \mathbf{0}$, and use Laplace transforms to compute the response of the system S defined by

$$S: \qquad \dot{\mathbf{x}}(t) = \mathbf{A}\mathbf{x}(t) + 1(t)\mathbf{f}$$

Show that, irrespective of what $\mathbf{f}$ might be, if degree $\psi <$ degree Δ, then the time dependence of $\mathbf{x}(t)$ involves only terms of the form $e^{\lambda_k t}$, $te^{\lambda_k t}$, . . . , $t^{m_k - 1}e^{\lambda_k t}$, where m_k is the multiplicity of the root λ_k of $\psi(\lambda)$. Therefore, it is the minimal polynomial ψ (and not the characteristic polynomial Δ) which characterizes the dynamics of S. (*Hint:* Use *6*.)

[1] This definition of multiplicity of an eigenvalue does not agree with that of all authors. The reader is cautioned to ascertain what each author means by the expression.

8 **Theorem** Let $\psi(\lambda) = \prod_{i=1}^{\sigma} (\lambda - \lambda_i)^{m_i}$ be the minimal polynomial of the matrix **A**. Let p and q be two arbitrary polynomials. Then

$$p(\mathbf{A}) = q(\mathbf{A}) \Leftrightarrow p \equiv q \qquad \text{mod } \psi$$

That is, the matrices $p(\mathbf{A})$ and $q(\mathbf{A})$ are equal if and only if the polynomials p and q have the same remainder when divided by the polynomial ψ.

Proof $\Leftarrow$ Divide p and q by ψ and let q_1 and q_2 be the quotients. Then, by assumption, the remainders are identical, say, r_1. Thus

$$\begin{aligned} p &= q_1\psi + r_1 \\ q &= q_2\psi + r_1 \end{aligned}$$

Hence, since the two equations above are identities in λ, we may substitute **A** for λ and get

$$\begin{aligned} p(\mathbf{A}) &= q_1(\mathbf{A})\psi(\mathbf{A}) + r_1(\mathbf{A}) = r_1(\mathbf{A}) \\ q(\mathbf{A}) &= q_2(\mathbf{A})\psi(\mathbf{A}) + r_1(\mathbf{A}) = r_1(\mathbf{A}) \end{aligned}$$

Hence $p(\mathbf{A}) = q(\mathbf{A})$.

$\Rightarrow$ By assumption, $(p - q)(\mathbf{A}) = \mathbf{0}$. Hence from the definition of ψ the polynomial $p - q$ is a multiple of ψ; hence, $p \equiv q \pmod{\psi}$.

9 **Corollary** With the notation of *8*,

$$p(\mathbf{A}) = q(\mathbf{A}) \Leftrightarrow p^{(l)}(\lambda_k) = q^{(l)}(\lambda_k) \qquad \text{for } l = 0, 1, \ldots, m_k - 1, \quad k = 1, 2, \ldots, \sigma$$

Proof $\Rightarrow$ By Theorem *8*, $p(\mathbf{A}) = q(\mathbf{A})$ implies that, for all λ's,

10 $$p(\lambda) = q_1(\lambda)\psi(\lambda) + r_1(\lambda)$$

11 $$q(\lambda) = q_2(\lambda)\psi(\lambda) + r_1(\lambda)$$

Since, for $k = 1, 2, \ldots, \sigma$, λ_k is a zero of order m_k of ψ, we have

$$(q_i\psi)^{(l)}\Big|_{\lambda=\lambda_k} = \sum_{\alpha=0}^{l} \binom{l}{\alpha} q_i^{(\alpha)}\psi^{(l-\alpha)}\Big|_{\lambda=\lambda_k} = 0 \qquad i = 1, 2$$

Therefore, from *10* and *11* we have

$$p^{(l)}(\lambda_k) = q^{(l)}(\lambda_k) \qquad k = 1, 2, \ldots, \sigma, \quad l = 0, 1, \ldots, m_k - 1$$

since they are both equal to $r_1^{(l)}(\lambda_k)$.

$\Leftarrow$ Since $p^{(l)}(\lambda_k) = q^{(l)}(\lambda_k)$ for $k = 1, 2, \ldots, \sigma$ and

$$l = 0, 1, \ldots, m_k - 1$$

and since p and q are polynomials, the polynomial $p - q$ is a multiple of $\prod_{k=1}^{\sigma} (\lambda - \lambda_k)^{m_k}$, hence of ψ. Therefore, $p - q = b\psi$, where b is a polynomial. Hence, $p(\mathbf{A}) - q(\mathbf{A}) = b(\mathbf{A})\psi(\mathbf{A})$ and

$$p(\mathbf{A}) = q(\mathbf{A}) \qquad \text{since } \psi(\mathbf{A}) = \mathbf{0}$$

3 *The index of an eigenvalue*

Now consider $\mathbf{A}$ as a linear transformation. From this point of view we have defined *an eigenvalue of* $\mathbf{A}$ as a value of λ for which the L.T. $\mathbf{A} - \lambda\mathbf{I}$ on $\mathcal{C}^n$ to $\mathcal{C}^n$ is not one-one. An eigenvalue thus defined is necessarily a zero of $\Delta(\lambda) = \det(\mathbf{A} - \lambda\mathbf{I})$.

1 **Definition** Let us denote by $\mathfrak{N}_i^k$ the null space of the L.T. $(\mathbf{A} - \lambda_i\mathbf{I})^k$. In symbols,

$$\mathfrak{N}_i^k \triangleq \{\mathbf{x} | (\mathbf{A} - \lambda_i\mathbf{I})^k\mathbf{x} = \mathbf{0}\}$$

Clearly,

$$\mathfrak{N}_i^1 \subset \mathfrak{N}_i^2 \subset \mathfrak{N}_i^3 \subset \cdots$$

2 **Definition** The *index of the eigenvalue* λ_i is the least integer such that

$$\mathfrak{N}_i^k = \mathfrak{N}_i^{k+1}$$

The index of λ_i will be denoted by ν_i. $\mathfrak{N}_i^{\nu_i}$, which we write $\mathfrak{N}_i$, is called the *generalized null space of* $\mathbf{A}$ *associated with the ith eigenvalue.*

3 *Example* Let $\mathbf{A} = \begin{bmatrix} 1 & 1 & 0 \\ 0 & 1 & 1 \\ 0 & 0 & 2 \end{bmatrix} \qquad \Delta(\lambda) = -(\lambda - 1)^2(\lambda - 2)$

Consider the eigenvalue $\lambda_1 = 1$. Since

$$\mathbf{A} - \mathbf{I} = \begin{bmatrix} 0 & 1 & 0 \\ 0 & 0 & 1 \\ 0 & 0 & 1 \end{bmatrix} \qquad (\mathbf{A} - \mathbf{I})^2 = \begin{bmatrix} 0 & 0 & 1 \\ 0 & 0 & 1 \\ 0 & 0 & 1 \end{bmatrix} = (\mathbf{A} - \mathbf{I})^3$$

and

$$\begin{aligned} \mathfrak{N}_1^1 &= \{\mathbf{x} | (\mathbf{A} - \mathbf{I})\mathbf{x} = \mathbf{0}\} = \{\mathbf{x} | x_2 = x_3 = 0\} \\ \mathfrak{N}_1^2 &= \{\mathbf{x} | (\mathbf{A} - \mathbf{I})^2\mathbf{x} = \mathbf{0}\} = \{\mathbf{x} | x_3 = 0\} \\ \mathfrak{N}_1^3 &= \mathfrak{N}_1^2 \end{aligned}$$

hence, $\nu_1 = 2$. Similarly, the index of $\lambda_2 = 2$ is $\nu_2 = 1$ with

$$\mathfrak{N}_2 = \{\mathbf{x} | x_1 = x_2 = x_3\}$$

4 *Exercise* For any integer k, $\mathfrak{N}_i^k$ is an *invariant subspace* of $\mathbf{A}$, that is,

$$\mathbf{x} \in \mathfrak{N}_i^k \text{ implies } \mathbf{Ax} \in \mathfrak{N}_i^k$$

Hint: Observe that $\mathbf{A}$ and $(\mathbf{A} - \lambda_i\mathbf{I})^k$ commute; hence,

$$(\mathbf{A} - \lambda_i\mathbf{I})^k\mathbf{Ax} = \mathbf{A}(\mathbf{A} - \lambda_i\mathbf{I})^k\mathbf{x}$$

5 *Exercise* If ν_i is the index of the eigenvalue λ_i, then for $l = 1, 2, \ldots,$

$$\mathfrak{N}_i^{\nu_i+l} = \mathfrak{N}_i^{\nu_i}$$

Hint: By definition it is true for $l = 1$. Next use induction, i.e., assume $\mathfrak{N}_i^{\nu_i+l-1} = \mathfrak{N}_i^{\nu_i}$ and show that $\mathfrak{N}_i^{\nu_i+l} = \mathfrak{N}_i^{\nu_i}$. Use the relation

$$(\mathbf{A} - \lambda_i\mathbf{I})^{\nu_i+l} = (\mathbf{A} - \lambda_i\mathbf{I})^{\nu_i+l-1}(\mathbf{A} - \lambda_i\mathbf{I}) \quad \lhd$$

The interesting fact is that the multiplicity m_i of λ_i is equal to the index ν_i of λ_i. In order to prove this fact, we establish the necessary and sufficient conditions for two polynomials in the L.T. $\mathbf{A}$ to be equal.

6 **Theorem** Let the L.T. $\mathbf{A}$ have the eigenvalues $\lambda_1, \lambda_2, \ldots, \lambda_\sigma$ with indices $\nu_1, \nu_2, \ldots, \nu_\sigma$, respectively. Let p and q be two polynomials. Then $p(\mathbf{A}) = q(\mathbf{A})$ if and only if, for $k = 1, 2, \ldots, \sigma$,

7 $$p^{(l)}(\lambda_k) = q^{(l)}(\lambda_k) \qquad l = 0, 1, 2, \ldots, \nu_k - 1$$

Proof Let z be the polynomial $p - q$. Then the theorem is equivalent to

$$z(\mathbf{A}) = \mathbf{0} \Leftrightarrow z^{(l)}(\lambda_k) = 0 \qquad l = 0, 1, \ldots, \nu_k - 1; k = 1, \ldots, \sigma$$

$\Rightarrow$ We may delete from $z(\lambda)$ any factor of the form $(\lambda - \mu)$, where μ is not an eigenvalue because the operator $(\mathbf{A} - \mu\mathbf{I})$ is one-one; hence, for any $\mathbf{y}$, $(\mathbf{A} - \mu\mathbf{I})\mathbf{y}$ is zero if and only if $\mathbf{y} = \mathbf{0}$. Let the resulting polynomial be r_1; then $r_1(\mathbf{A}) = \mathbf{0}$ by assumption. Let $r_1(\lambda) = \prod_{i=1}^{\sigma} (\lambda - \lambda_i)^{\gamma_i}$.

If λ_i is any eigenvalue of $\mathbf{A}$, there is a $\mathbf{y} \neq \mathbf{0}$ such that $\mathbf{Ay} = \lambda_i\mathbf{y}$; hence, $r_1(\mathbf{A})\mathbf{y} = r_1(\lambda_i)\mathbf{y} = \mathbf{0}$ because $r_1(\mathbf{A}) = \mathbf{0}$. Therefore, $r_1(\lambda_i) = 0$, that is, r_1 has a zero at each eigenvalue of $\mathbf{A}$. In other words, $\gamma_i \geq 1$ for each i. It remains to show that the multiplicity γ_i of each zero λ_i is at least as great as the index ν_i. Suppose it were not; that is, $\gamma_i < \nu_i$. Then there would be a vector $\mathbf{v}$ such that $\mathbf{0} \neq \mathbf{v} \in \mathfrak{N}_i^{\gamma_i+1}$ but $\mathbf{v} \notin \mathfrak{N}_i^{\gamma_i}$; thus, $(\mathbf{A} - \lambda_i\mathbf{I})^{\gamma_i+1}\mathbf{v} = \mathbf{0}$ and the vector $\mathbf{w}$ defined by $\mathbf{w} = (\mathbf{A} - \lambda_i\mathbf{I})^{\gamma_i}\mathbf{v}$ is nonzero. Let $r_1(\lambda) = (\lambda - \lambda_i)^{\gamma_i}t(\lambda)$; clearly the polynomial $t(\lambda)$ is such that $t(\lambda_i) \neq 0$, since γ_i is the multiplicity of the zero λ_i of r_1. Observe that $\mathbf{Aw} = \lambda_i\mathbf{w}$; hence,

$$\mathbf{0} = r(\mathbf{A})\mathbf{v} = t(\mathbf{A})(\mathbf{A} - \lambda_i\mathbf{I})^{\gamma_i}\mathbf{v} = t(\mathbf{A})\mathbf{w} = t(\lambda_i)\mathbf{w}$$

which is a contradiction since $t(\lambda_i) \neq 0$ and $\mathbf{w} \neq \mathbf{0}$. Therefore, $\gamma_i \geq \nu_i$ for each i, that is, $r_i(\lambda)$ has at λ_i a zero of multiplicity at least equal to ν_i. Hence $z(\lambda)$, which is a multiple of $r_i(\lambda)$, has the same property.

$\Leftarrow$ Let $\mathbf{u}_1, \mathbf{u}_2, \ldots, \mathbf{u}_n$ be a basis for the space For each $i = 1, 2, \ldots, n$, $\mathbf{u}_i, \mathbf{A}\mathbf{u}_i, \mathbf{A}^2\mathbf{u}_i, \ldots, \mathbf{A}^n\mathbf{u}_i$, being $n + 1$ vectors in an n-dimensional space, are linearly dependent. Hence, for

$$i = 1, 2, \ldots, n$$

there exists a polynomial $p_i(\lambda)$ such that $p_i(\mathbf{A})\mathbf{u}_i = \mathbf{0}$. Let $m(\lambda)$ be the least common multiple of the n polynomials $p_1, p_2, \ldots, p_n$. $m(\mathbf{A}) = \mathbf{0}$ because, for any vector $\mathbf{x}$, $m(\mathbf{A})\mathbf{x} = \mathbf{0}$, since $m(\mathbf{A})\mathbf{u}_i = \mathbf{0}$ for $i = 1, 2, \ldots, n$. As in the first part of the proof, if we may delete from $m(\lambda)$ all factors of the form $(\lambda - \mu)$, where μ is not an eigenvalue of $\mathbf{A}$, we get a new polynomial m_1 such that $m_1(\mathbf{A}) = \mathbf{0}$. Again as above, the last equality implies that $m_1(\lambda) = \prod_{i=1}^{\sigma} (\lambda - \lambda_i)^{\gamma_i}$ with $\gamma_i \geq \nu_i$ for each i. Let $m_2(\lambda) = \prod_{i=1}^{\sigma} (\lambda - \lambda_i)^{\nu_i}$. We still have $m_2(\mathbf{A}) = \mathbf{0}$ because the fact that $\mathfrak{N}_i^{\nu_i+k} = \mathfrak{N}_i^{\nu_i}$, $k = 1, 2, \ldots$, implies that the null space of $(\mathbf{A} - \lambda_i\mathbf{I})^{\nu_i+k}$ is the same as that of $(\mathbf{A} - \lambda_i\mathbf{I})^{\nu_i}$. Hence the null space of $m_2(\mathbf{A})$ is the same as that of $m_1(\mathbf{A})$, namely, the whole space. By assumption, the polynomial $z(\lambda)$ has a zero of order ν_k at λ_k for $k = 1, 2, \ldots \sigma$ and $m_2(\lambda) = \prod_{i=1}^{\sigma} (\lambda - \lambda_i)^{\nu_i}$. Therefore, $z(\lambda)$ is a multiple of $m_2(\lambda)$ and $m_2(\mathbf{A}) = \mathbf{0}$ implies $z(\mathbf{A}) = \mathbf{0}$. Q.E.D.

8 **Corollary** Let $\mathbf{A}$ be an L.T. in the n-dimensional space with eigenvalues $\lambda_1, \lambda_2, \ldots, \lambda_\sigma$ with respective multiplicities $m_1, m_2, \ldots, m_\sigma$ and indices $\nu_1, \nu_2, \ldots, \nu_s$. Then

$$m_i = \nu_i \qquad i = 1, 2, \ldots, \sigma$$

In other words, the multiplicity of an eigenvalue is equal to its index.

4 *Definition of a function of a matrix*[1]

1 **Definition** Let $\mathbf{A}$ be a matrix with eigenvalues $\lambda_1, \lambda_2, \ldots, \lambda_\sigma$. Let f be a function analytic in an open set containing $\lambda_1, \lambda_2, \ldots, \lambda_\sigma$. Let $p(\lambda)$ be a polynomial such that

[1] In the following we may interchangeably replace the word *matrix* by *linear transformation on* $\mathcal{C}^n$.

2 $$p^{(l)}(\lambda_k) = f^{(l)}(\lambda_k) \qquad l = 0, 1, 2, \ldots, m_k - 1; k = 1, 2, \ldots, \sigma$$

Then we define the function f of the matrix $\mathbf{A}$ as

$$f(\mathbf{A}) = p(\mathbf{A})$$

3 *Comment* This definition makes sense, since Theorem *D.2.8* guarantees that the result is independent of the choice of the polynomial p.

4 **Corollary** Let f, g be two functions analytic in an open set containing the λ_i's; then

$$f(\mathbf{A}) + g(\mathbf{A}) = h(\mathbf{A}) \qquad \text{where } h = f + g$$
$$f(\mathbf{A})g(\mathbf{A}) = k(\mathbf{A}) \qquad \text{where } k = fg$$

5 **Corollary** If $f(\lambda) = \sum_{k=0}^{N} a_k\lambda^k$, then $f(\mathbf{A}) = \sum_{k=0}^{N} a_k\mathbf{A}^k$.

6 **Corollary** $f(\mathbf{A}) = \mathbf{0}$ if and only if $f^{(l)}(\lambda_k) = 0$, $l = 0, 1, 2, \ldots, m_k - 1$; $k = 1, 2, \ldots, \sigma$.

7 *Exercise* Prove Corollaries *4* to *6*. (*Hint:* Refer to the definition.)

8 *Exercise* Show that for any matrix $\mathbf{A}$

$$e^{j\mathbf{A}} = \cos \mathbf{A} + j \sin \mathbf{A}$$
$$\sin 2\mathbf{A} = 2 \sin \mathbf{A} \cos \mathbf{A}$$
$$\cos^2 \mathbf{A} + \sin^2 \mathbf{A} = \mathbf{I}$$
$$\exp [\mathbf{A}(t_1 + t_2)] = [\exp (\mathbf{A}t_1)][\exp (\mathbf{A}t_2)] \qquad \forall t_1,t_2$$
$$\sin \mathbf{A}(t_1 + t_2) = \sin \mathbf{A}t_1 \cos \mathbf{A}t_2 + \cos \mathbf{A}t_1 \sin \mathbf{A}t_2 \qquad \forall t_1,t_2$$

5 *Geometric structure of the L.T.* A

Let $\mathbf{A}$ have $\lambda_1, \lambda_2, \ldots, \lambda_\sigma$ as eigenvalues with multiplicities $m_1, m_2, \ldots, m_\sigma$, respectively. Consider the polynomials $\varphi_k(\lambda)$ defined as follows: for each k, $1 \leq k \leq \sigma$,

1 $$\varphi_k(\lambda_k) = 1$$
$$\varphi_k^{(l)}(\lambda_k) = 0 \qquad l = 1, 2, \ldots, m_k - 1$$
$$\varphi_k^{(l)}(\lambda_j) = 0 \qquad l = 0, 1, 2, \ldots, m_j - 1; \; j = 1, 2, \ldots, k - 1, k + 1, \ldots \sigma$$

These polynomials are easily obtained from the partial-fraction expansion of $1/\psi(\lambda)$, where $\psi(\lambda)$ is the minimal polynomial of $\mathbf{A}$. Let us first expand in partial fractions and then collect terms:

2 $$\frac{1}{\psi(\lambda)} = \sum_{k=1}^{\sigma} \sum_{l=0}^{m_k-1} \frac{c_{kl}}{(\lambda - \lambda_k)^{l+1}} = \sum_{k=1}^{\sigma} \frac{n_k(\lambda)}{(\lambda - \lambda_k)^{m_k}}$$

Here the $n_k(\lambda)$ are polynomials in λ and of degree $[n_k(\lambda)] < m_k$. Then it is easy to show that

3 $$\varphi_k(\lambda) = \frac{\psi(\lambda) n_k(\lambda)}{(\lambda - \lambda_k)^{m_k}}$$

The polynomials $\varphi_k(\lambda)$ given by *3* satisfy the above conditions because

4 $$\sum_{k=1}^{\sigma} \varphi_k(\lambda) = 1 \qquad \text{for all } \lambda\text{'s}$$

5 $\varphi_k(\lambda)$ contains the factors $(\lambda - \lambda_j)^{m_j}$ for $j \neq k$ and $1 \leq j \leq \sigma$.

6 **Theorem** Let the polynomials φ_k, $k = 1, 2, \ldots, \sigma$, be defined by *3* and let $\varphi_k(\mathbf{A}) \triangleq \mathbf{E}_k$; then

(I) $\mathbf{E}_k \cdot \mathbf{E}_k = \mathbf{E}_k$ that is, $\mathbf{E}_k$ is a projection (*C.12.9*)

(II) $\mathbf{E}_k \cdot \mathbf{E}_j = \mathbf{0}$ $j \neq k$; $j,k = 1, 2, \ldots, \sigma$

(III) $\sum_{k=1}^{\sigma} \mathbf{E}_k = \mathbf{I}$

(IV) $\mathcal{C}^n = \mathcal{R}(\mathbf{E}_1) \oplus \mathbf{R}(\mathbf{E}_2) \oplus \cdots \oplus \mathbf{R}(\mathbf{E}_\sigma)$ that is, the whole space $\mathcal{C}^n$ is the direct sum of the ranges of $\mathbf{E}_1, \mathbf{E}_2, \ldots, \mathbf{E}_\sigma$

Proof (I) follows from $\varphi_k^2(\mathbf{A}) = \varphi_k(\mathbf{A})$ by Theorem *D.2.8*. (II) follows from Corollary *D.4.6* and the fact that the polynomial $\varphi_k(\lambda)\varphi_j(\lambda)$ has a zero of order $\geq m_i$ at each λ_i. (III) follows from *4*. (IV) Since $\varphi_k(\mathbf{A}) = \mathbf{E}_k$ is a polynomial in $\mathbf{A}$, $\mathbf{E}_k$ and $\mathbf{A}$ commute; therefore, $\mathcal{R}(\mathbf{E}_k)$, the range of $\mathbf{E}_k$, is an invariant subspace of $\mathbf{A}$. Therefore by (II) and (III), we conclude that the whole space is the direct sum of the ranges of $\mathbf{E}_1, \mathbf{E}_2, \ldots, \mathbf{E}_\sigma$; hence (IV).

7 *Comment* The decomposition of any $\mathbf{x}$ in components $\mathbf{x}_k$ such that $\mathbf{x}_k \in \mathcal{R}(\mathbf{E}_k)$ is given by

8 $$\mathbf{x} = \sum_{k=1}^{\sigma} \mathbf{E}_k\mathbf{x} = \sum_{k=1}^{\sigma} \mathbf{x}_k \qquad \text{where } \mathbf{x}_k \triangleq \mathbf{E}_k\mathbf{x}$$

The decomposition is unique because of (IV).

9 *Exercise* Show that (I) $\sum_{k=1}^{\sigma} \varphi_k(\lambda) = 1$ for all λ's, (II) $\varphi_j(\lambda)\varphi_k(\lambda)$ is a multiple of $\psi(\lambda)$ if $j \neq k$, and (III) $[\varphi_k(\lambda)]^2 \equiv \varphi_k(\lambda) \pmod{\psi}$.

10 *Exercise* Use the results of *9* to prove Theorem *6* by invoking only the definition of $\psi(\lambda)$ and Theorem *D.2.8*.

11 *Exercise* Show that the range of $\mathbf{E}_k$ is an invariant subspace of $\mathbf{A}$.

12 **Theorem** The range of $\mathbf{E}_k$ is the generalized null space $\mathfrak{N}_k$ associated with the eigenvalue λ_k. Symbolically, for $k = 1, 2, \ldots, \sigma$,

13 $$\mathcal{R}(\mathbf{E}_k) = \mathfrak{N}_k = \{\mathbf{x} | (\mathbf{A} - \lambda_k\mathbf{I})^{m_k}\mathbf{x} = \mathbf{0}\}$$

14 *Comment* From (II) and (III) of Theorem *6* the whole space $\mathcal{C}^n$ may be considered as the direct sum of the ranges of $\mathbf{E}_1, \mathbf{E}_2, \ldots, \mathbf{E}_\sigma$. With the present theorem the whole space $\mathcal{C}^n$ is the direct sum of the generalized null spaces[1]

15 $$\mathcal{C}^n = \mathcal{R}(\mathbf{E}_1) \oplus \cdots \oplus \mathcal{R}(\mathbf{E}_\sigma)$$

and

$$\mathcal{C}^n = \mathfrak{N}_1 \oplus \mathfrak{N}_2 \oplus \cdots \oplus \mathfrak{N}_\sigma$$

Proof (I) We prove that

16 $$\mathcal{R}(\mathbf{E}_k) \subset \mathfrak{N}_k$$

Suppose $\mathbf{0} \neq \mathbf{x} \in \mathcal{R}(\mathbf{E}_k)$; then, for some $\mathbf{y} \neq \mathbf{0}$,

$$\mathbf{x} = \mathbf{E}_k\mathbf{y} = \varphi_k(\mathbf{A})\mathbf{y}$$

and $$(\mathbf{A} - \lambda_k\mathbf{I})^{m_k}\mathbf{x} = (\mathbf{A} - \lambda_k\mathbf{I})^{m_k}\varphi_k(\mathbf{A})\mathbf{y} = n_k(\mathbf{A})\psi(\mathbf{A})\mathbf{y} = \mathbf{0}$$

where in the last step we used *3*. Therefore, $\mathbf{x} \in \mathfrak{N}_k$.

(II) We prove next that $j \neq k$ implies $\mathfrak{N}_j \cap \mathfrak{N}_k = \{\mathbf{0}\}$.[2] This result, together with *6*(IV) and *16*, implies that $\mathcal{R}(\mathbf{E}_k) = \mathfrak{N}_k$. Let $\mathbf{x} \in \mathfrak{N}_k \cap \mathfrak{N}_j$. Let $\rho_k(\lambda) = (\lambda - \lambda_k)^{m_k}$ and $\rho_j(\lambda) = (\lambda - \lambda_j)^{m_j}$. Hence, $\rho_k(\mathbf{A})\mathbf{x} = \mathbf{0}$ and $\rho_j(\mathbf{A})\mathbf{x} = \mathbf{0}$. Since ρ_k and ρ_j are relatively prime polynomials, there exist polynomials g_j and g_k such that, for all λ's,

$$g_k(\lambda)\rho_k(\lambda) + g_j(\lambda)\rho_j(\lambda) = 1$$

Hence $$g_k(\mathbf{A})\rho_k(\mathbf{A})\mathbf{x} + g_j(\mathbf{A})\rho_j(\mathbf{A})\mathbf{x} = \mathbf{x}$$

Since the left-hand side is equal to zero, $\mathbf{x} = \mathbf{0}$. Hence the only vector common to both $\mathfrak{N}_k$ and $\mathfrak{N}_j$ is the zero vector. Q.E.D.

17 **Corollary** The dimension of $\mathcal{R}(\mathbf{E}_k)$ (and of $\mathfrak{N}_k$ by Theorem *12*) is r_k, the order of λ_k as a zero of $\Delta(\lambda)$. [The proof follows directly from considering the system of n homogeneous equations $(\mathbf{A} - \lambda_k\mathbf{I})\mathbf{x} = \mathbf{0}$. The number of linearly independent solutions is equal to the order of the zero λ_k of det $(\mathbf{A} - \lambda\mathbf{I})$.] ◁

[1] Compare this result with Theorem *C.19.7*.

[2] That is, the only vector common to $\mathfrak{N}_j$ and $\mathfrak{N}_k$ is the zero vector $\mathbf{0}$.

The partition of the whole space $\mathcal{C}^n$ into invariant subspaces in Eq. *15* is of great importance. From *C.13.5* it follows that if we take as a basis for the whole space the union of bases of these invariant subspaces, the matrix representation of **A** will consist of diagonal blocks:

18
$$\begin{bmatrix} \mathbf{M}_1 & \mathbf{0} & \cdot & \cdot & \mathbf{0} \\ \mathbf{0} & \mathbf{M}_2 & & & \cdot \\ \cdot & & \mathbf{M}_3 & & \cdot \\ \cdot & & & \ddots & \cdot \\ \mathbf{0} & \cdots & \cdots & \cdots & \mathbf{M}_k \end{bmatrix}$$

The submatrix $\mathbf{M}_i$ represents **A** in the subspace $\mathfrak{N}_i$. The Jordan canonical form is obtained by picking in each $\mathfrak{N}_i$ a special basis. It can be shown that each $\mathbf{M}_i$ is then such that its only nonzero elements are on the main diagonal and on the diagonal next to the main one (either above it or below it).

19 *Exercise* Show that

$$\mathcal{C}^n = \mathfrak{N}_k \oplus \mathcal{R}[(\mathbf{A} - \lambda_k \mathbf{I})^{m_k}]$$

(*Hint:* Observe that $\mathfrak{N}_k = \mathcal{R}[\varphi_k(\mathbf{A})]$ and that, in the notation of the proof of *12*, φ_k and ρ_k are relatively prime.)

20 *Exercise* By a suitable choice of basis, **A** may be diagonalized if and only if all roots of $\psi(\lambda)$ are simple. This is equivalent to saying **A** is a simple L.T. if and only if all roots of $\psi(\lambda)$ are simple (see *C.18.9*).

21 *Exercise* The Jordan canonical form can be expressed as follows: there is a basis $\{\mathbf{e}_1, \mathbf{e}_2, \ldots, \mathbf{e}_n\}$ for the space $\mathcal{C}^n$, a set of integers

$$1 = n_1 < n_2 < \cdots < n_k \leq n$$

and an enumeration (with possible repetitions) $\mu_1, \mu_2, \ldots, \mu_k$ of the σ eigenvalues[1] of **A** such that

$$\mathbf{A}\mathbf{e}_i = \mu_j \mathbf{e}_i + \mathbf{e}_{i-1} \qquad \text{for } n_j < i < n_{j+1}$$

and
$$\mathbf{A}\mathbf{e}_{n_j} = \mu_j \mathbf{e}_{n_j}$$

The diagonal blocks will be of order $n_2 - n_1, n_3 - n_2, \ldots, n - n_k$. By *C.10.4*, this description of the Jordan form corresponds to that which has the 1's just above the main diagonal. The reason for this rather cryptic statement is the large number of cases covered by the description. Let us illustrate it by examples:

[1] That is, each one of the μ_i is a member of the set $\{\lambda_1, \lambda_2, \ldots, \lambda_\sigma\}$. Furthermore, repetitions may occur in the enumeration because k may be larger than σ.

$$\begin{bmatrix} \lambda_1 & 0 & 0 \\ 0 & \lambda_1 & 0 \\ 0 & 0 & \lambda_1 \end{bmatrix}$$

$\mathcal{C}^3 = \mathfrak{N}(\mathbf{A} - \lambda_1\mathbf{I})$; $\nu_1 = 1$, $\dim \mathfrak{N}(\mathbf{A} - \lambda_1\mathbf{I}) = 3$

$n_i = i$, $i = 1, 2, 3$; $\mu_1 = \lambda_1$, $\mu_2 = \lambda_1$, $\mu_3 = \lambda_1$

This case illustrates the parenthetical statement "with possible repetitions."

$$\left[\begin{array}{cc|c} \lambda_1 & 0 & 0 \\ 0 & \lambda_1 & 0 \\ \hline 0 & 0 & \lambda_2 \end{array}\right]$$

$\mathcal{C}^3 = \mathfrak{N}(\mathbf{A} - \lambda_1\mathbf{I}) \oplus \mathfrak{N}(\mathbf{A} - \lambda_2\mathbf{I})$; $\nu_1 = 1$, $\nu_2 = 1$
$\dim \mathfrak{N}(\mathbf{A} - \lambda_1\mathbf{I}) = 2$, $\dim \mathfrak{N}(\mathbf{A} - \lambda_2\mathbf{I}) = 1$
$n_i = i$, $i = 1, 2, 3$; $\mu_1 = \lambda_1$, $\mu_2 = \lambda_1$, $\mu_3 = \lambda_2$

$$\left[\begin{array}{cc|c} \lambda_1 & 1 & 0 \\ 0 & \lambda_1 & 0 \\ \hline 0 & 0 & \lambda_2 \end{array}\right]$$

$\mathcal{C}^3 = \mathfrak{N}_1^2 \oplus \mathfrak{N}(\mathbf{A} - \lambda_2\mathbf{I})$; $\nu_1 = 2$, $\nu_2 = 1$
$\dim \mathfrak{N}_1^2 = 2$; $\dim \mathfrak{N}(\mathbf{A} - \lambda_2\mathbf{I}) = 1$, $n_1 = 1$, $n_2 = 3$
$\mu_1 = \lambda_1$, $\mu_2 = \lambda_2$

$$\left[\begin{array}{cc|cc} \lambda_1 & 1 & 0 & 0 \\ 0 & \lambda_1 & 0 & 0 \\ \hline 0 & 0 & \lambda_1 & 1 \\ 0 & 0 & 0 & \lambda_1 \end{array}\right]$$

$\mathcal{C}^4 = \mathfrak{N}_1^2$; $\nu_1 = 2$; $\dim \mathfrak{N}_1^2 = 4$
$n_1 = 1$, $n_2 = 3$; $\mu_1 = \lambda_1$, $\mu_2 = \lambda_1$

$$\left[\begin{array}{ccc|c|cc} \lambda_1 & 1 & 0 & 0 & 0 & 0 \\ 0 & \lambda_1 & 1 & 0 & 0 & 0 \\ 0 & 0 & \lambda_1 & 0 & 0 & 0 \\ \hline 0 & 0 & 0 & \lambda_1 & 0 & 0 \\ \hline 0 & 0 & 0 & 0 & \lambda_2 & 1 \\ 0 & 0 & 0 & 0 & 0 & \lambda_2 \end{array}\right]$$

$\mathcal{C}^6 = \mathfrak{N}_1^3 \oplus \mathfrak{N}_2^2$; $\nu_1 = 3$, $\nu_2 = 2$
$\dim \mathfrak{N}_1 = 4$, $\dim \mathfrak{N}_2 = 2$
$n_1 = 3$, $n_2 = 4$, $n_3 = 6$; $\mu_1 = \lambda_1$,
$\mu_2 = \lambda_1$, $\mu_3 = \lambda_2$

22 *Exercise* (*Cayley-Hamilton theorem*) Let $\mathbf{A}$ be a matrix and Δ its characteristic polynomial. Then $\Delta(\mathbf{A}) = \mathbf{0}$. (*Hint:* Observe that each basis vector used in the representation *18* belongs to one of the $\mathfrak{N}_k$'s.)

6 *The fundamental formula*

1 **Theorem** If f is analytic in an open set containing $\lambda_1, \lambda_2, \ldots, \lambda_\sigma$, then

2 $$f(\mathbf{A}) = \sum_{k=1}^{\sigma} \sum_{l=0}^{m_k-1} \frac{(\mathbf{A} - \lambda_k\mathbf{I})^l}{l!} f^{(l)}(\lambda_k)\mathbf{E}_k$$

where $f(\mathbf{A})$ is defined by *D.4.1* and $\mathbf{E}_k$ is defined in *D.5.3* and *D.5.6*.
Proof Consider the interpolating polynomial

$$q(\mu) = \sum_{k=1}^{\sigma} \sum_{l=0}^{m_k-1} \frac{(\mu - \lambda_k)^l}{l!} f^{(l)}(\lambda_k)\varphi_k(\mu)$$

Making use of the properties *D.5.1* of the polynomials φ_k, it is easy to see that the conditions *D.4.2* are satisfied, namely,

$$q^{(l)}(\lambda_k) = f^{(l)}(\lambda_k) \qquad l = 0, 1, \ldots, m_k - 1; k = 1, 2, \ldots, \sigma$$

Hence, by Definition *D.4.1*,

$$q(\mathbf{A}) = f(\mathbf{A})$$

and *2* follows.

3 **Corollary: the fundamental formula for $f(\mathbf{A})$** Let

4 $$\mathbf{Z}_{kl} = (1/l!)(\mathbf{A} - \lambda_k\mathbf{I})^l\mathbf{E}_k$$

where $\mathbf{E}_k \triangleq \varphi_k(\mathbf{A})$ and the polynomial φ_k is defined by *D.5.3*. Then

5 $$f(\mathbf{A}) = \sum_{k=1}^{\sigma} \sum_{l=0}^{m_k-1} f^{(l)}(\lambda_k)\mathbf{Z}_{kl}$$

6 *Comment* Observe that the $\mathbf{Z}_{kl}$, called the *components* of the matrix **A**, depend only on **A** and not on f. This formula exhibits the fact that $f(\mathbf{A})$ depends only on the behavior of f at the eigenvalues of **A**. If λ_k is a simple root of $\psi(\lambda) = 0$, then $f(\mathbf{A})$ depends only on the value of f at λ_k. If λ_k is a zero of order m_k of ψ, then $f(\mathbf{A})$ depends only on the value of f and its $(m_k - 1)$ first derivatives at λ_k.

7 *Exercise* Show directly from *4* that $\mathbf{Z}_{k0} = \varphi_k(\mathbf{A})$, where φ_k is the polynomial defined by *D.5.3*.

8 *Exercise* If $j \neq k$, then for any $0 \leq l \leq m_k - 1$, $0 \leq m \leq m_j - 1$,

$$\mathbf{Z}_{kl}\mathbf{Z}_{jm} = \mathbf{0}$$

9 *Exercise* Show that

$$\mathbf{A} = \sum_{k=1}^{\sigma} (\lambda_k\mathbf{E}_k + \mathbf{Z}_{k1})$$

10 *Exercise* For any $k = 1, 2, \ldots$, if $j \geq m_k$, then

$$\mathbf{Z}_{kj} = \mathbf{0}$$

11 *Exercise* Let **A** have n distinct real eigenvalues. Let $\varepsilon_1, \varepsilon_2, \ldots, \varepsilon_n$ be an arbitrary sequence of 1 and -1's. Show that $\mathbf{X} \triangleq \sum_{i=1}^{n} \epsilon_i \sqrt{\lambda_i}\, \mathbf{E}_i$ satisfies the equation $\mathbf{X}^2 = \mathbf{A}$. In other words, any such **X** is a possible choice for $\sqrt{\mathbf{A}}$.

12 *Exercise* Show that

$$\exp(\mathbf{A}t) = \sum_{k=1}^{\sigma} \sum_{l=0}^{m_k-1} t^l e^{\lambda_k t}\mathbf{Z}_{kl}$$

13 *Exercise* Let f be an entire function (i.e., analytic in the whole complex plane). Show that

$$\frac{d}{dt} f(\mathbf{A}t) = \mathbf{A}f'(\mathbf{A}t) \qquad \forall t$$

where $\mathbf{A}$ is a matrix of constant elements and f' is the derivative of f. In particular, show that

$$\frac{d}{dt}[\exp(\mathbf{A}t)] = \mathbf{A}\exp(\mathbf{A}t)$$

14 *Exercise* Show that if $\mathbf{A}$ is skew symmetric (i.e., the transpose of $\mathbf{A}$, say, $\mathbf{A}'$, is $-\mathbf{A}$) then $\exp(\mathbf{A}t)$ is orthogonal (that is, $[\exp(\mathbf{A}t)]'[\exp(\mathbf{A}t)] = \mathbf{I}$). Give an interpretation of this fact by considering the differential equation $\dot{\mathbf{x}} = \mathbf{A}\mathbf{x}$ and studying the properties of its solutions.

15 *Exercise* If $\mathbf{Z}_{kl}\mathbf{x} \neq \mathbf{0}$, then $\mathbf{Z}_{kj}\mathbf{x} \neq \mathbf{0}$ for all $j < l$.

16 *Exercise* Application to Markov chains. Let $\mathbf{P} = (p_{ij})$ be a Markov matrix: $p_{ij} \geq 0$, $\forall i, j$; $\sum_{j=1}^{\infty} p_{ij} = 1$, $\forall i$. p_{ij} is the probability of going from state i to state j in one transition.

(I) Let $\mathbf{p} = (p_1, p_2, \ldots, p_n)$ and $\mathbf{q} = (q_1, q_2, \ldots, q_n)$, where $\mathbf{q} = \mathbf{pP}$. (Interpretation: p_i is the probability of being in state i; q_j is the probability of being in state j after one transition.) Use the fact that $p_i \geq 0$ and $\sum_{i=1}^{n} p_i = 1$ to prove that all eigenvalues of $\mathbf{P}$ are in the circle $|\lambda| \leq 1$ and that those which are on the boundary have index 1. (II) Show that $\lim_{N\to\infty} \sum_{0}^{N} \mathbf{P}^k/(N+1)$ converges to a matrix $\mathbf{E}$. (III) $\mathbf{E}^2 = \mathbf{E} = \mathbf{EP} = \mathbf{PE}$. (IV) $\mathcal{R}(\mathbf{E})$ is an invariant subspace of $\mathbf{P}$, in fact the null space associated with the eigenvalue 1.

(V)
$$\mathcal{C}^n = \mathcal{R}(\mathbf{E}) \oplus \mathcal{R}(\mathbf{I} - \mathbf{P})$$

(VI) For every $\mathbf{a}$ in $\mathcal{R}(\mathbf{E})$, there is a unique $\mathbf{x}$ such that $\mathbf{Px} = \mathbf{x}$ and $\mathbf{Ex} = \mathbf{a}$.

17 *Exercise* Show that

$$\mathbf{A}^N = \sum_{k=1}^{\sigma} \sum_{l=0}^{m_k-1} N(N-1)\cdots(N-l+1)\lambda_k^{N-l}\mathbf{Z}_{kl}$$

7 *Alternative expressions for $f(\mathbf{A})$*

Frequently, a function of a matrix is defined as follows. Let f be analytic in the disk $|\lambda| < R$, where $R > \max_{1\leq k\leq\sigma} |\lambda_k|$. Let the power-series representation of f over this disk be

$$f(\lambda) = \sum_{k=0}^{\infty} a_k\lambda^k$$

Then,

1 $$\sum_{k=0}^{\infty} a_k \mathbf{A}^k$$

is often taken as the meaning of $f(\mathbf{A})$.

This definition leads to exactly the same matrix $f(\mathbf{A})$ as the preceding one because of the following

2 **Theorem** Let f and each member of the sequence $\{f_N\}$ be analytic functions in an open set containing $\lambda_1, \lambda_2, \ldots, \lambda_\sigma$. Then, as $N \to \infty$

$$f_N(\mathbf{A}) \to f(\mathbf{A}) \Leftrightarrow f_N^{(l)}(\lambda_k) \to f^{(l)}(\lambda_k) \qquad 0 \leq l \leq m_k - 1$$

and $k = 1, 2, \ldots, \sigma$.

3 *Comment* If we take $f_N(\lambda) = \sum_{k=0}^{N} a_k \lambda^k$, then the above condition is fulfilled and Definition *D.4.1* gives the same result as *1*. Definition *D.4.1* has the advantage that it will suggest a computational procedure and that it is closely related to the geometrical structure of the L.T. $\mathbf{A}$.

Proof $\Leftarrow$ This follows immediately from *D.6.5* once it is observed that the $\mathbf{Z}_{kl}$'s depend only on $\mathbf{A}$ and not on f.

$\Rightarrow$ Suppose now that $f_N(\mathbf{A}) \to f(\mathbf{A})$; we shall prove that $f_N^{(l)}(\lambda_k) \to f^{(l)}(\lambda_k)$. Pick an arbitrary eigenvalue, say, λ_k. Then pick $\mathbf{0} \neq \mathbf{x} \in \mathfrak{N}_k$ such that $\mathbf{x} \notin \mathfrak{N}_k^{m_k-1}$. Hence $\mathbf{E}_k\mathbf{x} = \mathbf{x}$ and $\mathbf{E}_j\mathbf{x} = \mathbf{0}$ if $j \neq k$. Let $\mathbf{y}_\beta = (\mathbf{A} - \lambda_k\mathbf{I})^\beta\mathbf{x}$, $\beta = 0, 1, 2, \ldots, m_k - 1$. To simplify the notation, let us put $\mu_k \triangleq m_k - 1$. From *D.6.2*, $f_N(\mathbf{A})\mathbf{y}_{\mu_k} = f_N(\lambda_k)\mathbf{y}_{\mu_k}$. Hence, since $f_N(\mathbf{A}) \to f(\mathbf{A})$, $f_N(\lambda_k) \to f(\lambda_k)$. Similarly, from *D.5.2*, $f_N(\mathbf{A})\mathbf{y}_{\mu_k-1} = f_N(\lambda_k)\mathbf{y}_{\mu_k-1} + f'_N(\lambda_k)\mathbf{y}_{\mu_k}$; consequently $f'_N(\lambda_k) \to f'(\lambda_k)$. Proceeding in this way, we establish $f_N^{(l)}(\lambda_k) \to f^{(l)}(\lambda_k)$ for

$$l = 2, 3, \ldots, m_k - 1 \quad \text{Q.E.D.}$$

4 **Theorem** Let f be analytic in a domain D containing the eigenvalues λ_i of $\mathbf{A}$. Let C, the boundary of D, consist of a finite number of closed rectifiable Jordan curves; then

5 $$f(\mathbf{A}) = \frac{1}{2\pi j} \oint_C f(\lambda)(\lambda\mathbf{I} - \mathbf{A})^{-1}\, d\lambda$$

Proof Let $\rho(x) = 1/(\lambda - x)$. From *D.6.2* applied to the function ρ

$$(\lambda\mathbf{I} - \mathbf{A})^{-1} = \rho(\mathbf{A}) = \sum_{k=1}^{\sigma} \sum_{l=1}^{m_k-1} \frac{(\mathbf{A} - \lambda_k\mathbf{I})^l}{(\lambda - \lambda_k)^{l+1}} \mathbf{E}_k$$

Hence, in view of the analyticity of $f(\lambda)$ in the domain D,

$$\frac{1}{2\mu j}\oint_C f(\lambda)(\lambda\mathbf{I}-\mathbf{A})^{-1}\,d\lambda = \sum_{k=1}^{\sigma}\sum_{l=0}^{m_k-1}\frac{(\mathbf{A}-\lambda_k\mathbf{I})^l}{l!}f^{(l)}(\lambda_k)\mathbf{E}_k$$

The right-hand side according to *D.6.2* is the matrix $f(\mathbf{A})$; hence the theorem is established.

8 *Practical computation of* $f(\mathbf{A})$

The problem is the following: given an $n \times n$ matrix $\mathbf{A}$ whose elements are real or complex numbers and given a function, say, $f(x)$, find a convenient way for computing $f(\mathbf{A})$. This is an important problem in practice, as is shown by the following examples.

1 *Example* In the theory of linear time-invariant systems described by the system of differential equations $\dot{\mathbf{x}} = \mathbf{A}\mathbf{x} + \mathbf{a}(t)$, one is led to consider the matrix $e^{\mathbf{A}t}$. In this case $f(x) = e^{xt}$.

2 *Example* In the theory of linear time-invariant discrete systems described by $\mathbf{x}_{k+1} = \mathbf{A}\mathbf{x}_k + \mathbf{a}_k$, one is led to consider $\mathbf{A}^N$. In this case $f(x) = x^N$.

3 *Example* In the theory of stationary Markov chains one is particularly interested in the limiting behavior of $\mathbf{P}^N$ as $N \to \infty$, where $\mathbf{P}$ is the transition probability matrix of the chain.

Let $\lambda_1, \lambda_2, \ldots, \lambda_\sigma$ be the eigenvalues of $\mathbf{A}$: different eigenvalues are given a different label, so that $i \neq j$ implies $\lambda_i \neq \lambda_j$. If all the n eigenvalues of $\mathbf{A}$ are distinct, then $\sigma = n$; otherwise, $\sigma < n$.

Let $m_1, m_2, \ldots, m_\sigma$ be the multiplicity of the eigenvalues; that is, λ_i is a zero of order m_i of the *minimal* polynomial of $\mathbf{A}$.[1]

The interpolation method

This method is based directly on the definition of $f(\mathbf{A})$. We recall that if $p(\lambda) = \sum_{k=0}^{N} a_k\lambda^k$ is a polynomial, we define $p(\mathbf{A})$ as the matrix $\sum_{k=0}^{N} a_k\mathbf{A}^k$ with $\mathbf{A}^0 = \mathbf{I}$. Now, by Definition *D.4.1*,

4 $$f(\mathbf{A}) = p(\mathbf{A})$$

[1] As will be pointed out later, no error will be caused if one takes m_i to be the order of λ_i as a zero of the *characteristic* polynomial $\Delta(\lambda) = \det(\mathbf{A} - \lambda\mathbf{I})$. The computations, however, may be more complicated than they need be (see Remark *27*).

provided the interpolating polynomial p satisfies the following conditions:

5 $$f(\lambda_k) = p(\lambda_k) \qquad k = 1, 2, \ldots, \sigma$$

6 $$f^{(l)}(\lambda_k) = p^{(l)}(\lambda_k) \qquad k = 1, 2, \ldots, \sigma \qquad l = 1, 2, \ldots, m_k - 1$$

Thus, the computation of $f(\mathbf{A})$ consists of two steps: (I) the determination of an interpolating polynomial p satisfying conditions *5* and *6*, and (II) the evaluation of $p(\mathbf{A})$.

7 *Comment* Note the following special case: If the minimal polynomial $\psi(\lambda)$ has distinct roots, the interpolating polynomial is the well-known Lagrange interpolating polynomial

8 $$p(\lambda) = \sum_{k=1}^{\sigma} \frac{(\lambda - \lambda_1)(\lambda - \lambda_2) \cdots (\lambda - \lambda_{k-1})(\lambda - \lambda_{k+1}) \cdots (\lambda - \lambda_\sigma)}{(\lambda_k - \lambda_1)(\lambda_k - \lambda_2) \cdots (\lambda_k - \lambda_{k-1})(\lambda_k - \lambda_{k+1}) \cdots (\lambda_k - \lambda_\sigma)} f(\lambda_k)$$

Consequently

9 $$f(\mathbf{A}) = \sum_{k=1}^{\sigma} \left[\frac{f(\lambda_k)}{\prod\limits_{\substack{i=1 \\ i \neq k}}^{\sigma} (\lambda_k - \lambda_i)} \prod_{\substack{i=1 \\ i \neq k}}^{\sigma} (\mathbf{A} - \lambda_i \mathbf{I}) \right]$$

10 *Example* Compute exp $(\mathbf{A}t)$ for $\mathbf{A} = \begin{bmatrix} 1 & 1 & 0 \\ 0 & 1 & 0 \\ 0 & 0 & 2 \end{bmatrix}$.

Here $\lambda_1 = 1$, $m_1 = 2$; $\lambda_2 = 2$, $m_2 = 1$.

$$f(x) = \exp(xt) \qquad f'(x) = t \exp(xt)$$

Let $p(\lambda) = \alpha_0 + \alpha_1\lambda + \alpha_2\lambda^2$. Then $p'(\lambda) = \alpha_1 + 2\alpha_2\lambda$. The conditions *5* and *6* read

$$\begin{aligned} p(1) &= f(1) & \alpha_0 + \alpha_1 + \alpha_2 &= e^t \\ p'(1) &= f'(1) & \alpha_1 + 2\alpha_2 &= te^t \\ p(2) &= f(2) & \alpha_0 + 2\alpha_1 + 4\alpha_2 &= e^{2t} \end{aligned}$$

Solving the system, we get

$$\begin{aligned} \alpha_0 &= -2te^t + e^{2t} \\ \alpha_1 &= 2e^t + 3te^t - 2e^{2t} \\ \alpha_2 &= -e^t - te^t + e^{2t} \end{aligned}$$

With $\mathbf{A}^2 = \begin{bmatrix} 1 & 2 & 0 \\ 0 & 1 & 0 \\ 0 & 0 & 4 \end{bmatrix}$ we get

$$f(\mathbf{A}) = \alpha_0 \begin{bmatrix} 1 & 0 & 0 \\ 0 & 1 & 0 \\ 0 & 0 & 1 \end{bmatrix} + \alpha_1 \begin{bmatrix} 1 & 1 & 0 \\ 0 & 1 & 0 \\ 0 & 0 & 2 \end{bmatrix} + \alpha_2 \begin{bmatrix} 1 & 2 & 0 \\ 0 & 1 & 0 \\ 0 & 0 & 4 \end{bmatrix}$$

$$= \begin{bmatrix} e^t & te^t & 0 \\ 0 & e^t & 0 \\ 0 & 0 & e^{2t} \end{bmatrix} = e^{\mathbf{A}t}$$

11 *Comment* In general, we have

12 $$\exp(\mathbf{A}t) = \sum_{k=0}^{m-1} \alpha_k(t)\mathbf{A}^k$$

where m is the degree of the minimal polynomial of $\mathbf{A}$. The algebraic equations which determine the α_k's are of the form

$$\mathbf{M}\boldsymbol{\alpha} = \boldsymbol{\phi}(t)$$

where $\boldsymbol{\alpha} = (\alpha_0, \ldots, \alpha_{m-1})$; $\boldsymbol{\phi}$ is a vector whose elements are of the form $t^l e^{\lambda_k t}$, $k = 1, 2, \ldots, \sigma$; $l = 0, \ldots, m_k - 1$. The components of $\boldsymbol{\phi}$ form a set of functions of t which are linearly independent over any interval of positive length. Since the matrix $\mathbf{M}$ is a Vandermonde determinant in the case of simple eigenvalues and is closely related to a Vandermonde determinant when the eigenvalues are multiple, it is easy to show that $\mathbf{M}$ is always nonsingular. Therefore, the vector $\boldsymbol{\alpha}$ has components that are linearly independent linear combinations of the components of $\boldsymbol{\phi}$. Hence the

13 **Assertion** The components $\alpha_k(t)$ of $\boldsymbol{\alpha}$ form a set of functions of time that are linearly independent over any interval of positive length.

Method based on the fundamental formula

The second method is based on the fundamental formula, *D.6.5*, which states that, for any function f which is analytic at the eigenvalues $\lambda_1, \lambda_2, \ldots, \lambda_\sigma$ of $\mathbf{A}$,

14 $$f(\mathbf{A}) = \sum_{k=1}^{\sigma} \sum_{l=0}^{m_k-1} f^{(l)}(\lambda_k)\mathbf{Z}_{kl}$$

where the matrices $\mathbf{Z}_{kl}$, called the *components* of $\mathbf{A}$, depend exclusively on $\mathbf{A}$.

The matrices $\mathbf{Z}_{kl}$, although expressible as polynomials in $\mathbf{A}$, may be computed directly by inserting in the fundamental formula *14* appropriate trial functions f. The technique is best illustrated by an example.

15 *Example* Consider $\mathbf{A} = \begin{bmatrix} 1 & 1 & 0 \\ 0 & 1 & 1 \\ 0 & 0 & 2 \end{bmatrix}$, $\psi(\lambda) = (\lambda - 1)^2(\lambda - 2)$.

For any f analytic at $\lambda = 1$ and $\lambda = 2$, *14* states that

$$f(\mathbf{A}) = f(1)\mathbf{Z}_{10} + f'(1)\mathbf{Z}_{11} + f(2)\mathbf{Z}_{20}$$

The component matrices will be determined by taking special cases for f.

(I) Let $f(\lambda) = 1$ for all λ

16 $$\mathbf{Z}_{10} + \mathbf{Z}_{20} = \mathbf{I}$$

(II) Let $f(\lambda) = \lambda - 1$; hence

17 $$\mathbf{Z}_{11} + \mathbf{Z}_{20} = \mathbf{A} - \mathbf{I}$$

(III) Let $f(\lambda) = (\lambda - 1)^2$; hence

18 $$\mathbf{Z}_{20} = (\mathbf{A} - \mathbf{I})^2 = \begin{bmatrix} 0 & 0 & 1 \\ 0 & 0 & 1 \\ 0 & 0 & 1 \end{bmatrix}$$

16 to *18* constitute a system of equations in the $\mathbf{Z}_{ki}$. Observe that the coefficient matrix of the system of equations that has to be solved to find the $\mathbf{Z}_{ki}$ is in the triangular form; hence, the solution is obtained by successive substitutions from the bottom up.

$$f(\mathbf{A}) = f(1)\begin{bmatrix} 1 & 0 & -1 \\ 0 & 1 & -1 \\ 0 & 0 & 0 \end{bmatrix} + f'(1)\begin{bmatrix} 0 & 1 & -1 \\ 0 & 0 & 0 \\ 0 & 0 & 0 \end{bmatrix} + f(2)\begin{bmatrix} 0 & 0 & 1 \\ 0 & 0 & 1 \\ 0 & 0 & 1 \end{bmatrix}$$

In particular, if $f(\lambda) = e^{\lambda t}$, then $f'(\lambda) = te^{\lambda t}$ and

$$e^{\mathbf{A}t} = e^t\begin{bmatrix} 1 & 0 & -1 \\ 0 & 1 & -1 \\ 0 & 0 & 0 \end{bmatrix} + te^t\begin{bmatrix} 0 & 1 & -1 \\ 0 & 0 & 0 \\ 0 & 0 & 0 \end{bmatrix} + e^{2t}\begin{bmatrix} 0 & 0 & 1 \\ 0 & 0 & 1 \\ 0 & 0 & 1 \end{bmatrix}$$
$$= \begin{bmatrix} e^t & te^t & -e^t - te^t + e^{2t} \\ 0 & e^t & -e^t + e^{2t} \\ 0 & 0 & e^{2t} \end{bmatrix}$$

19 *Comment* The trial functions, written in the reverse of the order in which they appear above, are $(\lambda - 1)^2$, $(\lambda - 1)$, 1. It is easily recognized that they may be obtained by taking the minimum polynomial ψ and successively canceling one factor from ψ, then another from the resulting quotient, etc. It is clear that this technique will work in

general and will always result in a linear system of matrix equations for the $\mathbf{Z}_{ki}$ in which the matrix of the coefficients is triangular.

20 **Remark** If **A** has real elements, then both the characteristic and the minimal polynomial have real coefficients. Let the roots of the minimal polynomial be ordered so that $\lambda_1, \lambda_2, \ldots, \lambda_r$ are the real roots and $\lambda_{r+1}, \lambda_{r+1}^*, \ldots, \lambda_\sigma, \lambda_\sigma^*$ are the complex root pairs.

If the function $f(\cdot)$ has the property that $f(\lambda^*) = f(\lambda)^*$ for all λ, then cancellations occur and the fundamental formula *14* becomes

21
$$f(\mathbf{A}) = \sum_{k=1}^{r} \sum_{i=0}^{m_i-1} f^{(i)}(\lambda_k)\mathbf{Z}_{ki} + \sum_{k=r+1}^{\sigma} \sum_{i=0}^{m_i-1} \{\text{Re}\,[f^{(i)}(\lambda_k)]\mathbf{R}_{ki} + \text{Im}\,[f^{(i)}(\lambda_k)]\mathbf{X}_{ki}\}$$

where, for each complex root λ_k, we split $2\mathbf{Z}_{ki}$ into its real and imaginary parts,

$$2\mathbf{Z}_{ki} = \mathbf{R}_{ki} - j\mathbf{X}_{ki}$$

Consequently, *21* involves only real numbers with the corresponding increased convenience of the numerical work.

22 *Example* Evaluate $e^{\mathbf{A}t}$, where **A** is a 3×3 matrix with a real eigenvalue λ_1 and two complex conjugate eigenvalues λ_2 and λ_2^*. Let $\lambda_2 = \sigma_2 + j\omega_2$. Then from *21*

$$\exp(\mathbf{A}t) = e^{\lambda_1 t}\mathbf{Z}_{10} + e^{\sigma_2 t}[\mathbf{R}_{20}\cos\omega_{2 0}t + \mathbf{X}_{20}\sin\omega_2 t]$$

Apply *14* to $\varphi(\lambda) \triangleq (\lambda - \lambda_2)(\lambda - \lambda_2^*) = \lambda^2 - 2\sigma_2\lambda + |\lambda_2|^2$; hence

23
$$(\lambda_1^2 - 2\sigma_2\lambda_1 + |\lambda_2|^2)\mathbf{Z}_{10} = \mathbf{A}^2 - 2\sigma_2\mathbf{A} + |\lambda_2|^2\mathbf{I}$$

This equation determines $\mathbf{Z}_{10}$. The remaining unknown matrices $\mathbf{R}_{20}$ and $\mathbf{X}_{20}$ are obtained by substituting for f the functions $\varphi_1(\lambda) = \lambda - \sigma_2$, $\varphi_2(\lambda) = 1$ for all λ's in *21*:

24
$$(\lambda_1 - \sigma_2)\mathbf{Z}_{10} + \omega_2\mathbf{X}_{20} = \mathbf{A} - \sigma_2\mathbf{I}$$

25
$$\mathbf{Z}_{10} + \mathbf{R}_{20} = \mathbf{I}$$

The set of equations *23* to *25* can easily be solved for the three matrices $\mathbf{Z}_{10}$, $\mathbf{R}_{20}$, $\mathbf{X}_{20}$.

26 **Remark** There are convenient *numerical checks* on the evaluation of the $\mathbf{Z}_{kl}$'s:

(I) Put $f(\lambda) = \lambda$ in *14*. The right-hand side should give **A**.

(II) Put $f(\lambda) \equiv 1$ in *14*; then $\sum_{k=1}^{\sigma} \mathbf{Z}_{k0} = \mathbf{I}$.

(III) For $k = 1, 2, \ldots, \sigma$, $\mathbf{Z}_{k0}^2 = \mathbf{Z}_{k0}$.

(IV) If $k \neq j$, $\mathbf{Z}_{k0}\mathbf{Z}_{j0} = \mathbf{0}$.

$$\mathbf{Z}_{kl}\mathbf{Z}_{jm} = \mathbf{0} \qquad 0 \le l \le m_k - 1;\, 0 \le m \le m_j - 1$$

27 **Remark** It may happen that we know only the characteristic polynomial Δ and not the minimal polynomial ψ. We assert that if we apply the fundamental formula *14* in which we use for the m_k's the order of the λ_k's as zeros of Δ (instead of their order as zeros of ψ), the same matrix $f(\mathbf{A})$ will be obtained.

This assertion follows from two considerations. First, the order of λ_k as a zero of Δ is never smaller than its order as a zero of ψ; second, $\mathbf{Z}_{kl}$ for $l \geq m_k$ is the zero matrix. This is immediate since by *D.6.4* and *D.6.2* it is implied that $\mathbf{Z}_{kl} = q_{kl}(\mathbf{A})$, where the polynomial q_{kl} is given by

$$q_{kl}(\lambda) = \frac{1}{l!}(\lambda - \lambda_k)^l \varphi_k(\lambda) = \frac{\psi(\lambda) n_k(\lambda)}{l!}(\lambda - \lambda_k)^{l-m_k}$$

Thus q_{kl} is a multiple of ψ whenever $l \geq m_k$. Hence from the definition of the minimum polynomial, $\mathbf{Z}_{kl} = \mathbf{0}$ for $l \geq m_k$.

Therefore, when the system of equations for the $\mathbf{Z}_{kl}$'s is solved, one may find some of them to be identically zero.

REFERENCES AND SUGGESTED READING

1 Dunford, N., and J. T. Schwarz: "Linear Operators. Part I: General Theory," Interscience Publishers, Inc., New York, 1958.

2 Fantappie: Sulle Funzioni di Una Matriza, *Anais Acad. Brasil. Cienc.*, vol. 26, no. 1, pp. 25–33, 1954.

3 Frazer, R. A., W. J. Duncan, and A. R. Collar: "Elementary Matrices," Cambridge University Press, New York, 1938.

4 Gantmacher, F. R.: "The Theory of Matrices," vols. 1 and 2, Chelsea Publishing Company, New York, 1959.

5 Hamburger, H. L., and M. E. Grimshaw: "Linear Transformations in *n* Dimensional Space," Cambridge University Press, New York, 1951.

6 Wedderburn, J. H. M.: Lectures on Matrices, 1934, *Amer. Math. Soc. Colloq. Publ.*, vol. 17.

Glossary of Symbols

Following is a list of principal symbols appearing in the book. For convenience, the symbols are grouped into three categories; each symbol is followed by a brief description and, where necessary, the location of its first use or formal definition.

I. General conventions

1. Lowercase boldface letters denote vectors, e.g., $\boldsymbol{\alpha}$, $\mathbf{u}$, $\mathbf{x}$.
2. Capital boldface letters denote matrices, e.g., $\mathbf{A}$, $\mathbf{B}$, $\mathbf{M}$.
3. Greek and italic type are used for scalar-valued variables, functions and operators, e.g., α, $u(t)$, L.
4. Capital script letters denote systems and sets (spaces), e.g., $\mathcal{A}$, $\mathcal{R}$.
5. Capital letters are used to denote the Laplace transforms; e.g., $H(s)$ is the Laplace transform of $h(t)$.
6. f or $f(\cdot)$ denotes a function, with the dot standing for an undesignated variable; $f(t)$ denotes the value of f at time t. The latter convention is not adhered to strictly, so that in many instances $f(t)$ is used in the same sense as f or $f(\cdot)$ (*1.3.3*).
7. $f_{[t_0,t_1]}$ denotes a segment of f over the closed interval $[t_0,t_1]$, that is, the set of pairs $\{(t,f(t)),\ t_0 \leq t \leq t_1\}$. Likewise for open and semi-closed intervals (*1.3.3*).
8. Braces denote a set or a family; e.g., $\{\mathbf{x}\}$ is a set with generic element $\mathbf{x}$. $\{\mathbf{x}|P\}$ is a set of $\mathbf{x}$'s having property P. $\{u_i\}^n$ is the sequence $u_0, \ldots, u_n$.
9. A bar over a function indicates that its range is a function space, e.g., $\bar{\mathbf{A}}(\boldsymbol{\alpha};\mathbf{u})$ (*1.6.2*).
10. A dot over a time function indicates its time derivative; e.g., $\dot{\mathbf{x}}$. $\mathbf{x}^{(n)}$ or $\mathbf{x}^{(n)}(t)$ stands for the nth derivative of $\mathbf{x}(t)$.
11. Superscript asterisk (*) denotes (I) the complex conjugate of a number, e.g., ξ^*; (II) the adjoint of an operator, e.g., L^*; (III) the conjugate transpose of a matrix, e.g., $\mathbf{A}^*$; or (IV) the terminal state, e.g., $\boldsymbol{\alpha}^*$.
12. Superscript minus one ($^{-1}$) denotes the inverse of a system or a transformation or a matrix, e.g., $\mathcal{A}^{-1}$, L^{-1}, $\mathbf{A}^{-1}$.
13. Superscript dagger (†) denotes the pseudo inverse of a matrix, e.g., $\mathbf{A}^\dagger$.
14. Superscript perpendicular ($^\perp$) denotes the orthogonal complement of a subspace, e.g., $\mathcal{M}^\perp$.

15. Superscript tilde (˜) denotes (I) the z transform, e.g., $\tilde{F}(z)$, or (II) the suppressed output vector, e.g., $\tilde{\mathbf{y}}$.

II. General symbols and abbreviations

$\triangleq$	equals by definition; denotes
$\Rightarrow$	implies (*2.1.11*)
$\Leftarrow$	is implied by (*2.1.11*)
$\Leftrightarrow$	implies and is implied by; if and only if (*2.1.13*)
$\ni$	such that (*B.4.6 et seq.*)
$\forall$	for all (*1.3.2*)
$\exists$	there exists (*1.3.2*)
$\supset$	contains (*2.5.1*)
$\subset$	is a subset of; is contained in
$\in$	is an element of
$\notin$	does not belong to
$\cup$	union
$\cap$	intersection
$\{x\|P\}$	set of x's having property P
$\simeq$	state equivalent (*2.2.1*)
	approximately equals
$\sim$	asymptotically equals
$\not\simeq$	not state equivalent (*2.2.3*)
$\equiv$	system equivalence (*2.5.8*)
$\doteq$	zero-state equivalent (*2.9.1*)
$\overset{*}{=}$	conditionally equivalent (*2.7.5*)
$\times$	cartesian product (*1.4.17*)
	direct product (*1.10.2*)
$+$	parallel combination (*1.10.30*)
$\oplus$	direct sum (*C.7.7*)
$\overset{\perp}{\oplus}$	direct sum of orthogonal subspaces
$*$	convolution (*B.3.9*)
$\langle\cdot,\cdot\rangle$	scalar product (dots stand for undesignated variables) (*C.5.1*)
$\cdot\rangle\langle\cdot$	dyad (*5.4.15*)
(a,b)	open interval $a < t < b$
$(a,b]$	semiclosed interval $a < t \leq b$
$[a,b]$	closed interval $a \leq t \leq b$
$t-$	instant of time immediately preceding t
$t+$	instant of time immediately following t
$\\|\cdot\\|$	norm (*C.16.1*, *C.16.9*, *C.16.11*)
$\\|\cdot\\|_1$	l_1 norm (*C.16.7*)
$\\|\cdot\\|_2$	l_2 norm (*C.16.5*, *C.16.16*)
$\\|\cdot\\|_\infty$	l_∞ norm (*C.16.16*)
$\measuredangle\, s$	argument of the complex number s

$\mathfrak{F}$	Fourier transform (*B.2.15*)
$\mathfrak{L}$	Laplace transform (*B.2.2*)
$\mathfrak{L}_{II}$	bilateral Laplace transform (*B.2.11*)
$\mathcal{C}^n$	space of ordered n-tuples of complex numbers
$\mathcal{R}^n$	space of ordered n-tuples of real numbers
C^n	space of functions possessing continuous derivatives of order n
$\inf_x$	infimum over x (greatest lower bound)
$\sup_x$	supremum over x (least upper bound) (p. 401 n)
lim inf	limit inferior
lim sup	limit superior
sgn	signum function: sgn $x = 1$ for $x > 0$, sgn $x = -1$ for $x < 0$, sgn $x = 0$ for $x = 0$ (*7.7.10 et seq.*)
$\max(m,n)$	$\max(m,n) = m$ if $m \leq n$, $\max(m,n) = n$ if $n > m$ (*3.8.22*)
$\max_i$	maximum over i
det	determinant
tr	trace
diag	diagonal matrix
L.T.	linear transformation
i.s.L.	in the sense of Lyapunov
b.i.b.o	bounded-input bounded-output
R.H.S.	right-hand side
L.H.S.	left-hand side
Q.E.D.	*quod erat demonstrandum*
et seq.	*et sequens*
$\lhd$	end of discussion

III. Symbols with special meaning (arranged in alphabetical order)

$\mathcal{A}$	system, abstract object (*1.4.17*)
$\mathcal{A}\mathcal{B}$	tandem combination (product) of $\mathcal{A}$ and $\mathcal{B}$, with $\mathcal{B}$ followed by $\mathcal{A}$ (*1.10.23*)
$\mathbf{A}(\boldsymbol{\alpha};\mathbf{u}_{(t_0,t]})$	response of $\mathcal{A}$ at time t to input $\mathbf{u}_{(t_0,t]}$, with $\mathcal{A}$ in state $\boldsymbol{\alpha}$ at time t_0 (*1.6.2*)
$\mathbf{A}(\boldsymbol{\alpha};\mathbf{u})$	abbreviation of above (*1.6.2*)
$\bar{\mathbf{A}}(\boldsymbol{\alpha};\mathbf{u}_{(t_0,t]})$	response segment of $\mathcal{A}$ to input $\mathbf{u}_{(t_0,t]}$, with $\mathcal{A}$ in state $\boldsymbol{\alpha}$ at time t_0 (*1.6.2*)
$\bar{\mathbf{A}}(\boldsymbol{\alpha};\mathbf{u})$	abbreviation of above (*1.6.2*)
$\mathbf{A}(\boldsymbol{\theta};\mathbf{u})$	zero-state response of $\mathcal{A}$ to input $\mathbf{u}$ (*2.8.18*)
$\mathbf{A}(\mathbf{u})$	abbreviation of above (*2.8.18*)
$\mathbf{A}(\boldsymbol{\alpha};\mathbf{0})$	zero-input response of $\mathcal{A}$ starting in state $\boldsymbol{\alpha}$ (*2.8.23*)

$\mathcal{A}(\mathbf{u},\mathbf{y})$ input-output relation (*1.4.17*)

$\mathcal{A}^{-1}$ inverse of $\mathcal{A}$ (*2.10.6*)

$\mathcal{A}_l^{-1}$ left-constrained inverse of $\mathcal{A}$ (*2.10.20*)

$\mathbf{a}(t)$ equivalent input (*5.2.20*)

$\mathcal{C}^n$ space of ordered n-tuples of complex numbers

C^n space of functions possessing continuous derivatives of order n

$\mathcal{D}$ space of testing functions (p. 522)

$\mathcal{D}_K$ space of testing functions whose support is contained in the finite interval K (p. 523)

$\mathcal{D}$ system of the differential operator type (*3.3.1*)

D unit delay operator (*3.8.36*)

δ_{ij} Kronecker delta: $\delta_{ij} = 1$ for $i \neq j$, $\delta_{ii} = 0$ (*C.4.12*)

$\delta(t)$ delta function (unit impulse) occurring at $t = 0$ (*A.2.1*)

$\delta^{(n)}(t)$ delta function of order n (*A.2.5*)

$\mathbf{E}$ projection (*C.12.14*)

$\mathbf{E}_k$ projection on $\mathcal{N}_k$ (*D.5.6, D.5.13*)

$\exp \mathbf{A}t$ exponential of the matrix $\mathbf{A}t$ (*5.2.3, D.6.12*)

$\mathcal{F}$ Fourier transform (*B.2.15*)

$\mathbf{G}(t,\tau)$ Green's dyadic (*5.2.24*)

$\mathbf{\Gamma}$ matrix relating two sets of basis functions (*3.2.9*)

$h(t,\xi)$ impulse response [zero-state response at time t to $\delta(t - \xi)$] (*3.6.5, 6.3.2*)

$h(t)$ impulse response of a time-invariant system with $\xi = 0$ (*3.6.5*)

$\mathbf{h}(t)$ state impulse response (*3.7.31*)

$H(s)$ transfer function (*3.6.17, 9.2.1*)

$\mathbf{I}$ identity matrix, unitor (*1.9.8*)

$L(p)$ constant-coefficient differential operator (*3.8.11*)

$L(p,t)$ variable-coefficient differential operator (*4.5.58, 6.2.47*)

$\dfrac{1}{L(p)}$ system characterized by $L(p)y = u$ ($u \triangleq$ input, $y \triangleq$ output) (*4.3.1 et seq.*)

$\dfrac{1}{L(p)} \cdot M(p)$ system characterized by $L(p)y = M(p)u$ (*4.5.9 et seq.*)

$\mathcal{L}$ Laplace transform

$\mathcal{L}^{-1}$ inverse Laplace transform

$\mathcal{L}_{\text{II}}$ bilateral Laplace transform

λ_i ith eigenvalue of $\mathbf{A}$ (*5.4.5*)

$\mathcal{M}$ subspace of a linear vector space (*C.5.12*)

$\mathcal{N}(\mathcal{A})$ null space of the linear transformation $\mathcal{A}$ (*C.9.7*)

$\mathcal{N}(\mathbf{A})$ null space of the linear transformation represented by matrix $\mathbf{A}$ (*C.9.7*)

$\mathfrak{N}_k^l$	null space of $(\mathbf{A} - \lambda_k\mathbf{I})^l$ (*5.5.2, D.3.1*)
$\mathfrak{N}_k$	generalized null space of $\mathbf{A}$ associated with its kth eigenvalue (*5.5.5, D.3.2*)
$\mathbf{0}$	zero vector, zero matrix, null function (*2.8.1*)
o	of the order of (*7.4.1 et seq.*)
$1(t)$	unit step function (*A.2.1 et seq.*)
p	derivative in the distribution sense (*3.8.3, A.2.3, A.5.4 et seq.*)
$\phi_i(t,t_0)$	ith basis function (zero-input response) starting at t_0 (*3.6.32*)
$\phi_i(t)$	ith basis function starting at $t_0 = 0$ (*3.6.39*)
$\mathbf{\Phi}(t,t_0)$	state transition matrix (*6.2.8*)
$\mathbf{\Phi}(t)$	state transition matrix with $t = 0$ (*3.7.7, 5.2.8*)
$\hat{\mathbf{\Phi}}(s)$	Laplace transform of $\mathbf{\Phi}(t)$ (*5.3.2 et seq.*)
$\mathfrak{R}$	system of the reciprocal differential operator type (*3.3.1*)
$\mathfrak{R}(\mathfrak{A})$	range of the linear transformation $\mathfrak{A}$ (*C.9.6*)
$\mathfrak{R}(\mathbf{A})$	range of the linear transformation represented by the matrix $\mathbf{A}$ (*C.9.6*)
$R[\mathbf{u}(t)]$	range of the variable $\mathbf{u}(t)$ (at fixed t) (*1.4.3, 1.4.17*)
$R[\mathbf{u}]$	range of the time function $\mathbf{u}$ (*1.4.3, 1.4.17*)
$R_{[t_0,t_1]}$	space of input-output pairs over $[t_0,t_1]$ (*1.4.17*)
Re	real part of
Σ	state space (*1.6.2*)
$\Sigma(t,t_0)$	state space of a free interconnection at time t (*1.10.15*)
$\Sigma(t)$	state space of an interconnection at time t (*1.10.21*)
$\mathbf{s}(t)$	state at time t (*1.7.1*)
$\mathbf{s}(\mathbf{s}(t_0);\mathbf{u}_{(t_0,t]})$	state at time t when the input $\mathbf{u}_{(t_0,t]}$ is applied, with the system initially in state $\mathbf{s}(t_0)$ (*1.8.2*)
$\mathcal{S}[\mathbf{x}_1, \ldots, \mathbf{x}_n]$	subspace spanned by $\mathbf{x}_1, \ldots, \mathbf{x}_n$ (*A.4.2 et seq.*)
T	range of t (*1.3.1*)
T_δ	translation operator (*3.2.2*)
$<T,\phi>$	value of the distribution T at ϕ (*A.4.1*)
$\boldsymbol{\theta}$	zero state (*2.8.2*)
$\mathbf{u}$	input vector (*1.4.17*)
$\mathbf{uu}'$	$\mathbf{u}$ followed by $\mathbf{u}'$ (*1.6.16*)
u_j^i	jth component of input to $\mathfrak{A}_i$ (*2.6.1*)
$1(t)$	unit step function starting at $t = 0$ $[1(t) = 0$ for $t < 0$, $1(t) = 1$ for $t \geq 0]$ (*A.2.1 et seq.*)
$(\mathbf{u},\mathbf{y})$	input-output pair (*1.4.17*)
$\mathbf{x}(t)$	state at time t of a differential system (*3.6.2 et seq., 5.2.1*)
$\mathbf{x}(t;\mathbf{x}_0,t_0)$	state at time t given that the state at t_0 is $\mathbf{x}_0$ and no input is applied during $(t_0,t]$ (*5.2.11*)

$\mathbf{x}(t;\mathbf{x}_0,t_0;\mathbf{a}_{(t_0,t]})$ state at time t when the input $\mathbf{a}_{(t_0,t]}$ is applied and the initial state (at t_0) is $\mathbf{x}_0$ (*5.2.21*)

$\mathbf{y}$ output vector (*1.4.17*)

$\tilde{\mathbf{y}}$ suppressed output vector (p. 11)

$\mathbf{Z}_{kl}$ components of the matrix $\mathbf{A}$ (*5.3.1*)

$Z(s,\boldsymbol{\alpha})$ Laplace transform of the zero-input response starting in state $\boldsymbol{\alpha}$ (*4.7.6*)

Index

Page references in **boldface** type indicate definitions